Legumes for Soil Health and Sustainable Management

Ram Swaroop Meena • Anup Das
Gulab Singh Yadav • Rattan Lal
Editors

Legumes for Soil Health and Sustainable Management

Springer

Editors
Ram Swaroop Meena
Department of Agronomy
Institute of Agricultural Sciences (BHU)
Varanasi, UP, India

Anup Das
Division of Crop Production
ICAR Research Complex for NEH Region
Umiam, India

Gulab Singh Yadav
Division of Crop Production
ICAR Research Complex for NEH Region
Lembucherra, India

Rattan Lal
Carbon Management and Sequestration
Center, SENR/FAES
The Ohio State University
Columbus, OH, USA

ISBN 978-981-13-0252-7 ISBN 978-981-13-0253-4 (eBook)
https://doi.org/10.1007/978-981-13-0253-4

Library of Congress Control Number: 2018945195

Printed on acid-free paper

This Springer imprint is published by the registered company Springer Nature Singapore Pte Ltd.
The registered company address is: 152 Beach Road, #21-01/04 Gateway East, Singapore 189721, Singapore

Contents

About the Editors

Dr. Ram Swaroop Meena is born in a farmer's family in VOP. Harsana, Teh. Laxmangarh, Dist. Alwar, Rajasthan, India, on May 19, 1977. Dr. Meena had his schooling in the same village and graduated in Agriculture in 2003 from the Sri Karan Narendra Agriculture University, Jobner, Jaipur, Rajasthan. Dr. Meena has obtained his master's and the doctorate degrees in Agronomy from the Swami Keshwanand Rajasthan Agricultural University, Bikaner (Rajasthan), and securing first division in all the classes with triple NET, JRF, and SRF from the Indian Council of Agricultural Research, and RGNF award from the University Grants Commission. Dr. Meena has been awarded the Raman Fellowship from MHRD, GOI. He has completed his postdoctoral research on the soil carbon sequestration under Prof. Rattan Lal, Distinguished Scientist and Director, Carbon Management and Sequestration Center, Ohio State University, USA. He is working on soil sustainability, crop productivity, and resources use efficiency, under the current climatic era. Dr. Meena has supervised 17 PG and 4 Ph.D. students, and he has 8 years research and teaching experience at UG/PG/Ph.D. levels. He is working externally funded 3 running projects including DST (GOI), and involves many academic and administrative activities going on at institute/university level. Dr. Meena has published more than 90 research and review papers in peer-reviewed reputed journals, and contributed in the edited books with 25 book chapters at national and international levels. He has published 3 books on the national and another 3 on the international level. He has worked as an expert in the NCERT and MHRD, GOI, to develop 2 books for school education at XI and XII standards. Dr. Meena has received several awards, viz., Young

Scientist, Young Faculty, Global Research, Excellence in Research, Honourable Faculty Award, *etc*. Dr. Meena is a member of 9 reputed national and international societies, and is working as a general secretary, editor, member of editorial boards in twelve national, international peer-reviewed reputed journals, and attended several national and international conferences in the country and abroad. Dr. Meena is contributing to the agricultural extension activates on farmers' level as associate coordinator in trainings, meetings, workshops, and farmers fair.

Dr. Anup Das Principal Scientist (Agronomy) and Head, Division of Crop Production, Indian Council of Agricultural Research (ICAR), Research Complex for North Eastern Hill Region, Umiam, Meghalaya, has obtained his Master's degree from Tamil Nadu Agricultural University, Coimbatore, and Ph.D. from Indian Agricultural Research Institute (IARI), New Delhi. He has undergone special attachment research training on "Conservation agriculture for soil carbon sequestration" under Prof. Rattan Lal, Distinguished Scientist and Director, Carbon Management and Sequestration Center, Ohio State University, USA. Under SAARC-ACIAR programme, Dr. Das was trained on simulation modeling studies for enhancing crop and water productivity of rice-based cropping system in South Asia. During his 15 years' research career, Dr. Das has handled 15 externally funded projects, including Agricultural Model Intercomparison and Improvement Project (AgMIP), National Agricultural Innovation Project (NAIP), National Mission on Sustaining Himalayan Ecosystem, Task Force-6, Lal Bahadur Shastri Young Scientist Award challenge project (ICAR), Farmers Participatory Action Research Programme (FPARP), Network Project on Climate Change (NPCC), National Initiative on Climate Resilient Agriculture (NICRA), etc., and 12 institutional projects as Principal Investigator, Co-Principal Investigator, etc. Dr. Das has 4 years postgraduate teaching experience in College of Post-Graduate Studies (CPGS), Central Agricultural University (CAU), Umiam, Meghalaya, and guided and co-guided 17 masters and one Ph.D. student of CPGS, Umiam, and IARI, New Delhi. He is also involved in teaching program of three certificate courses on Organic Farming, Integrated Watershed Management, and Water Harvesting and Management of Indira Gandhi National Open University. He has in his credit 115 research papers, 15 books, 27 technical bulletins, and 52 book chapters. He is reviewer and editor of about 25 national and international journals. Dr. Das is also member of National Level Monitoring Team of National Food Security Mission, member of monitoring team of NICRA activities in Krishi Vigyan Kendra, and member secretary, NMSHE Task Force-6 activities in North East India. Dr. Anup Das is presently the Secretary, Indian Association of Hill Farming and Chief Editor of *Innovative Farming* – an international quarterly open access journal. He is recipient of many national awards and recognitions like NAAS Associate Fellow, Lal Bahadur Shastri Outstanding Young Scientist Award, Swami Sahajanand Saraswati Outstanding Extension Scientist Award, Outstanding Interdisciplinary Team Research Award, Fakhruddin Ali Ahmed Award for contribution to tribal farming system, Dr. HK Jain-CAU Award for excellence in agricultural research in north eastern states, and Dr. KG Tejwani Award for natural resource management, among others.

Dr. Gulab Singh Yadav is born in a farmer's family on July 1, 1982, at village-Amlonipur, Bareilly, Uttar Pradesh, India. He has obtained B.Sc. (Ag.) Honours from the Chandra Shekhar Azad University of Agriculture and Technology Kanpur, UP, India, M.Sc. (Ag) Agronomy and Ph.D. (Agronomy) from Indian Agricultural Research Institute (IARI), New Delhi, India, with first division and recipient of UP Rajaya Krishi Utpadan Mandi Parishad Scholarship, JRF and SRF-ICAR, B.Sc. (Ag.) Honours, M.Sc. (Ag) Agronomy, and Ph.D. (Agronomy), respectively. He had completed the postdoctoral research under the supervision of Prof. Rattan Lal, a renowned soil scientist from Carbon Management and Sequestration Center, Ohio State University, USA. Dr. Yadav has joined prestigious Agricultural Research Service (ARS)-ICAR, Government of India, on April 27, 2011 and presently serving as a Scientist (Agronomy) at Indian Council of Agricultural Research (ICAR), Research Complex for North Eastern Hill Region, Tripura Centre, Lembucherra, West Tripura, India.

During his 7 years research career, Dr. Yadav has involved in 6 externally funded projects, including National Agricultural Innovation Project (NAIP), National Mission on Sustaining Himalayan Ecosystem, Task Force 6, National Initiative on Climate Resilient Agriculture (NICRA), etc., and 4 institutional projects as Principal and Co-Principal Investigator. He has in his credit more than 50 research papers in international and national journals and several technical bulletins, book chapters, review, and popular articles. He is reviewer of about 10 national and international journals. He is recipient of best Research Scientist Award–2016 of ICAR Research Complex for North Eastern Hill Region, Tripura Centre, West Tripura, India, and DBT Associate–2016, Government of India. He is also a recipient of prestigious D.N. Borthakur Award (team award) for farming system research in North East India. Presently, he is a life member of several professional scientific societies.

Dr. Rattan Lal is a Distinguished University Professor of Soil Science and Director of the Carbon Management and Sequestration Center, the Ohio State University, and an Adjunct Professor of University of Iceland. With completion of education from Punjab Agricultural University, Ludhiana (B.Sc, 1963), Indian Agricultural Research Institute, New Delhi (M.Sc, 1965), and the Ohio State University, Columbus (Ph.D, 1968), he served as Sr. Research Fellow with the University of Sydney, Australia (1968–1969), Soil Physicist at IITA, Ibadan, Nigeria (1969–1987), and Professor of Soil Science at OSU (1987–to date). His current research focus is on climate-resilient agriculture, soil carbon sequestration, sustainable intensification, enhancing use efficiency of agroecosystems, and sustainable management of soil resources of the tropics. He is a fellow of the American Society of Agronomy (ASA, 1985), Soil Science Society of America (SSSA, 1986), Third World Academy of Sciences (1992), American Association for the Advancement of Sciences (1996), Soil and Water Conservation Society (SWCS, 1997), Indian Natl. Academy of Agricultural Sciences (1998), and Rothamsted, UK (2013). He received the Hugh Hammond Bennett Award of the SWCS, 2005; Borlaug Award (2005) and Liebig Award (2006) of the

International Union of Soil Sciences; M.S. Swaminathan Award (India) of 2009; COMLAND Award (Germany, 2009); and Sustained Achievement Award, Renewable Natural Resources Foundation (2017). He received honorary degree of Doctor of Science from Punjab Agricultural University (2001); the Norwegian University of Life Sciences, Aas (2005); Alecu Russo Balti State University, Moldova (2010); Technical University of Dresden Germany (2015); and University of Lleida, Spain (2017), and included in Thomson Reuters 2014–2017 list of most cited researchers and included in the World's Most Influential Scientific Minds (2002–2013). He was President of the World Association of the Soil and Water Conservation (1987–1990), the International Soil Tillage Research Organization (1988–1991), the Soil Science Society of America (2005–2007), and President of the International Union of Soil Sciences, Vienna, Austria (2017–2018). He was a member of the Federal Advisory Committee on National Assessment of Climate Change–NCADAC (2010–2014), member of the SERDP Scientific Advisory Board of the Department of Defense (2011), Senior Science Advisor to the Global Soil Forum of IASS, Potsdam, Germany (2010–2012), member Steering Committee of the Global Soil Week, IASS, Potsdam, Germany (2012–2016), member Advisory Board of FACCE-JPI of the European Council (2013–2016), Chair Advisory Committee to UNU-FLORES, Dresden, Germany (2014–2020), and member of the SPI of UNCCD, Bonn, Germany (2018–2020). Prof. Lal was a lead author of IPCC (1998–2000) and was awarded Nobel Peace Prize Certificate by IPCC in 2007 and Global Dryland Champion of the United Nations Convention to Combat Desertification, Bonn, Germany, in 2013. He has mentored 110 graduate students and 54 postdoctoral researchers, and hosted 169 visiting scholars from around the world. He has authored/co-authored more than 2227 research publications, including 868 refereed journal articles and 506 book chapters, has written 20 books, and edited/co-edited 68 books.

Legumes and Sustainable Use of Soils

1

Ram Swaroop Meena and Rattan Lal

Contents

R. S. Meena (✉)
Department of Agronomy, Institute of Agricultural Sciences (BHU), Varanasi, UP, India

R. Lal
Carbon Management and Sequestration Center, SENR/FAES, The Ohio State University, Columbus, OH, USA

R. S. Meena et al. (eds.), *Legumes for Soil Health and Sustainable Management*,
https://doi.org/10.1007/978-981-13-0253-4_1

Abstract

Sustainable use of soils is among the global challenges of the twenty-first century. In addition, growing food for 7.55 billion people (10.6% are prone to hunger and 26.7% to malnutrition and hidden hunger) in 2017 is a further challenge and threat to environment and sustainable production. Thus, food and nutritional demands must be met without degrading the natural environment. While the Haber-Bosch industrial process is producing about 100 Tg of reactive nitrogen (N) per annum globally, it creates enormous environmental problems. On a global basis, the supplied natural biological nitrogen fixation (BNF) is 110 Tg N per year on land and 140 Tg N per year in the ocean. Reducing the amount of N production is possible with the addition of legumes in the cropping systems. Soil sustainability implies the ecological balance, enhancement of soil functions, and biodiversity. Therefore, recommended technologies of crops and cropping systems which promote soil sustainability must be promoted. Legumes are known for their positive impacts, such as BNF, weed suppression, erosion control as cover crop, soil health improvement, and most importantly toward the eradication of malnutrition in third-world countries. Therefore, these crops can contribute to achieving the objectives of sustainable food and environmental security. Hence, inclusion of legumes in cropping system is inevitable in advancing soil sustainability and food and nutritional security (1 kg of grain legumes contain 180–430 g of protein) without compromising the long-term soil fertility potential. Rational soil management practices must involve legume-based rotations and intercropping considerations for restoring soil health, and soil sustainability should be given due emphasis.

Keywords

Food and nutritional security · Legumes · Soil health · Management

Abbreviations

IYP	International Year of Pulses
N	Nitrogen
UN	United Nations
IYS	International Year of Soils
SOC	Soil organic carbon
TG	Teragram
N_2O	Nitrous oxide
NO_3	Nitrate
P	Phosphorus
C	Carbon
SOM	Soil organic matter
LER	Land equivalent ratio

BMPs	Best management practices
BNF	Biological nitrogen fixation
SMB	Soil microbial biomass

1.1 Introduction

A challenge for today's agriculture is to ensure food and nutritional security for the rapidly growing global population. The global population is expected to reach 8.55 billion by 2030, 9.77 billion by 2050, and 11.18 billion by 2100 (UN 2017), along with the concomitant increase in global food demand. It is estimated that about 795 million of the 7.55 billion global population, or one person in nine, are prone to hunger (UN 2015). Of these, about 782 million are living in developing countries, representing 12.5% of the population of the regional population (UN 2015), signifying the key role of agricultural improvement. Therefore, the agriculture is under tremendous pressure to (1) manage food and nutritional security for the growing population, (2) mitigate climate change and its adverse effects, and (3) improve soil quality and the natural resource base (Lal 2015; Meena et al. 2015a). Hence, there is a need to incorporate legumes in cropping systems for soil sustainability and to achieve a balanced food and nutritional security (Nees et al. 2010). Further, legumes are of paramount importance in soil nourishment and eradication of malnutrition, especially in developing countries.

Legumes are members of the Leguminosae family. It comprises 650 genera and 18,000 species and is one of the 3rd largest angiosperms family on the earth, the others being Asteraceae and Orchidaceae (ILDIS 2017). Nutritional importance of legumes is second only to that of the cereals. One kg of legumes contains 180–430 g protein, 126–660 g carbohydrates, 7–400 g oil, 900–2400 mg calcium, 28–115 mg iron, 3–99 mg thiamine, 13–39 mg riboflavin, and 7–64 g dietic fibers (Berihun and Molla 2017). In addition, they are also rich in zinc, copper, manganese, folic acid, thiamine, vitamin C, lysine, tryptophan, threonine, methionine, cysteine, phosphorus, potassium, lipids, and fatty acids (Schmidt and Weidema 2015). In addition to forage and food uses, legumes can also be used to make tortillas, chips, doughnuts, bread, spreads, and types of snacks or in liquid form of yogurt and milk as value addition products (Meena et al. 2016). In the developing world, legumes are economic sources of proteins compared to animal proteins. Legumes are commercially cultivated on ~180 Mha (13–15% of the global arable land area) and account for ~26% of the world crop production (Medeot et al. 2010). Pulses alone contribute 34% of the dietary protein needs of the global population (Schuster 2013).

The Green Revolution during the 1960s enhanced the agronomic productivity of cereals with the use of short-statured high-yielding irrigation and fertilizer-responsive crop varieties. However, it created some negative effects on soil health and environment (Lal 2006, 2014). The most prominent negative effects are soil degradation, decline in total factor productivity, depletion of groundwater table, decline in soil fertility, losses in biodiversity, soil salinization, and development of weed resistance. In addition, especially in developing countries, nitrogen (N) fertilizers are being used extensively since the Green Revolution era due to the higher subsidy on these fertilizers, provided for increasing food production (Meena et al. 2015b). These practices further exacerbate the soil degradation and cause yield stagnation (Swarup et al. 2000; Lal 2010). Therefore, food security and soil sustainability are being jeopardized, especially in developing countries (Hati et al. 2006; Lal 2010).

Some food legumes, harvested for dry grains, are called pulses. The 68th General Assembly of the United Nations (UN) declared year 2016 as the "International Year of Pulses (IYP)" to raise awareness and to celebrate the role of pulses in human nutrition and welfare. The main objectives of the IYP were to (1) promote the worth and use of pulses throughout the food system, their benefits for soil fertility and sustainability, and (2) raise awareness about pulses and their role in sustainable food production and healthy diets (UN 2016). Likewise, the UN assembly also declared 2015 as the "International Year of Soils (IYS)" to enhance the understanding about the role of "healthy soils for a healthy life." The main objectives of the IYS were to (1) raise awareness in public and in policy makers on the soil and human life, (2) promote successful policies and plans for sustainable management of soil (Verma et al. 2015a), and (3) promote investment on soil sustainability for society and different land users (UN 2015). Consequently, the International Union of Soil Sciences (IUSS) has declared 2015–2024 as the International Decade of Soils (IUSS 2016).

Presently, there are growing concerns, among policy makers and the general public, regarding sustainability. The Brundtland Commission (1987) defined sustainability as "meeting the needs of the present without compromising the ability of future generations to meet their needs." Sustainability also implies a set of soil management policies addressed to the food productivity and environmental sustainability (Lal 2010; Verma et al. 2015a), maintaining crop productivity and agricultural stability in the long term (Ghosh et al. 2006). There are four aspects of sustaining soil functions: environmental, social, economic, and institutional (Hayati et al. 2010).

Despite the large number of ways and frameworks proposed for sustainability assessment, there are still some major shortcomings in techniques of assessing soil sustainability. First, the multidimensional nature of crops has not been sufficiently addressed (Hani 2007; Rossing et al. 2007; Yadav et al. 2017a), and many researchers still report an imbalance of indicators that represent the environmental, social, and economic dimensions of soil sustainability (Binder and Wiek 2007). Second, little information exists on the role of legume in soil sustainability (Stern 2006).

Modern agriculture is relying on the intensive use of inorganic fertilizers and irrigation for increasing crop yields, and the importance of legumes has been undermined (Van Werf and Petit 2002; Meena et al. 2014). Although the technology developed during the Green Revolution maximized outputs, input use efficiency was overlooked. Therefore, adverse impacts of intensive agriculture started to manifest itself in the production systems, with yield plateau or fall in productivity and widespread degradation of soil health and the environment, especially in the developing world (Binder et al. 2010). Furthermore, the intensive use of fertilizers may not sustain yields over the long term (Fustec et al. 2010; Bues et al. 2013). In this context, legumes, either as mono-crops or in a cropping sequence, have great potential to enhance soil functions in agricultural systems for sustaining productivity.

When used in rotation with nonleguminous crops, leguminous crops can restore soil organic carbon (SOC) stocks and minimize disease and pest incidences (Deakin and Broughton 2009). Furthermore, leguminous crops have biological N-fixing (BNF) capacity which improves soil health along with positive effects on the environment through reduction in the use of fertilizers and N losses in agricultural fields (Graham and Vance-Carroll 2015; Dhakal et al. 2015). Nitrogen is essential to all living bodies for the synthesis of proteins, nucleic acids, and other N-based compounds (Loreau and Hector 2001). Rhizobium is a species of rhizobacteria group that acts as symbiotically atmospheric N-fixer in legume plants. The *rhizobacteria* spp. infects the legume roots, leading to the development of nodules where the N-fixation takes place (Siddique et al. 2008). Atmospheric N-fixed by agriculturally important legumes is about 50–65 Tg annually, with 5–6 Tg fixed by other legumes in ecosystems (IAASTD 2008). Thus, with the growing threat of the deterioration of soil functions and environmental quality, the involvement of legumes in the agricultural system is an option to advance soil sustainability (Leterme and Carmenza Munoz 2002). The objective of this chapter is to analyze the diversified uses of legumes in agriculture and describe their future scope to achieve soil sustainability.

1.2 Prospects of Legumes in Developing Countries

Important global issues of the twenty-first century are the growing demand for healthy and nutritious food, declining crop productivity, intensifying the use of agrochemicals, and the attendant risks of growing pollution, accelerating soil erosion, declining soil fertility and organic carbon reserves, decreasing biodiversity, and declining soil sustainability (Fig. 1.1).

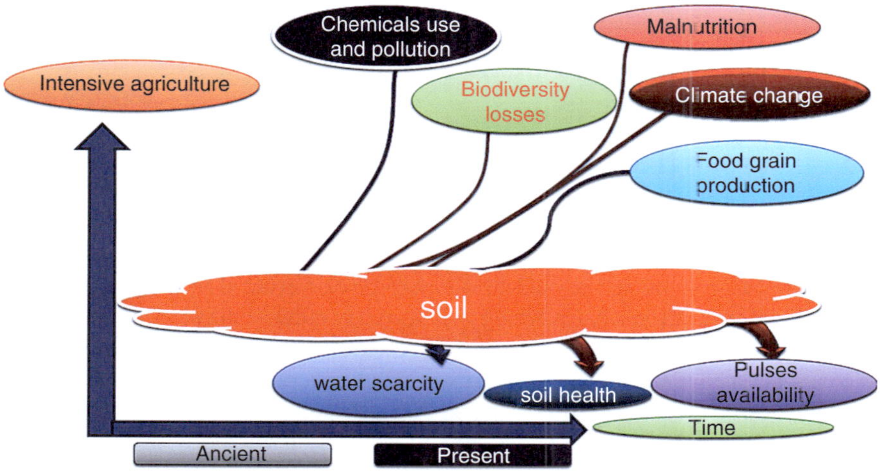

Fig. 1.1 Global issues in agriculture

Under these situations legumes can be used as an option for maintaining soil sustainability and improving the dietary composition in developing countries (Meena et al. 2015b). In most of the developing countries, legumes are substitutes of meat to the poor sector of the population. Legumes are also sources of protein and other essential minerals and amino acids (Leterme and Carmenza Munoz 2002). Thus, legumes have multipurpose use and are consumed either in processed forms or directly as food in developing countries (Berihun and Molla 2017). The use of pulses is also changing in developed countries where pulses are progressively considered as healthy foods compared with animal protein. However, the food price index of the developing countries is much higher than that of the developed countries (Reddy 2016).

The progress in proteinaceous food in developing countries can be drastically improved, there exists a deficiency of 77, 49, and 21% in meat, egg, and fish production, respectively. While the production of pulses increased by 1.5%/year between 1980 and 2012, the availability of pulses at present has stagnated at around 6.5 kg/capita per year in the developing world (FAO 2016). Yet, the per capita availability of pulses in India was 9.5 kg/year during the1960s (Joshi and ParthasarathyRao 2016). Therefore, dietary protein shares in the developing countries remain relatively lower compared to those of developed countries (USDA 2016). Furthermore, the per capita consumption of pulses in the developing countries has declined especially in Pakistan, Bangladesh, Sri Lanka, and those in sub-Saharan Africa (FAO 2016). With most of the agricultural development and research being focused on increasing cereal productivity in these areas, and with rapid increase in population, the per capita availability of pulses has declined in the developing countries. Furthermore, the yields and production of crops have also stagnated in developing countries due to degradation in soil health and decrease in SOC stocks. Therefore, inclusion of legume crops in the rotation cycle is an important option because of their role in BNF and soil health-restoring capacity (Nygren and Leblanc 2015). Legume crops play a diverse role in agricultural systems and the food sector. Thus, research on legume crops can have a significant impact on nutritional security and soil health on a long-term basis (Nees et al. 2010; Dhakal et al. 2015).

1.3 Current Need for Soil Sustainability

The rapidly increasing population is impacting food and nutritional security because of decreasing quality of water and soil resources which are already under great stress. The adverse impact on soil health may be exacerbated by soil degradation, pollution, global warming, declining soil fertility, industrialization, urbanization, and the rising price of fertilizers. Rapid increases in agricultural expansion and intensification have raised serious questions about agricultural sustainability (Verma et al. 2015a). Therefore, soil sustainability research and assessment must be strengthened to enhance awareness that soil is a truly living body with biological, physical, and chemical properties and functions which are essential to sound agroecosystem services for human well-being (Chen et al. 2003). There is an increasing

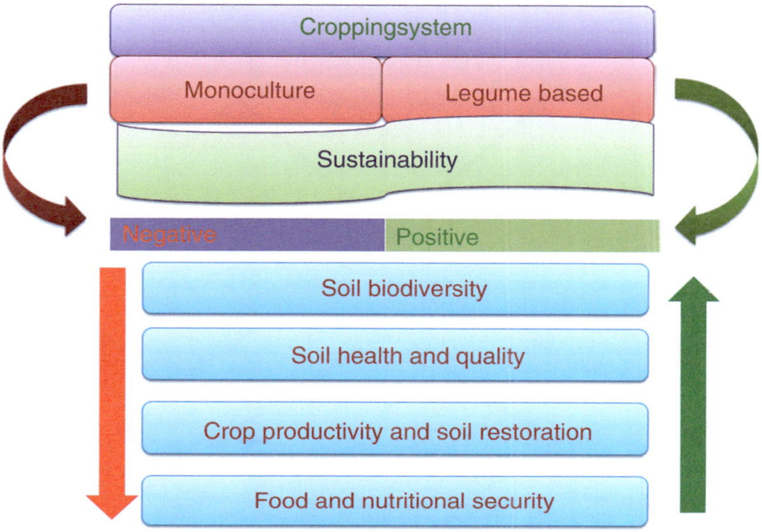

Fig. 1.2 Response of legume-based cropping system

recognition of the need for better soil preservation for a healthy environment, through innovative approaches in relation to food supply (Lal 2016; Meena et al. 2017a). The negative consequences of population pressure, environmental damage, soil degradation, and land constraints have been exacerbated by an inadequate understanding of the biological processes to optimize nutrient cycling, thereby reducing the use of agricultural inputs and increasing the use efficiency of inputs (Nees et al. 2010). Inclusion of legumes in crop rotation and intercropping can alleviate these constraints. Notable responses of including legume in a cropping system and their effects are outlined in Fig. 1.2.

The important benefits of legumes include soil restoration, increase in the SOC stock, improvement in N pool by BNF, and positive effects on the yield of succeeding crop. Thus, there is an enormous need to focus on the benefits of legumes and their role in the soil sustainability (Dhakal et al. 2016).

1.4 Role of Legumes in Soil Sustainability

Legumes are soil-amendment crops with strong benefits on soil health and must be an integral component of the farming systems (Hauggaard-Nielsen et al. 2007). Results from most short-term studies are encouraging and indicate that the legumes must be included within cropping system for soil health management (Binder and Wiek 2007). Legumes have positive effects (Fig. 1.3) on soil processes such as benefiting agroecosystems, agricultural productivity, soil conservation, soil biology, SOC and N stocks, soil chemical and physical properties, BNF, nitrous oxide (N_2O) emission, and nitrate (NO_3^-) leaching by reducing the need for chemical fertilizers

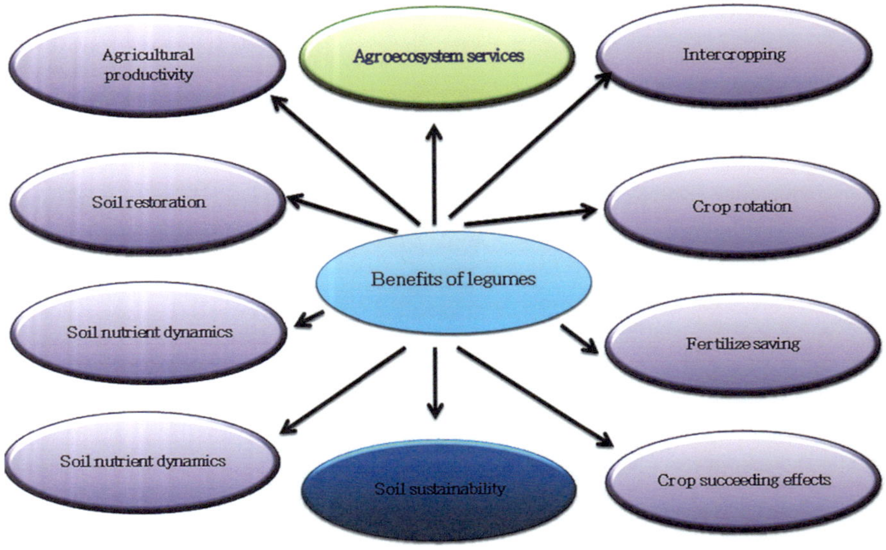

Fig. 1.3 Role of legumes in agriculture

(Crozat and Fustec 2006; Nees et al. 2010). Above all, legumes are now utilized as soil nourishment agents. However, these benefits on soil health must be quantified, and their mechanisms understood. Thus, incorporating legumes as a part of cropping systems is pertinent to better soil health and productivity (Binder et al. 2010; Dhakal et al. 2016).

1.4.1 Agroecosystem

Legumes contribute indirect/direct benefits to agroecosystems and society (Table 1.1). These benefits include (1) cultural services of productive cropping systems; (2) provisioning services of fiber, feed, food, and protein; (3) supporting services of water and nutrient cycling, soil formation, and the production of oxygen; and (4) regulating services of moderating climate change (Chen et al. 2003; Haslmayr et al. 2016). Intensive agriculture affects a range of ecosystem services (Lal 2013; Duru et al. 2015). Intermediary components in agriculturally sustainable agroecosystem are soil fauna, nutrient cycling, and soil biota (Chen et al. 2003; Kureh and Kamara 2005). Inclusion of legumes in the cropping systems can strengthen ecosystem services by promoting beneficial soil microorganisms and improving soil biodiversity. (Meena et al. 2015c). Also, legumes provide benefits to neighboring plants, and they also improve N mineralization and water relations, ensure protection from pests, and reduce soil erosion (Williams and Hedlund 2014). Legumes have a direct effect on crop productivity and influence crop diversity and community structure in the agroecosystem (Lal 2013). Furthermore, the extensive

Table 1.1 Advantage of including legumes in agroecosystems

Advantage of legumes	References
Food and nutritional security	Chen et al. (2003); Nees et al. (2010)
Human health safeguard	Leterme and Carmenza Munoz (2002); Binder and Wiek (2007); Lithourgidis et al. (2011)
Ecological balance	Binder and Wiek (2007); Lal (2013)
Effective agroecosystem	Chen et al. (2003); Haslmayr et al. (2016)
Control soil erosion, increase water and nutrient use efficiency	Lopez and Mundt (2000); Ratnadass et al. (2013); Lithourgidis et al. (2011); Williams and Hedlund (2014)
Increase agricultural productivity	Vance (2001); Deutsch et al. (2006); Meena et al. (2015b)
Sustain soil functions	Deakin and Broughton (2009); Derpsch (2015); Lal (2016)
Improve soil physical, chemical, and biological properties	Schoenholtz et al. (2000); Dexter (2004); Srinivasarao et al. (2012); Lal (2015)
Sustaining long-term soil health and quality	Padilla and Pugnaire (2006); Nees et al. (2010); Binder et al. (2010)
Biological nitrogen fixing (BNF)	Graham and Vance-Carroll (2015)
Reduce chemical fertilizer application for safer environment	Crozat and Fustec (2006); Nees et al. (2010)
Increase SOC stock	Lal (2015, 2016); Hauggaard-Nielsen et al. (2007)
Increase N stock	Hauggaard-Nielsen and Jensen (2005); Loreau and Hector (2001); Corre-Hellou et al. (2006)
Sequestrate soil carbon	Lal (2013); Duru et al. (2015)
Improve soil biota	Chen et al. (2003); Siddique et al. (2008); Deakin and Broughton (2009); Chintala et al. (2013)
Nutrient cycling and BNF capacity	Chalka and Nepalia (2006); Ghanbari et al. (2010)
This lower C/N ratio results in faster breakdown of legume residues	Carter (2002); Kureh and Kamara (2005)
Cut fertilizer costs	Fustec et al. (2010); Peyraud et al. (2009)

root system of legumes and secretion of exudates have a beneficial effect on soil health by improving the nutrient dynamics, structural stability, and quality of soil (Padilla and Pugnaire 2006). Legume crops also help in the cycling of major elements such as N, P, and C (Nees et al. 2010).

A range of grain legumes are growing in rotation with different crops worldwide. The most common cultivated grain legumes are faba bean (*Vicia faba* ssp. minor L.), dry bean (*Phaseolus vulgaris* L.), and pea (*Pisum sativum* L.), in the northern hemisphere. Legumes grown as grain food are cowpea (*Vigna unguiculata* L.), chickpea (*Cicer arietinum* L.), and dry bean. Peanut (*Arachis hypogea* L.), pigeon pea (*Cajanus cajan* L.), and lentil (*Lens esculenta* L.) are also used for food proposes. N harvest induces growth of legumes such as pea and soybean which are often of a high yield. For example, N accumulated in the seed is ~80% in soybean (Crépon 2006) and ~70% in pea (Crozat and Fustec 2006). Legumes are also agents for ecological facilitation processes in agroecosystems (Mahieu et al. 2007) because they contribute in N economy through BNF (Buragohain et al. 2017) and increase crop productivity and soil-N stock (Rochon et al. 2004; Crozat and Fustec 2006). Several legume crops (viz., lupin (*Lupinus angustifolius* L.), vetches (*Vicia genera* and *Vicia sativa* L.), velvet bean (*Mucuna pruriens* Bak.), fenugreek (*Trigonella foenum-graecum* L.), clovers (*Trifolium* sp.), *Crotalaria spectabilis*, or *Sesbania rostrata*) are also being used as green manure. They improve the nutrient cycling, soil organic matter (SOM), and nutrient-supplying capacity of soils. Therefore, compared with a continuous cereal-grown field, higher levels of available N are recorded in legume cultivated fields after harvesting, which indicates that inclusion of legume crops in rotation may lead to a positive N stock for the succeeding crop (Hauggaard-Nielsen and Jensen 2005). In addition, legume crops also enhance the resource use efficiency (Carlsson and Huss-Danell 2003; Jensen 2006).

Forage legumes have more potential of high biomass production over a range of climates. Four main forage legumes (clover (*Trifolium subterraneum* L.), white clover (*Trifolium repens* L.), alfalfa (*Medicago sativa* L.), and red clover (*Trifolium pratense* L.)) together cover a wide range of semiarid agroecosystem of the world (Soussana and Machado 2000; Corre-Hellou et al. 2006). Most widespread is the white clover and is used in agriculture. The bird's-foot trefoil (*Lotus corniculatus* L.) is also abundantly grown in temperate climates of northern countries (Loreau and Hector 2001; Corre-Hellou et al. 2007). Some experiments conducted using mixtures of leguminous crops with cereal crops have shown that positive effect is higher on soil ecology in low-input systems than others and give more stable yields in stressed soil environments (Høgh-Jensen 2006; Rasmunssen et al. 2007).

1.4.2 Agricultural Productivity

Achieving a sustainable productivity is a great challenge for the developing countries and is highly significant for alleviating the soil-related constraints (Verma et al. 2015a). Nonetheless, rational uses of agrochemicals are warranted as intensive use of agrochemicals which in turn has a deteriorating impact on the soil health (Vance 2001).

Furthermore, N is an abundant element on the globe and is also the major limiting element to the food production (Vance 2001). The N nutrition is an economic insurance against low yield and has been applied extensively for higher productivity (Deutsch et al. 2006), in spite of the fact that it is a major contributor to environmental contaminations (Foley et al. 2011). The BNF is an option for eco-friendly source of N, to optimize the application of fertilizer and to sustain agricultural productivity (Wichern et al. 2007). Most of the fixed N in legumes is harvested as a grain yield, and yet legumes deposit significant amounts of N in the soil and thereby improve soil N stock (Garg and Geetanjali 2007; Dhakal et al. 2015).

Legume-based crop rotations are better than nonleguminous with respect to environmental security. Yet, there are several factors that affect the choice of a rotation for high agricultural productivity (Khan et al. 2007). Important factors for consideration to identify suitable crop rotations for a particular area are soil properties (viz., SOC, soil-plant-water relations, soil pH, etc.) as these properties also affect the diversity, abundance, and agricultural productivity. Mono-cropping has negative effects on soil process (Mahieu et al. 2007; Derpsch 2015). In addition, soil microbial status is also significantly affected by crop rotation (Horrigan et al. 2002). Residues of legumes are a food base to those soil biota that improve nutrient mineralization, soil functions, and productivity. Thus, the role of legumes to improve the crop productivity potential is a promising strategy for addressing the challenges of low productivity. Legumes have a versatile potential for the perfection of agriculture productivity. Moreover, natural resources are used efficiently while also facilitating the soil quality and environment sustainability (Meena et al. 2016). The helpful legume-based agricultural practices are planned to improve productivity, agroecosystem, and diversity under unfavorable and harsh conditions (Chen et al. 2003). Among practices, legume-based crop rotation is one of the more effective ways to reduce the impact of diseases, diversify the cropping system, suppress weeds, and recover soil functions (Deutsch et al. 2006) and as a result increase agricultural productivity.

1.4.3 Intercropping

Growing two or more crops simultaneously, using a particular row arrangement on the same piece of land during a single growing season, is called intercropping (Brintha and Seran 2009). Intercropping is practiced to meet several ecological goals such as promoting species interaction, enabling natural regulation mechanisms, increasing biological diversity, and reducing farmer's risks against climatic aberrations (Lithourgidis et al. 2011; Kumar et al. 2016). Normally, intercrop components are from different families and species, with the main crop being of primary importance like food production and secondary crop for providing additional benefits (e.g., N-fixation, biomass, soil improvement, economic, etc.). The main aim of legume-based intercropping system is to produce more yields from the same field and improve natural resources' efficiency compared to mono-cropping (Inal et al. 2007). The mixture of a leguminous with a nonleguminous species can have yield advantages over monoculture. This is because leguminous crops improve soil functions through the symbiosis of legume-rhizobia (Fustec et al. 2010). Furthermore, intercropping with legumes provides many additional benefits to soil health and quality by reducing soil erosion, improving soil processes (Hauggaard-Nielsen et al. 2007), increasing moisture retention (Ghanbari et al. 2010), maintaining soil fertility (Hauggaard-Nielsen et al. 2009), increasing nutrient cycling, enhancing soil conservation (Lithourgidis et al. 2011) and BNF (Chalka and Nepalia 2006), saving labor (Hauggaard-Nielsen et al. 2007), and suppressing weed infestation (Liang et al. 2010). Legume-based intercropping systems are viable methods to introduce N into low-input agroecosystems and reduce reliance on mineral forms of N fertilizer (Seran and Brintha 2010) and also reduce the environmental damage through NO_3 leaching and N_2O emission. Legume-based intercropping is based on the principles of eco-friendly agriculture (Ashoka et al. 2017) as is practiced with the minimal use of inorganic N application (Dusa 2009). The legumes normally used as an intercrop (or mixed crop) in sustainable agricultural production included clover

(*Trifolium pratense* L.), pea (*Pisum sativum* L.), lentil (*Lens culinaris* L.) faba bean (*Vicia faba* ssp. minor L.), dry bean (*Phaseolus vulgaris*), pea, cowpea (*Vigna unguiculata* L.), chickpea (*Cicer arietinum* L.), peanut (*Arachis hypogea* L.), pigeon pea (*Cajanus cajan* L.), etc. (Blaser et al. 2006; Brintha and Seran 2009). Under favorable environmental conditions, legume crops add more N into the system, apparently leading to high yields of the main crops (Ghaley et al. 2005).

Several aspects to be considered for the success of the intercropping system (Brintha and Seran 2009) are light interception, crop species, and nutrient requirements. The potential of legume-cereal-based intercropping is to make N dependent on crop density (Abera et al. 2005). The choice of suitable crops depends on the plant growth habit, light, water, fertilizer, and land requirements (Seran and Brintha 2010). The associated nonlegume crops may benefit through N-fixation from leguminous crops (Sullivan 2003; Garg and Geetanjali 2007; Seran and Brintha 2010). The N-fixation or transfer occurs through plant root excretion, leaf fall, and N leached from leaves (Samba et al. 2007; Addo-Quaye et al. 2011). Further, N-fixed by a legume crop component may be available to the associated nonlegume crops in the current growing season (Adeniyan et al. 2007; Dahmardeh et al. 2010), known as direct N availability. Barbosae et al. (2008) reported that 25% of fixed N by component cowpea crop was transferred to maize crop. An important objective for legume-based intercropping is to ensure that an increased productivity per unit area is obtained compared to mono-cropping (Raji 2007). Addo-Quaye et al. (2011) and Muoneke et al. (2007) reported land equivalent ratio (LER) of 1.03–1.60 in the intercropping system indicating the efficient utilization of land resource by growing legume-based intercropping (Ram and Meena 2014).

1.4.4 Crop Rotations

Crop rotation is also a strategy for increasing productivity and sustainability in insensitive crop production systems, such as thorough inclusion of legumes in cereal production systems (Keeler et al. 2009). Crop rotations are an innovative technique to enhance productivity and improve soil sustainability. Benefits of a well-managed crop rotation are breaking plant pest cycles, improving yields, reducing chemical fertilizer inputs, improving soil fertility, increasing biological diversity of the agroecosystem, and controlling soil erosion (Ratnadass et al. 2013; Lithourgidis et al. 2011). Leguminous plants provide a high biomass in rotation that improves the SOC stock and maintains a high amount of active C in soil, which are important factors to soil health management (Hauggaard-Nielsen et al. 2007). Legume-based crop rotations that include grass produce more biomass and contribute to improving the SOC stock and can be adapted to any legume-based cropping system (Ram and Meena 2014). The legume-based green manure crops are also part of the crop rotation in many sustainable land development systems (Lithourgidis et al. 2011). Leguminous cover crops are also grown in a rotation primarily to improve soil fertility and prevent soil erosion (Lopez and Mundt 2000; Schipanski and Drinkwater 2012).

Legume cover crops are included in the system because of their N-fixation and high biomass production ability (Ramirez-Restrepo and Barry 2005). When the legumes are used strategically in a crop rotation, it can provide N to the succeeding crop. The quantity of N-fixed by this association between legumes and bacteria varies according to plant species and varieties, the crop management, climate, the soil type, and duration of the crop (Chu et al. 2004; Spehn et al. 2002). The quantity of N that a legume crop provides to succeeding crops depends on the quantity of N-fixed; the maturity of the legume crop, when it is incorporated into the soil, whether as a whole plant or only as the plant's root system that remains in the field; and the ecological conditions that govern the speed of decomposition (Addo-Quaye et al. 2011).

1.4.5 Soil Conservation

Conservation practices are increasingly being practiced in several regions on a global scale, primarily in response to improved recognition of soil degradation, quality, health, and sustainability. Legume crops are an ideal plant type for two components of conservation: crop rotation and soil cover (Mundt 2002; Srinivasarao et al. 2012). Crop residues of cultivated legume crops are a significant factor in the cropping system through their positive effects on soil biological, chemical, and physical functions as well as on soil water holding capacity and soil quality (Grandy et al. 2002; Mousavi et al. 2009). Kumar and Goh (2000) reported that the rotation of different crops with variable rooting patterns combined with minimal soil disturbance in zero-till systems promotes a more extensive network of root channels for macro- and micropores in the soil. This helps in water infiltration to deeper soil depths. The advantages of legumes in soil conservation have been reviewed by Siddique et al. (2008) in detail and are summarized as follows. The major advantage of legumes is the BNF to the soil and no exhaustion of existing soil N pools due to their capability to meet their own N needs through BNF. Leguminous crops add a high quality of SOM because of their high N/C ratio, deep root system, high water infiltration, and nutrient cycling in the soil (Dhakal et al. 2016). Leguminous crops provide soil protection during those periods when the main crop is not grown. Legume crops protect the soil from raindrops and reduce risks of soil erosion. Surface runoff is reduced by the legume crops by increasing soil water infiltration rate. A leguminous crop not only does the above ground soil protection but also helps in soil protection below ground by water infiltrating to the profile and thus keeps it in place.

1.4.6 Fertilizer Savings

Legumes have little fertilizer requirement except the starter dose, and the effect of the previous legume crop enables farmers to reduce fertilizer application in the succeeding crop. A food legume may save about 170–220 kg/ha of N fertilizer compared to an oilseed or nonlegume crop besides saving some 40–70 kg of N to the following crop (Fustec et al. 2010; Varma et al. 2017). Without any fertilizer application, clovers save 160–310 kg/ha of N fertilizer through supply of the BNF capacity (Bues et al. 2013), including 30–60 kg/ha N for the succeeding crop. Based on 4–5-year crop rotations, Dusa (2009) reported that 30 and 90 kg/ha of N fertilizers can be saved by forage and grain legume rotations compared to rotations with nonlegume crops. However, N fertilizer savings depend on site-specific conditions (Peyraud et al. 2009). More N savings can be achieved in the succeeding crops after legumes but are not optimally utilized in any of the present farming practices. As much as 30–45 kg/ha fertilizer are saved owing to inclusion of legumes in the cropping system (Loreau and Hector 2001; Mazvimavi et al. 2008; Addo-Quaye et al. 2011). When the fertilizer costs are higher than the product prices, there are limits on fertilizer use. Under these conditions, farmers can reduce the fertilizer needs by adopting suitable environmental schemes, organic farming, and other practices with leguminous crop to more fully utilize the N reserves. Legumes can decrease the energy footprint of cropping systems by reducing the need for N application and restoring the health of agroecosystem (Giller et al. 2009; Verma et al. 2015b). A large proportion of the N advantage of legumes comes from the crop residue and roots. The crop biomass slowly decomposes within few years to make available N and improve long-term soil fertility. Replacing fallow period with grain CropScan also decreases saline seeps and nitrate leaching (Alpmann et al. 2013). The possible advantages of leguminous cover crops to provide N is increasingly becoming more economical. The cost of N fertilizers has rather increased over the past four decades. At a time when the chemical inputs are no longer viable, farmers need to consider legume as an option to sustain soil fertility. Realizing that the price of natural gas-based N fertilizers will continue to increase, BNF will become an essential aspect of all agricultural systems.

1.4.7 Restore Polluted Soil

There are 3 million cases of pesticide poisoning every year and up to 220,000 deaths, mostly in developing nations (Prell and Poole 2006). The Green Revolution era witnessed incredible gains in global food production, but because of the intensive and indiscriminate usage of pesticides, the soil biodiversity was adversely affected (Khan et al. 2009; Jin et al. 2013). The term "pesticides" covers a diverse range of chemical compounds including fungicides, herbicides, nematicides, insecticides, rodenticides, molluscicides, plant growth regulators, as well as other chemicals used in agriculture (Klaassen and Watkins 2003). The adverse impacts of soil pollution (Table 1.2) in the agricultural system (soil, micro- and macroflora and fauna, and water bodies at farms) are well known (Franzluebbers 2002; Meena et al. 2017b).

The adoption of agricultural best management practices (BMPs) indicates that remediation aimed at recovery of polluted sites is at the core of achieving soil sustainability in the built agroecosystems (Deakin and Broughton 2009; Lal 2011). However, the remedial action in itself is required to be sustainable in a broad

Table 1.2 Some soil pollutants which degrade soil quality and health

Sources of soil pollutant	References
Pedogenic processes	Herawati et al. (2000)
Industrial operations such as metal forging, combustions of fossil fuels, smelting, etc.	Khan et al. (2009)
Mining areas	Wang et al. (2004)
Acid mine drainage	Williams et al. (2009)
Fly ash from energy sector	Liu et al. (2005)
Amending agricultural land with biosolids	Obbard et al. (1994)
Agricultural fields are irrigated with treated industrial wastewater	Sinha et al. (2006)
Industrial products like herbicides, pesticides, etc.	Klaasen and Watkins (2003)
Excessive drawing of groundwater leads to the release of arsenic	Mandal et al. (1996)

context, taking into account the ecological, the economic, as well as the social domains of sustainability (Schädler et al. 2011). In addition to controlling erosion, cover crops and their decaying residues reduce pollution by preventing runoff of nutrients and pesticides into the surface water (Fester et al. 2014). Also, cover crops may allow earlier field access and improve traction during wet spring. Rhizobiales, belonging to the Alphaproteobacteria class, are gram-negative bacteria of agronomic importance because some species have N-fixing symbiotic relationships with leguminous plants (Sato et al. 2005; Dadhich et al. 2015). Rhizobia invade the roots of legumes and form nodules to fix atmospheric N into ammonia, which is then provided to the host plants. This activity allows the plants to grow in the absence of an external N source (Bardos 2003; Bardos et al. 2011). In symbiotic N-fixation, hydrogen is a by-product of the process and has recently been discovered to be an ordinary element with novel bioactive characteristics that improve plant tolerance to abiotic factors (i.e., heavy metal toxicity and oxidative stress) (Cui et al. 2013). Recently, rhizobia are used for the elimination of various types of soil pollutants from the soil environment, ranging from pesticides and aromatic to linear hydrocarbons, chlorinated compounds, phenolic compounds (Jin et al. 2013), and others. Kaiya et al. (2012) observed that rhizobium is an important proliferating member of the degrading microcosm in polluted soil. Nevertheless, the pathways involved in the degradation and the bacterial catabolic enzymes of these compounds are not fully understood. Furthermore, rhizobia are also a prospective controlling tool for hazardous metal bioremediation (Fester et al. 2014). Potential mechanisms that are involved are (Fig. 1.4) (1) bioactive metabolites (Jin et al. 2013); (2) adsorption and accumulation of heavy metals and microbial secretion of enzymes to lessen their toxicity by altering the metals and increasing the bioavailability and complexation of metals; these actions can also directly/indirectly aid phytoremediation (Teng et al. 2015); and (3) volatilization of heavy metals by microbial action and their transformed products facilitate bioremediation, although this process is yet to be reported in rhizobia (Hao et al. 2012). However, abundant production of crop biomass in agroecosystems, microbial symbionts, is important to phytoremediation (Hao et al. 2012).

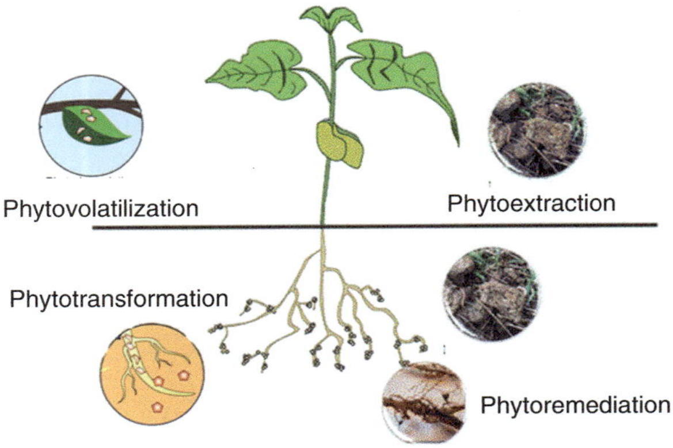

Fig. 1.4 The role of legume rhizobia in biodegradation of soil pollutants

The N-fixation and plant growth-promoting behavior of rhizobia improve the bioavailability of contaminants, crop biomass, the intake, soil fertility, and the ability to degrade organic pollutants and indirectly help photostabilization and translocation of pollution agents from soil to plant (Teng et al. 2015). These qualities could help rhizobia overcome the problems associated with phytoremediation and achieve higher working efficiency (Glick 2010). Thus, the symbiosis between rhizobia and legumes improves the removal rate for soil pollutants (Hao et al. 2012; Meena et al. 2014). In contrast to rhizospheric microorganisms (symbiotic and nonsymbiotic), the stable and balanced endophytic association between rhizobia and host plants provides a sustainable way to improve the soil processes (Li et al. 2013).

1.4.8 Soil Microbial Biomass

The soil is a most complex biological system of the earth (Gans et al. 2005) and may contain a million taxa in a 10 g sample (Young and Crawford 2004). Soil biogenesis is involved in primary processes such as nutrient cycling and soil formation and is the foundation stone of the biosphere. Soil microorganisms have an essential link between plant productivity and soil nutrient availability as they are indirectly/directly engaged in the nutrients cycling through the conversion of inorganic and organic forms of nutrients (Meena et al. 2015c).

Legumes are one of the important components to increase soil microbial biomass (SMB) in soils (Siddique et al. 2008; Deakin and Broughton 2009). Legumes play an important role in SMB and energetic key processes such as nutrient cycling and SOM decomposition and, thus, improve crop productivity and soil sustainability (Knight and Dick 2004; Lal 2012). The instability of SMB ensuring a number of key agroecological processes in soil could destructively alter agricultural productivity and soil sustainability (Devare et al. 2007). The relationship between the soil biota and legumes and their implications on diverse soil functions have positive effects on soil sustainability (Leterme and Carmenza Munoz 2002). The SMB is increased by legume-based rotations, with significant improvement in soil microbial community structure and soil health (Chintala et al. 2013). Some microorganisms which interact physically with leguminous crops in the rhizospheric zone can also improve crop productivity positively by enhancing plant growth and development (Deakin and Broughton 2009). The SMB is analogous to the eye of the needle through which all SOM must pass (Prell and Poole 2006) and, thus, is extensively used as a biological indicator in the evaluation of soil sustainability (Leterme and Carmenza Munoz 2002). The SOM is an instant sink for nutrients, organisms, and C. SMB also increases nutrient intake in crops during symbiotic associations. It contributes to soil physical structure, chemical processes, and pesticide degradation and suppresses soil pathogens (Liang et al. 2010). SMB and the microbial dynamics are pertinent indicators of changes in soil sustainability due to changes in soil properties. The SMB mainly occurs on the surface layer and varies according to soil profiles. SMB is the living component of the soil.

1.4.9 Soil Physical Properties

Soil physical property-related indicators are rapid and low-cost options. Important soil physical properties are bulk density, porosity, aggregate stability, and texture. These properties are also associated with water-related processes including aeration, runoff, erosion, water holding capacity, and infiltration rate (Dexter 2004). Therefore, soil is physically deprived when it has allowed water infiltration, poor cohesion, low aeration, low root density, and enhanced surface runoff and is consequently difficult for farm operations (Schoenholtz et al. 2000). Soil texture is one of the important factors affecting the relationship between gases and water, but it is independent of the soil management and is stable for a long time. Consequently, total porosity and bulk density represent the effects of soil use and management on the air/water relationships in a better manner (Beutler et al. 2002). The soil porosity is classified as textural, depending on the size of soil particles, and structural, depending on macrostructure and bio-pores (Dexter 2004), which may alter the characteristic soil-water retention (pF curve) based on soil particles (Meena et al. 2015c).

Legume crops have a potential to improve physical properties of soil by being a soil conditioner and improving the physical properties (Srinivasarao et al. 2012). Leguminous cover crops have a positive effect on soil physical properties mainly due to the production capacity of huge biomass which provides substrata for soil biological activity and SOM (Lal 2015). Furthermore, leguminous cover crops are grown to protect the soil from loss of plant nutrients and erosion, while green manure crops are grown for the purpose of improving soil physical properties. Leguminous crops are traditionally incorporated into the soil for the green manuring purpose. In effect, the leguminous crop residue management strategies offer advantages often attributed to green manuring. Moreover, some crops can physically modify the types of soil profile (Lithourgidis et al. 2011). Legumes also affect soil structure by their influence on aggregation (Mousavi et al. 2009). Leguminous cover crops can increase or maintain a good soil C/N ratio (Schädler et al. 2011) and increase in maintaining SOC stock (Dexter 2004). Legume crops frequently result in better infiltration of water, due to direct effects of the crop residue in soil formation and aggregation (Mousavi et al. 2009).

1.4.10 Soil Chemical Properties

Soil chemical properties for sustainability are connected with the ability to supply nutrients for crop and retaining/denaturing harmful chemical compounds or elements to the agroecosystem. Soil cation exchange capacity (CEC), pH, nutrient levels, and SOC concentration are the major chemical components used toward the assessment of soil fertility, especially when given the soil capacity for supporting high crop productivity (Schoenholtz et al. 2000; Meena et al. 2015b). Soil chemical properties have been associated with leguminous crops, and thus, the particulars of a soil property are easily interpreted and allow a speedy enhancement of the soil chemical properties by N-fixation and root biomass. These soil properties are also useful in considering the soil's capacity for sustaining agricultural production and productivity, maintaining nutrient cycling through legumes and sustainability, and improving SOM and crop biomass (Kelly et al. 2009a, 2009b). Soil chemical properties are used for assessment of available nutrients for crops and are based on globally well-known benefits of leguminous crops. Among them, legumes have a positive effect on pH, available nutrients, and SOC stock (Mugwe et al. 2004). Legume-based rotation induces changes in the bulk or the pH of the rhizosphere zone of soil under legumes. Root exudation of legumes and exchange or release of organic acids on the epidermal cell of root surfaces may enhance P availability (Bado et al. 2004; Meena et al. 2017b). On the poorly buffered acid sandy soils of the Sahel, such pH effects on P availability are important to small landholders (Buerkert et al. 2000). In addition, changes in pH are widely known to affect the growth and activity of microorganisms (Juma and Averbeke 2005; Lopez and Mundt (2000), which are also important components of diseases and nutrient cycling processes.

Leguminous green manure is a well-known generator of SOM. Green manure, apart from increasing soil N, releases P, maintains and renews the SOC, and improves soil chemical characteristics (Ogunwole et al. 2010). Incorporation of legume residues is beneficial to the soil for increasing SOC concentration which is not only important to agricultural productivity but also to sequestration of C from atmospheric CO_2 (Lal 2015; Turnbull and Bowman 2002). The potential benefits of incorporating leguminous crop into soil cannot be overemphasized (Sharma and Behera 2009). Observed that when leguminous cover crops are used as green manure and incorporated into the soil, their residues increase availability of N, P, K, and trace elements to the succeeding crops due to the lowering of the soil pH brought about by the CO_2 produced in the process of decomposition (Pikul et al. 2008). The green manure biomass either is incorporated from fast-growing legume crops grown in situ or transferred from another place for incorporation into the soil. The quantity of available nutrients provided by legume biomass is determined by the decomposition rate and nutrient concentrations depending on climatic situations, plant part, soil type, management regimes, and plant density (Adeboye et al. 2005). The rate of biological decomposition and consequent release of nutrient varies with ambient conditions such as temperature, moisture, aeration, pH, and other factors affecting nutrient availability (Liang et al. 2010). Incorporation of legumes before flowering makes available young legume biomass for quick decomposition. In general, the material of young plants decomposes faster than the mature plants and supplies N, P, and K into soils (Bado et al. 2004). Decomposition and mineralization of nutrients in the incorporated legume biomass influence the chemical properties with positive effects on nutrient status and fertility of the soil (Melero et al. 2007).

1.4.11 Soil Carbon Stock

Sequestration of SOC is one of the important determinants of soil fertility, productivity, and quality (Carter 2002; Yadav et al. 2017b). Crop residues increase C sequestration through decomposition of their residues, as presented in Fig. 1.5.

Increase in SOC stock improves soil tilth and workability, stabilizes soil aggregates, increases soil water holding (SWH) and aeration, enhances buffering capacities, and improves availability of nutrients through breakdown of residues (Lal 2015). The SOC stock depends on soil types (Wilhelm et al. 2004), crop and residue management (Webb et al. 2003), fertilizer N input (Giller 2001), and frequency and type of cropping system (Schimel et al. 1994). Structural degradation of soil is a serious problem in intensive agroecosystems due to the depletion of SOC stock (Mann et al. 2002). Kureh and Kamara (2005) observed that a decrease in SOC content paralleled the decrease in stability and that soil aggregate stability decreases in agricultural land compared with grassland. In the agricultural fields, legume crops contribute positively to the SOC stock, soil tilth, soil fertility, and overall soil sustainability (Lal

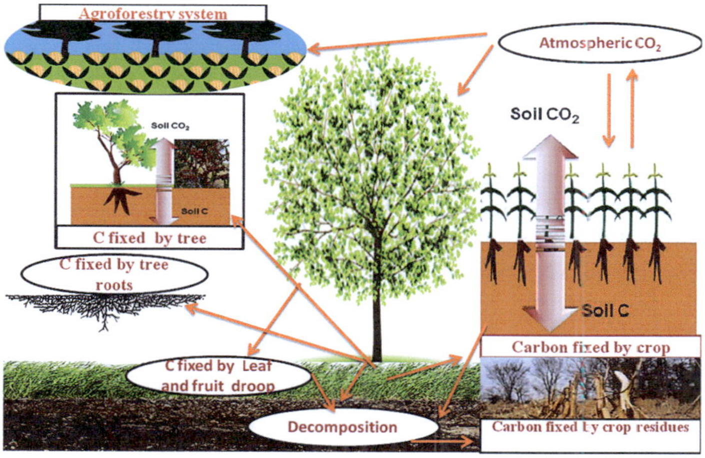

Fig. 1.5 Mechanisms of soil carbon sequestration

2015). The continuous crop production potentials of soils are directly related to their SOC content (Grandy et al. 2002; Lal 2006). In general, crop performance is positively related to the SOM content (Franzluebbers 2002; Lado and Ben-Hui 2004). The SOM influenced physical characteristics of the soil, and those are associated with soil structure and soil aggregate stability (Carter 2002). SOM compounds bind the primary soil particles in the aggregate, chemically and physically, and this, in turn, increases the aggregates stability and limits their breakdown during the wetting process (Doran 2000). Moreover, aggregate stability has tremendous effect on soil water holding capacity (SWHC), water infiltration, and aeration as well as penetration resistance and bulk density (Wilhelm et al. 2004). Legume-based cropping systems improve aggregate stability and extend the nutrient residence time in soil by reducing the mineralization rate. Biomass production can be increased by legume-based biculture, a mixture of legume with nonlegume species. Carbon sequestration varies among different leguminous cover crops depending on the total biomass production, decomposition rates, and conversion of liable C to soil recalcitrant C (McLauchlin and Hobbie 2004; Dhakal et al. 2015). Evaluation of these variations in field conditions is rather difficult due to high heterogeneous characteristics of the soil (Lal 2004a) and the difficulty in sampling to represent soil variability. Accumulation of SOC is a long-term slow biotic process influenced by abiotic factors (Jarenyama et al. 2000) and is often non-detectable within 1 or 2 years even with a considerable quantity of biomass input to the soil (Lal 2004b).

1.4.12 Soil N Pool

N is critical for the crop growth, second only to water and light. However, most crops depend on the intake of soil N to meet their needs; most notably the legumes, certain clades, are capable of N-fixation with the symbiotic relationship with rhizobia. The BNF benefits not only the legumes but also improves yield in succeeding crops (Carranca 2013), in agroforestry systems (Nygren and Leblanc 2015), and in grain-legume-cereal intercropping system (Chapagain and Riseman 2014, 2015). The N quantity made available to cereal crop derives from the breakdown of legume-biomass residues; furthermore, research results from mixed-cropping practices suggest that the crop may acquire N directly from companion crops through interplant N transfer (Yong et al. 2015; Varma et al. 2017) or the movement of N from "N-donor" to "N-receiver"(Moyer-Henry et al. 2006). However, the association of N tends to move from crop containing relatively high N (i.e., legumes) to those with a greater N demand (nonlegume) (Carranca 2013). N transfer is a highly variable mechanism and can provide anywhere from 10% to 85% of a receiver crop's N demand (Paynel et al. 2008). By adopting sustainable agricultural practices, N requirement can be met via N-fixation instead of from synthetic fertilizers (He et al. 2009; Chalk et al. 2014). Nonetheless, in order to take advantage of the biologically fixed N in the ecosystem, a higher understanding is needed of the factors that influence the effectiveness of crop-to-crop N transfer. An approach to boost N supply in cropping systems is the inclusion of N-fixing leguminous crops, which can provide N advantages to the second crops through N transfer. Nevertheless, a good understanding of the factors and mechanisms that manage N transfer is significant in the array to determine possible areas to improve this association. While chemical N fertilizers are expensive or have an environmental cost, cropping systems depend on the N-fixed by legumes and are more sustainable. Problems of high N fertilizer cost and availability are common in developing countries. The quantity of biologically fixed N/year by legumes varies significantly from zero to several hundred kg N/ha (Fustec et al. 2010; Carlsson and Huss-Danell 2014).

1.4.13 Crop Succeeding Effects

The selection of right legume for cultivation in the rotation decides the amount of N being sequestered into the soil, as presented in Fig. 1.6 (Briggs et al. 2005).

Crop rotation can improve biomass production and eventually the soil C and N sequestration, principally the rotations of legumes with nonlegumes, and C and N can be enhanced with succeeding effects (Table 1.3). This can be done by the use of appropriate crop rotations (Lal 2010; Ram and Meena 2014). The legume-based intercropping may increase N stock into the soil, and this total N may not be converted into available form during the current growing season, improving soil fertility to benefit a succeeding crop (Akinnifesi et al. 2007; Lithourgidis et al. 2011). Lopez and Mundt (2000) observed that Vogel's tephrosia (*Tephrosia vogelii*), velvet bean (*Mucuna pruriens*), and sunn hemp (*Crotalaria juncea*) often resulted in maize yields of 4–7 Mg/ha even with no additional N fertilizer application in the succeeding crop. Yusuf et al. (2009) reported that to maximize contribution of legume N to a succeeding crop, it is essential to maximize the total N amount in the leguminous crop, the amount of legume N mineralized, the efficiency of utilization of this mineral N, and the quantity of N derived from N-fixation.

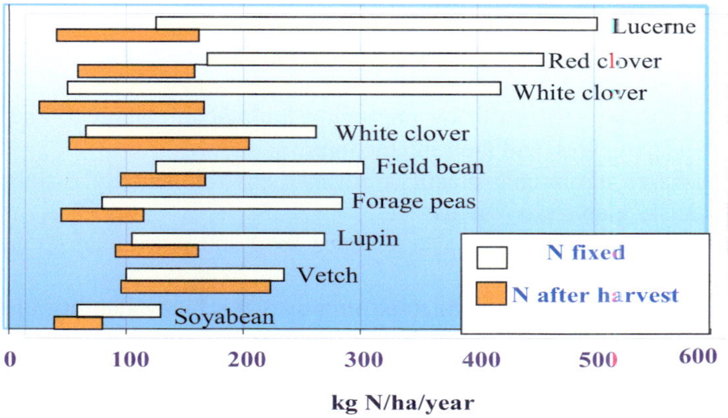

Fig. 1.6 Ranges of the quantities of N-fixed and that remaining after harvest (Data Source: Briggs et al. 2005)

Table 1.3 Effects of legumes on the maize yield in succeeding planting

Maize yield increases in pulses rotation as succeeding planting	References
48% higher yield soybean planting with recommended dose of fertilizer	Yusuf et al. (2009)
30% in 1-year rotation with soybean	Akinnifesi et al. (2007)
80% higher after 2-year cropping of soybean	Akinnifesi et al. (2007)
30% more yield recorded in cowpea rotation	Lithourgidis et al. (2011)
50% when rotated with *Sesbania* spp.	Kureh and Kamara (2005)
280% in the *Gliricidia* (*Gliricidia sepium*)-maize system in a long-term experiment	Kureh and Kamara (2005)

1.5 Future Outlook of Legumes

The review of literature presented herein ascertains that legume-based cropping holds a vast potential to advance soil sustainability. The multitude gaps of knowledge related to the advantages of legumes and their long-term positive effects on soil functions and behavior in various types of soils as well as understanding the management practices are in increasing demand. The capability of legumes to improve soil properties (e.g., physical, chemical, and biological) makes them integral to achieving the sustainability goals. Given the massive prospects of legumes as a soil amendment, the uncertainties outlined above need to be addressed objectively. An urgent need is to understand the future demand and the role of legumes in soil sustainability and food and nutritional security.

1.6 Conclusion

The synthesis presented supports the following conclusions:

1. Inclusion of legumes in crop rotations as an important strategy toward advancing soil sustainability.
2. Legumes restore soil health by improving biological, chemical, and physical properties, processes, and their interactions.
3. Legume-based agronomic practices advance environmentally sustainable and economically viable crop yields.
4. Improvements in soil health also advance food and nutritional security while improving the environment.
5. Bioremediation behavior of rhizobia improves the bioavailability of contaminants and the ability to degrade organic pollutants and thereby indirectly helps photostabilization and translocation of pollution agents from soil to plant.
6. Long-term and coordinated research efforts are needed to assess the effects of legume on different soils under diverse and changing climatic conditions on oil health, agronomic productivity, and environment quality.

References

Abera T, Feyissa D, Yusuf H (2005) Effects of inorganic and organic fertilizers on grain yield of maize-climbing bean intercropping and soil fertility in western Oromiya, Ethiopia. Tropentag 2005 Stuttgart-Hohenheim, October 11–13, 2005. Conference on international Agricultural Research for Development

Addo-Quaye AA, Darkwa AA, Ocloo GK (2011) Yield and productivity of component crops in a maize-soybean intercropping system as affected by time of planting and spatial arrangement. J Agric Biol Sci 6(9):50–57

Adeboye MKA, Iwuafor ENO, Agbenin JO (2005) Rotation effects of grain and herbaceous legumes on maize yield and chemical properties of an Alfisol in the northern Guinea Savanna, Nigeria. Niger J Soil Res 6:22–31

Adeniyan ON, Akande SR, Balogun MO, Saka JO (2007) Evaluation of crop yield of African yam bean, maize and kenaf under intercropping systems. Am Eurasian J Agric Environ Sci 2(1):99–102

Akinnifesi FK, Makumba W, Sileshi G, Ajayi OC, Mweta D (2007) Synergistic effect of inorganic N and P fertilizers and organic inputs from *Gliricidia sepium* on productivity of intercropped maize in Southern Malawi. Plant Soil 294:203–217

Alpmann D, Braun J, Schäfer BC (2013) AnalyseeinerBefragungunterer folgreichenKörnerleguminosenanbauernimkonventionellenLandbau. ErsteErg ebnisseausdemForschungsprojektLeguAN. In: DLG Wintertagung, ImFokus: HeimischeKörnerleguminosenvomAnbaubiszurNutzung. Berlin, p 20

Ashoka P, Meena RS, Kumar S, Yadav GS, Layek J (2017) Green nanotechnology is a key for eco-friendly agriculture. J Clean Prod 142:4440–4441

Bado BV, Sedogo MP, Lompo F (2004) Long term effects of mineral fertilizers, phosphate rock, dolomite and manure on the characteristics of an ultisol and maize yield in Burkina Faso. In: Bationo A (ed) Managing nutrient cycles to sustain soil fertility in sub-Saharan Africa. Academy Science Publishers, Nairobi, p 608

Barbosae S, Lima PI, Silva PS, de Oliveira OF, de Sousa RP (2008) Planting times of cowpea intercropped with corn in the weed control. Revista Caatinga (Mossoró, Brasil) 21(1):113–119

Briggs, DG (2005) Assessing and managing stands to meet quality objectives. In: Harrington CA. Schoenholz SH (ed) Productivity of western forests: a forest products focus. USDA forest service Pacific northwest research station, Portland, OR, General Technical Report PNW-GTR-642, pp 141–152

Bardos P (2003) A review of the contaminated land rehabilitation network for environmental technologies in Europe (CLARINET). Part 2: working group findings. Land Contamination Reclamation 11(1):15–30

Bardos P, Bone B, Boyle R, Ellis D, Evans F, Harries ND, Smith JWN (2011) Applying sustainable development principles to contaminated land management using the SuRF-UK framework. Remediation 21(2):1–138

Berihun T, Molla E (2017) Study on the diversity and use of wild edible plants in Bullen District Northwest Ethiopia. Aust J Bot 2017:1–10. https://doi.org/10.1155/2017/8383468

Beutler AN, Centurion JF, Souza ZM, Andrioli I, Roque CG (2002) Water retention in two oxisols under different uses. Revista Brasileira de Ciênciado Solo 26:829–834

Binder CR, Wiek A (2007) The role of transdisciplinary processes in sustainability assessment of agricultural systems. In: Hani J, Pinte'r L, Herren HR (eds) From common principles to common practice. Proceedings and outputs of the first symposium of the international forum on assessing sustainability in agriculture (INFASA). International Institute of Sustainable Development and Swiss College of Agriculture, Bern, pp 33–48

Binder CR, Feola G, Steinberger JK (2010) Considering the normative, systemic and procedural dimensions in indicator based sustainability assessments in agriculture. Environ Impact Assess Rev 30(2):71–81

Blaser BC, Gibson LR, Singer JW, Jannink JL (2006) Optimizing seeding rates for winter cereal grains and frost-seeded red clover intercrops. Agron J 98:1041–1049

Brintha I, Seran TH (2009) Effect of paired row planting of radish (*Raphanus sativus* L.) intercropped with vegetable amaranths (*Amaranths tricolor* L.) on yield components of radish in sandy regosol. J Agric Sci 4:19–28

Buerkert A, Bationo A, Dossa K (2000) Mechanisms of residue mulch induced cereal growth increases in West Africa. Soil Sci Soc Am J 64:346–358

Bues A, Preissel S, Reckling M, Zander P, Kuhlman T, Topp K, Watson CA, Lindström K, Stoddard FL, Murphy-Bokern D (2013) The environmental role of protein crops in the new Common Agricultural Policy. European Parliament, Directorate General for Internal Policies, Policy Department B: Structural and Cohesion Policies, Agricultural and Rural Development IP/B/AGRI/IC/2012–067, 112 pp.

Buragohain S, Sharma B, Nath JD, Gogaoi N, Meena RS, Lal R (2017) Impact of ten years of biofertilizer use on soil quality and rice yield on an inceptisol in Assam. India Soil Res. https://doi.org/10.1071/SR17001

Carlsson G, Huss-Danell K (2003) Nitrogen fixation in perennial forage legumes in the field. Plant Soil 253:353–372

Carlsson G, Huss-Danell K (2014) Does nitrogen transfer between plants confound 15N-based quantifications of N2 fixation? Plant Soil 374:345–358

Carranca C (2013) Legumes: properties and symbiosis. In: Camisão AH, Pedroso CC (eds) Symbiosis: evolution, biology and ecological effects. Animal science, issues and professions. Nova Science Publishers, New York. ISBN 978-1-62257-211-3

Carter MR (2002) Soil quality for sustainable land management: organic matter and aggregation interactions that maintain soil function. Agron J 94:38–47

Chalk PM, Peoples MB, Mcneill AM et al (2014) Methodologies for estimating nitrogen transfer between legumes and companion species in agro-ecosystems: a review of 15N-enriched techniques. Soil Biol Biochem 73:10–21

Chalka MK, Nepalia V (2006) Nutrient uptake appraisal of maize intercropped with legumes and associated weeds under the influence of weed control. Indian J Agric Res 40:86–91

Chapagain T, Riseman A (2014) Barley–pea intercropping: effects on land productivity, carbon and nitrogen transformations. Field Crop Res 166:18–25. https://doi.org/10.1016/j.fcr.2014.06.014

Chapagain T, Riseman A (2015) Nitrogen and carbon transformations, water use efficiency and ecosystem productivity in monocultures and wheat-bean intercropping systems. Nutr Cycl Agroecosyst 101:107–121. https://doi.org/10.1007/s10705-014-9647-4

Chen G, Zhu H, Zhang Y (2003) Soil activities and carbon and nitrogen fixation. Res Microbiol 154:393–398

Chintala R, Mollinedo J, Schumacher TE, Malo DD, Julson JL (2013) Effect of biochars on chemical properties of acidic soil. Arch Agron Soil Sci 60:393–404

Chu GX, Shen QR, Cao JL (2004) Nitrogen fixation and N transfer from peanut to rice cultivated in aerobic soil in intercropping system and its effect on soil N-fertility. Plant and Soil 263:17–27

Corre-Hellou G, Fustec J, Crozat Y (2006) Interspecific competition for soil N and its interaction with N2 fixation, leaf expansion and crop growth in pea-barley intercrops. Plant Soil 282:195–208

Corre-Hellou G, Brisson N, Launay M, Fustec J, Crozat Y (2007) Effect of root depth penetration on soil N sharing and dry matter in pea-barley intercrops given different soil N supplies. Field Crop Res 103:76–85

Crépon K (2006) Protein supply in Europe and the challenge to increase grain legumes production: a contribution to sustainable agriculture. In: Grain legumes and the environment: how to assess benefits and impacts. Proceedings of the AEP workshop, (ed.) AEP, 18–19 November 2004, Zürich, Switzerland, pp 13–16

Crozat Y, Fustec J (2006) Assessing the role of grain legumes in crop rotation: some agronomic concepts that can help. In: Grain legumes and the environment: how to assess benefits and impacts. Proceedings of the AEP workshop, (ed.) AEP, 18–19 November 2004, Zürich, Switzerland, pp 55–60

Cui WT, Gao CY, Fang P, Lin GQ, Shen WB (2013) Alleviation of cadmium toxicity in *Medicago sativa* by hydrogen-rich water. J Hazard Mater 260:715–724

Dadhich RK, Meena RS, Reager ML, Kansotia BC (2015) Response of bio-regulators to yield and quality of Indian mustard (*Brassica juncea* L. Czernj. and Cosson) under different irrigation environments. J App and Nat Sci 7(1):52–57

Dahmardeh M, Ghanbari A, Syahsar BA, Ramrodi M (2010) The role of intercropping maize (*Zea mays* L.) and cowpea (*Vigna unguiculata* L.) on yield and soil chemical properties. Afr J Agric Res 5(8):631–636

Deakin WJ, Broughton WJ (2009) Symbiotic use of pathogenic strategies: Rhizobial protein secretion systems. Appl Soil Ecol 7:312–320

Derpsch R (2015) The extent of conservation agriculture adoption worldwide: implications and impact. In: Proceedings of the Third World Congress on Conservation Agriculture: Linking Production, Livelihoods and Conservation, Nairobi, Kenya, 3–7 October 2015

Deutsch B, Kahle P, Voss M (2006) Assessing the source of nitrate pollution in water using stable N and O isotopes. Agron Sustain Dev 26:263–267

Devare M, Londono RLM, Thies JE (2007) Neither transgenic Bt maize (MON863) nor tefluthrin insecticide adversely affect soil microbial activity or biomass: a 3-year field analysis. Soil Biol Biochem 39:2038–2047

Dexter AR (2004) Soil physical quality part I. Theory, effects of soil texture, density, and organic matter, and effects on root growth. Geoderma 120:201–214

Dhakal Y, Meena RS, De N, Verma SK, Singh A (2015) Growth, yield and nutrient content of mung bean (*Vigna radiata* L.) in response to INM in eastern Uttar Pradesh, India. Bangladesh J Bot 44(3):479–482

Dhakal Y, Meena RS, Kumar S (2016) Effect of INM on nodulation, yield, quality and available nutrient status in soil after harvest of green gram. Legum Res 39(4):590–594

Doran JW (2000) Soil health and global sustainability: translating science into practice. Agric Ecosyst Environ 88:119–127

Duru M, Therond O, Martin G, Martin-Clouaire R, Magne MA, Justes E, Journet EP, Aubertot JN, Savary S, Bergez JE, Sarthou JP (2015) How to implement biodiversity-based agriculture to enhance ecosystem services: a review. Agron Sustain Dev 35:1259–1281

Dusa EM (2009) Researches regarding the productivity of oat-lentil intercropping in the organic agriculture system. Res J Agric Sci 41(1):22–26

FAO (2016) Hunger. In: Poverty and climate change: the challenges today and tomorrow

Fester T, Giebler J, Wick LY, Schlosser D, Kästner M (2014) Plant-microbe interactions as drivers of ecosystem functions relevant for the biodegradation of organic contaminants. Curr Opin Biotechnol 27:168–175

Foley JA, Ramankutty N, Brauman KA, Cassidy ES, Gerber JS, Johnston M (2011) Solutions for a cultivated planet. Nature 478:337–342

Franzluebbers AJ (2002) Water infiltration and soil structure related to organic matter and its stratification with depth. Soil Tillage Res 66:197–205

Fustec J, Lesuffleur F, Mahieu S, Cliquet JB (2010) Nitrogen rhizodeposition of legumes: a review. Agron Sustain Dev 30:57–66

Gans J, Wolinsky M, Dunbar J (2005) Computational improvements reveal great bacterial diversity and high metal toxicity in soil. Science 309:1387–1390

Garg N, Geetanjali (2007) Symbiotic nitrogen fixation in legume nodules: process and signaling. A review. Agron Sustain Dev 27:59–68

Ghaley BB, Hauggaard-Nielsen H, Hogh-Jensen H, Jensen ES (2005) Intercropping of wheat and pea as influenced by nitrogen fertilization. Nutr Cycling Agroecosyst 73:201–212

Ghanbari A, Dahmardeh M, Siahsar BA, Ramroudi M (2010) Effect of maize (L.) – cowpea (*Vigna unguiculata* L.) intercropping on light distribution, soil temperature and soil moisture in arid environment. J. Food Agric Environ 8:102–108

Ghosh PK, Manna MC, Dayal D, Wanjari RH (2006) Carbon sequestration potential and sustainable yield index for groundnut- and fallow-based cropping systems. J Agric Sci 144:249–259

Giller KE (2001) Nitrogen fixation in tropical cropping systems, 2nd edn. CABI, Wallingford, 423 pp
Giller KE, Witter E, Corbeels M, Tittonell P (2009) Conservation agriculture and smallholder farming in Africa: the heretics' view. Field Crop Res 114:23–34
Glick BR (2010) Using soil bacteria to facilitate phytoremediation. Biotechnol Adv 28:367–374
Graham PH, Vance-Carroll P (2015) Legumes: importance and constraints to greater use. Plant Physiol 131:872–877
Grandy AS, Porter GA, Erich MS (2002) Organic amendment and rotation crop effects on the recovery of soil organic matter and aggregation in potato cropping systems. Soil Sci Soc Am J 66:1311–1319
Hani F (2007) Globla agriculture in need of sustainability assessment. In: Hani J, Pinte'r L, Herren HR (eds) From common principles to common practice, proceedings and outputs of the first symposium of the international forum on assessing sustainability in agriculture (INFASA). International Institute of Sustainable Development and Swiss College of Agriculture, Bern, pp 3–17
Hao XL, Lin YB, Johnstone L, Baltrus DA, Miller SJ, Wei GW et al (2012) Draft genome sequence of plant growth-promoting rhizobium *Mesorhizobium amorphae*, isolated from zinc-lead mine tailings. J Bacteriol 194:736–737
Haslmayr HP, Geitner C, Sutor G, Knoll A, Baumgarten A (2016) Soil function evaluation in Austria—development, concepts and examples. Geoderma 264:379–387
Hati KM, Swarup A, Singh D, Misra AK, Ghosh PK (2006) Long-term continuous cropping, fertilization and manuring effects on physical properties and organic carbon content of a sandy loam soil. Aust J Soil Res 44:487–495
Hauggaard-Nielsen H, Jensen ES (2005) Facilitative root interactions in intercrops. Plant Soil 274:237–250
Hauggaard-Nielsen H, Jornsgaard B, Kinane J, Jensen ES (2007) Grain legume–cereal intercropping: the practical application of diversity, competition and facilitation in arable and organic cropping systems. Renew Agric Food Syst 23:3–12
Hauggaard-Nielsen H, Gooding M, Ambus P, Corre-Hellou G, Crozat Y, Dahlmann C, Dibet A, von Fragstein P, Pristeri A, Monti M, Jensen ES (2009) Pea–barley intercropping for efficient symbiotic N_2–fixation, soil N acquisition and use of other nutrients in European organic cropping systems. Field Crop Res 113:64–71
Hayati D, Ranjbar Z, Karami E (2010) Measuring agricultural sustainability. In: E. Lichtfouse (ed.), Biodiversity, biofuels, agroforestry and conservation agriculture, 73 Sustainable Agriculture Reviews 5, pp 73–100
He X, Xu M, Qiu GY, Zhou J (2009) Use of 15N stable isotope to quantify nitrogen transfer between mycorrhizal plants. J Plant Ecol 2:107–118
Herawati N, Suzuki S, Hayashi K, Rivai IF, Koyama H (2000) Cadmium, copper, and zinc levels in rice and soil of Japan, Indonesia, and China by soil type. Bull Environ ContamToxicol 64:33–39
Høgh-Jensen H (2006) The nitrogen transfer between plants: an important but difficult flux to quantify. Plant Soil 282:1–5
Horrigan L, Lawrence RS, Walker P (2002) How sustainable agriculture can address the environmental and human health harms of industrial agriculture. Environ Health Perspect 110:445–456
IAASTD (2008) Agriculture at a crossroads: global report/International Assessment of Agricultural Knowledge, Science and Technology for Development (IAASTD). Press, Island
ILDIS (2017) International legume database and information service, World Database of Legumes. http://www.ildis.org/ dated 02/01/2017
Inal A, Gunes A, Zhang F, Cacmak I (2007) Peanut/maize inter-cropping induced changes in rhizosphere and nutrient concentrations in shoots. Plant Physiol Biochem 45:350–356
IUSS (2016) Bulletin of the International Union of Soil Sciences (IUSS). http://www.iuss.org/files/iuss-bulletin129_72dpi.pdf
Jarenyama P, Hesterman OB, Waddington SR, Harwood RR (2000) Relay-intercropping of sunn hemp and cowpea into a smallholder maize system in Zimbabwe. Agron J 92:239–244

Jensen ES (2006) Grain legume functions in crop rotations. In: Grain legumes and the environment: how to assess benefits and impacts. Proceedings of the AEP workshop, (Ed.) AEP, 18–19 November 2004, Zürich, Switzerland, pp 49–54
Jin QJ, Zhu KK, Cui WT, Xie YJ, Han B, Shen WB (2013) Hydrogen gas acts as a novel bioactive molecule in enhancing plant tolerance to paraquat-induced oxidative stress via the modulation of heme oxygenase-1 signalling system. Plant Cell Environ 36:956–969
Joshi PK, ParthasarathyRao P (2016) Global pulses scenario: status and outlook. Ann N Y Acad Sci:1–12
Juma KA, Van Averbeke (2005) Response of Muxe to N and P availability in pots. In: Tenywa JS, Adipala E, Nampala P, Tussiime G, Okori P, Kyanmuhangire W (eds). African Crop Science. Proceedings, vol. 7, Kampala, Uganda, 5–9 Dec 2005 pp 1179–1182
Kaiya S, Utsunomiya S, Suzuki S, Yoshida N, Futamata H, Yamada T et al (2012) Isolation and functional gene analyses of aromatic hydrocarbon-degrading bacteria from a polychlorinated-dioxin-dechlorinating process. Microbes Environ 27:127–135
Keeler BL, Hobbie SE, Kellogg LE (2009) Effects of long-term nitrogen addition on microbial enzyme activity in eight forested and grassland sites: implications for litter and soil organic matter decomposition. Ecosystems 12:1–15
Kelly A, Redmond L, King F (2009a) GHG MACC brief: a provisional collation of some GHG MACC curves in circulation. Dublin, AP EnvEcon Ltd
Kelly B, Allan C, Wilson BP (2009b) Soil indicators and their use by farmers in the Billabong Catchment, southern New South Wales. Aust J Soil Res 47:234–242
Khan DF, Herridge DF, Peoples MB, Shah SH, Khan T, Madani MS, Ibrar M (2007) Use of isotopic and non-isotopic techniques to quantify below-ground nitrogen in faba bean and chickpea. Soil Environ 26:42–47
Khan MS, Zaidi A, Wani PA, Oves M (2009) Role of plant growth promoting rhizobacteria in the remediation of metal contaminated soils. Environ Chem Lett 7:1–19
Klaassen CD, Watkins JB (2003) Casarett and Doull's essentials of toxicology. McGraw-Hill, New York
Knight TR, Dick RP (2004) Differentiating microbial and stabilized β-glucosidase activity relative to soil quality. Soil Biol Biochem 36:2089–2096
Kumar K, Goh KM (2000) Crop residues and management practices: effects on soil quality, soil nitrogen dynamics, crop yield and nitrogen recovery. Adv Agron 68:198–279
Kumar S, Sheoran S, Kumar SK, Kumar P, Meena RS (2016) Drought: a challenge for Indian farmers in context to climate change and variability. Progress Res Int J 11:6243–6246
Kureh I, Kamara AY (2005) Effects of sole cropping, intercropping and rotation with legume trap-crops on Striga control and maize grain yield in farmers' fields in the Northern Guinea Savanna. In: Badu-Apraku B, Fakorede MAB, Lum, AF, Menkir A, Ouedraogo, M (eds), Demand-driven technologies for sustainable maize production in West and Central Africa. Fifth Biennial West and Central Africa Regional Maize Workshop, 3–6 May 2005, IITABenin. pp 169–179
Lado M, Ben-Hui M (2004) Organic matter and aggregate size interactions in infiltration, seal formation and soil loss. Soil Sci Soc Am J 68:935–942
Lal R (2004a) Carbon sequestration in soils of Central Asia. Land Degr Dev 15:563–572
Lal R (2004b) Soil carbon sequestration to mitigate climate change. Geoderma 123:1–22
Lal R (2006) Enhancing crop yields in developing countries through restoration of soil organic carbon pool in agricultural lands. Land Degrad Dev 17:197–206
Lal R (2010) Soil quality and ethics: the human dimension. In: Lal R Stewart BA (eds) Food security and soil quality, Advances in soil science. Taylor & Francis (CRC Press), Boca Raton, pp 301–308
Lal R (2011) Sequestering carbon in soils of agro-ecosystems. Food Policy 36 S33–S39
Lal R (2012) Climate change and soil degradation mitigation by sustainable management of soils and other natural resources. Agric Res 1:199–212
Lal R (2013) Intensive agriculture and the soil carbon pool. J Crop Improve 27:735–751. https://doi.org/10.1080/15427528.2013.845053
Lal R (2014) Societal value of soil carbon. J Soil Water Conserv 69:186A–192A

Lal R (2015) Restoring soil quality to mitigate soil degradation. Sustainability 7:5875–5895. https://doi.org/10.3390/su7055875
Lal R (2016) Soil health and carbon management. Food and Energy Secur 5(4):212–222. https://doi.org/10.1002/fes3.96
Leterme P, Carmenza Munoz L (2002) Factors influencing pulse consumption in Latin America. Br J Nutr 88(Suppl 3):251–254
Li Y, Liang F, Zhu YF, Wang FP (2013) Phytoremediation of a PCB contaminated soil by alfalfa and tall fescue single and mixed plants cultivation. J Soils Sediments 13:925–931
Liang B, Lehmann J, Sohi SP, Thies JE, O'Neill B, Trujillo L, Gaunt J, Solomon D, Grossman J, Neves EG, FJ L˜a (2010) Black carbon affects the cycling of non-black carbon in soil. Org Geochem 41:206–213
Lithourgidis AS, Dordas CA, Damalas CA, Vlachostergios DN (2011) Annual intercrops: an alternative pathway for sustainable agriculture. Austr J Crop Sci 5:396–410
Liu XM, Wu JJ, Xu JM (2006) Characterizing the risk assessment of heavy metals and sampling uncertainty analysis in paddy field by geostatistics and GIS. Environ Pollut 141:257–264
Lopez CG, Mundt CC (2000) Using mixing ability analysis from two way cultivar mixtures to predict the performance of cultivars in complex mixtures. Field Crop Res 68(2):121–132
Loreau M, Hector A (2001) Partitioning selection and complementarity in biodiversity experiments. Nature 412:72–76
Mahieu S, Fustec J, Faure ML, Corre-Hellou G, Crozat Y (2007) Comparison of two 15N labelling methods for assessing nitrogen rhizodeposition of pea. Plant Soil 295:193–205
Mandal BK, Chowdhury TR, Samanta G, Basu GK, Chowdhury PP, Chanda CR, Lodh D, Karan NK, Dhar RK, Tamili DK, Das D, Saha KC, Chakraborti D (1996) Arsenic in groundwater in seven districts in West Bengal, India – the biggest calamity in the world. Curr Sci 70:976–986
Mann L, Tolbert V, Cushman J (2002) Potential environmental effects on inter row runoff and infiltration. Soil Sci Soc Am J 89:149–166
Mazvimavi K, Twomlow S, Belder P, Hove L (2008) An assessment of the sustainable uptake of conservation farming in Zimbabwe. Global theme on agroecosystems report no. 19. International Crops Research Institute for the Semi-Arid Tropics, Bulawayo, Zimbabwe, pp 51
McLauchlin KK, Hobbie SE (2004) Comparison of labile soil organic matter fractionation techniques. Soil Sci Soc Am J 68:1616–1625
Medeot DB, Paulucci NS, Albornoz AI, Fumero MV, Bueno MA, Garcia MB, Woelke MR, Okon Y, Dardanelli MS (2010) Plant growth promoting Rhizobacteria improving the legume–rhizobia Symbiosis. In: Khan MS, Musarrat J, Zaidi A (eds) Microbes for legume improvement. Springer, Vienna, pp 473–494
Meena VS, Maurya BR, Meena RS, Meena SK, Singh NP, Malik VK (2014) Microbial dynamics as influenced by concentrate manure and inorganic fertilizer in alluvium soil of Varanasi, India. African J Microb Res 8(1):257–263
Meena RS, Meena VS, Meena SK, Verma JP (2015a) Towards the plant stress mitigate the agricultural productivity: a book review. J Clean Prod 102:552–553
Meena RS, Yadav RS, Meena H, Kumar S, Meena YK, Singh A (2015b) Towards the current need to enhance legume productivity and soil sustainability worldwide: a book review. J Clean Prod 104:513–515
Meena RS, Meena VS, Meena SK, Verma JP (2015c) The needs of healthy soils for a healthy world. J Clean Prod 102:560–561
Meena RS, Bohra JS, Singh SP, Meena V, Verma JP, Verma SK, Sihag SK (2016) Towards the prime response of manure to enhance nutrient use efficiency and soil sustainability a current need: a book review. J Clean Prod 112:1258–1260
Meena RS, Gogai N, Kumar S (2017a) Alarming issues on agricultural crop production and environmental stresses. J Clea Prod 142:3357–3359
Meena RS, Meena PD, Yadav GS, Yadav SS (2017b) Phosphate solubilizing microorganisms, principles and application of MicrophosTechnology. J Clean Prod 145:157–158

Melero S, Madejon E, Ruiz JC, Herencia JF (2007) Chemical and biochemical properties of a clay soil under dryland agriculture system as affected by organic fertilization. Eur J Agron 26:327–334
Mousavi S, Yousefi-Moghadam S, Mostafazadeh-Fard B, Hemmat A, Yazdani MR (2009) Effect of puddling intensity on physical properties of a silty clay soil under laboratory and field conditions. Paddy Water Environ 7(1):45–54
Moyer-Henry KA, Burton JW, Israel DW, Rufty TW (2006) Nitrogen transfer between plants: a 15N natural abundance study with crop and weed species. Plant Soil 282:7–20
Mugwe J, Mugendi D, Okoba B, Tuwei P, O'Neill M (2004) Soil conservation and fertility improvement using leguminous Shrubs in central highlands of Kenya: NARFP Case study. In: Bationo A (ed) Managing nutrient cycles to sustain soil fertility in sub-Saharan Africa. Academy Science Publishers, Nairobi, p 608
Mundt (2002) Use of multiline cultivars and cultivar mixtures for disease management. Annu Rev Phytopathol 40:381–410. https://doi.org/10.1146/annurev.phyto.40.011402.113723
Muoneke CO, Ogwuche MAO, Kalu BA (2007) Effect of maize planting density on the performance of maize/soybean intercropping system in a guinea savannah agroecosystem. Afr J Agric Res 2(12):667–677
Nees B, Anderberg S, Olsson L (2010) Structuring problems in sustainability science: the multi-level DPSIR framework. Geoforum 41(3):479–488
Nygren P, Leblanc HA (2015) Dinitrogen fixation by legume shade trees and direct transfer of fixed N to associated cacao in a tropical agroforestry system. Tree Physiol 00:1–14
Obbard JP, Sauerbeck D, Jones KC (1994) Dehydrogenase activity of the microbial biomass in soils from a field experiment amended with heavy metal contaminated sewage sludges. Sci Total Environ 142:157–162
Ogunwole JO, Iwuafor ENO, Eche NM, Diels J (2010) Effect of organic and inorganic soil amendments on soil physical and chemical properties in a West Africa Savanna agroecosystem. Trop Subtrop Agroecosyst 12:247–255
Padilla FM, Pugnaire FI (2006) The role of nurse plants in restoration of degraded environments. Front Ecol Environ 4:196–202
Paynel F, Lesuffleur F, Bigot J et al (2008) A study of 15N transfer between legumes and grasses. Agron Sustain Dev 28:281–290
Peyraud JL, Le Gall A, Lüscher A (2009) Potential food production from forage legume-based systems in Europe: an overview. Ir J Agric Food Res 48:115–135
Pikul JL, Johnson JMF Jr, Schumacher TE, Vigil M, Riedell WE (2008) Change in surface soil carbon under rotated corn in eastern South Dakota. Soil Sci Soc Am J 72:1738–1744
Prell J, Poole P (2006) Metabolic changes of rhizobia in legume nodules. Trends Biotechnol 14:161–168
Raji JA (2007) Intercropping soybean and maize in a derived savanna ecology. Afr J Biotechnol 6(16):1885–1887
Ram K, Meena RS (2014) Evaluation of pearl millet and mung bean intercropping systems in arid region of Rajasthan (India). Bangladesh J Bot 43(3):367–370
Ramirez-Restrepo C, Barry T (2005) Alternative temperate forages containing secondary compounds for improving sustainable productivity in grazing ruminants. Anim Feed Sci Technol 120:179–201
Rasmunssen J, Eriksen J, Jensen ES, Esbensen KH, Høgh-Jensen H (2007) In situ carbon and nitrogen dynamics in rye-grass clover mixtures: transfers, deposition and leaching. Soil Biol Biochem 39:804–815
Ratnadass A, Blanchard E, Lecompte P (2013) Ecological interactions within the biodiversity of cultivated systems. In: Cultivating biodiversity to transform agriculture. Hainzelain ed., ed Quae, Cirad, pp 141–179
Reddy A (2016) Sustainable indicators of food, nutritional and health outcomes in India. Yojna, December 2016

Rochon JJ, Doyle CJ, Greef JM, Hopkins A, Molle G, Sitzia M, Scholefield D, Smith CJ (2004) Grazing legumes in Europe: a review of their status, management, benefits, research needs and future prospects. Grass Forage Sci 59:197–214
Rossing WAH, Zander P, Josien E, Groot JCJ, Meyer BC, Knierim A (2007) Integrative modelling approaches for analysis of impact of multifunctional agriculture: a review for France, Germany and the Netherlands. Agric Ecosyst Environ 120(1):41–57
Samba T, Coulibay BS, Koné A, Bagayoko M, Kouyaté Z (2007) Increasing the productivity and sustainability of millet based cropping systems in the Sahelian zones of West Africa. In: Bationo A (eds), Advances in integrated soil fertility management in sub-Saharan Africa: challenges and opportunities, pp 567–574
Sato Y, Monincová M, Chaloupková R, Prokop Z, Ohtsubo Y, Minamisawa K et al (2005) Two rhizobial strains, Mesorhizobium loti MAFF303099 and Bradyrhizobium japonicum USDA110, encode haloalkane dehalogenases with novel structures and substrate specificities. Appl Environ Microbiol 71:4372–4379
Schädler S, Morio M, Bartke S, Rohr-Zänker R, Finkel M (2011) Designing sustainable and economically attractive brownfield revitalization options using an integrated assessment model. J Environ Manag 92:827–837
Schimel DS, Braswell BH, Holland EA (1994) Climatic, edaphic and biotic controls over storage and turnover of carbon in soils. Glob Biogeochem Cycles 8:279–293
Schipanski ME, Drinkwater LE (2012) Nitrogen fixation in annual and perennial legume–grass mixtures across a fertility gradient. Plant Soil 357:147–159
Schmidt J, Weidema BP (2015) Shift in the marginal supply of vegetable oil. Int J LCA 13(3):235–239
Schoenholtz SH, VamMiegroet H, Burger JA (2000) A review of chemical and physical properties as indicators of forest soil quality: challenges and opportunities. For Ecol Manag 138:335–356
Schuster GI (2013) Cultivated, primarily as food source. In: Nutritional Aspects of Legumes, vol. I. Encyclopedia of Life Support System (EOLSS). www.eolss.net/samplechapters/c10/e5-02-02.pdf
Seran TH, Brintha I (2010) Review on maize based intercropping. J Agro 9(3):135–145
Sharma AR, Behera UK (2009) Nitrogen contribution through Sesbania green manure and dual – purpose legumes in maize–wheat cropping systems: agronomic and economic considerations. Plant Soil 325:289–304
Siddique KHM, Johansen C, Kumar Rao JVDK, Ali M (2008) Legumes in sustainable cropping systems. In: Kharkwal MC (ed) Food legumes for nutritional security and sustainable agriculture. Proceedings of the fourth international food legumes research conference (IFLRC-IV), October 18–22, 2005, New Delhi, India, vol 1. Indian Society of Genetics and Plant Breeding, New Delhi, pp 787–819
Sinha S, Gupta AK, Bhatt K, Pandey K, Rai UN, Singh KP (2006) Distribution of metals in the edible plants grown at Jajman, Kanpur (Indian) receiving treated tannery wastewater: relation with physico-chemical properties of the soil. Environ Monit Assess 115:1–22
Soussana JF, Machado O (2000) Modelling the temperate grasses and legumes in cut mixtures. In: Lemaire et al (eds) Grassland ecophysiology and grazing ecology. CAB International, London, pp 169–190
Spehn E, Scherer Lorenzen M, Schmid B, Hector A, Caldeira M, Dimitrakopoulos P, Finn J, Jumpponen A, O'donnovan G, Pereira J (2002) The role of legumes as a component of biodiversity in a cross-European study of grassland biomass nitrogen. Oikos 98:205–218
Srinivasarao C, Venkateswarlu B, Lal R (2012) Long-term effects of soil fertility management on carbon sequestration in a rice-lentil cropping system of the Indo-Gangetic plains. Soil Sci Soc Am J 76(1):167–178
Stern N (2006) Stern review: economic of climate change. United Kingdom's Treasury, London
Sullivan P (2003) Intercropping principles and production practices. Appropriate Technology Transfer for Rural Areas Publication. http://www.attra.ncat.org

Swarup A, Manne MC, Sigh GB (2000) Impact of land use and management practices on organic carbon dynamics in soils of India. In: Lal R, Kimble JM, Stewart BA (eds) Global climate change and tropical ecosystems. Lewis Publisher, Boca Raton, pp 261–281
Teng Y, Wang X, Li L, Li Z, Luo Y (2015) Rhizobia and their bio-partners as novel drivers for functional remediation in contaminated soils. Front Plant Sci 6(32):1–11
Turnbull J, Bowman WD (2002) Variable effects of nitrogen additions on the stability and turnover of soil carbon. Nature 419:915–917
UN (2015) United Nation Department of Economics and Social Welfare, Population Division, Revision of World Population Prospects. https://esa.un.org/unpd/wpp. dated 10/17/2016
UN (2016) United Nations proclaims 2016 as "International year of pulses". Global pulse confederation. Retrieved 24 January 2016, (A/RES/68/231)
USDA (2016) National nutrient database for standard reference release 28, released September 2015, slightly revised May 2016
Van der Werf HMG, Petit J (2002) Evaluation of the environmental impact of agriculture at the farm level: a comparison and analysis of 12 indicator-based methods. Agric Ecosyst Environ 93(1–3):131–145
Vance CP (2001) Symbiotic nitrogen fixation and phosphorus acquisition: plant nutrition in a world of declining renewable resources. Plant Physiol 127:390–397
Varma D, Meena RS, Kumar S, Kumar E (2017) Response of mung bean to NPK and lime under the conditions of Vindhyan Region of Uttar Pradesh. Leg Res 40(3):542–545
Verma JP, Meena VS, Kumar A, Meena RS (2015a) Issues and challenges about sustainable agriculture production for management of natural resources to sustain soil fertility and health: A book review. J Clean Prod 107:793–794
Verma JP, Jaiswal DK, Meena VS, Meena RS (2015b) Current need of organic farming for enhancing sustainable agriculture. J Clean Prod 102:545–547
Wang C, Shen Z, Li X, Luo C, Chen Y, Yang H (2004) Heavy metal contamination of agricultural soils and stream sediments near a copper mine in Tongling, People's Republic of China. Bull Environ ContamToxicol 73:862–869
Webb J, Bellamy P, Loveland PJ, Goodlass G (2003) Crop residue returns and equilibrium soil organic carbon in England and Wales. Soil Sci Soc Am J 67:928–936
Wichern F, Mayer J, Joergensen RG, Müller T (2007) Release of C and N from roots of peas and oats and their availability to soil microorganisms. Soil Biol Biochem 39:2829–2839
Wilhelm WW, Johnson JMF, Hatfield JL, Voorhees WB, Linden DR (2004) Crop and soil productivity response to corn residue removal: a literature review. Agron J 96:1–17
Williams A, Hedlund K (2014) Indicators and trade-offs of ecosystem services in agricultural soils along a landscape heterogeneity gradient. Appl Soil Ecol 77:1–8
Williams PN, Lei M et al (2009) Occurrence and partitioning of cadmium, arsenic and lead in mine impacted paddy rice-Hunan, China. Environ Sci Technol 43:637–642
Yadav GS, Lal R, Meena RS, Datta M, Babu S, Das LJ, Saha P (2017a) Energy budgeting for designing sustainable and environmentally clean/safer cropping systems for rainfed rice fallow lands in India. J Clean Prod 158:29–37
Yadav GS, Lal R, Meena RS, Babu S, Das A, Bhomik SN, Datta M, Layak J, Saha P (2017b) Conservation tillage and nutrient management effects on productivity and soil carbon sequestration under double cropping of Rice in north eastern region of India. Ecol Indi http://www.sciencedirect.com/science/article/pii/S1470160X17305617
Yong T, Liu X, Yang F et al (2015) Characteristics of nitrogen uptake, use and transfer in a wheat-maize-soybean relay intercropping system. Plant Prod Sci 18:388–397
Young IM, Crawford JW (2004) Interactions and self-organization in the soil-microbe complex. Science:304, 1634–1637
Yusuf AA, Iwuafor ENO, Olufajo OO, Abaidoo RC, Sanginga N (2009) Effect of crop rotation and nitrogen fertilization on yield and nitrogen efficiency in maize in the northern Guinea Savanna of Nigeria. Afr J Agric Res 4(10):913–921

Cereal-Legume Cropping System in Indian Himalayan Region for Food and Environmental Sustainability

2

Anup Das, M. Thoithoi Devi, Subhash Babu, Meraj Ansari, Jayanta Layek, S. N. Bhowmick, Gulab Singh Yadav, and Raghavendra Singh

Contents

A. Das (✉)
Division of Crop Production, ICAR Research Complex for NEH Region, Umiam, India

M. Thoithoi Devi · S. Babu · J. Layek
ICAR Research Complex for NEH Region, Tripura Centre, Agartala, Tripura, India

M. Ansari
ICAR Research Complex for NEH Region, Manipur Centre, Imphal, India

S. N. Bhowmick
ICAR Research Complex for NEH Region, Tripura Centre, Agartala, India

G. S. Yadav
Division of Crop Production, ICAR Research Complex for NEH Region, Lembucherra, India

R. Singh
ICAR Research Complex for NEH Region, Sikkim Centre, Gangtok, India

R. S. Meena et al. (eds.), *Legumes for Soil Health and Sustainable Management*,
https://doi.org/10.1007/978-981-13-0253-4_2

Abstract

The Indian Himalayan Region (IHR) is extended from Jammu and Kashmir to the northeastern part of the country and shows a great differentiation in climatic, edaphic, geological, vegetation, and other features due to complex variegation of agroecosystems which leads to diverse agroecological zones. Agriculture is the important source of livelihood of the region, and rice (*Oryza sativa* L.), wheat (*Triticum aestivum* L.), and maize (*Zea mays* L.) are the main crops of the entire IHR. Unsustainable agricultural practices, such as monocropping, conventional tillage, indiscriminate use of fertilizers and pesticides, etc., apart from land degradation and cropland scarcity have serious implications for livelihood security in IHR. Under such scenario, there is a need to diversify cropping pattern to make the entire agricultural system sustainable and environmentally secure. Inclusions of legumes in cereal-based cropping system either as intercrop or in sequence of crop rotation are the most promising options for diversified sustainability of the system and enhance the cropping intensity. Diverse habitat of IHR favors the growth and development of an amazing variety of legumes and other crops which make this region the rich hub for agricultural crop diversity specifically the legume crops. Broad bean (*Vicia faba*), horse gram (*Macrotyloma uniflorum*), field pea (*Pisum sativum*), black gram (*Vigna mungo*), adzuki bean (*Vigna angularis*), cowpea (*Vigna unguiculata*), soybean (*Glycine max*), lentil (*Lens esculenta*), green gram (*Vigna radiata*), beans (*Phaseolus* sp.), lathyrus (*Lathyrus sativus*), pigeon pea (*Cajanus cajan* L), etc. are some of the legumes cultivated by the farming communities in IHR. Rice bean [*Vigna umbellata* (Thunb.) Ohwi and Ohashi and mucuna/velvet bean [*Mucuna pruriens* (L.) DC.] are some of the specific legumes grown abundantly in the eastern IHR which has immense food and natural resource conservation values. Albeit the legume species provides food, fuel, fodder, etc. and has multifarious roles in agriculture and natural resource conservation, their ability to fix atmospheric nitrogen in root nodules and subsequently contributions to the soil fertility give them the unique identity. Legume-based systems improve several aspects of soil fertility, such as soil organic carbon (SOC) and humus content and nitrogen and phosphorus availability, suppress weed growth through smothering effects, increase production per unit area, enhance land use efficiency, reduce runoff and soil loss, etc.

Inclusion of legume provides sustainability to nonlegume cereal component by enriching soil fertility and increasing system productivity and returns. Significant reductions in the release of greenhouse gases, viz., carbon di-oxide, nitrous oxide etc., are a logical consequence of reduced fertilizer and energy use in arable systems with legumes. Pulses are considered the key crops for intensification of rice and maize-fallows of IHR due to their short-duration, hardy, and low-input requiring nature, hence offers a tremendous opportunity to utilize residual soil moisture.

Keywords

Cereal · Cropping system · Indian Himalayan Region · Legume · Land degradation · Sustainability

Abbreviations

B:C	Benefit/cost
BNF	Biological nitrogen fixation
CA	Conservation agriculture
CI	Cropping intensity
CO_2	Carbon dioxide
CT	Conventional tillage
DHA	Dehydrogenase activity
FP	Farmers' practice
EHR	Eastern Himalayan region
IHR	Indian Himalayan Region
LER	Land equivalent ratio
LUE	Land use efficiency
MEY	Maize equivalent yield
mt	Million tons
MT	Minimum tillage
N_2O	Nitrous oxide
NER	North East Region
NT	No-till
PEM	Protein-energy-malnutrition
RDA	Recommended dietary allowances
SMBC	Soil microbial biomass carbon
SOC	Soil organic carbon
SOM	Soil organic matter
USD	US dollar
WUE	Water use efficiency

2.1 Introduction

The Indian Himalayan Region (IHR) with width of 250–300 km across stretches over 2500 km beginning from Arunachal Pradesh in the east to Jammu and Kashmir in the west and spreads between 21°57′–37°5′ N latitudes and 72°40′–97°25′ E longitudes (Bhatt et al. 1999). This great chain of mountains in Indian territory extends all along the northern border of the country which extends from the eastern border of Pakistan in the west to the western border of Myanmar in the east covering partially/fully 12 states of India, viz., Jammu and Kashmir, Himachal Pradesh, Uttaranchal, and Sivaliks of Punjab and Haryana in the west, West Bengal in the east, and Sikkim, Arunachal Pradesh, Nagaland, Manipur, Mizoram, Tripura, Meghalaya, and hills of Assam in the northeast (Samal et al. 2005). In continuation of various mountainous countries, the IHR is differentiated by their climatic, edaphic, geological characters, vegetation, cropping patterns, crop rotations, and other features due to complex variegation of agroecosystems. It resulted in representation of diverse agroecological zones in the Himalayas which provides the myriad microhabitats. This diverse habitat favors the growth and development of an amazing variety of legumes and other crops over thousands of years by the hill farmers which make this region the rich hub for agricultural crop diversity specifically the legume crops. The hill and mountain areas of the Himalayas are ecologically fragile and economically not developed well with several problems imposing severe limitations on resource productivity level (Bhatta and Vetaas 2016). Moreover, agriculture is mostly subsistence in nature and is the important source of livelihood of the region. Unsustainable agricultural practices, land degradation, and cropland scarcity have serious implications for food security and Himalayan livelihoods (Partap 1999; Das et al. 2014a Meena et al. 2017a). Crisis area studies, conducted by the Mountain Farming Systems Programme of International Centre for Integrated Mountain Development (ICMOD), documented evidences of unsustainable mountain agriculture in IHR (Shrestha 1992; Pandey 1992; Jodha and Shrestha 1994). Many of the indicators used to illustrate this are derived from the farmers' responses to the lack of adequate cropland management. For example, marginal rainfed croplands in some areas are abandoned because farmers do not perceive that the benefits of cultivation outweigh the costs involved (Partap 1999).

The North East Region (NER) of India also called as the Eastern Himalayan region (EHR) comprises of eight states, namely, Arunachal Pradesh, Assam, Manipur, Meghalaya, Mizoram, Nagaland, Sikkim, and Tripura, and covers about 8.3% geographical area and has about 4% populations of the country (Ngachan et al. 2010). Out of the total 26.27 m ha geographical area of NER of India, around 77% is hills and senile plateau, while only 12% net area is under cultivation (Das et al. 2016). Rainfed, monocropping, and subsistence-type agriculture are the characteristic features of NER. Paddy is grown as the major crop of the region followed by maize. Food crops account for more than 80% of the gross cropped area, and cereals occupy about 70% of that (Gupta et al. 1998). Crop diversification by involving legumes is the need of the hour for the development of sustainable farming systems in this region. Inclusion of legumes in cropping system either as intercrop or in sequence of crop rotation is one of the most promising options for diversified sustainability of the system (Saha et al. 2011; Meena et al. 2015a), and it will create an opportunity to enhance the cropping intensity (CI) manifolds. Out of 23.1 m ha area under pulses in the country, the NER has only 255.99 thousand ha (1.1%). This region contributes only 216.6 thousand tons (1.3%) to the country's total pulse production (2014–2015) of 17.2 million tons (mt). With this production level, the per capita pulse availability in NER is hardly 12.5 g against 46 g at national level. Considering the recommended per capita dietary pulse intake of 50 g, the pulse production in this region needs to be increased by almost four times to make this region self-sufficient in pulses (Das et al. 2016). The fact that the productivity of the pulses in NER (848 kg/ha) is higher than that of country's (743 kg/ha) suggests that this region suits well to requirement of pulses production. The NER has a much wider spectrum of pulses grown than any other regions of the world, and these include pigeon pea [*Cajanus cajan* (L.) Millsp.], pea (*Pisum sativum* L.), rice bean (*Vigna umbellata* Thunb), French bean/rajmash (*Phaseolus vulgaris* L), soybean (*Glycine max*), green gram (*Vigna radiata* L), black gram (*Vigna mungo* L), lentil (*Lens culinaris* Medik), broad bean (*Vicia faba* L), lablab bean [*Lablab purpureus* (L.)], and cowpea [*Vigna unguiculata* (L.) Walp]. Rice bean (*Vigna umbellata*), a highly photosensitive short-day plant, is a very important rainy (*kharif*) season pulse of this region and is an integral component of *jhum*. Recently, identified photo-insensitive genotypes of rice bean have offered prospects of cultivation of this crop in spring/summer season also.

There are many species and varieties of legumes that are cultivated by the farming communities of IHR like broad bean, horse gram (*Macrotyloma uniflorum*), field pea, black gram, adzuki bean (*Vigna angularis*), cowpea, soybean, lentil), green gram, rice bean, etc. Besides, several species and varieties of beans (*Phaseolus* sp.) are exclusive to higher Himalayas. Legume crops are of multipurpose in value and play significant role in providing agricultural, food, nutritional, and livelihood security to the hill farmers. They have been closely interlinked with cereals in a way that in agriculture legumes complement cereals in terms of cropping pattern and crop cycle (Ram and Meena 2014) and provide rich protein and a variety of minerals and nutrients to a cereal-based diet (FAO 1982). Pulses are described as "poor men's meat" (FAO 1982), due to its high protein content. The significance of legume species is not limited only to the use as fuel, fodder, etc. but extended to traditional rituals and ceremonies also. Cultivation of pulses are also an effective means of rehabilitating degraded soils and can contribute significantly to achieving the twin objectives of increasing productivity as well as improving the sustainability of cereal-based cropping systems (Yadav et al. 1998). Naturally the atmospheric nitrogen (N) fixation ability of legumes enriches the soil with available N and not only meets the N requirement of associated crop but also the subsequent crops (Dhakal et al. 2016) and further reduces the use of synthetic fertilizer and organic manures in agriculture. The latter though have been traditionally used in hill agroecosystem and are less available due to dwindling forest cover and decrease in domesticated animal population (Semwal and Maikhuri 1996). Increase in yield of subsequent crops raised after pulses to the tune of about 20–40% has been reported (Joshi 1998). The importance of legumes is much more vital in degraded hill and mountain ecosystem of IHR either in the form of N contribution or through leaf fall. It has been estimated that pigeon pea, soybean, cowpea, groundnut (*Glycine max*), field pea, and rice bean can produce 2.0, 1.2, 2.0, 0.5, 0.48, and 1.1 Mg/ha of leaf fall, respectively, which directly contributes to soil fertility and helps to ameliorate soil acidity (Hazarika et al. 2006). Pulse crops by virtue of their inherent capacity to add N to the soil and recharging of soil micronutrients through its taproot system are the crop of choice for organic production system in the IHR specially NER. Most of the indigenous legumes of the Himalayas have high ecological and economic potential and grow well in harsh environmental conditions (rainfed, low temperature, low fertility, moisture stress, etc.) with little external inputs (Maikhuri et al. 1996). Of late, a decline in interest of hill and mountain farmers toward traditional/indigenous legume crop has been observed as a result of climatic, cultural, and socioeconomic changes. This decline is considered as a big challenge to the indigenous legume crops and their wild relatives, and consequently the subsistence farming system of the region is under threat for sustenance.

2.2 Pulse Scenario in India

India is the largest producer and consumer of pulses in the world accounting for about 29% of the world area and 19% of the world's production (Singh et al. 2015). Ironically, the country's pulse production has been hovering around 14–15 million toones (mt), coming from a near-stagnated area of 22–23 m ha, since 1990–1991 (Singh et al. 2013). The pulse area, production, and productivity increased by 21, 104, and 68% in 2014–2015 as compared to 1950–1951 (Fig. 2.1) (Tiwari and Shivhare 2016). However, record pulse production of 22 mt was achieved in India during 2016–2017. This was mainly due to special emphasis on pulses by providing minimum support price, making available critical inputs to farmers along with large-scale awareness and demonstrations across the country. The low productivity in pulses may be attributed to cultivation under nutrient- and moisture-starved

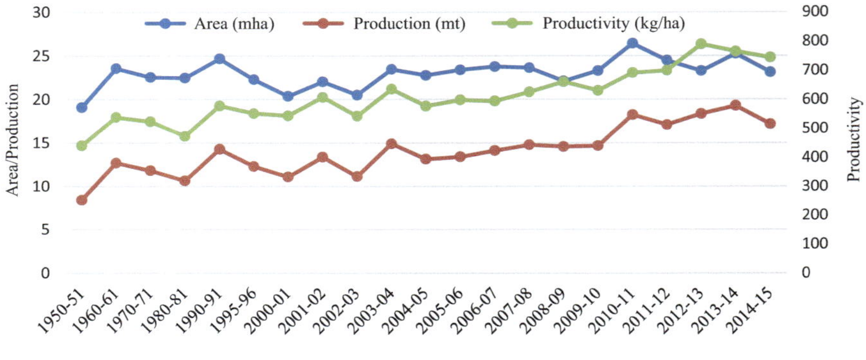

Fig. 2.1 Trends in area, production, and productivity of total pulses in India

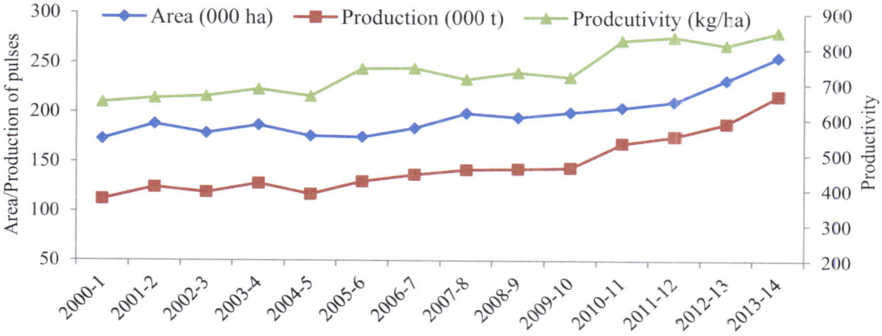

Fig. 2.2 Trends in area, production, and productivity of total pulses in NER (eastern Himalayas)

conditions, priority to cereals and cash crops, non-availability of location-specific pest and disease-resistant varieties, and lack of organized marketing infrastructure (Meena et al. 2015a).

2.2.1 Pulse Scenario in Eastern, Western, and Central Himalayas

The area under pulse crops increased sharply in the NER from 112 thousand ha in 2000–2001 to about 217 thousand ha in 2013–2014 (Fig. 2.2). The productivity of pulses during this period also enhanced from 647 to 848 hg/ha. Assam has the highest pulse area of 149.00 thousand hectares and contributed 104.00 thousand tons of pulse production during 2013–2014, followed by Nagaland and Manipur which accounted 37.80 and 30.38 thousand hectares area and contributed 42.5 and 28.65 thousand tons production, respectively (Das et al. 2016). Among the western and central Himalayan states, during 2014–2015, maximum area and production of pulses are in Uttar Pradesh, while the highest productivity of 1252 kg/ha is found in Himachal Pradesh which is 68% higher than the national average (743 kg/ha), and the lowest productivity is of Jammu and Kashmir (292 kg/ha, 61% lower than the national average) (Fig. 2.3) (Tiwari and Shivhare 2016).

2.2.2 Major Pulses of IHR

The major pulses grown in IHR are green gram, black gram, pigeon pea, rice bean, cowpea, etc. in *kharif* and French bean, chickpea (*Cicer arietinum*), lentil, khesari (*Lathyrus sativus*), and field pea in *rabi* season. In hills, various other beans such as faba bean/broad bean, adzuki bean, moth bean (*V. aconitifolia*), and lablab bean or *sem* are also grown as pulse. In *jhum* areas of the NER, rice bean is still a predominant pulse crop. Important pulses grown in IHR along with their growing season and major uses are presented in Tables 2.1 and 2.2.

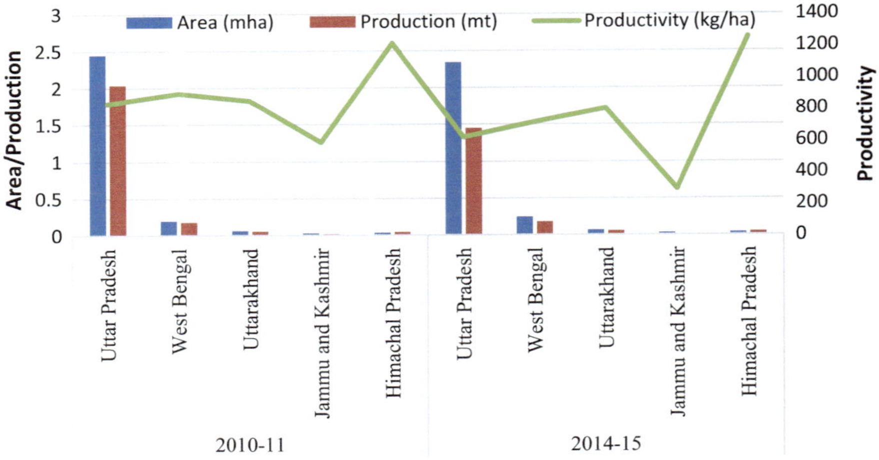

Fig. 2.3 Trends in area, production, and productivity of total pulses in western and central Himalayan states

Table 2.1 Pulse/grain legumes grown in Eastern Himalayan region

Botanical name	Local name	Growth period	Uses
Cajanus cajan	Arhar	June–Dec	Cooked as "dal," consumed during jaundice, dried stem used as fuel
Cicer arietinum	Chana	Oct–Feb	Cooked as "dal," whole plant as vegetable, and sprouted seeds consumed raw
Lathyrus sativus	*Soshta batura* (white seeded)	Oct–Feb	Highly nutritious, consumed with vegetables or with chana dal, always taken as mixed legume diet, tender pods and leaves as vegetables, rough "chapatti," whole plant used as fodder, manure
	Kalo Batura (black seeded)	Sep–Feb	
	Pahelo batura (brown mottled), khesari		
Lens esculenta	Masoor	Oct–Feb	Cooked as a dal, light meal taken during severe jaundice, indigestion, and loss of appetite
Macrotyloma uniflorum	Gahat	June–Sep	Consume to withstand extreme cold, lowering high cholesterol
		July–Oct	
Phaseolus lunatus	*Lachen tibi, Singtamey simi, Ghew bori*	Aug–Oct	Sprouted seed used as a breakfast, dal, tender pod as a vegetable
Phaseolus vulgaris	Kalo Mantulal simi/Alpatre simi/Harey simi	July–Nov	Sprouted seed used as a breakfast, dal, tender pod as a vegetable

(continued)

Table 2.1 (continued)

Botanical name	Local name	Growth period	Uses
Pisum sativum	Matar/Kerau	Oct–Feb	Cooked as "dal" or as mixed vegetable of tender pods, seeds, and leaves
Vicia faba	Bakulla	Oct–Jan	Young pods as vegetable. Dried bean fried, roasted, or mixed with pea, gram, mung, germinated to form soup, eaten with rice
Vigna mungo sp. niger	Kalo dal (black seeded)	Aug–Nov	Medicinal (at constipation, weakness), cooked as pulse meal or "dal" or mixed with radish, cabbage, salt, and lettuce leaves
Vigna mungo sp. viridis	*Pahenlo dal* (green seeded)	Aug–Nov	Cooked as dal and as a medicine (for low blood pressure)
Vigna radiata	*Mung dal*	Aug–Nov	Cooked dal
Vigna umbellata	Rato maysum dal (red-seeded)	July–Oct	As a dal, raw plants or tender pods. Seed flour mixed with paddy straw and water to form feed meal
	Kalo maysum dal (black-seeded)	July–Oct	As a dal, whole plant as energetic cattle feed
	Seto maysum dal (white-seeded)	July–Oct	Dry seed or cooked, mixed and boiled with pea and urd. Tender pods with young seed consumed for lowering cholesterol
	Tulo maysum dal	Aug–Nov	Mixed with chana and pea, palak, carrot, cooked as mixed eaten during extreme cold fever
Vigna unguiculata	Tuney bori	July–Oct	Grains are cooked or mixed with local herbs. Whole plant used as mulch or cover crops, also consumed during cold fever
Vigna sinensis	Thangre	July–Oct	Fodder/green manuring

Source: Das et al. (2016)

Table 2.2 Pulse/grain legumes grown in central and western Himalayan states

State	Pulse/grain legume
Jammu and Kashmir	Field pea (*Pisum sativum*), chickpea (*Cicer arietinum*), lentil, horse gram (*Macrotyloma uniflorum*), *Phaseolus* sp., etc.
Himachal Pradesh	Urd bean (*Vigna mungo*), lentil (*Lens esculenta*), field pea (*Pisum sativum*), horse gram (*Macrotyloma uniflorum*), rice bean (*Vigna umbellate*), etc.
Uttar Pradesh	Chickpea (*Cicer arietinum)*, lentil (*Lens esculenta*), field pea (*Pisum sativum*), pigeon pea (*Cajanus cajan*), urd bean (*Vigna mungo*), mung bean (*Vigna radiate)*, broad bean (*Vicia faba*) (grown sporadically), etc.
Uttarakhand	Mung bean (*Vigna radiate)*, urd bean (*Vigna mungo*), lentil (*Lens esculenta*), horse gram (*Macrotyloma uniflorum*), chickpea (*Cicer arietinum*), rice bean (*Vigna umbellate*), etc.

Table 2.3 Biochemical constituents of some indigenous legumes of Eastern Himalaya

Legume	Biochemical constituents	References
Broad bean	Starch (62–65.3%), protein (20–25%), dietary fiber (20.4–26.8%),	Sharma et al. (2003)
Winged bean	Protein (40–50%)	Sharma et al. (2003)
Jack bean	Protein (23%- 34%), carbohydrate (55%), crude fiber (2.55 ± 0.15%), ash content (3.45 ± 0.96%)	Akpapunam and Sefa-Dedeh (1997); Abitogun and Olasehinde (2012)
Rice bean	Carbohydrate (58.2–72.0%), crude protein (18.3–32.2%), ash (3.5–4.9%), soluble ether extract (0.1–0.5%), crude fiber (3.6–5.5%)	Buergelt et al. (2009)
Pigeon pea	Protein (21.0%), starch (48.4%), soluble sugar (5.1%), crude fiber (8.2%), fat (2.3%) in green seed Protein (18.8%), starch (53.0%), soluble sugar (3.1%), crude fiber (6.6%), fat (1.9%) in mature seed.	Saxena et al. (2010)
French bean	Carbohydrate (59.12%), protein (22.1%)	Ganie et al. (2014)
Cowpea	Carbohydrate (56–66%), protein (22–24%), crude fiber (5.9–7.3%), ash (3.4–3.9%), fat (1.3–1.5%)	Gómez (2004)
Peas	Carbohydrate (61.55%), protein (21.87%), fiber (1.47%), oil (1.58%), ash (2.70%)	Mishra et al. (2010)

2.2.2.1 Indigenous Pulses of EHR

There are some unique pulses grown in the EHR from time immemorial due to their ability to adapt in diverse agroclimatic conditions of the region. Some of the indigenous legumes are adzuki bean), broad bean, moth bean, local khesari, winged bean (*Psophocarpus tetragonolobus*), tree beans (*Parkia roxburghii*), jack bean (*Canavalia ensiformis*), sword bean (*Canavalia gladiata*), etc. *Vigna vexillata* is another less known potential pulse cum tuber crop of this region resembling cowpea and produces both edible seeds and tubers. Bakthul bean (*Sesbania grandiflora*) is also an indigenous, underutilized potential leguminous crop generally grown in NER (Das and Ghosh 2012). Indigenous legumes are superior in nutritional value over many of the more commonly grown pulses (Table 2.3). Rice bean is grown predominantly under the rainfed conditions in a mixed farming system under the shifting cultivation or in kitchen gardens (Yadav et al. 2017a, b). It is a multipurpose legume which can be used as food, fodder, green manure, cover crop, etc. In view of enormous variability, high yield (1.5–2.0 Mg/ha), tolerance to disease and pest, and wider adaptability, this underexploited legume grown predominantly in mixed farming system has great potential to become an important pulse crop of the region. Due to high content of L-dopa, broad bean can potentially be included in dietary strategies to manage Parkinson's disease (Kempster et al. 1993). Winged bean is also an important pulse crop of the region used both as pulse and vegetables as a favorite supplementary delicacy. Winged bean has very high contents of crude fat (1.7%), protein (50.7%), potassium (8.9 mg/g), calcium (8.06 mg/g), and magnesium (5.72 mg/g) in full mature seed and high total soluble sugar (488.80 mg/g), nonreducing sugar (415.95 mg/g), and starch (420.60 mg/g) in tuber (Ningombam et al. 2012). Mucuna/velvet bean [*Mucuna pruriens* (L.) DC.] is an important pulse crop of the IHR having medicinal potential and plays a significant role in soil conservation (Kala, 2005).

Pahenlo dal (unique green-seeded urd bean) is one of the most extensively grown pulses in Sikkim as *kharif*/pre-*rabi* crop. Local collections of *Pahenlo dal* are semi spreading types, and pods are less hairy than black-seeded urd bean. *Pahenlo dal* possesses higher amount of protein as compared to *Kalo dal* (black-seeded urd beans) and better in taste. The seeds of *Pahenlo dal* are thin and longer than the *Kalo dal*. SKMPD-3 is most promising variety of *Pahenlo dal* developed by ICAR Research Complex for NEH Region, Sikkim Centre, Sikkim through local selections. *Kholar bean* (a variant of rajmash) is very popular in Nagaland and mostly cultivated in *jhum* fields.

2.2.3 Cropping System in Central Himalayas

Crops and cropping systems in the central Himalayas are diverse due to large agro-ecological and cultural diversity, which has led to variable cropping patterns. About 80% of people of the central Himalayas practice subsistence agriculture (Maikhuri et al. 2001). Land holdings are small, with fragmented and terraced slopes covering 85% of total agricultural land which is rainfed, while the valley area, which covers 15% of agricultural land, is irrigated. In the central Himalayas, intercrop combinations traditionally involve cereals with millet, millet with legumes, and legumes with legumes. Intercropping finger millet (non-legume) with legumes often results in higher resource use efficiency compared to sole cropping (Chandra 2007; Ram and Meena 2014). The intercropping species that differ in sowing and harvesting times, and their maximum demands on environmental resources, extends the duration of resource use (Chandra et al. 2011b). Finger millet (*Eleusine coracana*) is known for its high mineral contents (Chethan and Malleshi 2007), and legumes fix nitrogen symbiotically. Such complementarities between crops in resource use are particularly important in low input subsistence farming systems such as those in the central Himalayas. Finger millet and black gram are commonly intercropped, and these species are the most important staple food crops in the Garhwal area of the central Himalaya.

2.3 Prospect of Legumes in IHR

Farmers and people of mountain ecosystem of IHR are resource poor and reside in ecologically sensitive region. Land degradation, low productivity, poor nutrition and health, low income, and vulnerability to climatic variabilities and extremes are some of the features of the Himalayan ecosystem (Meena et al. 2017a). Thus, inclusion of food legumes in production systems can play multiple roles and services at

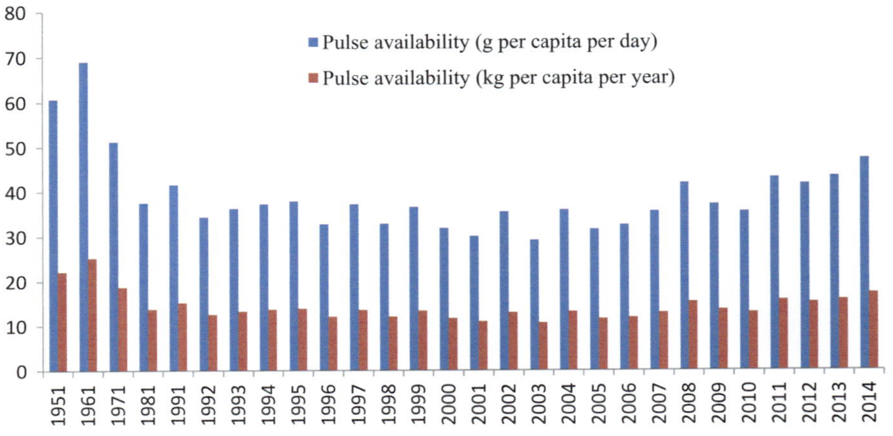

Fig. 2.4 Per capita availability of pulses in India. (*Source*: Agricultural Statistics at a glance-2014)

food system level, both for human and animal consumption, as a source of plant proteins (Tharanathan and Mahadevamma 2003); at production system level, through atmospheric N fixation and due to their ability in mitigating greenhouse gases emissions (Lemke et al. 2007; Meena et al. 2015b); and at cropping system levels, through crop diversification breaking the cycles of pests and diseases and contributing to balance the deficit in plant protein production (Peoples et al. 2009).

The recommended dietary allowances (RDA) of pulses for adult male and female are 60 g and 55 g per day against availability of 46 g/day in 2016 (Fig. 2.4). The per capita availability is much lower in IHR. The deficiency of protein in human diet often leads to protein-energy malnutrition (PEM) causing various forms of anemia. Legumes are a significant source of protein, dietary fiber, carbohydrates, vitamins, and dietary minerals. 100 g legumes provide energy 321–570 Kcal, protein 21–28 g, fat 0.8–48.0 g, carbohydrate 21–63.4 g, and total dietary fiber 9.0–22.7 g. Lentil is very rich in pantothenic acid (2.12 mg/100 g), vitamin B_6 (0.54 mg/100 g), folate (479 mg/100 g), etc. Cowpea is also rich in vitamin B_6 (0.44 mg/100 g), folate (546 mg/100 g), etc., and vitamin C content of soybean is highest (6 mg/100 g) among pulses (Table 2.4).

2.4 Land Degradation in IHR

Land serves as the storage bin for water and nutrients required for growth and development of different crops. Soil degradation is a great challenge in India specially in IHR with respect to food and environmental security. According to the NBSS&LUP (2004), out of the total geographical area of 328.37 m ha of India, approximately 146.8 m ha (44.7%) is under various kinds of degradation. Out of the total degraded area of the country, ~ 42.9 m ha (29.22%) is in IHR (Table 2.5). Degradation due to water erosion is the most serious problem in India, leading to loss of top fertile soil

Table 2.4 Nutritive value of some important legumes (per 100 g) of IHR

Pulses	Energy (Kcal)	Protein (g)	Fat (g)	Carbohydrate (g)	Total dietary fiber (%)	Thiamin (mg)	Riboflavin (mg)	Niacin (mg)	Iron (mg)	Zinc (mg)	Calcium (mg)
Chickpea	368	21.0	5.7	61.0	22.7	0.30	0.51	1.5	6.2	3.4	105
Pigeon pea	342	21.7	1.49	62.0	15.5	0.45	0.51	2.9	5.2	2.7	130
Urd bean	347	24.0	1.6	63.4	16.2	0.41	0.37	2.3	8.4	3.5	110
Mung bean	345	25.0	1.1	62.6	16.3	0.72	0.15	2.3	6.7	2.7	132
Lentil	346	27.2	1.0	60.0	11.5	0.45	0.49	2.6	7.5	4.7	56.0
Field pea	345	25.1	0.8	61.8	13.4	0.47	0.21	2.9	4.4	3.0	55.0
Rajmash	345	23.0	1.3	63.4	18.2	0.53	0.22	2.08	3.4	1.9	186
Cowpea	346	28.0	1.3	63.4	18.2	0.50	0.48	2.36	7.5	3.7	80.3
Horse gram	321	23.0	2.3	59.1	15.0	0.42	0.20	1.5	7.0	–	287
Moth bean	330	24.0	1.5	61.9	–	0.45	0.09	1.5	9.6	–	202
Peanut	570	25.0	48.0	21.0	9.0	0.60	0.30	12.9	2.0	3.3	62.0
Soybean	446	16.49	19.9	30.1	9.3	0.87	0.87	1.62	15.7	4.9	277
Khesari						0.39	0.41				

Source: *The Nutritive value of Indian Foods & the planning satisfactory Diets (ICMR)*; http://www.iipr.res.in/pdf/15.1_270615.pdf

Table 2.5 Extent of land degradation in IHR

State	Total degraded area (m ha)	% of degraded area to total ground area of the state
Manipur	1.9	42.6
Mizoram	1.9	89.2
Meghalaya	1.2	53.9
Assam	2.2	28.2
Arunachal Pradesh	4.6	53.8
Nagaland	1.0	60
Sikkim	0.2	33
Tripura	0.6	59.9
Himachal Pradesh	4.2	75
Jammu and Kashmir	7	31.6
Uttar Pradesh + Uttarakhand	15.3	52
West Bengal	2.8	31
Total	42.9	–

Source: NBSS & LUP-ICAR (2005) on 1:250,000 scale

and formation of gullies, landslides, and terrain deformation. The average soil erosion rate of India was about 16.4 Mg/ha per year, resulting in an annual total soil loss of 5.3 billion tons (Dhruvanarayan and Ram 1983;; Datta et al. 2017). About 29% of total eroded soil is lost to the sea, 61% eroded soil is translocated from one place to another, and the remaining 10% is deposited in reservoirs leading to imbalance in ecosystem.

India is losing a huge exchequer every year due to soil and land degradation. Costs of inaction on land degradation are higher than the costs of action, indicating the benefits that will accrue if sufficient conservation practices are undertaken. The total annual costs of land degradation by land use and cover change in 2009 as compared to 2001 in India are estimated to be about 5.35 billion USD with an annual per capita cost of 4.4 USD (Indian Ministry of Statistics and Programme Implementation, 2014**).** The ratio of action over inaction is in the range of 20–40% in humid regions in general and above 40% in subhumid and arid regions (Mythili and Jann Goedecke 2016). Total direct cost of soil degradation was estimated at INR 448.6 billion with cost of soil erosion in lost production at INR 361 USD (Sehgal and Abrol 1994). Annual per capita cost of soil degradation of Himalayan states are much higher than other states (Table 2.6) due to steep topography, high rainfall, cultivation practices, and anthropogenic activities like road constructions, urbanization, etc. Monetary losses amounting to 22% (*564.64 million USD*) for pulses due to water erosion have been reported by Sharda et al. (2010).

In NER, *jhum* cultivation (primitive form of slash and burn agriculture) is still practiced in approximately 0.756 million ha land area resulting in burning phytomass (including forest floors) of more than 8.5 mt annually (Choudhury et al. 2015a), and it accounted for 146.6 Mg/ha per year soil loss in the first year *jhum* cultivation, 170.2 Mg/ha/year in second year *jhum* cultivation, and 30.2 Mg/ha per year in

Table 2.6 Annual per capita cost (USD) of soil degradation in Himalayan vs. other states of India

States	Annual per capita cost (USD) of soil degradation
Arunachal Pradesh	76.6
Manipur	47.6
Meghalaya	42.5
Nagaland	46.9
Mizoram	175.1
Sikkim	47.0
Tripura	40.1
Jammu and Kashmir	20.0
Uttarakhand	20.3
Jharkand	6.6
Chhattisgarh	10.0
West Bengal	0.9
Punjab	0.3
Tamil Nadu	3.5
India	4.4

Sources: Indian Ministry of Statistics and Programme Implementation (2014); Mythili and Goedecke (2016)

abandoned *jhum* fallow (Darlong 1996). In case of cultivation in hill areas, it has been seen that soil loss is more in case of *jhumming* in comparison to others. Agricultural activities leading to soil/land degradation are low and imbalanced fertilization, excessive tillage and use of heavy machinery, crop residue burning and inadequate organic matter inputs, poor irrigation and water management, poor crop rotations, pesticide overuse and soil pollution (Verma et al. 2015a, b, c), etc. Apart from faulty agricultural activities that led to soil degradation, other human-induced land degradation activities include overgrazing, deforestation, and careless forest management; urban growth, industrialization, and mining; and natural (earthquakes, tsunamis, droughts landslides, volcanic eruptions, etc.) and social sources (land shortage, decline in per capita land availability, population increase, etc.) of land degradation, etc.

Some of the strategies to mitigate land degradation are soil erosion control measures, rainwater harvesting, terracing, and bio-fencing; cover cropping, agroforestry, grass water ways, intercropping, strip cropping, and contour farming; watershed approach; residue recycling and minimal soil disturbance; integrated nutrient management, application of organic manure, and reclamation of acid soils; vegetative barriers, mulching, and diversified cropping; agroforestry; conservation agriculture (CA); etc. Soil conservation measures, such as contour plowing, bunding, and use of strips and terraces, can decrease erosion and slow runoff water. Pulses are the candidate crop for all the above conservation measures for erosion control, fertility enhancement, and resource conservation (Dhakal et al. 2016). Pulses act as cover crop that reduces the impact of falling raindrops on the soil surface, increases infiltration of water into the soil, reduces runoff velocity resulting in reduction in loss of top soil and nutrients, etc. Intercropping of cowpea with maize (2:1) has been reported to decrease runoff and soil loss by 10% and 28%, respectively, compared to pure maize (Srinivasarao et al. 2014). When cereal crops like maize and sorghum are cultivated along with legumes like green gram (*Vigna radiata* L.), groundnut, black gram (*Vigna mungo* L.), and cowpea (*Vigna unguiculata*) as intercrops, sufficient ground cover is ensured, and hence, there is drastic reduction in erosion hazards (Rao and Khan 2003). Integrated application of NPK mineral fertilizers along with organic manure increases crop productivity, improves soil physical properties and organic carbon (SOC) content, and decreases soil loss by improving soil aggregation(Varma et al. 2017a, b). It has been reported that about 25% and 19% of gross C input contributed to higher SOC concentrations after 9 years of irrigated or after 30 years of rainfed soybean-wheat production, respectively (Kundu et al. 2007, Bhattacharyya et al. 2009). Increase in SOC pool due to restoration of salt-affected soils has been also reported by Bhojvaid and Timmer (1998).

2.5 Role of Legumes in Soil Sustainability

Inclusion of legumes in cereal-based cropping system substantially reduces the use of fertilizers and energy requirement and consequently lowers down the GHG emissions. The role of legumes in climate change mitigation has not been properly addressed. Legumes have the great potential to lower the emission of GHGs such as carbon dioxide (CO_2) and nitrous oxide (N_2O) compared to mineral N fertilization-based conventional agriculture (Reckling et al. 2014). Legumes also play an important role in the carbon sequestration in soils (IPCC 2007; Yadav et al. 2017a) and reduce the fossil energy inputs in the farming system (Jensen et al. 2012). The legume benefits in agronomic terms have been divided into "N effect" and "break crop effect" (Chalk 1998). The "N effect" component is due to N provision from BNF (Peoples et al. 2009). The N effect of legume is generally higher in situations of low N fertilizer application to subsequent crops (Preissel et al. 2015), whereas the benefits of "break crop effect" include improvements in soil properties such as soil organic matter (SOM) and structure (Hernanz et al. 2009), soil aggregation, water retention and available water (Angus et al. 2015), phosphorus (P) mobilization (Shen et al. 2011), reduction in GHG emission (Stagnari et al. 2017; Meena et al. 2017b), and less insect, disease, and weed problems (Robson et al. 2002).

Continuous cultivation along with conventional tillage systems may cause significant SOC losses through decomposition of humus (Christopher and Lal 2007). Shifting from pasture to field crops has been reported to result in loss of soil C stocks between 25% and 43% (Soussana et al. 2004). Inclusion of legumes in cropping/farming systems can improve multiple soil properties, such as SOC and humus content and N and P availability (Jensen et al. 2012). Besides, the

Table 2.7 N economy due to inclusion of pulses in sequential cropping

Preceding legume	Following cereal	Fertilizer N equivalent (kg N/ha)
Chickpea	Maize	60–70
	Pearl millet	40
Pigeon pea	Wheat	40
	Maize	20–49
	Pearl millet	30
Lentil	Pearl millet	40
	Maize	18–30
Peas	Pearl millet	40
	Maize	20–32
Green gram	Pearl millet	30
Lathyrus	Maize	36–48
Cowpea	Pearl millet	60
Pigeon pea	Sorghum	51
Mung bean	Rice	40
Chickpea	Rice	40
Rajmash	Rice	40
Fodder cowpea	Rice	40

Source: Lee and Wani (1989); Hazarika et al. (2006)

pulse and legume crop can fix the atmospheric N in soil and improve the soil health (Singh and Singh 2002), reduce the soil loss, conserve the soil and water, and suppress the weed growth through smothering effects (Konlan et al. 2013). The N economy due to inclusion of pulses in sequential cropping is given below in the Table 2.7.

2.5.1 Legume Effect on Soil Properties

Improvement in soil physicochemical and biological properties due to legume-based systems has been reported by Jensen et al. (2011, 2012). Ngangom et al. (2017a, 2017b) also reported substantial improvement in maize equivalent yield and soil physical, chemical, and biological properties due to inclusion of legumes in maize-based cropping system in Meghalaya plateau of EHR. Due to inclusion of legumes (green gram, French bean), SOC content is enhanced by 5.7–6.9%, available N by 5.6–11.0%, available P by 4.9–27.6%, available K by 10.3–11.2%, soil microbial biomass carbon (SMBC) by 4.7–8.4%, and dehydrogenase activity (DHA) by 11.0–32.3% as compared to maize-fallow. They also observed improvement in soil moisture stock and water holding capacity due to inclusion of pulses in the maize-based cropping system (Table 2.8; Fig. 2.5). Legumes can increase SOC content in several ways, by supplying biomass, organic C, and N to soil (Lemke et al. 2007; Meena et al. 2017c), as well as releasing hydrogen gas as by-product of BNF, that promote nodule development in the rhizosphere (La Favre and Focht 1983). Legumes also promote SOC content through breakdown of its residues

Table 2.8 Effect of cropping systems on maize equivalent yield (MEY), soil physical, chemical and biological properties after harvesting of winter crops

Cropping system	MEY (Mg/ha)	Bulk density (Mg/ m³)	SOC (%)	Available nutrients (kg/ha)			SMBC (μg/g soil)	DHA (μgTPF/ hr./g)
				N	P	K		
Maize-fallow	3.12	1.28	1.59	242.0	12.3	249.7	238.5	28.2
Maize-French bean (bunch type)	7.02	1.28	1.68	259.9	12.9	275.5	249.8	32.0
Maize-French bean (pole type)	8.39	1.27	1.72	255.5	12.0	277.6	258.6	37.3
Maize-black gram	6.55	1.24	1.70	268.6	15.7	248.4	255.8	31.3

Source: Ngangom et al. (2017a)

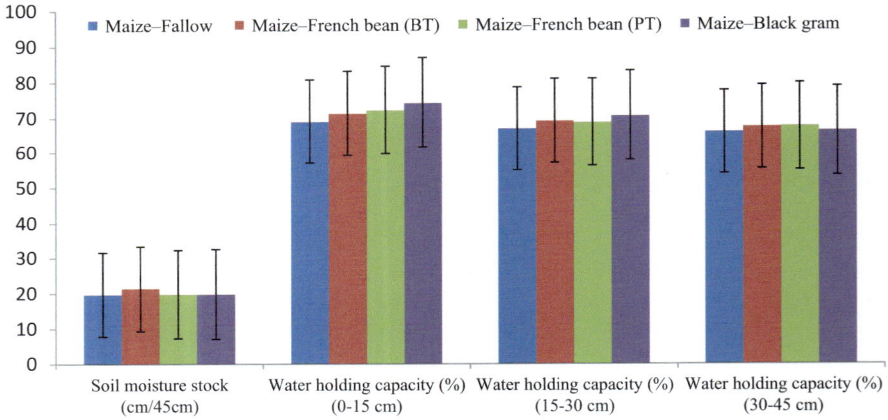

Fig. 2.5 Soil moisture stock and water holding capacity as affected by inclusion of pulses in maize-based system (Ngangom et al. 2017b)

(Köpke and Nemecek 2010; Alpmann et al. 2013). Minimun/no-till along with residue retention has been reported to reduce production costs substantially when combined with the diversification of the crop rotation by including a legume. Cost savings up to 21% has been reported when switched over from a conventional tillage, cereal-based rotation to a conservation tillage-based rotation involving legumes, compared to a cost savings of 12.5% only when the tillage system was changed, but the cereal-dominated rotation was maintained (Luetke-Entrup et al. 2003). Inclusion of pea in rotation with cereal exerted the most positive action to SOC content relative to other legumes (Hajduk et al. 2015). Legumes can reduce soil compaction by providing a continuous network of residual root channels and macropores in the subsoil, penetrating soil hardpans through its vigorous taproot systems (Jensen and Hauggaard-Nielsen 2003; Peoples et al. 2009). Intercropping of groundnut, French bean, rice bean [*Vigna umbellata* (Thunb.)], and cowpea has been reported to reduce soil loss and enhance soil properties and system productivity compared to mono-cropping of maize in sloping lands of Umiam, Meghalaya (Rajkhowa et al., 2016). Maize + soybean (2:2)-groundnut (237.6 kg/ha) and maize + soybean – French bean (232.2 kg/ha) had higher soil available N after three cropping cycles than rice + soybean (4:2)- tomato (225.8 kg/ha) and rice + soybean-rapeseed (220.5 kg/ha) in mid hills of EHR (Das et al. 2014a, b).

2.5.2 Efficient Utilization of P

Legumes solubilize soil phosphates through root exudates, and their deep taproot system contributes to efficient recycling and utilization of nutrients (Nuruzzaman et al. 2005, Jensen and Hauggaard- Nielsen 2003; Yadav et al. 2017b). Roots of leguminous crops releases organic acids like carboxylic acids that solubilize phos-vthat are generally not available to plants. The release of organic acids depends on the soil pH and P concentrations. More organic acids are generally released when soil P concentration is low, and depending upon species of legumes grown up to eight acids are released through root exudations (Egle et al. 2003). Release of organic acids through root exudations benefits the P uptake by the cereal component in mixture (Li et al. 2007) and also cereals crops grown as succeeding crop after legume crops (Nuruzzaman et al. 2005). Increase in P availability in soil and root rhizosphere due to acidification in maize-cowpea intercropping system than in sole cropping has been reported (Latati et al. 2016).

Significant reductions in GHG emission, viz., CO_2 and N_2O, due to low fertilizer especially N fertilizer and energy use in crop production with legumes have been reported (Jensen et al. 2011). However, CO_2 emissions from roots of legumes having nodules can be higher than those of other crops. This C emission is considered not to impact CO_2 concentrations in the atmosphere since it has been captured through recent photosynthesis (Jensen et al. 2011). Legume crops have the potential to reduce N_2O losses when grown in legume-cereal intercropping systems (Pappa et al. 2011). Furthermore, the inhibiting effect on denitrification in the root zone of some plants by rhizobia and other bacteria can contribute to reduction in N_2O emissions (Henry et al. 2008; Verma et al. 2015a, b, c). N_2O emissions to the atmosphere were also reported to be lower from legume systems than those of N-fertilized nonlegume crops and pastures (Dusenbury et al. 2008; Jensen et al. 2011).

2.6 Scope of Legume in Existing Cropping Systems

Pulses are considered the key crops for intensification of rice-fallows (Ghosh et al. 2016). Pulses being short duration, hardy, and low-input requiring in nature offer a tremendous opportunity to utilize residual soil moisture in rice-fallow (Ghosh et al. 2016). Further, given unique characteristics of biological nitrogen fixation (BNF),

Table 2.9 Cropping system in different Himalayan agroclimatic region

Agroclimatic zones	States represented	Annual rainfall (mm)	Cropping system
Western Himalayan region (dry temperate, terraced farming, low biodiversity)	Jammu and Kashmir, Himachal Pradesh, Uttar Pradesh	1650–2000	Rice-chickpea/lentil/field pea, maize-chickpea/field pea, ragi-chickpea/lentil/field pea, maize-urd bean/mung bean/wheat, pigeon pea-wheat, mung bean/urd bean-mustard, common bean-potato
Central Himalayan region (subtropical in Southern foothills, warm temperate in middle Himalayan valleys, cool temperate in middle Himalayas)	Uttarakhand	490–1570	Rice-pea, sorghum (grain/fodder)-chickpea, fallow-chickpea, sorghum + pigeon pea-fallow, pearl millet + pigeon pea-fallow, rice/maize-chickpea/lentil, moth bean/mung bean/urd bean-wheat, rice/maize-chickpea/lentil/field pea, moth bean/mung bean/urd bean-wheat, pearl millet-chickpea
Eastern Himalayan region (moist subtropical, shifting cultivation, high biodiversity)	Manipur, Meghalaya, Nagaland, Arunachal Pradesh, Sikkim, Tripura, Mizoram, hilly tracts of Assam, West Bengal	1840–3530	Rice-urd bean/mung bean, rice-lathyrus, maize-maize-urd bean, maize-pigeon pea/horse gram, maize-chickpea/lentil/field pea, jute-urd bean-chickpea/lentil, soybean/groundnut-rapeseed/mustard

Source: Singh et al. (2005); Pattanayak and Das (2017)

deep rooted, potential to establish with surface broadcast in standing rice fields and soil fertility restoration property pulses can be best fitted in rice-fallows (Ali et al. 2014). Further, inclusion of pulses in rice-fallows is a low-cost approach that can improve the farm income of the resource poor hill farmers. In spite of a few point of preference over other crops, a number of abiotic factors related to soil and water largely limit the production potential of pulses in rice-fallows. In addition, biotic factors also cause severe loss to the pulse crops in rice-fallows (Bandyopadhyay et al. 2015). Consequently, strategic management options needs to be worked out to cope up with the challenging environments of rice-fallows (Das et al. 2016). Legumes due to their competitive nature and environmental and socioeconomic benefits are suitable for introduction in modern cropping systems to increase crop diversity and reduce use of fertilizer and other external inputs (Stagnari et al. 2017). Cropping systems involving pulses in Himalayan agroclimatic region are given in Table 2.9.

2.6.1 Potential Future Pulse-Based Cropping System for IHR

Most of the IHR states are deficient in food grain production and are expected to increase in deficiency by 2050 due to increasing population growth. The CI of the IHR states is also low. However, some of the states like Tripura have already achieved 186% CI, thus showing the way for the other states to follow. With the inclusion of pulses in the existing system, adequate conservation measures, efficient use of natural resources, improved seeds, planting materials, and policy supports, it is possible to achieve at least 200% CI even under rainfed conditions. When maize is planted by last week of April or first week of May, it is harvested by August, which is the peak period of monsoon in Northeast India. The crops like French bean, black gram, rapeseed, etc. can be cultivated with residual moisture. A profitable crop yield is achieved when timely sowing of *rabi* crops done along with adequate nutrient and pest management is practiced. Similarly, after rice harvest in October for upland and November for lowland, pulses like pea, lentil, and oilseeds like rapeseed can be grown with residual moisture (Das et al. 2016). In case of upland, lifesaving irrigation is required during vegetative and flowering stages. With the change in lifestyle and increase in income, demand for rice will be reduced; instead consumption of pulses, fruits, and vegetables will increase. Thus, diversification of rice- and maize-based cropping system with pulses and vegetables is the need of the hour. Also, many farmers and entrepreneurs are looking for commercial livestock and poultry farming. However, high cost of feed is the limiting factor for promotion of these sectors. Thus, cultivation of food-feed crops for dual purpose should be emphasized. Crops like maize, soybean, sweet potato, colocasia, elephant foot yam, dioscorea, etc. are having high-yield potential and feed value. 300% CI with tomato/potato/French bean in pre-*kharif* (Jan-May), okra in *kharif*, and French bean/rajmash in pre-*rabi* were achieved under raised beds in lowland (Das et al. 2014a; Varma et al. 2017a, b). Similarly, rice-pulses-French bean/buckwheat is possible in lowland with one or two irrigation facilities for pre-*rabi* crops. Among the maize-based cropping systems, maize (green cobs)-groundnut-mustard was the most profitable cropping system producing (11.90 Mg/ha) maize equivalent yield with highest economic return of Rs 13,550/ha under upland conditions in mid-altitudes followed by French bean (green pods)-French bean grain (Munda et al. 1999). When maize is taken as green cob, it can be harvested by the onset of monsoon, and regular rice or maize can be taken during rainy season. After rice/maize, pulses or oilseeds can be cultivated. Year 2015 and then 2016 are reported to be the warmest year in history. The NER also follows almost similar trend. There is widening of gap between day and night temperatures in many places. Rain fall is becoming more erratic and unpredictable with more of extreme events. Droughts and flood are becoming more frequent in the region (Das et al. 2009; Meena et al. 2016). Drought of 2006, 2009, and 2014 and floods of recent past are the major reason for worry to achieve sustainable food production in IHR. Thus, there is need for crop diversification and adoption of a farming system approach, so that in the event of drought or floods some or other components will survive and farmers will continue to get their livelihood support. Inclusion of tree components like leguminous *Parkia roxburghii* in farming system will add to the resilience of the system and assured income. Vertical intensification, terrace farming, kitchen gardening, family farming, hydroponics, aeroponics, etc. are some of the future cropping systems and crop production approaches to adopt climate change and make agriculture a profitable enterprise (Pattanayak and Das 2017).

The fact that NER of India has been identified as hub for organic food production, emphasis has to be given on high-value crops having more market demand. Crop rotations, including a leguminous fertility building crops and cash crops, are the main principles for nutrient management within organic production systems. Organic rotations are divided into phases: one phase improves the level of soil N, and another depletes it. The N building and depleting phases must be in balance, with opportunities for fertility building for long-term sustenance farming (Meena et al. 2015a; Berry et al. 2003). Leguminous cover crops such as soybean, rice bean, velvet bean, etc. are suitable intercrops with maize to utilize the land efficiently, increase ground cover while providing the N through biological fixation, and add organic manure to soil. Intercropping of cereals and legumes not only sustain productivity but also offers the opportunity to increase the use of symbiotically fixed N (Das et al. 2014a). Organic production of rice, pea, lentil, French bean, soybean, groundnut, etc. has been reported to sustain soil health and crop productivity in NER India (Das et al. 2014a). Some important futuristic cropping systems of Northeast India are given in Table 2.10.

2.6.2 Legume Cropping in Western and Central Himalayas

The farmers of western and central Himalayas have been practicing low-input agriculture with major emphasis on conserving crop biodiversity both at species and intraspecies level (Bisht et al. 2006; Meena et al. 2015c). The region has two distinct crop seasons upon which cropping patterns are evolved, viz., the winter or rabi season (from October to March) and the summer or *kharif* season (from April to October). The agriculture is predominantly rainfed (85%), and area under irrigation is only about 15% (Maikhuri et al. 1996). Rice and wheat are the most important crops in irrigated farms, whereas indigenous crops like *Amaranthus viridis*, *Eleusine coracana*, *Panicum miliaceum*, *Hordeum vulgare*, and *Setaria italica* and various legume crops like *Vigna radiata*, *Macrotyloma uniflorum*, *V. angularis*, *V. unguiculata*, *Pisum arvense*, *Glycine max*, etc. are raised in rainfed farms. The farm area under rainfed conditions is generally divided into two equal halves locally called as "Mullasari" and "Mallasari." Grain legumes are basically rainy season crop raised under rainfed conditions. Some leguminous crops like pea and lentil are cultivated during post-rainy (*rabi*) season. Some of these pulses are cultivated on the field bunds of rice fields in irrigated land, whereas few are grown only in kitchen gardens. In greater Himalayan region, legumes are mostly mixed with traditional nonlegumes like millets (*Eleusine*, *Echinochloa*, sorghum, and maize) and pseudocereals

Table 2.10 Present and futuristic cropping systems with pulses in Northeast India

State	Existing cropping system	Futuristic cropping system with pulses
Sikkim	Maize-fallow	Maize-*pahenlo* dal-buckwheat
	Maize-rice	Maize-vegetables pea
	Maize-rapeseed	Maize-*pahenlo* dal-rapeseed
	Maize-buckwheat	Maize + bean-rapeseed
Manipur	Rice-fallow	Rice-pea/lentil
	Vegetable-fallow	Rice-broad bean
	Rice-rice	Maize-broad bean
Meghalaya	Rice-fallow	Rice-pea/lentil/rapeseed
	Maize-fallow	Maize-French bean/black gram/rapeseed
	Rice-vegetables	French bean/carrot/potato-okra-French bean/rajmash
	Ginger/turmeric-fallow	Maize + soybean/groundnut-potato/tomato/French bean
		Broccoli-potato-rajmash/French bean
		Ginger/turmeric + soybean-rice bean
Tripura	Rice-rice	Rice-pea/lentil/rapeseed
	Rice-fallow	Rice-rice-pea/lentil
	Vegetable-fallow	Maize-rice-pea/lentil
		Maize-black gram/green gram
Nagaland	Rice-fallow	Rice-pea/lentil (lowlands)
	Maize-fallow	Maize-pulses (French bean, black gram)
Arunachal Pradesh	Rice-fallow	Rice-pea/lentil (lowland)
	Maize-fallow	Maize + soybean-French bean/black gram
Mizoram	Rice-fallow	Rice-pea/lentil (lowland)
	Maize-fallow	Maize + soybean-French bean/black gram/green gram

Source: Pattanayak and Das (2017); Das et al. (2016)

(*Amaranthus*, *buckwheat*, etc.). As many as 10–12 crops are grown together by the hill farmers with the objective to obtain maximum and diverse yield on per unit area basis (Shiva and Vanaja 1993; Ghosh and Dhyani 2004). Inclusion of nonleguminous crops like sorghum, maize, millets, etc. with legumes like French bean, cowpea, etc. provides support for climbing to the legume component and minimizes risks of disease and weed problems and also the harmful impacts of intensive cultivation of cereals on soil fertility. Cultivation of legume crops is much simple and requires less labor and attention than other crops. Rainy season legumes hardly require any irrigation, and rainwater is sufficient to meet its water requirement of the crop. At vegetative stage, when the crop roots grasp the soil firmly, a local farm implement called "Maaua" is operated in the field for soil loosening. Green and succulent pods of some legumes like *French bean* and cowpea are harvested early for use as green vegetables, but grains get ready for harvest around 125–135 days (Dhanai et al. 2016).

2.7 Cereal + Legume Intercropping

In intercropping, the crops are arranged in definite rows. Sowing of both main and intercrops may be done simultaneously or in staggered manner. Similarly, harvesting time may also differ. Intercropping ensures desired plant stand due to definite row arrangement and, hence, facilitates easy cultural operation, spraying of pesticides and harvesting, and higher returns. The major principles of intercropping are the varied and contrasting maturities, height and rooting pattern, nutrition needs, etc. so that the component crops in intercropping complement each other rather than competing for the resources such as water, light, nutrient, space, etc. and reduce risks of climatic adversities. Intercropping is particularly more productive under rainfed conditions due to climatic anomalies and poor availability of farming resources like credit, inputs, etc. Pulses are intercropped with cereals, oilseeds, millets, vegetables, etc. for efficient use of natural resources and assured income by small and marginal farmers (Ghosh et al. 2016, Das et al., 2016; Meena et al. 2017c).

Intercropping is a common practice, not only in the central Himalayas, India, but worldwide because it minimizes the risk of crop failure due to adverse effects of pests, improves the use of limited resources, reduces soil erosion, increases yield stability, and is cost-effective (Chandra et al. 2013). Several scientists have been working with cereal-legume intercropping systems (Dwivedi et al. 2015) and proved its success compared to the monocrops. Common crop combinations in intercropping systems of this region are cereal + legume, particularly maize + cowpea, maize + soybean, maize + pigeon pea, maize + groundnuts, maize + beans, sorghum + cowpea, millet + groundnuts, and rice + pulses (Dwivedi et al. 2015). Intercropping of cereals and legumes would be valuable because the component crops can utilize different sources of N (Chu et al., 2004). The cereal may be more competitive than the legume for soil mineral N, but the legume can fix N symbiotically if effective strains of *Rhizobium* are present in the soil. Intercropping is much less risky than monocropping considering that if one crop of a mixture fails, the component crops may still be harvested. Biological nitrogen fixation is the major source of nitrogen in legume-cereal mixed cropping systems when nitrogen fertilizer is limited (Fujita et al., 1992). Moreover, because inorganic fertilizers have much environmental damage such as nitrate pollution, legumes grown in intercropping are regarded as a sustainable and alternative way of introducing N into lower-input agroecosystems (Fustec et al., 2010). In addition, roots of the legume component can decompose and release nitrogen into the soil where it made available to subsequent crops. Intercrops of maize with legumes (e.g., maize + cow pea) can also substantially increase forage quantity and quality and decrease the requirements for protein supplements compared with maize sole crops (Javanmard et al. 2009). The amount of nitrogen fixed by the legume component in cereal-legume intercropping systems depends on several factors, such as species, plant morphology, density of component crops, rooting ability, type of management, and competitive abilities of the component crops (Dwivedi et al. 2015).

Table 2.11 Land equivalent ratio (LER) as affected by intercropping

System	LER	Location	References
Maize + soybean (2:2)	1.41	Northwest Himalayas	Pandey et al. (1999)
Maize + green gram (1:1)	1.54	West Bengal	Patra et al. (1999)
Maize + grain legume	1.22–1.54	Uttarakhand	Khola et al. (1999)
Maize + field pea (1:1)	1.51	Uttarakhand	Devi (2014)
Maize + field pea (2:2)	1.60	Uttarakhand	Devi (2014)
Maize + cowpea (1:2)	1.63	Eastern Himalayas	Choudhary et al. (2014)
Maize + French bean (1:2)	1.75	Eastern Himalayas	Choudhary et al. (2014)
Maize + black gram (1:2)	1.66	Eastern Himalayas	Choudhary et al. (2014)

Intercropping with legumes has a positive impact on symbiosis for nitrogen fixation and increasing soil fertility. The infertile land requires more nitrogen for proper plant growth and better yield, and thus the demand for soil nitrogen in Himalayan rainfed agriculture is increasing day-by-day. Unique characteristics like high protein content (2–3 times more than cereals), nitrogen-fixing ability, soil ameliorative properties, and ability to thrive better under unfavorable conditions make pulses an integral component of agriculture and cuisine in central Himalayas (Chandra et al. 2013). When nonleguminous crops like maize and millets are with legumes, they provide support for climbing to the latter, minimize disease and weed problems, and alleviate the negative impacts of continuous cereal cultivation on soil fertility. Five years study on a sloping land (35% slope) in eastern Himalayas (1000 m ASL) revealed that the average maize equivalent yield (MEY) was the highest for rice-lentil system (16.2 Mg/ha) followed by maize-French bean (14.6 Mg/ha) and maize-rapeseed system (9.01 Mg/ha), while the lowest MEY was recorded in monocropped farmers' practice (3.08 Mg/ha). Soil organic carbon (SOC) stock in the fifth year of study at 0–30 cm was much higher under maize-French bean system (61.4 Mg/ha) compared to monocropped maize (55.4 Mg/ha) (Rajkhowa et al. 2016). Land equivalent ratio (LER) indicates the area which will be required to produce the same productivity as that of an intercropping system. Intercropping of pulses with cereals like maize is reported to give LER of 1.22 to as high as 1.75 in different locations of IHR (Table 2.11).

2.7.1 Effect of Legumes on Succeeding/Associated Crops

Legumes improve soil fertility and increase yield of the subsequent or associated crops; the extent of such benefits however largely depends upon the total plant biomass production, N_2 fixation, amount of N added to the soil through roots and nodules, and the leaf fall which are governed by agroclimatic and management practices to a great extent (Wani et al. 1994). Various research findings suggest that the carry-over of N for succeeding crops may be 60–120 kg in berseem (*Trifolium alexandrinum*), 75 kg in Indian clover (*Medicago sativa*), 75 kg in cluster bean

(*Cyamopsis tetragonoloba*), 35–60 kg in fodder cowpea (*Vigna sinensis*), 68 kg in chickpea (*Cicer arietinum*), 55 kg in black gram (*Vigna radiata*), 54–58 kg in groundnut (*Arachis hypogaea*), 50–51 kg in soybean (*Glycine max*), 50 kg in Lathyrus, and 36–42 kg per ha in pigeon pea (*Cajanus cajan*) (Das and Ghosh 2012). Legumes with indeterminate growth are more efficient in N fixation capacity than determinate types.

Fodder legumes in general are more potent in increasing the productivity of succeeding cereals. Through a symbiotic association with legumes, *Rhizobium* bacteria can convert atmospheric N into an organic form in the root nodules of crops. The accumulation of N depends on the length of the growing season, local climate and soil conditions. If a legume is grown as a green manure crop, biomass N produced can (in some cases) supply the entire N requirement for the subsequent crop. The proportion of N from N fixation in crops ranges from zero – usually where environmental stresses are severe and prevent nodulation – to 98% in crops grown under ideal conditions. The amount of N fixed has been recorded as up to 450 kg/ha/crop in the tropics. Legume residues contain P, potassium, and other nutrients that are recycled in relatively available forms for subsequent crops (Meena et al. 2014).

2.8 Role of Legume in Improving Input Use Efficiency

Legumes are more efficient in using scarce natural resources like water, nutrients, etc. than cereals. Inclusion of legumes in cereal-based cropping systems enhances equivalent yield, production efficiency, land use efficiency (LUE), energy productivity (Fig. 2.6), water use efficiency (WUE), water productivity, benefit/cost (B:C) ratio, employment generation, etc. (Das et al. 2014b, c; Ngangom et al. 2017c).

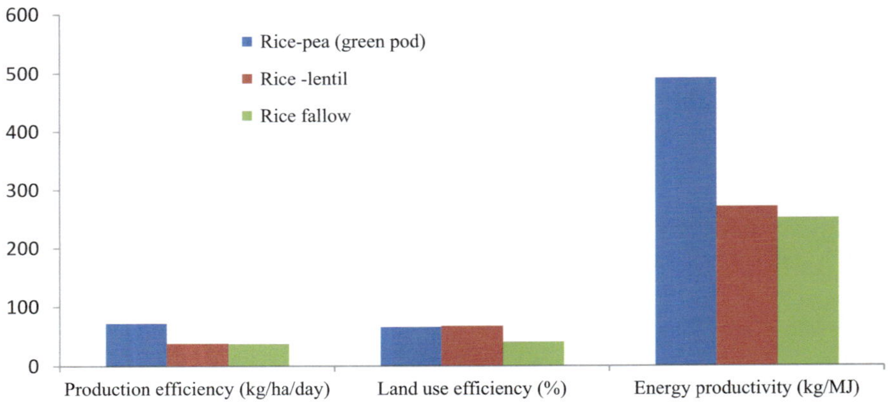

Source: Das et al. (2014b)

Fig. 2.6 Production efficiency, land use efficiency, and energy productivity as affected by inclusion of pulses in rice-fallow. (Source: Das et al. 2014b, c)

Table 2.12 Equivalent yield, WUE, water productivity, B:C ratio, and employment generation as affected by inclusion of pulses in rice and maize-fallow

Crop sequence	Rice/ maizeequivalent yield (Mg/ha)	WUE (kg/ ha.mm)	Waterproductivity (kg/m^3 water)	B:Cratio	Employment (man-days/ ha. year)
Rice-based cropping systems					
Rice-pea	10.72	19.69	0.64	5.78	230
Rice-lentil	6.42	12.34	0.34	2.24	210
Rice-fallow	3.26	5.99	0.18	2.17	100
Maize-based cropping systems					
Maize-French bean	7.07	11.58	0.32	3.88	150
Maize-rapeseed	3.84	6.98	0.23	2.30	110
Maize-fallow	2.64	6.23	0.16	2.39	60

Source: Das et al. (2014b)

The rice equivalent yield (REY) obtained with rice-pea sequence under CA was 229% higher compared to farmers' practice (FP) of rice monocropping at mid-altitude of Meghalaya in EHR. Similarly, maize-French bean under CA recorded 168% higher maize equivalent yield (MEY) compared to that of maize monocropping in Meghalaya hills (Das et al. 2014a). The enhancement in net return with maize-French bean and rice-pea under CA over respective FP were as high as 465 and 360%, respectively. Similarly, water use efficiency (WUE) enhanced by 228% under rice-pea sequence relative to monocropping of rice in mid-altitude of Meghalaya (Table 2.12). In the same study, the water productivity achieved with rice-pea (INR 9.57 m^{-3} water) and maize-French bean (INR 6.59 m^{-3} water) sequences under CA was significantly higher compared to that under FP of monocropping rice (INR 2.66 m^{-3} water) and maize (INR 1.80 m^{-3} water). Thus, the study indicated opportunity of CA cereal-pulse system in EHR for conserving natural resources and enhancing productivity (Das et al., 2014a). Production efficiency, land use efficiency, and energy use efficiency have been reported to increase when pea or lentil included as winter crop after lowland rice in HER (Das et al. 2014c).

Replacing traditional maize-wheat system with baby corn + French bean-pea-summer squash increased maize grain equivalent yield (190.8%), production efficiency (190.5%), productivity of the system (182.1%), gross returns (121.0%), net returns (176.4%), B:C ratio (64.1%), and profitability of the system (176.4%) in Palampur, Himachal Pradesh (Ramesh et al. 2016). They also reported that integrated nutrient management (75% RDF + 25% through FYM) also significantly increased maize grain equivalent yield by 4.3% over recommended dose of fertilizers. Padhi et al. (2010) found significant increase in grain yield of finger millet taken after cowpea (20 and 37%) and cluster bean (11 and 37%) grown for fodder or incorporation after harvest of green pods during *kharif* season compared to single crop of finger millet, respectively. In a study conducted at Pantnagar, Uttarakhand, Devi and Singh (2015) reported that intercropping of baby corn with field pea, besides increasing total production per unit area, also suppresses weed growth (33.78% and 30.68% higher weed control efficiency than sole baby corn and sole pea, respectively) and

Table 2.13 Effect of cropping sequence on equivalent yield, energy use efficiency, and income

Cropping system	System productivity (kg/ha per day)	Net energy return (x 000, MJ/ha)	Energy use efficiency (output/input ratio)	Energy productivity (g/ MJ)	Water use efficiency (kg/ha per mm)
Maize	4.0	11.49	10.10	0.317	4.74
Soybean	3.53	6.07	6.96	0.427	4.77
Groundnut	4.23	4.76	5.51	0.606	7.19
1:2 maize/ soybean	5.35	14.70	9.47	0.308	6.22
1:5 maize/ soybean	5.61	13.55	10.80	0.406	6.47
1:2 maize/ groundnut	5.75	14.24	9.07	0.326	6.77
1:5 maize/ groundnut	6.22	12.64	9.92	0.439	7.26

Source: Modified from Choudhury and Kumar (2016)

severity of diseases. Appropriate legume intercrop and planting geometry (row ratio) allow better light interception to the crop canopy and improve performance of cereal-based cropping system. In a study on eastern Himalayan region (Table 2.13), soybean and groundnut intercropping in different row ratios with maize resulted in improvement of water use efficiency by 83.2%, harvest monetary benefit by 87.5%, benefit/cost ratio by 92.3%, and energy productivity by 38.5% with 1:5 row ratios of maize/groundnut over solitary maize (Meena et al. 2015a; Choudhury and Kumar 2016). Choudhary et al. (2014) reported that the amount of N, P, and K removed by weeds was less in intercropping system of maize with pulses (cowpea, French bean, black gram) as compared with sole maize plot. Water use efficiency improved by 83.2%, harvest monetary benefit by 87.5%, benefit/cost ratio by 92.3%, and energy productivity by 38.5% with maize-groundnut intercropping (1:5) over solitary maize (Choudhary and Kumar 2016). They also reported that increased in interception of solar radiation in the bottom of the canopy ranged from 19% to 33% in an intercropping system of maize with groundnut and soybean.

2.8.1 Weed Smothering Efficiency of Pulses

Crops like rice and maize which are grown during rainy season suffer from excessive weed problem due to favorable climatic and microenvironmental conditions, causing a huge loss to productivity and income of farmers (Choudhury et al. 2015b; Dadhich et al. 2015). Maize being a wide-spaced crop and mostly cultivated during rainy season suffers from severe weed competition leading to 30–70% yield loss (Hugar and Palled, 2008). Intercropping of legumes provides the opportunities to suppress the weeds in addition to increasing system productivity (Banik et al. 2006). In a study on maize + legume intercropping system in Basar (660 m ASL), Arunachal Pradesh by Choudhary et al. (2014) reported that weed density were higher under sole maize followed by intercropping of 1:1 row proportion of maize with groundnut or maize with soybean. Soybean sole crop (66.7%) followed by 1:5 maize/soybean intercropping (56.4%) registered higher weed smothering efficiency over sole maize. Thus, intercropping of maize with groundnut or soybean in eastern Himalayan region enhanced system productivity and suppressed weeds. The maize-growing areas in the country are mostly under maize-legume systems. However, from region to region the differences lie mainly in the varieties and legume species. Grain legumes are mostly grown as intercrops, in sequence or rotations with maize in mid-altitude subhumid (common beans and soybean), highlands (faba bean and chickpea), dry land (common bean, pigeon pea, cowpea, and groundnut), and low-altitude subhumid (cowpea) agroclimatic conditions (Dwivedi et al., 2016). Intercropping of finger millet with black gram at a density not exceeding 75% of a sole black gram culture may improve overall yields and income from mountain agriculture system (Chandra et al., 2013). Intercropping of soybean with rice (4:2) has been reported to enhance system productivity and soil fertility in mid-altitude subtropical EHR (Das et al. 2014d).

2.9 Opportunity for Pulses Under Conservation Agriculture (CA) in NER

Excessive tillage may result in short-term increase in fertility but degrades soils in the medium and long run through structural degradation, loss of organic matter, erosion, and falling biodiversity. In order to keep production system in different land situations sustainable, CA based on minimum/no-till (NT) system, optimum residue management, and crop rotation is an alternative to conciliate agriculture with its environment and overcome the imposed constraints of the climate change and continuous escalation of inputs costs. Pulses do not need fine seed bed; they perform well even on rough seed bed with good aeration. Thus, there is good scope for adoption of conservation tillage in pulses (Kumar et al. 2016). Land configuration like raised and sunken beds or modified furrow-irrigated raised beds is also having good scope in poorly drained soils as pulses can be cultivated on the raised beds and water-loving crops like rice can be cultivated in the sunken beds (Das et al. 2014a). Conventionally in NER, after *kharif* rice, fields remain fallow in lowland, mainly due to excess moisture owing to seepage from surrounding hillocks in mid altitudes. A simple drainage around the rice fields/plots with appropriate outlets at physiological maturity creates the desirable situations for cultivation of pulse crops like pea and lentil (Das et al. 2012). Das et al. (2011) reported that among various field pea varieties tried, Prakash, Vikash, IPFD 99–13 recorded maximum green pod yield followed by IPFD 1–10-(3.29 Mg/ha), IPFD −99-25, and HUDP. Among the lentil varieties tested, DPL-15, DPL 62, IPL81, and IPL406, 1303 were found potential for NT cultivation in rice-fallows (Layek et al. 2014). It is also concluded that minimum tillage (MT) in rice followed by NT in pea/lentil was optimum for higher productivity of succeeding pea/lentil compared to conventional tillage (CT) in

Table 2.14 Yield attributes and yields of lentil as influenced by rice residue management practices under upland and lowland conditions

Ecosystems/residue management practices	Upland				Lowland			
	Pods/ plant	Seeds/ pod	1000 seed wt. (g)	Seed yield (Mg/ha)	Pods/ plant	Seeds/ pod	1000 seed wt. (g)	Seed yield (Mg/ha)
Residue retention	20.26	1.84	25.01	0.72	65.37	2.39	25.05	1.64
20 cm standing stubble	18.64	1.51	24.35	0.67	55.75	2.00	24.38	1.51
Residue removal	17.29	1.46	24.31	0.54	44.26	1.95	24.32	1.44
SEm±	0.53	0.05	0.08	0.03	1.01	0.05	0.09	0.06
CD ($p = 0.05$)	1.56	0.15	0.22	0.07	2.96	0.16	0.26	0.18

Source: Das et al. (2016)

rice-NT in pea/lentil and NT in rice-NT in pea/lentil under lowland conditions (Das et al. 2013). With appropriate agronomic interventions and varietal screening, pea and lentil could be popularized at mid-altitude for food and nutritional security of small and marginal farmers of the region (Ghosh et al. 2011). In a field study in Sikkim, Singh et al. (2015) reported significantly higher green pod yield (5.89 Mg/ha), gross returns (1.93×10^3 \$/ha; 1 \$ = 65 INR), net returns (1.48×10^3 \$ha), and B:C ratio (3.27) with NT over MT and CT. Similarly, NT required 44 and 28.3% less energy as compared to CT and MT, respectively. Inclusion of pulses, like pea and lentil, enhances farmers' income and employment substantially compared to rice-fallow or other systems. Farmer's net income enhanced by 446.2 \$/ha and 330.8 \$/ha over rice-fallow system due to inclusion of pea (vegetable) and lentil, respectively, in the system following NT practice (Das et al. 2012). Another field study in Meghalaya indicated that in situ retention of previous rice crop residues gave significantly higher grain yield of succeeding lentil (DPL 62) followed by 20 cm stubble under upland condition (Table 2.14). Similarly, yield enhancement of lentil due to in situ retention of previous rice crop residues was 14% higher compared to residue removal (1.44 Mg/ha) in lowland (NASF Annual Report 2015).

There is a good scope for expansion of small-seeded varieties of pulses like lentil, lathyrus, etc. under *utera* (*paira* cropping) cultivation in lowland rainfed rice-fallow land in NER (Das et al. 2014a). Field experiments conducted at ICAR Research Complex for NEH Region, Umiam, Meghalaya, on *utera* cultivation of lentil with rice recorded seed yield of 400 to 600 kg/ha (Das et al. 2014a). The *utera* cultivation can be more effective by using short-duration and high-yielding varieties of rice to vacate the field early giving sufficient time for pulse crop in the field.

Maize which is the second most important crop of NER is generally grown during *kharif* season in uplands, sloping lands, and shifting cultivated area. The early and timely sown maize (April/May) is harvested during July/August and allows cultivation of pulse crops like French bean, black gram, and green gram. Retention of standing stalk of previous maize crop (0.75– 1 m) followed by sowing of subsequent French bean crop by opening a narrow furrow in between two lines of previous maize crop using a NT seed drill or furrow opener is a recommended practice for conserving moisture, saving time, and enhancing CI in NER (Das et al. 2011). As maize is grown with a wider line spacing (60 cm), two lines of French bean/ pulses (30 cm spacing) are accommodated in between two lines of previous maize. This method saves about 50% cost in staking besides conserving soil moisture, saving time, and improving soil quality. Similarly, black gram and green gram are grown under NT tillage with conserved moisture in maize-fallow producing a seed yield of 1–1.5 Mg/ha (Das et al. 2011).

2.10 Strategies for Enhancing Area and Productivity of Legumes

Pulse production and productivity in India have been stagnant for the last few decades even after several efforts and program. Due to their low productivity, pulses are grown as residual/alternate crops with minimal inputs mostly on marginal lands in IHR after taking care of food/income needs from high-input high-productivity crops like rice, maize, and other cereals by the most farmers (Das et al., 2016). Pulses are mostly grown as rainfed crops with negligible fertilizer or other agro-inputs leading to thirsty and hungry conditions (Meena et al. 2016). Thus, poor yield of pulses is mainly related to poor management practices owing to their relatively low status (crop of secondary importance) in the cropping system (Ghosh et al., 2016). Further, pulse crops are adversely affected by a large number of biotic (insect-pest, diseases, weed) and abiotic (water, nutrient, low temperature, etc.) stresses, which are responsible for instability and poor yields to a great extent. Some of the strategies to enhance area and productivity of legumes in IHR are as follows.

2.10.1 Horizontal Inclusion of Pulses in Cropping System

Cropping system is broadly grouped into sequential cropping and intercropping. Sequential cropping may be a regular rotation of different crops in which the crops follow a definite order of appearance on the land or it may consist of only one crop grown year after year on the same area. To enhance the pulse production, it must be included in cropping system either in sequential (horizontal) or intercropping (vertical) (Fig. 2.7). The prominent sequential cropping systems involving different pulses are indicated below.

Rice-lentil/field pea/lathyrus In recent years, development of early- to medium-maturing varieties of lentil suitable for planting up to mid of December (HUL-57) with yield potential of 1.0–1.2 Mg/ha has enabled farmers to adopt rice-lentil system instead of rice monocropping especially in the valley areas, where residual moisture is available (Ansari et al. 2015a). Rice/maize-lentil cropping system has been successfully demonstrated in Manipur Hills of EHR (Ansari et al. 2015b). Introduction of short-duration pea varieties is also another option to enhance pulse production. In Manipur, vegetable/garden pea after rainy season rice has been found

Fig. 2.7 Diversion of monocropping rice to rice-lentil sequential cropping system at Sekmai Hijam Khunou village, Thoubal district, Manipur

to be promising in terms of productivity and income. Lathyrus has also very good potential after rice as relay crop. *Makhyatmubi* is a local cultivar of pea which is very popular in Manipur, and it has very high potential in the valleys under rice-pea system (Ansari et al. 2015a).

Pigeon Pea-Mustard With the advent of short-duration cultivars of pigeon pea (UPAS-120, Manak and Pusa 992), it can be grown in *jhum* cultivated areas of EHR, where pigeon pea can mature in 130–150 days and rapeseed-mustard can be taken as sequence crop. This will provide desired stability and sustainability to productivity of fragile *jhum* land system by sustaining soil fertility due to leaf fall and N fixation by pigeon pea (Ansari et al., 2015b). But there are some issues, which need to be addressed for wider adoptability and profitability from this system. Presently, most of the short-duration varieties of pigeon pea available for cultivation are affected by sterility mosaic, fusarium wilt, and *Phytophthora* blight diseases and have tendency to prolong maturity with the late monsoon. Therefore, development and adoption of suitable varieties, which could mature by early November with high-yielding potential, are required (Das et al. 2016).

Rice/Maize-Urd Bean/Mung Bean Cultivation of *rabi*/summer urd bean and mung bean is getting momentum due to availability of short-duration suitable high-yield potential cultivars. Summer urd bean after rice and urd bean/mung bean/rajma after maize are having high potential in terrace and hill agriculture areas (Das et al. 2016). Babu et al. (2016) reported that maize (green cobs)-urd bean (*Pahenlo dal*)-buckwheat system was the most resource-efficient system and recorded significantly higher system productivity (8.89 Mg/ha), net return (4.67×10^3 \$/ha), B:C ratio (2.59), and employment generation (285 man-days/ha) over other cropping systems. With regard to the energetics, this system recorded 188.9, 192, 25.8, and 6.5% higher gross energy return, net energy return, energy use efficiency, and energy productivity, respectively, over maize-fallow rotation.

Rice-Broad Bean Broad bean is very popular and choicest food legume in Manipur due to its high market demand and price. It is cultivated across the Manipur after harvesting of cereals. The local broad bean is a long-duration crop. Due to introduction of high-yielding short-duration cultivars, maize-broad bean system is becoming in Manipur hills of EHR (Ansari et al. 2015a).

Rice Bean/Soybean-Mustard Rice bean is an indigenous legume crop of EHR and extensively grown in terraces, hill slope, and foot hills (except low land areas). Local cultivars have high-yield potential, but they are long duration in nature. Inclusion of high-yielding short-medium-duration cultivars of rice bean will provide enough opportunity to take mustard in sequential cropping (Das et al. 2016).

2.10.2 Vertical Inclusion of Pulses in Cropping System

Intercropping of pulses with cereals is more successful in terrace and *jhum* cultivated areas, where there is least possibility of water stagnation. Intercropping of pulses with maize is more popular in terrace cultivation, and mixed cropping of pulses with rice/maize is more popular in *jhum* cultivated areas of Manipur. Maize + rice bean (Fig. 2.8), maize + soybean, and maize + Urd bean/mung bean (Fig. 2.9) are more remunerative in *kharif* season. Perennial pigeon pea in intercropping system is generally grown on bunds with cereals, oilseeds, and pulses. In *rabi* season, mustard + lentil/lathyrus/pea, *Makhyatmubi* (local pea variety) + vegetables (Fig. 2.10), and broad bean + vegetables are more remunerative intercropping systems.

Fig. 2.8 Maize + rice bean grown in 2: 4 ratios in terrace land at ICAR Langol experimental farm, Manipur, maize harvested during pod initiation stage, and maize stalk act as support for trailing

Fig. 2.9 Maize + soybean (left) and maize + urd bean (right) intercropping systems

Fig. 2.10 *Makhyatmubi* (local pea variety) + cabbage and maize + pigeon pea intercropping at Thoubal, Manipur

2.10.3 Cultivation of Pulses in Field Bund

In IHR, due to difficult terrain, steep slope, erosion hazard, etc., in most of the cases, up to 10% of the cultivated land is left for field bunds, terrace risers, and other conservation measures. In these bunds, pulses like pigeon pea, soybean, green gram, black gram, etc. can be grown. This practice besides increasing pulse production and income will help in conserving soil, water, and other natural resources (Das et al. 2016). Similar practice is also popular in other Himalayan countries like Nepal (Gharti et al. 2014).

2.10.4 Inclusion of Pulses in Shifting Cultivation

The continuance of *jhum* in the NER is closely linked to ecological, socioeconomic, cultural, and land tenure systems of tribal communities. Since the community owns the lands, the village council or elders divide the *jhum* land among families for their subsistence on a rotational basis in most *jhum* dominated areas. The dry broadcast method involves sowing in the month of March/April and harvesting in August/September. Wet sowing is done in the month of May/June and harvested during October/November. Shifting cultivation is widely practiced in hilly and sloping areas, and settled terrace farming is done in foothill or low slope areas, mostly near the adjacent rivers, streams, or other water sources. Depending on the slope, wet broadcast on bunded fields or dry broadcast on unbunded fields is practiced. In Nagaland, alder (*Alnus nepalensis*)-based *jhum* farming is common, where alder trees are maintained in the field by regular lopping and crops are cultivated in interspaces. Rice bean, soybean, pigeon pea, etc. are cultivated in *jhum* fields as mixed crop with rice, maize, and millets in the IHR (Das et al. 2016). There is tremendous opportunity to enhance pulses production especially in *Jhum* areas, where farmers are growing crops in mixed cropping on rotational basis. Some potential pulses for *jhum* areas are pigeon pea, rice bean, soybean, mung bean/urd bean, broad bean, Makhyatmubi, winged bean, cowpea, lima bean, and tree bean (*Parkia roxburghii* indigenous to Himalayan states like Manipur, Nagaland, etc.) either as sole cropping or intercropping or agroforestry system. One of the ways to increase yield in this area is adoption of intercropping system which is cultivation of two or more crop species in the same field with definite row proportions. In intercropping system, crops are complementary in terms of growth pattern, aboveground canopy, rooting system, and their water and nutrient demand (Singh et al. 2008).

2.10.5 Farm Mechanization

Large proportions of cultivated area in IHR are hilly and mountainous, and farm mechanization is nonexistent. As compared to other parts of the country, low-cost farm machineries and implements for the hilly terrains should be made available for the farmers.

2.10.6 Transfer of Technology

Transfer of technology programs (farmers' trainings, frontline demonstrations, sensitization programs, etc.) have to be organized to familiarize the farmers with the improved production technology of pulse crops and hammer home the point that yield will increase substantially if these crops are grown as per the recommended package of practices. Such programs are very important to bridge the gap between actual and potential yield of legume crops.

2.11 Future Perspectives

The population of India will grow to about 1.69 billion by 2050, and the total pulse requirement will be 32 million tons. Similarly the IHR population will also grow due to poor and marginal nature of the farmers, low level of education, and dependence on agriculture as source of livelihood. To meet the required level of pulse production, an annual growth rate of 2.2% is required (Singh and Pratap, 2015). EHR alone has about 1.5 million hectare area under rice-fallow and a substantial area under maize fallows. Bringing these areas under pulses (at least 25% area) will contribute significantly to improving pulses scenario of the IHR and country (Das et al. 2014a,c). Reducing postharvest losses to the tune of 10–15% from the current losses will also add to ~ 1 million tons of pulses. Since pulses are the most important constituents of vegetarian diets for supplementing proteins, target-oriented breeding strategies are required for improving the content and quality of proteins in these. The availability of critical inputs like quality seeds, manure, biofertilizers, micronutrient, biopesticides, etc. and lifesaving irrigation through micro-irrigation and rainwater harvesting and its efficient utilization will further lead toward reducing the yield gaps in pulses. Efforts are to be focused on development of bio-intensive eco-friendly integrated pest management module for managing the major insect pests and diseases which restricts adequate benefit from pulses. To combat abiotic stresses such as drought, water logging, etc., wild gene pool is being explored to introgress resistance/tolerance genes through pre-breeding activities (Singh and Pratap, 2015). IHR being vulnerable to various kinds of land degradation due high rainfall, monocropping, faulty cultivation practices, steep slopes, and low use of manure and fertilizers, legumes, and pulses will have to play a vital role for providing food security and soil sustainability in the region (Das et al. 2016). There is an urgent need to initiate a National Mission on Pulses specific to IHR for food, nutritional, and environmental security of the fragile ecosystem of the world. It is rightly said that "a good health of Himalayas will ensure a good health of the country and possibly the world." Thus, all stakeholders has to join hand to save the Himalayan Ecosystem and agriculture through dissemination and demonstration of sustainable best practices like inclusion of legumes and pulses in cropping systems among many other approaches.

2.12 Conclusion

India is the largest pulse producer and consumer of the world, but ironically our pulse demand is not fulfilled by the home-grown pulses rather depend on the import which create heavy burden on the government exchequer and also raise the question on the various stakeholder involved in agriculture. The answer lies within the system only, but it requires the concerted effort of all, right from the farmers to policy makers. The potential of the IHR in legume production is still untapped as the entire region provides the opportunity in the rice/maize-fallow area for pulse production. The entire region consider pulses as the crop of secondary importance due to which it does not attract much of the farmer's crop management attention. There is the need of mass awareness campaign in the entire region on the multifarious benefits of legumes such as enriching the soil with the available nitrogen, conserves soil and nutrients by reducing runoff and erosion, a source of plant proteins, mitigating greenhouse gases emissions, breaking the cycles of pests and diseases through crop diversification, and contributing to balance the deficit in plant protein production. In the present climate variability and degrading natural resource scenario, pulses should be given the enough attention at the field level in the entire Indian Himalayan Region for livelihood security and environmental sustainability.

References

Abitogun AS, Olasehinde EF (2012) Nutritional evaluation of seed and characterization of crude jack bean (*Canavalia ensiformis*) oil. IOSR J Appl Chem 1(6):36–40

Agricultural Statistics at a glance (2014) Government of India, Ministry of Agriculture, Department of Agriculture and Cooperation, Directorate of Economics and Statistics

Akpapunam MA, Sefa-Dedeh S (1997) Jack bean (*Canavalia ensiformis*): Nutrition related aspects and needed nutrition research. Plant Foods Hum Nutr 50(2):93–99

Ali M, Ghosh PK, Hazra KK (2014) Chapter 7: Resource conservation technologies in rice fallow. In: Ghosh PK, Kumar N, Venkatesh MS, Hazra KK, Nadarajan N (eds) Resource conservation technology in pulses. Scientific Publishers, Jodhpur, pp 83–88. ISBN: 978-81-7233-885-5

Alpmann D, Braun J. Schäfer BC (2013) AnalyseeinerBefragunguntererfolgreichenKörnerleguminosenanbauernimkonventionellenLandbau. ErsteErgebnisseausdemForschungsprojekt

Angus JF, Kirkegaard JA, Hunt JR, Ryan MH, Ohlander L, Peoples MB (2015) Break crops and rotations for wheat. Crop Pasture Sci 66:523–552

Ansari MA, Prakash N, Baishya LK, Sanatombi Kh, Bungbungcha Ch, Singh TB (2015a). Participatory evaluation of lentil in rice-fallow with residual in-situ soil moisture under different tillage practices. Compendium of National Seminar on Sustaining Hill Agriculture in Changing Climate. Indian Association of Hill Farming and ICAR Research Complex for NEH Region, Meghalaya, pp 342–344. 5–7 December, 2015. Agartala, Tripura

Ansari MA, Prakash N, Sanatombi Kh (2015b) Enhancing food and nutritional security under rice-fallow cropping system for sustainable rice-lentil production system in North Eastern India. National Seminar on advances of life science, Manipur University, March 23–24, 2015, pp 97–98

Babu S, Singh R, Avasthe RK, Yadav GS, Rajkhowa DJ (2016) Intensification of maize (*Zea mays*) –based cropping sequence in rainfed ecosystem of Sikkim Himalayas for improving system productivity, profitability, employment generation and energy-use efficiency under organic management condition. Indian J Agric Sci 86(6):778–784

Bandyopadhyay KK, Sahoo RN, Singh R, Pradhan S, Singh S, Krishna G, Pargal S, Mahapatra SK (2015) Characterization and crop planning of Rabi fallows using remote sensing and GIS. Curr Sci 108(11):2051–2062

Banik P, Midya A, Sarkar BK, Ghose SS (2006) Wheat and chickpea intercropping systems in an additive series experiment: advantages and weed smothering. Eur J Agron 24:325–332

Berry PM, Stockdale EA, Sylvester-Bradley R, Philipps L, Smith KA, LordEI WCA, Fortune S (2003) N, P and K budgets for crop rotations on nine organic farms in the UK. Soil Use Manage 19:112–118

Bhatt DK, Joshi VK, Arora RK (1999) Conodont biostratigraphy of the lower Triassic in Spiti Himalaya, India. Geol Soc India 54(2):153–167

Bhatta KP, Vetaas OR (2016) Does tree canopy closure moderate the effect of climate warming on plant species composition of temperate Himalayan oak forest? J Vegetation Sci 27(5):948–957

Bhattacharyya R, Prakash V, Pandey SC, Kundu S, Srivastva AK, Gupta HS (2009) Effect of fertilization on carbon sequestration in soybean-wheat rotation under two contrasting soils and management practices in the Indian Himalayas. Aust J Soil Res 47:592–601

Bhojvaid PP, Timmer VR (1998) Soil dynamics in an age sequence of *Prosopis juliflora* planted for sodic soil restoration in India. Forest Ecol Manage 106:181–193

Bisht IS, Rao KS, Bhandari DC, Sunil N, Maikhuri RK, Dhillon BS (2006) A sustainable site for in situ (on-farm) management of plant diversity in traditional agroecosystems of western Himalaya in Uttaranchal state: a case study. Genetic Res Crop Evol 53:1333–1350

Buergelt D, Oppen MV, Yadavendra JP (2009) Quality parameters in relation to consumer's preferences in rice bean. Presentation at the International Conference on Grain Legumes: Quality Improvement, Value Addition and Trade, February 14–16, 2009, Kanpur, India

Chalk PM (1998) Dynamics of biologically fixed N in legume-cereal rotations: a review. Austr J Agric Res 49:303–316

Chandra A (2007) Traditional agrodiversity management in central Himalayan village ecosystem. Ph.D. thesis, University of Delhi, Delhi, India
Chandra A, Pardha SP, Maikhuri RK, Saxena KG, Rao KS (2011) Traditional agro diversity management: a case study of central Himalayan village ecosystem. J Mountain Sci 08:62–74
Chandra A, Kandari LS, Negi VS, Maikhuri RK, Rao KS (2013) Role of intercropping on production and land use efficiency in the Central Himalaya, India. Environ Int J Sci Technol 8:105–113
Chethan S, Malleshi NG (2007) Finger millet polyphenols: characterization and their nutraceutical potential. Am J Food Technol 2(7):582–592
Choudhary VK, Kumar PS (2016) Productivity, Water Use and Energy Profitability of Staggered Maize–Legume Intercropping in the Eastern Himalayan Region of India. Proc Natl Acad Sci, India, Section B: Biological Sciences 86(3):547–557
Choudhary VK, Dixit A, Kumar PS, Chauhan BS (2014) Productivity, Weed Dynamics, Nutrient Mining, and Monetary Advantage of Maize-Legume Intercropping in the Eastern Himalayan Region of India. Plant Prod Sci 17(4):342–352
Choudhury BU, Fiyaz Abdul R, Mohapatra KP, Ngachan SV (2015a) Impact of land uses, agro physical variables and altitudinal gradient on soil organic carbon concentration of North-Eastern Himalayan Region of India. Land Degrad Dev 27:1163. https://doi.org/10.1002/ldr.2338
Vijay Kumar Choudhary, Anil Dixit, Paramasivam Suresh Kumar, Bhagirath Singh Chauhan, (2015b) Productivity, Weed Dynamics, Nutrient Mining, and Monetary Advantage of Maize-Legume Intercropping in the Eastern Himalayan Region of India. Plant Production Science 17(4):342–352
Christopher SF, Lal R (2007) N management affects carbon sequestration in North American cropland soils. Crit Rev Plant Sci 26:45–64
Chu GX, Shen QR, Cao JL (2004) Nitrogen fixation and N transfer from peanut to rice cultivated in aerobic soil in intercropping system and its effect on soil N-fertility. Plant Soil 263:17–27
Dadhich RK, Meena RS, Reager ML, Kansotia BC (2015) Response of bio-regulators to yield and quality of Indian mustard (*Brassica juncea* L. Czernj. and Cosson) under different irrigation environments. J App and Nat Sci 7(1):52–57
Darlong P (1996) Soil erosion in the North East: Past, present and future. In: Zahid Hussain (ed)
Das A, Ghosh PK, Choudhury BU, Patel DP, Munda GC, Ngachan SV, Chowdhury P (2009) Climate change in northeast India: recent facts and events –worry for agricultural management, ISPRS Archives XXXVIII-8/W3 Workshop Proceedings: Impact of Climate Change on Agriculture, 32–37.
Das A, Ghosh PK (2012) Role of legumes in sustainable agriculture and food security: an Indian perspective. Outlook Agric 41:4. https://doi.org/10.5367/oa.2012.0109
Das A, Patel DP, Munda GC, Ghosh PK, Ngachan SV, Choudhury BU, GI Ramkrushna, Saha R, Rajkhowa DJ, Panwar AS, Rajesh K, Manoj K, Juri B (2011) Conservation agriculture in rice and maize based cropping systems: package and practices for NEH Region. Research Bulletin no. 75. ICAR Research Complex for NEH Region, Umiam, Meghalaya, p 36
Das A, Ramkrushna GI, Patel DP, Choudhury BU, Munda GC, Rajkhowa DJ, Ngachan SV (2012) Zero tillage pea, lentil and toria cultivation in rice fallow for diversification and resource conservation in hills. ICAR Research Complex for NEH Region, Umiam-793 103, Meghalaya
Das A, Ramkrushna GI, Ngachan SV, Munda GC (2013) Resource conservation technologies in pulse based cropping systems in NEH region. In: Ghosh PK, Kumar N, Venkatesh MS, Hazra KK, Nadarajan (eds) Resource conservation technology in pulses. Scientific Publishers, Jodhpur, pp 43–57
Das A, Patel DP, Munda GC, Ramkrushna GI, Kumar M, Ngachan SV (2014a) Improving productivity, water and energy use efficiency in lowland rice (*Oryza sativa*) through appropriate establishment methods and nutrient management practices in the mid-altitude of north-east India. Expl Agric 50(3):353–375
Das A, Ramkrushna GI, Choudhury BU, Ngachan SV, Tripathi AK, Singh RK, Patel DP, Tomar JMS, Mohapatra KP, Layek J, Munda GC (2014b) Conservation agriculture in rice and maize based cropping systems for enhancing crop and water productivity - participatory technology demonstration in north east India. Indian J Soil Conserv 42(1):196–203

Das A, Patel DP, Ramkrushna GI, Munda GC, Ngachan SV, Buragohain J, Kumar M, Naropongla (2014c) Crop diversification, crop and energy productivity under raised and sunken beds: results from a seven-year study in a high rainfall organic production system. Biol Agric Hort 30(2):73–87

Das A, Patel DP, Kumar M, Ramkrushna GI, Ngachan SV, Layek J, Lyngdoh M (2014d) Influence of cropping systems and organic amendments on productivity and soil health at mid altitude of North East India. Indian J Agric Sci 84(12):1525–1530

Das A, Babu S, Yadav GS, Ansari MA, Singh R, Baishya LK, Rajkhowa DJ, Ngachan SV (2016) Status and strategies for pulses production for food and nutritional security in north-eastern region of India. Indian J Agron 61(4th IAC Special issue):129–143

Datta R, Kelkar A, Baraniya D, Molaei A, Moulick A, Meena RS, Formanek P (2017) Enzymatic degradation of lignin in soil: a review. Sustain MDPI (9):1163. https://doi.org/10.3390/su9071163, 1–18

Devi MT (2014) Studies on planting pattern and weed management in field pea + baby corn intercropping system. Part of Ph.D. thesis submitted to G.B. Pant University of Agriculture and Technology, Pantnagar

Devi MT, Singh VK (2015) Composition of weeds and yield as affected by different weed management and planting patterns in field pea and baby corn intercropping system. Ann Agric Res 36(3):279–289

Dhakal Y, Meena RS, Kumar S (2016) Effect of INM on nodulation, yield, quality and available nutrient status in soil after harvest of green gram. Legum Res 39(4):590–594

Dhanai R, Negi RS, Singh S (2016) Rural transformation by agriculture diversification and innovation adoption: a study from Rudraprayag district, Garhwal Himalaya, India. Int J Eng G Tech 2:90–103

Dhruvanarayan VVN, Ram B (1983) Estimation of soil erosion in India. J Irrig Drain Eng 109:419–434

Dusenbury MP, Engle RE, Miller RP, Lemke RL, Wallander R (2008) Nitrous oxide emissions from a northern great plains soil as influenced by N management and cropping systems. J Environ Qual 37:542–550

Dwivedi A, Dev I, Kumar V, Yadav RS, Yadav M, Gupta D, Singh A, Tomar S (2015) Potential role of maize-legume intercropping systems to improve soil fertility status under smallholder farming systems for sustainable agriculture in India. Int J Life Sci Biotechnol Pharma Res 4(3):14–157

Dwivedi A, Singh A, Naresh RK, Kumar M Kumar V, Bankoti P, Sharma DK, Thaneshwar SA, Singh O (2016) Towards sustainable intensification of maize (*Zea mays* L.) + legume intercropping systems; experiences; challenges and opportunities in India; a critical review. J Pure Appl Microbiol 10(1):725–740

Egle K, Romer W, Keller H (2003) Exudation of low molecular weight organic acids by *Lupinus albus* L., *Lupinus angustifolius* L. and *Lupinus luteus* L. as affected by phosphorus supply. Agronomie 23:511–518

FAO (1982) Legumes in human nutrition. FAO, Rome

Fujita K, Ofosu-Budu KG, Ogata S (1992) Biological nitrogen fixation in mixed legume-cereal cropping systems. Plant Soil, 141:155–176

Fustec J, Lesuffleur F, Mahieu S, Cliquet JB (2010) Nitrogen rhizodeposition of legumes. A review. Agron Sustain Dev 30:57–66

Ganie MA, Akhter F, Bhat MA, Najar GR (2014) Growth, yield and quality of French bean (*phaseolus vulgaris* l.) As influenced by sulphur and boron application on inceptisols of Kashmir. The Bioscan 9(2):513–518

Gharti DB, Darai R, Subedi S, Sarker A, Kumar S (2014) Grain legumes in Nepal: present scenario and future prospects. World Agric Res 2(5):216–222

Ghosh P, Dhyani PP (2004) Baranaaja: the traditional mixed cropping system of the Central Himalaya. Outlook Agric 33:261–266

Ghosh PK, Das A, Munda GC, Patel DP, Saha R (2011) Prospects and opportunities for conservation agriculture in rice based cropping systems of North East India. Conserv Agric News Lett (PACA) 18:4–5

Ghosh PK, Hazra KK, Nath CP, Das A, Acharya CL (2016) Scope, constraints and challenges of intensifying rice (*Oryza sativa*) through pulses. Indian J Agron 61(4th IAC Special issue):122–148

Gómez C (2004) Cowpea: post-harvest operations. Food and Agriculture Organization of the United Nations (FAO), Rome

Gupta HS, Sarma BK, Sahay G (1998) Present status and prospects of pulse production in North Eastern Hill region. Indian J Hill Farming 11(1&2):77–87

Hajduk E, Właśniewski S, Szpunar-Krok E (2015) Influence of legume crops on content of organic carbon in sandy soil. Soil Sci Ann 66:52–56

Hazarika UK, Munda GC, Bujarbaruah KM, Das A, Patel DP, Prasad K, Kumar R, Panwar AS, Tomar JMS, Bordoloi JS, Sharma M, Gogoi G (2006) Nutrient management in organic farming. Technical Bulletin No. 30.ICAR Research Complex for NEH Region, Meghalaya

Henry S, Texier S, Hallet S, Bru D, Dambreville C, Chèneby D, Bizouard F, Germon JC, Philippot L (2008) Disentangling the rhizosphere effect on nitrate reducers and denitrifiers: insight into the role of root exudates. Environ Microbiol 10:3082–3092

Hernanz JL, Sanchez-Giron V, Navarrete L (2009) Soil carbon sequestration and stratification in a cereal/leguminous crop rotation with three tillage systems in semiarid conditions. Agric Ecosyst Environ 133:114–122

Hugar HY, Palled YB (2008) Effect of intercropping vegetables on maize and associated weeds in maize-vegetable intercropping systems. Karnataka J Agric Sci 21(2):159–161

Indian Ministry of Statistics and Programme Implementation (2014) Gross state domestic products. http://mospi.nic.in/Mospi_New/upload/State_wise_SDP_2004-05_14mar12.pdf. Accessed on Oct 2017

IPCC (2007) Climate change: synthesis report. Summary for policymakers. Intergovernmental Panel on Climate Change (IPCC)

Javanmard A, Nasab ADM, Javanshir A, Moghaddam M, Janmohammadi H (2009) Forage yield and quality in intercropping of maize with different legumes as double cropped. J Food Agric Environ 7:163–166

Jensen ES, Hauggaard-Nielsen H (2003) How can increased use of biological N2 fixation in agriculture benefit the environment? Plant Soil 252:177–186

Jensen ES, Peoples MB, Boddey RM, Gresshoff PM, Hauggaard-Nielsen H, Alves BJR, Morrison MJ (2011) Legumes for mitigation of climate change and the provision of feedstock for biofuels and biorefineries. A review. Agron Sust Dev 32:329–364

Jensen ES, Peoples MB, Boddey RM, Gresshoff PM, Hauggaard-Nielsen H, Alves BJ, Morrison MJ (2012) Legumes for mitigation of climate change and the provision of feedstock for biofuels and biorefineries. A review. Agron Sustain Dev 32:329–364

Jodha NS, Shrestha S (1994) Towards sustainable and more productive mountain farming. In: Proceedings of the international symposium on mountain environment and development: constraints and opportunities. ICIMOD, Kathmandu

Joshi PK (1998) Performance of grain legume in the Indo-Gangetic plain. In: JVDK KR, Johansen C (eds) Residual effect of legume in rice–wheat cropping system of Indo-Gangetic plains. International Crop Research Institutes for Semi-Arid Tropics (ICRISAT), Oxford & IBH Publishing Co. Pvt. Ltd, New Delhi

Kala CP (2005) Ethnomedicinal botany of the Apatani in the Eastern Himalayan region of India. J Ethnobiol Ethnomed 1:11. https://doi.org/10.1186/1746-4269-1-11

Kempster PA, Bogetic RC, Secombe JW, Martin HD, Blazs NDH, Wahlqvist ML (1993) Motor effects of broad beans (*Vicia faba*) in Parkinson's disease: single dose study. Asia Pac J Clin Nutr 2:85–89

Khola OPS, Dube RK, Sharma NK (1999) Conservation and production ability of maize (*Zea mays*) legume intercropping systems under varying dates of sowing. Indian J Agron 44(1):40–46

Konlan S, Sarkodie-addo J, Asare E, Kombiok MJ (2013) Groundnut (Arachis hypogaea L.) varietal response to spacing in the Guinea savanna agro-ecological zone of Ghana: growth and yield. Afr J Agric Res 8(22):2769–2777

Köpke U, Nemecek T (2010) Ecological services of faba bean. Field Crops Res 115:217–233

Kumar S, Sheoran S, Kumar SK, Kumar P, Meena RS (2016) Drought: a challenge for Indian farmers in context to climate change and variability. Progr Res Int J 11:6243–6246

Kundu S, Bhattacharyya R, Ved-Prakash GBN, Gupta HS (2007) Carbon sequestration and relationship between carbon addition and storage under rainfed soybean–wheat rotation in sandy loam soil of the Indian Himalayas. Soil Tillage Res 92:87–95

La Favre JS, Focht DD (1983) Conservation in soil of H_2 liberated from N_2 fixation by H up-nodules. Appl Environ Microb 46:304–311

Latati M, Bargaz A, Belarbi B, Lazali M, Benlahrech S, Tellah S (2016) The intercropping common bean with maize improves the rhizobial efficiency, resource use and grain yield under low phosphorus availability. Eur J Agron 72:80–90

Layek J, Chowdhury S, Ramkrushna GI, Das A (2014) Evaluation of different lentil cultivars in lowland rice fallow under no-till system for enhancing cropping intensity and productivity. Indian J Hill Farm 27(2):4–9

Lee KK, Wani SP (1989) Proceedings of Colloquium on soil fertility and fertilizer management in semi-arid and tropical India. IFDC Muscle Sholas, AL, pp 89–108

Lemke RL, Zhong Z, Campbell CA, Zentner RP (2007) Can pulse crops play a role in mitigating greenhouse gases from North American agriculture? Agron J 99:1719–1725

Li L, Zhang L, Zhang F (2007) Crop mixtures and the mechanisms of overyielding. Encyclopedia of Biodiversity, vol 2, https://doi.org/10.1016/B978-0-12-384719-5.00363-4

Luetke-Entrup N, Pahl H, Albrecht R (2003) Fruchtfolgewert von Körnerleguminosen. In: UFOP-Praxisinformation. Union zurFörderung von Oel- und Proteinpflanzen (UFOP), Berlin, p 20

Maikhuri RK, Rao KS, Saxena KG (1996) Traditional crop diversity for sustainable development of Central Himalayan agroecosystem. Int J Sust Dev World Ecol 3:8–31

Maikhuri RK, Rao KS, Semwal RL (2001) Changing scenario of Himalayan agroecosystems: loss of agrobiodiversity, an indicator of environmental change in Central Himalayas. Environmentalist 21:23–39

Meena VK, Kaushik MK, Meena RS (2014) Response of cluster bean [*Cyamopsis tetragonoloba* (L.)] to growth substances in sub-tropical climate of Rajasthan. Indian J Ecol 41(1):190–191

Meena RS, Yadav RS, Meena H, Kumar S, Meena YK, Singh A (2015a) Towards the current need to enhance legume productivity and soil sustainability worldwide: A book review. J Clean Prod 104:513–515

Meena RS, Meena VS, Meena SK, Verma JP (2015b) Towards the plant stress mitigate the agricultural productivity: a book review. J Clean Prod 102:552–553

Meena VS, Maurya BR, Meena RS (2015c) Residual Impact of Well-Grow Formulation and NPK on Growth and Yield of Wheat (*Triticum aestivum* L.). Bangladesh J Bot 44(1):143–146

Meena RS, Bohra JS, Singh SP, Meena VS, Verma JP, Verma SK, Shiiag SK (2016) Towards the prime response of manure to enhance nutrient use efficiency and soil sustainability a current need: A book review. J Clean Prod 112:1258–1260

Meena RS, Gogaoi N, Kumar S (2017a) Alarming issues on agricultural crop production and environmental stresses. J Clean Prod 142:3357–3359

Meena RS, Meena PD, Yadav GS, Yadav SS (2017b) Phosphate Solubilizing Microorganisms, Principles and Application of Microphos Technology. J Clean Prod 145:157–158

Meena RS, Kumar V, Yadav GS, Mitran T (2017c) Response and interaction of *Bradyrhizobium japonicum* and *Arbuscular mycorrhizal* fungi in the soybean rhizosphere: a review. Plant Growth Reg, Accepted in press

Mishra MK, Dubey RK, Rao SK (2010) Nutritional composition of field pea (*Pisum sativum* var arvense). Legum Res 33(2):146–147

Munda GC, Hazarika UK, Saxena DC, Raj S, Patel DP (1999) Performance of cropping systems under mid altitude rainfed dry terraces of Meghalaya. Indian J Hill Farmg 12(1 & 2):106–110

Mythili G, Jann Goedecke E (2016) Economics of Land Degradation in India. In: Nkonya et al (eds) In: Economics of Land Degradation and Improvement – A Global Assessment for Sustainable Development. https://doi.org/10.1007/978-3-319-19168-3_15
NASF Annual Report (2015) National Agricultural Science Fund Project (ICAR, New Delhi) on "Mitigating abiotic stresses and enhancing resource-use efficiency in pulses in rice fallows through innovative resource conservation practices". Annual Report, 2015. ICAR Research Complex for NEH Region, Umiam, Meghalaya
National Bureau of Soil Survey & Land Use Planning (NBSS&LUP) (2004) Soil Map (1:1 Million Scale). NBSS&LUP, Nagpur
National Bureau of Soil Survey and Land Use Planning (NBSS&LUP) (2005) Annual report 2005. Nagpur; NBSS&LUP, Nagpur
Ngachan SV, Anupam M, Kadirvel G, Das A, Kanchan S (2010) Conservation of natural resources for sustainable hill agriculture. ICAR Research Complex for NEH Region, Umiam, p 428
Ngangom B, Das A, Ramkrushna GI, Layek J (2017a) Soil properties and nutrient uptake as influenced by mulching in maize (*Zea mays*)-based cropping systems. Indian J Agron 62(2):34–38
Ngangom B, Das A, Savita, Krishnappa R (2017b) Soil physical properties and productivity as influenced by soil moisture conservation measures under maize based cropping system in arid soils of North East India. Int J Curr Microbiol App Sci 6(3):428–436
Ngangom B, Das A, Ramkrushna GI, Layek J (2017c) Soil properties and nutrient uptake as influenced by mulching in maize (Zea mays) based cropping systems. Indian J Agron 62(2):219–133
Ningombam RD, Singh PK, Salam JS (2012) Proximate composition and nutritional evaluation of underutilized legume, *Psophocarpus tetragonolobus* grown in Manipur, Northeast India. Am J Food Technol **7**:487–493
Nuruzzaman M, Lambers L, Bolland MDA, Veneklaas EJ (2005) Phosphorus uptake by grain legumes and subsequently grown wheat at different levels of residual phosphorus fertiliser. Austr J Agric Res 56:1041–1047
Padhi AK, Jena BK, Panigrahi RK (2010) Performance of double cropping systems involving legume vegetables and finger millet under rainfed condition. Indian J Agric Res 44(1):14–19
Pandey KK (1992) Sustainability of the environmental resource base and development priorities of a mountain community. ICI MOD Occasional Paper No. 19. Kathmandu
Pandey AK, Prakash V, Singh R, Mani VP (1999) Effect of intercropping pattern of maize (*Zea mays*) and soybean (*Glycine max* L. Merril) on yield and economics under mid-hills of N-W Himalayas. Ann Agric Res 20(3):354–359
Pappa VA, Rees RM, Walker RL, Baddeley JA, Watson CA (2011) Nitrous oxide emissions and nitrate leaching in an arable rotation resulting from the presence of an intercrop. Agric Ecosyst Environ 141:153–161
Partap T (1999) Sustainable land management in marginal mountain areas of the Himalayan region. Mountain Res Dev:251–260
Patra BC, Mandal BB, Mandal BK, Padhi AK (1999) Suitability of maize (*Zea mays*) based intercropping system. Indian J Agric Sci 69(11):759–762
Pattanayak A, Anup Das (2017) Present and future cropping system of North East India in changing climatic conditions. In: SOUVENIR, conference on "Farmers First for conserving soil and water resources in North Eastern Region (FFCSWR 2017), February 09–11, 2017. Indian Association of Soil and Water Conservationists, Dehradun, Uttarakhand, pp 44–48
Peoples MB, Hauggaard-Nielsen H, Jensen ES (2009) The potential environmental benefits and risks derived from legumes in rotations. In: N fixation in crop production. American Society of Agronomy, Crop Science Society of America, Soil Science Society of America, pp 349–385
Preissel S, Reckling M, Schläfke N, Zander P (2015) Magnitude and farm economic value of grain legume pre-crop benefits in Europe: a review. Field Crop Res 175:64–79
Rajkhowa DJ, Anup D, Das SK, Sen A, Tripathi AK, Mohanty AK, Hazarika S, Saikia US, Krishnappa R, Aochen C, Kumar A, Ngachan SV (2016) National Initiative on Climate Resilient Agriculture (NICRA) Research Highlights from North Eastern Hill Region (2010–11 to 2014–15). ICAR Research Complex for NEH Region, Umiam, p 105
Ram K, Meena RS (2014) Evaluation of pearl millet and Mung bean intercropping systems in arid region of Rajasthan (India). Bangladesh J Bot 43(3):367–370

Ramesh SC, Negi RS, Rana SS (2016) Innovative farming practices under the changing climate in the North Western Himalayan region, India. International Agronomy Congress:1149–1150

Rao JV, Khan IA (2003) Research gaps in intercropping systems under rainfed conditions in India, an on farm survey. Hyderabad, CRIDA

Reckling M, Preissel S, Zander P, Topp CFE, Watson CA, Murphy-Bokern D, Stoddard FL (2014) Effects of legume cropping on farming and food systems. Legume Futures Report 1:6. Available from www.legumefutures.de

Robson MC, Fowler SM, Lampkin NH, Leifert C, Leitch M, Robinson D et al (2002) The agronomic and economic potential of break crops for ley/arable rotations in temperate organic agriculture. Adv Agron 77:369–427

Saha SN, Shroff JC, Patel RH, Usadadiya VP (2011) Influence of intercropping and weed management practices on weed and yields of maize. Int J Sci Nat 2:47–50

Samal PK, Palni LMS, Dhyani PP (2005) Status and trends in research and development projects in the mountains: a situational analysis in the Indian Himalaya. Int J Sust Dev World Ecol 12(4):479–488

Saxena KB, Kumar RV, Sultana R (2010) Quality nutrition through pigeon pea—a review. Sci Res 2(11):1335–1344

Sehgal J, Abrol IP (1994) Soil degradation in India: status and impact. Oxford and IBH, New Delhi, p 80

Semwal RL, Maikhuri RK (1996) Structure and functioning of traditional hill agro–ecosystems of Garhwal Himalaya. Biol Agric Hort 13:267–289

Sharda VN, Dogra P, Prakash C (2010) Assessment of production losses due to water erosion in rainfed areas of India. J Soil Water Conserv 65(2):79–91

Sharma BK, Das A, Bujarbaruah KM (2003) Underutilized life support crop species: production and research in North Eastern Hill region of India. ICAR Research Complex foe North Eastern Hill Region, Umiam

Shen J, Yuan L, Zhang J, Li H, Bai Z, al CX (2011) Phosphorus dynamics: from soil to plant. Plant Physiol 156:997–1005

Shiva V, Vanaja RP (1993) Cultivating diversity: biodiversity conservation and seed politics. Research foundation for Science Technology and Natural Resource Policy. Natraj Publishers, Dehradun, 130pp

Shrestha S (1992) Mountain agriculture: indicators of unsustainability and options for reversal, MFS Discussion Paper No. 32. ICIMOD, Kathmandu

Singh NP, Pratap (2015) Achieving self-sufficiency in pulses production. In: Dixit GP, Jagdish S, Singh NP (eds) Pulses: challenges and Opportunities under changing climatic scenario. 29 September to October 1, 2014. JNKV Jabalpur, Indian Society of Pulses Research and Development. ICAR – Indian Institute of Pulses Research, Kanpur, India, pp 1–13

Singh B, Singh Y (2002) Concepts in nutrient management. In: Singh G, Kollar JS, Sekhan HS (eds) Recent advances in agronomy. Indian Society of Agronomy, New Delhi, pp 92–109

Singh G, Sekhon HS, Kolar JS (2005) Pulses. Agrotech Publishing Academy, Udaipur, p 592

Singh U, Saad AA, Singh SR (2008) Production potential, biological feasibility and economic viability of maize (*Zea mays*) based intercropping system under rainfed conditions. Indian J Agric Sci 78(12):1023–1027

Singh AK, Manibhushan BBP, Singh KM, Upadhyaya A (2013) An analysis of oilseeds and pulses scenario in Eastern India during 2050–2051. J Agric Sci 5(1):241–249

Singh R, Babu S, Avasthe RK, Yadav GS, Rajkhowa DJ (2015) Influence of tillage and organic nutrient management practices on productivity, profitability and energetics of vegetable pea (*Pisum sativum* L.) in rice-vegetable pea sequence under hilly ecosystems of north-east India. Res Crops 16(4):683–688

Soussana JF, Loiseau P, Vuichard N, Ceschia E, Balesdent J, Chevallier T, Arrouays D (2004) Carbon cycling and sequestration opportunities in temperate grasslands. Soil Use Manage 20:219–230

Srinivasarao CH, Venkateswarlu B, Lal R (2014) Long-term manuring and fertilizer effects on depletion of soil organic stocks under Pearl millet-cluster bean-castor rotation in Western India. Land Degrad Dev 25:173–183

Stagnari F, Maggio A, Galieni A, Pisante M (2017) Multiple benefits of legumes for agriculture sustainability: an overview. Chem Biol Technol Agric 4(2):1–13

Tharanathan RN, Mahadevamma S (2003) Grain legumes—a boon to human nutrition. Trends Food Sci Tech 14:507–518

Tiwari AK,Shivhare AK (2016) Pulses in India: retrospect and prospects. Goverment of India, Ministry of Agriculture & Farmers Welfare (DAC&FW), Directorate of Pulses Development, Vindhyachal Bhavan, Bhopal, M.P.- 462004. Publication No.: DPD/Pub.1/Vol. 2

Varma D, Meena RS, Kumar S (2017a) Response of mung bean to fertility and lime levels under soil acidity in an alley cropping system in Vindhyan Region, India. Int J Chem Stu 5(2):384–389

Varma D, Meena RS, Kumar S, Kumar E (2017b) Response of mung bean to NPK and lime under the conditions of Vindhyan Region of Uttar Pradesh. Legum Res 40(3):542–545

Verma JP, Jaiswal DK, Meena VS, Meena RS (2015a) Current need of organic farming for enhancing sustainable agriculture. J Clean Prod 102:545–547

Verma JP, Meena VS, Kumar A, Meena RS (2015b) Issues and challenges about sustainable agriculture production for management of natural resources to sustain soil fertility and health: a book review. J Clean Prod 107:793–794

Verma SK, Singh SB, Prasad SK, Meena RN, Meena RS (2015c) Influence of irrigation regimes and weed management practices on water use and nutrient uptake in wheat (*Triticum aestivum* L. Emend. Fiori and Paol.). Bangladesh J Bot 44(3):437–442

Wani SP, Rupela OP, Lee KK (1994) BNF technology for sustainable agriculture. In: Transactions of the 15th World Congress of Soil Science, 10–16 July 1994. The International Society of Soil Science and The Mexican Society of Soil Science, Acapulco, Mexico, pp 245–262

Yadav RL, Dwivedi BS, Gangwar KS, Prasad K (1998) Overview and prospects for enhancing residual benefits of legumes in rice and wheat cropping systems in India. In: Kumar Rao JVDK, Johansen C (eds) Residual effects of legumes in rice and wheat cropping systems of the Indo-Gangetic Plain. ICRISAT, Oxford &IBH Publishing Co. Pvt. Ltd, New Delhi, pp 207–226

Yadav GS, Lal R, Meena RS, Babu S, Das A, Bhomik SN, Datta M, Layak J, Saha P (2017a) Conservation Tillage and Nutrient Management Effects on Productivity and Soil Carbon Sequestration Under Double Cropping of Rice in North Eastern Region of India. Ecol Indi. http://www.sciencedirect.com/science/article/pii/S1470160X17305617

Yadav GS, Lal R, Meena RS, Datta M, Babu S, Das LJ, Saha P (2017b) Energy budgeting for designing sustainable and environmentally clean/safer cropping systems for rainfed rice fallow lands in India. J Clean Prod 158:29–37

Grain Legumes for Resource Conservation and Agricultural Sustainability in South Asia

3

Narendra Kumar, K. K. Hazra, C. P. Nath, C. S. Praharaj, and U. Singh

Contents

N. Kumar (✉) · K. K. Hazra · C. P. Nath · C. S. Praharaj · U. Singh
ICAR-Indian Institute of Pulses Research, Kanpur, India

R. S. Meena et al. (eds.), *Legumes for Soil Health and Sustainable Management*,
https://doi.org/10.1007/978-981-13-0253-4_3

Abstract
Degradation of natural resources is a major environmental concern that threatens the agroecosystem health and food security in South Asian countries. About 1.8 billion people (24% of world population) are living in this region in an area of 5.03 km^2. The higher population pressure on agricultural land (7 person ha^{-1}) has further threatened the existing resources to a great extent. Thus, conserving natural resource base is essential to feed the burgeoning population. Continuous practice of cereal-cereal rotation including rice-wheat in Indo-Gangetic plains have emerged several soil- and environmental-related issues. Diversification of cereal-cereal cropping systems is warranted to mitigate those issues and to adapt to the changing climatic condition and to enhance the resource-use efficiency on a sustainable basis. Grain legumes are the suitable candidate crop for diversification because of its inherent capacity to build up soil health and in conserving natural resources. There exists a large scope to introduce pulses as the second crop in 22.2 million hectare areas of rice fallows in India, Bangladesh, and Nepal. System intensification with inclusion of mungbean in summer fallows of rice-wheat cropping system could add an additional pulse crops area of 1.0 m ha in Indo-Gangetic plains. Several alternative grain legume inclusive crop rotations have been identified for the different agro-zones that certainly could play an important role in popularizing the conservation of agriculture in cereal-dominated production systems of South Asia. Endowed with an inherent potential biological N-fixation (30–150 kg N ha^{-1}), of the deep root system, the root exudates mediated P-solubilization, and nutrient-rich residues of grain legumes improve the soil fertility and enhance the soil profile nutrient cycling. Crop diversification with grain legumes has additional benefits associated with improving water productivity, reducing input cost, and minimizing incidence of diseases and pests. Besides this, the low application rate of the N fertilizer to grain legumes has the advantage of reducing greenhouse gas emissions and groundwater pollution. Thus, grain legumes would play a crucial role in resource conservation, ecosystem balance, and in the sustainability of agricultural systems of South Asia.

Keywords
Conservation agriculture · Crop diversification · Nutrient cycling · Resource conservation · Rice fallow · South Asia · Sustainability

Abbreviations

@	At the rate of
AM	Arbuscular mycorrhiza
BNF	Biological nitrogen fixation
C	Carbon
CA	Conservation agriculture

CEY	Chickpea equivalent yield
cm	Centimeter
DNA	Deoxyribonucleic acid
DTPA	Diethylenetriaminepentaacetic acid
g	Gram
GHG	Greenhouse gas
IGP	Indo-Gangetic plains
ka	Hectare
kg	Kilogram
kj	Kilojoule
km	Kilometer
m	Million
mg	Milligram
mm	Millimeter
N	Nitrogen
NW	Northwest
P	Phosphorus
PEY	Pigeonpea equivalent yield
RCTs	Resource conservation technologies
SOC	Soil organic carbon
SOM	Soil organic matter
t	Tonne
Tg	Teragram
VAM	Vesicular-arbuscular mycorrhiza
w/w	Weight/weight

3.1 Introduction

Grain legumes are considered as the second most important group of food crops after cereals. Developing nations contribute about three-fourth to the global grain legume production, and the remaining one-fourth comes from developed nations. In 2014, the global grain legume production was 85.2 million tonnes from an area of 77.5 million hectares with an average yield of 909 kg ha^{-1}. India not only contributes maximum (around 25–28%) to the total global grain legume production but also largest consumer in the world. The total grain legume production in South Asia is 22.3 million tonnes from an area of 33.7 million hectares and with a yield of 660 kg ha^{-1}. India, Pakistan, Bangladesh, Nepal, and Afghanistan are the major grain legume-producing countries in South Asia with a relative share of 89.7%, 3.1%, 1.8%, 1.4%, and 0.3%, respectively (Table 3.1) (FAOSTAT 2014). The chickpea (*Cicer arietinum* L.) contributed the maximum (about 34.1%) to South Asia grain legume production followed by dry beans (30.6%), pigeonpea [*Cajanus cajan* (L.) Millsp.] (16.1%), lentil (*Lens culinaris* Medikus) (6.9%), and dry peas (3.7%) (Table 3.2).

Table 3.1 Area, production, and yield of grain legumes in South Asian countries and world (2014)

Countries	Area ('000 ha)	Production ('000 t)	Yield (kg ha^{-1})
India	30309.0	19980.0	659
Pakistan	1358.6	682.2	502
Nepal	292.5	310.5	1062
Bangladesh	277.3	395.2	1425
Afghanistan	79.8	60.0	752
Sri Lanka	17.4	21.7	1251
Maldives	0.16	0.100	870
Total, South Asia	33708.4	22257.7	660
Total, world	85191.5	77473.1	909

Source: FAOSTAT 2014

Table 3.2 Area, production, and yield of major grain legumes in South Asia (2014)

Crop	Area (million hectare)	Production (million tonnes)	Yield (kg ha^{-1})
Chickpea	11.49	10.55	919
Beans, dry	10.3	4.56	442
Pigeonpea	5.62	3.31	589
Lentil	2.32	1.58	681
Peas, dry	1.26	0.84	667
Cowpeas, dry	0.011	0.015	1327
Total grain legumes	33.71	22.26	660

Source: FAOSTAT 2014

Grain legumes, belonging to the family Fabaceae, are a wonderful gift of nature to mankind. They provide not only nutritious food, feed, and fodder but also help in conserving the natural resources and maintain ecological balance. The cultivated legumes can be classified as grain legumes [chickpea, lentil, peas (*Pisum sativum* L.), grass pea (*Lathyrus sativus* L.), faba bean (*Vicia faba* L.), French bean (*Phaseolus vulgaris* L.), pigeonpea, urdbean (*Vigna mungo* (L.) Hepper), mungbean (*Vigna radiata* (L.) R. Wilczek), cowpea (*Vigna unguiculata* (L.) Walp.), moth bean (*Vigna aconitifolia* (Jacq.) Marechal), horse gram (*Macrotyloma uniflorum* (Lam.) Verdc.), etc.], oilseed legumes [soybean (*Glycine max* (L.) Merr.) and groundnut (*Arachis hypogaea* L.)], vegetable legumes [peas, faba bean, cowpea, French bean, cluster bean (*Cyamopsis tetragonoloba* (L.) Taub.), winged bean (*Psophocarpus tetragonolobus* (L.) D.C.), etc.], forage legumes [cowpea, berseem (*Trifolium alexandrinum* L.), lucerne (*Medicago sativa* L.), cluster bean, etc.], and range legumes [stylo (*Stylosanthes* spp.), siratro (*Macroptilium atropurpureum* (DC.) Urb.), etc.].

Grain legumes are the rich source of dietary protein. Other than proteins, these are an imperative source of the 15 essential minerals required by human beings. The sustenance estimations of seeds of grain legumes are high giving out 1040 to 1430 kJ calories per 100 g (Singh 2015). The protein content (20–28%) in grain legumes is twofold than that of most cereals and, however, is generally inadequate in sulfur-containing amino acids such as methionine and cystine. Proteins of cereal grains are poor in lysine but are higher in sulfur-containing amino acids. Therefore, the combination of grain legumes and cereals provides all essential amino acids (comparable with milk protein) in right proportionate required in a balanced human diet (Singh 2015; Meena et al. 2015a). The utilization of grain legumes minimizes the danger of assortment of chronic degenerative sicknesses, for example, cancer, obesity, diabetes, and cardiovascular diseases (Patterson et al. 2009). Grain legumes are likewise rich in dietary fiber, complex sugar, starch, and various vitamins and minerals, viz., folate, potassium, selenium, and zinc. In addition, a wide assortment of non-nutritive bioactive constituents, for example, catalyst inhibitors, phytic acid, lectins, phytosterols, phenols, and saponins, are available in grain legumes which have well-being defensive impacts (Champ 2002). The antioxidant and DNA protective properties of phytic acid (Phillippy 2003), antioxidant, and other vital physiological properties of phenolic compounds (Yeh and Yen 2003) and hypocholesterolemic impacts and anticancer action of saponins (Shi et al. 2004) are well documented.

Degradation of natural resources is an important environmental issue that debilitates the biological system well-being and in the sustenance of security around the world. The overexploitation of natural resources (soil and water) leads to a decrease in response to applied agricultural inputs, for example, tillage, fertilizer, water, inter-cultivation, and pesticides. Prior to the 1960s, different kinds of grain legumes were the important parts of crop rotations in the South Asia. The "green revolution" dating from the 1960s has increased the area and production of food grain crops, mainly rice and wheat which met the food demand of the ever-increasing population of South Asia. This has brought an imbalance in production between the different groups of crops, particularly the grain legumes which relegated from fertile areas to poor and marginal lands under rainfed farming. The maximum gain in area and production due to the green revolution in South Asia is recorded in wheat which increased from 25.9 million a hectare and 22.9 million tonnes during 1965 to 50.8 million a hectare and 140.9 million tonnes during 2014, respectively. However, the area and production of grain legumes during the 1960s to 2009 remained almost stagnant. The continuous practice of the exhaustive rice and the wheat production system over the last six decades has resulted in many problems which have restricted the ability of these resources to produce the level matching future food grain requirement of the country. The deteriorating production and sustainability of the cereal-based systems are evident from either stagnation or decline in the yield and factor productivity of rice and wheat due to an undesirable decline in soil physical environment and excessive mining of essential plant nutrients from soil (Yadav et al. 1998a; Timsina and Connor 2001). The overexploitation of groundwater for the irrigation of rice and wheat has crossed the natural ability to recharge itself in many parts of South Asia. In recent years, a growing deceleration in total factor productivity and deterioration of soil health under the cereal-based cropping system has necessitated for diversification of existing cropping systems (Yadav 1998; Buragohain et al. 2017). Thus, enhancing and sustaining the natural resource base is of paramount importance.

Endowed with a unique ability for biological N fixation (BNF), having a deep root system, low water requirements, and capacity to withstand drought, grain legumes constitute an important component of crop diversification and help to alleviate the detrimental effect of monoculture of cereal-based cropping systems. Thus, grain legumes can be an important source to reverse the detrimental effect of rice and wheat production systems and contribute to accomplishing the twin objectives of enhancing system productivity and sustainability of cereal-based cropping systems in South Asia (Yadav et al. 1998). Thus, the present chapter deals with the role of grain legumes in enhancing resource-use efficiency and sustainability of existing cereal-based cropping systems in South Asia (Tables 3.1 and 3.2).

3.2 Crop Diversification with Grain Legumes

The lack of crop diversity is one of the fundamental causes for several soil, environment, and pest problems, which are now increasingly being evident from different agro-regions of South Asia (Chauhan et al. 2012; Congreves et al. 2015). Indeed, with advancement in irrigation facilities and higher accessibility to the farm inputs, a large number of farmers have inclined toward cereals, which in turn aggravated the second-generation problems (Kerr et al. 2007). Therefore, in order to address the negative issues associated with continuous cereal-based systems, the diversification/intensification of conventional cropping systems could be a strategic option in achieving the production sustainability (Njeru 2016). Crop diversification is essentially an important component of profitable and sustainable agriculture (Hatfield and Karlen 1994). Inclusion of pulses/grain legumes in the cropping system is an age-old practice (Ghosh et al. 2007). Endowed with the inherent potential of a deep root system, BNF, and most importantly complementary with cereals and other non-legume crops, grain legumes could essentially serve in a key role in crop diversification/intensification in different production systems (Hazra et al. 2014; Meena et al. 2017). Presently, conservation agriculture (CA) is increasingly being advocated for cereal-based cropping systems of South Asia (Jat et al. 2014). Among the three major principles of CA, crop rotation particularly with grain legumes has been strongly recommended as a "missing ingredient" for restoration of soil health and resource conservation (Snapp et al. 2002). In general, grain legumes are thought to be less profitable than cereal crops. In view of this, von Richthofen et al. (2006) anticipated that grain legume inclusive rotations could fetch similar or higher economic returns when compared with non-legume-based rotations.

To date, several short-term as well as long-term effects of grain legume crop/s in different cropping systems have been documented, which are mostly optimistic in terms of productivity and soil fertility (Ganeshamurthy 2009). Long-term inclusion of grain legumes like mungbean, pigeonpea, and chickpea in the conventional maize-wheat system of subtropical Indo-Gangetic plains (IGP) can improve soil health, particularly the soil organic carbon (SOC) (Venkatesh et al. 2013). They also specified that grain legume crops can improve both the soil labile and non-labile fractions of soil organic C. Likewise, Ghosh et al. (2012) reported similar results from lowland ecosystem of Indo-Gangetic plains (IGP), where the intensification of the rice-wheat cropping system with cultivation of the mungbean during summer (April–May) has resulted in an improved SOC. A similar positive effect of growing summer mungbean in rice-wheat system on SOC was previously reported in *mollisols* and *inceptisols* of IGP (Saraf and Patil 1995; Ghosh and Sharma 1996). Likewise, every year and alternate year, substituting of wheat crop with chickpea in the rice-wheat system has been recommended to minimize the fertilizer input and irrigation requirement and at the same time maintain the total system productivity in subtropical IGP conditions (Hazra et al. 2014; Verma et al. 2015). In fact, the favorable effect of grain legumes in crop rotation is mainly associated with an improved soil health and higher accessibility of plant-available nutrients in the soil (Hazra et al. 2014; Ghosh et al. 2006). The effect of grain legumes was found to be more prominent where nutrient-rich grain legume residues are returned to the soil for an extensive period. Apart from the above ground crop residues, the progressive decomposition of leftover legume roots can enrich the soil N-pool equivalent to ~40 kg N ha^{-1} (Singh et al. 2005).

Besides this, inclusion of grain legumes in cropping systems is the need of the hour for several regional interests. For instance, in Northwestern India, the continuous practice of intensively irrigated rice-wheat cropping systems led to over-exhaustion of groundwater and developed secondary salinization and deteriorated the soil's physical conditions, which eventually made the system unsustainable and less productive over the years. To cope up with this situation, short-duration pigeonpea is now being advocated in the place of rice for this region to curtail the demand for irrigation water (Kumar et al. 2016). Presently, the demand for grain legumes is rising rapidly; and to meet the growing demand, more area should be dedicated for grain legume cultivation. The mungbean cultivation in summer fallows of irrigated cereal-based cropping system offers an immense scope to practice an "ecologically intensive" cropping system (Sharma et al. 2000b; Venkatesh et al. 2015; Dhakal et al. 2016). The short-duration heat-tolerant mungbean crop can be easily accommodated after the winter crop in rice (*Oryza sativa* L.)-wheat (*Triticum aestivum* L), maize (*Zea mays* L.)-wheat, rice-potato (*Solanum tuberosum* L.), rice-chickpea, and rice-mustard (*Brassica* spp.) cropping systems of IGP, where assured that irrigation facilities were available. Likewise, in IGP conditions, chickpea is emerging as a potential alternative for wheat. Cropping systems such as rice-rajmash-summer mungbean, rice-lentil, rice-field pea, rice-wheat-mungbean in lowland, pigeonpea-wheat, and maize-wheat-mungbean in upland have been found promising under irrigated conditions (IIPR 2009). Nonetheless, the appropriate agro-techniques and short-duration varieties have to be developed for a higher adoption of grain legumes in these new niches. Apart from these, a large area of South Asia including India and Bangladesh is kept fallow after growing of the rainfed rice crop due to lack of irrigation facilities. In these rice fallows, grain legumes like lentil, chickpea, lathyrus, and mungbean (peninsular India) can be grown as secondary crop with appropriate soil moisture conservation practices and manipulating cultivation practice of both rice and pulse crops (Ghosh et al. 2016; Kumar et al. 2016a; Ali et al. 2014). This way, in the near future, certainly grain legumes would play a very important role toward utilizing untrapped niches and in improving sustainability of the cereal-dominated cropping systems.

Given the advantage of higher economic return, soil health maintenance, weed control, and less risk, intercropping is very popular in the areas of smallholding farmers (Ghosh et al. 2007). Grain legumes are the important candidate crop for intercropping, and several grain legumes inclusive of intercropping systems have been found highly productive and less competitive with the component crop. Some of the potential grain legume inclusive intercrop systems are pigeonpea + sorghum (*Sorghum bicolor* L.), pigeonpea + maize, soybean + pigeonpea (Ghosh et al. 2006), sugarcane (*Saccharum officinarum*) + lentil, potato + rajmash, chickpea + mustard, linseed (*Linum usitatissimum* L.) + lentil (IIPR 2009), etc. Intercropping of early pigeonpea with groundnut (5:2), chickpea + wheat/barley (2:1), and chickpea + mustard (6,2) is some of the potential grain legume-based intercropping system under irrigated conditions (Ali 1992; Ali 2004; Ali et al. 2012). Similarly, spring planted sunflower + mungbean/urdbean intercropping also became popular in some parts of Northern India (Ali et al. 1998). Mungbean (variety PDM 11 and PDM 84–143) and urdbean (variety DPU 88–31) intercropping with spring-planted sugarcane has been found promising under irrigated condition (Panwar et al. 1990). In the dry lands of subtropical India, pearl millet is grown extensively, and different grain legumes, viz., green gram, black gram, castor (*Ricinus communis* L.), cowpea, and groundnut, can be accommodated as intercrops (Ghosh et al. 2007; Ram and Meena 2014).

3.3 Grain Legumes for Restoration of Soil Health

3.3.1 Biological Nitrogen Fixation

Grain legumes have been utilized since time immemorial as an essential N source across the world. Mainly in South Asia, the winter grain legumes like lentil, chickpea, faba bean, and field pea are the major source of protein in human diets. Grain legumes are used in agribusiness since they improve the profitability and sustainability of agriculture. The most important being the ability to fix N by biological fixation. After carbon and water, N is the most imperative constraining elements for the development of plant and yield of crops (Vance 1997; Peoples et al. 1995). Young (1992) stated that the three groups of microorganisms are able to fix N_2 in beneficial interaction with plants: the nodule-forming organisms (*Rhizobium*, *Azorhizobium*, *Allorhizobium*, *Mesorhizobium*, *Sinorhizobium*, and *Bradyrhizobium*), the actinomycetes (Frankia), and the *Cyanobacteria* (*Nostoc*, *Anabaena*).

The contribution of N is the most ordinarily watched essential advantage of leguminous crops. Part of fixed N by leguminous crops is utilized by the succeeding crop; thereby the N fertilizer requirement decreases in the succeeding crop (Reeves 1994). Grain legumes are moderately high in protein content. This can be specifically credited to the capacity of grain legumes to supply the vast majority of its own N needs with the assistance of advantageous *Rhizobia* microbes living in their roots. Grain legumes fix 30–150 kg N/ha depending upon the rhizobial population, host crop, management level, and ecological conditions. The N fixation amount by legumes is quantified to a great extent by the hereditary capability of the crops and by plant accessible N rate in the soil. *Rhizobium* and *Bradyrhizobium* are responsible for symbiotic N fixation in grain legumes. The *Rhizobium* is quickly developing acid-releasing microbes, whereas the *Bradyrhizobium* is a slow grower that doesn't deliver acid to the soil (Brady and Weil 2002). Soil determinants, for example, temperature, moisture, and pH further decide the N fixation limit of grain legume crops. The amount of N released by the leguminous crops in the soil is adequate to make desired yield level of succeeding non-leguminous crops, while higher N requiring crop, for instance, the corn, by and large need supplemental N. Frye et al. (1988) suggested in such crops, N rates could be cut down evidently while keeping up the expected crop yields. Moreover, Peoples et al. (1995) reported that in agriculture systems about 90 to 140 Tg N year^{-1} is supplied through biological N fixation (BNF). However, more checks on these values are necessary; most evidences proved that the BNF contributes more N than addition of synthetic N fertilizers for plant development. Usually, BNF provides 50–60% of the N uptake by grain legumes, 55–60% of the N in N-fixing trees, and 70–80% of the N uptake by leguminous pastures.

3.3.2 Nutrient Recycling

Grain legumes being hardy and having low input requirement offer a tremendous opportunity toward effective utilization of resources. Further, given the unique characteristics of the BNF, it has potential to establish itself with surface broadcast and soil fertility restoration property, and grain legumes can be best suited for resource savings. The N-sparing and synergistic effects of grain legumes are well recognized. The intrinsic N-fixing capacity of the grain legume crops enables them to meet the large proportion of their N requirement and also helps in economizing N in succeeding non-legume crops. In sequential crop involving grain legumes, the preceding grain legume crop may contribute 18–70 kg N ha^{-1} to soil, and thereby a considerable amount of N can be saved in succeeding crops. In rice-wheat rotation the growing of short-duration mungbean in summer may bring up N economy up to 40–60 kg N ha^{-1} in the succeeding rice crop. The N economy due to preceding pigeonpea over sorghum was found to be 51 kg N equivalent ha^{-1}. The effect of *rabi* (winter) grain legumes on yield and N economy in following rice revealed that chickpea, rajmash, and lentil exhibited a most favorable effect in economizing N to the extent of 40 kg ha^{-1} (IIPR 2009).

Grain legumes can recycle soil profile nutrients because of their deep root systems bringing about a more proficient utilization/recycling of applied nutrients and reduce the loss of soil nutrients especially nitrate underneath the root zone. The quantity of C and N provided by roots of grain legumes crop can be critical for enhancing the soil's organic matter (Sainju et al. 2005; Verma et al. 2015a) and thus

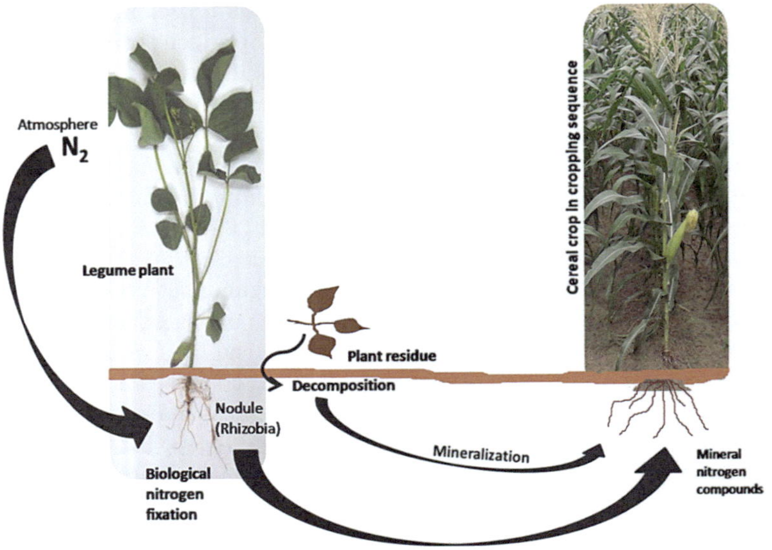

Fig. 3.1 Benefits of grain legumes inclusive cropping systems

improving nutrient recovery. The healthy and profuse root framework may have a predominant role in soil C and N cycles and may impact soil organic C and N levels than the above ground plant biomass. The legume crops have been accounted for to decrease the potential for NO_3 draining from farms (Staver and Brinsfield 1990). Meisinger et al. (1998) reported that grain legumes minimize the concentration of NO_3 by 20 to 80% in leachate over control of (non-leguminous crops). Grain legume crops store the inorganic soil N in between two principal crop seasons in an organic form, thereby reducing the NO_3 leaching. The N is along these lines discharged to the succeeding crop (Fig. 3.1). The association of grain legumes with vesicular-arbuscular mycorrhiza (VAM) helps in increasing the availability of nutrients and water to crop plants. Grain legumes add organic matter through leaf fall, root biomass, and easily degradable crop residues. Grain legumes also release organic acids into the soil (Fageria et al. 2002), in this manner mobilizing inaccessible soil nutrients. It is notable that acidification of the rhizosphere can solubilize a few low solvent macronutrients and micronutrients. For example, root instigated rhizosphere chemical changes have been accounted for to expand accessibility of P to pigeonpea (Ae et al. 1990). Roots of this plant discharge piscidic acid, which chelates Fe and in this manner free a portion of the firmly bound soil P. Henceforth, pigeonpea is effectively developed in P inadequate tropical soils. Grain legumes having a high biomass and more root exudates may contribute a noteworthy amount of C stock in the subsurface layer, along these lines increasing the C sequestration.

3.3.3 Soil Health Improvement

3.3.3.1 Soil Physical Properties

Grain legumes are true component crops in the cereal-dominating cropping systems of South Asia for enhancing soil physicochemical and biological properties. Wilhelm et al. (2004) revealed that the significance of soil organic matter (SOM) in enhancing soil productivity and sustainability is notable. Legume crops enhance SOM which balance out soil aggregates, make the soil easily cultivable, and increase air circulation, soil water holding, and buffering limits. Further, SOM breakdown delivers accessible nutrients to plants. SOM ties the primary soil particles in the aggregates, physically and chemically, and thus increases the stability of the soil aggregates and limits their breakdown amid the wetting procedure (Lado and Ben-Hui 2004). According to Tisdall and Oades (1982), roots and hyphae are the major binding agents for macroaggregates (>0.25 mm), while humic compounds promote microaggregate (<0.25 mm) formation. Frey et al. (1999) reported that the length of fungal hyphae ranged from 19 to 292 mg g^{-1} soil and was 1.9 to 2.5 times higher in grain legume system than over no legume system. Due to more rhizosphere activity and rhizo-deposition, macroaggregates are gradually bound together by temporary (i.e., fungal hyphae and roots) and transient binding agents (i.e., microbial and plant-derived polysaccharides). Crop rotations that included grain legumes are generally beneficial to aggregate stability and formation of a favorable soil structure. The fungi present in the grain legume crop rhizosphere produce a glycoprotein called "glomalin." The sticky part of glomalin entraps soil mineral, organic matter, and debris to form stable soil aggregates. Hence, the microbial activity of rhizosphere is directly responsible for the improved soil structure in crop rotations involving grain legumes. In a long-term rotational experiment, a higher percentage of soil aggregates exceeding 0.25 mm was recorded where the preceding crop was a legume (Sharma et al. 2000a; Meena et al. 2014). The narrow C/N ratios of grain legume residues fasten their decomposition and improve SOM, thereby impacting soil aggregations and lessen soil bulk density. Ganeshamurthy et al. (2006) reported that inclusion of mungbean in rice-wheat system resulted in lower bulk density and higher hydraulic conductivity. The improvement in overall soil physical parameters under grain legume inclusive cropping systems is also recorded in two sets of long-term study in sandy loam (*Typic Ustochrept*) soil of the Indo-Gangetic Plain (Kumar et al. 2012).

3.3.3.2 Soil Chemical Properties

Chemical properties impacted by grain legume crops are the soil pH, nutrient accessibility, exchange capacity, etc. Grain legume crops have the ability to reduce the pH of soil in the rhizosphere and make microenvironment favorable for nutrient availability. Since grain legumes acquire a greater part of their N requirement from the air as diatomic N rather than from the soil as NO_3, their net effect lowers the pH of

Table 3.3 Effect of cropping systems and nutrient management on soil fertility

Treatments	Soil organic C (%)	Avail. N (kg ha^{-1})	Avail. P_2O_5 (kg ha^{-1})	Avail. K_2O (kg ha^{-1})
R-W	0.35c	258.9c	18.1c	222.9c
R-C	0.38b	272.5b	20.7ab	237.9b
R-W-R-C	0.37bc	266.6b	19.2b	238.0b
R-W-M	0.42a	286.3a	21.1a	262.2a

R rice, *W* wheat, *C* chickpea, *M* mungbean

the soil. Among grain legumes, chickpea reduces the pH most followed by pea and pigeonpea (Singh et al. 2009). It was also reported by Singh et al. (2009) that significant amount of organic residues is added through grain legume crops to the soil in the form of root biomass and leaf litters. Roots and leaf litters being rich in N facilitate fast decomposition of crop residues in soil and increase microbial activity. The grain legume crop residues may change relatively unavailable nutrients P in organic forms to available P to succeeding crops. For example, lupine can retain more P than most other grain legume crops from soil testing low in phosphorus (Braum and Helmke 1995). On decay, natural P in the cover crop tissues could give a labile sort of P to succeeding crops. Soil biological and chemical properties are intimately related in controlling of soil tilth. Soil microorganisms assume a significant role in keeping up soil quality on account of their involvement in nutrient recycling through the breaking down of organic matter and nutrients stock. Inclusion of legume in cropping system not only economizes the N requirement of cropping system but also helps in the efficient utilization of native phosphorus due to secretion of certain acids that help in solubilization of various forms of phosphorus. This capacity of the legumes makes them efficient in native utilization of phosphorus present in different forms. Increased availability of P is a result of P acquisition from insoluble phosphates through root exudates. Chickpea has the ability to access P normally which is not available to other crops by mobilizing sparingly soluble Ca-P by acidification of rhizosphere through its citric acid root exudates in Vertisols, and pigeonpea have been characterized for dissolution of Fe-P in Alfisol (Ae et al. 1991). Long-term incorporation of grain legumes in rice-wheat and maize-wheat systems altogether enhances SOC and accessible N, P, K, S, and DTPA-extractable Zn and in this manner expanded the nutrient take-up by cereal component crops. Long-term cultivation of mungbean enriched the SOC by 12.0 and 12.5% in maize-wheat and rice-wheat rotation, respectively, proposing the significance of fallow management for SOC management in tropics (Table 3.3). Considering the relative efficiency in SOC management, the crop rotation was found in the order of maize-wheat-mungbean > pigeonpea-wheat > maize-wheat-maize-chickpea > maize-wheat (Venkatesh et al. 2013).

3.3.3.3 Soil Biological Properties

Living soil organisms contribute less than 0.5 percent (w/w) of the total soil mass, yet they play a significant role in agroecological sustainability by influencing several soil properties and processes that directly or indirectly affect crop yields. Many researchers have demonstrated that soil microbial activities are related to the soil physicochemical properties. Soil microbial community structure, size, and functions are highly dynamic and are greatly influenced by soil properties, crop management, and nature of crops grown. Grain legumes may give great natural conditions for the growth and development of soil microorganisms. The soil microbial biomass is the living portion of the soil that includes basically microorganisms and parasites, including soil microfauna and green growth (Kumar and Goh 2000). The enzymatic activity in the soil is generally a product of magnitude of microbial population in soil, being gotten from intracellular, from cell-related, or from free mixes. The symbiotic association of *Rhizobium* and arbuscular mycorrhizal (AM) with roots of grain legumes increases N and phosphorus availability in soil for plant use. This is attributed due to fixation of atmospheric N by root nodulating Rhizobium bacteria and through enzymatic activities of the AM fungi. The grain legume crops boost the dehydrogenase, urease, protease, phosphatase, and β-glucosidase reactions in the soil. Inclusion of grain legumes in rice-wheat and maize-wheat systems has shown altogether enhanced soil biological properties (soil microbial biomass C and dehydrogenase activity) in a long-term study (Venkatesh et al. 2013; Dhakal et al. 2015). Thus, realizing the significance of soil microorganisms in terms of agroecological sustainability and crop productivity, nowadays, and soil biological parameters are included as important indicators toward determining the soil health.

3.4 Grain Legumes for Water Economy

Presently, the increasing water scarcity is a potential threat for crop production. Given the higher water use efficiency, grain legumes are always a preferred choice under water-limited conditions (Siddique et al. 2008). Water requirement of grain legumes is lower than cereals (Table 3.4). Global water consumption by cereals is reported to be about 60% as against 4% in grain legumes. Grain legumes have the ability to use water more efficiently than other crops due to their morphological and physiological features. Due to their deep root system, grain legumes are able to extract moisture from a deeper layer of soil profile thereby having the ability to thrive well under rainfed situations. Thus, including grain legumes in the cropping system could substantially reduce the irrigation water requirement and thereby

Table 3.4 Water requirement of grain legumes and cereal crops

Kharif/summer crop	Water requirement (cm)	Rabi crops	Water requirement (cm)
Urdbean (summer)	22–30	Chickpea	12–21
Mungbean (summer)	20–35	Lentil	10–12
Urdbean (kharif)	6–12	Field pea	12–14
Mungbean (kharif)	12–15	Rajmash	20–25
Pigeonpea	16–23	Wheat	30–45
Rice	100–220		
Maize	25–40		

curtails the production cost. A higher emphasis should be directed to design the climate resilient and low-input sustainable cropping systems involving grain legumes as a key component. In parallel, identification of stress-tolerant genotypes can further improve the water productivity. Pala et al. (2007) have compared wheat-wheat, wheat-chickpea, and wheat-lentil production systems in Mediterranean water-limited condition. They found that wheat-lentil production system had higher water productivity followed by wheat-chickpea and was least in case of wheat-wheat system. On the same line, Timsina and Connor (2001) suggested that grain legumes can be included in rice-wheat cropping system of IGP for upscaling water and nutrient use efficiency of the rotation. Besides this, it was found that adoption of the drip irrigation in pigeonpea and pigeonpea-based intercropping system improves the water productivity (Verma et al. 2015b; Praharaj et al. 2017). In general under subtropical climates like in IGP, *rabi* grain legumes like chickpea and lentil need only one irrigation, whereas wheat crop required five to six irrigations. Therefore, the problem of groundwater depletion commonly observed in rice-wheat regions of Indo-Gangetic plains could be reversed by replacing one of the cereal crops by pulse crop. Crop management strategies like laser land leveling and ridge furrow planting enable the crop to efficiently utilize the rainfall water and thus further improve the water use efficiency in dryland areas.

3.5 Weed Smothering Effects of Grain Legumes

Weeds are always a major constraint in the agricultural production system. It is perceived that existence of some weeds in fields can be helpful to the crop as it gives sustenance and living space to a wide range of agriculturally important organisms. However, weed population above critical threshold limits can adversely affect the crop productivity and quality of produce. Therefore, weed control has become imperative in arable crop production. The major emphasis of weed management strategies is to maintain weed populations below threshold level through a scope of cultivation methodologies all through the turn, which implies that immediate control activities inside the individual product have a more noteworthy guarantee to get success. It is critical to contemplate weeds as a component of the biodiversity of the agroecosystem, so weed control strategies should include their management rather than eradication. There is an extensive variety of weed control techniques, and consideration of grain legumes in crop rotation is a wise option to maintain weed population below the threshold level. Grain legumes manage the weeds in a particular situation as (i) crop rotation/diversification, (ii) cover crops, (iii) intercrop, and (iv) crop residues and through allelopathy.

3.5.1 Crop Rotation

Grain legumes in a crop rotation are key determinants for the levels of weeds in a system and affect the relative dominance of various weed species. The advantages of crop rotation rely upon the determination of crops and their order in a rotation.

Constant development of a solitary crop or crops having same cultivation practices permits certain weed species to end up plainly predominant in the framework, and, after some time, these weed species turn out to be difficult to control. For example, *Phalaris minor* Retz. has become a menace in the cereal-dominating cropping systems of South Asia (Brar 2002). Likewise, Chauhan and Johnson (2010) reported that weedy rice *(Oryza sativa* L.) is turning into a major weed issue in rice monoculture in Southeast and South Asia. Hence, crop rotation is critical with crop having an alternate developmental behavior. The fallow period between two main crops can be utilized by grain legumes to smother weeds. Replacing one rice crop in rice-rice-rice or rice-rice system with a grain legume in the dry season may altogether help in diminishing the seed bank of weedy rice in the soil. Extremely reassuring outcomes have been found in decreasing the weedy rice seed bank in a rice-rice-rice systems in Vietnam (Chauhan et al. 2010) when rice in dry season was supplanted by mungbean. In India, from a long-term experiments reported that (Kumar and Singh 2009; Hazra et al. 2012) incorporation of grain legume can minimize the pervasion of *Phalaris minor* and *Avena fatua* (wild oat) in winter crops. A crop like mungbean and urdbean which develop quickly and can contend with the weeds ought to be incorporated into the crop rotation either as sole crop or intercropping (Kumar et al. 2013; Meena et al. 2015b).

3.5.2 Intercropping

At present, intercropping is experienced where growers look for the highest consolidated yield of at least two crops for each unit of land or hazard not meeting the farmer's income. Intercropping infers growing of at least two crops of various development propensities at the same time on a similar land, which provides early cover of canopy and seedbed utilize bringing about decreased weed development by rivalry for various natural resources among crops component. Intercrops can be more powerful than sole crops in utilizing resources and stifling of the weed development due to existence of a complementary relationship in resource utilization (nutrients, water, and light) and facilitative interaction between intercrop plants (Liebman et al. 2001). Intercropping provides the sustainability of agricultural production systems especially under rainfed situations. Intercropping of short growth length, speedy developing, and early developing grain legume crop with longer growth habit and wide-spaced crops resulted in early land cover and smothering rising weeds adequately. The intercropping of corn with legumes prompted a higher soil cover and diminished light accessibility for weeds, which brought about a decrease in weed populations and dry matter contrasted with sole crops (Kumar et al. 2010b; Meena 2013). Reduction in development of weeds by grain legume crops is more prominent at a low-efficiency site than at a high-profitability site. In contrast to sole cropping, a grain legume-added intercropping system diminishes relative cover of weeds by 41% and lessened the population of *Senecio vulgaris* L. by 58% and enhances crop yield by 10% (Baumann et al. 2000). In pigeonpea-based intercropping (Table 3.5), fast-growing early-maturing grain legumes (cowpea and mungbean) decreases weed population by 30 to 40% compared to 22% by sorghum

Table 3.5 Weed smothering efficiency of important cropping system

Intercropping systems	Weed smothering efficiency (%)
Pigeonpea + urdbean	32.82
Pigeonpea + mungbean	31.01
Pigeonpea + cowpea	39.06
Pigeonpea + sesame	36.6
Pigeonpea + pearl millet	50.8
Maize + urdbean	17.3
Maize + pigeonpea	16.4

Source: Ali (1988)

(Ali 1988). In central and peninsular India, pigeonpea + sorghum has been found to be the most productive system on Vertisols, whereas on alfisols and entisols, pigeonpea + pearl millet proved to be the ideal system. It was found that weeds caused 79.93% reduction in pigeonpea grain yield if weeds were allowed to grow till harvest; however, grain yield losses were only 38.19% in pigeonpea + soybean intercropping system (Ali and Singh 1997). A similar effect was also reported in chickpea +wheat (Banik et al. 2006) and chickpea + mustard (Kaur et al. 2014) intercropping systems. Improvement in yield and weed suppression has also been demonstrated in many environments for cereal-grain legume intercrops.

3.5.3 Cover Crop

Grain legumes for cover crops have turned into a suitable alternative for sustainable farming in the light of its importance in soil health and crop yield improvement besides controlling weeds. Cover crops smother weeds by giving a physical obstacle; however, cover crops likewise trap light and regulate soil temperature, both of which act as a germination barrier for some small seeded seasonal weed species. Various grain legumes (urdbean, cowpeas, mungbean, and horse gram) have been found to stifle and reduce the population of different weeds due to competition for resources or allelopathic effect. Constant soil cover with grain legumes decreases light absorption by weed seeds and rivals the weeds for space and other resources. Norris and Kogan (2000) reported that the utilization of the cowpeas as cover crop advances the parasitic, bacterial, and mycorrhizal groups that might be unfavorable for weed growth and development but favorable for crop plants. Hence, cover crops control weeds and thus minimize the use of herbicides in agriculture. Weed control is generally the best under thick cover crops like cowpea, horse gram, and peas and, further, when cover crops are maintained for the longest manageable time. Cover crops may also influence weed diversity and dominance through modification of nutrient cycling, especially N cycling.

3.5.4 Pulse Crop Residues and Allelopathy

Surface crop residues can influence seed germination by means of physical and biochemical changes in the seed zone of soil layer. The two fundamental physical impacts of soil surface crop residue are decreasing light and shielding of the soil surface. Shielding of the soil surface has impacts on both soil temperature and moisture (Varma et al. 2017). Grain legumes may likewise add to weed control by delaying weed seed germination and decreasing weed population and weed growth and, thereby, minimize the crop yield loss due to weeds. Additionally, the surface cover in combination with grain legume crops changes the chemical environment of the soil zone around weed seed by means of allelopathy. According to Liebman and Davis (2000), the allelopathic reaction of crop residue decomposition has seen more articulated consequences in germination of small weed seeds. Various grain legume crops (lentil, cowpeas, and lupins) have been found to stifle and decrease the many weeds due to crop-weeds competition or allelopathic properties. Soybean crop residues have an ability to smother weeds and help improve the performance of summer squash (*Cucurbita pepo* L.) and tomato (*Solanum lycopersicum* L.) (Barker and Bhowmik 2001). Also, water-soluble extracts from lentil crop residues are lethal to stinkweed (*Thlaspi arvense* L.), downy brome (*Bromus tectorum* L.), and flixweed (*Descurainia sophia* [L.] Webb.) rather than wheat and confirmed that residues of these crops might be utilized for the specific control of weeds in wheat and to minimize the dependence on herbicides (Moyer and Huang 1997).

3.6 Grain Legumes in Conservation Agriculture

The basic principles of conservation agriculture (CA) such as least disturbance of soil, rational retention of adequate crop residue on the soil surface, and sensible crop rotation for improving livelihood and ecological security are well met while bringing grain legumes in production systems. Grain legume crops are considered as hardy crops which can thrive better than many other crops under adverse conditions and thus have immense value in CA. Inclusion of grain legumes in cereal-based crop rotations enhances input-use efficiencies and hence is considered as one of the best in resource conservation technology. Some of the CA-related values of grain legumes are as follows:

3.6.1 Reduced Tillage

Unlike cereals, grain legumes need rough and well-aerated seedbed. Grain legumes like lentil, lathyrus, urdbean, and mungbean are amenable for surface seeding in rice fallows under relay (*paira*) cropping. Chickpea and lentil also perform well under no-till (Table 3.6). The deep and strong roots of pigeonpea are capable of breaking hard pan in subsoil.

Table 3.6 Effect of tillage practices on chickpea yield

Treatment	Grain yield (kg ha^{-1})	Increase over conventional tillage (%)
No-till dibbling +mulching	1660	28.2
No-till drill + mulching	1589	22.7
Conventional tillage	1295	
CD (P = 0.05)	115	

Table 3.7 Grain yield of rice and wheat as influenced by residue management in rice-wheat-mungbean system

	Yield (Kg ha^{-1})		
Residue incorporation	Rice	Wheat	Sustainability Yield index
Rice-wheat-mungbean	3507	5082	0.79
Rice-wheat	3327	4902	0.74
Rice	3089	4855	0.74
No residue	2839	4434	0.67

3.6.2 Water Saving

Water requirements of grain legumes are much less than cereals and commercial crops; hence, they have a comparative advantage in rainfed/dryland areas. The water requirement of rice crop is 900–2500 mm, wheat 400–450 mm, and sugarcane 1400–2500 mm; however, grain legumes need only 250–300 mm of water. On account of their unique morphological and physiological features, they are capable of utilizing water more efficiently. Further, their deep root system enables them to draw soil moisture from deeper layers of the soil profile. In general under subtropical climate like in Indo-Gangetic plains, *rabi* grain legumes like chickpea and lentil need only one irrigation that too many times meet through winter rains. With 1 ha-mm of water, 12.5 kg chickpea could be produced as against 7.0 kg wheat and 2.5 kg rice.

3.6.3 Crop Residues

Grain legumes not only provide an excellent cover to soil surface due to their dense canopy but also leave substantial amount of easily decomposable crop residues. Incorporation of the mungbean residue further improved yield of rice and wheat as well as the sustainability index over and above rice + wheat residues (Table 3.7; Fig. 3.2). The low harvest index in grain legumes eventually provides large amount of crop residues. A good crop of chickpea may provide 8–10 tonnes of straw (crop residue). Winter grain legumes and pigeonpea shed a large number of their leaves at maturity (2–3 t ha^{-1} dry leaves) which provides a thin soil cover (Table 3.8).

Fig. 3.2 Pigeonpea leaf litters at podding and after harvest

Table 3.8 Leaf litter fall and nutrient contribution through leaf litter

Crop	Leaf litters (t ha^{-1})	N (kg ha^{-1})	P (kg ha^{-1})	K (kg ha^{-1})
Chickpea	1.1–1.7	7–14	3–5.5	8–20
Lentil	1.3–1.6	8–10	3.5–4.5	12.5–19
Pigeonpea	1.3–2.8	8–16	2.5–5	13.5–24

3.6.4 Crop Diversity

Crop diversity in cropping system is one of the important principles of CA. Grain legumes on account of their numerous virtues like short duration, low input requirements, biological N_2 fixation, nutrient recycling due to deep root system, and reversing the adverse effect of continued cereal-based production system are considered ideal for crop diversification. Grain legumes may find a place in the existing crop rotation as an intercrop, catch crop, substitute of low yielding cereals/millets, green manure crop, and alley crop. A sizeable shift of chickpea from the Indo-Gangetic plains to central and southern regions and cultivation of spring/ summer urdbean and mungbean as well as short-duration pigeonpea in the irrigated belt of north India is a glaring example of crop diversification. Some of the prominent crop rotations are rice-wheat-mungbean, rice/maize/bajra-chickpea/lentil/grass pea, pigeonpea (short duration, 140–155 days)-wheat, and rice-rice-urdbean/mungbean.

Thus, the inclusion of grain legumes in cereal-based crop rotations under CA may be considered as one of the RCTs which will reverse the negative effect of cereal-cereal rotation systems in this region.

3.7 Higher Productivity and Sustainability

Declining yield and factor productivity of major cereal-based cropping systems are a major concern since the last two decades in South Asia especially in IGP. This has forced the researchers and farmers to look for an alternative crop which can solve dual problems of declining factor productivity and system sustainability. Grain

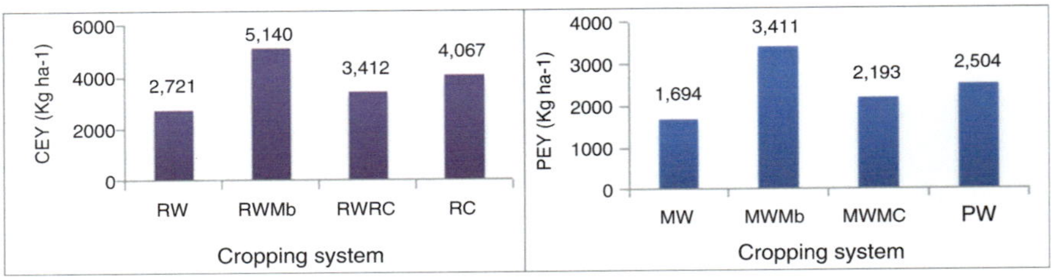

Fig. 3.3 Long-term effect of rice- and maize-based cropping system on system productivity; R, rice; W, wheat; Mb, mungbean; C, chickpea; M, maize; P, pigeonpea. (Source: IIPR 2012–2013)

legumes due to the intrinsic nature of fixation of atmospheric N and improving soil health can be fitted well under such situations in cereal-based systems. The lower cost of cultivation and higher market price resulted in higher system productivity and profitability from the grain legume-based cropping system in comparison to the cereal-based system.

In a long-term study at IGP, this revealed higher system productivity and sustainability in pulse inclusive of diversified cropping system over cereal-cereal system. The increase in system productivity is attributed due to increase in yield of component crops. Among different cropping systems, rice-wheat-mungbean gave a maximum production of 5, 140 kg ha^{-1} chickpea equivalent yield (CEY). This was followed by rice-chickpea and lowest under rice-wheat. In another set of a long-term study in which maize-wheat system performance was compared with pigeonpea- wheat, maize-chickpea, and maize-wheat-mungbean. Among these systems, maximum productivity of 3,411 kg ha^{-1} pigeonpea equivalent yield (PEY) was recorded in maize-wheat-mungbean followed by pigeonpea-wheat and least under maize-wheat (Fig. 3.3). However, results of another field trial on resource conservation revealed that rice-chickpea-mungbean and rice-chickpea performed better than the rice-wheat system in terms of system productivity, economics, and sustainability (IIPR 2012–2013). Similar results were also reported by Ali and Kumar (2006) and Ghosh et al. (2012).

Growing grain legumes as intercrop also a way to increase total productivity per unit are notable in time. Intercropping of short-duration (60–75 days) grain legumes (mungbean and urdbean) is most popular among farmers in IGP of South Asia. The special feature of this system is that the productivity of the base crop, i.e., pigeonpea, remains unaffected and an additional 400–500 kg ha^{-1} of mungbean or urdbean or 600–800 kg ha^{-1} of sorghum can be obtained without any additional inputs. Intercropping of winter grain legumes like chickpea and lentil with oilseeds is common in rainfed areas. Literatures reveal that high productivity and monetary returns can be obtained from chickpea + mustard, lentil + linseed, and wheat + lentil intercropping systems (Ali and Mishra 1996; Singh and Rathi 2003). Similarly, the horse gram can be also intercropped with early pigeonpea and maize in mid hills of Himalaya (Kumar et al. 2010a). Further, under rainfed wheat+ chickpea was found more remunerative than wheat + mustard, but in irrigated conditions, wheat+ mustard proved more profitable over wheat + chickpea. Lentil and linseed make a perfect combination for intercropping as compared to other *rabi* crops in rainfed conditions. Many other intercropping systems were also reported by several workers Ahlawat et al. (2005), Kumar et al. (2006, 2008, and 2010a, b). It has also been observed that growing one row of mungbean gives about half tonne/ha additional yield of mungbean without affecting the sugarcane yield. A further increase in mungbean rows 2–3 makes the systems nonprofitable. It has been also found that mungbean is more suitable than urdbean (Yadav et al. 1987 and Panwar et al. 1990). Another study has proved synergistic effects of urdbean and mungbean on cane yield in spring-planted crop with additional yield of 0.4–0.5 tonnes per hectare of these legumes (Lal et al. 1999; Varma and Meena 2016). Similarly, lentil is suitable for intercropping with autumn-planted sugarcane.

3.8 Grain Legumes in Rice Fallows

Growing rice is the predominant activity for farmers during the *kharif* season in most parts of South Asian countries. It is grown in both irrigated and rainfed conditions under various cropping systems. About 22.2 million hectares of land in South Asia remains fallow after rice harvest during rabi/winter season (Gumma et al. 2016) due to number of biotic, abiotic, and socioeconomic constraints. Out of total rice fallow area, 88% lies in India followed by 8.6% cent in Bangladesh. Despite ample opportunities, rice fallow systems did not get enough attention in the past. A number of abiotic factors related to soil and water lead to low productivity of grain legumes in rice fallows during the past several years. Low moisture content in soil after rice harvest followed by a fast decline in soil moisture with the advancement of *rabi* season results in mid- and terminal drought at flowering and pod-filling stages which adversely affect the productivity of grain legumes. Due to anaerobic conditions in rice cultivation, many of the organisms including rhizobia would not be able to survive. Besides the inherent constraints, rice fallows also affect seed germination, seedling emergence, and crop establishment due to disruption of the soil structure, soil water deficit, poor aeration, and mechanical impedance of the seed zone. This hostile environment creates a potential threat to microbial activity, nutrient availability, root growth (root is mostly confined to the top soil layer), and water and nutrients uptake; thus, subsoil resources in rice fallows remain unutilized.

Pulses with properties like low input requirements, short duration, ability to establish even with surface broadcast in standing rice fields (para/utera cropping), and soil fertility restoration are ideal crops for the rice fallow agroecosystem. They have the ability to fix atmospheric N and thus improve/restore soil fertility of sick soils which developed due to continuous cultivation of the rice crop. If this area is brought under cultivation, it may benefit millions of poor and small farmers solely dependent on agriculture for their livelihood. Productivity and profitability from grain legumes in rice fallows can be improved with suitable crop management technique and even by utilizing residual soil moisture. By adopting improved technologies like resource conservation, short duration, disease resistance, improved varieties, timely sowing, plant population, biofertilizers inoculation, fertilizer application, timely weed management practices, need-based plant protection measures coupled with proper irrigation schedule (lifesaving) would definitely increase the yield of grain legumes in rice fallow agroecological situation (Kumar et al. 2016a, b). Further, resource conservation technologies which deal with soil moisture conservation, organic matter buildup, improvement in soil structure, and microbial population could be an appropriate approach to address these problems in rice fallow. Therefore, if crop residues are retained on the soil surface in combination with suitable planting techniques (no-till planting or paira cropping), it may alleviate terminal drought/heat stress in pulses by conserving and regulating the soil moisture (Kumar et al. 2013a; 2016c). A minimum soil traffic by adoption of a suitable technology involving no-till and minimum soil disturbance and management of crop residues (conservation tillage) could lead to favorable effect on soil properties that further conserve the soil moisture to a longer period for plant use (Kumar et al. 2014; Meena et al. 2015c).

Conservation tillage with proper crop residue management is reported to reduce soil water evaporation, soil sealing, and crusting (Kumar et al. 2016a). It is also evident that hydraulic conductivity under straw retained in no-till drill is many times higher than that of a conventional tillage. In fact higher yield of lentil after wet season (rainy season) rice and with conservation tillage was also reported by Bandyopadhyay et al. (2016) under rainfed areas of eastern India. This will also reduce cost of cultivation through savings in labor, time, and farm power and improve input-use efficiency. Traditionally, seeds of pulses (lentil, lathyrus, mungbean, and urdbean) are broadcasted in the standing rice (para/utera cropping) field without any tillage. Under such situation, 20–30 cm rice stubble needs to be maintained in the field to get an advantage similar to that of conservation tillage. In areas where grain legumes are sown after a harvest of rice with land preparation, zero-till seeding may be advocated as it facilitates advance planting by 7 to 10 days and saves energy and labor. Rice ratoons are also a major problem in growing grain legumes under rice fallows. It is observed that a large quantity of residual soil moisture is lost by nonproductive rice ratoons. Besides, the farm pond concept need to be advocated for the harvesting of excess rainfall and to use during the critical growth period of grain legumes for life saving irrigation through micro-irrigation system. To contain the growth of rice ratoons, postemergence herbicides like quizalofop-ethyl @ 100 g ha^{-1} can be used at 3–4 leave stage (Kumar et al. 2013b). Under soil moisture stress, movement of plant nutrients from soil is a limiting factor for plant growth and yield. Under such situations, foliar nutrition of 2 percent urea and micronutrients may be used to mitigate the effect of soil moisture stress to certain extent (Kumar et al. 2014a).

3.9 Grain Legumes for Ecosystem Services

Agriculture is one of the major sources of the greenhouse gas (GHG) emission and soil water pollution and also the major consumer of fossil energy. The projected environmental change will certainly impact on productivity and sustainability of

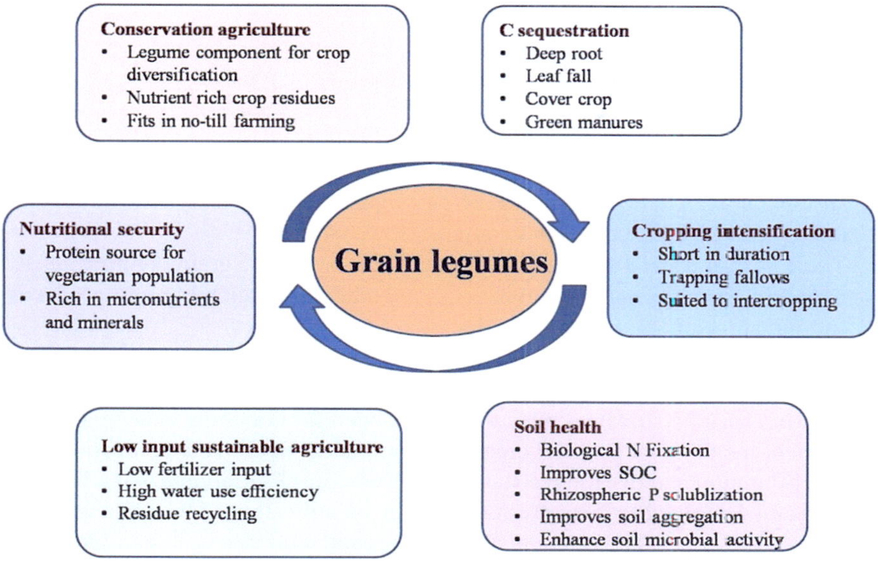

Fig. 3.4 The multipronged benefits of grain legumes in soil-plant systems and human nutrition

agricultural production systems (Keatinge et al. 2014) in the near future. A higher reliance on cereal-based rotations may lead to higher agronomic and ecological risks in the background of global climate change (Ebert 2014). According to the Newton et al. (2011), to improve the crop resilience to biotic and abiotic stress, there is a need to increase the heterogeneity (both temporal and special) into the cropping system. Grain legumes in the cropping system can play a vital role in ecosystem services (Fig. 3.4). Inclusion of grain legumes in intensive cereal-based crop rotations curtails the rate of N fertilizers, subsequently reduces the energy use, and GHG emission per unit cropping area (Nemecek et al. 2008). Likewise, Fuhrer (2006) (Meena and Yadav 2015) specified that grain legumes can minimize the use of fossil energy as well as reduce the N losses. Based on a comparative assessment of N fixation by legume and industrial fertilizer manufacturing, Crews and Peoples (2004) concluded that the ecological impact of grain legume N fixation is positive. The low C:N ratio of legume residues increases the retention of soil C and N and improves environmental quality (Drinkwater et al. 1998). However, legumes cultivation sometimes favors higher N_2O emissions. The main processes involved in the N_2O emission in legumes are rhizobial denitrification within the nodule, nitrification/denitrification of biologically fixed N, and decomposition of N-rich legume residue. Added to this, the altered N dynamics with the symbiotic N fixation may cause N losses like NO_3^- leaching. Intercropping of grain legumes in cereals can reduce nitrate leaching (Yadav 1981). Sugarcane + urdbean and pigeonpea + maize resulted in a low nitrate N leaching as compared to the sole cropping of sugarcane and maize (Yadav 1982). In addition, the inclusion of short-duration summer grain legumes reduces the fallow period between two crops (rice-wheat) and thus reduces C loss during hot summer and enhances C sequestration of a system.

3.10 Way Forward

The research emphasis on following aspects needs to be focused:

- The quantification of intensification effects due to inclusion of different grain legumes in existing cropping systems on the system productivity, profitability, sustainability, soil health, and insect-pest dynamics in system mode needs to be attempted under different agroecological regions.
- There is also a need to quantify the beneficial effect of grain legumes on soil microbial diversity and their dynamics, soil nutrients availability, soil productivity, and agroecological sustainability.
- Intensification approaches are highly location specific which depend upon the existing resources and socioeconomic and climatic conditions. Therefore, location-specific strategies need to be developed for obtaining beneficial advantages of inclusion of grain legumes as a component of crop diversification using short-duration and disease resistance varieties against different driving forces.
- Concerted research efforts are required for identification of a climate-resilient grain legume-based production system (cropping sequence or intercropping) for different agroecosystems. Further, possibilities need to be explored to identify water-saving technology such as micro-irrigation, resource conservation technology, and conservation tillage. In situ soil, the moisture conservation strategy needs to be strengthened for mitigation of mid-season and terminal drought in *rabi* season grain legumes especially under rice fallows. Farm pond water harvesting can be promoted to harvest excessive runoff water during rainy season for use in rabi season grain legumes as lifesaving irrigation through the micro-irrigation system.
- Conservation agriculture is increasingly being practiced in many parts of South Asia. Diversification or intensification of the rice-wheat system under CA in the Indo-Gangetic Plain through popularization of short-duration varieties of pigeonpea, chickpea, lentils, and summer mungbean needs to be promoted as the key to sustainability.
- Widespread micronutrient deficiency is observed in South Asia which adversely affects the yields of grain legumes in this region. Thus, to minimize the micronutrient deficiency in grain legumes, emphasis should be given for evaluation and development of micronutrient fortified customized fertilizers for different agroclimatic zones. Further, to improve grain quality, agronomic bio-fortification strategies need be promoted to ensure nutritional security in this region.
- Suitable resource conservation technologies need to be developed for grain legumes to mitigate the ill effect of increasing ambient temperature and CO_2 concentration and soil temperature.
- Yield loss in grain legumes due to weeds is more than in cereal crops. Mostly, application of preemergence herbicides followed by manual weeding is used in

grain legumes to control weeds in all seasons. But due to high cost and unavailability of labor on time, later flush of diversified weeds must be controlled through postemergence herbicides to realize higher yield of grain legumes as it is possible in cereal crops. Unfortunately, none of the postemergence herbicides available in the market are effective for controlling weeds in grain legumes especially during the rabi season. Thus, a new generation highly effective postemergence herbicides need to be identified for effective weed control in grain legumes. Further, development of genotypes tolerant to postemergence herbicide (imazethapyr, metribuzin, and glyphosate) is the need of the hour.

- In the last few decades, research evidences have shown that some of the grain legumes can be grown during nonconventional seasons. Thus, research should be initiated for strengthening of agro-techniques for popularization of the *rabi* rajmash in northern plains, spring rajmash in NW Himalaya, and pre-*rabi* pigeonpea in flooded areas under irrigated conditions.
- Grain legumes are mostly grown under rainfed conditions. Many biotic and abiotic stresses are affecting the yields of grain legumes in this region. Therefore, research efforts can be thrust on developing abiotic stresses (water logging, heat, and salinity), tolerant pulse genotypes, and their improved management for minimizing the ill effects of climatic variability. Further, transplanting of long duration pigeonpea can be popularized as a contingent crop measure to assure optimum plant population.
- Grain legume production systems are poorly mechanized. Sowing is commonly done through broadcasting of seeds and harvesting through manual labors. Thus, for increasing profitability and for enhancing the yield, farm machineries should be developed for grain legume production systems. Popularizing the machine harvestable varieties further can facilitate large-scale cultivation of grain legumes. The varieties having higher podding height above the ground and effect plant type even at maturity will be an obvious choice for mechanical harvesting. Therefore, genotypes suitable for mechanical harvesting need to be developed in different grain legume crops.
- Grain legumes can be best fitted under organic farming in comparison to cereal crops due to its low requirement of inputs like nutrients and water. The profuse development of roots and nodules is observed in grain legumes under application of organic manure. Thus, suitable agro-techniques and cropping systems should be developed for grain legumes for organic farming.
- New niches (intensification or diversification) should be identified for inclusion of grain legumes particularly in irrigated cereal-based agroecosystems for increasing pulses acreage and production in South Asia.
- For making grain legume cultivation profitable, the development of a market regulation mechanism for fluctuating prices and supportive policy is to be developed to make grain legume cultivation a profitable enterprise.

References

Ae N, Arihara J, Okada K, Yoshihara T, Johansen C (1990) Phosphorus uptake by pigeonpea and its role in cropping systems of the Indian subcontinent. Science (Washington, DC) 248:477–480

Ae N, Arihara J, Okada K (1991) Phosphorus nutrition of grain legumes in the semi arid tropics. ICRISAT, Hyderabad, p 33

Ahlawat IPS, Gangaiah B, Singh O (2005) Production potential of chickpea (*Cicer arietinum*)-based intercropping systems under irrigated conditions. Indian J Agron 50:27–30

Ali M (1988) Weed suppressing ability and productivity of short duration legumes intercropped with pigeonpea under rainfed conditions. Trop Pest Manage 34:384–387

Ali M (1992) Effect of summer legumes on productivity and nitrogen economy of succeeding rice (*Oryza sativa*) in sequential cropping. Indian J Agric Sci 62:466–467

Ali M, Kumar S (2006) Prospects of mungbean in rice-wheat cropping systems in Indo-Gangetic Plains of India. In: S. Shanmugasundaram (ed) Improving income and nutrition by incorporating mungbean in cereal fallows in the Indo-Gangetic Plains of South Asia DFID Mungbean Project for 2002–2004. Proceedings of the final workshop and planning meeting, Punjab Agricultural University, Ludhiana, Punjab, India, 27–31 May 2004, pp 246–254

Ali M, Mishra JP (1996) Technology for late-sown gram. Indian Farming 46(9):67–71

Ali M, Singh KK (1997) Management of pulse crops under intercropping systems. In: Asthana AN, Ali M (eds) Recent advances in pulses research. Indian Society of Pulses Research and Development, IIPR, Kanpur, pp 489–508

Ali M, Mishra JP, Singh KK (1998) Genotypic compatibility and spatial arrangements in spring sunflower and green gram (*Phaseolus radiatus*) intercropping. Indian J Agric Sci 6:636–637

Ali M, Kumar N, Ghosh PK (2012) Milestones on agronomic research in pulses in India. Indian J Agron 57(3rd IAC Special Issue):52–57

Ali M, Ghosh PK, Hazra KK (2014) Resource conservation technologies in rice fallow. In: Ghosh PK, Kumar N, Venkatesh MS, Hazra KK, Nadarajan N (eds) Resource conservation technology in pulses. Scientific Publisher, Jodhpur, pp 83–89

Bandyopadhyay PK, Singh KC, Mondal K, Nath R, Ghosh PK, Kumar N, Basu PS, Singh SS (2016) Effects of stubble length of rice in mitigating soil moisture stress and on yield of lentil (*Lens culinaris* Medik) in rice-lentil relay crop. Agr Water Manag 173:91–102

Banik P, Midya A, Sarkar BK, Ghose SS (2006) Wheat and chickpea intercropping systems in an additive series experiment: advantages and weed smothering. Eur J Agron 24:325–332

Barker AV, Bhowmik PC (2001) Weed control with crop residues in vegetable cropping systems. J Crop Prod 4:163–184

Baumann DT, Kropff MJ, Bastiaans L (2000) Intercropping leeks to suppress weeds. Weed Res 40:359–374

Brady NC, Weil RR (2002) The nature and properties of soils, 13th edn. Upper Saddle River, Prentice Hall, p 960

Brar LS (2002) Current status of herbicide resistance in Punjab and its management strategies. In: Proceedings of international workshop on herbicide resistance and zero tillage in rice-wheat cropping system. CCSHAU, Hisar, pp 6–10

Braum SM, Helmke PA (1995) White lupin utilizes soil phosphorus that is unavailable to soybean. Plant Soil 176:95–100

Buragohain S, Sharma B, Nath JD, Gogaoi N, Meena RS, Lal R (2017) Impact of ten years of biofertilizer use on soil quality and rice yield on an Inceptisol in Assam, India. Soil Res. https://doi.org/10.1071/SR17001

Champ MM (2002) Non-nutrient bioactive substances of pulses. British J Nutr 88:S307–S319

Chauhan BS, Johnson DE (2010) Weedy rice (*Oryza sativa* L.) I. Grain characteristics and growth response to competition of weedy rice variants from five Asian countries. Weed Sci 58:374–380

Chauhan BS, Migo T, Westerman PR, Johnson DE (2010) Post-dispersal predation of weed seeds in rice fields. Weed Res 50:553–560

Chauhan BS, Mahajan G, Sardana V, Timsina J, Jat ML (2012) Productivity and sustainability of the rice–wheat cropping system in the indo–Gangetic Plains of the Indian subcontinent: problems, opportunities, and strategies. Adv Agron 117:315–369

Congreves KA, Hayes A, Verhallen EA, Van Eerd LL (2015) Long-term impact of tillage and crop rotation on soil health at four temperate agroecosystems. Soil Till Res 152:17–28

Crews TE, Peoples MB (2004) Legume versus fertilizer sources of nitrogen: ecological tradeoffs and human needs. Agric Ecosyst Environ 102:279–297

Dhakal Y, Meena RS, De N, Verma SK, Singh A (2015) Growth, yield and nutrient content of mungbean (*Vigna radiata* L.) in response to INM in eastern Uttar Pradesh, India. Bangladesh J Bot 44(3):479–482

Dhakal Y, Meena RS, Kumar S (2016) Effect of INM on nodulation, yield, quality and available nutrient status in soil after harvest of green gram. Legum Res 39(4):590–594

Drinkwater LE, Wagoner P, Sarrantonio M (1998) Legume-based cropping systems have reduced carbon and nitrogen losses. Nature 396(6708):262–265

Ebert AW (2014) Potential of underutilized traditional vegetables and legume crops to contribute to food and nutritional security, income and more sustainable production systems. Sustain 6:319–335

Fageria NK, Baligar VC, Clark RB (2002) Micronutrients in crop production. Adv Agron 77:185–268

FAOSTAT (2014) www.fao.org/faostat/en/#data/QC.March 8& 24, 2017

Frey SD, Elliott ET, Paustain K (1999) Bacterial and fungal abundance and biomass in conventional and no-tillage agroecosystem along two climatic gradients. Soil Biol Biochem 31:573–585

Frye WW, Blevins RL, Smith MS, Corak SJ, Varco JJ (1988) Role of annual legume cover crops in efficient use of water and nitrogen. In: Hargrove WL (ed) Cropping strategies for efficient use of water and nitrogen, vol 51. American Society of Agronomy, Madison., Spec. Publ. No., pp 129–154

Fuhrer J (2006) Environmental aspects of the nitrogen cycle in legume-based cropping systems. In: AEP (ed) Grain legumes and the environment: how to assess benefits and impacts? Zurich, November 18–19, 2004. AEP and FAL, pp 85–91

Ganeshamurthy AN (2009) Soil changes following long–term cultivation of pulses. J Agric Sci 147:699–706

Ganeshamurthy AN, Ali M, Srinivasarao C (2006) Role of pulses in sustaining soil health and crop production. Indian J Fert 1(12):29–40

Ghosh PK, Sharma KC (1996) Direct and residual effect of green manuring in rice wheat rotation. Crop Res 1:133–136

Ghosh PK, Mohanty M, Bandyopadhyay KK, Painuli DK, Misra AK (2006) Growth, competition, yields advantage and economics in soybean/pigeonpea intercropping system in semi–arid tropics of India: II Effect of nutrient management. Field Crops Res 96:90–97

Ghosh PK, Bandyopadhyay KK, Wanjari RH, Manna MC, Misra AK, Mohanty M, Rao AS (2007) Legume effect for enhancing productivity and nutrient use-efficiency in major cropping systems–an Indian perspective: a review. J Sustain Agric 30:59–86

Ghosh PK, Venkatesh MS, Hazra KK, Kumar N (2012) Long-term effect of pulses and nutrient management on soil organic carbon dynamics and sustainability on an inceptisol of Indo-Gangetic plains of India. Exp Agric 48:473–487

Ghosh PK, Hazra KK, Nath CP, Das A, Acharya CA (2016) Scope, constraints and challenges of intensifying rice (*Oryza sativa*) fallows through pulses. Indian J Agron 61(4th IAC Special Issue):S122–S128

Gumma MK, Thenkabail PS, Teluguntla P, Rao MN, Mohammed IA, Whitbread AM (2016) Mapping rice fallow cropland areas for short-season grain legumes intensification in South Asia using MODIS 250 m time-series data. Int J Digit Earth 9:981–1003

Hatfield JL, Karlen DL (1994) Sustainable agriculture systems. Lewis Publication, Boca Raton

Hazra KK, Kumar N, Venkatesh MS, Ghosh PK (2012) Inclusion of pulses in rice-wheat system can reduce *Phalaris minor* population. Pulses Newsletter 23(3):6

Hazra KK, Venkatesh MS, Ghosh PK, Ganeshamurthy AN, Kumar N, Nadarajan N, Singh AB (2014) Long–term effect of pulse crops inclusion on soil–plant nutrient dynamics in puddled rice (*Oryza sativa* L.)–wheat (*Triticum aestivum* L.) cropping system on an Inceptisol of Indo–Gangetic plain zone of India. Nutr Cycl Agroecosyst 100:95–110

IIPR (2009) 25 years of pulses research at IIPR. Indian Institute of Pulses Research, Kanpur

IIPR (2012–2013) Annual Report. Indian Institute of Pulses Research, Kanpur – 208 024 (Uttar Pradesh), India

Jat RK, Sapkota TB, Singh RG, Jat ML, Kumar M, Gupta RK (2014) Seven years of conservation agriculture in a rice–wheat rotation of eastern Gangetic Plains of South Asia: yield trends and economic profitability. Field Crops Res 164:199–210

Kaur R, Kumar A, Sharma BC, Brijnandan KP, Sharma N (2014) Weed indices in chickpea + mustard intercropping system. Indian J Weed Sci 46:333–335

Keatinge JDH, Ledesma DR, Keatinge FJD, Hughes J'A (2014) Projecting annual air temperature changes to 2025 and beyond: implications for vegetable horticulture worldwide. J Agric Sci 52:38–57

Kerr RB, Snapp S, Chirwa M, Shumba L, Msachi R (2007) Participatory research on legume diversification with Malawian smallholder farmers for improved human nutrition and soil fertility. Exp Agric 43:437–453

Kumar K, Goh KM (2000) Crop residues and management practices: effects on soil quality, soil nitrogen dynamics, crop yield, and nitrogen recovery. Adv Agron 68:197–319

Kumar N, Singh KK (2009) Weed management in pulses: why, when and how. Indian Farming 60(4):9–12

Kumar R, Ali M, Arya RL, Mishra JP (2006) Enhancing productivity and profitability of chickpea *(Cicer arietinum)* + Indian mustard *(Brassica juncea)* intercropping system. Indian J Agron 51:27–30

Kumar N, Prakash V, Meena BL, Gopinath KA, Srivastva AK (2008) Evaluation of *toria* (*Brassica campestris*) and lentil (*Lens culinaris*) varieties in intercropping system with wheat (*Triticum aestivum*) under rainfed conditions. Indian J Agron 53:47–50

Kumar N, Prakash V, Mina BL, Gopinath KA, Srivastva AK (2010a) Effect of sowing methods, growth retardant and intercropping on horse gram (*Microtyloma uniflorum*) productivity. Indian J Agric Sci 80: 335–337

Kumar R, Gopal R, Jat ML, Gupta RK (2010b) Conservation agriculture based strategies for sustainable weed management in maize (*Zea mays*). Training Manual, Maize for Freshers. Directorate of Maize Research, New Delhi, India

Kumar N, Singh MK, Ghosh PK, Hazra KK, Venkatesh MS, Nadarajan N (2012) Resource conservation technology in pulse based cropping systems. Technical Bulletin 1/2012. Indian Institute of Pulses Research, Kanpur, pp. 35 pages

Kumar N, Basu PS, Ghosh PK, Venkatesh MS, Hazra KK, Praharaj CA, SenthilKumar M, Yadav A, Singh S (2013a) RCT for enhancing pulses productivity in rice fallows. Pulses Newsletter 24(3):6

Kumar N, Ghosh PK, Hazra KK, Venkatesh MS, Basu PS, Yadav SK (2013b) Management of rice ratoon in rice-fallow pulses system. Pulses Newsletter 24(4):3

Kumar N, Hazra KK, Singh MK, Venkatesh MS, Kumar L, Singh J, Nadarajan N (2013c) Weed management techniques in pulse crops, Technical bulletin 10/2013. Indian Institute of Pulses Research, Kanpur, p 47

Kumar N, Singh SS, Praharaj CS, Yadav A, Yadav SL, Singh S (2014a) Effect of nutrient management system under rice-fallow condition. Pulses Newsletter 25(4):6

Kumar N, Singh SS, Ghosh PK, Basu PS, Singh MK, Venkatesh MS, Hazra KK, Praharaj CS, SenthilKumar M, Yadav SK, Singh S, Yadav A (2014b) Enhancing chickpea (*Cicer arietinum*) productivity through conservation practices under rainfed rice-fallow regions of India. "Extended summaries of voluntary papers" of National Symposium on "agricultural diversification for sustainable livelihood and environmental security" November 18–20, 2014, Ludhiana, Punjab, pp 734–735

Kumar N, Hazra KK, Nadarajan N, Singh S (2016a) Constraints and prospects of growing pulses in rice fallows of India. Indian Farming 66(6):13–16

Kumar N, Hazra KK, Nath CP, Praharaj CS, Singh U, Singh SS (2016b) Pulses in irrigated ecosystem: Problems and prospects. Indian J Agron (4th IAC Special Issue):S262–S268

Kumar N, Singh SS, Ghosh PK, Praharaj CS, Basu PS, Hazra KK, Senthilkumar M, Singh MK (2016c) Mitigating abiotic stresses in pulses under rice fallows in India. Extended summaries Vol. 1: 4th international agronomy congress, sec: abiotic and biotic (weeds) stress management, Nov. 22–26, 2016, New Delhi, India, pp 370–371

Lado M, Ben-Hui M (2004) Organic matter and aggregate size interactions in infiltration, seal formation and soil loss. Soil Sci Soc Am J 68:935–942

Lal M, Singh AK, Ali M, Kumar R (1999) Agronomic evaluation of intercropped mungbean/urdbean genotypes in spring planted sugarcane. Annual report. IIPR, Kanpur

Liebman M, Davis AS (2000) Integration of soil, crop and weed management in low-eternal input farming systems. Weed Res 40:27–47

Liebman M, Mohler CL, Staver CP (2001) Ecological Management of Agricultural Weeds. Cambridge University Press, Cambridge

Meena RS (2013) Response to different nutrient sources on green gram (*Vigna radiata* L.) Productivity. Indian J Ecol 40(2):353–355

Meena RS, Yadav RS (2015) Yield and profitability of groundnut *(Arachis hypogaea* L) as influenced by sowing dates and nutrient levels with different varieties. Legum Res 38(6):791–797

Meena VS, Maurya BR, Meena RS, Meena SK, Singh NP, Malik VK (2014) Microbial dynamics as influenced by concentrate manure and inorganic fertilizer in alluvium soil of Varanasi, India. African J Microb Res 8(1):257–263

Meena RS, Yadav RS, Meena H, Kumar S, Meena YK, Singh A (2015a) Towards the current need to enhance legume productivity and soil sustainability worldwide: a book review. J Clean Prod 104:513–515

Meena RS, Dhakal Y, Bohra JS, Singh SP, Singh MK, Sanodiya P (2015b) Influence of bioinorganic combinations on yield, quality and economics of Mungbean. Am J Exp Agric 8(3):159–166

Meena RS, Meena VS, Meena SK, Verma JP (2015c) Towards the plant stress mitigate the agricultural productivity: a book review. J Clean Prod 102:552–553

Meena RS, Gogaoi N, Kumar S (2017) Alarming issues on agricultural crop production and environmental stresses. J Clean Prod 142:3357–3359

Meisinger JJ, Hargrove WL, Mikkelsen RL, Williams JR, Benson VW (1998) Effects of cover crops on ground water utility. In: Hargrove WL (ed) Cover crop for clean water. Soil and Water Conservation Society, Ankeny, pp 9–11

Moyer JR, Huang HC (1997) Effect of aqueous extracts of crop residues on germination and seedling growth of ten weed species. Bot Bull Acad Sinica 38:131–139

Nemecek T, von Richthofen JS, Dubois G, Casta P, Charles R, Pahl H (2008) Environmental impacts of introducing grain legumes into European crop rotations. Eur J Agron 28:380–393

Newton AC, Johnson SN, Gregory PJ (2011) Implications of climate change for diseases, crop yields and food security. Euphytica 179:3–18

Njeru EM (2016) Crop diversification: a potential strategy to mitigate food insecurity by smallholders in sub-Saharan Africa. J Agric Food Syst Community Dev 3(4):63–69

Norris RF, Kogan M (2000) Interactions between weeds, arthropod pests, and their natural enemies in managed ecosystems. Weed Sci 48:94–158

Pala M, Ryan J, Zhang H, Singh M, Harris HC (2007) Water-use efficiency of wheat-based rotation systems in a Mediterranean environment. Agric Water Manag 93(3):136–144

Panwar BS, Singh BV, Sharma JC (1990) Feasibility of intercropping in autumn planted sugarcane. Indian Sugar 39:755–756

Patterson CA, Maskus H, Dupasquier C (2009) Pulse crops for health. AACC International Inc. Cereals Foods World 54(3):108–112

Peoples MB, Herridge DF, Ladha JK (1995) Biological nitrogen fixation: an efficient source of nitrogen for sustainable agricultural production. Plant Soil 174:3–28

Phillippy BQ (2003) Inositol phosphates in food. Adv Food Nutr Res 45:1–60

Praharaj CS, Singh U, Singh SS, Kumar N (2017) Micro-irrigation in rainfed pigeonpea–upscaling productivity under eastern Gangetic Plains with suitable land configuration, population management and supplementary fertigation at critical stages. Curr Sci 112(1):95–107
Ram K, Meena RS (2014) Evaluation of pearl millet and Mungbean intercropping systems in arid region of Rajasthan (India). Bangladesh J Bot 43(3):367–370
Reeves DW (1994) Cover crops and rotations. In: Hatifield JT, Stewart BA (eds) Crops residue management. Lewis Publishers, Boca Rotan, pp 125–172
Sainju UM, Singh BP, Whitehead WF (2005) Tillage, cover crops, and nitrogen fertilization effects on cotton and sorghum root biomass, carbon, and nitrogen. Agron J 97:1279–1290
Saraf CS, Patil RR (1995) Fertiliser use in pulse based cropping systems. Fert News 40:55
Sharma S, Upadhyay RG, Sharma CR (2000a) Effect of rhizobium inoculation and nitrogen on growth, dry matter accumulation and yield of black gram. Legum Res 23:64–66
Sharma SN, Prasad R, Singh S, Singh P (2000b) On-farm trials of the effect of introducing a summer green manure of mungbean on the productivity of a rice–wheat cropping system. J Agric Sci 134:169–172
Shi J, Arunasalam K, Yeung D, Kakuda Y, Mittal G, Jiang Y (2004) Saponins from edible legumes: chemistry, processing, and health benefits. J Med Food **7**:67–78
Siddique KHM, Johansen C, Kumar Rao JVDK, Ali M (2008) Legumes in sustainable cropping systems. In: Kharkwal MC (ed) Food legumes for nutritional security and sustainable agriculture proceedings of the fourth international food legumes research conference (IFLRC-IV), October 18–22, 2005, New Delhi, India, vol 1. Indian Society of Genetics and Plant Breeding, New Delhi, pp 787–819
Singh J (2015) Improving nutritional quality of pulses. In: Dixit JP, Singh J and Singh NP (eds) Pulses: challenges & opportunities under changing climatic scenario (Proceedings of National Conference on Pulses: Challenges & Opportunities Under Changing Climatic Scenario, September 29–October 1, 2014, JNKVV Jabalpur) Indian Society of Pulses Research and Development, ICAR-Indian Council of Pulses Research, Kanpur-208 024, UP, India, pp 285–308
Singh KK, Rathi KS (2003) Dry matter production and productivity as influenced by staggered sowing of mustard intercropped at different row ratios of chickpea. J Agron Crop Sci 189:169–175
Singh Y, Singh B, Timsina J (2005) Crop residue management for nutrient cycling and improving soil productivity in rice–based cropping systems in the tropics. Adv Agron 85:269–407
Singh KK, Ali M, Venkatesh MS (2009) Pulses in cropping systems. IIPR, Kanpur, p 39
Snapp SS, Kanyama-Phiri GY, Kamanga B, Gilbert R, Wellard K (2002) Farmer and researcher partnerships in Malawi: developing soil fertility technologies for the near-term and far-term. Exp Agric 38:411–431
Staver KW, Brinsfield RB (1990) Patterns of soil nitrate availability in corn production systems: implications for reducing groundwater contamination. J Soil Water Cons 45:318–323
Timsina J, Connor DJ (2001) Productivity management of rice-wheat cropping systems: issues and challenges. Field Crops Res 69:93–132
Tisdall JM, Oades J (1982) Organic Matter and Water-Stable Aggregates in Soils. J Soil Sci 33:141–163.
Vance CP (1997) Enhanced agricultural sustainability through biological nitrogen fixation. In: Legocki A, Bothe H, Puhler A (eds) Biological fixation of nitrogen for ecology and sustainable agriculture. Springer, Berlin, pp 179–186
Varma D, Meena RS (2016) Mungbean yield and nutrient uptake performance in response of NPK and lime levels under acid soil in Vindhyan region, India. J App Nat Sci 8(2):860–863
Varma D, Meena RS, Kumar S (2017) Response of mungbean to fertility and lime levels under soil acidity in an alley cropping system in Vindhyan region, India. Int J Chem Stu 5(2):384–389
Venkatesh MS, Hazra KK, Ghosh PK, Prahraj CS, Kumar N (2013) Long–term effect of pulses and nutrient management on soil carbon sequestration in Indo–Gangetic plains of India. Can J Soil Sci 93:127–136

Venkatesh MS, Hazra KK, Singh J, Nadarajan N (2015) Introducing summer mungbean in cereal based production system. Indian Farming 65(1):12–13

Verma JP, Jaiswal DK, Meena VS, Meena RS (2015a) Current need of organic farming for enhancing sustainable agriculture. J Clean Prod 102:545–547

Verma JP, Meena VS, Kumar A, Meena RS (2015b) Issues and challenges about sustainable agriculture production for management of natural resources to sustain soil fertility and health: a book review. J Clean Prod 107:793–794

Verma SK, Singh SB, Prasad SK, Meena RN, Meena RS (2015c) Influence of irrigation regimes and weed management practices on water use and nutrient uptake in wheat (*Triticum aestivum* L. Emend. Fiori and Paol.). Bangladesh J Bot 44(3):437–442

Von Richthofen JS, Pahl H, Nemecek T, Odermatt O, Charles R, Casta P, Sombrero A, Lafarga A, Dubois G (2006) Economic interest of grain legumes in European crop rotations GL-Pro Report, WP3

Wilhelm WW, Johnson JMF, Hatfield JL, Voorhees WB, Linden DR (2004) Crop and soil productivity response to corn residue removal: a literature review. Agron J 96(1):17

Yadav RL (1981) Intercropping pigeonpea to conserve fertilizer nitrogen in maize and produce residual effect on sugarcane. Exp Agric 17:311–315

Yadav RL (1982) Minimising nitrate nitrogen leaching by parallel multiple cropping in long duration row crops. Exp Agric 18:37–42

Yadav RL (1998) Factor productivity trend in a rice-wheat cropping system under long-term use of chemical fertilizers. Exp Agric 34:1–18

Yadav RL, Prasad SR, Singh K (1987) Fertilizer requirement and row arrangements of pulses in sugarcane based intercropping systems. Indian J Agron 32:80–84

Yadav RL Dwivedi BS, Gangwar KS, Prasad K (1998) Overview and prospects for enhancing residual benefits of legumes in rice and wheat cropping systems in India. In: Kumar Rao JVDK, Johansen C, Rego TJ (eds) Residual effect of legumes in rice and wheat cropping systems of the Indo-Gangetic plains. International Crops Research Institute for Semi Arid Tropics, Patancheru, India. Oxford & IBH Publishing Co. Pvt. Ltd., New Delhi, pp 207–225

Yeh CT, Yen GC (2003) Effects of phenolic acids on human phenol sulfo- transferases in relation to their antioxidant activity. J Agric Food Chem 51:1474–1479

Young JPW (1992) Phylogenic classification of nitrogen-fixing organisms. In: Stacey G, Burris RH, Evans H (eds) Biological nitrogen fixation. Chapman & Hall, New York, pp 43–86

Role of Legumes in Soil Carbon Sequestration

4

Sandeep Kumar, Ram Swaroop Meena, Rattan Lal, Gulab Singh Yadav, Tarik Mitran, Babu Lal Meena, Mohan Lal Dotaniya, and Ayman EL-Sabagh

Contents

S. Kumar
Department of Agronomy, CCS Haryana Agricultural University, Hisar, Haryana, India

R. S. Meena (✉)
Department of Agronomy, Institute of Agricultural Sciences (BHU), Varanasi, UP, India

R. Lal
Carbon Management and Sequestration Center, SENR/FAES, The Ohio State University, Columbus, OH, USA

G. S. Yadav
Division of Crop Production, ICAR Research Complex for NEH Region, Lembucherra, India

T. Mitran
School of Environment and Natural Resources, The Ohio State University, Columbus, OH, USA

B. L. Meena
ICAR – Central Soil Salinity Research Institute, Karnal, India

M. L. Dotaniya
ICAR – Indian Institute of Soil Science, Bhopal, India

A. EL-Sabagh
Department of Agronomy, Faculty of Agriculture, Kafrelshekh University, Kafr El-Shaikh, Egypt

R. S. Meena et al. (eds.), *Legumes for Soil Health and Sustainable Management*,
https://doi.org/10.1007/978-981-13-0253-4_4

Abstract

The soil organic carbon (SOC) pool is the key indicator of soil health and quality which in turn plays a vital role to soil sustainability. The continuous uses of unsustainable agricultural approaches have depleted most of the SOC pool of global agricultural lands. Promoting cultivation of leguminous crops, grasses, shrubs, and trees offers multiple advantages, e.g., augmenting crop and soil productivity and adapting to climate change by increasing resilience of agroecosystems. As per model-based prediction by World Bank, the cumulative soil carbon (C) sequestration of pulses in Asia and Africa is expected to be 33.0 and 35.12 Mg ha^{-1}, respectively, by 2030. Legumes have the potential to reduce the CO_2 emitted during the manufacturing of chemical nitrogenous fertilizers through their biological nitrogen fixation (BNF) capacity. Therefore, the main advantage of using legumes is to ensure that the BNF which in turn reduces the amount of nitrogen (N) fertilizer required for the succeeding crop. A meta-analysis study suggested that the legumes have the capacity to store 30% higher soil organic carbon (SOC) when compared to other species; this is because of their N-fixing ability. The leguminous vegetation improves soil health and soil C content as per the nature of the specific crop. The C sequestration potential and the amount of organic C returned by leguminous species to soil depend largely on specific legume species, growth behavior, root morphology and physiology, leaf morphology, climatic conditions, structure and aggregation, prevailing cropping system, and agronomic interventions during crop growth period. The aboveground plant biomass (e.g., plant leaves, branches, stem, foliage, fruits, wood, litter-fall) and the belowground plant biomass (e.g., dead roots, carbonaceous substances from root exudates, rhizospheric deposition, and legume-promoted microbial biomass C) directly contribute to the SOC pool.

Keywords
Legumes · Soil C sequestration · Soil quality · Soil microbes · Climate change

Abbreviations

ANPP	Aboveground net primary productivity
C	Carbon
DM	Dry matter
DW	Dry weight
GT	Giga tone
SMBC	Soil microbial biomass carbon
SOC	Soil organic carbon
SOM	Soil organic matter
WSOC	Water-soluble organic carbon

4.1 Introduction

The challenges of the soil organic carbon (SOC) sequestration and preservation are manifold, and some are caused by human-induced factors such as low adoption rates of sustainable soil management practices (FAO and ITPS 2015). After carbon (C) enters the soil in the form of organic material from soil fauna and flora, it may persist in the soil for decades, centuries, or even millennia. The soil organic matter (SOM) has a direct linear relation with SOC, as SOM contains about 58% organic C (Kumar et al. 2006). The SOC is the biggest C pool in the terrestrial biosphere and is more than double of the atmospheric C and the biotic pools combined (Jobbagy and Jackson 2000; Liang et al. 2016).

Legume-based cover cropping is one of the important options for soil C sequestration in agriculture and plays a critical role to mitigate climate change (Lal 2015; Meena et al. 2015c). Leguminous cover crops supply organic C to the soil by decomposing their residues. Continued crop production potentials of soils are directly related to their SOC contents (Lal 1998; Mann et al. 2002). Within limits, crop productivity is positively related to the SOC content (Doran 2002) because of improvement in soil physical, chemical, and biological properties and consequently crop yields (Franzluebbers 2002). The soil physical characteristics influenced by SOC are those associated with soil aggregation and aggregate stability (Six et al. 2002). The management of legumes residues determines the soil aggregation and thus influences the soil C sequestration (Franzluebbers 2002). Six et al. (2002) advocated that sun hemp (*Crotalaria juncea*) crop have a highly dense and extensive root system which accounts in the formation of macroaggregates. The retention of legumes residue promotes soil macroaggregation and SOC storage (Kushwaha et al. 2001). Naturally, the legumes fix greater amount of nitrogen (N) through the process of biological nitrogen fixation (BNF) which, in turn, contributes to C sequestration with an average rate of 0.88 Mg ha^{-1} $year^{-1}$ (Diekow et al. 2005), greater than the grasses (0.71 Mg ha^{-1} $year^{-1}$) (Martin et al. 2012). The capability of alfalfa (*Medicago sativa*) to formation of stable soil aggregates is much higher (Six et al. 2002). SOC enhances the stability of soil aggregates and structure because SOM remains physically protected in the core of soil aggregates. The stability of soil aggregation decides the soil water contents, gaseous exchange between soil and atmosphere, soil microbial communities, and nutrient cycling (Sexstone et al. 1985). The soil structure is comprised of primary soil particles, macro- and microaggregates acting as physical units of aggregates. The turnover of legumes residue in soil is the base of soil aggregation, which ensures the availability of C to the soil microbial community as a source of metabolic energy, leading to improvement in soil biological diversity and stimulating biodegradation of harmful soil contaminants (Grego and Lagomarsino 2008). This is closely associated with the increased soil aggregation and consequently reduced decomposition of organic C by soil microbes (Conceicao et al. 2013).

The crop diversity with legumes, higher is amount and value of the crop residues being incorporated into the soil, which is available for soil microbes as feed or a source of energy. Accordingly, the activities of microbial communities are augmented, which in turn promote the formation of macroaggregation, where SOM remain safe to microbial attack and the subsequent decomposition and can accrue (Tiemann et al. 2015; Yadav et al. 2017). For example, in mega-aggregates the concentration of soil C increased up to 33% in soils under diverse planting with legumes as compared to the soil under monocropping of corn (Tiemann et al. 2015).

Legume-based cropping systems can increase biomass input and extend the biomass residence time in soil by reducing the mineralization rate (Meena 2013). Biomass production can also be increased by legume-based bicultural system, a mixture of legume with nonlegume species. The magnitude of soil C sequestration varies among different leguminous cover crops depending exclusively on total biomass production, decomposition rates, and conversion of liable C to soil recalcitrant C (McLauchlan and Hobbie 2004; Lal 2004a, b; Benbi and Brar 2009). Accumulation of SOC is a long-term biotic process influenced by abiotic factors (Jarenyama et al. 2000; Lal 2004a, b). Thus, it is an important to optimize the management skills through legume-based cultivation practices (Curtin et al. 2000). The objective of this chapter is to deliberate the rate and magnitude of the soil C sequestration potential of legumes. By changing the land-use pattern following a sustainable way through introducing leguminous crops, shrubs, and tree species in the existing system, the annual C sequestration rate could be increased by 20–75 g C m^{-2}, and soil organic C may reach a new equilibrium in the interior, several years onward (Liu et al. 2013). Legumes are an imperative component of ecological sustainability and nutritional security and a valuable source of food proteins. This chapter will be helpful to researchers and policy makers to know the role of legumes in the alarming issues of climate change (Meena et al. 2017a) and food and nutritional security, especially after the Paris Agreement of December 2015, which appeals for action to store and increase the sink capacity of SOC pool for soil sustainability and world food security.

4.2 Soil C Sequestration Through Legumes

The intensive cultivation without the inclusion of legumes may result in the common problem of reduced SOC pool (Jarenyama et al. 2000; Lal 2010). The purpose to tap the C sequestration potential of soil and the cultivation of plants belonging to Papilionaceae family needs to be endorsed in the agricultural system (FAO and ITPS 2015). The population explosion and economical emergencies, especially in the developing world, the significance of advanced legume-based approaches, and their impacts on soil and ecological dimensions need to be considered more now than in the past (Reeves 1997; Jarenyama et al. 2000). More importantly, care should be taken while selecting the appropriate legume crops in farming practice. Some approaches can accelerate the economic production but still are C exhaustive in nature and, so, increase the CO_2 emission from soil into the atmosphere (Chan 2008). Likewise, Chaer et al. (2011) confirmed that the apparent profits of growing perennial legumes toward capturing atmospheric CO_2 and gathering it into the soil additionally their capability to offer a strong sink to store the CO_2 in their biomass.

Although the actual quantity of soil C sequestration in the different horizons highly varies with the soil management or legume-based farming practices along with the climatic conditions, soil properties, and the native C level of that sites (Whitbread et al. 2000), the C capturing capacity of soil can be enhanced and improved via legume-based farming practices and perhaps those that restored soil fertility and health. Promoting legume cultivation offers multiple advantages: augmenting crop and soil productivity (Meena et al. 2015b), adapting to climate change resilience, sequestering atmospheric C, and dropping of concentration of greenhouse gases (GHGs) emission from the atmosphere (Curtin et al. 2000; FAO and ITPS 2015). Legume crops residue is one of the main sources of C in agricultural soils and produces a considerable amount of residues which in turn favor the accumulation of humus and in consequent soil's C pool upon incorporation into the soil (Hajduk et al. 2015; Dhakal et al. 2016). Unlike cereals and other crops, the cultivation of legumes substantially enlarges the soil C sequestration over the other nonlegume crops because of their ability for larger belowground biomass production and leaf-shedding ability (Ganeshamurthy 2009; Abberton 2010). The increased SOC content in the legume-based rotation had also been confirmed in several scientific studies (Podleoeny 2005; Rochester 2011; Rutkowska and Pikua 2013; Hajduk et al. 2015). The legume-based farming practices are improved via adopting an ecologically sustained strategy with high diversity, mixed farming, and a sensible crop rotation (Ram and Meena 2014). While enclosing legume, in leguminous tree-based agroforestry systems, and adding of legume shrubs into the silvopastoral system are found to be good concerning sustainable soil C sequestration (Curtin et al. 2000; Chan 2008).

It is also important whether the forage legume is annual or perennial in growth habit. The potential effects of forage legumes seem to be higher with perennial legume species (Young et al. 2009). These are associated with two things, firstly, the potential of legumes for higher C inputs because of their ability to efficiently utilize rainfall and their extended growing period (Peoples and Baldock 2001) and, secondly, decreased C losses from organic compounds over annual legumes as perennial legumes help in maintaining lower water contents in soil, thus, reducing the soil microbial activities, associated decomposition, and consequently the release of organic C through microbial respiration (Young et al. 2009). Several scientific reports suggested that raising of perennial species of legumes such as alfalfa, tamarind (*Desmanthus virgatus*), and purple bean (*Macroptilium atropurpureum*) enhanced an abundance of SOC pool as compared to the SOC returned by the annual crop (Whitbread et al. 2000: Young et al. 2009).

Besides the nature and growth habits of legume crops, the nutrient management in legumes and associated pasture are a key factor in the direction of building SOC pool which directly affects the plant biomass production and consequently the contribution of plants in SOC accumulation (Chan et al. 2011). Once the ways of C sequestration are recognized, the adoptions of proper agronomic practices are crucial to preserve and improve the SOC pool (Lal 2009). Thus, any crop management intervention that increases crop production and the retention of crop root and shoots residues response to lessen C losses from soil directly or indirectly and increase soil C sequestration (Hoyle et al. 2011). Williams and Donald (1957) observed that the augmenting C sequestration in the soil is directly correlated to the plant biomass production and hereafter to soil fertility. The enhancing soil fertility is probably the utmost actual approach for increasing the C sink capacity of the soil (Abberton 2010; Verma et al. 2015). The soil C dynamics rest on the balance between C inputs (crop residues and other organic substances) and outputs (CO_2 evaluation from SOC decomposition) (Wang 2015). The amount of C to be stored in soil also varies based on the total quantity and quality of residues being added in the soil. This suitably explains the reasons of dissimilarities in C sequestration among different crop rotations (Wamisho 2013). Apart from producing a large amount of aboveground plant biomass, the increase in belowground plant biomass, i.e., plant roots, also has a greater significance in C sequestration (Gregorich et al. 2001; Kundu et al. 2007; Kane 2015).

According to Adeboye et al. (2006), the starter dose of 20 kg N ha^{-1} in three legume plants, i.e., soybean (*Glycine max*), cowpea (*Vigna unguiculata*), and cavalcade centro (*Centrosema pascuorum*), enhances the organic C buildup in the soil. It is reported that BNF is responsible for a vigorous plant growth which in turn assimilated more CO_2 from the atmosphere through the process of photosynthesis. The assimilated C in plants returns to the soil upon their incorporation and subsequent decomposition. Lange et al. (2015) stated that the increase in the total N content and SOC was interrelated; the decreased root biomass production by legumes produced the negative effect of SOC. Conversely, they advocated most pronounced effects of the legume on soil C sequestration in the topmost 5 cm soil, but not in top 30 cm soil when considering as a whole. Although crop species have a vital role in retaining amount and quality of SOC reserves apart from the diversity of crop residue (Mandal et al. 2007). The total quantity of crop residues being added to the soil system is considerably higher in soils under legume cultivation after cereal crops than in uncultivated fallow soils (Sombrero and de Benito 2010). The residues of perennial legumes are considered as good in quality and produced in large quantity, which are more effective in supplying a considerable amount of soil C for the long term (World Bank 2012). Gregorich et al. (2001) concluded that the quality of residues had an imperative role in the direction of maintaining or increasing soil C in agroecosystem (Dhakal et al. 2016). They also advocated that the soil under legume-based system have a tendency to be more preservative of residue C inputs, mostly from roots and their exudates than that of soils from monoculture. Apart from the amount of C inputs, the C sequestration efficiency (i.e., conversation rate of C inputs) also decides the contribution of incorporated organic C inputs into the soils (Liang et al. 2016).

4.3 Legume Improves Carbon in Soils

The rate of organic C production and sequestration by the legumes or their mixtures with grasses is correlated with soil texture which is reported higher in heavy soils than light soils (McLauchlan 2006; Gawel 2011). Kong et al. (2009) laboratory experiments evidenced the inverse correlations between clay content and the decomposition rate of SOC. The beneficial effects of growing legume plants in silty soils already rich in organic matter sequester more C compared to the growing of cereals such as oats (Hajduk et al. 2015). In the same direction, Gami et al. (2009) also confirmed the synergistic relationship between silt + clay soil and soil C sequestration. In contrast, legume cultivation in sandy-textured soils did not have a significant effect on soil C storage after crop harvesting (7.39 g C kg^{-1} dry matter), as compared to the initial organic content (7.76 g C kg^{-1} dry matter) (Hajduk et al. 2015). Wang et al. (2010) in an experiment on fine sandy (Dadhich et al. 2014) and gravelly loamy soils with six winter and summer cover crops observed that the C sequestration capability of faba bean (*Vicia faba*) was the maximum which added 597 g C m^{-2}, whereas white clover (*Trifolium repens*) accumulated the lowest quantity of C (149 g m^{-2}). Among summer crops, the maximum amount of C (481 g m^{-2}) was returned by sun hemp into the gravelly loamy soil, whereas the lowest amount was added by castor bean (*Ricinus communis*) with a value of 102 g m^{-2} at 30 °C. At the end of the completion of one cycle of summer and winter cover crops, the soil C pool increased by 39.1 and 13.8% in fine sandy and gravelly loamy soil, respectively, over the respective soils. A long-term study conducted by Rutkowska and Pikula (2013) reported that on loamy- and sandy-textured soils, the crop rotation with legumes/cover crops is the principal factor toward stabilizing the SOC pool.

4.4 Leguminous Cover Cropping Enhances C Pool

Cover crops are referred to as the growing of crops like legumes aimed toward protection and improvement of soil in between the periods of regular crop production (Nair et al. 2015). The growing of legumes as cover crop in rotation improves soil structure and quality by accelerating organic C storage in soil strata through production of large biomass (World Bank 2012; Nair et al. 2015). Alternating their extensive well-developed deep and shallow root systems helps in improving SOC pool and prevents soil erosion (Nair et al. 2015). Santos et al. (2011) from an experiment of over 17 years concluded that the forage-based rotation of alfalfa and rye grass (*Lolium perenne*) sequesters SOC considerably more over cover crop-based rotation with cereal crops. They established that the roots of forage-based or cover-based rotation had more predominant role in the direction of building SOC pool and residues. Nair et al. (2015) and Ram and Meena (2014) advocated that legume-based cover crop accumulated more C into the soil over the grass-based cover crop rotation with nonlegume crops. The adoption of velvet beans (*Mucuna pruriens*) as cover crop resulted in a higher yield of dry matter by 11.9 Mg ha^{-1} annually in shoot residues, which was significantly higher than the dry matter produced by the conventionally unfertilized monoculture of maize (*Zea mays*). The replacement of reed meadows field by alfalfa because of the higher demand of forage for live pool resulted in increased SOC levels (Zhang et al. 2009).

4.5 Legumes Crops in Rotation Improve C Pool

The selection of legume crops for cultivation in the rotation decides the amount of C being sequestered into the soil (Abdurahman et al. 1998). Crop rotation can improve biomass production and eventually the soil C sequestration, principally the rotations of legumes with nonlegumes, and SOC can be enriched by the use of appropriate crop rotations (Lal 2010). Legume-based cropping systems and SOC sequestration capacity are presented in Table 4.1.

The C sequestration potential and the amount of organic C being added into the soil strata by the leguminous crop greatly vary with the selection of appropriate legume. The growth habit, canopy structure, quantity and quality of residues left on the soil surface, root physiology and pattern, number of leaves being produced, climatic stimuli, soil aggregation, existing cropping system, and agronomic interventions during the crop cycle improve SOC pool (Chan and Heenan 1996; Gregorich et al. 2001; World Bank 2012; Nair et al. 2015). This is because of the higher conversation efficiency from residue C to soil C by legumes in rotation over the monoculture cereal crops such as wheat (*Triticum aestivum*) (Campbell et al. 1999). The legume-based rotations are more efficient in converting biomass to SOC in comparison to the grass-based rotation. Inclusion of legumes in rotation has the potential of guaranteeing the in situ availability of N which in turn plays a vital role in generating higher biomass C. It also promotes the release of C via root exudation into the rhizospheric zone (Hajduk et al. 2015). The N fixed by the root nodules of legumes also accelerates the C sequestration potential of succeeding crops in the rotation, more likely because of the improved microbial activities and biomass production by successive crop (Nair et al. 2015). The N provided by the legumes enhances the N utilization efficiency and produces more root biomass and, thus, leads to C inputs in soil (Conceicao et al. 2013; Meena et al. 2013).

Table 4.1 Carbon sequestration in different soybean-based cropping systems

Crops rotation	C/N ratio	Rate of C sequestration (Mg ha^{-1} C per year)
Soybean-corn	11:70	0.68
Soybean-soybean	11:74	0.69
Soybean-sunflower	11:77	0.57
Soybean-oilseed radish	11:75	0.54
Soybean-sun hemp	11:63	0.79

Data source: Junior et al. (2016)

Scientific studies also reported improved organic C status under the rotation of forage legumes with grain crops after several years of cropping compared with that of a monoculture of grain crop (Campbell and Zentner 1993; Gregorich et al. 2001). In the same direction, Campbell et al. (1999) reported that the wheat-lentil (*Lens culinaris*) rotation sequestered more C pool compared with that of C sequestered by wheat crop alone in monoculture. Venkatesh et al. (2013) and Meena et al. (2015b) reported increased SOC pool due to the addition of pulse in upland maize-based system and most importantly in top 0–20 cm soil layer. Rochester (2011) also reported that the higher C sequestration takes place in the subsoil rather than that of surface soil (up to 30 cm). In soil, fresh plant residue, roots, living microbes, and macrofauna account <15% of the total SOC pool, whereas stable humus, partially decomposed plant residues, and indecomposable organic C, i.e., charcoal, represent the balance (Hoyle et al. 2011: Jensen et al. 2012). The distribution of the soil C pool relied on the soil deepness. The passive C fractions dominate in subsurface soil horizons over the active C pool, whereas in the surface soil, the active C has more storage over passive C pool (Venkatesh et al. 2013). Blair and Crocker (2000) scrutinized the effects of different rotations including legumes and uncultivated fallow in a long-term experiment and stated that introduction of legumes in rotation promoted the buildup of liable C reserve in the soil profile more than those with the continuous wheat and a long-uncultivated fallow period.

The soil C reserve had reported 130–134 Mg ha^{-1} in top 70 cm soil when maize was grown in rotation with alfalfa which was considerably greater than the maize under monoculture, as it stored only 109–115 Mg ha^{-1} C in soil (Jensen et al. 2012). On the contrary, in the absence of appropriate crop rotation, plant biomass production and associated SOC decrease because of a higher incidence of biotic factors, i.e., weeds, insects, and diseases (Wang et al. 2010). Besides this, increase in cropping intensity or frequency by cultivating bare fallow land in the crop rotation is an alternative way toward enhancing biomass production and consequently the soil C sequestration (Wang et al. 2010). In subhumid to arid climates, the cotton (*Gossypium* spp.)-legume system performed better with respect to soil C sequestration than the other cropping systems, while C sequestration under the rotation of sorghum (*Sorghum bicolor*)-legume was better in semiarid regions, but in dry areas, the same rotation failed to sequester soil C (Chaudhury et al. 2016; Ram and Meena 2014).

The introduction of soybean in rotation returns higher organic C in soil than that added by cowpea (*Vigna unguiculata*) because soybean produced a 35% higher stoverbiomass than the cowpea (Akinnifesi et al. 2007). The higher C inputs through plant roots and crop residues may lead to greater organic C under soybean rotation. Pigeon pea (*Cajanus cajan*) also produced the greater amount of dry leaves (3.0 Mg ha^{-1}), which was about twice that of the dry leaves returned by the cowpea (1.14 Mg ha^{-1}) (Abdurahman et al. 1998). Newaj and Yadav (1994) also found higher SOC under the pulse-based cropping system, but the highest improvement was noticed in rotation involving pigeon pea because of its long growing duration. The SOC content was found to be increased by 10.7% over the initial content under the legume-based system (Newaj and Yadav 1994). The inclusion of cowpea and green gram (*Vignaradiata*) in maize-wheat cropping system after wheat harvest increased SOC by 6.3% due to their leguminous biomass (Singh and Sandhu 1980; Meena 2013). In the same way, the average amount of total C input under pigeon pea-wheat (49.04 Mg C ha^{-1}) and maize-wheat-green gram (48.68 Mg C ha^{-1}) system was higher as compared to that without legume inclusion in existing maize-wheat system (37.06 Mg C ha^{-1}) (Venkatesh et al. 2013). This significant improvement in total SOC under maize-wheat-green gram and pigeon pea-wheat was 11 and 10% over the existing conventional maize-wheat system, respectively. The monocropping of cotton in black soils failed to sequester the organic C in soil, but when it was rotated with pigeon pea, the sequestration of organic and inorganic C increased by 0.75 and 1.12%, respectively (Chaudhury et al. 2016). In Asia, the inclusion of the green gram as a grain legume in the existing short fallow of rice (*Oryza sativa*)-wheat system after wheat harvest produced a total dry matter of 4.5 tons ha^{-1} and increased soil C (Yaqub et al. 2010; Meena et al. 2015a, b, c). According to Hajduk et al. (2015), pea (*Pisum sativum*) sequesters and deposits more C into the soil column compared to lupin (*Lupinus albifrons*) (7.58 vs. 7.23 g C kg^{-1}dry matter) which had a small effect on C storage in soil. This was closely associated with the narrowing of lupin leaves as it assimilates less photosynthate.

4.6 Green Manuring with Legumes

The critical determinants of soil C pool are the total quantity of organic matter input into the soil from plant residues and the duration for which they remain in soil that varies with their decomposition rate (Paustian et al. 1995). In general, farmers left the soil bare or fallow between the two crops in the predominant cropping system like after wheat harvest in rice-wheat cropping system in the Indo-Gangetic plains of Southern Asia (Kane 2015; Curtin et al. 2000). It means no biomass is produced in 3–4 months of the year on that land meaning that of considerably lower C inputs into the soil. Hence, it is difficult to balance the C losses that occur during the year. The introduction of legumes as green manure between two existing crops in the cropping system ensures a good biomass production which in turn offsets C losses and confirms net gain of C in the soil (Tiemann et al. 2015). In green manuring, accumulated standing biomass is directly incorporated into the soil system (Zentner et al. 1996). Biederbeck et al. (1996) stated that green manuring with annual legumes in place of partial fallow may add dry matter to the tune of 3 Mg ha^{-1}. Furthermore, they have reported that rotation in which legumes was involved added 49% higher stubble C, 133% higher stubble N, and about 60% extra SOC below 30 cm of soil. Likewise, the introducing vetch (*Vicia sativa*) along with soybean as a winter green manure crop in rotations before the maize crop increased the SOC sequestration by about 10 Mg C ha^{-1} after 13 years of experimentations (Sisti et al. 2004). The rate of crop residue decomposition varies negatively with its C/N ratio and lignin content as lignin is resistant to microbial degradation (Silver and Miya 2001; Dhakal et al. 2016). In the same way, Curtin et al. (2000) advocated that replacement of partial fallow with legume green manuring had a little impact (~25%) on soil C storage because green manuring crops have narrow C/N ratio (12–13: 1) and low in lignin content; this is why they decompose quickly (Zentner et al. 1996; Curtin et al.

2000). The growing of sun hemp as green manure improved the SOC pool by 0.92% but reduced soil inorganic carbon (SIC) by 0.64% (Chaudhury et al. 2016). The advantages of sun hemp in capturing atmospheric C and subsequent storing in soil may be credited to a lower C/N ratio (23: 1) of its residue, which helps in enhancing the N availability, required for hasty residue conversion into particulate C fraction of C pool (Junior et al. 2016). Legume-based green manuring contributes to GHGs emission reduction in two ways, *first*, by converting plant C into SOC and, *second*, by reducing the requirement of nitrogenous fertilizers consequently in lowering of N_2O emissions (Biederbeck et al. 1996; Curtin et al. 2000; Nair et al. 2015).

4.7 Legume-Based Pasture Development

In several parts of the world, the mixed pastures based on ryegrass (*Lolium multiflorum*) and clovers (*Trifolium* spp.) are grown extensively, but still scientific studies on the effects of pasture mixture on soil C sequestration are very rare (Daly et al. 1996; Chan et al. 2011). The mixture of legume species with pasture increases the production of below- and aboveground biomass which in turn ensures a higher soil C pool (World Bank 2012; FAO 2016). The evidence from many research experiments in different climatic conditions and regions confirmed the SOC acquisition ability of legume-based pastures (Wright et al. 2004; Chan et al. 2011). Arrouays et al. (2001) reported that the growing of legume mixture replacing grass leys resulted increase 10 Mg C ha^{-1} into the soil at the end of 20 years of study period. Yang and Kay (2001) reported significantly greater mean SOC reserve under continuous alfalfa in top 40 cm soil in comparison to the continuous corn. Mortensen et al. (2004) also quantified the similar legume effects of yellow-flowering alfalfa (*Medicago sativa*) on building SOC pool when inter-seeded with grasses in range lands and stated that the mean rate of SOC deposition in upper 1.0 m soil layer was 1.56, 0.65, and 0.33 Mg C ha^{-1} per year after 4, 14, and 36 years of inclusion of alfalfa, respectively. In contrast, alfalfa would be expected to have more capability than the clover to enhance the C accumulation in >30 cm soil depth because of its extensive deep root system (Peoples and Baldock 2001). The rotation of wheat-sweet clover (*Melilotus officinalis*) accelerated the SOC contents at different soil depths and also had the highest size of SOC (Liu et al. 2003a, b). Liu et al. (2003a) advocated that adoption of improved crop rotation can upsurge the organic C pool and improve quality and structure of black soils, thus, capture higher atmospheric CO_2 and consequently mitigate the greenhouse effects. The introduction of perennial legume peanut (*Arachis pintoi*) in pastures increased C sequestration potential by 7.8 Mg ha^{-1} every year in top 1.0 m soil profile over the growing of pure grass, regardless of legume's contribution with only 20% of the total root biomass (Fisher et al. 1994). Further, the C balance should be favorably changed by enhancing the chemical complexity, quantity, and quality of carbonic substance being added in soil to compensate the C losses from microbial attack or decomposition (Kane 2015; Yadav et al. 2017). It is supposed that carbonic compounds in the legume green manure crops are resistant enough to microbial decomposition and may also enhance the diversity and complexity of soil C for its long-term stabilization (Wickings et al. 2012).

4.8 Leguminous Oilseeds Improve Soil C Pool

The soils under two important oilseed crops, i.e., soybean and groundnut (*Arachis hypogaea*), are believed to be the important sites in the global C cycle (Bhattacharyya et al. 2009). Most of the scientific studies proved the positive response of oilseed legumes in the way of soil C sequestration (Aziz et al. 2014). In subtropical regions' soybean-wheat cropping system, the C contribution from harvestable aboveground biomass of soybean was observed to be 22% higher than wheat, which resulted in 18% of annual total C being assimilated more than wheat (Kundu et al. 2001; Meena et al. 2015a, b, c). The long-term experiments are one of the effective pathways to track the effects of continuous soybean monocropping as well as the crop residue retention on soil surface change in SOC levels (Leigh and Johnston 1994). In the same direction, the results of a long-term (30 years) experiment revealed that under rainfed soybean-wheat system, the SOC level increased because of the greater C returned by the system than that of the quantity needed to sustain SOM content (Kundu et al. 2007). They found that soybean and wheat annually produced 3.54 and 3.32 $Mgha^{-1}$ aboveground biomass, respectively, and the average share of both the crops in harvestable aboveground biomass was 29 and 24%, respectively. In soybean, about 13% of gross biomass (below- and aboveground plant) was alone contributed by root exudates and rhizospheric deposition (Kemper et al. 1998), even though, in most of the crops, these values ranged from 5 to 20% (Kundu et al. 2007). Kundu et al. (1997) also reported that the soybean returned belowground root biomass to the soil which was 31% of the aboveground biomass. The contribution of different parts of soybean like root nodules, roots, and leaves in the gross C returned by the crop to the soil was 32, 34, and 38%, respectively. In soybean, rhizospheric deposition of C was the main contributor in total C inputs. The winter and summer cover crops are the foremost choice to improve the C turnover trough advancement in soybean cultivation (Amado and Bayer 2008). A multi-locational trials in different agroecological zones are reported that the increased total organic C under soybean-wheat system was up by 46% over the cereal-based rotation of corn-black oat (*Avena strigosa*) (Amad et al. 2006), although the rotations generated almost equal quantity of crop residues, most likely because of the quick breakdown of soybean residues which thereafter accelerated the dry matter production capacity of wheat in sequence (Weber and Mielniczuk 2009). Bhattacharyya et al. (2009) also found higher trends of C pool returned into the soil from roots and rhizospheric deposition of soybean with a value of 620 and 362 kg C ha^{-1} $year^{-1}$ in irrigated and rainfed conditions, respectively (Meena et al. 2017b).

A 15-year experiment with corn-soybean cropping system registered the highest productivity and returned huge crop residues to the soil system in comparison to the monocropping of either corn or soybean (Drinkwater et al. 1998). Further, Drinkwater et al. (1998) advocated that the application of crop residue of having a narrow C/N ratio to improve soil fertility in the corn-soybean system has the potential to increase the soil C sequestration by 13–30 Tg year^{-1}, which is equal to 1–2% of projected C emissions from fossil fuel combustions. Based on the assumption of 50% residue retention, the global cumulative C sequestration for soybean varies from 0.5 MT to 0.7MT by 2030. Junior et al. (2016) found the increased total C content when pigeon pea, sun hemp, and soybean were grown as a sequential crop after corn. Among these, soybean/corn rotation and soybean monocropping sequestered the significantly greater amount of C in soil followed by the sun hemp. The soybean monocropping increased soil C in soil as particulate C fraction over the monocropping of corn. It was because of the lower C and N balance of soybean residues than that of corn, which in turn, contributed to the rapid C addition into the soil system as a medium particulate fraction. In this study, the mean C deposition rate in soil under these crop sequences was 0.63 Mg C ha^{-1} annually, which was lower than that observed (0.85 Mg C ha^{-1} annually) by Sa and Lal (2009) and much higher (0.35 Mg C ha^{-1} annually) than that reported by Bayer et al. (2006). The incorporation of surface litter and aboveground biomass of groundnut into the soil caused considerable variation in SOC, which quickly shrank once by coming in contact with favorable temperature and soil moisture for their breakdown (Capriel 1991). Ghosh et al. (2006) in their 6-year experiment stated that the total annual C input was comparatively more under groundnut-based [0.24 (r = 0.80, $P \leq 0.01$)] crop rotation than that of fallow-based [0.36 (r = 0.80, $P \leq 0.01$)] system; moreover, the rate of C loss was also higher. Among the groundnut-based system, the gross annual C production to some extent was higher in groundnut-mustard (*Brassica nigra*) rotation over the groundnut-wheat, groundnut-sunflower (*Helianthus annuus*), groundnut-chickpea (*Cicer arietinum*), and groundnut-groundnut system, even though the contribution of rhizospheric deposition (Meena et al. 2017b) and root biomass in total C inputs in soil was higher under the groundnut-wheat system.

4.9 Leguminous Shrubs Enhance the Soil C Pool

The conversion of pasture and forest to cropland has been responsible for 59 and 24% loss of stored soil C globally (He et al. 2011; Guan et al. 2016). The impacts of the shrub encroachment on the change in SOC pool are in direct or indirect control of several biotic (shrub types) and abiotic (soil texture and climate) factors (Kulmatiski and Beard 2013). Since the last decade, an increase in dominance and abundance of legume shrubs has been reported in the world's grasslands (Kulmatiski and Beard 2013). The replacement of grasslands by shrub-intruded grasslands can be responsible for some ecological consequences, like altering biodiversity, the pattern of SOC, and the local C balance (Jackson et al. 2002). The gain or loss in net C in soil stratum heavily depends on the type of shrub in the encroachment of either grassed species or legume closely associated with their aboveground net primary production (ANPP). Li et al. (2016) reported that the rate of increase in SOC contents was higher in leguminous shrub-based grassland than that of nonlegume shrubs with an effect of 0.32 and 0.15%, respectively. Legume-based shrub planting exhibits the amazing capacity of countering atmospheric CO_2 via soil C sequestration and, thus, plays a counter role in the global C cycle (Liu et al. 2010). The proliferation of woody leguminous shrub can influence the C pool at regional or global level. In this context, the SOC pool could improve, if legume shrubs are more productive over herbaceous plants (Wheeler et al. 2007).

Guan et al. (2016) evaluated the effect of bush clover (*Lespedeza davurica*), milk vetch (*Astragalus adsurgens*), and alfalfa on the gross amount of SOC returned in upper 2 m soil layer over 7 years. Their report revealed that the level of SOC significantly increased in soil under legume crops/shrubs whereas decreased to some extent under the fallow uncultivated soil in top 2 m soil profile at the end of 7 years of the experiment. This increment in SOC concentration in soil under cultivation of bush clover, milk vetch, and alfalfa was 19.9, 14.6, and 24.1 Mg C ha^{-1}, respectively, and reduced in bare soil by 4.2 Mg C ha^{-1}. The increased SOC content in deeper soil layer under bush clover, milk vetch, and alfalfa might be due to improved turnover and proliferation of finer roots deeper in soil or might be due to the downward movement of soluble C from upper to lower soil strata with water after heavy irrigation or rainfall (Guan et al. 2016). The projected, root biomass C in milk vetch 20 Mg C ha^{-1} was observed more than the aboveground biomass C production of 15 Mg C ha^{-1} in 7 years of study (Fan et al. 2015). Similarly, in bush clover, the root biomass C production was 15 Mg C ha^{-1} (32 Mg ha^{-1} root biomass) having an aboveground biomass production of 42 Mg ha^{-1} (Guan et al. 2016). The higher SOC deposition in top 30 cm soil by bush clover was attributable to the greater turnover of legume residues, litter, carboxylic acids, and other carbonic substances along with root proliferation in the surface layer (Zhou et al. 2006). The litter-fall and its decay are crucial for the formation of organic matter, mineralization, and terrestrial C sequestration (Zhou et al. 2012). The decomposition of deposited litter on soil surface depends on various factors like quality of litter precipitation, solar radiation, soil water content, biological activities, etc. (Austin and Vivanco 2006). The litter of leguminous shrubs, i.e., bladder senna (*Colutea arborescens*) and tree medic (*Medicago strasseri*), is low in lignin content which are prone to quick degradation of organic C (Alegre et al. 2004).

Su et al. (2005) stated that planting of littleleaf pea shrub (*Caragana microphylla*) substantially improved the soil C status in upper 0–5 and 5–20 cm soil layers in sand dunes of semiarid regions. Liu et al. (2013) reported increase in 21.84, 30.0, and 39.09% C content under littleleaf pea shrub at the age of 5, 21, and 34 years, respectively, compared to the grassland ecosystem in top 20 cm soil stratum. The result might be linked with the inclusion of organic layer and preplanting disturbance (Laganiere et al. 2010). The ANPP of the young shrub is less, and the total amount of C returned in soil is not adequate to offset the C pool (Liu et al. 2013). With the advancement in age of the plant, the ANPP proportionally increases along with newly changed microclimate and greater C protection which stimulate the gathering of C in the soil stratum (Six et al. 2002). The legume shrubs colutea (*Colutea cilicica*) and bladder senna (*Colutea arborescens*) produced the organic C with a value of 0.69 Mg and 0.66 Mg ha^{-1} $year^{-1}$, respectively, in a semiarid climate (Alegre et al. 2004). Another shrub legume tree medic (*Medicago arborea*) generated 1.75 Mg organic C ha^{-1} annually (Alegre et al. 2004), presented in Table 4.2.

Table 4.2 SOC increase through different legumes

Legumes	Soil depth (cm)	SOC increase (Mg ha^{-1})	References
Colutea cilicica		0.69	Alegre et al. (2004)
Colutea arborescens		0.66	Alegre et al. (2004)
Medicago arborea		1.75	Alegre et al. (2004)
Lespedeza davurica	200	19.9	Guan et al. (2016)
Astragalus adsurgens	200	14.6	Guan et al. (2016)
Medicago sativa	200	24.1	Guan et al. (2016)
Vicia sativa		10.0	Sisti et al. (2004)
Medicago sativa	100	0.33–1.56	Mortensen et al. (2004)
Arachis pintoi	100	7.8	Fisher et al. (1994)

Soil texture seems to be key factor influencing the SOC buildup under shrub encroachment (Li et al. 2016). The results of the structural equation model (SEM) also showed that the soil texture had the greater potential to change the rate of SOC turnover over time accounted for the shrub encroachment (Six et al. 2000). As of a fact, shrubs are made up of a larger amount of ANPP in coarse-textured soils than that of fine-textured soils with same cover (Liang et al. 1989). It means, if shrubs are included within the grassland system in the sandy-/coarse-textured soils, they will exhibit a greater capacity of biomass production, SOC increment, and, thus, C sequestration. As of another fact, the bulk capacity of coarse-textured soils have a tendency to be high, so a positive correlation was present between the soil bulk density and the change in mean percent SOC content. Wheeler et al. (2007) stated that the inclusion of C_3 legume shrubs into the grassland predominated by the C_4 species resulted in an increase in the SOC level by about 23% in coarse-textured soils than the clay loam soils under similar environmental conditions. The soil N content also significantly contributed in regulating the effect of shrub encroachment in building of SOC pool (Kirkby et al. 2013). It is supposed that an amount of 1 g N is capable of storing 7–13 g of C as per the global meta-analysis (Johnson and Curtis 2001; Yang et al. 2011). Shrub encroachment apparently stimulates the soil N accumulation and consequently secures their role in accumulating SOC contents (Jackson et al. 2002). The increased returns of roots of along with the root exudation greatly decide the accumulation of C in plant and consequently in the SOC contents (Sun et al. 2001; Shahzad et al. 2015). The minor roots share about 33% of annual ANPP and that directly affect the C dynamics and sequestration (Zhou et al. 2012).

4.10 Legume Trees Improve Soil C Pool

Along with the food, feed, fiber, fuel, and fodder, legume trees are also important to enhance the soil fertility and soil C sequestration presented in Fig. 4.1.

The soil under the agricultural system is supposed to be an excellent source of C sink (Albrecht and Kandji 2003; Ibrahim et al. 2010; Abberton 2010). The soils

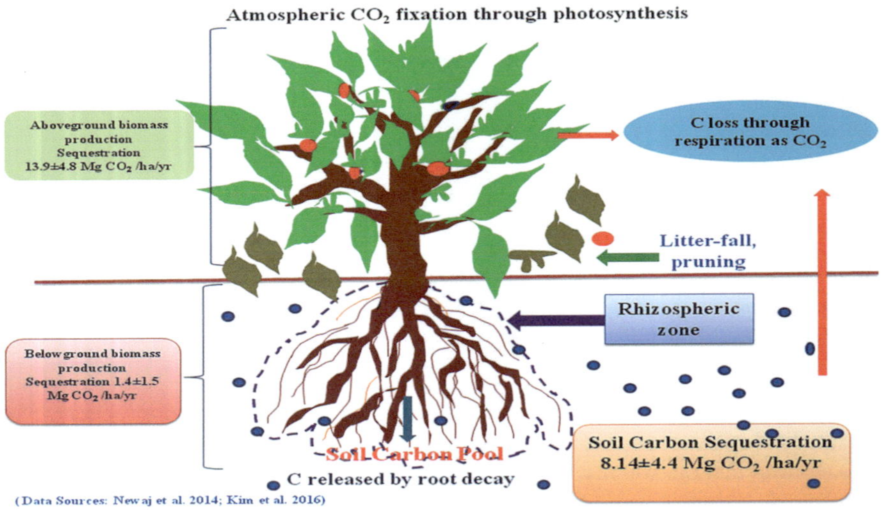

Fig. 4.1 C sequestration mechanism of leguminous trees

Table 4.3 SOC increase through different legume tree species

Trees	Soil depth (cm)	SOC increase (Mg ha^{-1})	References
Leucaena leucocephala	0–10	8.34	Onim et al. (1990)
Cajanus cajan	0–30	0.73	Onim et al. (1990)
Acacia auriculiformis	0–10	3.41	Dreschel et al. (1991)
Cassia siamea	0–10	5.20	Dreschel et al. 1991
Sesbania sesban	0–30	3.10	Onim et al. (1990)
Crotalaria grahamiana	0–20	1.69	Impala (2001)

under tree plantation have greater potential to maintain a higher concentration of SOC pool in their profile, as presented in Table 4.3.

The tree-based land-use systems could be sustainable in many ways to capture and accumulate the atmospheric CO_2 because of their high C capturing capacity, cost-effectiveness, and accompanying soil benefits (Dhruw et al. 2009; Wani and Qaisar 2014). Based on the information collected from various sites, the SOC pool in global soils ranged from 6.9 to 302 Mg ha^{-1} (Nair et al. 2009). In spite of huge variation in these values, the difference between various systems, soil, and ecological provenances, the tree plantation have a significant place in enhancing soil C sequestration. So, the sink capacity of a soil can be improved by introducing the legume trees in the system (Albrecht and Kandji 2003). For a reason that the legume trees have high potentiality to establish and survive under the harsh conditions, also they have the capacity to generate high biomass C (Macedo et al. 2006, 2008;

Meena et al. 2015b). It is anticipated that after some time of establishment, the biomass produced by legume trees will be amalgamated into the soil.

The multipurpose legume trees have greater potential to accrue the huge amount of aboveground biomass and to convert it toward organic C through regular foliage pruning to use as green manure or to feed animals, qualitative and quantitative leaf litter-fall, and root exudation and depositions in the rhizosphere (Kumar et al. 2001; Bertin et al. 2003; Becker et al. 2016). The litter-fall pruned plant parts and biomass of dead trees accumulate on the soil surface which is recycled efficiently to direct contribution to SOC deposition (Ibrahim et al. 2010). Fornara et al. (2009) identified that finer root particles having a diameter of less than 2 mm enhance total annual primary productivity and significantly contribute toward the terrestrial ecosystem and global C pools. In the tree species, the root densities are low in upper 30 cm of soil depth, and root nodules are noticed very rarely. Even the root nodulation increases below 4 m depth of soil (Virginia et al. 1986). In general, it was found that the C constitutes nearly about 50% of the dry weight of branches and 30% of plant foliage parts. The belowground plant parts such as living and dead root biomass, other plant parts, soil microbes, and C deposited in soil strata share about 66% (2/3rd) of total C sequestration (Pinho et al. 2012). These added materials are an important substrate for the diverse microbial population and their activities (Bertin et al. 2003). Beside these complex processes, the contribution of legume trees in enhancing soil C sequestration is an interesting and needful subject of discussion at current to lower down the atmospheric CO_2 (Bertin et al. 2003).

The leguminous trees store the C in their woody biomass, deeper in soil and instable soil. The soil C flux in 0–25 cm upper soil profile could be ascribed SOC redistribution in the course of cultivation practices (Kaongaa and Smith 2009). The lower subsurface soil horizons (25–200 cm) accumulated 75% of C flux most likely owing to the soil C density, leaching of SOC from upper horizons to lower one with rainwater, deep-rooted behavior of legume trees and associated an additional root C turnover, SOM deposition through cultivation practices and increased resistance of SOM fluxes to microbial breakdown deeper in soil because of unfavorable physico-chemical environments for microbial functionaries that restricts decomposition (Sanchez 1995; Kaongaa and Smith 2009). The deep-rooted leguminous trees store the C in their woody biomass, deeper in soil and stable soil aggregates and enhance the potential of subsoil as C sink by promoting plant growth (Rosenstock et al. 2014).

The need of promoting the legume tree plantation for sustainable intensification of the ecological system and soil C dynamics is of increasing interests (Abberton 2010). Under legume tree plantation, the soil C sequestration is linked with the total biomass production and consequently the soil fertility. The soil fertility is one of the chief ways for enhancing the sink capacity of the soil C (Abberton 2010; Verma et al. 2015), although as earlier discussed the quantity of C accumulated as ANPP and the potential of soil to capture and store the C highly depends on the type, age, and population density of the tree, climatic parameters, soil texture and aggregation, agronomic intervention, and so on (Ibrahim et al. 2010). The capacity of an ecosystem to capture the atmospheric CO_2 is also regulated by soil biological activities and the C use efficiency (Becker et al. 2016). The climatic parameters like relative humidity and temperature greatly influenced the soil microbial activities and associated organic matter decomposition and, thus, the soil C sequestration (Pinho et al. 2012).

Numerous studies have shown that the plantation of the leguminous tree accumulates a large amount of C in their biomass as well as soil system (Pregitzer and Euskirchen, 2004; Macedo et al. 2006). The legume tree species also have an imperative role toward increasing C concentration in soils of dryland (Nair et al. 2015). The acacia (*Acacia nilotica*) and prosopis (*Prosopis* sp.) plantations reported to increase C pool by 2 Mg ha^{-1} in soils of subtropical regions (Geesin et al. 2000). The inclusion of leguminous tree species senegal (*Faidherbia albida*) and gliricidia (*Gliricidia sepium*) increases the soil C pool in the agricultural system of semiarid regions (Kalinda et al. 2015; Kaonga and Coleman 2008). Balieiro et al. (2008) reported the beneficial effects of incorporation of leguminous tree guachapele (*Pseudosamanea guachapele*) with nonlegume trees eucalyptus (*Eucalyptus globulus*) on the soil C accumulation. They evaluated that the soil under pure plantation of guachapele and eucalyptus derived 27 and 19% SOC from tree species, whereas this value increased up to 40% when both these were mixed. This difference in SOC accumulation in legume and nonlegume trees might be associated with the quantity of litter being produced.

The Calliandra (*Calliandra calothyrsus*) a small leguminous tree or large shrub when used as live fence sequestered 2 Mg C m^{-2} of live fence (3–4 Mg dry matter m^{-2}) at the age of 10 months, which epitomize C equal to 20 Mg ha^{-1} in an average fence width of 1 m. Likewise, the live fence of 4 years gliricidia produced a gross annual C up to 35 and 50 Mg ha^{-1} (7 and 9.5 Mg m^{-2} total biomass) when pruned at 4- and 6-month intervals, respectively. Kang et al. (1999) also stated that gliricidia and subabul (*Leucaena leucocephala*) added 15% more SOC equivalent to 2 Mg C ha^{-1} under 12-year hedgerow intercropping than that added by the sole crops in Alfisol. Under the hedgerow intercropping system, subabul increased the SOC content in 0–15 cm soil depth by 1.23 and 0.94% in hedgerow and alleys, respectively, over the continuous cropping after 5 years (Kang 1997). Lasco and Suson (1999) in their experiment reported that the subabul added on an average 16 Mg C ha^{-1} during 6 years. They have reported increases in ANPP from 4 Mg in first year to 64 Mg in sixth year, while about 25% of the aboveground biomass C was estimated in woody debris. The increase in annual SOC deposition in top 15 cm soil stratum by the legume tree subabul substantially improved by 0.08, 0.26, and 0.76 Mg C ha^{-1} at the age of 14, 20, and 38 years of planting, respectively, over the grass pasture in tropical regions (Radrizzani et al. 2011). The introduction of subabul, cassia tree (*Senna siamea*) and pasture grass specie (*Imperata cylindrica*) were found to be more efficient in enhancing the SOC content in upper 20 cm soil (Gaiser et al. 2012), whereas legume tree guava (*Inga edulis* sp.) returned 12% more organic C (0.23 Mg C ha^{-1}) into the soil profile after 5 years of the hedgerow intercropping (Alegre and Rao 1996). But Rhoades et al. (1998) stated guava returned 20 Mg ha^{-1} additional C in upper 15 cm furrow slice than that of open pasture. Another 10 m height flowering leguminous tree coral bean (*Erythrina bertroana*) recorded the same trend of 30 and 55 Mg C ha^{-1} annually, respectively, in the same research. de Jong et al. (1995) also reported the C sequestration potential of live fence trees of 24–36 Mg C ha^{-1} in 25–30-year cycle. However, there is a need to consider the entire C cycle of the trees for the better assessment of C sequestration potential of the system (Albrecht and Kandji 2003).

4.11 Legume Root Biomass Role in Soil Carbon Sequestration

Unless legume crops were harvested, all underground biomass turnover was present for being assimilated into the SOM (Nair et al. 2015). The underground plant portion and root production directly contribute into the soil system and by itself have greatest potential to incorporate directly into the SOM pool (Sanderman et al. 2010; Sainju 2016). Several scientific studies had reported the correlations between plant root extension and the distribution of SOC within soil depth. The SOC contributed by plant root system remains present in the soil for a longer time in comparison to the aboveground residues associated with slow decomposition of roots in soil (Kuzyako and Domanski 2000; Rasse et al. 2005; Blagodatskaya et al. 2014). Legume-based crop rotation accounts to soil C sequestration because of increased rate of SOC accumulations at different soil depths as per the rooting habit of different species (Nair et al. 2015). The crop rotation with deep-rooted legumes builds the C pool deeper in soil via rhizospheric deposition, and root biomass extended deep in the soil. Gregorich et al. (2001) and Meena et al. (2017b) compared the legume-based rotation (maize-oat-alfalfa-alfalfa) with monoculture maize in over 35 years under fertilized and non-fertilized conditions. They concluded that the C sequestration in legume-maize rotation was 20 Mg ha^{-1} higher as compared to the monoculture of maize at the end of experiment. And more importantly, they stated that SOC in legume-based rotation below plow layer (top 15 cm layer) appeared to be in a biologically resistant form (e.g., more aromatic C content), signifying the role of deep-rooted legumes toward enhancing long-term C sequestration in deeper soil. Similarly, under legume-based rotation, the size of C_3-C pool beneath the plow layer was 40% higher over the monoculture.

The increased atmospheric CO_2 possibly induces the legumes root biomass production which in due course leads C returns to the soil. This is associated with the greater root turnover along with exudation of numerous carbonic substances which in turn contribute to greater soil C. The C sequestration through the root exudates and release of organic compounds from plant root during plant growth cycle and also the slaughtering of minor roots and their hairs by root elongation is well documented (Kemper et al. 1998; Rasse et al. 2005; Kumar et al. 2006). Legume roots exhibit the property to synthesize, store, and release various types of compounds (Kumar et al. 2006). The term "rhizo-deposition" is used to describe the release of carbonic substances from the plant roots as root exudates (Kuzyako and Domanski 2000). These secreted organic substances comprehend water-soluble and complex compounds like mucilage, root border cells, simple and complex sugars, vitamins, organic acids, amino acids, phenolics, polypeptides, polysaccharides, extracellular enzymes ,and other secondary plant metabolites. These compounds accumulate in the rhizospheric zone of plant roots (Paul and Clark 1996).

The transfer of photosynthates in the form of organic compounds in the direction of root's surrounding areas through roots occurs by the mechanism of root exudation, respiration, lysis, and root death (Kemper et al. 1998). The mechanisms of root death can be categorized into three states: (i) *apoptotic death* (occurring of C re-translocation to other plant-growing regions), (ii) *non-apoptotic death* (sewerage of principal roots – roots excised, rests metabolically energetic, and gradually exhausts inner soluble C pool); and (iii) *non-apoptotic death* (mechanically injured roots – metabolic activities terminate instantly and whole C contents go into the soil system) (Kumar et al. 2006). The all soluble C substance existing in the root may enter into the rhizospheric zones; on the other hand, the root exudates are still dominated by solutes having lower molecular weights in the cytoplasm present within the cell membrane (Kuzyakov et al. 2001). The repetitive discharge of carbonic substances from plant roots is further categorized into two classes: *first*, exudates which are discharged with a particular intention which is strongly regulated by the plant and, *second*, exudates which are lost on account of passive diffusion over which plants have limited or no control. The roots also liberate the CO_2 into the soil ecosystem as a result of carbohydrate respiration. The C in soluble form is discharged by non-metabolically active plant roots which are termed as "lysis," whereas metabolically active plant roots account for CO_2 respiration and exudation (Rasse et al. 2005).

The rhizospheric depositions also have a great role in sequestering of the C. The root-derived exudate accelerates the working capacity of soil micro-/macroflora and fauna by providing the C as a source of energy (Paul and Clark 1996). The C released in rhizospheric zones assist proliferation of the growth and dynamics of microbial populations in the surface rhizosphere, endorhizosphere, and outside of the roots, i.e., ectorhizosphere. As soon as the releases of the C substance in the rhizosphere provinces occur, the soil bacteria attack and consume the substrate and proliferate quickly regarding growth, activity, and number which upon death contribute to SOC pool (Kumar et al. 2006). Blagodatskaya et al. (2014) also confirmed the higher microbial diversity and profusion along with rapid growth inside the rhizospheric environment of legumes when compared to the fallow soil. Now clearly, this was because of the abundant C availability released by the roots in the rhizosphere. The phenolic compounds secreted by the legume roots are supposed to be involved in the genetic triggering of the legume-*Rhizobium* association (Nair et al. 2015). The root-derived chemical promotes the growth and development of plant-fungal symbiotic association and provides the powerful signals that change the mycorrhizal association with the host plants. The dense root system of certain legumes and improved microbial functionaries can encourage the soil aggregation (Bronick and Lal 2005). Plant root system alters the soil structure using their physiology, biochemistry, and functions. The size and number of root nodules per plant also significantly account toward the SOC pools. Root nodules support the bacteroids inhabiting in the rhizosphere. Henceforth, these supposed that the legume cultivation is a potential way to increase soil C sequestration. (Libault 2014).

4.12 Legumes Improve Soil Microbial Biomass C (SMBC)

The soil under legume cultivation behaves in a similar pattern as to that of an organically managed field; this is also the reason behind the increased organic C content and associated soil microbial activities under legumes cultivation presented in Fig. 4.2.

Many researchers reported the increased N availability and metabolic activities of microbes and below- and aboveground biomass production in the legume-based system (Robinson et al. 1996; Adeboye et al. 2006; Diepeningen et al. 2006). Soil microbial biomass carbon (SMBC) comprises of 2–5% of total soil C (Smith and Paul 1990). The biological activities and subsequent SMBC in soil system under legume crops presented a higher C than nonleguminous crop. This was supported by the increased amount of biomass C, diversity of added plant residue, and biomass ninhydrin under legume cultivation over nonleguminous (Robinson et al. 1996;). Crop rotation with legumes significantly altered the status of both water-soluble organic carbon (WSOC) and SMBC (Adeboye et al. 2006; Ganeshamurthy 2009).

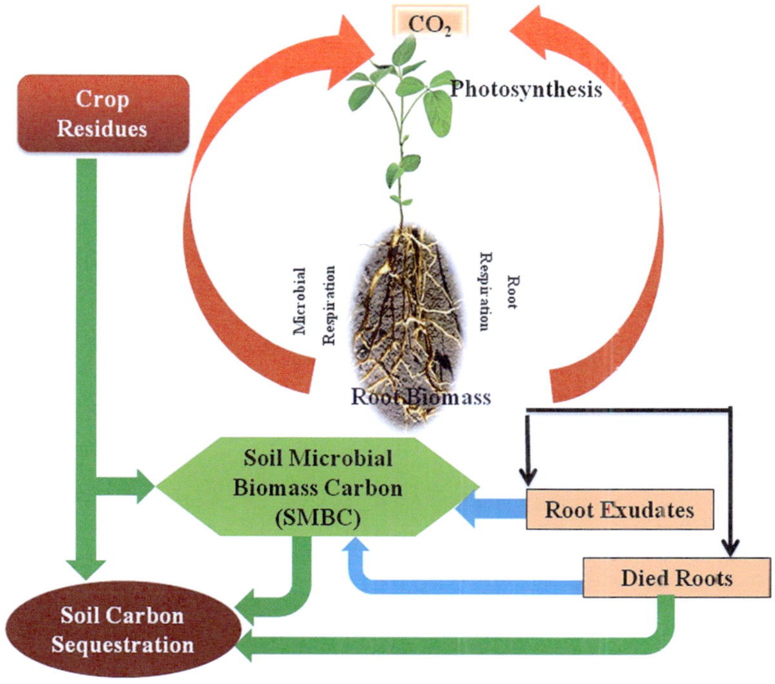

Fig. 4.2 Soil microbial biomass carbon and sequestration through legumes

They found that crop rotation may not always affect the total SMBC but may affect physiologically distinct sub-component of microbial biomass. Similarly, the soybean rotation returned the greater SMBC and WSOC, whereas the fallow rotation produced the least values of these parameters. These SMBC and WSOC were linearly correlated with SOC and soil pH but not with the total N content (Moore et al. 2000). The crop rotation of maize-wheat-green gram and pigeon pea-wheat significantly improved the SMBC by 10 and 15% over the existing conventional maize-wheat system (Venkatesh et al. 2013). Adeboye et al. (2006) in their study found that any rotation with legume has a higher proportion of organic C as compared to the fallow rotation. This might be due to the low C/N ratio or high quality of residues of legumes returned into the soil which is more favorable for growth and survival of soil microbial populations and hereafter greater biomass C as a proportion of SOC.

4.13 Conclusion and Future Perspective

Legumes are an imperative component of ecological sustainability and nutritional security and a valuable source of food proteins. Most of the global soils have become depleted in SOC pool over the years under intensive agriculture while a strong link between nutritional security and the C pool in terrestrial ecosystems, notably the SOC pool. Legume-based cropping systems crops are some of the key components to significant improvement in SOC pool in stressed soils. Legumes have positive effects on soil processes such as benefiting agroecosystem, agricultural productivity, soil conservation, soil biology, SOC and N pool, soil chemical and physical properties, BNF, nitrous oxide (N_2O) emission, and nitrate (NO_3^-) leaching by reducing the need for chemical fertilizers. The SOC pool, its amount, and depth-distribution improve along with involvement of legumes in the systems as a component of soil quality and a source of numerous ecosystem services. This will be helpful to researchers and policy makers to form decisions on the alarming issues of climate change and food and nutritional security, especially after the Paris Agreement of December 2015, which appeals for action to store and increase the sink capacity of SOC pool for soil sustainability and world food security.

References

Abberton (2010) Enhancing the role of legumes: potential and obstacles. In: Grassland carbon sequestration: management, policy and economics. Proceedings of the workshop on the role of grassland carbon sequestration in the mitigation of climate change. Rome, April 2009

Abdurahman MD, Seeling B, Rego TJ, Reddy BB (1998) Organic matter inputs by selected cropping systems on a vertisol in the semi-arid tropics of India. Ann Arid Zone 37:363–371

Adeboye MKA, Iwuafor ENO, Agbenin JO (2006) The effects of crop rotation and nitrogen fertilization on soil chemical and microbial properties in a Guinea Savanna Alfisol of Nigeria. Plant Soil 281:97–107. https://doi.org/10.1007/s11104-005-3828-5

Akinnifesi FK, Makumba W, Sileshi G, Ajayi OC, Mweta D (2007) Synergistic effect of inorganic N and P fertilizers and organic inputs from *Gliricidia sepium* on productivity of intercropped maize in southern Malawi. Plant Soil 294:203–217

Albrecht A, Kandji ST (2003) Carbon sequestration in tropical agroforestry systems. Agric Ecosyst Environ 99:15–27. https://doi.org/10.1016/S0167-8809(03)00138-5

Alegre JC, Rao MR (1996) Soil and water conservation by contour hedging in the humid tropics of Peru. Agric Ecosyst Environ 57:17–25

Alegre J, Alonso-Bl'azquez N, de Andr'es EF, Tenorio JL, Ayerbe L (2004) Revegetation and reclamation of soils using wild leguminous shrubs in cold semiarid Mediterranean conditions: Litterfall and carbon and nitrogen returns under two aridity regimes. Plant Soil 263:203–212

Amad TJC, Bayer C, Conceicao PC, Spagnollo E, de CBHC, Veiga M (2006) Potential of carbon accumulation in no-till soils with intensive use and cover crops in southern Brazil. J Environ Qual 35:1599–1607. https://doi.org/10.2134/jeq2005.0233

Amado TJC, Bayer C (2008) Revised carbon sequestration rates in tropical and subtropical soils under no-tillage in Brazil

Arrouays D, Deslais W, Badeau V (2001) The carbon content of top soil and its geographical distribution in France. Soil Use Manag 17:7–11

Austin AT, Vivanco L (2006) Plant litter decomposition in a semi-arid ecosystem controlled by photo-degradation. Nature 442:555–558

Aziz I, Tariq M, Islam KR (2014) Impact of sole cropping and multiple cropping on soil humified carbon fraction. Pak J Bot 46(6):2157–2162

Bayer C, Martin-neto L, Mielniczuk J, Pavinato A, Diekow J (2006) Carbon sequestration in two Brazilian Cerrado soils under no-till. Soil Tillage Res 86:237–245. https://doi.org/10.1016/j.still.2005.02.023

Becker JN, Gutlein A, Cornejo NS, Kiese R, Hertel D, Kuzyakov Y (2016) Legume and non-legume trees increase soil carbon sequestration in Savanna. Ecosystems 20:989. https://doi.org/10.1007/s10021-016-0087-7

Benbi DK, Brar JS (2009) A 25-year record of carbon sequestration and soil properties in intensive agriculture. Agron Sustain Dev/Springer Verlag/EDP Sciences/INRA 29(2):257–265

Bertin C, Yang X, Weston LA (2003) The role of root exudates and allelochemicals in the rhizosphere. Plant Soil 256(1):67–83

Bhattacharyya R, Prakash V, Kundu S, Pandey SC, Srivastva AK, Gupta HS (2009) Effect of fertilisation on carbon sequestration in soybean–wheat rotation under two contrasting soils and management practices in the Indian Himalayas. Aust J Soil Res 47:592–601

Biederbeck VO, Bouman OT, Campbell CA, Bailey LD, Winkleman GE (1996) Nitrogen benefits of four green manure legumes in dryland cropping systems. Can J Plant Sci 76:307–515

Blagodatskaya E, Blagodatsky S, Anderson TH, Kuzyakov Y (2014) Microbial growth and carbon use efficiency in the rhizosphere and root-free soil. PLoS One 9(4):e93282. https://doi.org/10.1371/journal.pone.0093282

Blair N, Crocker GJ (2000) Crop rotation effects on soil carbon and physical fertility of two Australian soils. Aust J Soil Res 38:71–84

Bronick CJ, Lal R (2005) Manuring and rotation effects on soil organic carbon concentration for different aggregate size fractions on two soils in north eastern Ohio, USA. Soil Tillage Res 81:239–252

Campbell CA, Zentner RP (1993) Soil organic matter as influenced by crop rotation and fertilization. Soil SciSoc Am J 57:1034–1040

Campbell CA, Zentner RP, Liang BC, Roloff G, Gregorich EG, Blomert B (1999) Organic C accumulation in soil over 30 years in semiarid south western Saskatchewan: effect of crop rotations and fertilizers. Can J Soil Sci 80:179–192

Capriel P (1991) Inference of graduated management intensity on humus chemistry. Bayerisches land writs chaftlichesJahrbuch 68:367–369

Chaer GM, Resende AS, Campello EFC, de FSM, Boddey RM (2011) Nitrogen-fixing legume tree species for the reclamation of severely degraded lands in Brazil. Tree Physiol 31:139–149. https://doi.org/10.1093/treephys/tpq116

Chan Y (2008) Increasing soil organic carbon of agricultural land. Primefact 735
Chan KY, Heenan DP (1996) The influence of crop rotation on soil structure and soil physical properties under conventional tillage. Soil Tillage Res 37:113–125
Chan KY, Conyers MK, Li GD, Helyar KR, Poile G, Oates A, Barchia IM (2011) Soil carbon dynamics under different cropping and pasture management in temperate Australia: results of three long-term experiments. Soil Res 49:320–328
Chaudhury ST, Bhattacharyya SPW, Pal DK, SahrawatKL NA, Chandran P, Venugopalan MV, Telpandel B (2016) Use and cropping effects on carbon in black soils of semi-arid tropical India. Curr Sci 110(9):1652–1698
Conceicao PC, Diekow J, Bayer C (2013) Combined role of no-tillage and cropping systems in soil carbon stocks and stabilization. Soil Tillage Res 129:40–47. https://doi.org/10.1016/j.still.2013.01.006
Curtin D, Wang H, Selles F, Zentner RP, Biederbeck VO, Campbell CA (2000) Legume green manure as partial fallow replacement in semiarid Saskatchewan: effect on carbon fluxes. Can J Soil Sci 80:499–505
Dadhich RK, Reager ML, Kansoti BC, Meena RS (2014) Efficacy of growth substances on mustard (*Brassica juncea* L.) under hyper arid environmental condition of Rajasthan. Ecoscan 8(3–4):269–272
Daly MJ, Hunter RM, Green GN, Hunt L (1996) A comparison of multi-species pasture with ryegrass–white clover pasture under dryland conditions. Proc N Z Grassland Assoc 58:53–58
de Balieiro FC, Pereira MG, José B, Alves R, de RAS, Franco AA (2008) Soil carbon and nitrogen in pasture soil reforested with eucalyptus and guachapele. R Bras Ci Solo 32:1253–1260
de Jong BHJ, Montoya-Gomez G, Nelson K, Soto-Pinto L, Taylor J, Tipper R (1995) Community forest management and carbon sequestration: a feasibility study from Chiapas, Mexico. Interciencia 20:409–416
Dhakal Y, Meena RS, Kumar S (2016) Effect of INM on nodulation, yield, quality and available nutrient status in soil after harvest of green gram. Leg Res 39(4):590–594
Dhruw SK, Singh LJ, Singh AK (2009) Storage and sequestration of carbon by leguminous and non-leguminous trees on red lateritic soil of Chhattisgarh. Indian Forester 135(4):531–538
Diekow J, Mileniczuk J, Knicker H, Bayer C, Dick DP, Kögel-Knabner I (2005) Soil C and N stocks affected by cropping systems and nitrogen fertilization in a southern Brasil Acrisol managed under no-tillage for 17 years. Soil Tillage Res 81:87–95
Diepeningen AD, de Vos OJ, Korthals GW, Van Bruggen AHC (2006) Effects of organic versus conventional management on chemical and biological parameters in agricultural soils. Appl Soil Ecol 31:120–135
Doran JW (2002) Soil health and global sustainability: translating science into practice. Agric Ecosyst Environ 88:119–127
Dreschel P, Glaser B, Zech W (1991) Effect of four multipurpose tree fallow in central Togo. Agrofor Syst 16:193–202
Drinkwater LE, Wagoner P, Sarrantonio M (1998) Legume-based cropping systems have reduced carbon and nitrogen losses. Nature 396:262
Fan JW, Du YL, Turner NC, Wang BR, Fang Y, Xi Y, Guo XR, Li FM (2015) Changes in root morphology and physiology to limited phosphorus and moisture in a locally-selected cultivar and an introduced cultivar of *Medicago sativa* L. growing in alkaline soil. Plant Soil 392:215–226
FAO (2016) Hunger, poverty and climate change: the challenges today and tomorrow
FAO, ITPS (Intergovernmental Technical Panel on Soils) (2015) Status of the World's soil resources (SWSR) – main report. FAO, Rome
Fisher MJ, Rao IM, Ayarza MA, Lascono CE, Sanz JI, Thomas RJ, Vera RR (1994) Carbon storage by introduced deep-rooted grasses in South American savannas. Nature 371:236–238
Fornara DA, Tilman D, Hobbie SE (2009) Linkages between plant functional composition, fine root processes and potential soil N mineralization rates. J Ecology 97:48–56
Franzluebbers AJ (2002) Water infiltration and soil structure related to organic matter and its stratification with depth. Soil Tillage Res 66:197–205

Gaiser T, Stahr K, Bernard M, Kang BT (2012) Changes in soil organic carbon fractions in a tropical Acrisol as influenced by the addition of different residue materials. Agrofor Syst 86:185–195. https://doi.org/10.1007/s10457-011-9417-0
Gami SK, Lauren JG, Duxbury JM (2009) Influence of soil texture and cultivation on carbon and nitrogen levels in soils of the eastern Indo-Gangetic Plains. Geoderma 153:304–311
Ganeshamurthy AN (2009) Soil changes following long-term cultivation of pulses. J Agric Sci 147:699–706
Gawel E (2011) The role of fine-grained legume plants in a farm. Water Environ Rural Areas 11(3):73–91
Geesin D, Felker P, Bingham RL (2000) Influence of mesquite (*Prosopisglandulosa*) on soil nitrogen and carbon development: implications for global carbon sequestration. J Arid Environ 46:157–180
Ghosh PK, Manna MC, Dayanand R (2006) Carbon sequestration potential and sustainable yield index for groundnut- and fallow-based cropping systems. J Agric Sci 144:249–259. https://doi.org/10.1017/S0021859606006046
Grego S, Lagomarsino A (2008) Soil organic matter in the sustainable agriculture: source or sink of carbon? In: Soil carbon sequestration under organic farming in the Mediterranean environment by Sara Marinari and Fabio Caporali, pp 39–51. ISBN: 978-81-7895-327-4
Gregorich EG, Drury CF, Baldock JA (2001) Changes in soil carbon under long-term maize in monoculture and legume-based rotation. Can J Soil Sci 81:21–31
Guan XK, Turner NC, Song L, Gu YJ, Wang TC, Li FM (2016) Soil carbon sequestration by three perennial legume pastures is greater in deeper soil layers than in the surface soil. Bio Geosci 13:527–534. https://doi.org/10.5194/bg-13-527-2016
Hajduk E, Wlaoeniewski S, Szpunar-Krok E (2015) Influence of legume crops on content of organic carbon in sandy soil. Soil Sci Annu 66(2):52–56. https://doi.org/10.1515/ssa-2015-0019
He Z, Huang C, Zheng H, Zhou J, Pang J, Li X, Wang L (2011) Holocene loess and its deposition dynamics in the upper reaches of the Huaihe River. J Geogr Sci 21:561–573
Hoyle FC, Baldock JA, Murphy DV (2011) Soil organic carbon – role in Australian farming systems. In: Tow P, Cooper I, Partridge I, Birch C (eds) Rainfed farming systems. Springer, Heidelberg
Ibrahim M, Guerra L, Casasola F, Neely C (2010) Importance of silvopastoral systems for mitigation of climate change and harnessing of environmental benefits. In: Proceedings of the Workshop on the role of grassland carbon sequestration in the mitigation of climate change. Rome, April 2009
Impala (2001) First annual report of IMPALA project, covering the period October 2000–December 2001, Nairobi, Kenya, 68 pp
Jackson RB, Banner JL, Jobbagy EG, Pockman WT, Wall DH (2002) Ecosystem carbon loss with woody plant invasion of grasslands. Nature 418:623–626. https://doi.org/10.1038/nature00910
Jarenyama P, Hesterman OB, Waddington SR, Harwood RR (2000) Relay-intercropping of sunnhemp and cowpea into a smallholder maize system in Zimbabwe. Agron J 92:239–244
Jensen E, Peoples M, Robert B, Peter G, Hauggaard-Nielsen et al (2012) Legumes for mitigation of climate change and the provision of feedstock for biofuels and biorefineries: a review. Agron Sustain Dev 32(2):329–364. https://doi.org/10.1007/s13593-011-0056-7
Jobbagy EG, Jackson RB (2000) The vertical distribution of soil organic carbon and its relation to climate and vegetation. Ecol Appl 10:423–436
Johnson DW, Curtis PS (2001) Effects of forest management on soil C and N storage: meta-analysis. For Ecol Manag 140:227–238. https://doi.org/10.1016/S0378-1127(00)00282-6
Junior GFS, Corá JE, Lal R (2016) Soil aggregation according to the dynamics of carbon and nitrogen in soil under different cropping systems. Pesqagropec Bras Brasília 51(9):1652–1659. https://doi.org/10.1590/S0100-204X2016000900065
Kalinda C, Mburu D, Ngamau K, Chisala LA, Zulu D, Kihoro J (2015) Short term leguminous trees-tillage interactions and their effect on soil-water content in a semi-arid agroforestry parkland. Open J Fores 5:668–677. https://doi.org/10.4236/ojf.2015.57059

Kane D (2015) Carbon sequestration potential on agricultural lands: a review of current science and available practices. In: association with: National Sustainable Agriculture Coalition Breakthrough Strategies and Solutions, LLC

Kang BT (1997) Alley cropping – soil productivity and nutrient recycling. For Ecol Manag 91:75–82

Kang BT, Caveness FE, Tian G, Kolawole GO (1999) Long-term alley cropping with four species on an Alfisol in Southwest Nigeria—effect on crop performance, soil chemical properties and nematode population. Nutr Cycl Agroecosyst 54:145–155

Kaonga ML, Coleman K (2008) Modelling soil organic carbon turnover in improved fallows in eastern Zambia using the RothC-26.3 model. For Ecol Manag 256(5):1160–1166. https://doi.org/10.1016/j.foreco.2008.06.017

Kaongaa ML, Smith TPB (2009) Carbon pools in tree biomass and the soil in improved fallows in eastern Zambia. Agrofor Syst 76:37–51. https://doi.org/10.1007/s10457-008-9185-7

Kemper WD, Alberts EE, Foy CD, Clark RB, Ritchie JC, Zobel RW (1998) Arenchyma, acid tolerance and associative N_2 fixation enhance carbon sequestration in soil. In: Lal R, Kimble JM, Follet RF, Stewart BA (eds) Management of carbon sequestration in soil. CRC Press, Boca Raton/New York, pp 221–234

Kirkby CA et al (2013) Carbon-nutrient stoichiometry to increase soil carbon sequestration. Soil Biol Biochem 60:77–86. https://doi.org/10.1016/j.soilbio.2013.01.011

Kong XB, Dao TH, Qin J, Qin H, Li C, Zhang F (2009) Effects of soil texture and land use interactions on organic carbon in soils in North China cities' urban fringe. Geoderma 154:86–92

Kulmatiski A, Beard KH (2013) Woody plant encroachment facilitated by increased precipitation intensity. Nat Clim Chang 3:833–837. https://doi.org/10.1038/Nclimate1904

Kumar BM, George SJ, Suresh TK (2001) Fodder grass productivity and soil fertility changes under four grass + tree associations in Kerala. India Agrofor Syst 52:91–106

Kumar R, Pandey S, Pandey A (2006) Plant roots and carbon sequestration. Curr Sci 9(7):885–890

Kundu S, Singh M, Tripathi AK, Manna MC, Takkar PN (1997) Time-course of di nitrogen fixation in soybean grown on TypicHaplusterts of Madhya Pradesh. J Indian Soc Soil Sci 45:274–278

Kundu S, Singh M, Saha JK, Biswas A, Tripathi AK, Acharya CL (2001) Relationship between C addition and storage in a vertisol under soybean-wheat cropping system in sub-tropical Central India. J Plant Nut Soil Sci 164:483–486

Kundu S, Bhattacharyya R, Prakash V, Ghosh BN, Gupta HS (2007) Carbon sequestration and relationship between carbon addition and storage under rainfed soybean–wheat rotation in a sandy loam soil of the Indian Himalayas. Soil Tillage Res 92:87–95. https://doi.org/10.1016/j.still.2006.01.009

Kushwaha CP, Tripathi SK, Singh KP (2001) Soil organic matter and water-stable aggregates under different tillage and residue conditions in a tropical drylandagroecosystem. Appl Soil Ecol 16:229–241

Kuzyako Y, Domanski G (2000) Carbon input by plants into the soil. Rev J Plant Nutr Soil Sci 163:421–431

Kuzyakov Y, Ehrensberger H, Stahr K (2001) Carbon partitioning and below-ground translocation by *Loliumperenne*. Soil Biol Biochem 33:61–74

Laganiere J, Angers DA, Pare D (2010) Carbon accumulation in agricultural soils after afforestation: a meta-analysis. Glob Change Biol 16:439

Lal R (1998) Soil erosion impact on agronomic productivity and environmental quality. Crit Rev Plant Sci 17:319–464

Lal R (2004a) Soil carbon sequestration impacts on global climate change and food security. Science 304:1623–1627

Lal R (2004b) Soil carbon sequestration in India. Climatic Change 65:277–296

Lal R (2009) Soil carbon sequestration for climate change mitigation and food security. Souvenir, Platinum Jubilee symposium on soil science in meeting the challenges to food security and environmental quality. Indian Society of Soil Science, New Delhi, pp 39–46

Lal R (2010) Enhancing eco-efficiency in agro-ecosystems through soil carbon sequestration.Crop Sci 50 (Suppl 1):S–120

Lal R (2015) Restoring soil quality to mitigate soil degradation. Sustainability 7:5875–5895. https://doi.org/10.3390/su7055875

Lange M et al (2015) Plant diversity increases soil microbial activity and soil carbon storage. Nat Commun 6:67–07. https://doi.org/10.1038/ncomms7707

Lasco RD, Suson PD (1999) A *Leucaena leucocephala*-based indigenous fallow system in Central Philippines: the Naalad system. Int Tree Crops J 10:161–174

Leigh RA, Johnston AE (1994) Long-term experiments in agricultural and ecological sciences. CAB International, Wallingford

Li H et al (2016) Effects of shrub encroachment on soil organic carbon in global grasslands. Sci Rep 6:28974. https://doi.org/10.1038/srep28974

Liang YM, Hazlett DL, Lauenroth WK (1989) Biomass dynamics and water use efficiencies of five plant communities in the short grass steppe. Oecologia 80:148–153

Liang F et al (2016) Three-decade long fertilization-induced soil organic carbon sequestration depends on edaphic characteristics in six typical croplands. Sci Rep 6:30350. https://doi.org/10.1038/srep30350

Libault M (2014) The carbon-nitrogen balance of the nodule and its regulation under elevated carbon di oxide concentration. Biomed Res Int. https://doi.org/10.1155/2014/507946

Liu X, Han X, Song C, Herbert SJ, Xing B (2003a) Soil organic carbon dynamics in black soils of China under different agricultural management systems. Commun Soil Sci Plant Anal 34(7–8):973–984. https://doi.org/10.1081/CSS-120019103

Liu XB, Han XZ, Herbert SJ, Xing B (2003b) Dynamics of soil organic carbon under different agricultural management systems in the black soil of China. Commun Soil Sci Plant Anal 34:973–984

Liu X, Liu FM, Liu DQ, Sun GJ (2010) Soil organic carbon, carbon fractions and nutrients as affected by land use in semi-arid region of loess plateau of China. Pedosphere 20:146

Liu JB, Zhang YQ, Wu B, Qin SG, Jia X, Feng W (2013) Changes in soil carbon, nitrogen and phosphorus along a chronosequence of *Caragana microphylla* plantation, north western China. Pol J Environ Stud 23(2):385–391

Macedo MO, Campello EFC, Andrade AG, de FSM (2006) Establishment of legume trees on heaps of blast furnace slag. Floresta Ambiente 13:20–25

Macedo MO, Resende AS, Garcia PC, Boddey RM, Jantalia CP Urquiaga S, Campello EFC, Franco AA (2008) Changes in soil C and N stocks and nutrient dynamics 13 years after recovery of degraded land using leguminous nitrogen-fixing trees. For Ecol Manag 255:1516–1524. https://doi.org/10.1016/j.foreco.2007.11.007

Mandal B, Majumdar B, Bandyopadhya PK, Hazra GC, Gangopadhyay A, Samantaray RN, Mishra AK, Chaudhury J, Saha MN, Kundu S (2007) The potential of cropping systems and soil amendments for carbon sequestration in soils under long-term experiments in subtropical India. Glob Chang Biol 13:357–369

Mann L, Tolbert V, Cushman J (2002) Potential environmental effects of corn (*Zea mays* L.) Stover removal with emphasis on soil organic matter and erosion. Rev Agric Ecosyst Environ 89:149–166

Martin M, Dos R, Angersd A, Coraj E (2012) Carbohydrate composition and water-stable aggregation of an Oxisol as affected by crop sequence under no-till. Soil Sci Soc Am J 76:475–484. https://doi.org/10.2136/sssaj2011.0110

McLauchlan KK (2006) Effects of soil texture on soil carbon and nitrogen dynamics after cessation of agriculture. Geoderma 136:289–299. https://doi.org/10.1016/j.geoderma.2006.03.053

McLauchlan KK, Hobbie S (2004) Comparison of labile soil organic matter fractionation techniques. Soil SciSoc Am J. 68:1616–1625

Meena RS (2013) Response to different nutrient sources on green gram (*Vignaradiata* L.) Productivity. Indian J Ecol 40(2):353–355

Meena VS, Maurya BR, Verma R, Meena RS, Jatav GK, Meena SK, Meena R, Meena SK (2013) Soil microbial population and selected enzyme activities as influenced by concentrate manure and inorganic fertilizer in alluvium soil of Varanasi. Biosci 8(3):931–935

Meena RS, Meena VS, Meena SK, Verma JP (2015a) Towards the plant stress mitigate the agricultural productivity: a book review. J Clean Prod 102:552–553

Meena RS, Yadav RS, Meena H, Kumar S, Meena YK, Singh A (2015b) Towards the current need to enhance legume productivity and soil sustainability worldwide: a book review. J Clean Prod 104:513–515

Meena VS, Maurya BR, Meena RS (2015c) Residual impact of well-grow formulation and NPK on growth and yield of wheat (*Triticumaestivum* L.). Bangladesh J Bot 44(1):143–146

Meena RS, Gogaoi N, Kumar S (2017a) Alarming issues on agricultural crop production and environmental stresses. J Clean Prod 142:3357–3359

Meena RS, Kumar V, Yadav GS, Mitran T (2017b) Response and interaction of *Bradyrhizobiumjaponicum* and *Arbuscularmycorrhizal*fungi in the soybean rhizosphere: a review. Plant Growth Reg 84(2):207–223

Moore JM, Klose S, Tabatabai MA (2000) Soil microbial biomass carbon and nitrogen as affected by cropping systems. Biol Fertil Soil 3:200–210

Mortensen MC, Schuman GE, Ingram LJ (2004) Carbon sequestration in rangelands inter seeded with yellow-flowering alfalfa (*Medicago sativa* ssp. *Falcate*). Environ Manag 33:S475–S481

Nair PKR, Kumar BM, Nair VD (2009) Agroforestry as a strategy for carbon sequestration. J Plant Nutr Soil Sci 172(1):10–23

Nair R, Mehta CR, Sharma S (2015) Carbon sequestration in soils-a review. Agric Rev 36(2):81–99

Newaj R, Yadav DS (1994) Changes in physicochemical properties of soil under intensive cropping systems. Indian J Agron 3:373–378

Onim JFM, Otieno K, Dzowela B (1990) The role of sesbania as multipurpose trees in small-scale farms in western Kenya. In: Macklin B, Evans DO (eds), Perennial Sesbania species in agroforestry systems, eds . Proceedings of a workshop held in Nairobi, March 27–31, 1989. The Nitrogen Fixing Tree Association, Hawaii, USA, pp. 167–179

Paul EA, Clark FE (1996) Soil microbiology and biochemistry, 2nd edn. Academic Press, San Diego

Paustian K, Robertson GP, Elliott ET (1995) Management impacts on carbon storage and gas fluxes (CO_2, CH_4) in mid-latitude cropland. In: Lal R et al (eds) Soil management and the greenhouse effect. Lewis Publishers, Boca Raton, pp 69–83

Peoples MB, Baldock JA (2001) Nitrogen dynamics of pastures: nitrogen fixation inputs, the impact of legumes on soil nitrogen fertility and the contributions of fixed nitrogen to Australian farming systems.Aust J Exp Agric. 41: 327–346

Pinho RC, Miller RP, Alfaia SS (2012) Agroforestry and the improvement of soil fertility: a view from Amazonia. Appl Environ Soil Sci 2012:1–11. https://doi.org/10.1155/2012/616383

Podleoeny J (2005) Legumes in Poland – perspectives for growing and seed utilization. Acta Agrophysica 6(1):213–224

Pregitzer KS, Euskirchen ES (2004) Carbon cycle and storage in world forest: biome patterns related to forest age. Glob Chang Biol 10:2052–2077

Radrizzani A, Shelton HM, Dalzell SA, Kirchhof G (2011) Soil organic carbon and total nitrogen under *Leucaena leucocephala* pastures in Queensland. Crop Pasture Sci 62:337–345

Ram K, Meena RS (2014) Evaluation of pearl millet and Mungbean intercropping systems in arid region of Rajasthan (India). Bangladesh J Bot 43(3):367–370

Rasse DP, Rumpel C, Dignac MF (2005) Is soil carbon mostly root carbon? Mechanisms for a specific stabilisation. Plant Soil 269:341–356

Reeves DW (1997) The role of soil organic matter in maintaining soil quality in continuous cropping systems. Soil Tillage Res 43:131–167

Rhoades CC, Eckert GE, Coleman DC (1998) Effect of pasture trees on soil nitrogen and organic matter: implications for tropical montane forest restoration. Restor Ecol 6(3):262–270

Robinson CA, Cruse RM, Ghaffarzadeh M (1996) Cropping systems and nitrogen effects on mollisol organic carbon. Soc Sci Am J 6:264–269

Rochester IJ (2011) Sequestering carbon in minimum-tilled clay soils used for irrigated cotton and grain production. Soil Tillage Res 112:1–7

Rosenstock TS, Tully KL, Arias-Navarro C, Neufeldt H, Butterbach-Bahl K, Verchot LV (2014) Agroforestry with N_2-fixing trees: sustainable development's friend or foe? Curr Opin Environ Sustain 6:15–21

Rutkowska A, Pikua D (2013) Effect of crop rotation and nitrogen fertilization on the quality and quantity of soil organic matter. In: Hernandez Soriano MC (ed) Soil processes and current trends in quality assessment. In Tech, Rijeka, pp 249–267

Sa JCM, Lal R (2009) Stratification ratio of soil organic matter pools as an indicator of carbon sequestration in a tillage chronosequence on a Brazilian Oxisol. Soil Tillage Res 103:46–56

Sainju UM (2016) A global meta-analysis on the impact of management practices on net global warming potential and greenhouse gas intensity from cropland soils. PLoS One 11(2):e0148527. https://doi.org/10.1371/journal.pone.0148527

Sanchez PA (1995) Science in agroforestry. Agrofor Syst 30:5–55. https://doi.org/10.1007/BF00708912

Sanderman J, Farquharson R, Baldock J (2010) Soil carbon sequestration potential: a review for Australian agriculture. A report prepared for the Department of Climate Change and Energy Efficiency CSIRO National Research Flagships

Santos N, Dieckow J, Cimelio, Bayer Molin R, Nerildefavarettovolneipaulettiandjonatasthiagopiva (2011) Forages cover crops and related and root additions in no-till rotations. To C sequestration in a subtropical Ferralsol. Soil Till Res 111:208–218

Sexstone AJ, Revsbech NP, Parkin TB, Tiedje JM (1985) Direct measurement of oxygen profiles and denitrification rates in soil aggregates. Soil Sci Soc Am J 49(3):645–651

Shahzad T, Chenu C, Genet P, Barot S, Perveen N, Mougin C, Fontaine S (2015) Contribution of exudates, arbuscular mycorrhizal fungi and litter depositions to the rhizosphere priming effect induced by grassland species. Soil Biol Biochem 80:146–155

Silver WL, Miya RK (2001) Global patterns in root decomposition: comparisons of climate and litter quality effects. Oecologia 129:407–419

Singh G, Sandhu HS (1980) Studies on multiple cropping II. Effect of crop rotation on physical and chemical properties of soils. Indian Agron J 25:57–67

Sisti CPJ, Dos Santos HP, Kochhann RA, Alves BJR, Urquiaga S, Boddey RM (2004) Change in carbon and nitrogen stocks in soil under 13 years of conventional or zero tillage in southern Brazil. Soil Tillage Res 76:39–58

Six JP, Elliott KET, Combrink C (2000) Soil structure and organic matter distribution of aggregate-size classes and aggregate-associated carbon. Soil Sci Soc Am J 64:681–689

Six J, Callewaert P, Lenders S, Degryze S, Morris SJ, Gregorich EG, Paul EA, Paustian K (2002) Measuring and understanding carbon storage in afforested soils by physical fractionation. Soil Sci Soc Am J 66:1981

Smith JL, Paul EA (1990) The significance of soil microbial biomass estimations. In: Bollag JM, Stotzky G (eds) Soil biochemistry, vol 6. Marcel Decker, New York, pp 357–396

Sombrero A, de Benito A (2010) Carbon accumulation in soil. Ten-year study of conservation tillage and crop rotation in a semi-arid area of Castile-Leon, Spain. Soil Tillage Res 107:64–70

Su Y, Zhang T, Li Y, Wang F (2005) Changes in soil properties after establishment of *Artemisia halodendron* and *Caragana microphylla* on shifting sand dunes in semiarid Horqin Sandy land, Northern China. Environ Manag 36:272–281

Sun Q, Han J, Gui R, Liu Z, Liu G (2001) Biomass in *Lespedeza dahurica* S. Grassland of China 4:21–26

Tiemann LK, Grandy AS, Atkinson EE, Marin-Spiotta E, McDaniel MD (2015) Crop rotational diversity enhances belowground communities and functions in an agroecosystem. Ecol Lett 18(8):761–771

Venkatesh MS, Hazra KK, Ghosh PK, Praharaj CS, Kumar N (2013) Long-term effect of pulses and nutrient management on soil carbon sequestration in indo-Gangetic plains of India. Can J Soil Sci 93:127–136

Verma JP, Meena VS, Kumar A, Meena RS (2015) Issues and challenges about sustainable agriculture production for management of natural resources to sustain soil fertility and health: a book review. J Clean Prod 107:793–794

Virginia RA, Jenkins MB, Jarrell WM (1986) Depth of root symbiont occurrence in soil. Biol Fertil Soil 2(3):127–130
Wamisho K (2013) A meta regression analysis of soil organic carbon sequestration in corn belt states: implication for cellulosic biofuel production, 2013 Annual Meeting, August 4–6, 2013, Washington, DC, 149741, Agricultural and Applied Economics Association
Wang J (2015) Contributions of wheat and maize residues to soil organic carbon under long-term rotation in North China. Sci Rep 5:11409. https://doi.org/10.1038/srep11409
Wang Q, Li Y, Ashok A (2010) Growing cover crops to improve biomass accumulation and carbon sequestration: a phytotron study. J Environ Protection 1(2):145–149
Wani NR, Qaisar KN (2014) Carbon per cent in different components of tree species and soil organic carbon pool under these tree species in Kashmir valley. Curr World Environ 9(1):174–181
Weber MA, Mielniczuk J (2009) Estoque e disponibilidade de nitrogênio no solo emexperimento de longa duração. Revista Brasileira de Ciênciado Solo 33:429–437. https://doi.org/10.1590/S0100-06832009000200020
Wheeler CW, Archer SR, Asner GP, McMurtry CR (2007) Climatic/edaphic controls on soil carbon/nitrogen response to shrub encroachment in desert grassland. Ecol Appl 17:1911–1192
Whitbread AM, Blair GJ, Lefroy RDB (2000) Managing legume leys, residues and fertilisers to enhance the sustainability of wheat cropping systems in Australia. Soil physical fertility and carbon. Soil Tillage Res 54:77–89
Wickings K, Grandy AS, Reed SC, Cleveland CC (2012) The origin of litter chemical complexity during decomposition (N Johnson, ed.). Ecol Lett. 15(10):1180–1188
Williams CH, Donald CM (1957) Changes in organic matter and pH in a podzolic soil as influenced by subterranean clover and superphosphate. Aust J Agric Res 8:179–189
World Bank (2012) Carbon sequestration in agricultural soils. Economic and sector work report number 67395-GLB. https://doi.org/10.1017/S0021859609990104
Wright AL, Hons FM, Rouquette FM Jr (2004) Long-term management impacts on soil carbon and nitrogen dynamics of grazed Bermuda grass pastures. Soil Biol Biochem 36:1806–1816
Yadav GS, Lal R, Meena RS, Babu S, Das A, Bhomik SN, Datta M, Layak J, Saha P (2017) Conservation tillage and nutrient management effects on productivity and soil carbon sequestration under double cropping of Rice in north eastern region of India. Ecol India. http://www.sciencedirect.com/science/article/pii/S1470160X17305617
Yang XM, Kay BD (2001) Rotation and tillage effects on soil organic carbon sequestration in a typic Hapludalf in southern Ontario. Soil Till Res 59:107–114
Yang Y, Luo Y, Finzi AC (2011) Carbon and nitrogen dynamics during forest stand development: a global synthesis. New Phytol 190:977–989. https://doi.org/10.1111/j.1469-8137.2011.03645
Yaqub M, Mahmoo T, Akhtar M, Iqbal MM, Ali S (2010) Induction of mungbean [*Vigna Radiata*(L.) Wilczek] as a grain legume in the annual rice-wheat double cropping system. Pak J Bot 42(5):3125–3135
Young RR, Wilson B, Harden S, Bernardi A (2009) Accumulation of soil carbon under zero tillage cropping and perennial vegetation on the Liverpool Plains, eastern Australia. Aust J Soil Res 47:273–285
Zentner RP, Campbell CA, Biederbeck VO, Selles F (1996) Indianhead black lentil as green manure for wheat rotations in the Brown soil zone. Can J Plant Sci 76:417–422
Zhang T, Wang Y, Wang X, Wang Q, Han J (2009) Organic carbon and nitrogen stocks in reed meadows soils converted to alfalfa fields. Soil Till Res 105(1):143–148
Zhou Z, Sun OJ, Huang J, Liu L, Liu P, Han X (2006) Soil carbon and nitrogen stores and storage potential as affected by land-use in an agro-pastoral ecotone of northern China. Biogeochemistry 82:127–138
Zhou Y, Pei Z, Su J, Zhang J, Zheng Y et al (2012) Comparing soil organic carbon dynamics in perennial grasses and shrubs in a saline-alkaline arid region, north western China. PLoS One 7(8):e42927. https://doi.org/10.1371/journal.pone.0042927

Perennial and Intercrop Legumes as Energy Crops for Biogas Production

5

P. Walter Stinner, Arno Deuker, Tina Schmalfuß, Christopher Brock, Nadja Rensberg, Velina Denysenko, Paul Trainer, Kurt Möller, Joachim Zang, Leandro Janke, Wilson Mozena Leandro, Katja Oehmichen, Denny Popp, and Jaqueline Daniel-Gromke

Contents

P. Walter Stinner (✉) · T. Schmalfuß · N. Rensberg · V. Denysenko · P. Trainer · L. Janke
K. Oehmichen · J. Daniel-Gromke
Deutsches Biomasseforschungszentrum gemeinnützige GmbH – DBFZ, Leipzig, Germany
e-mail: walter.stinner@dbfz.de; w.stinner@posteo.de; tina.schmalfuss@dbfz.de; nadja.rensberg@dbfz.de; velina.denysenko@dbfz.de; paul.trainer@dbfz.de; leandro.janke@dbfz.de; katja.oehmichen@dbfz.de; jaqueline.daniel-gromke@dbfz.de

A. Deuker
STEAG New Energies GmbH, Kompetenzzentrum Biomassebeschaffung und Stoffstrommanagement, Saarbrücken, Germany
e-mail: arno.deuker@steag.com

C. Brock
Forschungsring für Biologisch-Dynamische Wirtschaftsweise e.V., Darmstadt, Germany
e-mail: brock@forschungsring.de

K. Möller
Landwirtschaftliches Technologiezentrum, Augustenberg, Karlsruhe, Germany
e-mail: kurt.moeller@uni-hohenheim.de

J. Zang
Instituto Federal de Goiás – IFG, Goiânia, Brazil

W. M. Leandro
Universidade Federal de Goiás – UFG, Goiânia, Brazil

D. Popp
Helmholtz Centre for Environmental Research – UFZ, Leipzig, Germany
e-mail: denny.popp@ufz.de

R. S. Meena et al. (eds.), *Legumes for Soil Health and Sustainable Management*, https://doi.org/10.1007/978-981-13-0253-4_5

Abstract

Biogas generation opens new possibilities for the use of legume growths and thus for legume cropping, as quality restrictions are lower than for other forms of utilization. The German example of energy cropping shows the possibility of up to 20% of forage legume integration for bioenergy production into the cropping systems. This can allow an improvement of sustainability and resilience, especially with regard to reduced external inputs, improved humus, energy and nutrient balances, reduced greenhouse gas emissions, and the general positive impact of forage and intercrop legumes on cropping systems.

One hectare of main crop forage legumes delivers the energy for the operation of 12–17 ha of arable land. Crop residues and intercrop legumes mean additional energy yield.

For organic cropping systems, the anaerobic digestion of forage legumes allows higher nitrogen efficiency combined with reduced greenhouse gas emissions and reduced nitrate leaching risk. The higher N-efficiency results in higher yields and higher raw protein contents of nonlegume crops in crop rotations.

Without regarding the additional positive impacts, forage legumes as energy crops are only competitive under special conditions (e.g., cultivation on marginal sites). Economic internalization of positive external effects would promise legume cropping for this issue. For the further bioeconomy, the legume-based green biorefineries can deliver energy, colors (e.g., based on chlorophyll and carotins), vitamins, proteins, fibers, fatty acids, etc. In this way, forage legumes can get (again) key elements of sustainable economies.

Keywords
Legume-based green biorefineries · Biogas · Post fossile cropping · Energy balance · Climate friendly cropping · Nitrogen efficiency · Cropping systems · Organic cropping

Abbreviations

BNF	Biological di-nitrogen fixation
C	Carbon
CH_4	Methane
$CH_{4,STP}$	Methane at standard temperature and pressure (0 °C; 1013.25 bar)
CO_2eq.	Carbon dioxide equivalents
CCM	Corn-cob-mix
DM	Dry matter
FM	Fresh matter
KTBL	Kuratorium für Technik und Bauwesen in der Landwirtschaft
N	Nitrogen
N_2O	Nitrous oxide
SOM	Soil organic matter
STP	Standard temperature and pressure (0 °C, 1013.25 bar)
VS	Volatile Solids

5.1 Introduction

The role of forage and intercrop legumes has decreased in the modern conventional agriculture, which bases on fossil fuels for the generation of nitrogen (N) fertilizers and pesticides (e.g., for weed control), as well as for the operation of farm machines. For fodder issues, the combination of corn silage and soy meal enables bigger herds and enhanced animal yields with lower costs in specialized farms.

As shown in Sect. 5.3 "Energy Potentials of Legume Mixtures in Organic Farming Systems" (Deuker et al. 2011; Stinner 2011), legume crops can be a good energy source in sustainable cropping systems (Meena et al. 2015a). They can be digested in biogas plants, which deliver a mobile organic fertilizer and biogas, a multipurpose energy carrier that can be used for on-demand power and heat supply, for vehicle operation, as well as for private and industry purposes. Biogas as a form of bioenergy can contribute to sustainable energy systems and even to a sustainable agriculture in a post-fossil energy future. Crop cultivation only makes sense with a value creating use. As shown in the following, the bioeconomy development can open new possibilities for legume cropping.

Bioenergy use is not only a traditional form of renewable energy provision with a current share of about 61% of the global renewable energy provision or ca. 15% of the total global primary energy consumption (see Fig. 5.1).

It also plays an important role within the sector of renewable energies in industrialized countries. As shown in Fig. 5.2, the share of bioenergy in Germany, a highly industrialized country in transition to an energy supply primarily based on renewable resources, amounts to about 70% as measured against the total share of renewable energies in German primary energy consumption in 2016. A difference between global and German energy consumption refers to the end energy consumption. Energy from biomass for cooking purposes is less important in an industrialized country, whereas bioenergy plays a big role for heat and electricity production.

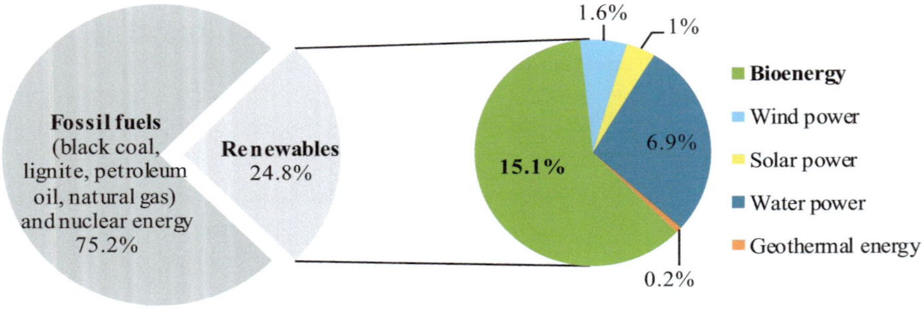

Fig. 5.1 Share of renewable energies in global primary energy consumption (556 EJ) in 2016; based on Witt et al. (2017); own diagram

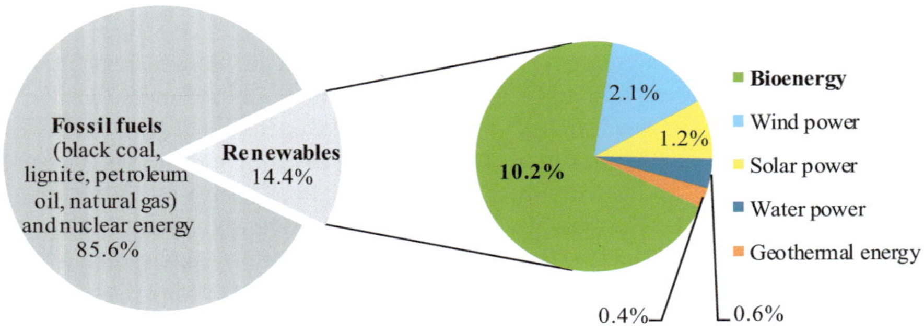

Fig. 5.2 Share of renewable energies in German primary energy consumption (13 EJ) in 2016; based on Lenz et al. (2017); own diagram

Especially the biogas technology is an important component for future (i) agriculture, (ii) energy, and (iii) bio-based economy systems. The big advantages for the (i) agriculture are the delivery of digestate, an efficient mobile organic fertilizer with humus reproduction effect, and the substrate flexibility, which allows the use of different organic wastes and residues as well as diverse energy crops, even green growths, including legumes, weeds, and so on.

In the energy system (ii), biogas can be used as fuel for electricity generation, for the transport and for the heat and industry sector, eventually after upgrading to biomethane, which is equivalent to natural gas. The gas grid opens long-term storage options for methane, which enables a flexible utilization even over seasons. For on-site electricity production, the biogas can be stored for a limited period of time, which allows a flexible electricity production. Hereby, fluctuating residual energy needs can be balanced, which are caused by a fluctuating demand and by a fluctuating renewable energy production by wind and solar power. From the agricultural point of view, the fuel option to operate machines is of very high importance. Biomethane can fuel tractors and other farm machines, which implies an own agricultural energy source and independence from the global fossil oil supply chains. Furthermore, the fuel cell suitability of methane in combination with decentral electric engines for several works enables the further development of strong innovation cycles for agriculture with doubled energy efficiency and reduced weight by saving of heavy mechanical power transfer.

For material usage in a future bio-based economy (iii), biogas technology opens pathways for fiber supplied by the digestate, for carboxylate generation (from adapted hydrolysis step), and for industrial products from biogas or methane. Especially the carboxylate platform opens various opportunities for the transfer of the economy into a fossil-carbon(C)-free future. There are options for plant protection agents like caproic acid, for bulk chemicals, for liquid fuels, and for the integration of surplus fluctuating power from wind and solar power plants.

Due to these aspects, by the end of 2016, about 8700 biogas plants and biogas upgrading plants (upgrading biogas to biomethane) have been in operation and predominantly contribute to electricity and heat generation in Germany. Most of these biogas plants are operated by agricultural holdings (see Fig. 5.3). The main biomass resources used for biogas production are animal excrements such as manure and dung (Meena et al. 2013), as well as nonlegume energy crops (see Fig. 5.4).

Because of restrictions concerning potentials, availability, degradability, and logistic issues of organic residues and wastes, the highest share of substrate input for biogas production accounts for energy crops (see Figs. 5.4 and 5.5). Considering the energy contents of different biomass resources, about 80% of the energy supply results from energy crops. Due to the legal financial support of liquid manure utilization, it is one of the substrates with >40% (mass-based) in biogas plants contributing to the on-site electricity generation.

For biomethane production, energy crops are the most prevalent biomass resources as well. The share of energy crops is higher than in plants with on-site cogeneration (compare Figs. 5.4 and 5.5). The reason for this is a higher performance (substrate input, installed capacity) due to the economies of scale of

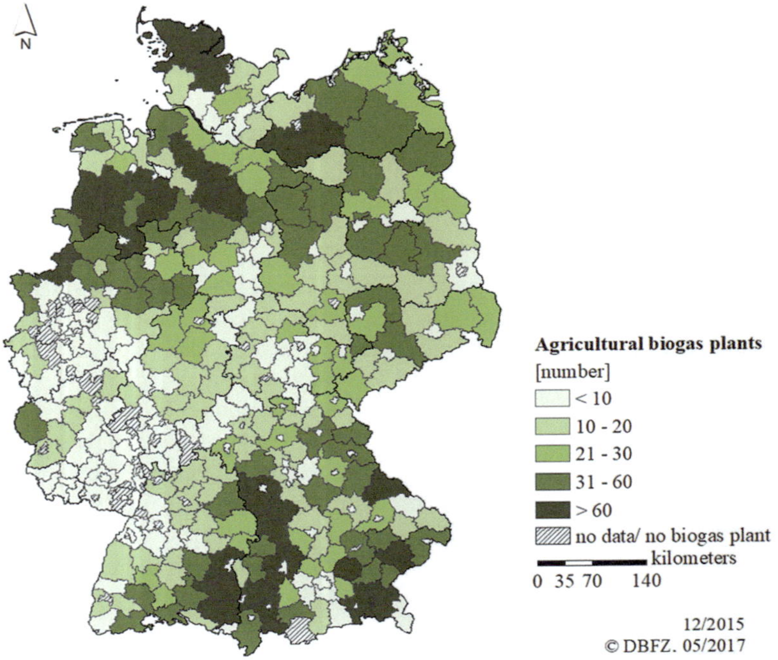

Fig. 5.3 Distribution of biogas plants referred to districts in Germany in 2015; own diagram

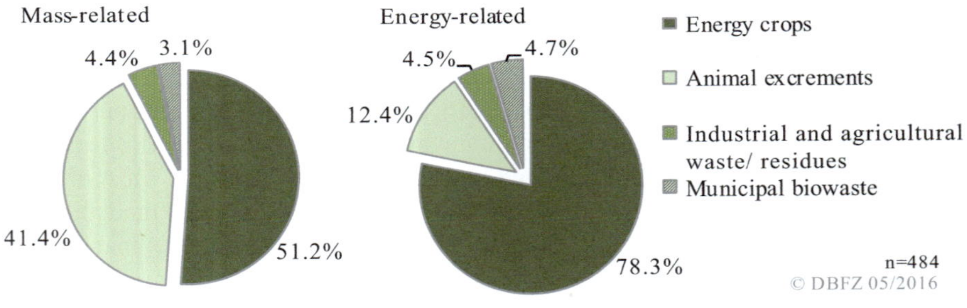

Fig. 5.4 Biomass resources in biogas plants (on-site electricity generation) referring to mass content and energy content; based on DBFZ operators' survey 2016, reference year 2015; own diagram

upgrading technologies on the one hand and the high transportability of energy crops on the other hand, as compared to liquid manure.

Among energy crops, mainly annual nonlegume crops are used. Corn silage (*Zea mays*, whole crop silage) is by far the most common energy crop used for biogas production (73% referring to mass content). Grass silage (different species, e.g., *Lolium perenne* or *Phleum pratense*) contributes to the total amount of energy crops

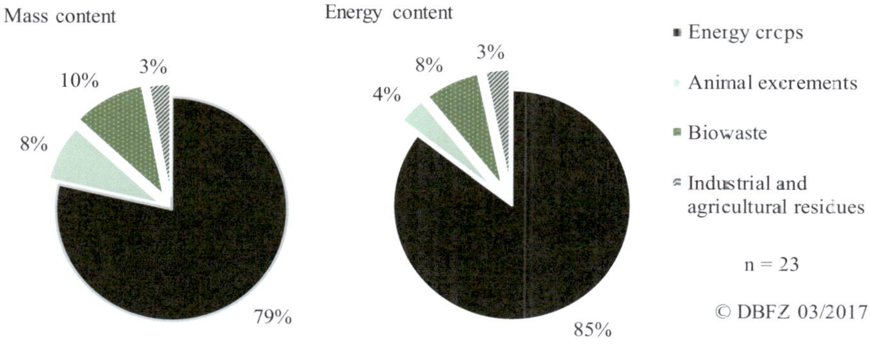

Fig. 5.5 Biomass resources in biomethane production units referring to mass content and energy content; based on DBFZ operators' survey 2016, reference year 2015; own diagram

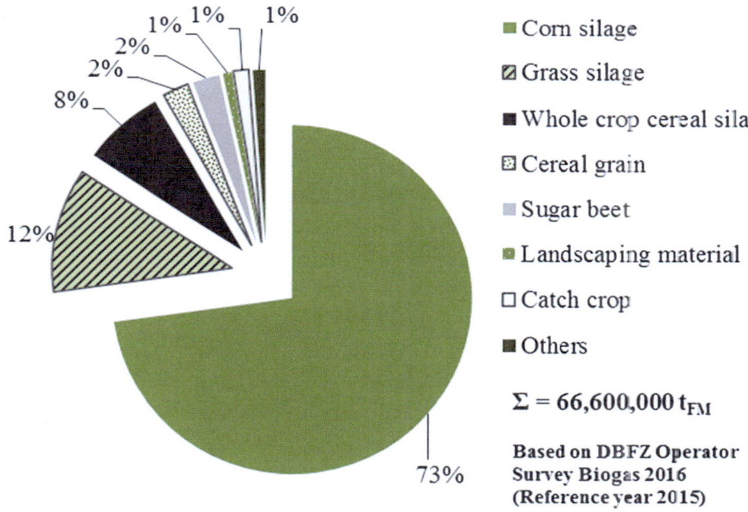

Fig. 5.6 Distribution of energy crops in biogas plants referring to mass content; based on DBFZ operators' survey 2016, reference year 2015; own diagram

used in biogas plants with a share of about 12%, whereas whole crop cereal silage (different species, e.g., *Secale cereale*; 8%), cereal grain (different species, e.g., *Secale cereal*; 1%), and sugar beet (*Beta vulgaris* ssp.; 2%) are of minor importance (see Fig. 5.6). Currently, legumes hardly play a role for biogas and biomethane production.

The energy main crop area for biogas and biomethane production was estimated according to the amount of electricity production based on biogas, by reference to the substrate distribution according to plant operators' surveys by DBFZ as well as by means of a middle methane yield per hectare. If 70% of the determined grass land is not directly cultivated for biogas production but rather results from the third/fourth grass cut (so-called autumn cut) and that land for catch crops and landscaping material can be excluded, the cultivation area for biogas and biomethane production accounts for about 1.5 million ha in 2015.

The following subchapters show the options and advantages of legumes as bioenergy crops as well as the advantages for a biogas use of legumes, which are cropped for soil fertility reasons in (organic) farming systems (Verma et al. 2015). Thus, there are positive effects on N-efficiency, on biological di-nitrogen fixation (BNF), on yields and qualities of crops and on environmental parameters, esp. on the reduction of direct and indirect greenhouse gas emissions and on nitrate leaching. Reasons for the present lack of utilization of legume energy crops for biogas production are the higher complexity and the more difficult economic conditions as discussed in Sect. 5.4 "Economical Aspects and Competitiveness of Legume Energy Crops."

The objectives of this chapter "Perennial and Intercrop Legumes as Energy Crops for Biogas Production" are to show (i) the options and economic restrictions of utilization of legumes for bioenergy and bioeconomy, (ii) the resulting chances and advantages for cropping systems, and (iii) the impact for future sustainable agricultural systems in post-fossil energy centuries.

5.2 Gas Yields of Different Forage Crops

According to the data of Keymer (2017), the degradability and methane yields of different forage legumes and legume mixtures were calculated and compared to typical nonlegume crops for biogas production. The results clearly show that heterogeneous biomass, such as green growths, is adequate for biogas production (see Table 5.1). Methane yields, which are related to the organic matter, depend on the degradability of the whole biomass, including cellulose, hemicellulose, sugars, fats, starch, and proteins, and on methane concentration as second factor rather than on particularly high contents of fats (like plant oil or biodiesel yields) or of sugars/starch (like first-generation ethanol). As related to the methane yield, the advantages of intensive nonlegume substrates are only based on the higher degradability of sugar and starch in comparison to lignocellulose. With 346 $m^3$of methane (CH_4) at standard temperature and pressure (STP; 0 °C; 1013.25 bar) per ton of volatile solids (VS), a mixture of peas *(Pisum sativum)* and vetch (*Vicia sativa*) in budding growth phase is in the range of corn silage (340 $m^3CH_{4,STP}*t_{VS}^{-1}$), the most common energy crop for biogas production in Germany (see above in the Introduction).

Even though the data suggest the advantage of high degradability and of correlated high specific methane yields, it should be mentioned that younger growths with higher degradability normally imply a higher number of cuts, resulting in relatively higher harvest costs (€$*t^{-1}$ of dry matter (DM)). Additionally, some perennial legumes, esp. red clover (*Trifolium pratense*) and alfalfa (*Medicago sativa*), need to reach the end of flowering development to improve their wintering stability for overcoming to the next year. This is largely because from flowering on, the reserves of the lower parts of the plant (esp. the roofs) are refilled.

Table 5.1 Degradability and methane yields of common nonlegume energy crops compared to legumes; modified data Keymer (2017); own table

Substrate	Degradability % degraded	Methane yield of organic matter $m^3CH_{4,STP}*t_{VS}^{-1}$	Methane yield of fresh matter $m^3CH_{4,STP}*t_{FM}^{-1}$
Common nonlegumes			
Corn silage	84	340	104
CCM (*Zea mays*, corn-cob-mix)	98	380	244
Sorghum silage (*Sorghum* sp.)	79	320	78
Whole crop cereal silage	79	330	101
Green rye silage (*Secale cereale*)	81	320	74
Sugar beet silage	96	360	78
Cereal seeds	97	380	322
Grass silage	79	320	101
Legumes and legume mixtures			
Clover grass silage (mix of different species, e.g., *Lolium perenne* or *Phleum pratense* + *Trifolium pratense* and/or *Trifolium repens*)	76	320	88
Clover/alfalfa silage (*Trifolium pratense* or *Trifolium repens* or *Medicago sativa* or a mixture of different species)	67	290	77
Egyptian clover (*Trifolium alexandrinum*), green, begin of flowering	69	306	33
Alfalfa grass (*Medicago sativa* mixed with a grass species, e.g., *Lolium perenne* or *Phleum pretense*), 1st cut, begin of flowering	73	310	47
Persian clover (*Trifolium resupinatum*), green, begin of flowering	71	308	31
Clover grass 1st cut, begin of budding	76	325	43
Clover grass 1st cut, flowering	70	289	45
Clover grass 2nd cut, begin of flowering	68	284	43
Clover grass 2nd cut, budding	71	302	43
Clover grass 2nd cut, before budding	74	319	42
Clover grass 3rd cut	71	302	43
Red clover, green, 1st cut, begin of flowering	68	288	52
Red clover, green, 1st cut, budding	72	308	47
Red clover, green, 2nd cut, begin of flowering	66	278	45
Red clover, green, 2nd cut, budding	69	296	42
Red clover, green, 2nd cut, before budding	72	316	39

(continued)

Table 5.1 (continued)

Substrate	Degradability % degraded	Methane yield of organic matter $m^3CH_{4,STP}*t_{VS}^{-1}$	Methane yield of fresh matter $m^3CH_{4,STP}*t_{FM}^{-1}$
Alfalfa, green, 1st cut, begin of flowering	67	287	52
Alfalfa, green, 1st cut, end of flowering	63	267	50
Alfalfa, green, 1st cut, budding	70	308	48
Alfalfa, green, 2nd cut, begin of flowering	66	286	52
Alfalfa, green, 2nd cut, end of flowering	61	266	48
Alfalfa, green, 2nd cut, budding	69	305	49
Alfalfa grass, 1st cut, begin of budding	75	323	43
Alfalfa grass, 1st cut, opening of buds	72	305	47
Alfalfa grass, 1st cut, flowering	68	284	51
Alfalfa grass, 2nd cut, begin of budding	72	314	44
Substrate	Degradability % degraded	Methane yield of organic matter $m^3CH_{4,STP}*t_{VS}^{-1}$	Methane yield of fresh matter $m^3CH_{4,STP}*t_{FM}^{-1}$
Alfalfa grass, 2nd cut, opening of buds	69	300	48
Alfalfa grass, 2nd cut, flowering	66	283	51
Forage pea, before mid of flowering	73	317	39
Field bean (*Vicia faba*), green	65	277	45
Legume-sunflower mix (*Helianthus annuus* mixed with a legume, e.g., *Phaseolus vulgaris*), green	71	300	46
Vetch-rye mixture, green	74	306	48
Pea-vetch mixture, begin of flowering	72	316	41
Pea-vetch mixture, end of flowering	69	293	47
Pea-vetch mixture, budding	77	346	36
Legume mixture, begin of flowering	71	319	39
Legume mixture, end of flowering	64	268	49
Legume mixture, budding	74	324	33
Sainfoin (*Onobrychis* sp.), 1st cut, mid of flowering	63	267	51
Sainfoin, 1st cut, before flowering	66	290	49
Sainfoin, 2nd cut, before flowering	68	292	48

FM fresh matter

The biogas yields are subject to the same interrelations than the like energy concentrations are in the area of animal feeding. Due to the lignifying processes and the stronger microfibril creation, the elder growths have a lower degradability, which results in lower methane yields as well as in lower energy concentrations (Datta et al. 2017). As an important aspect, it should be pointed out that energy concentration and quick degradability are relatively less important in the biogas process comparing to ruminant fodder as shown below. In comparing the quality, more detailed differences between biogas plants and animals must be taken into consideration concerning the influences of the growths on the usability in biogas or forage sector:

- Animals must maintain a high relative energy need – they need higher fodder qualities to reach adequate yields. Furthermore, (i) the uptake of fodder is under more strict limitation, and (ii) only the surplus of energy to the maintain need can be used for the generation of animal products.
- Animals normally have less than 24 h for the digestion process due to the limited capacity of the digestive system. In contrast, biogas plants generally have more than 40 days of retention time for digesting energy crops.
- For animals, there is a strong influence of taste on fodder uptake. Typically, silages with bad taste (butyric or even higher concentrations of acetic acid) result from growths with low DM and sugar concentration or strong buffer. Legume growths have a high buffering capacity due to proteins or cations.
- Proteins are only important for animal husbandry. A biogas plant has no need for special amino acids, and the need for N is much lower because the product gas is (nearly) free of N. The degraded protein N stays in the process. In opposite to this, animal product production needs proteins and special amino acids.
- The higher harvest costs of (younger, easier degradable) material with a higher number of cuts are given for biogas plants as well as for animals in stables. However, grazing husbandry systems can cut the intercorrelation of higher digestibility with higher harvesting costs.

5.3 Energy Potentials of Legume Mixtures in Organic Farming Systems

Within a multidisciplinary project (Möller et al. 2006a, b; Stinner 2011, Deuker 2013), among others, the attainable energy yields from digestion of clover grass and cover crops of two organic crop rotations were studied. The trials were installed in 2001 and ended in summer 2005. The results of the years 2003 and 2004 were considered for this paper due to installation effects in 2001 and 2002 and a not completed investigation in 2005. The field trials were carried out at the experimental farm of the University of Gießen "Gladbacherhof" (50° 23′ 44″ North, 08° 14′ 46″ East, temperature Ø 9.3 °C, precipitation Ø 670 mm).

5.3.1 Crop Rotations

5.3.1.1 Stockless Organic Farming

- Clover grass (seed amounts: *Trifolium pratense* at 7.5 kg*ha^{-1}, *Medicago sativa* at 7.5 kg*ha^{-1}, *Lolium perenne* at 5 kg*ha^{-1}, *Phleum pratense* at 5 kg*ha^{-1}, and *Festuca pratensis* at 5 kg*ha^{-1})
- Potatoes (*Solanum tuberosum*)
- Winter wheat (*Triticum aestivum*)

- Grain peas (*Pisum sativum*)
- Winter wheat
- Spring wheat (*Triticum aestivum*) with undersown clover grass

5.3.1.2 Organic Dairy Farming

Grassland on 30% of farmland area, arable land on 70% of farmland, stocking rate 0.8 livestock units per hectare farmland

- Clover grass (forage)
- Clover grass (forage)
- Winter wheat
- Potatoes
- Winter rye (*Secale cereale*)
- Grain peas
- Spelt (*Triticum spelta*)
- Spring wheat with undersown clover grass

Cover crops (common vetch: *Vicia sativa* at 90 kg*ha^{-1} and oil radish: *Raphanus sativus* at 5 kg*ha^{-1}) were cultivated after winter cereals and peas in both crop rotations.

5.3.2 Calculation of Attainable Energy Yields

The energy yield from digestion of clover grass and cover crops was calculated with measured crop yields of the years 2003/2004 and with the help of data from literature on the rate of methane (Deuker et al. 2008, 2011). Firstly, KTBL (Kuratorium für Technik und Bauwesen in der Landwirtschaft) values were used (KTBL 2005) as values for methane yields (the dataset of KTBL has been established in collaboration with leading German biogas researchers in order to determine a scientifically based calculation instrument to be used by enterprises and authorities). When no values were found for special substrates, the energy yield was estimated with values of similar materials (e.g., mean of different crucifer cover crops for oil radish). The energy yield is expressed in standard cubic meters of methane per hectare. One $m^3CH_{4,STP}$ is equivalent to about 1 l of diesel fuel. Silage losses were estimated to be 15% of DM (Thaysen 2012).

5.3.3 Substrates for Anaerobic Digestion

Stockless Organic Farming The total DM yields of clover grass (the sum of four cuts) varied between 14.5 and 15.8 t_{DM}*ha^{-1}. The cut of undersown clover grass following spring wheat in autumn reached 1.0 to 2.0 t_{DM}*ha^{-1}. The cover crop yields ranged from 1.8 to 3.1 DM*ha^{-1}. The corresponding VS yields referring to 1 hectare farmland are shown in Table 5.2.

Table 5.2 Energy yields of legume mixtures in different organic farming systems

		Yields			Silage losses	Methane potential			Energy yield		
		$VS*ha^{-1}$			%	$m^3CH_{4,STP}*t_{VS}^{-1}$			$m^3CH_{4,STP}ha^{-1}$		
Farming system	Substrate	From	To	Ø		From	To	Ø	From	To	Ø
Stockless organic	Clover grass	13.0	14.1	13.6	15	186	390	307	2055	4674	3549
	Intercrops	1.9	2.0	2.0	15	231	376	325	373	639	553
Dairy organic	Intercrops	2.0	2.1	2.1	15	231	376	325	393	671	580

Table 5.3 Substrate-related methane production price using cover crops: trial yields

Cost position	Price €*ha^{-1}
Seeding	128.70
Harvest and silage average [22.9 t_{FM}*ha^{-1}]	175.06
Feeding biogas plant [19.4 t_{FM}*ha^{-1} (15% silage losses)]	51.32
Spreading biogas slurry [18.0 t*ha^{-1} (1.5 t*ha^{-1} biogas produced)]	57.07
Total	412.15
Methane yield [m^{3}CH$_4$*ha^{-1}]	574
Substrate-related methane production price [€*m^{3}CH$_4$$^{-1}$]	0.72

Organic Dairy Farming Cover crop yields varied from 1.9 to 4.0 t_{DM}*ha^{-1}. The VS yields of all digestible by-products (t_{VS}*ha^{-1}) are shown in Table 5.2.

5.3.4 Energy Content of Substrates

The range of given data on the yields of methane is relatively large (Deuker et al. 2008; Keymer 2006), while different methods of determination were used (calculation, batch digestion trial, continuous digestion trial). Specific factors such as particle size, duration of digestion, and pretreatment (green material, silage) influence the methane yield (Table 5.2).

5.3.5 Energy Yields

The energy yield in the stockless organic rotation amounted to about 3550 m^{3}CH$_{4,STP}$*ha^{-1} for clover grass and to 550 m^{3*}CH$_{4,STP}$*ha^{-1} for intercrops, whereas in the organic dairy rotation, the clover grass has an energy yield of about 580 m^{3}CH$_{4,STP}$*ha^{-1} (see Table 5.2).

5.4 Economical Aspects and Competitiveness of Legume Energy Crops

5.4.1 Fertile Arable Sites

5.4.1.1 Agricultural Residues and By-Products

The calculated prices for methane production with the data from the organic farming biogas trial (Möller et al. 2006a, b) are shown in Table 5.3. General and specific assumptions used for the calculation are listed in Appendix A and B.

As a result, an average methane production price from intercrops is calculated at 0.72 €*m^{3}CH$_4$$^{-1}$. From the same trial, the clover grass yields were used to calculate the biogas potential and energy price. The results are shown in Table 5.4.

Table 5.4 Substrate-related methane production price using clover grass: trial results

Cost position	Price €*ha^{-1}
Seeding	260.74
Harvest and silage average [41.8 t_{FM}*ha^{-1}]	677.77
Feeding biogas plant [35.53 t_{FM}*ha^{-1} (15% silage losses)]	91.18
Spreading biogas slurry [26.9 t*ha^{-1} (8.7 t*ha^{-1} biogas produced)]	91.92
Total	1121.60
Methane yield [m^{3}CH$_4$*ha^{-1}]	3423
Substrate-related methane production price [€*m^{3}CH$_4$$^{-1}$]	0.33

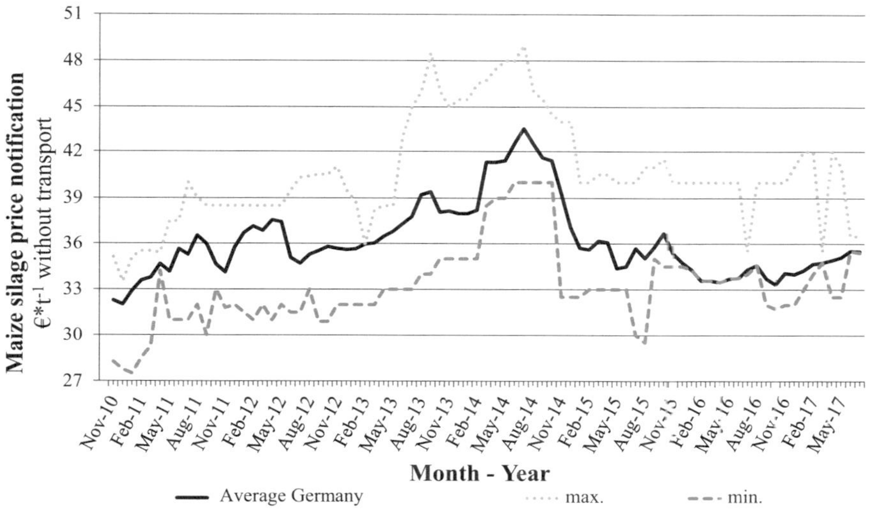

Fig. 5.7 Corn silage price notifications of the Brandenburger BauernZeitung (2010–2017), average for Germany, and minimum/maximum for selected regions (Schleswig-Holstein, Mecklenburg-West Pomerania, Lower Saxony East, Lower Saxony West, Saxony-Anhalt, Thuringia, Saxony); own diagram

5.4.1.2 Corn Silage

Due to the new developments in the agricultural market and the rise of energy (fuel) prices, corn silage prices tend to fluctuate a lot in the last years (Fig. 5.7). In spite of a shift of average German corn silage yields from 39 Mg*ha^{-1} in 2010 to 46 Mg*ha^{-1} in 2011 (DMK 2012a, b), in that period the price increased substantially. Production price components are detailed in Table 5.5.

Methane yields per ha of corn silage were calculated on the basis of corn silage yields according to Destatis, Limburg-Weilburg district, 2003–2004 (same region as project results).

Table 5.5 Substrate-related methane production price using maize, literature data

Cost position	Price €*ha^{-1}
Seeding	554.31
Harvest and silage average [47.9 t_{FM}*ha^{-1}]	324.71
Feeding biogas plant [42.2 t_{FM}*ha^{-1} (12% silage losses)]	108.17
Spreading biogas slurry [31.1 t*ha^{-1} (11.1 t*ha^{-1} gas produced)]	103.55
Total	1090.73
Methane yield [m^{3}CH$_4$*ha^{-1}]	4373
Substrate-related methane production price [€*m^{3}CH$_4$$^{-1}$]	0.25

5.4.2 Marginal Sites and Organic Cropping Systems

The above-shown results are the common sense, showing higher costs for legume energy crops compared to corn silage on fertile arable sites if the additional positive effects on following crops, saving of N-fertilizer, etc., are not included.

For organic cropping systems, dependent on the positive impact of legumes, the calculation is different if there is less need for legume growth for animal nutrition than for legume cropping for crop rotation issues (organic farming with low amount of roughage-consuming animals). In this case, the seeding costs belong to the cropping system, and the harvest costs must be reduced for the costs of mulching. Additional benefits from higher N-efficiency (Stinner 2011; Chap. 5; Verma et al. 2015) may be included but are waived in the following conservative calculation (Table 5.6).

The result of 0.21 €*m^{3}CH$_4$$^{-1}$ shows a clear competitiveness of legumes as energy crops in organic farming when compared with conventional corn silage. Even though this calculation is simplified and disregards the above named advantages of enhanced N-efficiency on the one hand and the higher effort for plant operation, esp. for stirring plus the lower organic loading rate (due to the higher ammonia contents) on the other hand, this shows feasible options, which are realized in some cases (e.g., joint biogas plant of a group of organic farms in Hallerndorf).

Marginal sites are characterized by different constraints. For this reason, to compare corn silage and fodder-legume silage for biogas production, the following calculations are just rough examples and shall show the need for individual calculation in such cases.

On marginal sites, the yield level can be reduced due to the limitation of vegetation period, water, and temperature level. Furthermore, there can be risks of failure by drought, for corn additionally by spring frost, spring cold (which enhances the risk of strong weed competition), and wild boars. Both aspects influence the relative preference. Table 5.7 shows the influence of a 30% yield reduction of corn silage by marginal conditions, and Table 5.8 shows the additional effect of a 50% failure risk once in 5 years.

Table 5.6 Substrate-related methane production price using clover grass in organic system with low stocking rate

Cost position	Price €*ha^{-1}
Harvest and silage average [41.8 t_{FM}*ha^{-1}]	527.77
Feeding biogas plant [35.53 t_{FM}*ha^{-1} (15% silage losses)]	91.18
Spreading biogas slurry [26.9 t*ha^{-1} (8.7 t*ha^{-1} biogas produced)]	91.92
Total	710.87
Methane yield [m^{3}CH$_4$*ha^{-1}]	3423
Substrate-related methane production price in organic system with low stocking rate [€*m^{3}CH$_4$$^{-1}$]	0.21

Table 5.7 Substrate-related methane production price for corn silage on marginal site (30% lower yield than on fertile site)

Cost position	Price €*ha^{-1}
Seeding	554.31
Harvest and silage average [33.5 t_{FM}*ha^{-1}]	250.00
Feeding biogas plant [29.5 t_{FM}*ha^{-1} (12% silage losses)]	75.71
Spreading biogas slurry [21.8 t*ha^{-1} (7.7 t*ha^{-1} gas produced)]	72.49
Total	951.51
Methane yield [m^{3}CH$_4$*ha^{-1}]	3061
Substrate-related methane production price [€*m^{3}CH$_4$$^{-1}$]	0.31

Table 5.8 Substrate-related methane production price for corn silage on marginal site (30% lower yield than on fertile site) and failure risk of 50% once in 5 years, costs in failure year

Cost position	Price €*ha^{-1}
Seeding	554.31
Harvest and silage average [16.8 t_{FM}*ha^{-1}]	180.00
Feeding biogas plant [15 t_{FM}*ha^{-1} (12% silage losses)]	37.86
Spreading biogas slurry [10.9 t*ha^{-1} (3.9 t*ha^{-1} gas produced in previous year)][a]	72.49
Total	844.66
Methane yield [m^{3}CH$_4$*ha^{-1}]	1531
Substrate-related methane production price in failure year [€*m^{3}CH$_4$$^{-1}$]	0.55
Substrate-related methane production price in average [€*m^{3}CH4^{-1}][b]	0.36

[a]Fertilization due to expected yields = digestate from previous year; over-expected yield in failure year means over-calculated fertilization, resulting in nutrient losses. For the following year, this means less digestate fertilizer than needed for expected yield; costs for additional need of external fertilizer are not included

[b]Additional costs for substrate organization in year of failure are not included

Table 5.9 Substrate-related methane production price of clover grass under marginal conditions (30% yield reduction)

Cost position	Price €*ha^{-1}
Seeding	260.74
Harvest and silage average [29.3 t_{FM}*ha^{-1}]	474.44
Feeding biogas plant [24.87 t_{FM}*ha^{-1} (15% silage losses)]	63.83
Spreading biogas slurry [18.8 t*ha^{-1} (6.1 t*ha^{-1} biogas produced)]	62.51
Total	861.52
Methane yield [m^{3}CH$_4$*ha^{-1}]	2396
Substrate-related methane production price [€*m^{3}CH$_4$$^{-1}$]	0.36

Forage legume costs show more resilience against marginal conditions, e.g., smaller yields due to drought or a short vegetation period imply a reduction of the number of cuts (e.g., two or three cuts instead of three or four cuts). Additionally, the fix costs of seeding and field preparation are lower, especially when established as nurse crop

For three main reasons, corn silage as energy crop has higher cost risks on marginal lands than forage legume mixtures do. (i) As corn silage has relatively high costs for seeding including field preparation, a reduction of yield implies a strong enhancement of substrate-related costs. (ii) There is only one harvest, which is the main impact of good cost efficiency under good conditions. When yields are low, cost savings for harvest are under-proportionate. (iii) Corn has a higher risk for failure, esp. by spring cold and spring frost. The latter does not play a role for legume forage mixtures.

As shown in Table 5.9, the substrate-related methane costs of forage legumes are only a little bit higher on marginal sites than on fertile sites. They become equal to the costs of corn silage if there is a risk of failure every 5 years. In calculations for energy crops in Eastern Europe (Russia, Ukraine, Belarus), even economic advantages of perennial forage legumes (established as underseed) against annual nonlegume energy crops were found (Thrän & Pfeiffer 2012).

If the seeding costs are covered by the cropping system for having the advantage of forage legumes, there is a further reduction of 0.12 €*m^{3}CH$_4$$^{-1}$, resulting in a strong economic preferability compared to corn silage under such conditions.

5.5 Legume Energy Crop Effects in Stockless Organic Farming Systems

In organic farming, clover/grass ley provides N to the whole cropping system via symbiotic BNF, and furthermore, it controls certain weeds. When ruminants are not present, a common practice in organic farming is to leave the biomass from clover/grass ley in the field for their residual fertility effect. In a field experiment, a whole cropping system was implemented with a typical crop rotation for such farming systems, and it was evaluated whether the use of N could be improved by harvesting and processing biomass from clover/grass ley biomass, crop residues, and cover crops in a biogas digester and using the effluents as a fertilizer, as compared to common practice (Stinner et al. 2008; Dhakal et al. 2016).

The results of the field experiments indicate that digestion of clover/grass ley, crop residues, and cover crops can increase the crop DM and N-yields as well as the N-content of wheat grains in organic stockless systems. Harvesting and digestion of residues and their reallocation after digestion resulted in a better and a more even allocation of N within the whole crop rotation, in a higher N-input via BNF and in lower N-losses due to emissions. In comparison, the BNF was reduced by about 20% when the aboveground biomass was left in the field for mulching (Stinner et al. 2008). As compared to animal husbandry, by digesting, the N remains almost completely within the system in case of storing the effluents in a closed system which leads to a high intensity in nutrient cycling. Moreover, it was pointed out that harvesting and digesting of this biomass probably results in a higher N-availability of digested manures in comparison to the same amounts of undigested biomass (Stinner et al. 2008). In conclusion, clover/grass ley biomass, crop residues, and cover crops represent a large unexploited energy potential and could be used by anaerobic digestion to produce biogas and, moreover, to establish and improve more efficient organic stockless systems.

5.6 Legume Intercropping for Biogas Production in Sugarcane Production Systems

According to FAO (2013), sugarcane (*Saccharum officinarum*) is the most produced agricultural commodity in the world with an annual production of $1.8*10^9 t_{FM}*a^{-1}$, followed by other grasses and legumes ($1.1*10^9 t_{FM}*a^{-1}$), maize (*Zea mays*; $1.0*10^9 t_{FM}*a^{-1}$), and rice (*Oryza sativa*; $0.7*10^9 t_{FM}*a^{-1}$). It is cultivated as a perennial crop with 4–6 years of cultivation time, mostly in tropical and subtropical regions of Brazil, India, China, and Thailand. Sugarcane is mainly used for sugar, bioethanol, and electricity production. Typically, it is planted again after clearing for several times, creating decades of mono-cropping in large mono-cropping landscapes.

Despite the advantages of intercropping during sugarcane cultivation, such as increase in soil fertility, N-fixation, protection against erosion, and improvement of biological diversity (also reducing pests by reducing infection chains or enhancement of anti-phytopathogenic potential), the intercropping is not considered a common practice during sugarcane cultivation due to the additional effort without direct economic use of the intercrop (Suman et al. 2006; Li et al. 2013; Lithourgidis et al. 2011).

However, the anaerobic digestion of sugarcane residues opens the possibility of co-digestion with legume-based intercrop. The large amounts of organic residues of cane processing (i.e., vinasse, filter cake, or parts of bagasse and straw) allow the construction of biogas plants up to the size of multimegawatts (Janke 2017). In this case, it may open a feasible way to integrate the cropping of different legumes in different forms of cropping periods into the sugarcane cultivation systems. For this option, combining the use of several sugarcane residues with the use of legume intercrops in one process, the uniquely feedstock flexibility of the biogas process is a basic requirement.

Additionally, anaerobic digestion of sugarcane residues co-digested with legumes would provide advantages both for the process kinetics and digestate quality. Depending on the harvest time, legumes have a typical N-content of about 3–4% DM, resulting in a C/N ratio of between 10 and 15. This high N-content can balance the low N-content of sugarcane straw (C/N ratio of 83) and bagasse (C/N ratio of 116). For the other residues (i.e., vinasse and filter cake), co-digestion with legumes would be an important source of alkalinity for a more stable anaerobic digestion process (Janke et al. 2016; Yadav et al. 2017).

5.7 Optimizing of Nitrogen Effects (N-Harvest, N-Losses, N-Utilization in Crop Rotation)

The data on system effects of biogas fermentation in organic cropping systems that were obtained in the two-field trials between 2002 and 2005 were evaluated concerning nutrient uptake, nutrient cycle, and N-losses in five different manuring systems such as "stable manure" or "fermentation of slurry and crop residues including external substrates" (for more information, see Möller et al. 2006a, b; Meena et al. 2015a, b).

In both systems, with and without livestock, no differences in yields and nutrient uptake of the legume crops were observed (Moller et al. 2006a, b).

Within the livestock system, the nonlegume crops of the slurry system had no higher yields than those of the stable manure system, but their N-uptake was 6% higher. By including crop residues in the fermentation process, the N-uptake within the system increased by about 10%, and simultaneously, the nitrate leaching potential was reduced by 10%. An additional 10% increase in N-yield was achieved by adding external substrates. However, they did not significantly increase DM yields.

In the stockless system, the fermentation did not affect the yields and N-uptake of legumes, potatoes, and summer wheat. As opposed to this, a significant increase of 10% in yields and N-content of winter wheat was measured. Moreover, the nitrate leaching potential was significantly decreased by 10%, and the N_2O emissions could be even reduced by 40%. The addition of external substrates did not lead to further increases in yields or N-uptake.

In conclusion, the comparison of the systems with and without fermentation of slurry did not show effects on yields, nitrate leaching, and N-uptake. Adding crop residues led to a decreased nitrate leaching potential as well as to an increased N-uptake, whereas the inclusion of external substrates did not affect the N-uptake.

5.8 Reduction of Nitrous Oxide Emissions from Legume Growths in Stockless Cropping Systems by Biogas Use

Nitrogen is a major plant nutrient required for high yields of agricultural crops and is therefore very important concerning its efficient utilization, especially as fertilizer in organic farming systems. An inadequate synchrony of crop N-demand and soil N-supply leads to N-losses from agricultural systems and can thereby pose a threat to sustainability (Möller and Stinner 2009; Dhakal et al. 2016). Furthermore, agriculture soils are main sources for N_2O emissions (Mosier 2001) and accounted for nearly 70% of the anthropogenic N_2O emissions in Germany (Statistisches Bundesamt 2005; UBA 2017). These emissions increase after amending soils with organic manures having a narrow C/N ratio (Möller and Stinner 2009). The prerequisite for N_2O formation is the availability of reactive N as ammonium or nitrate. Nitrification of ammonium in the presence of oxygen or the denitrification of nitrate in an anoxic environment could trigger N_2O emissions. Therefore, anaerobic digestion of manure in biogas plants impedes nitrification of ammonia and enables to reduce these emissions during manure storage. Since oxygen is absent in the headspace of the plant's gastight cover, N_2O emissions will be eliminated if the manure is digested (Oenema et al. 2005). With regard to this, investigations on whether different manuring options with and without digestion of residues (field residues, stable wastes) affect the gaseous N-losses showed that emissions are much higher when the legume clover/grass ley is cut, chopped, and left on the field instead of harvesting and removing it from the aboveground biomass (Möller and Stinner 2009; Meena et al. 2015a, b). The N_2O emissions increased significantly in organic manuring when liquid effluents of the biogas digester are applied, when green manures with a narrow C/N ratio (like cover crops) are incorporated, and when the aboveground biomass of clover/grass ley is left on the field to decompose (mulching). In contrast to mulching and incorporating biomass as green manure, the N_2O emissions decreased by harvesting as well as by digesting crop residues and clover/grass ley during winter. The subsequent reallocation of the effluents within the same crop rotation, on the other hand, increased emissions after addition to the soil. However, considering the balance of the whole arable organic cropping system, N_2O emissions could be significantly decreased by 38%. In conclusion, anaerobic digestion of field residues, especially of legumes such as clover/grass ley, in biogas plants has high potential to lower the N_2O emissions of agricultural soils (Möller and Stinner 2009).

5.9 Effect of Perennial and Intercrop Legume Mixtures on Soil Organic Matter Compared to Nonlegume Energy Crops

With around 10% of the cropping area in Germany for biogas purposes (see Sect. 5.1 "Introduction"), and a further worldwide use of crops for biofuel production like ethanol, biodiesel, and plant oil, energy crops are an important part of the cropping systems. Due to the substrate flexibility of the biogas technology, this opens new possibilities for higher amounts of perennial and intercrop legumes in the cropping systems.

There is comprehensive evidence in the literature that the inclusion of legumes into crop rotations results in higher levels of soil organic matter (SOM) compared to rotations without legumes (Gregorich et al. 2001; Rauhe 1969; Schjønning et al. 2007; Ram and Meena 2014). The positive impact of legumes on SOM is driven by the complex impact of high C-input through high amounts of plant residues remaining in soils, BNF, provision of a source of N for the retention of substrate C, and the absence of disturbance of the soil under the crop. The BNF means an N-supply to the crops, e.g., from the mineralization of SOM or from mobile fertilizers generated from the growths, is substituting for external N-sources. In a system-related perspective, the inclusion of legumes into crop rotations further decreases the area of nonlegume crops and provides fodder for farm animals (or feedstock for digestion), which in turn produces a high-quality manure with regard to SOM buildup (Leithold et al. 2015).

Published results consistently show that perennial fodder legumes (and their mixtures with grasses) leave higher amounts of residues in the soil than all annual crops. In their meta-analysis of data from Canada, Bolinder et al. (2007) calculated an aboveground biomass/root ratio for fodder legumes that was less than half the one for most annual crops in their study. Root amounts were, thus, higher with fodder legumes than with annual crops at the same amount of aboveground biomass. This is in line with results from long-term field experiments in the former German Democratic Republic, as reported by Michel (1988) and Körschens (1982). However, C-input with grain legumes was reported to be considerably lower than with perennial fodder legumes (Bolinder et al. 2007). In their study, Plaza-Bonilla et al. (2016) conclude that crop rotations with grain legumes but without perennial fodder legumes need the inclusion of catch crops to compensate for organic matter losses. Schulz et al. (2015) even found decreasing SOM levels under an organic crop rotation with grain legumes and catch crops in a long-term field experiment in Germany, while a competing treatment with 33% perennial fodder legumes maintained SOM levels in the observation period. The cropping of catch crops naturally increases organic matter and C-input to soils. Therefore, a regular inclusion of catch crops into crop rotations increases SOM levels compared to rotations without catch crops on a global average, as shown by Poeplau and Don (2015) in a meta-analysis on that issue. The authors included the "crop functional type" – legume or nonlegume – as a factor in their evaluation and did not find a significant impact. The reason for this observation is likely to be that the organic matter input to soils must not differ between catch crops and crop mixtures with and without legumes. Furthermore, it was shown that organic matter input is the main driver for SOM level changes at a site (e.g., Kong et al. 2005). However, buildup of SOM is not only dependent on C but also on the availability of N for the soil microbes (Kirkby et al. 2013; Meena et al. 2014a, b). From a stoichiometric point of view, the C/N ratio of a soil indicates the share of C from plant residues (and other organic inputs) that can be retained in the soil in the long run, and a higher N-availability to microbes increases the amount of potential C retained in the soil (Schimel and Weintraub 2003). This process adds to the explanation of the observed positive impact of legumes on SOM. While nonlegumes withdraw N from the soil, legumes add N to the soil via BNF. The N-input through BNF can result in a net N-gain in the soil even in harvested stands. Mineral fertilization of nonlegumes does not seem to decrease the uptake of soil N by the plants considerably but leads to a higher overall N-uptake (Stevens et al. 2005). Further, excess mineral N is more easily lost from the system than organically bound N due to the higher mobility.

The combined effect of organic C- and N-inputs can even be used to explain observed differences between perennial fodder legumes, grain legumes, and catch crops with regard to SOM formation. Total organic C- and N-inputs with perennial fodder legumes usually are much higher than those with all annual legume and non-legume crops. The consistently observed positive impact of perennial fodder legumes on SOM levels can therefore be assumed to result from this situation. Grain legumes provide much lower organic C- and N-inputs. Due to the comparably high share of total plant N in the aboveground biomass, the harvest of grain legumes may even result in a net N-loss, especially if the whole plant is exported including the straw. Catch crops are not harvested, and C- and N-inputs with legume and nonlegume crops must not differ between crop groups. Further, both groups provide an N-input into the soil, as nonlegume catch crops usually receive N-fertilizer or take up excess N remaining in the soil solution after the preceding crops, while legumes benefit from BNF. The impact of fertilizer N thus could explain the missing differentiation between legume and nonlegume catch crops in their impact on SOM levels reported by Poeplau and Don (2015) and Meena et al. (2014a, b).

Apart from C- and N-inputs, the absence of soil disturbance under perennial legumes might contribute to SOM formation. According to Balesdent et al. (2003), the effect of tillage events on SOM is mainly due to the disruption of soil structure, which makes organic matter available for microbes that has been physically protected before the perturbation.

However, legumes do not play an important role as energy crops (besides plant oil from soybean (*Glycine max*)). Especially for biogas use, this is due to economic reasons and additionally due to the intensive biogas production in regions, which are overloaded with nutrients from intensive animal husbandry, which is reasoned by the fact that animal manure is a very good substrate and the addition of energy crops allows better economic results by allowing the advantage to use the economy of scales due to the transportability of energy crops (see Fig. 5 3).

Concluding, it shall be stressed that the inclusion of legumes in crop rotations as a feedstock for biogas production is a viable option for the sustainable management of SOM. The contribution to SOM formation is one important ecosystem service provided by legumes, adding to the value of these crops in rotations.

5.10 Greenhouse Gas Emissions Reduction by Legumes Including the N-Harvest Effect

As shown in the Introduction, energy crops are an important source for biogas production in Germany. Nonlegume crops like corn silage, grass silage, whole crop grain silage, and other nonlegume crops are the most significant ones for this purpose, whereas legume crops are of rather little importance. The reasons therefore are high yields per ha (Vetter et al. 2014), especially high yields per single harvest as there is typically only one harvest per year. However, legume forage mixtures must be cut three to five times a year to obtain the same or a slightly less DM yield. The high yield and the low effort imply high net energy yields. Consequently, this enables a high potential for mitigation of greenhouse gas emissions in the course of replacing fossil fuels.

However, the BNF of legumes allows an additional yield. In organic farms, it is the basic input; in conventional farms, it can compensate inorganic N-fertilizers if the digestate is used as fertilizer for the nonlegume cash crops. Since the production of chemical N-fertilizers is highly energy intensive and emits greenhouse gases caused by fossil energy consumption as well as by N_2O generation, these emissions could be avoided by producing an effective organic fertilizer with the N gained by BNF. Accordingly, this production can be a further add-on effect for the reduction of greenhouse gas emissions.

The obtained data on the BNF of legume crops as biogas substrates were recalculated and compared to data on the process of producing mineral N-fertilizer concerning energy savings and greenhouse gases (Stinner 2015; Dhakal et al. 2016). Besides an additional possible methane yield that can be obtained, the results also show a significant potential for energy saving and mitigation of greenhouse gases by substituting mineral fertilizers by digestate. Possible net savings in nonrenewable energies and greenhouse gas emissions of 5.28 kg of carbon dioxide equivalents (CO_2eq.) per kg of replaced mineral fertilizer were determined (Stinner 2015). The value is based on the difference of the total greenhouse gases caused by mineral fertilizer (9.01 kgCO_2eq.) and of those caused by the replacement of 1 kg inorganic N by digestate (3.73 kgCO_2eq.). Furthermore, when mineral fertilizer is replaced with BNF in a conventional system, the utilization of clover grass leys as substrate can lead to a reduction of primary energy consumption by more than 6.4 MWh*ha^{-1} and to a mitigation of greenhouse gas emissions of at least 2 tCO_2eq.*ha^{-1} per year. Using legume intercrop mixtures enables a yearly energy reduction of 0.84 MWh*ha^{-1} and emission saving of 200 kgCO_2eq.*ha^{-1} (Stinner 2015).

In conclusion, the results show a strong prior underestimation of legumes as energy crops. The effect of BNF should be included in their balance evaluations concerning possible energy yields and greenhouse gas savings. The amount is remarkable if mineral N is replaced by BNF of legume crops for biogas production in case of using the digestate as fertilizer for the nonlegume crops.

5.11 Anaerobic Digestion of Coumarin-Rich Perennial Legumes like *Melilotus* spp.

Sweet clover (*Melilotus* spp.) is an upright and tall-growing legume native to Europe and Asia (Turkington et al. 1978). It is a valuable legume as it fixes more N than other legumes such as alfalfa and red as well as white clover (*Trifolium repens*; Goplen 1980). Sweet clover is modest regarding growth conditions and, once established, tolerates wet as well as dry conditions. It can be cultivated on a variety of clay and loam soils even if sandy as long as soil pH is 6.5 or higher (Smith and Gorz 1965). Sweet clover forms deep tap roots making it highly valuable for soil improvement. Crops, which formerly could not be cultivated due to low soil quality, can be established after sweet clover cultivation (Smith and Gorz 1965).

Besides the usage for improving degraded soils, sweet clover is a valuable forage crop. However, the utilization as pasture crop is controversial as there are conflicting reports about the capability of cattle to adapt to the bitter taste of sweet clover (as summarized in Bull 2013) which is due to the plant secondary metabolite coumarin (Smith and Brink 1938). In contrast to grazing, sweet clover biomass has a good hay and silage quality (Smith and Gorz 1965). However, a dimeric form of coumarin, dicoumarol, might be formed in spoiled silage causing the so-called sweet clover disease (Roderick 1931). Dicoumarol causes fatal internal bleedings due to anticoagulating properties (Overman et al. 1942).

Since the usage as forage crop is limited, energetic utilization in terms of biogas production from sweet clover biomass gained recent attention (Bull 2013; Popp et al. 2015; Meena et al. 2017). On sandy soils, sweet clover is beneficial in crop rotations with maize, a common biogas substrate, if it precedes the maize cultivation. Sweet clover allows high maize yields without a need for N-fertilization (Bull 2013). When sweet clover is seeded in spring and harvested in autumn of the first year, sweet clover can be harvested again before cultivation of maize as catch crop, or sweet clover can be used as manure crop for cultivating maize as main crop in the following year.

Ensuing from such a crop rotation with maize, also the sweet clover biomass might be used as biogas substrate. However, as coumarin adversely affects the activity of ruminal microorganisms (Cansunar et al. 1990; Moniello et al. 1996), the anaerobic digestion process might be impaired. In continuous lab-scale anaerobic digestion experiments, an inhibitory effect of coumarin was evident by a pronounced decrease of the biogas yield and a substantial increase of volatile fatty acid concentrations (Popp et al. 2015). Concentrations of volatile fatty acids are used as indicator of the process stability as they are central intermediates of the anaerobic digestion process. Acid concentrations reached critical levels due to coumarin, which presumably would result in complete process breakdown in full-scale systems. However, in the lab-scale systems, the biogas process was able to adapt to coumarin, which was correlated to reorganization of the bacterial community composition and to the anaerobic degradation of coumarin during biogas production. In contrast to the bacterial community, the methanogenic community composition remained stable. As a consequence of the adaptation, the biogas production resumed like before the inhibition (Popp et al. 2015). Hence, coumarin acts as inhibitor of the anaerobic digestion process leading to transient process instabilities.

In general, coumarin-rich substrates like sweet clover should be utilized for biogas production with great care. In order to avoid any process inhibition, the microbial communities should first be adapted to coumarin by feeding small amounts of coumarin-rich substrates. Once the microbial community is adapted, coumarin-rich substrates can be anaerobically digested in high dosage without compromising the process performance. With low-coumarin cultivars, process instabilities might be prevented, but as they yield less biomass (Goplen 1971, 1981), they are not favorable for biogas production in terms of economic efficiency.

5.12 Options for Legume Energy Crops in Conventional Farming Systems

As shown by Stinner and Rensberg (2011), forage legumes (and other fodder crops like fodder beets) have strongly decreased in the recent decades due to the strongly enhanced competitiveness of corn silage. Section 5.2 "Gas Yields of Different Forage Crops" shows the relative easier utilization of fodder crops with lower energy content such as legume forage mixture growths as biogas substrate than as fodder for animals. This means biogas production generally allows a higher degree of freedom for cropping systems than other forms of utilization of agricultural products. This implies special chances for forage legume mixtures to become again important parts of the cropping systems.

The potential of the combination of (legume) intercrops and the use of field residues and of forage legumes as main crops should be high enough to exchange nonlegume energy crops, as derived from Sect. 5.3 "Energy Potentials of Legume Mixtures in Organic Farming Systems." The exchange of nonlegume energy crops for biogas generation in Germany (which can be transferred to other regions) opens up a big opportunity to enhance the sustainability of the whole farming systems (Meena et al. 2015a, b).

As there are currently about 1.5 million ha of energy crops for biogas production in Germany in 2016 (FNR 2016; Chap. 2), there is a potential of around 12% of the arable land, which can be transferred to forage legumes. If the crops for liquid biofuels, esp. biodiesel and bioethanol, are also replaced by biomethane (FNR 2017), there are 2.41 million ha (> 20% of arable land) of energy crops, which are in principle available for forage legume cropping in Germany.

A prerequisite for higher shares of legume energy crops is the economic competitiveness. As shown in Sect. 5.4 "Economical Aspects and Competitiveness of Legume Energy Crops," economic competitiveness of forage legumes compared to the usual energy crop corn silage (see Sect. 5.1 "Introduction") for biogas production can be provided on marginal sites and under organic farming systems with low stocking rates. If forage legumes as energy crops shall be established in a bigger extent to become a key element of more sustainable but high productive future farming systems, there is a need for supporting measures.

This support can be provided both on the basis of energy policy and on that of agricultural policy. In the area of energy policy, the German Renewable Energy Sources Act seems to be a good measure. Due to the numbers shown in Sect. 5.4 "Economical Aspects and Competitiveness of Legume Energy Crops," 0.08 €*$m^3CH_4^{-1}$, equivalent to around 0.02 €*kWh^{-1} of electric energy, would cover the cost difference on typical conventional farms on fertile sites. As a side effect, this would be a support for organic agriculture, which is strongly dependent on forage legume cropping for soil fertility as well as for pest and weed management reasons.

In the area of agriculture policy, a support of ca. 280 €*ha^{-1} of forage legume cropping or 56 €*ha^{-1} of arable land with a minimum of 20% of forage legumes with adequate utilization would equalize the costs to corn silage energy cropping. This can be done as a measure for supporting organic agriculture as well as other forms of sustainable cropping.

5.13 Conclusion

The shown aspects indicate the potential of the combination of legumes, especially forage and intercrop legumes, with its utilization for biogas production. As a conclusion, biogas utilization of legumes gives the following options:

1. It allows a higher grade of freedom as compared with fodder use. For example, the biogas process can be adapted to toxic substances like coumarin, allowing the use of legume species with special advantages for cropping systems. The longer retention time in biogas plants and the lower own energy demand compared to ruminants allow lower energy concentrations (yield reduction effect is lower than in ruminant husbandry).
2. The farm-own energy need for cropping measures (ca. 200–300 l of diesel fuel per ha in average) can be covered by a small percentage of main crop forage legumes, as 1 ha of forage legumes generates an amount of energy equivalent to nearly 3500 l of diesel fuel. Methane-operated tractors are not common, but available.
3. The German example of energy cropping (currently mostly nonlegumes) shows the possibility of up to 20% of forage legume integration for bioenergy production into the cropping systems without restrictions for food and feed production.
4. For organic cropping systems, especially under stockless conditions or with low stocking rates, the anaerobic digestion of forage legumes allows not only the generation of farm-own energy for several purposes but especially higher N-efficiency combined with reduced greenhouse gas emissions and reduced nitrate leaching risk. The higher N-efficiency results in higher yields and higher raw protein contents of nonlegume crops in crop rotations.

5. However, the use of forage legumes currently does not play a decisive role for bioenergy production. The reasons are mainly the costs, which are around 0.08 € higher per m^3 of methane or 0.02 € higher per kWh of electricity if forage legumes are cropped instead of corn silage on fertile sites under conventional conditions. On marginal sites or under organic farming conditions, competitiveness between forage legumes and corn silage can be given.
6. If forage and intercrop legume cropping shall gain importance due to a more sustainable agriculture, the biogas utilization option will allow the combination of sustainability and productivity. Certainly, the establishment of such a system needs support. This support can be organized in the area of energy policy, in the area of agricultural policy, or in a combination of both. In energy policy, the option is a higher feed-in tariff for electricity or methane based on legumes, maybe also for other substrates with special high sustainability advantages according to the German Renewable Energy Sources Act of 2014. In agriculture policy, support for organic agriculture or for forage and intercrop legumes integration with adequate utilization would enable the use for fodder and biogas issues.
7. A support of ca. 280 €*ha^{-1} of forage legume cropping or 56 €*ha^{-1} of arable land with a minimum of 20% of forage legumes with adequate utilization would equalize the costs to corn silage energy cropping. This can be done as a measure for supporting organic agriculture as well as for other forms of sustainable cropping.

5.14 Future Outlook

The agriculture system of the future has to be resilient, productive, free of fossil C-need, and environment-friendly. Therefore, the cropping systems need to work with lower input, using synergy effects to combine resilience, long-term soil fertility, environmental friendliness, and high productivity.

Cropping of different perennial and intercrop forage legume mixtures is a productive brick for these issues on the cropping side due to their role for N-input, for soil fertility enhancement (humus reproduction, erosion avoiding, soil regeneration period), for weed and pest control, as well as for biodiversity protection (fodder base and habitat for lots of different species).

Adapted green biorefineries with biogas technique as central element may become the counterpart of this cropping element. The value creation of such utilization can allow the cropping in an extent, which is necessary for the sustainability of the cropping system. Such biorefinery can provide important products for arable, horticulture, and grassland cropping systems. The first product is methane as fuel with high energy density in compressed form. It can be used to operate piston engines of farm machines as a currently available, but not common technique. In the future, it can operate fuel cells with factor 1.5–2 enhanced efficiency and with the opening of the innovation window of electric-operated farm machines. Especially when combined with autonomous operation, this enables strongly reduced weight of farm machines, which is necessary to reduce soil compaction. As further products for cropping issues, plant strengthening/plant protection agents (e.g. capronic acid, fermentation product from carboxylate platform); new fertilizers (e.g. organic CULTAN-able N or N-phosphate fertilizers), allowing higher nutrient efficiencies and reduced environmental emissions; and peat-free substrates for horticulture (based on digestate fibers or terra preta substrate mixtures from digestate and charcoal, ideally after joint composting) may be developed from such biorefinery, leading to sustainable cropping.

For the further bioeconomy, the legume-based green biorefineries can deliver energy, colors (e.g., based on chlorophyll and carotins), vitamins, proteins, fibers, fatty acids, etc. In this way, forage legumes can become (again) key elements of sustainable economies.

Appendices

Appendix A: General Assumptions for Calculation of Production Prices

Field acreage 5 ha, distance field to farm 5 km according to Vetter (2009).

Labor prices are set with the scheme from KTBL Betriebsplanung 2010/2011 (2010) for permanent employees with maximum ancillary labor costs (49%) to 14.91 €.

Biogas mass is assumed at 1318 kg*m^{-3}.

Specific methane yield was calculated with the dataset from Deuker et al. (2008).

Appendix B: Assumptions for Calculation of Production Prices from Intercrops

Seeding: tractor 67 kW, working width 6 m, 108 kg*ha^{-1}, seed cost 100 €*ha^{-1} (Schäfer 2012).

Whole crop silage yield and silage conservation:

Yield: 2 × 6.2 m, 250 kW harvester
Transport: trailer 50 m^3, 18 t; 160 kW
Compacting: wheel loader, 13.5 t, 105 kW; light goods bucket, 4 m^3
Feeding biogas plant: wheel loader 67 kW cutting bucket 2 m^3
Spreading digestate: slurry tanker with trailing distribution boom 24 m^3; 12 m; 160 kW

References

Balesdent J, Chenu C, Balabane M (2003) Relationship of soil organic matter dynamics to physical protection and tillage. Soil Till Res 53:215–230

Bolinder MA, Janzen HH, Gregorich EG, Angers DA, VandenBygaart AJ (2007) An approach for estimating net primary productivity and annual carbon inputs to soil for common agricultural crops in Canada. Agric Ecosys Environ 118:29–42

Brandenburger BauernZeitung (2010–2017) Data from issue 2010 (42) to 2017 (02). Deutscher Bauernverlag GmbH, Berlin

Bull II (2013) Untersuchungen zum Anbau und zur Verwertung von Steinklee. Dissertation, Universität Rostock

Cansunar E, Richardson AJ, Wallace G, Stewart CS (1990) Effect of coumarin on glucose-uptake by anaerobic rumen fungi in the presence and absence of Methanobrevibacter smithii. FEMS Microbiol Lett 70(2):157–160

Datta R, Kelkar A, Baraniya D, Molaei A, Moulick A, Meena RS, Formanek P (2017) Enzymatic degradation of lignin in soil: a review. Sustain MDPI 9(7):1163. https://doi.org/10.3390/su9071163. 1–18

Deuker A (2013) Energieerzeugung in landwirtschaftlichen Biogasanlagen: Potenziale und Grenzen. Sustainable biogas production in agriculture – potential and limitations. Including three journal articles. Verlag Dr. Köster, Berlin

Deuker A, Stinner W, Leithold G (2008) Biogas energy potentials from agricultural by-products: examples from organic farming in Germany compared with energy maize. In: 16th European biomass conference and exhibition: from research to industry and markets, Valencia, Spain, 2–6 June 2008, pp 274–284

Deuker A, Stinner W, Leithold G (2011) Biogas energy from agricultural by-products: energy yields and effects on organic farming systems compared with energy maize cropping. In: Behnassi M, Draggan S, Yaya S (eds) Global food insecurity. Springer, Dordrecht, pp 269–279. https://doi.org/10.1007/978-94-007-0890-7_19

Dhakal Y, Meena RS, Kumar S (2016) Effect of INM on nodulation, yield, quality and available nutrient status in soil after harvest of green gram. Leg res 39(4):590–594

DMK (2012a) Hohe Erträge und Erntemengen beim Mais. Deutsches Maiskomitee eV. (DMK) http://www.maiskomitee.de/web/intranet/news.aspx?news=b49dbc47-b873-4e2d-9856-8305e7c8624c. Accessed 25 Jan 2012

DMK (2012b) Flächenerträge.Deutsches Maiskomitee e.V. (DMK) http://www.maiskomitee.de/web/public/Fakten.aspx/Statistik/Deutschland/Fl%C3%A4chenertr%C3%A4ge. Accessed 10 Aug 2017

FAO (2013) Food and agricultural commodities production [WWW Document]. FAOSTAT. faostat.fao.org. Accessed 25 May 2015

FNR (2016) Cultivation of renewable resources in Germany. Fachagentur Nachwachsende Rohstoffe eV. (FNR). https://mediathek.fnr.de/downloadable/download/sample/sample_id/1406/. Accessed 03 Aug 2017

FNR (2017) Biofuels in comparison. In: Bioenergy in Germany Fact and Figures 2016. Fachagentur Nachwachsende Rohstoffe e.V. (FNR). http://www.fnr.de/fileadmin/allgemein/pdf/broschueren/Broschuere_Basisdaten_Bioenergie_englisch_2017.pdf. Accessed 03 Aug 2017

Goplen BP (1971) Polara, a low coumarin cultivar of sweet clover. Can J Plant Sci 51:249–251

Goplen BP (1980) Sweet clover production and agronomy. Can Vet J 21(5):149–151

Goplen BP (1981) Norgold – a low coumarin yellow blossom sweet clover. Can J Plant Sci 51:1019–1021

Gregorich EG, Drury CF, Baldock JA (2001) Changes in soil carbon under long-term maize in monoculture and legume-based rotation. Can J Soil Sci 81:21–31

Janke L (2017) Optimization of anaerobic digestion of sugarcane waste for biogas production in Brazil. Dissertation, Department of Waste Management and Material Flow Rostock University (in press)

Janke L, Leite AF, Batista K, Silva W, Nikolausz M, Nelles M, Stinner W (2016) Enhancing biogas production from vinasse in sugarcane biorefineries: effects of urea and trace elements supplementation on process performance and stability. Bioresour Technol 217:10–20. https://doi.org/10.1016/j.biortech.2016.01.110

Keymer U (2006) Biogasausbeuten verschiedener Substrate. http://www.lfl.bayern.de/ilb/technik/10225/index.php?sel_list=24%2Cb&strsearch=&pos=left. Accessed 10 May 2007

Keymer U (2017) Biogasausbeuten verschiedener Substrate. http://www lfl.bayern.de/iba/energie/049711/. Accessed 28 Feb 2017

Kirkby CA, Richardson AE, Wade LJ, Batten GD, Blanchard C, Kirkegaard JA (2013) Carbon-nutrient stoichiometry to increase soil carbon sequestration. Soil Biol Biochem 60:77–86

Kong AYY, Six J, Bryant DC, Denison RF, van Kessel C (2005) The relationship between carbon input, aggregation, and soil organic carbon stabilization in sustainable cropping systems. Soil Sci Soc Am J 69:1078–1085. https://doi.org/10.2136/sssaj2004.0215

Körschens M (1982) Der Einfluss unterschiedlicher Fruchtarten auf den Ct-Gehalt des Bodens. [In German]. Arch Acker- u. Pflanzenbau u. Bodenkd 26:711–716

KTBL (2005) Gasausbeute in landwirtschaftlichen Biogasanlagen. Kuratorium für Technik und Bauwesen in der Landwirtschaft e.V. (KTBL), Darmstadt

KTBL (2010) Betriebsplanung Landwirtschaft 2010/11. Kuratorium für Technik und Bauwesen in der Landwirtschaft e.V. (KTBL), Darmstadt

Leithold G, Hülsbergen K-J, Brock C (2015) Organic matter returns to soils must be higher under organic compared to conventional farming. J Plant Nutr Soil Sci 178:4–12

Lenz V, Müller-Langer F, Denysenko V, Daniel-Gromke J, Rensberg N, Rönsch C, Janczik S, Kaltschmitt M (2017) Erneuerbare Energien. BWK 69(5):54–77

Li X, Mu Y, Cheng Y, Liu X, Nian H (2013) Effects of intercropping sugarcane and soybean on growth, rhizosphere soil microbes, nitrogen and phosphorus availability. Acta Physiol Plant 35(4):1113–1119. https://doi.org/10.1007/s11738-012-1148-y

Lithourgidis AS, Dordas CA, Damalas CA, Vlachostergios DN (2011) Annual intercrops: an alternative pathway for sustainable agriculture. Aust J Crop Sci 5(4):396–410. http://searchinformitcomau/documentSummary;dn=281409060336481;res=IELHSS ISSN: 1835–2693. Accessed 20 Mar 17

Meena VS, Maurya BR, Verma R, Meena RS, Jatav GK, Meena SK, Meena R, Meena SK (2013) Soil microbial population and selected enzyme activities as influenced by concentrate manure and inorganic fertilizer in alluvium soil of Varanasi. The Bioscience 8(3):931–935

Meena VS, Maurya BR, Meena RS, Meena SK, Singh NP, Malik VK (2014a) Microbial dynamics as influenced by concentrate manure and inorganic fertilizer in alluvium soil of Varanasi, India. Afr J Microb Res 8(1):257–263

Meena RS, Yadav RS, Meena VS (2014b) Response of groundnut (*Arachis hypogaea* L.) varieties to sowing dates and NP fertilizers under western dry zone of India. Bangladesh J Bot 43(2):169–173

Meena RS, Yadav RS, Meena H, Kumar S, Meena YK, Singh A (2015a) Towards the current need to enhance legume productivity and soil sustainability worldwide: a book review. J Clean Prod 104:513–515

Meena VS, Maurya BR, Meena RS (2015b) Residual impact of well-grow formulation and NPK on growth and yield of wheat (*Triticum aestivum* L.). Bangladesh J Bot 44(1):143–146

Meena HR, Meena RS, Lal R, Singh GS, Mitran T, Layek J, Patil SB, Kumar S, Verma T (2017) Response of sowing dates and bioregulators on yield of clusterbean under current climate in alley cropping system in eastern U.P. Indian Leg Res, Accepted in press

Michel D (1988) Beitrag zur Quantifizierung der Stickstoff- und Humusreproduktionsleistung der Luzerne aus einem Fruchtfolge-Düngungsversuch /On the evaluation of data from a test on crop rotation and fertilization in the case of lucernes with respect to their nitrogen and humus production efficiency. Tag-Ber Akad Landwirtsch-Wiss DDR 269:367–374

Möller K, Stinner W (2009) Effects of different manuring systems with and without biogas digestion on soil mineral nitrogen content and on gaseous nitrogen losses (ammonia, nitrous oxides). Eur J Agron 30:1–16. https://doi.org/10.1016/j.eja.2008.06.003

Möller K, Stinner W, Leithold G (2006a) Biogas in organic agriculture: effects on yields, nutrient uptake and environmental parameters of the cropping system. Presented at Joint Organic Congress, Odense, Denmark, 30–31 May 2006. http://orgprints.org/7223/

Möller K, Leithold G, Michels J, Schnell S, Stinner W, Weiske A (eds) (2006b) Auswirkung der Fermentation biogener Rückstände in Biogasanlagen auf Flächenproduktivität und Umweltverträglichkeit im Ökologischen Landbau – Pflanzenbauliche, ökonomische und ökologische Gesamtbewertung im Rahmen typischer Fruchtfolgen viehhaltender und viehloser ökologisch wirtschaftender Betriebe. Endbericht Deutsche Bundesstiftung Umwelt, Osnabrück

Moniello G, Richardson AJ, Duncan SH, Stewart CS (1996) Effects of coumarin and sparteine on attachment to cellulose and cellulolysis by Neocallimastix frontalis RE1. Appl Environ Microbiol 62(12):4666–4668

Mosier AR (2001) Exchange of gaseous nitrogen compounds between agricultural systems and the atmosphere. Plant Soil 228:17–27

Oenema O, Wrage N, Velthof GL, Groenigen JW, van Dolfing J, Kuikman PJ (2005) Trends in global nitrous oxide emissions from animal production systems. Nutr Cycl Agroecosyst 72:51–65

Overman RS, Stahmann MA, Sullivan WR, Huebner CF, Campbell HA, Link KP (1942) Studies on the hemorrhagic sweet clover disease: VII. The effect of 3,3′-methylenebis(4-hydroxycoumarin) on the prothrombin time of the plasma of various animals. J Biol Chem 142:941–955

Plaza-Bonilla D, Nolot J-M, Passot S, Raffaillac D, Justes E (2016) Grain legume-based rotations managed under conventional tillage need cover crops to mitigate soil organic matter losses. Soil Till Res 156:33–43

Poeplau C, Don A (2015) Carbon sequestration in agricultural soils via cultivation of cover crops – a meta-analysis. Agric Ecosys Environ 200:33–41

Popp D, Schrader S, Kleinsteuber S, Harms H, Sträuber H (2015) Biogas production from coumarin-rich plants - inhibition by coumarin and recovery by adaptation of the bacterial community. FEMS Microbiol Ecol 91(9):fiv103

Ram K, Meena RS (2014) Evaluation of pearl millet and mungbean intercropping systems in arid region of Rajasthan (India). Bangladesh J Bot 43(3):367–370

Rauhe K (1969) Der Einfluß des Futterbaues sowie der organischen und mineralischen Düngung auf den C- und N-Gehalt des Bodens im Fruchtfolgedüngungsversuch Seehausen/Certain humus losses which require humus substitution would occurwith the cultivation of fellow crops and grain, according to the conditions of soil and climate. [In German]. Arch Acker- u Pflanzenbau u Bodenkd 13:455–462

Roderick LM (1931) A problem in the coagulation of the blood. Am J Phys 96(2):413–425

Schäfer (2012) Saatgut. http://www.roman-schaffer.com/saatgut/Saatgut2012.pdf. Accessed 25 Jan 2012

Schimel JP, Weintraub MN (2003) The implications of exoenzyme activity on microbial carbon and nitrogen limitation in soil: a theoretical model. Soil Biol Biochem 35:549–563

Schjønning P, Munkholm LJ, Elmholt S, Olesen JE (2007) Organic matter and soil tilth in arable farming: management makes a difference within 5–6 years. Agric Ecosys Environ 122:157–172

Schulz F, Brock C, Schmidt H, Franz K-P, Leithold G (2015) Development of soil organic matter stocks under different farm types and tillage systems in the organic arable farming experiment Gladbacherhof. Arch Agron Soil Sci 60:313–326

Smith WK, Brink RA (1938) Relation of bitterness to the toxic principle in sweet clover. J Agr Res 57:145–154

Smith WK, Gorz HJ (1965) Sweet clover improvement. In: Norman AG (ed) Advances in agronomy, vol 17. Academic Press Inc, New York

Statistisches Bundesamt (2005) Statistisches Jahrbuch 2004 für die Bundesrepublik Deutschland

Stevens WB, Hoeft RG, Mulvaney RL (2005) Fate of Nitrogen-15 in a long-term nitrogen rate study: II nitrogen uptake efficiency. Agron J 97:1046–1053

Stinner W (2011) Auswirkungen der Biogaserzeugung in einem ökologischen Marktfruchtbetrieb auf Ertragsbildung und Umweltparameter. Dissertation, Chair of organic agriculture, Giessen University, (ed) Verlag Dr. Köster, Berlin

Stinner W (2015) The use of legumes as a biogas substrate – potentials for saving energy and reducing greenhouse gas emissions through symbiotic nitrogen fixation. Energy Sust Soc 5(4). https://doi.org/10.1186/s13705-015-0034-z

Stinner W, Rensberg N (2011) Perspektive der Biogasgewinnung aus nachwachsenden Rohstoffen – Bewertung von Substratalternativen zu Silomais. http://forumue.de/wp-content/uploads/2015/04/DBFZ_Maisalternativen.pdf. Accessed 21 Apr 2017

Stinner W, Möller K, Leithold G (2008) Effects of biogas digestion of clover/grass-leys, cover crops and crop residues on nitrogen cycle and crop yield in organic stockless farming systems. Eur J Agron 29(2):125–134. https://doi.org/10.1016/j.eja.2008.04.006

Suman A, Lal M, Singh AK, Gaur A (2006) Microbial biomass turnover in Indian subtropical soils under different sugarcane intercropping systems. Agron J 98:698–704. https://doi.org/10.2134/agronj2005.0173

Thaysen J (2012) Wie hoch sind die Silierverluste? Bauernblatt 25 Aug 2012, pp 10–12

Thrän D, Pfeiffer D (eds) (2012) Brückenschlag nach Osteuropa – Biomassepotenziale und -nutzungsoptionen in Russland, Weißrussland und der Ukraine. Schriftenreihe des BMU-Förderprogramms "Energetische Biomassenutzung", vol 6, Leipzig http://www.biogasunden-ergie.de/downloads/scholwin_publication_4.pdf. Accessed 21 Apr 2017

Turkington RA, Cavers PB, Rempel E (1978) The biology of Canadian weeds. 29. Melilotus alba Desr. and M. officinalis (L.) Lam. Can J Plant Sci 58:523–537

UBA (2017) National trend tables for the German Atmospheric Emission Reporting 1990–2015. Umweltbundesamt, Dessau. https://www.umweltbundesamt.de/sites/default/files/medien/361/dokumente/2017_01_23_em_entwicklung_in_d_trendtabelle_thg_v1.0.xlsx. Accessed 21 Apr 2017

Verma JP, Jaiswal DK, Meena VS, Meena RS (2015) Current need of organic farming for enhancing sustainable agriculture. J Clean Prod 102:545–547

Vetter A (2009) Biogassubstrate – welche "Exoten" haben potential für die Zukunft. In Gülzower Fachgespräche Band 32 Tagungsband "Biogas in der Landwirtschaft – Stand und Perspektiven, pp 278–288

Vetter A, Eckner J, Strauß C, Nehring A (2014) Entwicklung und Optimierung von standortangepassten Anbausystemen für Energiepflanzen im Fruchtfolgeregime. Abschlussbericht zum Teilprojekt 1. (EVA II). https://doi.org/10.2314/GBV:826946658

Witt J, Magdowski A, Janzczik S, Kaltschmitt M (2017) Erneuerbare Energien – Globaler Stand 2016. BWK 69(07/08):6–28

Yadav GS, Lal R, Meena RS, Babu S, Das A, Bhomik SN, Datta M, Layak J, Saha P (2017) Conservation tillage and nutrient management effects on productivity and soil carbon sequestration under double cropping of rice in north eastern region of India. Ecol Ind http://www.sciencedirect.com/science/article/pii/S1470160X17305617

Symbiotic Tripartism in the Model Plant Family of Legumes and Soil Sustainability

6

Vinod Vijayakumar

Contents

Abstract

The demands of feeding a world population are expected to double by 2050. This is because 2.5 billion will be added to the urban population alone. This massive undertaking has posed many challenges toward agricultural productivity and increase in food quality, quantity, and production of protein-rich crops, but on the other hand, modern aggressive agricultural practices have rendered the current acreage of arable land and soil unsustainable to meet the demands of sustainable cropping systems. However, the beneficial role of legumes in cropping systems such as symbiotic nitrogen fixation, intercropping, and rotation of legumes with

V. Vijayakumar (✉)
College of Food, Agricultural and Environmental Sciences, Department of Food Science and Technology, The Ohio State University, Columbus, OH, USA

R. S. Meena et al. (eds.), *Legumes for Soil Health and Sustainable Management*,
https://doi.org/10.1007/978-981-13-0253-4_6

cereals offers credible potential for providing economically sustainable advantages for farming. The inherent capacity of legumes to form symbiotic associations with biological nitrogen-fixing (BNF) rhizobia and phosphorus-acquiring arbuscular mycorrhizal fungi (AMF), i.e., symbiotic tripartism, further advocates the use of legumes as cover crops, increasing soil fertility, rhizospheric processes, and sustainable (food/oil) crop production. Furthermore, it is estimated that BNF of legumes contribute to five to seven times less greenhouse gas (GHG) emissions per unit area compared to other crops, in addition to estimates of total global BNF of 122 T gN/year (=million tons of N), while AMF play a critical role in global carbon cycle, with estimates of the amount of total C fixed to be up to 20% which is *c.* 5 T PgC/year (=billion tons of C). In view of this importance of symbiotic tripartism in natural and managed ecosystems, this chapter emphasizes the genetic and symbiotic feature(s) of legumes in large-scale community and global food security programs and soil sustainability and management.

Keywords
Arbuscular mycorrhizal symbiosis · Common symbiosis pathway · Dual inoculation · Legume-rhizobia symbiosis · Plant mineral nutrition · Rhizosphere

Abbreviations

AMS	Arbuscular mycorrhizal symbiosis
CSP	Common SYM pathway
N	Nitrogen
P	Phosphorus
RNS	Root nodule symbiosis
SOM	Soil organic matter

6.1 Introduction

The last few decades have seen tremendous changes in agriculture and in food systems of the world. While global trade in agricultural commodities has increased, the interconnectivity among food resources in developed and developing countries has also contributed and exacerbated the challenges. These challenges are related to malnutrition, food security, environmental degradation, and large-scale soil sustainability which in turn make it harder to achieve the targets of Sustainable Development Goals (SDGs), eliminating poverty and hunger (http://www.fao.org/3/a-i6627e.pdf, FAO, 2017). Hence, the transition of existing agricultural practices (heavy fertilizer and pesticide use, monoculture crops, etc.) to more eco- and environmental-friendly management practices is the pressing need of large-scale globalization of sustainable agriculture and crop production. Worldwide, agricultural systems are struggling to produce high-quality food with minimal external inputs and help in reducing negative environmental impacts (Wittwer et al. 2017; Meena et al. 2017a). The conventional intensive agricultural systems which mainly rely on the heavy use of pesticides and mineral fertilizers and monoculture cropping systems have further rendered the use of arable land to a bare minimum. For example, the intensive application of large amount of relatively inexpensive nitrogenous compounds has led to an undesirable and unprecedented negative impact on aquatic and terrestrial ecosystems. This net negative impact is also illustrated by their increased potential accumulation as other forms of reactive nitrogen (N) and other nitrates in soils and waters, and as nitrogen oxides and nitrous oxides in the atmosphere (Sulieman and Tran 2015). The use of organic farming and conservation agriculture has provided promising avenues for increased ecological benefit in terms of sustaining soil quality, fertility, and biodiversity. However, challenges to the increase in acreage that are needed to produce substantial quantity of high-quality food along with global food security still largely remain unaddressed.

One way of mitigating lower crop yields and increasing soil sustainability is using cover crops in crop rotation schemes, by introducing nitrogen-fixing legume species for improving N nutrition and increasing soil N organic pool, which in turn will benefit the growth of succeeding main crops (Wittwer et al. 2017). A further advantage of legumes as cover crops is their capacity to form tripartite mutualistic symbiotic associations with various other soil-borne fungi and bacteria. Such symbiotic interactions play a key role in natural and managed ecosystems for plant nutrient use efficiency (NUE). An intriguing example to study in the context of plant NUE would be the differential regulation in N and phosphorus (P) uptake mechanisms and assimilation and its effects on nutrient mobility and metabolic activity in the soil root zone and beyond. More than 90% of all plant families studied (80% of species, including crops and fruiting trees; Strack et al. 2003) in both agricultural and natural environments form symbiotic associations of some kind or the other (mycorrhizal and root-nodule symbioses). So, understanding the extent to which the plant-microbe associations in the rhizosphere (volume of soil influenced by roots) affect plant productivity and survival is of paramount ecological importance. The year 2016 was declared as the "International Year of Pulses" by the UN general assembly, and this was in recognition of the contributions and major role that pulses (dry grain legumes) play in global food security, human well-being, and environmental sustainability challenges. Although pulses form a subgroup of legumes, other crop members of the plant family *Leguminosae* are often neglected in not only western diets (Messina 1999; Anand et al. 2015) but also in scientific interest (Ianneta et al. 2016), except for soybeans (Messina 2010). Hence, this chapter's objective is to provide and emphasize the genetic and environmentally sustainable symbiotic features of legumes in global food security and effective educational and research programs. Furthermore, there is increasing evidence from plant-microbe interaction studies that fungal phytopathogens may have thrived successfully by recruiting and exploiting gene networks of ancient symbiotic pathways (Martin et al. 2008). These studies provide a promising entry point in elucidating the molecular mechanisms of plant NUE in legumes, which can consequently be used as model plant systems for plant breeding and understanding key genetic mechanisms in soil sustainability. Finally, understanding the molecular and regulatory mechanisms is important. This will facilitate mechanistic understanding of fungal and rhizobacterial symbionts in governing plant growth and productivity in nutrient-rich/nutrient-scarce environments, thus, enabling better management of sustainable forest and agroecosystems in a biosphere which is increasingly threatened by global climate change. In the future fundamental efforts in ensuring food and nutritional security should be taken using naturally occurring phenomena so that the earth's biogeochemistry is not affected negatively any further.

6.1.1 Soil and Plant Mineral Nutrition

The foundation of agriculture world over is laid in soils. The agricultural soils are a major source of nutrients and in combination with water provide a distinct influence on human and the ability to sustain health, nutrition, and global food security (Lal 2012; Brevik and Sauer 2015). Hence, soils form a dynamic and diverse natural ecosystem. They help in the improved conservation and management of the agroecosystems. Conservation is one of the major issues facing the twenty-first century challenges for crop production and global food security (Brevik 2013). Minerals, soil organic matter (SOM), living organisms, gas, and water are the five ingredients that soil(s) are composed of. While soil minerals (classes, clay, silt, and sand) provide soils their texture, SOM is used as the best indicator of agricultural soil quality and forms a critical ingredient that comprises of various decomposed states of plant, animal, and microbial biomass (Needelman 2013). The natural cycling of nutrients from soils to plants and animals and back to soils as SOM helps maintain the balance of essential mineral nutrients required for plant growth and development. Of the 17 essential nutrient elements required for plant growth (Table 6.1, excluding the structural elements carbon, hydrogen, and oxygen), the primary macronutrients nitrogen, phosphorous, and potassium (N, P, and K, respectively) form the principle limiting factors affecting crop productivity (Parikh and James 2012).

Mineral deficiencies in soil and/or as agricultural input disrupt the plant metabolism, function, and, hence, its physiology and growth. Deprivation of an essential mineral element (N/P/K) leads to the generation of nutrient deficiency symptoms in a plant (Table 6.1), which in turn leads to the expression of metabolic disorders. These disorders are related to the roles played by the missing element making it indispensable for growth (van der Ploeg et al. 1999; Ohkama-Ohtsu and Wasaki 2010; White and Brown 2010). The intake of nutrients by the roots of higher plants is characterized by their selective ion intake and storage in specific tissues, cell types, and subcellular compartments (White and Brown 2010). A plant root system grows continuously throughout the year depending on the availability of water and mineral nutrients in the immediate microenvironment surrounding the root, called the "rhizosphere" (Fig. 6.1). In both natural and managed (agro)ecosystems, the (bio)availability, intake, and exchange of key nutrients in the rhizosphere determine the plant growth, productivity, and yield.

Table 6.1 The 14 essential mineral elements required for plant growth and development[a]

Element	Symbol	Primary form(s)	Required for[b]	Most probable deficiency symptom[b]
Primary macronutrients				
Nitrogen	N	NO_3^-, NH_4^+	Production of proteins, nucleic acids (DNA &RNA), and chlorophyll	Leaf chlorosis, stunted and slow growth
Phosphorus	P	$H_2PO_4^-$, HPO_4^{2-},PO_4^{3-}	Development of energy, sugars, and nucleic acids	Dark green plants, older leaves purplish or red
Potassium	K	K^+	Enzyme activity, photosynthesis, proteins synthesis, and sugar transport	Chlorotic leaf tips and leaf burns at margins
Secondary macronutrients				
Calcium	Ca	Ca^{2+}	Cell wall construction and regulation	Terminal bud leaves – hooked, turn brown, die back
Magnesium	Mg	Mg^{2+}	Chlorophyll synthesis and energy production (important as a cofactor)	Interveinal chlorosis, leaves red or with dead spots
Sulfur	S	SO_4^{2-}	Proper amino acid and protein synthesis	Light green young leaves, no chlorotic spotting or striping
Micronutrients				
Boron	B	BO_3^{3-}	Cell wall formation and reproductive tissue regeneration	Light green terminal buds, leaves – twisted, brittle, die back
Chlorine	Cl	Cl^-	Leaf turgor and photosynthesis	Chlorotic and necrotic lesions (spotting)
Copper	Cu	Cu^+, Cu^{2+}	Chlorophyll production, respiration, and protein synthesis	Chlorosis of young leaves, withered tips, and ultimately die
Iron	Fe	Fe^{3+}	Respiration and photosynthetic reactions	Interveinal chlorosis, distinct veins, and chlorotic areas
Manganese	Mn	Mn^{2+}	Chloroplast organellar biogenesis	Interveinal chlorosis, non-distinct veins, and spotty chlorosis
Molybdenum	Mo	MoO_4^{2-}	Proper enzyme activity and N fixation in legumes	Light green plants, pale and necrotic spotting leaves
Nickel	Ni	Ni^{2+}	Proper seed germination and beneficial in (N) metabolism	Interveinal chlorosis in younger leaves
Zinc	Zn	Zn^{2+}	Growth hormone production and internode elongation	Interveinal mid-leaf chlorosis, stunted growth

a and b, data compiled from references: Parikh and James 2012; McCauley et al. 2011, respectively

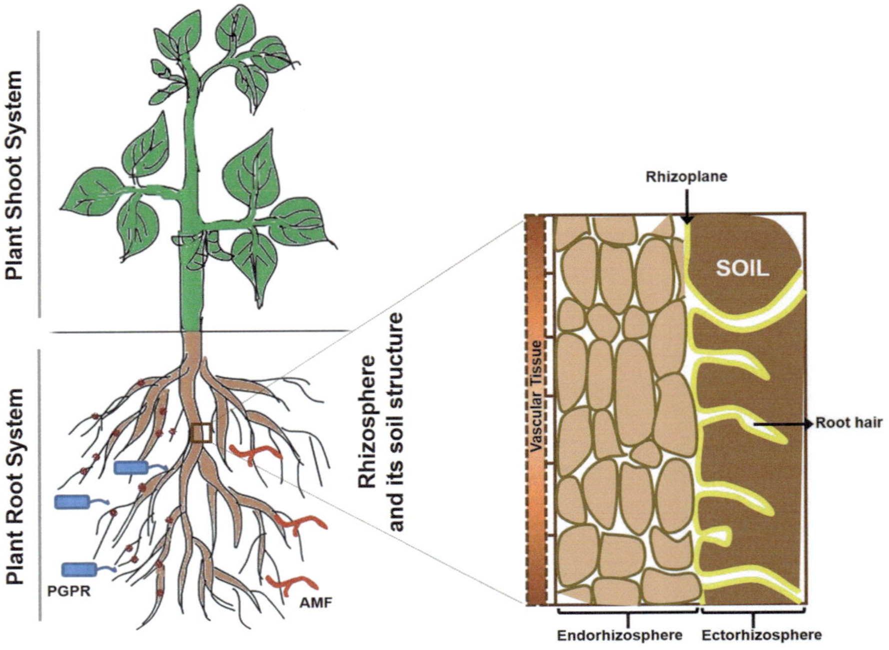

Fig. 6.1 Schematic representation of the soybean root system and its tripartite symbioses with plant growth-promoting rhizobia (PGPR) and AMF. Zoomed inset depicts a section of the root showing the organization of the rhizospheric soil structure (adapted from McNear 2013)

6.1.2 Rhizosphere, Rhizomicrobiome, and Rhizodeposits

The rhizosphere can be defined as the interface between plant roots and the volume of soil that is influenced by roots. This is where interactions among the multitudinous array of soil-dwelling microorganisms affect biogeochemical cycling besides plant growth, productivity, and tolerance to biotic and abiotic stress (Phillippot et al. 2013; Oburger and Schmidt 2016; Meena et al. 2018). The rhizospheric soil structure (McNear 2013) can be further separated into zones of soil around the outer roots and root hairs (ectorhizosphere), the soil adhering to the root surface (rhizoplane), and the inner root (endorhizosphere) where key plant-microbe interactions and early signaling events occur (Fig. 6.1). The microbial community that lives in the rhizosphere consists of a myriad of species: bacteria, fungi, oomycetes, nematodes, protozoa, archaea, arthropods, and viruses, commonly referred to as the "rhizosphere microbiome" or "rhizomicrobiome" (Mendes et al. 2013; Venturi and Keel 2016). Many abiotic (drought, salt, temp, soil constraints, etc.) factors govern the assembly of the rhizosphere microbiome and its community. This is done by exerting either a positive or a negative influence on the deposition of nutrients, exudates (small molecules and chemical signals), mucilage, etc., collectively referred to as the "rhizodeposits" (Philippot et al. 2013). The type, nature, and concentration of select rhizodeposits have been shown to determine microbial diversity and activity that precedes and/or influences the outcome of plant-microbe below-ground interactions (Mendes et al. 2014). Many large-scale studies have shown the species-specific effect of plants on the composition and shaping of the microbial rhizospheric community under agro- and natural ecosystems (Philippot et al. 2013; Mendes et al. 2014; and references there in). Further, the shaping and assembly of the rhizospheric community by the rhizodeposits is largely believed to be occurring via two general processes: (i) direct stimulation of rhizomicrobiome multiplication in the vicinity of the roots by rhizodeposits (nutrients, mucilage, etc.) and (ii) detection and rapid response to root exudates (low molecular weight compounds – small molecules and/or chemical signals) of either plants or microbes, resulting in cellular response(s) entailing regulatory signaling cascades (Venturi and Keel 2016). This intriguing complexity and the dynamic nature of rhizosphere and its interactions makes it a key factor, the understanding of which is critical to enhancing global plant productivity and ecosystem functioning (Philippot et al. 2013).

6.2 Legumes as Models of Sustainable Agriculture

Legumes represent the second major (*c.* 14% of total land under cultivation) agricultural crop worldwide and are widely acknowledged for their beneficial role in cropping systems. This is especially true for the warm-season legumes like soybean, cowpea, common bean, groundnut, and pigeon pea (Rubiales and Mikic 2015). Although many different efforts have been made (www.fao.org; Foyer et al. 2016, Varshney 2016) to bring pulses at the forefront and highlight their importance in cropping systems, improving soil characteristics, mitigation of greenhouse gases, biodiversity, and human health (Siddique et al. 2012; Courty et al. 2015; Kouris-Blazos and Belski 2016; Kaur et al. 2016; Stagnari et al. 2017), cultivation of (food) legumes still remains largely below that of other major cereal crops, e.g., rice, wheat, and maize (Siddique et al. 2012).

Hence, keeping in mind the significance of legumes in nature, their potential for conservation agriculture, soil sustainability, and expansion of crop land, a further emphasis is placed on the rhizosphere of leguminous plant(s), their rhizospheric communities, interactions and effects on plant mineral nutrition, and NUE. In comparison with major cereal crops, legumes are crucial for providing answers to global demands of a protein-rich diet even under circumstances of an ever-increasing threat of global climate change and aggressive agricultural practices. The major advantage of legumes in addition to contributing high-quality organic matter to the soil (high N: C ratio) is their inherent ability to utilize and fix low inputs of N by forming symbiotic associations with the soil microbiota (Siddique et al. 2012). Here, legumes play a special role in the symbiotic interactions of plants with microbes. This is because they can establish symbioses with both rhizobia and AMF. This fundamental ability of legumes to form functional interactions with soil-borne microbes for plant NUE and ability to adapt to minimal soil disturbance (no-till farming practices) and chemical fertilizer input(s) makes them best suited for the conservation agriculture practices of the future.

One of the most prevalent and commonly occurring plant-microbe interactions in nature worldwide is a tripartite mutualistic endosymbiotic association of leguminous plants with different endosymbiotic N-fixing gram-negative bacteria and various soil-borne phosphorus-acquiring AMF, forming legume-rhizobia symbiosis (LRS) and arbuscular mycorrhizal symbiosis(AMS), respectively (Parniske 2008; Oldroyd et al. 2009; Hause and Schaarschmidt 2009). In the rhizosphere, soil bacterium of the *Rhizobium* sp. interacts with roots of leguminous plants resulting in the formation of N-fixing root nodules (Beringer et al. 1979; Dhakal et al. 2016). On the other hand, some fungi colonize the plant roots by infecting them and forming symbiotic relationships, called mycorrhizas (Guissou 2009). The nutrient that gets most influenced by rhizobial infections is N by the process of symbiotic/biological N fixation. However, under mycorrhizal infections apart from an efficient P uptake, mycorrhizas are also known to increase the uptake of many other nutrients both directly and indirectly like zinc (Zn), copper (Cu), and possibly N by increasing the cation/anion ratio (Smith et al. 1992). Recently studies on plant nutrient research have also highlighted the importance of N uptake in plants especially under mycorrhizal symbioses (Kiers et al. 2011; Fellbaum et al. 2012; Courty et al. 2015). Furthermore, many recent studies have highlighted the importance of dual inoculations of AMF and rhizobia to be beneficial in terms of promoting plant growth in legumes like soybean and in inhibiting fungal diseases, e.g., soybean red crown rot (Wang et al. 2011; Gao et al. 2012; Meng et al. 2015). It has also been shown that colonization of plants by AMF leads to a potential synergistic interaction with P-solubilizing bacteria (PSB). This suggests functional recruitment of soil-bacterial species in the rhizosphere with the capacity to solubilize poorly available P (Ordonez et al. 2016; Meena et al. 2017b). The LRS forms one of the most prominent examples of plant growth-promoting rhizobacteria (PGPR) being suggestive of a direct correlation between plant-growth promotions through symbiotic N fixation. PGPRs are known to be efficient at metabolizing root exudates (carbohydrates) and in turn providing plants with N needed for amino acid synthesis (Berg 2009).

The tripartite symbiosis/associations of legumes with rhizobia and AMF in natural and managed ecosystems lead to the functional adaptation of any particular legume species to soils characterized by nutrient-limited conditions, e.g., N and P (Courty et al. 2015). Further, in meeting global demands of increased food production and food security, understanding the fundamental principles in efficient nutrient use by plants and their rhizospheric interactions is of prime importance. In the past, this has paved the way for the next generation of vastly improved agroecosystem management practices. In the future, similar studies may promise to further the development of sustainable agriculture with minimal chemical (pesticide and/or mineral fertilizer) inputs.

6.3 Symbiotic Plant-Beneficial Interactions

Rhizospheric microbiome and their interactions with roots help plants achieve important functions necessary for their growth, maintenance of health, and productivity. In all documented cases of improved plant NUE, plant growth promotion correlates directly with beneficial associations. A key feature in the establishment and maintaining of microbial symbioses (symbiotic plant-beneficial interactions) is the ability of both the micro- and macro-symbionts to participate in a mutualistic exchange of nutrients, i.e., the C and N cycles (Courty et al. 2015). While, the micro-symbiont depends on the ability of the plant to transfer organic carbon (C) or photosynthate, the macro-symbiont relies on the ability of microbes to take up/fix inorganic or organic forms of nutrients from soil and translocate them to the plant root. These interactions in nature may occur under different levels of functional associations wherein both the partners to a larger extent are mutually benefitted. Many different rhizospheric processes play a crucial role in the successful establishment of plant-root symbioses. Nutrient-limiting conditions have been shown to increase root hair formation and root exudation and subsequent recruitment of plant-beneficial microbes into the rhizospheric community of legumes, leading to the successful establishment of a tripartite mutualistic symbiosis (Scheublin et al. 2004; Brown et al. 2013; Giehl and von Wiren 2014; Ren et al. 2017).

6.3.1 Arbuscular Mycorrhizal Symbiosis (AMS)

Plant root associations with fungi of the glomeromycotan phyla are commonly referred to as arbuscular mycorrhizas. Hence, mycorrhizal ("fungus-root") associations are specialized symbiotic associations of soil fungi with plant roots (Brundrett 2002; Bucher 2007). Evidence based on paleobotanical and morphological studies in combination with DNA-based molecular analysis suggest that mycorrhizal fungi are probably the first terrestrial fungi to colonize land long before plants did (Brundrett 2002). The AMF are known to associate with a wide diversity of plants and together with ectomycorrhizas (EcMs) are believed to colonize greater than 70–90% of all land plants (van der Heijden et al. 2015). The mycorrhizal associations are typically characterized by an infection which begins outside the plant root epidermis as a local concentration of fungal infection units leading to the formation of "special types of appressoria called hyphopodia" (Parniske 2008). Upon successful infection, the fungus forms inter- and intracellular hyphae, penetrating and colonizing the root cortical cells. Branches of the intercellular hyphae penetrate the root cortical parenchyma cells branching profusely within. This in turn gives rise to intracellular specialized structures forming an interface, the main sites of nutrient exchange between the host plant and mycorrhizal fungus (Smith et al. 1992). In mutualistic mycorrhizal associations, emphasis is placed mainly on the translocations of P, N, and C at the interfaces of soil-fungus and fungus-plant (Bonfante and Genre 2010). A central aspect of AM associations is the formation of a functional symbiosome – the symbiotic interface where exchange of nutrients takes place.

During AMS, the plant acquires P_i and other nutrients through the fungus, in exchange for photosynthetic carbohydrates as carbon source. Interestingly, two recent reports (Jiang et al. 2017; Luginbuehl et al. 2017) have presented new evidence leading to a paradigm shift in the functional biochemistry of AMS. Their results suggest active participation of photobiont in providing the AMF with a robust source of C that fulfills their metabolic needs, i.e., an interaction wherein the plant supplies lipids as carbon source for sustenance of colonization by AMF. The evidence presented by authors Jiang et al. (2017) is of further interest for they show that plant fatty acids is also transferred in interactions with parasitic fungi and is required for pathogen colonization, *Golovinomyces cichoracerum*. These results pave ways for studies understanding the importance of fatty acid biosynthesis in plant-fungal interactions for they facilitate host invasion mechanisms in mutualistic mycorrhizal and pathogenic fungi alike.

Mycorrhizal associations have been shown to evolve in many ways to improve the overall fitness of both plant and fungal symbionts. This suggests an active mechanism of mutualism. It has also been shown that in systems managed by humans (agroecosystems), most mycorrhizal associations improve plant productivity, but in some cases the mycorrhizal associations may also be considered to be parasitic on plants especially when the net costs of symbiosis exceed its benefits (Johnson et al. 1997), culminating in the exploitative mycorrhizas of achlorophyllous, myco-heterotrophic plants (Merckx et al. 2009). However, considering the whole paradigm of mycorrhizal associations that is existing in nature, where mycorrhizas are formed in an enormously wide variety of host plants (angiosperms, gymnosperms, the pteridophyte sporophytes – all having roots, as well as gametophytes of some hepatics and the pteridophytes which do not) as obligate symbionts (Smith and Read, Mycorrhizal Symbiosis, 3rd edition, 2008), their continuum of mutualistic associations in terms of plant productivity and as efficient nutrient uptake mechanisms often outweighs those of its parasitic associations.

6.3.2 Legume-Rhizobia Symbiosis (LRS) or Root Nodule Symbiosis (RNS)

Rhizobia are members of soil bacteria called "diazotrophs" that are capable of assimilating and fixing atmospheric N gas into a more readily usable form such as ammonia (NH_3). It is directly usable for plant growth (Brewin 2010; Syntikov 2013). Although, there are two main groups of diazotrophs that possess a similar mechanism of biological (di)N fixation, the rhizobacterial species are given a precedence in text here. This is because of their ability to enter into symbiotic associations with legumes. The members of other group of diazotrophs are free-living bacteria classified as associative N fixers. The ability of rhizobacterial species to fix molecular N is regarded as one of the prominent processes that determine the biological productivity of the planet. The N cycle acts as a critical player of biogeochemical transformations of the earth (Syntikov 2013). Further, the propensity of legumes to get in associations and allow for the development of rhizobial populations in the rhizosphere is key to the agricultural productivity worldwide, for *Leguminosae/Fabaceae* is a large family consisting of 917 plant genera forming a major group of *angiosperms* (flowering plants) (http://www.theplantlist.org/browse/A/Leguminosae/). Thus, legumes are one of the most sought-after major food crops in sustainable agricultural cropping systems (Ohyama 2017).

The interaction between legumes and the rhizobia species belonging to the family of *Rhizobiaceae* (genera, *Azorhizobium*, *Bradyrhizobium*, *Mesorhizobium*, *Rhizobium*, and *Sinorhizobium*) results in the formation of a characteristic special structure, the nitrogen-fixing nodules on roots (Beringer 1979; Syntikov 2013). The establishment of bacterial infection and the development of root nodules (RN) occur as consequences of a series of complex morphophysiological changes in the cells of both partners requiring expression, regulation, and modulation of many genes and enzymes (Stougaard 2000). This process culminates in the establishment of a functional interaction, the legume-rhizobia symbiosis (LRS) or in other terms referring to the formation of a functional RN for biological N_2-fixation, the RNS. The sequential stages of RN development in LRS/RNS involves (i) preinfection stage, where exchange of signals takes place between the symbionts, leading to recognition and adsorption of beneficial bacteria to the root surface; (ii) bacterial infection, an invasion of bacteria in the root zone behind the root tip where root hairs are growing; and (iii) infection thread and symbiosome membrane (SM) formation and nodule organogenesis, infection of root hairs occurs through the formation of an infection thread, and its progression that takes the bacteria via root hairs into the root cortex distributing them into cells; later the SM, a regulated interface (physical barrier), is formed between the host plant and the infecting bacteria. This is where movement of solutes is regulated by the plant and finally culminates in the formation of a fully functional N-fixing root nodule (Clarke et al. 2014).

6.3.3 The Common Denominators of Symbiotic Tripartism and Early Signaling

The knowledge that AMS is extremely ancient and had existed more than 400 million years ago (Ma) (Parniske 2008), legumes 60 Ma (Lavin et al. 2005), and the LRS evolution about 58 Ma (Sprent 2007) suggests that evolution of symbiotic signaling mechanisms occurred first in the AMS and subsequently molecular players were recruited and adapted in the LRS. AMS is one of the oldest forms of symbiotic associations known to mankind. Its presumed ecological importance, mechanisms, and signaling events underlie the formation of a functional symbiosome which is poorly understood. But, it is just beginning to be unraveled at a remarkable pace (Bucher et al. 2014; Courty et al. 2015; Geurts et al. 2016; Genre and Russo 2016; Holmer et al. 2017). It is a well-known fact that in many plant-microbe interactions, detection and attraction of the symbiotic partner occur prior to a cell contact with the initiation of a molecular dialog. This leads to the physical stages of plant and fungal cell-to-cell contact and the beginning of a functional interaction.

The characteristic features of legumes forming symbiotic interactions with both rhizobia and AMF (symbiotic tripartism) have been explored scientifically quite extensively. Hence, the use of legumes as model plant species has greatly facilitated the identification of plant genes essential for symbiotic associations. For instance, recent work on the model legume species of *Medicago truncatula* (barrel clover), *Lotus japonicus* (bird's-foot trefoil), *Pisum sativum* (pea), and *Glycine max* (soybean) have led to the identification and characterization of several genes encoding common signaling components of symbiosis, involving both AMS and LRS (Parniske 2008), collectively referred to as the "common symbiotic (Sym) pathway, CSP" (Oldroyd et al. 2009; Genre and Russo 2016) or the "common symbiosis signaling pathway, CSSP" (Bucher et al. 2014). Many of these studies have revealed that the activation of the CSP/CSSP is by the production of symbiotic signals, the lipochitooligosaccharides (LCOs), which are similar in structure to LCOs produced in both rhizobia ("nod factors," NFs – rhizobial signaling molecules) as well as AMF ("Myc factors," MFs – mycorrhizal signaling molecules) (D'Haeze and Holsters 2002; Maillet et al. 2011; Nadal and Paszkowski 2013). In contrast, plants have been shown to produce novel classes of plant hormones like flavonoids and strigolactones. These are exuded by the plant roots as part of early signal exchange and recognition by rhizobia and AMF, respectively (Liu and Murray 2016; Akiyama et al. 2005; Besserer et al. 2006). However, the perception of these signals by initial plant receptors has been shown to be different. For example, in legumes (*Lotus japonicus*) the perception of NFs is by two LysM-type receptors, LjNFR1/LjNFR5 (Geurts et al. 2016). A conceptual model of the molecular players involved in the establishment and development of the LRS and AMS (Oldroyd et al. 2009; Ercolin and Reinhardt 2011; Genre and Russo 2016) is depicted in Fig. 6.2.

Briefly, it is shown that at least seven genes are involved and required for both AM and RN symbioses. These genes encode proteins which are involved either directly or indirectly in signal transduction network elicited by the development of arbuscules and nodules. The respective genes were identified to be a leucine-rich-repeat (LRR) receptor kinase (e.g., LjSYMRK) (Endre et al. 2002; Stracke et al. 2002; Yoshida and Parniske 2005), a cation channel (e.g., LjCASTOR and LjPOLLUX) (Imaizumi-Anraku et al. 2005), a putative nuclear pore component (e.g., LjNUP133 and LjNUP85) (Kanamori et al. 2006; Saito et al. 2007, respectively), and a calcium- and calmodulin-dependent protein kinase (e.g., LjCCaMK and LjCYCLOPS) (Lévy et al. 2004; Gleason et al. 2006; Tirichine et al. 2006). Further, functional genomic studies of *lotus* plants carrying mutations in genes of the "CSP/CSSP," e.g., SYMRK, showed upon rhizobial infection, a nod factor (NF)-dependent root hair deformation response which led to the identification of genes upstream and/or existence of a parallel pathway (Stracke et al. 2002). The one distinguishing feature of legumes in perceiving rhizobial-specific signal molecules was later exploited to identify two LysM-type serine/threonine receptor kinase genes, NFR1 and NFR5, the putative NF-receptor genes (Radutoiu et al. 2003). Thus, it appears though the two different symbioses share central components of the signaling pathways triggering their symbiotic associations, there seems to be a coexistence of other distinct and parallel signaling pathways specific for LR and/or

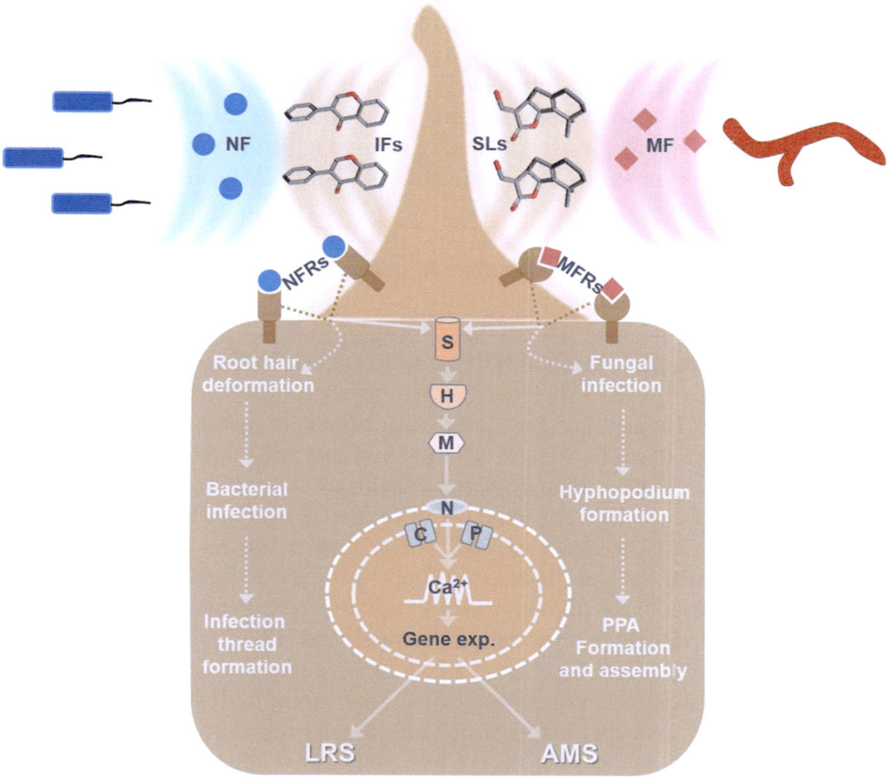

Fig. 6.2 Diagrammatic representation of a "common symbiosis signaling pathway" that governs both the *LRS* and *AMS*. Sensing of *N* and *P* limitation conditions in the soil leads to the formation and release of plant hormones, isoflavones (*IFs*) and strigolactones (*SLs*) leading to the activation of rhizobial and AM fungal response through the release of symbiotic signals nod factors (*NFs*) and Myc factors (*MFs*), respectively. Recognition of *NFs* and *MFs* activates a common signal transduction pathway, transducing characteristic cell-specific responses. The common components of the pathway are abbreviated as follows: *NFR* nod factor receptor, *MFR* Myc factor receptor, *S* receptor-like kinase SYMRK, *H* the biosynthetic enzyme HMGR1 (3-hydroxy-3-methylglutaryl CoA reductase-1), *M* mevalonate biosynthesis, *N* nucleoporins, *C* and *P* potassium channel of nuclear membranes (Castor/Pollux), Ca^{2+} induction of NF and MF specific calcium transients, and finally the Ca^{2+} signals are transduced by a Ca^{2+}/calmodulin-dependent protein kinase (CCaMK) and an interacting transcription factor (Cyclops)

AM symbioses. This led to the proposal of a model in which the rhizobial NFs and the putative AM fungal signals (Myc factors) activate signaling pathways having common components, the SYMRK proteins (Parniske 2008). Recent studies, however, have suggested the wider role for CSP/CSSP proteins in symbiotic and non-symbiotic signaling (Genre and Russo 2016). Furthermore, other genetic screens for elucidation of signaling events in the early (rhizodermal cells) to late (inner cortical cells) colonization stages of mycorrhizal infection and symbiosome development have led to the identification of many other key regulators (Gobbato et al. 2012; Gobbato 2015; Xue et al. 2015), chemical (e.g., lipid-derived) signals (Vijayakumar et al. 2016; Siebers et al. 2016), and fatty acid biosynthesis program (Jiang et al. 2017; Luginbeuhl et al. 2017), respectively. It remains to be seen as to what one may decipher from studies on symbiotic signaling in non-legumes or other molecular genetic studies in legume hosts, which may provide a picture of mycorrhizal associative processes of importance to sustainable agriculture and forest ecosystems.

6.4 Nitrogen and Phosphate: Acquisition, Uptake, and Assimilation

6.4.1 Symbiotic/Biological Nitrogen Fixation (S/BNF)

N is an important element and the single most important agricultural input that is critical for crop production. Recent data showed that the global total N fertilizer consumption derived from both cropland expansion and fertilizer application has increased *c.* tenfold, while in the same period (1961–2013) P fertilizer consumption increased *c.* fourfold (Lu and Tian 2017). The analysis of the effects of human-induced N/P imbalances and the effects of increasing the N/P ratio of atmospheric depositions suggests that P limitation will be met first (due to slow rate of P release from weathering compared to losses from erosion) and N limitation second in most regions of the world (Penuelas et al. 2013). Further, a recent study on mitigation of excessive eutrophication under conventional application of P fertilizers has provided evidence for a successful use of synthetic apatite nanoparticles as an effective alternative to commonly used soluble P fertilizers thus minimizing eutrophication (Liu and Lal 2014). Given that various forms of N (chemical and/or inorganic fertilizers, organic manure, etc.) have been a relatively inexpensive input, farmers worldwide have had an economic incentive in adding a little extra in their efforts for increasing crop production and yield (Ribaudo et al. 2011). However, the increased application of large amounts of nitrogenous compounds has had a number of undesirable impacts on aquatic and terrestrial ecosystems and atmospheric resources. The increasing use of inexpensive reactive N in agriculture has also increased its potential to be lost to the environment as other forms of reactive N, like ammonia (NH_3), ammonium (NH_4^+), nitrogen oxides (NO_x), nitrous oxide (N_2O), and nitrate (NO_3^-), giving rise to a serious concern relating to our soil and water pollution (Galloway et al. 2003; Masclaux-Daubresse et al. 2010). The direct effects of reactive N on species composition, diversity, dynamics, and the functioning of terrestrial, freshwater, and marine ecosystems have been previously reported (Matson and Vitousek 2006). In addition, human-induced increases in reactive N emissions into the environment have also been shown to contribute toward harmful changes to ecosystems: ozone-induced injury to crops and forests, eutrophication, greenhouse gas emissions, soil acidification, and NO_3^- contamination of drinking water aquifers (Ribaudo et al. 2011; Chen et al. 2011), all playing part in a larger consortium affecting global climate change. Furthermore, the consequences of excess N in the

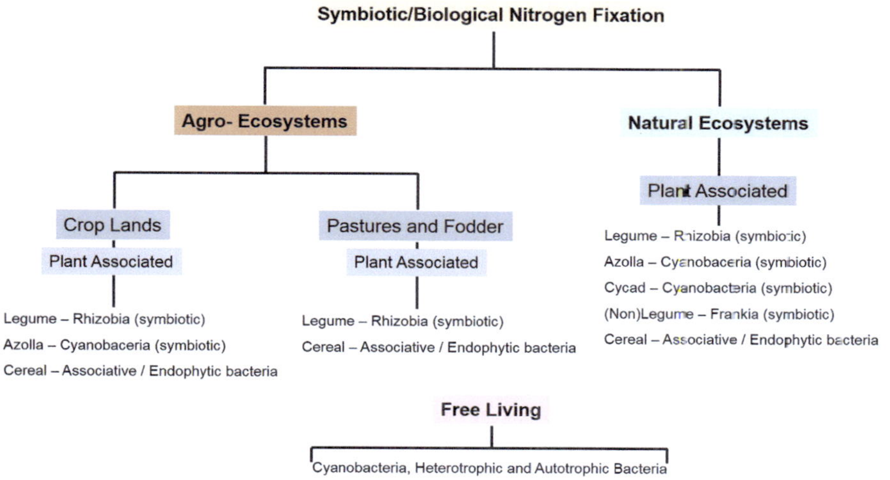

Fig. 6.3 Summary of biological nitrogen-fixing organisms found in natural and agroecosystems

environment is compounded by other factors like the loss of wetlands and the increase in impervious surfaces, such as asphalt roads and parking lots. Given these pressing conditions/factors of aggressive farming for increased food production, the biological nitrogen fixation (BNF) capability of the members of the family of *Rhizobiaceae* is critical for the management of both forest and agroecosystems. Driving the point further is the suggested addition of *c*. 40 million tons of N each year to agroecosystems by the activity of BNF in LRS (Clarke et al. 2014).

In temperate climates and agricultural soils, plants, fungi, and bacteria take up NH_4^+ and NO_3^- as the predominant sources of N, compared to amino acids and other N-based organic acids (Jackson et al. 2008). NO_3^-, the most important source of N available for crops (Wang et al. 2012), is present in mM concentrations in soil solution. However, NO_3^- being more mobile than NH_4^+ and requiring high energy costs of assimilation and reduction to NH_4^+ most often tends to be least favored as compared to NH_4^+ (Courty et al. 2015). Plant roots can transport both NO_3^- and NH_4^+ by the processes of mass flow and diffusion; however, symbiotic associations of plant roots with AMF tend to increase the surface area of root absorption to a far greater extent and beyond the immediate vicinity of root surfaces. Most natural ecosystems are characterized by nutrient-poor or nutrient-limiting conditions, and under such circumstances, the BNF allows for the incorporation of N_2 from the atmosphere directly into plants for their proper growth and development. In the absence of BNF by rhizobacteria, the amounts of NH_3 that could be readily assimilated by the plant are limited, and the anthropogenic addition of NH_3 and/or nitrate fertilizer compounds the problem of surface and groundwater pollution further (Lambers et al. 2008). Also, considering the high energy costs for soil bacteria to reduce NO_3^- toNH_4^+ and the prerequisites for the presence of other soil-borne bacteria which under intensive agricultural practices may not be amenable as sources of soil fertility. A summary of the BNF carried out by the specialized group of prokaryotes adapted from authors Herridge et al. (2008) and Wagner (2012) is provided in Fig. 6.3. The importance of symbiotic BNF can further be accentuated by the estimated 122 T gN (= million tons of N) of the total amount of biological nitrogen fixed annually and globally, which is equal to or more than the estimated annual production of N fertilizers by artificial Harber-Bosch process (Ohyama 2017), suggesting a massive role of BNF in global N cycling. While estimates of global N fixed in other symbiotic, associative, and free-living bacteria are very difficult, informed analyses suggest a total of 50–70 T gN to be fixed from biological agents in agricultural ecosystems (Herridge et al. 2008; Ohyama 2017). Further, the total amount of global BNF fixed by LRS has been calculated ("with some degree of confidence") to be 21 T gN annually, with soybean capable of fixing 16.4 T gN annually (Herridge et al. 2008). Taken together, this unequivocally suggests that BNF is "still the most important N source in agro-ecosystems" (Ianneta et al. 2016; Ohyama 2017).

6.4.2 Phosphate Acquisition and Uptake

Phosphorous, an essential macronutrient, occurs in the environment as inorganic orthophosphate (P_i) forming inert complexes with cations such as iron phosphate ($FePO_4$) and aluminum phosphate ($AlPO_4$) and in organic molecules like phytate, lecithin, etc. P_i is the only form directly accessible to plants, and their concentration rarely exceeds 10 μM in soil solutions. This poses a serious problem for plant growth and productivity, for P_i concentration in soil solution is low in the sub micromolar range ($\leq$ 10 μM) while the plant requirements are high in the millimolar (mM) range (Schachtman et al. 1998; Javot et al. 2007). An additional problem for efficient P_i uptake by plants is posed by its immobile nature in the soil compared to other mineral nutrients. Hence, an increased acquisition of P_i by the roots quickly generates a P_i depletion zone surrounding the epidermis and the root hairs, for the rate of P_i uptake exceeds the rate of P_i diffusion into the soil solution. The uptake mechanism in which plants directly acquire P_i from the soil environment by the root epidermis and root hairs is defined as the direct P_i uptake (DPU) pathway, and the mechanism where P_i is taken up from the soil solution by arbuscular mycorrhizal associations is defined as the mycorrhizal P_i uptake (MPU) pathway (Bucher 2007; Smith and Smith 2011). Plants forming symbiotic associations with fungi benefit from their efficient methods of P_i acquisition capacity. The fungal mycelia grow up to 100 times more than the plant root hairs, thus increasing the absorptive surface area of the plants from few centimeters surrounding their roots to many meters far away from the P_i depletion zone and beyond their normal reach. Following acquisition of P_i, (DPU and/or MPU pathways) the plant still needs to distribute the acquired P_i to different tissues throughout the plant, by the cycling of P_i between the roots and shoots via the conductive tissues, the xylem, and the phloem. Changes in P_i concentrations both within and outside their environment impact many cellular responses of plants. Hence, the P_i levels within the cytoplasm and intracellular

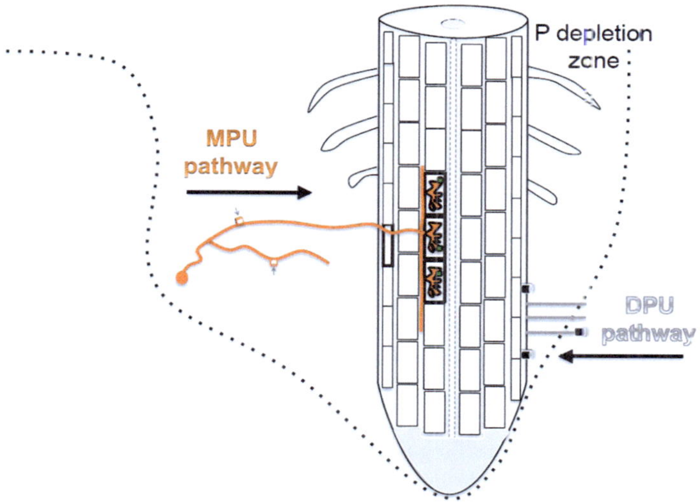

Fig. 6.4 Schematic representation of the direct- and mycorrhizal-phosphate uptake (DPU and MPU) pathways in arbuscular mycorrhizas

organelles need to be tightly controlled, and here plants have evolved and developed many strategies in regulating and maintaining P_i levels within strictly manageable limits. In cellular processes, P_i often needs to be transported across unfavorable electrochemical gradients, and active transporters are needed in this transporting of P_i across the membranes and into the root cells and other subcellular compartments. Through various expression patterns and substrate specificities, the plant P_i transporters (symporters and translocators) further interconnect the metabolism of subcellular compartments and tissues providing flexible ways of altering metabolic fluxes throughout the plant whenever changes in P_i are encountered (Javot et al. 2007). The existence of such efficient transport mechanisms suggests that most transport proteins in plant cells are energized by the creation of electrochemical gradient of protons across the plasma membrane. The formation of electrochemical gradients across the plasma membrane is due to the action of plasma membrane H^+ pumps fueled by ATP (H^+-ATPases). A diagrammatic representation of the direct and mycorrhizal phosphate uptake pathways and P_i efflux across the plasma membrane (adapted from Bucher 2007, and Smith and Smith 2011) are depicted in Fig. 6.4.

6.5 The Biological Triad, Potential for Soil Sustainability

The biological triad in this context refers to the symbiotic tripartism in model legumes and their beneficial rhizospheric processes by symbiotic bioactivity and soil sustainability. An improvement in nutrient management has been a longstanding goal of many developing and developed countries and in the USA; conservation programs within the United States Department of Agriculture (USDA) have aimed at providing financial and technical assistance to farmers in adopting recommended management practices (RMPs) to reduce N losses to the environment, including nutrient and manure management and planning (Lal 2012). However, N-related problems and barriers still persist, with large amounts of cropland not being properly farmed using good N management practices. This has led to the suggestion of "community-based natural resources management" as a useful strategy to mitigate and restore contaminated and degraded soils by functionally enhancing the ecosystem services (Lal 2012). While certain pioneering studies in urban agriculture for crops and fruit plants have provided novel answers to the challenges of increase in urban land acreage, productivity in urban vacant land and improvement in soil quality at previously degraded crop/urban sites (Zezza and Tasciotti 2010; Beniston and Lal 2012; de Souza et al. 2013; Beniston et al. 2016) the full-scale utility of such systems in smallholder farms and in agricultural practices is largely desired. A shift in urban land uses toward grain legumes and further exploring the genetic improvements in legumes may provide interesting answers to the contribution of legumes toward sustainable urban cropping systems (Anglade et al. 2016).

Legumes make a viable alternative to environment- friendly sustainable agriculture in both urban and agroecosystems thus playing a vital role in plant ecosystems worldwide. The establishment of symbiotic tripartism and the understanding of the factors that govern the association of the organisms of the three kingdoms (plantae, bacteria, and fungi) are fundamental to developing predictive models in sustainable agriculture. In the absence of symbiotic tripartism, the recruitment and assemblage of soil communities promoting plant growth are minimal (Nihorimbere et al. 2011; Rascovan et al. 2016). The processes of beneficial interactions of legumes with rhizobia and AMF improves soil fertility and plant productivity, which in turn has a profound influence on the rhizospheric processes favoring the practices of good agricultural management and long-term sustainability. Hence, in many cases dual inoculations (*Rhizobia* in combination with AMF and/or other beneficial bacteria) have served as simple experimental models to understand their interaction effects in soil-plant interface. Many studies have shown the synergistic beneficial effects of such dual inoculations on a range of host plants belonging to the family of *Leguminosae* (Table 6.2). These benefits in terms of plant biomass have provided solutions to a variety of sustainable agriculture demands. Interestingly, some studies have also shown a negative impact of dual inoculations on certain variety of host plants, suggesting the regulation of symbiotic interactions by both photo- and micro- bionts (Franzini et al. 2010; Sakamoto et al. 2013). However, the positive impacts of dual inoculation were observed on most other family members of *Leguminosae* (Table 6.2). The beneficial effects of dual inoculation on warm-season legumes such as soybean, cowpea, common bean, groundnut, and pigeon pea are heartening, for the use of soybean, for example, in multiple cropping systems (as cover crop; intercropping and rotation crop), has shown major benefits toward enhanced yields, NUE, and reduced incidence of disease, etc. (Meng et al. 2015; Foyer et al. 2016; Marzban et al. 2017). The increase in NUE of cereal crops in rotation with legumes is critical to the development of sustainable agriculture with less

Table 6.2 Overview of certain recent studies (2010–2017) shown to have reported synergistic effects of dual inoculations in the family of legumes/beans/peas

Rhizobial type and/or strain no.	AMF type	Host plant and/or plant variety	Study effect or conclusion	Reference(s)
Rhizobium tropici 899 (Rh I) and *Rhizobium* 912 (Rh II)	*Glomus intraradices* and *Glomus mossae*	*Phaseolus vulgaris* L. cvs Efequince (L-1); Romano bush (L-2); contender (L-3); Borlotto (L-4)	Host- and variety-dependent inhibition of nodule development and N_2-fixation	Franzini et al. (2010)
Rhizobium (AcM05)	*Glomus fasciculatum*	*Acacia mellifera*	Dual inoculation and N/P/– supplementation showed increased biomass, total N, and P content. Further, increased nodule nitrogenase activity, dry-matter yield, and total chlorophyll content in leaves were observed	Lalitha et al. (2011)
Rhizobium tropici (CIAT899)	*Glomus intraradices* (BEG 157)	*Phaseolus vulgaris* L., CocoT, and flamingo	Dual inoculation showed significantly increased plant-growth parameters, nodulation, N and P accumulation, and P-use efficiency	Tajini et al. (2012)
Mesorhizobium cicero	*Glomus intraradices*, *Glomus etunicatum* and *Glomus mossae*	*Cicer arietinum* L. cv.ILC482	Dual inoculation enhances N, P, zinc, iron, and copper content in plants	Tavasolee et al. (2013)
Rhizobium sp.	*Glomus fasciculatum*	*Cajanus cajan* L.	Increased N and P content; potential enhancing of N, P, and chlorophyll contents of *C. cajan*	Bhattacharjee and Sharma (2012)

(continued)

Table 6.2 (continued)

Rhizobial type and/or strain no.	AMF type	Host plant and/or plant variety	Study effect or conclusion	Reference(s)
Rhizobium leguminosarum	*Glomus etunicatum*	*Phaseolus vulgaris* var. contender	External NH_4^+ supply negatively affects AM colonization and BNF. However, the presence of AM at moderate NH_4^+ levels improves nodule growth and BNF	Mortimer et al. (2012)
Bradyrhizobium japonicum MN-S and TAL-102	*Glomus intraradices*	*Vigna radiata*	Dual inoculation showed potential for plant growth promotion and for enhancing crop production of *V. radiata*	Yasmeen et al. (2012)
Bradyrhizobium japonicum USDA110	*Gigaspora rosea*	*Glycine max* (L.) Merr. Enrei and Kanto100	AM fungal colonization has a stimulatory effect on nodulation. While autoregulation of host plants is initiated upon nodulation, this suppresses AM fungal colonization	Sakamoto et al. (2013)
*Bradyrhizobiumjaponicum*SH212	*Glomus mossae*	*Glycine max* L. cv. Dongnong no.42 and *Zea mays* L. cv. Dongnong no.48	Dual inoculation significantly increased N_2-fixation efficiency of soybean in the soybean/maize intercropping system and promoted N transfer from soybean to maize	Meng et al. (2015)
Rhizobium leguminosarum L.	*Glomus mosseae* L.	*Pisum sativum* L.	Dual inoculation showed potentially enhanced productivity, profitability, and NUE in pea productivity in Himalayan acid Alfisol	Bai et al. (2016)
Bradyrhizobium japonicum	*Glomus mosseae*	*Glycine max* L. Merr.	In P-deficient environments, plant interactions with *Rhizobia* and AMF lead to regulation of symbiotic partners best suited for NUE	Ding et al. 2016

Rhizobial type and/or strain no.	AMF type	Host plant and/or plant variety	Study effect or conclusion	Reference(s)
Rhizobium leguminosarum	*Glomus fasciculatum*, *Glomus mosseae*,and *Glomus aggregatum*	*Lens culinaris* NARC.2008–4	Dual inoculation showed enhanced (45%) seed yield per plant compared to 24% and 28% for VAM and *Rhizobial* single inoculations	Yaseen et al. (2016)
Bradyrhizobium diazoefficiens USDA 110	*Gigaspora rosea* MAFF520062	*Glycine max* (L.) Merr.	Dual inoculation under excess Zn, supported higher-shoot biomass by modulating Mn, Fe, P, and polyphenol levels	Ibiang et al. (2017)
Rhizobium leguminosarum	*Glomus mosseae*	*Phaseolus vulgaris* "Derakhshan" and *Zea mays* 'S.C. 704'	Intercropping of maize and soybean beneficial for the increase in leaf N and P concentrations and root growth of maize	Marzban et al. (2017)
*Mesorhizobiummediterraneum*UPM-Ca36	*Rhizophagus irregularis* BEG 140	*Cicer arietinum* L. cv. Chk 4198	Dual inoculation promotes significant increases in biomass and crude grain protein content under water-deficit conditions	Oliveira et al. (2017)
Rhizobia (*Ensifer* sp.)	*Glomus mosseae*	*Sesbania cannabina*	Tripartite symbiosis lead to the enhanced degradation of polycyclic aromatic hydrocarbons	Ren et al. (2017)
Bradyrhizobium japonicum	*Glomus macrocarpum* and *Acaulospora colombiana*	*Phaseolus lunatus* L. Walp	X-ray microanalytical studies showed 13 elements to be abundant under tripartite symbiotic conditions	Rodak et al. (2017)

(continued)

Table 6.2 (continued)

Rhizobial type and/or strain no.	AMF type	Host plant and/or plant variety	Study effect or conclusion	Reference(s)
Rhizobium and PSB	*Acaulospora scrobiculata* and *Rhizophagus irregularis*	*Glycine max* (L.) Merr. Var. JS 93–05, *Phaseolus mungo* Roxb. Var. IPU2–43, *Vigna radiata* (L.) R. Wilczek var. Samrat, *Cicer arietinum* L. var. DCP.92–3, *Lens culinaris* Medikus var. DPL-15, and *Pisum sativum* var. Vikas	Shade adversely affected growth, biomass, and yield of all plant species tested. However, application of *Rhizobium* and/or PSB in addition to AMF increased growth of all six pulses, irrespective of their shade or non-shade conditions	Shukla et al. (2017)

and lesser use of N fertilizers for productivity and environmental stability. For it has been predicted that in the years from 1961 to 2000, the NUE of cereals had decreased from *c*. 80% in 1961 to around 30% in the year 2000 (Erisman et al. 2008; Tilman et al. 2002), with more than 50–80% of applied N fertilizer to cereal crops lost as agricultural runoff (UNEP and WHRC 2007; Foyer et al. 2016).

The symbiotic tripartism in managed agroecosystems of legume cultivation also improves soil ecology and fertility in systems where crop rotation is being implemented. In combination with practices of no – or conservation – tilling (NT or CT) (Islam and Reeder 2014; Busari et al. 2015), the use of legumes as rotation crop provides the other crops in rotation the essential nutrients that are most required for their growth and productivity. The approach of rotation of legumes and intercropping of legumes with cereal crops or other non-legumes has tremendous benefits for yield, productivity, the NUE, reduced disease occurrence, and improved access to other mineral elements, e.g., P by the recruitment of other P-solubilizers (Foyer et al. 2016; Chapelle et al. 2016; Wittwer et al. 2017). Functional relationships of legumes with naturally occurring AMF combinations will further enhance uptake of N, P, and other microelements and confer plant protection, thus having a net positive impact on soil ecology and sustainability. This way substantially higher net yields could still be maintained without the use of additional source of mineral fertilizers, which we know now are detrimental to the cause of environmentally friendly sustainable agriculture.

The increased bioactivity in soils and rhizodeposition also led to the recruitment of other potential beneficial rhizosphere microbiome and probable factors of plant protection. The use of grain legumes has also shown reduced greenhouse gas emissions (Foyer et al. 2016), and in symbiotic tripartism, they also favor bioremediation of polluted soils. And fungi being primarily absorptive feeders depend largely on their ability to feed by secretion of extracellular enzymes (osmotrophy) and, hence, display an extensive biochemical capacity to degrade environmental pollutants (Harms et al. 2011). However, the combined capacities of fungi and bacteria in functional and associative collaborations may harbor other important features useful for bioremediation of polluted environments and soils.

6.6 Conclusions

Application of synthetic N and P fertilizer(s) in agricultural production systems has posed a major challenge to global cropland agroecosystems. For mitigation of anthropogenic inputs, the use of "cover crop functional groups (legumes versus non-legumes)" provides the best possible management strategy. Further, the inclusion of legumes in management and agricultural productivity strategies confers additional benefits due to their inherent capacity to form symbiotic relationships with rhizosphere-microbiomes. Both AMS and RNS play key roles in their ecosystems and exert influence on vast majority of soil microbial populations and root exudates that shape the rhizosphere. Symbiotic associations thus contribute immensely to C storage in soils and in altering the quality and quantity of SOM, hence soil sustainability.

Given the uniqueness and global importance of BNF and AMS in the biosphere, in addition to the limited fossil energy resources, and recurring health and environmental concerns regarding agricultural chemical inputs, the potential role of legumes in agricultural practices is beginning to be realized on a global scale. This promises to stimulate and promote a wider adoption of leguminous crops (grain legumes) for progressive replacements of largely industrialized cereal-crop systems and coping with environmental challenges and meeting the demands of SDGs of large-scale sustainable agriculture and crop production systems.

6.7 Future Perspectives

The rhizospheric soil assemblages and communities are diverse and require an increased understanding of N and P acquisition strategies under rhizobial and mycorrhizal associations (Courty et al. 2015). The identification of key molecular players in the two evolutionarily related symbiotic systems of AMS and LRS has provided valuable insights into the deciphering of symbiotic processes and their potential for sustainable agriculture through the use of legumes (Gobbato 2015; Kamel et al. 2016; Hartman et al. 2017). Further, a concerted push for protein-rich food sources in the quest for food security and the symbiotic capabilities of legumes places them at the forefront of sustainable cropping systems and agricultural sustainability in both the developed and developing nations (Foyer et al. 2016). The long-term SDGs of "Food and Agriculture Organization (FAO) of the United Nations" in achieving a successful transition to sustainable agricultural development can largely be supported by the capacity for BNF, P-acquisition, and NUE offered by grain legumes. Further, the enormous potential that a boost in grain legume production provides may serve as an important catalyst in pivoting large-scale efforts in reducing pesticide and mineral fertilizer use and environmental sustainability (Vance 2001; Parniske 2008; Foyer et al. 2016). Taken together, notable future investigations into the natural variations in AM and rhizobial function and plant responsiveness in terms of productivity, food security, and soil fertility may provide a much-needed complementarity in meeting the demands of sustainable agriculture and an increasing demand for global food security.

References

Akiyama K, Matsuzaki K, Hayashi H (2005) Plant sesquiterpenes induce hyphal branching in arbuscular mycorrhizal fungi. Nature 435:824–827

Anand SS, Hawkes C, de Souza RJ, Mente A, Dehghan M, Nugent R, Zulyniak MA, Weis T, Bernstein AM, Krauss RM, Kromhout D, Jenkins DJA, Malik V, Martinez-Gonzalez MA, Mozaffarian D, Yusuf S, Willett WC, Popkin BM (2015) Food consumption and its impact on cardiovascular disease: importance of solutions focused on the globalized food system. J Am Coll Cardiol 66:1590–1514

Anglade J, Medina MR, Billen G, Garnier J (2016) Organic market gardening around the Paris agglomeration: agro-environmental performance and capacity to meet urban requirements. Environ Sci Pollut Res. https://doi.org/10.1007/s11356-016-6544-1

Bai B, Suri VK, Kumar A, Choudhary AK (2016) Influence of dual inoculation of AM fungi and *Rhizobium* on growth indices, production economics, and nutrient use efficiencies in garden pea (*Pisum sativum* L.). Commun Soil Sci Plant Anal 47:941–954
Beniston JW, Lal R(2012) Improving soil quality for urban agriculture in the north central U.S. in carbon sequestration in urban ecosystems, Lal R, Augustin B (eds) Springer, Dordrecht, pp 279–314
Beniston JW, Lal R, Mercer KL (2016) Assessing and managing soil quality for urban agriculture in a degraded vacant lot soil. Land Degrad Develop 27:996–1006
Berg G (2009) Plant–microbe interactions promoting plant growth and health: perspectives for controlled use of microorganisms in agriculture. Appl Microbiol Biotechnol 84:11–18
Beringer JE, Brewin N, Johnston AWB, Schulman HM, Hopwood DA (1979) The rhizogium-legume symbiosis. Proc R Soc Lond 204:219–233
Besserer A, Puech-Pages V, Kiefer P, Gomex-Roldan V, Becard G, Sejalon-Delmas N (2006) Strigolactones stimulate arbuscular mycorrhizal fungi by activating mitochondria. PLoS Biol 4: 1239–1247
Bhattacharjee S, Sharma GD (2012) Effect of dual inoculation of arbuscular mycorrhiza and *Rhizobium* on the chlorophyll, nitrogen and phosphorus contents of pigeon pea (*Cajanus cajan* L.). Adv Microbiol 2:561–564
Bonfante P, Genre A (2010) Mechanisms underlying beneficial plant-fungus interactions in mycorrhizal symbiosis. Nat Commun 1:48
Brevik EC (2013) The potential impact of climate change on soil properties and processes and corresponding influence on food security. Agriculture 3:398–317
Brevik EC, Sauer TJ (2015) The past, present, and future of soils and human health studies. Soil 1:35–46
Brewin NJ (2010) Root nodules (Legume–Rhizobium Symbiosis). eLS. https://doi.org/10.1002/9780470015902.a0003720.pub2
Brown LK, George TS, Dupuy LX, White PJ (2013) A conceptual model of root hair ideotypes for future agricultural environments: what combination of traits should be targeted to cope with limited P availability? Ann Bot 112:317–330
Brundrett MC (2002) Coevolution of roots and mycorrhizas of land plants. New Phytol 154:275–304
Bucher M (2007) Functional biology of plant phosphate uptake at root and mycorrhiza interfaces. New Phytol 173:11–26
Bucher M, Hause B, Krajinski F, Kuester F (2014) Through the doors of perception to function in arbuscular mycorrhizal symbioses. New Phytol 204:833–840
Busari MA, Kukal SS, Kaur A, Bhatt R, Dulazi AA (2015) Conservation tillage impacts on soil, crop and the environment. Int Soil Water Conserv Res 3:119–129
Chapelle E, Mendes R, Bakker PA, Raaijmakers JM (2016) Fungal invasion of the rhizosphere microbiome. ISME J 10:265–268
Chen X, Cui Z, Vitousek PM, Cassman KG, Matson PA, Romheld V, Zhang F (2011) Integrated soil-crop system management for food security. Proc Natl Acad Sci 108:6399–6404
Clarke VC, Loughlin PC, Day DA, Smith PMC (2014) Transport processes of the legume symbiosome membrane. Front Plant Sci 5:699
Courty PE, Smith P, Koegel S, Redecker D, Wipf D (2015) Inorganic nitrogen uptake and transport in beneficial plant root-microbe interactions. CRC Crit Rev Plant Sci 34:4–16
D'Haeze W, Holsters M (2002) Nod factor structures, responses and perception during initiation of nodule development. Glycobiology 12:79–805
de Souza RG, da Silva DKA, de Mello CMA, Goto BT, da Silva FSB, Sampaino EVSB, Maia LC (2013) Arbuscular mycorrhizal fungi in revegetated mined dunes. Land Degrad Develop 24:147–155
Dhakal Y, Meena RS, Kumar S (2016) Effect of INM on nodulation, yield, quality and available nutrient status in soil after harvest of green gram. Leg res 39(4):590–594
Ding X, Zhang S, Wang R, Li S, Liao X (2016) AM fungi and rhizobium regulate nodule growth, phosphorous (P) uptake, and soluble sugar concentration of soybeans experiencing P deficiency. J Plant Nutr 39:1915–1925

Endre G, Kerszt A, Kevei Z, Milhacea S, Kalo P, Kiss GB (2002) A receptor kinase gene regulating symbiotic nodule development. Nature 417:962–966

Ercolin F, Reinhardt D (2011) Successful joint ventures of plants: arbuscular mycorrhiza and beyond. Trends Plant Sci 16:356–362

Erisman JW, Sutton MA, Galloway J, Klimont Z, Winiwarter W (2008) How a century of ammonia synthesis changed the world. Nat Geosci 1:636–639

Fellbaum CR, Gachomo EW, Beesetty Y, Choudhari S, Kiers ET, Bueking H (2012) Carbon availability triggers fungal nitrogen uptake and transport in arbuscular mycorrhizal symbiosis. Proc Natl Acad Sci U S A 109:2666–2671

Foyer CH, Lam HM, Nguyen HT, Siddique KH, Varshney RK, Colmer TD, Cowling W, Valliyodan B, Considine MJ (2016) Neglecting legumes has compromised human health and sustainable food production. Nat Plants 2:16112

Franzini VI, Azcon R, Mendes FL, Aroca R (2010) Interactions between *Glomus* species and *Rhizobium* strains affect the nutritional physiology of drought-stressed legume hosts. J Plant Physiol 167:614–619

Galloway JN, Aber JD, Erisman JW, Seitzinger SP, Howarth RW, Cowling EB, Cosby BJ (2003) The nitrogen cascade. Bioscience 53:341–356

Gao X, Lu X, Wu M, Zhang H, Pan R, Tian J, Li S, Liao H (2012) Co-inoculation with rhizobia and AMF inhibited soybean red crown rot: from field study to plant defense- related gene expression analysis. PLoS One 7:e33977

Genre A, Russo G (2016) Does a common pathway transduce symbiotic signals in plant–microbe interactions? Front Plant Sci 7:96

Geurts R, Xiao TT, Reinhold-Hurek B (2016) What does it take to evolve a nitrogen-fixing endosymbiosis? Trends Plant Sci 21:199–208

Giehl RFH, von Wiren N (2014) Root nutrient foraging. Plant Physiol 166:509–517

Gleason C, Chaudhuri S, Yang T, Munoz A, Poovaiah BW, Oldroyd GED (2006) Nodulation independent of rhizobia induced by a calcium-activated kinase lacking autoinhibition. Nature 441:1149–1152

Gobbato E (2015) Recent development in arbuscular mycorrhizal signaling. Curr Opin Plant Biol 26:1–7

Gobbato E, Marsh JF, Vernie T, Wang E, Maillet F, Kim J, Miller JB, Sun J, Bano SA, Ratet P (2012) GRAS-type transcription factor with a specific function in mycorrhizal signaling. Curr Biol 22:2236–2241

Guissou T (2009) Contribution of arbuscular mycorrhizal fungi to growth and nutrient uptake by jujube and tamarind seedlings in a phosphate (P-) deficient soil. Afr J Microbiol Res 3:297–204

Harms H, Schlosser D, Wick LY (2011) Untapped potential: exploiting fungi in bioremediation of hazardous chemicals. Nat Rev Microbiol 9:177–192

Hartman K, van der Heijden MGA, Roussely-Provent V, Valser JC, Schlaeppi K (2017) Deciphering composition and function of the root microbiome of a legume plant. Microbiome 5:2

Hause B, Schaarschmidt S (2009) The role of jasmonates in mutualistic symbioses between plants and soil-born microorganisms. Phytochemistry 70:1589–1599

Herridge DF, Peoples MB, Boddey RM (2008) Global inputs of biological nitrogen fixation in agricultural systems. Plant Soil 311:1–18

Holmer R, Rutten L, Kohlen W, van Velzen R, Geurts R (2017) Commonalities in symbiotic plant-microbe signaling. Adv Bot Res 82. ISSN 0065-2296

Ianneta PP, Young M, Bachinger J, Bergkvist G, Doltra J, Lopez-Bellido RJ, Monti M, Pappa VA, Reckling M, Topp CF, Walker RL, Rees RM, Watson CA, James EK, Squire GR, Begg GS (2016) Comparative nitrogen balance and productivity analysis of legume and non-legume supported cropping systems: the potential role of biological nitrogen fixation. Front Plant Sci 7:1700

Ibiang YB, Mitsumoto H, Sakamoto K (2017) *Bradyrhizobia* and arbuscular mycorrhizal fungi modulate manganese, iron, phosphorus, and polyphenols in soybean (*Glycine max* (L.) Merr.) under excess zinc. Environ Exp Bot 137:1–13

Imaizumi-Anraku H, Takeda N, Charpentier M, Perry J, Parniske M, Hayashi M (2005) Plastid proteins crucial for symbiotic fungal and bacterial entry into plant roots. Nature 433:527–531
Islam R, Reeder R (2014) No-till and conservation agriculture in the United States: an example from the David Brandt farm, Carroll, Ohio. Int Soil Water Conserv Res 2:97–07
Jackson LE, Burger M, Cavagnaro TR (2008) Roots, nitrogen transformations, and ecosystem services. Annu Rev Plant Biol 59:341–363
Javot H, Pumplin N, Harrison MJ (2007) Phosphate in the arbuscular mycorrhizal symbiosis: transport properties and regulatory roles. Plant Cell Environ 30:310–322
Jiang Y, Wang W, Xie Q, Liu N, Liu L, Wang D, Zhang X, Yang C, Chen X, Tang D, Wang E (2017) Plants transfer lipids to sustain colonization by mutualistic mycorrhizal and parasitic fungi. Science 356:1172–1175
Johnson NC, Graham JH, Smith FA (1997) Functioning of mycorrhizal associations along the mutualism-parasitism continuum. New Phytol 135:575–585
Kamel L, Keller-Pearson M, Roux C, Ane JM (2016) Biology and evolution of arbuscular mycorrhizal symbiosis in the light of genomics. New Phytol 213:531–536
Kanamori N, Madsen LH, Radutoiu S, Frantescu M, Esben MHQ, Miwa H, Downie JA, James EK, Felle HH, Haaning LL, Jensen TH, Sato S, Nakamura Y, Tabata S, Sandal N, Stougaard J (2006) Anucleoporin is required for induction of Ca2+ spiking in legume nodule development and essential for rhizobial and fungal symbiosis. Proc Natl Acad Sci U S A 103:359–364
Kaur S, Aggarwal R, Lal R (2016) Assessment and mitigation of greenhouse gas emissions from ground water irrigation. Irrig and Drain 65:762–770
Kiers ET, Duhamel M, Beesetty Y, Mensah JA, Franken O, Jansa J, Bucking H (2011) Reciprocal rewards stabilize cooperation in the mycorrhizal symbiosis. Science 333:880–882
Kouris-Blazos A, Belski R (2016) Health benefits of legumes and pulses with a focus on Australian sweet lupins. Asia Pac J Clin Nutr 25:1–17
Lal R (2012) Climate change and soil degradation mitigation by sustainable management of soils and other natural resources. Agric Res 1:199–212
Lalitha S, Rajeshwaran K, Senthil kumar P, Kumar S (2011) Role of AM fungi and rhizobial inoculation for reclamation of phosphorus deficient soil. Asian J Plant Sci 10:227–232
Lambers H, Raven JA, Shaver GR, Smith SE (2008) Plant nutrient-acquisition strategies change with soil age. Trends Ecol Evol 23:95–103
Lavin M, Herendeen PS, Wojciechowski MF (2005) Evolutionary rates analysis of *Leguminosae* implicates a rapid diversification of lineages during the tertiary. Syst Biol 54:574–594
Levy J, Bres C, Geurts R, Chalhoub B, Rosenberg C, Debelle F (2004) A putative Ca2+ and calmodulin-dependent protein kinase required for bacterial and fungal symbioses. Science 303:1361–1364
Liu C-W, Murray JD (2016) The role of flavonoids in nodulation host-range specificity: an update. Plants 5:33
Liu R, Lal R (2014) Synthetic apatite nanoparticles as a phosphorus fertilizer for soybean (*Glycine max*). Sci Rep 4:5686
Lu C, Tian H (2017) Global nitrogen and phosphorus fertilizer use for agriculture production in the past half century: shifted hot spots and nutrient imbalance. Earth Syst Sci Data 9:181–192
Luginbuehl LH, Menard GN, Kurup S, Van Erp H, Radhakrishnan GV, Breakspear A, Oldroyd GED, Eastmond PJ (2017) Fatty acids in arbuscular mycorrhizal fungi are synthesized by the host plant. Science 356:1175–1178
Maillet F, Poinsot V, Andre O, Puech-Pages V, Becard G, Denarie J (2011) Fungal lipochitooligosaccharide symbiotic signals in arbuscular mycorrhiza. Nature 469:58–64
Martin F, Aerts A, Ahren D, Brun A, Danchin EG, Duchaussoy F, Tunlid A, Tuskan G, Grigoriev IV (2008) The genome of *Laccaria bicolor* provides insights into mycorrhizal symbiosis. Nature 452:88–92
Marzban Z, Faryabi E, Torabian Z (2017) Effects of arbuscular mycorrhizal fungi and *Rhizobium* on ion content and root characteristics of green bean and maize under intercropping. Acta Agric Slov 109:79–88

Masclaux-Daubresse C, Daniele-Vedele F, Dechorgnat J, Chardon F, Gaufichon L, Suzuki A (2010) Nitrogen uptake, assimilation and remobilization in plants: challenges for sustainable and productive agriculture. Ann Bot 105:1141–1157

Matson PA, Vitousek PM (2006) Agricultural intensification: will land spared from farming be land spared for nature? Conserv Biol 20:709–710

McCauley A, Jones C, Jacobsen J (2011) Plant nutrient functions and deficiency and toxicity symptoms. Nutrient management module no. 9, CCA 1.5 NM CEU

McNear DH Jr (2013) The rhizosphere – roots, soil and everything in between. Nat Educ Knowl 4:1

Meena RS, Gogaoi N, Kumar S (2017a) Alarming issues on agricultural crop production and environmental stresses. J Clean Prod 142:3357–3359

Meena RS, Meena PD, Yadav GS, Yadav SS (2017) Phosphate solubilizing microorganisms, principles and application of microphostechnology. J Clean Prod 145:157–158

Meena RS, Vijayakumar V, Yadav GS, Mitran T (2018) Response and interaction of Bradyrhizobium japonicum and Arbuscular mycorrhizal fungi in the soybean rhizosphere. A Rev Plant Growth Regul 84:207–223

Mendes LW, Kuramee EE, Navarrete AA, van Veen JA, Tsai SM (2014) Taxonomical and functional microbial community selection in soybean rhizosphere. ISME J 8:1577–1587

Mendes R, Garbeva P, Raaijmakers JM (2013) The rhizosphere microbiome: significance of plant beneficial, plant pathogenic, and human pathogenic microorganisms. FEMS Microbiol Rev 37:634–663

Meng L, Zhang A, Wang F, Han X, Wang D, Li S (2015) Arbuscular mycorrhizalfungiandrhizobium facilitatenitrogenuptakeandtransfer insoybean/maizeintercropping system. Front Plant Sci 6:339

Merckx V, Bidartondo MI, Hynson NA (2009) Myco-heterotrophy: when fungi host plants. Ann Bot 104:1255–1261

Messina M (2010) A brief historical overview of the past two decades of soy and isoflavone research. J Nutr 140:1350S–4S

Messina MJ (1999) Legumes and soybeans: overview of their nutritional profiles and health effects. Am J Clin Nutr 70:439S–450S

Mortimer PE, Perez-Fernandez MA, Valentine MA (2012) Arbuscular mycorrhiza maintains nodule function during external NH4+ supply in *Phaseolus vulgaris* (L.). Mycorrhiza 22:237–245

Nadal M, Paszkowski U (2013) Polyphony in the rhizosphere: presymbiotic communication in arbuscular mycorrhizal symbiosis. Curr Opin Plant Biol 16:473–479

Needelman BA (2013) What are soils? Nat Educ Knowl 4:2

Nihorimbere V, Ongena M, Smargiassi M, Thonart P (2011) Beneficial effect of the rhizosphere microbial community for plant growth and health. Biotechnol Agron Soc Environ 15:327–337

Oburger E, Schmidt H (2016) New methods to unravel rhizosphere processes. Trends Plant Sci 21:243–255

Ohkama-Ohtsu N, Wasaki J (2010) Recent progress in plant nutrition research: cross- talk between nutrients, plant physiology and soil microorganisms. Plant Cell Physiol 51:1255–1264

Ohyama T (2017) The role of legume-rhizobium symbiosis in sustainable agriculture. In Sulieman S, Tran LSP (eds) Legume nitrogen fixation in soils with low phosphorus availability. https://doi.org/10.1007/978-3-319-55729-8_1

Oldroyd GED, Harrison MJ, Paszkowski U (2009) Reprogramming plant cells for endosymbiosis. Science 324:753–754

Oliveira RS, Carvalho P, Marques G, Ferreira L, Vosatka M, Freitas H (2017) Increased protein content of chickpea (*Cicer arietinum* L.) inoculated with arbuscular mycorrhizal fungi and nitrogen-fixing bacteria under water deficit conditions. J Sci Food Agric. DOI https://doi.org/10.1002/jsfa.8201

Ordonez YM, Fernandez BR, Lara LS, Rodriguez A, Uribe-Velez D, Sanders IR (2016) Bacteria with phosphate solubilizing capacity alter mycorrhizal fungal growth both inside and outside the root and in the presence of native microbial communities. PLoS One 11:e0154438

Parikh SJ, James BR (2012) Soil: the foundation of agriculture. Nat Educ Knowl 3:2

Parniske M (2008) Arbuscular mycorrhiza: the mother of plant root endosymbioses. Nat Rev Microbiol 6:763–775

Penuelas J, Poulter B, Sardans J, Ciais P, van der Velde M, Obersteiner M, Janssens IA (2013) Human-induced nitrogen–phosphorus imbalances alter natural and managed ecosystems across the globe. Nat Commun 4:2934

Phillipot L, Raaijmakers JM, Lemanceau P, van der Putten WH (2013) Going back to the roots: the microbial ecology of the rhizosphere. Nat Rev Microbiol 11:789–799

Radutoiu S, Madsen LH, Madsen EB, Felle HH, Sandal N, Stougaard J (2003) Plant recognition of symbiotic bacteria requires two LysM receptor-like kinases. Nature 425:585–592

Rascovan N, Carbonetto B, Perrig D, Diaz M, Canciani W, Abalo M, Alloat J, Gonzalez-Anta G, Vasquez MP (2016) Integrated analysis of root microbiomes of soybean and wheat from agricultural fields. Sci Rep 6:28084

Ren C-G, Kong C-C, Bian B, Liu W, Li Y, Liou Y-M, Zhi-Hong X (2017) Enhanced phytoremediation of soils contaminated with PAHs by arbuscular mycorrhiza and rhizobium. Int J Phytoremed. https://doi.org/10.1080/15226514.2017.1284755

Ribaudo M, Delgado J, Hansen L, Livingston M, Mosheim R, Williamson J (2011) Nitrogen in agricultural systems: implications for conservation policy. Economic Research Service/USDA/ERR-127

Rodak BW, Freitas DS, Bamberg SM, Carneiro MAC, Guilherme LRG (2017) X-ray microanalytical studies of mineral elements in the tripartite symbiosis between lima bean, N2-fixing bacteria and mycorrhizal fungi. J Microbiol Methods 132:14–20

Rubiales D, Mikic A (2015) Introduction: legumes in sustainable agriculture. CRC Crit Rev Plant Sci 34:2–3

Saito K, Yoshikawa M, Yano K, Miwa H, Hayashi M, Kawaguchi M (2007) NUCLEOPORIN85 is required for calcium spiking, fungal and bacterial symbioses, and seed production in *Lotus japonicus*. Plant Cell 19:610–624

Sakamoto K, Ogiwara N, Kaji T (2013) Involvement of autoregulation in the interaction between rhizobial nodulation and AM fungal colonization in soybean roots. Biol Fertil Soils 49:1141–1152

Schachtman DP, Reid RJ, Ayling SM (1998) Phosphorus uptake by plants: from soil to cell. Plant Physiol 11:447–453

Scheublin TA, Ridgway KP, Young JPW, van der Heijden MGA (2004) Nonlegumes, legumes, and root nodules harbor different arbuscular mycorrhizal fungal communities. Appl Environ Microbiol 70:6240–6246

Shukla A, Kumar A, Chaturvedi OP, Nagori T, Kumar N, Gupta A (2017) Efficacy of rhizobial and phosphate-solubilizing bacteria and arbuscular mycorrhizal fungi to ameliorate shade response on six pulse crops. Agrofor Syst. https://doi.org/10.1007/s10457-017-0070-0

Siddique KHM, Johansen C, Turner NC, Jeuffroy MH, Gan Y, Alghamdi SS (2012) Agron Sustain Dev 32:45–64

Siebers M, Brands M, Wewer V, Duan Y, Hoelzl G, Doermann P (2016) Lipids in plant–microbe interactions. Biochim Biophys Acta 1861:1379–1395

Smith SE, Robson AD, Abbott LK (1992) The involvement of mycorrhizas in assessment of genetically dependent efficiency of nutrient uptake and use. Plant Soil 146:169–179

Smith SE, Smith FA (2011) Roles of arbuscular mycorrhizas in plant nutrition and growth: new paradigms from cellular to ecosystem scales. Annu Rev Plant Biol 62:227–250

Smith, Read (2008) Mycorrhizal symbiosis, 3rd edn. Academic press, Boston. ISBN: 9780123705266

Sprent J (2007) Evolving ideas of legume evolution and diversity: a taxonomic perspective on the occurrence of nodulation. New Phytol 174:11–25

Stagnari F, Maggio A, Galieni A, Pisante M (2017) Multiple benefits of legumes for agriculture sustainability: an overview. Chem Biol Technol Agric 4:2

Stougaard J (2000) Regulators and regulation of legume root nodule development. Plant Physiol 124:531–540

Strack D, Fester T, Hause B, Schliemann W, Walter MH (2003) Arbuscular mycorrhiza: biological, chemical, and molecular aspects. J Chem Ecol 29:1955–1979

Stracke S, Kistner C, Yoshida S, Mulder L, Szczyglowski K, Parniske M (2002) A plant receptor-like kinase required for both bacterial and fungal symbiosis. Nature 417:959–962

Sulieman S, Tran L-SP (2015) Legume nitrogen fixation in a changing environment. https://doi.org/10.1007/978-3-319-06212-9_1

Syntikov DM (2013) How to increase the productivity of the soybean- rhizobial symbiosis. In: A comprehensive survey of international soybean research – genetics, physiology, agronomy and nitrogen relationships. https://doi.org/10.5772/51563

Tajini F, Trabelsi M, Drevon J-J (2012) Combined inoculation with *Glomus intraradices* and *Rhizobium tropici* CIAT899 increases phosphorus use efficiency for symbiotic nitrogen fixation in common bean (*Phaseolus vulgaris* L.). Saudi J Biol Sci 19:157–163

Tavasolee A, Aliasgarzard N, SalehiJoulani G, Mardi M, Asgharzadeh A (2013) Interactive effects of arbuscular mycorrhizal fungi and rhizobial strains on chickpea growth and nutrient content in plant. Afr J Biotechnol 10:7585–7591

Tilman D, Cassman KG, Matson PA, Naylor R, Polasky S (2002) Agricultural sustainability and intensive production practices. Nature 418:671–677

Tirichine L, Imaizumi-Anraku H, Yoshida S, Murakami Y, Madsen LH, Miwa H, Nakagawa T, Sandal N, Albrektsen AS, Kawaguchi M, Kawasaki S, Stougaard J (2006) Deregulation of a Ca2+/calmodulin-dependent kinase leads to spontaneous nodule development. Nature 441:1153–1156

UNEP and WHRC (2007) Reactive nitrogen in the environment: too much or too little of a good thing? (The United Nations environment program, 2007); www.unep.org/pdf/dtie/Reactive_Nitrogen.pdf

van der Heijden MGA, Martin FM, Selosse M-A, Sanders IR (2015) Mycorrhizal ecology and evolution: the past, the present, and the future. New Phytol 205:1406–1423

van der Ploeg RR, Bohm W, Kirkham MB (1999) On the origin of the theory of mineral nutrition of plants and the law of the minimum. Soil Sci Soc Am J 63:1055–1062

Vance CP (2001) Symbiotic nitrogen fixation and phosphorus acquisition. Plant nutrition in a world of declining renewable resources. Plant Physiol 127:390–397

Varshney RK (2016) Exciting journey of 10 years from genomes to fields and markets: some success stories of genomics-assisted breeding in chickpea, pigeon pea and groundnut. Plant Sci 242:98–07

Venturi V, Keel C (2016) Signaling in the rhizosphere. Trends Plant Sci 21:187–198

Vijayakumar V, Liebisch G, Buer B, Xue L, Gerlach N, Blau S, Schmitz J, Bucher M (2016) Integrated multi-omics analysis supports role of lysophosphatidylcholine and related glycerophospholipids in the *Lotus japonicus–Glomusintraradices* mycorrhizal symbiosis. Plant Cell Environ 39:393–315

Wagner SC (2012) Biological nitrogen fixation. Nat Educ Knowl 3:15

Wang X, Pan Q, Chen F, Yan X, Liao X (2011) Effects of co-inoculation with arbuscular mycorrhizal fungi and rhizobia on soybean growth as related to root architecture and availability of N and P. Mycorrhiza 21:173–181

Wang Y-Y, Hsu P-K, Tsay Y-F (2012) Uptake, allocation and signaling of nitrate. Trends Plant Sci 17:458–467

White PJ, Brown PH (2010) Plant nutrition for sustainable development and global health. Ann Bot 105:1073–1080

Wittwer RA, Dorn B, Jossi W, van der Heijden MG (2017) Cover crops support ecological intensification of arable cropping systems. Sci Rep 7:41911

Xue L, Cui H, Buer B, Vijayakumar V, Delaux P-M, Junkermann S, Bucher M (2015) Network of GRAS transcription factors involved in the control of arbuscule development in *Lotus japonicus*. Plant Physiol 167:854–871

Yaseen T, Ali K, Munsif F, Rab A, Ahmad M, Israr M, Bariach AK (2016) Influence of arbuscular mycorrhizal fungi, *Rhizobium* inoculation and rock phosphate on growth and quality of lentil. Pak J Bot 48:2101–2107

Yasmeen T, Hameed S, Tariq M, Ali S (2012) Significance of arbuscular mycorrhizal and bacterial symbionts in a tripartite association with *Vigna radiata*. Acta Physiol Plant 34:1519–1528
Yoshida S, Parniske M (2005) Regulation of plant symbiosis receptor kinase through serine and threonine phosphorylation. J Biol Chem 280:9203–9209
Zezza A, Tasciotti L (2010) Urban agriculture, poverty and food security: empirical evidence from a sample of developing countries. Food Policy 35:265–273

Rhizobium-Legume Symbioses: Heavy Metal Effects and Principal Approaches for Bioremediation of Contaminated Soil

7

Sara Lebrazi and Kawtar Fikri-Benbrahim

Contents

Abstract

Leguminous plants play a vital role in agriculture, economy, and even food security for the world's population. Indeed, they are considered as a major source of protein for human food worldwide, providing 22% protein, 32% fat, and 7% carbohydrates. They provide a bulk of soil organic matter (SOM) in agricultural soils and have a crucial role in the soil for long-term sustainability. This is due to their significant role in improving soil fertility and ability to form *Rhizobium*-legume symbiosis enabling atmospheric nitrogen (N) fixation. Recently, *Rhizobium*-legume symbioses have attracted attention for their biochemical and ecological capacity to degrade and remove organic pollutants. They are also known for their resistance to heavy metal which make them efficient tools for rehabilitating contaminated soils. However, high heavy metal concentrations in

S. Lebrazi · K. Fikri-Benbrahim (✉)
Sciences and Technology Faculty, Sidi Mohamed Ben Abdellah University, Fes, Morocco
e-mail: Kawtar.fikribenbrahim@usmba.ac.ma

R. S. Meena et al. (eds.), *Legumes for Soil Health and Sustainable Management*,
https://doi.org/10.1007/978-981-13-0253-4_7

soil may have an adverse effect on both *Rhizobium* and its host plant and also on their symbiotic properties. In fact, the repartition of heavy metals in soil is widespread, with an annual global heavy metal release estimated at 22.10^{-3} Tg of Cd, 939.10^{-3} Tg of Cu, 783.10^{-3} Tg of Pb, and 1.35 Tg of Zn. Moreover, consumption of agri-foods grown in heavy metal-polluted soils may have serious implications on human health. Recent data indicate that exposure to low levels of some heavy metals such as cadmium can have adverse health effects, mainly in the form of kidney damage, but also bone and fracture effects.

In this chapter, the harmful effects of heavy metals on the environment and humans are described along with their natural and anthropogenic origins. The disturbances induced by HM to host plants and to symbiotic N-fixing (SNF) bacteria are explained. The importance of phytoremediation as an alternative method of pollutant cleanup is highlighted. Then its main categories are elaborated along with the role of some legumes in phytoextraction and phytostabilization. Then, several approaches and strategies are aimed at improving the bioaccumulation potential and bioremediation of heavy metals based on the use of *Rhizobium*-legume symbiosis. These methods have been objectively discussed. The co-inoculation of plants with rhizobia and plant growth-promoting rhizobacteria (PGPRs) resistant to heavy metals presents important advantages for promoting the plant growth besides reinforcing the bacterial potency for heavy metal intake.

Furthermore, the use of bacterial genetic/molecular engineering approaches, particularly for the symbiotic association *Rhizobium*-legume, has proved to be an interesting and significant alternative. It offers a greater degradation capacity of various metal contaminants to promote contaminated soil remediation.

Keywords
Rhizobium · Legume · Symbiosis · Contamination · Heavy metals · Phytoremediation · Soil fertility

Abbreviations

ACC deaminase	1-Aminocyclopropane-1-carboxylate deaminase
BCF	Bioconcentration factor
CDF	Cation diffusion facilitator
EDTA	Ethylenediaminetetraacetic acid
EPS	Exopolysaccharide
HM	Heavy metal
HME-RND	Heavy metal efflux-resistance nodulation and cell division
IAA	Indole-3-acetic acid
MFP	Membrane fusion protein
MFS	Major facilitator superfamily
MTEs	Metal trace elements

OMF	Outer membrane factors
PGPR	Plant growth-promoting rhizobacteria
RND	Resistance nodulation and cell division
ROS	Reactive oxygen species
TF	Translocation factor
Tg	Teragrams

7.1 Introduction

Legumes are highly significant in socioeconomical and ecological systems. Besides being a source of dietary protein and enabling food and nutritional security (22% protein, 32% fat, and 7% carbohydrates) (Machehouri et al. 2017), they also play a key role in improving soil fertility due to *Rhizobium*-legume nitrogen (N)-fixing symbiosis (Mandal and Bhattacharyya 2012). The reduction of N_2 is carried out usually in the plant's roots in specialized organs called nodules (Kalloniati et al. 2015). This symbiotic relation is responsible for a substantial portion of conversion of the global flux of atmospheric N in an available form for plants and participates in the production of 50% of ~175 teragrams (Tg) of total biological N fixation (BNF). This process provides nearly half of the N used in agriculture per year (Ögütçü et al. 2008). In addition, the increase and expansion of the organic fertilizer's role would reduce the requirement for chemical fertilizers and reduce their negative environmental effects whose global annual demand is about 50 Tg (Machehouri et al. 2017). Hence, biofertilization plays a major role in addressing environmental pollution and degradation issues (Ögütçü et al. 2009; Malusa and Vassilev 2014; Mahanty et al. 2016). However, several edaphic factors such as soil's physicochemical composition and properties can affect, on one hand, the host plant's development and, on the other, the rhizobial efficiency. This consequently influences the N-fixing symbiosis activity (Kinkema et al. 2006; Mandal and Bhattacharyya 2012; Yadav et al. 2017a). Heavy metal soil contamination is among the critical limiting factors of *Rhizobium*-legume symbiosis. Indeed they are considered to be major inorganic environmental pollutants that can persist in the soil for a long period (Angelovičová and Fazekašová 2014). Moreover, they are continuously added to the soil by various anthropogenic activities. These activities involve the use of wastewater for irrigation in agricultural production. In fact, polluted water is currently used in about 20 million hectares of agricultural land. In several developing countries in Asia (China, Pakistan, India) and Africa (Ghana, Ethiopia, Kenya, and Morocco), undiluted sewage is frequently used for irrigation. This is because it is considered to be a rich source of nutrients compared with other water sources (Qureshi et al. 2016). The accumulation of heavy metals in the environment can have an ecotoxicological effect on plants and soil microorganisms, which present large consequences on the ecosystem functioning.

Some metals such as iron (Fe), molybdenum (Mo), and nickel (Ni) are essential for the growth of both rhizobia and their plant host. Others such as cadmium (Cd), lead (Pb), mercury (Hg), etc. don't seem to be beneficial and could be toxic at relatively low concentrations (Singh and Prasad 2015). However, all these metals can be toxic at a certain concentration (EL Hilali 2006). In fact, copper (Cu) can promote microbial growth at low concentrations, but at high concentrations, it may inhibit their growth (Wei et al. 2009; Meena et al. 2014). In contrast, Cd even at very low concentration is considered toxic for both plants and bacteria and can alter the nitrogenase activity. Consequently, this could adversely affect other metabolic activities of the microsymbiont (Ahmad et al. 2012). The exposure to high concentrations of heavy metals present in soil can have a detrimental effect on human health. This is because it can be readily transferred into the human body through suspended dust or by direct contact (Sun et al. 2010; El Aafi et al. 2015).

The development of new approaches based on the use of genetically engineered bacteria can be considered as an emerging technology; this later should be given more importance as an ecological and effective way for the elimination and optimization of contamination by toxic metals in soils (Joutey et al. 2013). Several genes have been studied for their beneficial effects of being introduced into rhizobacteria and plants to improve biodegradation and bioaccumulation of persistent toxic metal elements in contaminated soils. Knowledge of the physicochemical properties of genetically engineered bacteria such as their phenotypic, metabolic, and symbiotic potential, as well as their interaction with the environment, makes it possible to better predict the effectiveness of bioremediation (Rittmann et al. 2006; Azad et al. 2014). Several previous studies have highlighted the development of transgenic technologies and its applications in bioremediation under complex environmental conditions (Ezezika et Singer 2010; Singh et al. 2011; Delgadillo et al. 2015).

In this chapter, we will focus mainly on the latest findings in bioremediation studies using rhizobial bacteria. Then we would describe the various mechanisms and strategies involved in the rehabilitation of soil contaminated with metal pollutants. The last part of the chapter will evaluate the complementary and synergistic benefits of co-inoculation with multiple bacteria and also of the use of genetically modified strains for improved phytoremediation.

7.2 Metal Impacts on Human Health

Legumes play important roles in one's daily diet, thanks to their high protein content. However, various anthropogenic activities such as fertilizer utilization, irrigation with dusty or contaminated waters, etc. are the main causes of increased concentrations of heavy metals in the environment. Legumes exposed to polluted soil can take up heavy metals by absorption. This could threaten the environment and human health through ingestion, inhalation, and skin absorption (Liu et al. 2013). Some metals such as Cu, Cr, and Zn (trace elements) play an essential role in maintaining the human metabolism. Hence, Cu, for example, is of substantial importance to a human life; however, chronic intakes of this heavy metal can induce carcinogenic and non-carcinogenic risks when they exceed their safe threshold limits such as headache, neurologic involvement, and liver disease (US EPA 2000).

Chronic exposure to certain heavy metals even in low concentrations is known to produce major damage to human health and cause several types of cancer. In fact, just occupational exposure to dust and mists containing hexavalent Cr may increase the risk of lung cancer in human (Park et al. 2004). The consumption of food contaminated with Cd is associated with an increased risk of postmenopausal breast cancer (Itoh et al. 2014; Ashoka et al. 2017). The long-term exposure of arsenic (As) could also be the origin of several human health problems including respiratory, gastrointestinal, cardiovascular, hematological, renal, hepatic, developmental, reproductive, neurological, genotoxic, immunological, mutagenic, and carcinogenic effects (Liu et al. 2013).

7.3 Origin of Heavy Metal Contamination in Soil

Contamination of food plants by pollutants, particularly heavy metals, is a major societal concern. Indeed, more and more crops are grown in peri-urban areas and can, therefore, be impacted by industrial pollution. This is a global problem, and very few countries are spared. Over the past decades, the repartition of heavy metals in soil is widespread; actually it is estimated that the annual worldwide release of heavy metals is about 22.10^{-3} Tg of Cd, 939.10^{-3} Tg of Cu, 783.10^{-3} Tg of plomb, and 1.35 Tg of Zn (Singh et al. 2003; Oves et al. 2012). These contaminations can be of several origins. The main problem with heavy metals such as plomb, Cd, Cu, and Hg is that they cannot be biodegraded and therefore they persist for long periods in soil. Their presence in the soil can be natural, or they can have an anthropogenic origin (Fig. 7.1).

Heavy metals occur naturally in rocks and are released during their alteration to the geochemical background form. The natural concentration of heavy metals in soils depends on the nature of the rock, its location, and age. However, the major source of contamination is anthropogenic in origin. The main anthropogenic

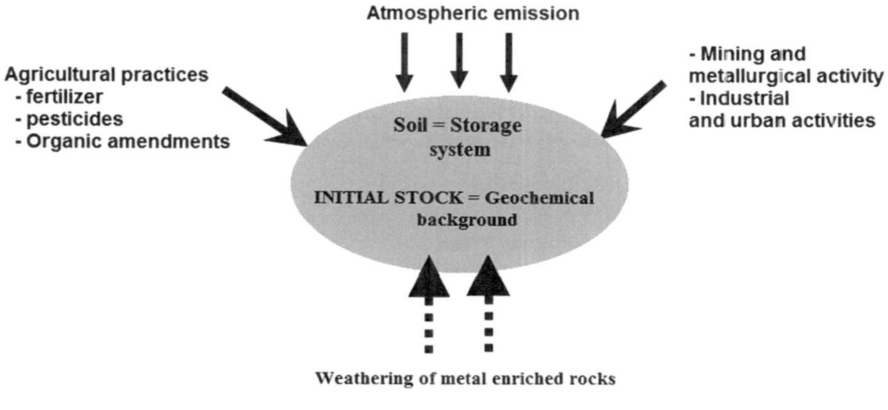

Fig. 7.1 Sources of heavy metals in soil. (Modified: Huynh 2009)

pollution factors responsible for increased metal fluxes are atmospheric pollution (Oves et al. 2012) resulting from industrial and urban activities (factory discharges, exhaust fumes, etc.). Some agricultural practices are responsible for the heavy metal introduction in the soil. In fact, despite the economic benefits of industrial fertilizers and pesticides in improving crop yields, negative effects on both the environment and humans health are much more harmful (Fantke et al. 2012; Meena et al. 2017a). This is especially for non-biodegradable heavy metals, such as Cd, Pb, Hg, Zn, and Cu, which can persist in soil for several decades. Moreover, the high fertilizer cost and the problems of recurrent droughts inducing unavailability or low water availability may lead to the use of wastewater for irrigation. This wastewater, rich in organic matter and fertilizers, contains undesirable chemical elements that can have high levels of metals (Ez-Zarhouny et al. 2015), such as hexavalent chromium contained in tannery effluents. These heavy metals accumulate in the soil and diffused by various phenomena, according to biogeochemical conditions, and then they are absorbed by plants which can eventually become contaminated (Mench et al. 2000).

Industrial pollution including smelting, metal forging, combustions of fossil fuels, etc. is another source of metal pollution (Khan et al. 2009). Also, inappropriate treatment of tailings and acid mine drainage is an especially important source of heavy metal pollution in agricultural fields surrounding the mining areas (Williams et al. 2009).

The major role of industrial and agricultural practices in soil contamination must be taken into account: it concerns a large part of territory. The heavy metal accumulation and transfer is, therefore, a risk to human health through contamination of the food chain and the environment as a whole.

7.4 Effect of Heavy Metal on *Rhizobium*-Legume Symbiosis

The metal trace element (MTE) bioaccumulation in plant's tissues can induce disturbances at different metabolic stages of plants including *Rhizobium*-legume symbiosis (Fig. 7.2).

Soil microorganisms seem to be the first organisms affected by heavy metal contamination. High concentration of heavy metals in soil can cause dramatic changes and damages in microbial composition and activities (Abd-Alla et al. 2014). Several metals such as Cu, Ni, Cd, As, and Zn have shown their ability to alter the growth, morphology, and many activities of multiple groups of microorganisms. In fact, heavy metal cations can inhibit sensitive enzyme activity by attachment onto cysteine residues, glutamic acid, or aspartic acid, which form part of the active sites of several enzymes. A large excess of metal cation can also compete with other soil cations that normally serve as an essential nutrient for the plant (e.g., Ca^{2+}, K^{+}, $Mg2^{+}$, etc.). The absorption of MTEs (metal trace elements) by plant's root surface results in inhibition or stimulation of soil's cation uptake, causing significant changes in plant metabolism (Singh et al. 2016; Datta et al. 2017).

These metals also induce a decrease in the plant's chlorophyll content, a decrease in photosynthesis following an alteration of electron transport, and a perturbation of

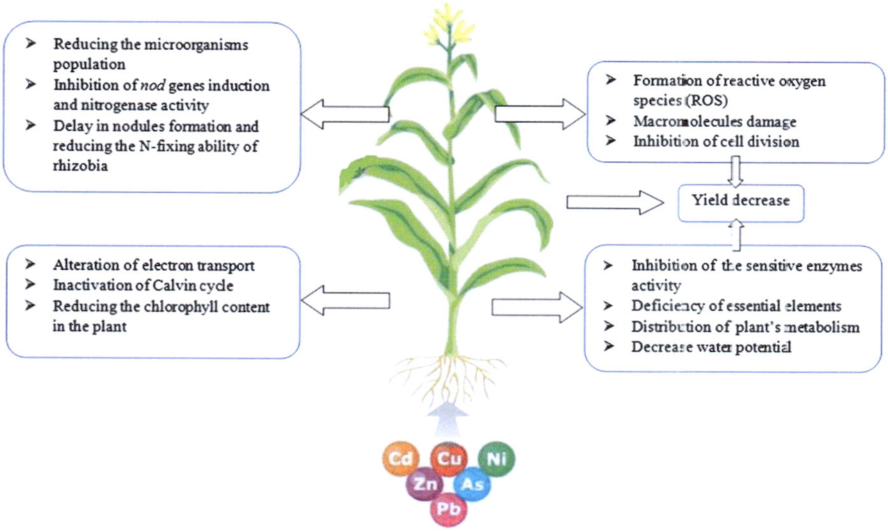

Fig. 7.2 Heavy metal toxicity to symbiotic N-fixing microorganism and host plants

the Calvin cycle enzymes (e.g., Rubisco disturbance, enzyme-catalyzing fixation of atmospheric CO_2 required for photosynthesis) (Sethy and Ghosh 2013).

Heavy metals can also alter the nucleic acid structure and prevent gene expression by the metal cation fixation on the phosphate groups of nucleic acids (Hengstler et al. 2003). Furthermore, the toxic effect induced by heavy metals can be expressed by interactions of metal ions with physiological ions, such as Cd^{2+} with Zn^{2}, with N^{+} or with Ca^{2+}, which cause a function suppression of the latter. Direct interaction of heavy metals with the sulfhydryl group (-SH) of functional proteins can lead to the destruction of their structure and the inhibition of their activities (Singh et al. 2016). It has also been reported that elevated metals concentrations might stimulate the formation of free radicals and reactive oxygen species (ROS) which consequently damage the macromolecules (Sharma and Dietz 2009; Dal Corso et al. 2013; Singh et al. 2016). For instance, some metals, such as Cd, Cu, Zn, Ni, Cr, etc., have been reported to enhance lipid peroxidation (Wani and Khan 2012) involved in oxidative stress by the Fenton reaction for both rhizobia and their host legumes. But despite the existence of some studies on the mechanisms and the factors involved in controlling responses to oxidative stress in rhizobia, only little is known about the effects of heavy metals (Balestrasse et al. 2001).

A high concentration of heavy metals may reduce the isoflavonoid exudate and therefore inhibit the induction of *nod* genes which subsequently cause a loss of N-fixing ability of rhizobia in association with some leguminous host (Ahmad et al. 2012; Meena et al. 2015a). In fact, increasing arsenic (As) concentration in the nutrient solution delayed nodule formation in the soybean (*Glycine max*) cv. Curringa plants inoculated with *Bradyrhizobium japonicum* CB1809 and decreased nodule number per plant at harvest. In addition, it was observed that inoculated plants exhibited poor root hairs and low shoot and root dry matter content, as the metal concentration in soil solution increased (Reichman 2007). Abnormally higher metal concentration can also limit absorption of water and nutrients by the plant and affect their health. However, when the metal is located within the plant tissues, it can interact directly with cellular components and disrupt metabolic activities. This can cause cell damage and in some cases can lead to a plant's death (Ahmad et al. 2012). Moreover, the addition of 50–200 mg kg^{-1} of Cd, Zn, Co, and Cu to soils for *Lablab purpureus* cultivation showed negative effects on nodulation, growth, and nitrogenase activity of plants in pot and field trials (Abd-Alla et al. 2014; Dhakal et al. 2016). Sepehri et al. (2006) revealed that 2 mg Cd/kg soil could affect the symbiotic properties of *S. meliloti* strains and consequently the *S. meliloti*-alfalfa symbiosis.

7.5 Mechanisms of Heavy Metal Resistance in Rhizobia

Even though *Rhizobium*-legume symbiosis can be affected by several environmental conditions including heavy metals, some rhizobial strains have shown their ability and efficiency to persist in polluted soil (Table 7.1) and hence can be used as an indicator organism to many toxic chemicals, such as heavy metals as referred by Botsford (1999). *Rhizobium leguminosarum* bv. trifolii isolated from soils treated with sludge showed their ability to induce nodulation in white clover (*Trifolium repens* L. Blanca) but not to fix N (Pereira et al. 2006). In the heavy metal-contaminated soil, the potential of *Rhizobium*-legume symbiotic association to fix N depends essentially on the resistance of rhizobial population to these metals. To survive under metal-stressed conditions, some microorganisms as rhizobia have developed a variety of adaptation mechanisms to reduce and to counteract the adverse effect of high concentrations of heavy metals. This is done while ensuring the maintenance of the biological role of essential ions (Lebrazi and Fikri-Benbrahim 2014; Meena et al. 2017b).

In general, there are mainly four adaptation mechanisms to Zn, Cd, and Pb; and microorganisms possess a combination of these mechanisms.

Immobilization of Heavy Metals on the Bacterial Wall by the Production of Exopolysaccharides (EPS) EPS are polymers, made up of polysaccharides and several other macromolecules, are excreted in the medium and are able to absorb heavy metals present in the environment. The production of EPS and their ability to sequester metal ions have been reported in some *Burkholderia fungorum* strains isolated from polluted rice fields to reduce the bioavailability of Cd (Zhang and Min 2010).

Intracellular Sequestration of Heavy Metals by the Production of Various Chelating Proteins The SmtA-type metallothioneins are intracellular proteins rich

Table 7.1 Metal(loid)'s resistance in some rhizobial strains isolated from contaminated soils

Rhizobium species	Metal(loid) resistance	References
Azorhizobium caulinodans	4–5 mM Cd	Zhengwei et al. (2005)
Bradyrhizobium sp.	5.1 mM Ni	Wani et al. (2007)
	21.4 mM Zn	
Bradyrhizobium sp. STM2464	15 mM Ni	Chaintreuil et al. (2007)
Mesorhizobium amorphae sp. CCNWGS0123	2.4 mM Cr	Xie et al. (2013)
	2.4 mM Cu	
	1.6 mM Zn	
	1.2 mM Ni	
Mesorhizobium metallidurans	16–32 mM Zn	Vidal et al. (2009)
	0.3–0.5 mM Cd	
Mesorhizobium sp. RC1 and RC4	7.7 mM Cr	Wani et al. (2009)
Rhizobium etli AY460185	5 mM Cd	Abou-Shanab et al. (2007)
	5 mM Cr	
	15 mM Ni	
	15 mM Pb	
	10 mM Zn	
Rhizobium halophytocola KT327204.1	7 mM Mn	Gupta et al. (2016)
	1 mM Ni	
	3 mM Pb	
	1 mM Fe	
Rhizobium leguminosarum bv. viciae E20-8	2 mM Cd	Figueira et al. (2005)
Rhizobium metallidurans sp. Nov	35 mM Zn	Grison et al. (2015)
	0.5 mM Cd	
Rhizobium sp. RP5	6 mM Ni	Wani et al. (2008a)
	28.8 mM Zn	
Rhizobium radiobacter LBA4213	2 mM Cd	EL Aafi et al. (2015)
	2 mM Cu	
	7 mM Pb	
	2.5 mM Zn	
Rhizobium sp. VMA301	2.8 mM As	Mandal et al. (2008)
Sinorhizobium medicae MA11	10 mM As	Pajuelo et al. (2008)
Sinorhizobium meliloti CCNWSX0020	1.8 mM Cu	Li et al. (2014)
Sinorhizobium sp. M14	250 mM As(V)	Drewniak et al. (2008)
	20 mM As(III)	

in cysteine and can produce metal-chelating agents called siderophores, which are involved in the acquisition of various heavy metals (Ahemad and Kibret 2014). It was described that *Synechococcus PCC 7942* was able to sequester Zn and Cd cations (Blindauer et al. 2002). Other BmtA bacterial metallothioneins have been described in *Anabaena PCC 7120*, *Pseudomonas aeruginosa*, *P. putida*, and *Escherichia coli* which are able to bind multiple Zn ions (Blindauer et al. 2002). The transcriptomic responses of *E. coli* on Zn or Cd stress highlights the induction of genes involved in the biosynthesis of cysteine that can sequester excess metal (Maynaud et al. 2014).

Extracellular Sequestration of Heavy Metals by Precipitation of Insoluble Forms *Cupriavidus* sp. and *Klebsiella aerogenes* can detoxify Cd by excreting sulfides to limit its influx into the cell (Aiking et al. 1982). *Citrobacter* strains use phosphates to precipitate Cd outside the cell preventing its entry (Macaskie et al. 2000). In *C. metallidurans* CH34 strain isolated from an industrial environment, the plasmid-borne czc operon involved in the efflux of metal cations outside the cell has been shown to be induced following a high concentration of metals.

This induction is then followed by a bioprecipitation process of metals in the form of bicarbonates or hydroxides to prevent the reentry of exported cations (Lodewyckx 2001).

Active Transport of Heavy Metals by Effluent Pumps Metallic cation transporters such as PIB-ATPases across the cell membrane, against the concentration gradient, using ATP hydrolysis as an energy source are used. This family is classified into four subgroups according to their specificity on the metals:

1. PIB-1 includes P-ATPases carrying Cu^+/Ag^+ such as CopA of *E. coli* (González-Guerrero and Argüello 2008).
2. PIB-2 includes P-ATPases carrying $Zn^2N^{+/}Cd^{2+}/Pb^{2+}$ such as ZntA of *E. coli.* and CadA of *Staphylococcus aureus* (Lu et al. 2016).
3. PIB-3 includes P-ATPases carrying $Cu^+/Ag^+/Cu^{2+}$ such as CopB of *Enterococcus hirae* (Schurig-Briccio and Gennis 2012).
4. PIB-4 includes P-ATPases carrying Co^{2+} as CoaT of *Synechocystis* sp. (Blasi et al. 2012).

Protein complexes type HME-RND (heavy metal efflux-resistance nodulation and cell division) are large protein complexes belonging to the family of RND proteins described in the transport of substrates involved in resistance to heavy metals (*Cupriavidus metallidurans*), in nodulation (*M. loti*), and cell division (*E. coli*) (Nies 2003). HME-RND consists of three proteins:

- RND protein constituting the pump located in the internal membrane which is coupled to two other membrane proteins
- The outer membrane protein (OMF, outer membrane factors)
- The membrane fusion protein (MFP) forming an antiportation cation/two protons, allowing the export of metallic ions outside the cell

Indeed, protons from the respiratory chain are expelled into the periplasmic space to serve as antiporter's co-carriers to metallic cations. Among the most highly described HME-RNDs, CzcABC protein complex encoded by *czc* operon is present on the plasmid pMOL30 of *C. metallidurans* CH34 which allows adaptation to high

Co, Ni, Zn, and Cd concentrations (Von Rozycki and Nies 2009; Buragohain et al. 2017). The protein of the CDF (cation diffusion facilitator) family is a secondary membrane transporter of proton/cation antiport type, using proton motor force or potassium gradient as a source of energy. These carriers were the first to be defined as specific to exclusion of metal cations. They have six TMs of which the first four appear to be conserved with a serine and aspartate residue. They also possess numerous metal cation-binding sites and numerous histidine residues at the C- and N-terminal ends and in the region between the transmembrane helices four and five which are essential for metal transport (Paulsen et al. 1997). The CzcD protein of CDF type described in *C. metallidurans* CH34 is involved in the export of Zn, Co, and Cd from cytoplasm to periplasm (Scherer and Nies 2009). The MFS family (major facilitator superfamily) are uniport, symport, or antiport membrane proteins that carry small-sized solutes in response to a chemiosmotic gradient, such as carbohydrates, Krebs cycle intermediates, antibiotics, amino acids, nucleosides, vitamins, as well as cations and anions, but not macromolecules (Tian et al. 2013). The best known example is NreB protein described in the strain 31A of *C. metallidurans*, which allows specific efflux of Ni (Grass et al. 2001). This protein possesses 12 TMs with a histidine-rich C-terminal domain, not essential to the protein function but necessary for the acquisition of complete resistance (Grass et al. 2001).

Currently, the description of metallic rhizobia, isolated from heavy metal-contaminated environments, associated with legumes, and involved in strategies of phytoremediation, is not adequately documented. Many studies on mechanisms of resistance to heavy metals are based on the study of the model bacterium in this field, *C. metallidurans* CH34 (von Rozycki and Nies 2009). Some researchers have evaluated the mechanisms of heavy metal adaptation in rhizobia. Peirera et al. (2006) showed that *R. leguminosarum* bv. viciae were able to withstand high concentrations of Cd by the intracellular synthesis of biomolecules (polysaccharides, thiols, and organic acids) which enable sequestering Cd and, thus, limit its toxicity and possible damage. In addition, they demonstrated that the extracellular production of surface lipopolysaccharides (LPS) helps to immobilize Cd on the cell wall and then limit its cell penetration. Moreover, production of glutathione, a tripeptide containing a thiol group, which is generally produced during oxidative stress has been demonstrated in the detoxification of metal cations such as Cd and Ni for *Rhizobium leguminosarum* bv. viciae and *Bradyrhizobium* sp. (Figueira et al. 2005; Bianucci et al. 2012).

Chaintreuil et al. (2007) and Meena et al. (2017c) described the first Ni-resistant *Bradyrhizobium* isolated from the nodules of *Serianthese calycina* on the Ni-contaminated soil of New Caledonia. They demonstrated that Ni resistance was due to two resistance determinants encoding efflux systems of the HME-RND type; Operon *cnr* (49% identity with CnrA of *C. metallidurans CH34*) and *nre* (61% identity with NreB of CH34). The presence of these two Ni resistance markers gives to this strain an advantage for growth in these contaminated environments and for the establishment of symbiosis with the legume.

Furthermore, Hao et al. (2012) and Li et al. (2012), respectively, sequenced genomes of *Mesorhizobium amorphae* CCNWGS0123 isclated from *Robinia pseudoacacia* nodules and *Ensifer meliloti* CCNWSX0020 isolated from *Medicago lupulina* nodules, from Zn-/Pb-contaminated mining sites, and identified numerous heavy metal resistance genes potentially involved in revegetation strategies. In *M. amorphae* CCNWGS0123, genes involved in Cu resistance have been identified such as those coding for efflux systems of HME-RND (CusAB) type, of PIB-ATPases (close to CopA PIB-ATPase), and of multicopper oxidases (Hao et al. 2012). In *E. meliloti* CCNWSX0020, they have also been identified such as those coding for efflux systems of PIB-ATPase type and multicopper oxidases (Li et al. 2012).

7.6 *Rhizobium*-Legume Symbiosis as a Tool of Phytoremediation

Phytoremediation is the use of plants that have constitutive and adaptive mechanisms to extract, sequestrate, or decontaminate terrestrial or aquatic environments (Fig. 7.3) (Lal 2009; Mandal and Bhattacharyya 2012). This technology is a natural and economic alternative to physical and chemical methods of cleanup and can be applied to both organic and inorganic pollutants. Phytoremediation consists of five main subgroups (Tekaya et al. 2014):

(a) Phytoextraction: It is an ideal method for reducing or removing pollutants from soil without adversely affecting the soil's properties. It is based on the use of plants that absorb MTEs in soils by root system and then transfer and accumulate them in their harvestable parts (Ensley 2000). Regarding this method, several studies have been carried out to evaluate its effectiveness by focusing on accumulative or hyperaccumulative plant species able to produce high biomass (Ensley 2000; Meena et al. 2015b). Unfortunately, the majority of hyperaccumulating plants have a major shortcoming to produce aerial biomass, to grow slowly, and their potential effectiveness for removing MTE from soil or substrate is very limited. The times are expressed at least in decades and even in centuries (Robinson et al. 1998).
(b) Phytodegradation or phytotransformation: It is based on the decomposition of organic products and contaminants taken up by plants through metabolic processes within the plant or the external breakdown of contaminants by enzymes produced and released by the plants. This process concerns complex organic molecules that are degraded into molecular contaminants in soils, sludges, sediments, and groundwater medium (Tangahu et al. 2011).
(c) Rhizofiltration: This requires intervention of roots to absorb metals from contaminated waters or wastewaters such as agricultural runoff, industrial discharge, or acid mine drainage.
(d) Phytostabilization: It consists of immobilization and reduction of toxicity and bioavailability of metals by plant roots in association with microorganisms. Thus, vegetation cover formed could limit the pollution spread by wind erosion, runoff, or leaching and has a barrier function to limit exchanges between soil and atmosphere or water. This technique is relatively common for older

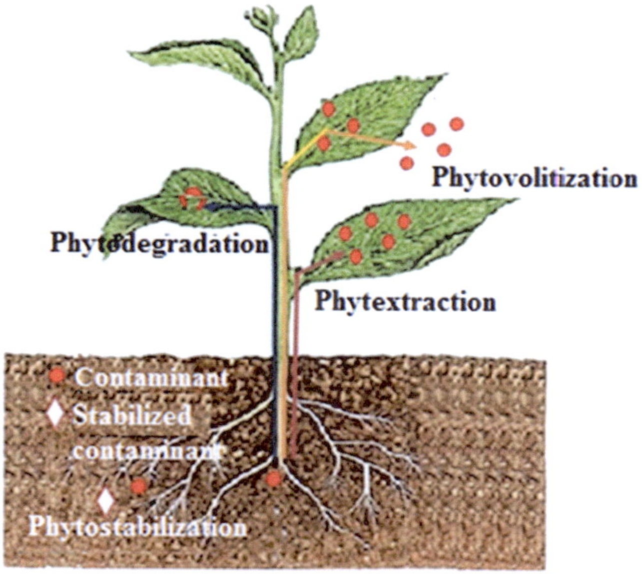

Fig. 7.3 Different forms of bioremediation

mines through varieties of plants resistant to metals such as *Agrostis tenuis* cv., *Festuca rubra* cv., *Goginan* cv., or *Parys* (Johnson et al. 1994). These resistances are often increased by symbiosis, colonizing roots, like arbuscular mycorrhizae via sequestration of metals in fungal hyphae. Arbuscular mycorrhizae can also excrete, via their hyphae, glycoproteins and/or glomulin complexing metals contained in the soil (Göhre and Paszkowski 2006).

(e) Phytovolatilization: This process is based on using plants to extract some metals from the soil and transform them into volatile elements followed by allowing volatilization, into the atmosphere, of the metal such as mercury or arsenic contained in plants. This method is not always satisfactory because of the release of toxic substances into the atmosphere (Vara Prasad and de Oliveira Freitas 2003).

The two most commonly considered branches are phytoextraction and phytostabilization.

The principal parameters which influence the plant suitability for phytoextraction are its growth rate, the amount of produced biomass, tolerance, and ability of this plant's shoots to accumulate high concentrations of metals (Ali et al. 2011). The establishment of these phytotechnologies requires a contact between the contaminant and the plant's roots to ensure the absorption of those contaminants. The metals absorbed by plants are subsequently transported to their shoots and leaves in spite of their possible toxicity. Moreover, a previous research attests that adding soil bacteria can have an essential role in accelerating this process and improving plant growth by sequestration of heavy metals (Ali et al. 2011; Mandal and Bhattacharyya 2012; Meena and Meena 2017). The phytoremediation is considered as an innovative and environmental friendly procedure that offers a real alternative to ecological and financial approaches for heavy metal removal in contrast to physicochemical processes, which are not only expensive but also harmful to the soil structure.

However, this technology is a relatively slow process because it can require several years to reduce the toxicity of metal contents in soil and to make it come up to an acceptable level. This is because most metallic hyperaccumulator plants are characterized by their small size, slow growth, and low biomass production (Khan 2005).

The use of legumes in metal removal from contaminated areas has several benefits on both ecological and agronomic scale. Almost all legumes known for their use in metal remediation could be divided into two application types: phytostabilization and phytoextraction. The main contribution of rhizobia on the two phytoremediation types is to improve plant growth (Hao et al. 2014). Rhizobia have demonstrated their capacity to be used as a powerful tool for heavy metal bioremediation. In particular, the symbiotic interaction between tolerant rhizobia and legumes should provide a high efficiency for phytoremediation (Carrasco et al. 2005; Ike et al. 2007) and N compounds contributing to the soil enrichment following the biological fixation of N in legume nodules. Furthermore, resistant rhizobia can directly enhance the pollutant intake and translocation from soil to plant and the organic pollutant degradation ability due to their N-fixing capacity and plant growth-promoting traits. These bacteria can also improve other soil biodegrading microorganism growth and facilitate their intervention in soil phytoremediation (Teng et al. 2015; Meena et al. 2015c). The positive effects of inoculation with rhizobia on some legume plants are outlined in Table 7.2.

7.7 Heavy Metal Bioavailability as a Key for Evaluating Phytoextraction Efficiency

Bioavailability can be defined as the fraction of a contaminant's total amount in the soil that is available or can become available for intake by organisms in a specified period (Petruzzelli et al. 2015). It presents a critical factor which can affect phytoextraction efficiency of target heavy metals because it is in direct relationship with soil and plant since the plant intake is mainly related to concentrations of metals and their available forms in the soil environment. This may explain the importance of bioavailability for phytoremediation strategy involved in cleaning sites contaminated by heavy metals. The heavy metals/metalloids in soil can be divided into three categories depending on their bioavailability: readily bioavailable (Cd, Zn, Ni, As, Se, and Cu), moderately bioavailable (Co, Fe, and Mn), and least bioavailable (Pb, U, and Cr) (Ali et al. 2013). Soil properties are considered like major factors that may affect metal bioavailability and which can vary continuously. Among these main parameters, the pH is, for example, one of the major parameters influencing the bioavailability of metal trace elements in soil solutions by regulating precipitation-dissolution, specific adsorption, and complexation processes and

Table 7.2 Effects of inoculations with rhizobia on legume plants grown in heavy metal (HM)-contaminated soils

Plant	Rhizobia inoculum	Heavy metals	Inoculation effect on plant	References
Cicer arietinum	*Mesorhizobium* sp. RC3	Cr	Increased dry matter, number of nodules, seed yield, grain protein, and root and shoot N, increased the Cr uptake in roots, shoots, and grains	Wani et al. (2008b)
Lens culinaris	*Rhizobium leguminosarum* RL9	Zn	Increased dry matter, numbers and dry mass of nodules, leghemoglobin, seed yield, and grain protein, decreased shoot Zn content	Wani et al. (2008c)
Leucaena leucocephala	*Mesorhizobium* sp. UFLA 01-765	Zn- Cd	Improved growth and nodule formation, increased shoot N accumulation and HM uptake	Rangel et al. (2016)
Lolium multiflorum Lam.	*Bradyrhizobium* sp. YL-6	Cd	Increased shoot and root dry weight, enhanced Cd root and shoot uptake	Guo and Chi (2014)
Medicago lupulina	*Sinorhizobium meliloti* CCNWSX0020	Cu	Enhanced root and shoot dry weight, increased plant growth, biomass, and Cu uptake	Fan et al. (2011) and Kong et al. (2015)
Medicago sativa	*Sinorhizobium meliloti*	Cd	Increased biomass and nutrient acquisition, enhanced absorption and translocation to the shoots of HM	Ghnaya et al. (2015)
Medicago sativa	*Sinorhizobium meliloti* Alf12	As- Pb	Increased growth and shoot N, greater HM accumulation	Carrasco et al. (2005)
Medicago sativa	*Sinorhizobium meliloti* S412	Cu	Greater HM accumulation in nodules and shoots	Nouairi et al. (2015)
Pisum sativum	*Rhizobium* sp. *RP5*	Ni- Zn	Increased dry matter, nodule numbers, root N, shoot N, leghemoglobin, seed yield, and grain protein, decreased shoot Ni and Zn content	Wani et al. (2008a)
Vigna radiata	*Bradyrhizobium* sp. RM8	Ni- Zn	Improved the nodule numbers, leghemoglobin, seed yield, grain protein, and root and shoot N, reduced the uptake of Zn and Ni	Wani et al. (2007)
Zea mays L.	*Rhizobium leguminosarum* TAL-102	Pb	Enhanced growth and dry biomass, increased Pb uptake by plant	Hadi and Bano (2010)

therefore their transfer to plants. Hence, low pH causes reduction of sorption processes due to the acid-catalyzed dissolution of oxides and their sorption sites, while the solubility decreases as pH increases by the formation of metal-organic complexes (Petruzzelli et al. 2015; Kumar et al. 2017).

Other influencing factors include temperature which is responsible for modification in the chemical nature and the mobility of organometal complexes and also their intake by plants. Ionic strength can reduce the sorption of heavy metals by soil surfaces due to increased competition from alkaline metals (Petruzzelli and Pezzarossa 2003).

Heavy metal phytoextraction efficiency can be determined by calculating the bioconcentration factor (BCF) and translocation factor (TF). BCF refers to the efficiency of a plant species in accumulating and concentrating a metal into its tissues from the surrounding environment (Ladislas et al. 2012) and is calculated by the following equation (Ashraf et al. 2012):

$$\mathrm{BCF} = C_{\mathrm{tissue}} / C_{\mathrm{soil}}$$

where C_{tissue} represents metal concentration in plant tissues and C_{soil} is concentration of the same metal in soil.

Translocation factor is defined by the plant's efficiency to take up metal in its roots and to translocate it in its shoots (Ahmadpour et al. 2012). This TF is determined as follows (Padmavathiamma and Li 2007):

$$\mathrm{TF} = C_{\mathrm{shoot}} / C_{\mathrm{root}}$$

where C_{shoot} is a metal concentration in plant shoots and C_{root} is a metal concentration in plant roots.

The efficient plants for phytoextraction are characterized by their ability to accumulate high levels of toxic metals in their shoots, and only plant species with both BCF and TF >1 are considered like hyperaccumulators which have the potential to be used for phytoextraction (Ahmadpour et al. 2012). It is reported that some species of legumes such as *Lupinus* and *Astragalus* may accumulate high concentrations of heavy metals in their shoots and can, therefore, be used for phytoextraction (Pastor et al. 2003; Hao et al. 2014).

The use of rhizobia can effectively increase the capacity of these host legumes to tolerate and accumulate toxic heavy metals in their shoots (Hao et al. 2014).

7.8 Plant Growth-Promoting Traits for Enhancing Plant Phytoremediation Ability

Bacteria have been reported to have the ability to reduce metal intake by plants (Ahemad and Kibret 2014; Babu et al. 2015). Immobilization of heavy metals in the soil upon bacterial inoculation is a very important property to increase plant growth and reduce metal(loid) intake by plants (Ahemad and Kibret 2014; Wang et al. 2017). Rhizobia are among the PGPR characterized by different mechanisms by which they can directly stimulate and improve their host plant's proliferation. These mechanisms include production of indole-3-acetic acid (IAA) phytohormone, solubilization of phosphate, secretion of siderophores allowing the mobilization of iron, and synthesis of some specific enzymes like 1-aminocyclopropane-1-carboxylate deaminase (ACC deaminase) which reduce the ethylene stress levels in plants. Due to the involvement of one or more of these mechanisms, PGPRs can reduce toxic effects of metal on the plant (Pereira and Castro 2014; Ma et al. 2016). Some plant hormones, such as IAA, induce an ATPase activation of the plasma membrane and increase cell division and elongation of coleoptiles (tissues protecting apex and first leaves) (Pazurkiewicz-Kocot et al. 2003), leading to intracellular processes regulation and plant physiology modification.

Heavy metal-tolerant plant species with metal accumulation capacity have been widely used in phytoremediation programs. Moreover, rhizobacteria producing IAA have been used to assist the phytoremediation of heavy metal-contaminated soil (Yu et al. 2014). In hydroponic culture, simultaneous addition of IAA (0.1 mM) and EDTA (0.2 mM) can induce a 28-fold increase in Pb intake in *Medicago sativa* leaves when compared to a sixfold increase with EDTA alone (0.2 mM) (López et al. 2005). In another study, *Bacillus subtilis* SJ-101, producing IAA, increased the growth of *Brassica juncea* in contaminated soil by increasing its tolerance to metals and also its accumulation ability by 1.5 times (Zaidi et al. 2006; Dadhich et al. 2015).

Siderophores are low molecular weight molecules, synthesized by microorganisms, and regarded as organic chelators with a very high affinity for Fe (Hider and Kong 2010). So, siderophore-producing bacteria can help plants to adapt to heavy metal stress and promote Fe absorption. Rhizobial species, including *R. meliloti*, *R. leguminosarum* bv. viciae, *R. leguminosarum* bv. phaseoli, *R. leguminosarum* bv. trifolii, *R. tropici*, *Sinorhizobium meliloti*, and *Bradyrhizobium*, are known for their ability to produce siderophores (Gómez-Sagasti and Marino 2015) and their susceptibility to Ni toxicity (He and Yang 2007).

1-Aminocyclopropane-1-carboxylate deaminase (ACC deaminase) is a pyridoxal phosphate-dependent enzyme largely available in diverse soil bacterial and fungal species. This enzyme can regulate and lower levels of ethylene by metabolizing ACC, an ethylene precursor. ACC deaminase regulates the production of ethylene in response to a multitude of biotic and abiotic stresses. Plants inoculated with the ACC deaminase bacteria can regulate their ethylene levels and consequently contribute to having a more extensive root system. Such root proliferation in contaminated soil can lead to an enhanced uptake of heavy metals (Arshad et al. 2007). Strains, such as *R. leguminosarum* bv. viciae, *R. hedysari*, *R. gallicum*, *B. japonicum*, *B. elkani*, *M. loti*, and *S. meliloti*, have been known to produce ACC deaminase (Duan et al. 2009; Gopalakrishnan et al. 2015).

Metals are generally stable and adhere strongly to soil particles, which make them hardly biodegradable and slightly available for intake by phytoextracting plants (Gamalero and Glick 2012; Ahemad 2015; Meena et al. 2017d). Therefore, they persist longer in the environment. Under heavy metal stress conditions, some metal-resistant phosphate-solubilizing bacteria (PSB) were considered as very promising agents to solubilize some insoluble and biologically unavailable metals by secreting organic acids which present some chelating properties and thus facilitate metal bioavailability for plant intake (Panhwar et al. 2013; Ahemad 2015).

Wang et al. (2017) indicated that metal(loid)-tolerant *Rhizobium tropici* KX008303 isolated from the metal(loid)-contaminated rhizosphere could reduce water-soluble content in the soil. In addition, this strain significantly increased the edible tissue biomass and reduced available Cd content in soil and Cd and As intake of edible tissues of *Brassica rapa* L. ssp. *Chinensis L.* and *Raphanus sativus L.* var. radculus pers. Their results also revealed that the metal uptake is accompanied by the production of siderophore and 71.1 μg ml^{-1} of IAA. Another recent study indicated that *Rhizobium radiobacter* EC1B increased the production of IAA with increasing concentrations of Cd and showed a general decrease in IAA production with increasing metal levels (Zn) with a significant improvement of the ACC deaminase activity and the production of siderophore (Moreira et al. 2016). *Phyllobacterium myrsinacearum* RC6b is specifically considered as an effective metal mobilizer for improving phytoremediation due to its capacity to mobilize high concentrations of Cd, Pb, and Zn in soils. *P. myrsinacearum* RC6b is characterized by its high degree of resistance to Cd (350 ppm), Zn (1000 ppm), and Pb (1200 ppm). Furthermore, *P. myrsinacearum* RC6b has multiple plant growth beneficial traits such as insoluble phosphate's solubilization and ACC deaminase's, IAA's, and siderophore's productions. RC6b can increase Cd and Zn concentrations in the shoot tissues by 57% and 34%, respectively (Ma et al. 2013).

Rhizobium strain RL9 isolated from the nodules of lentil (*Lens esculenta Moench*) grown in metal-contaminated soils presented a tolerance to high Pb concentrations up to 1600 ppm. This strain has plant growth-promoting traits; hence, it produced a good amount of IAA (33 ppm), and it was also positive for siderophore, HCN, and ammonia (Wani and Khan 2012). *Rhizobium* strain ND2 isolated from *Phaseolus vulgaris* root nodules demonstrated its ability to absorb and immobilize chromium and seemed to be an efficient plant growth-promoting bacteria, thanks to its ability to produce IAA, EPS, ammonia, protease, and catalase and to enhance root growth even under Cr (VI) stress (Karthik et al. 2016).

7.9 Use of PGPRs and Rhizobia as a Synergic Team for Efficient Phytoremediation

The use of multiple beneficial rhizobacteria has been exploited for more efficient and sustainable soil rehabilitation. Co-inoculation with multiple bacteria may have complementary and synergistic benefits to plants (Larimer et al. 2012; Teng et al. 2015).

Consortia inoculation allows the treatment of complex pollutants or mixtures of pollutants by several specialized microorganisms (Roane et al. 2001). The selection of rhizospheric microorganisms has important advantages in promoting plant growth and protecting it against the toxicity of pollutants (Kuiper et al. 2001; Yadav et al. 2017b). In fact, the co-inoculation with PGPRs may increase the potency of these bacteria for heavy metal intake. It was observed that *Sinorhizobium meliloti* strain CCNWSX0020, resistant to 1.4 mmol Cu^{2+}/L and isolated from *Medicago lupulina* plants growing in metal-contaminated soil, and *Pseudomonas brassicacearum* strain Zy-2-1 isolated from *Sphaerophysa salsula*'s root nodules increased plant growth and Cu content in plant's shoots and roots as the concentrations of Cu in the medium increased in comparison to *Sinorhizobium* inoculation alone. The higher level of Cu translocation from roots to shoots for the co-inoculated plants under stress conditions indicates that co-inoculation pattern improves metal extraction potential and consequently the translocation behavior of Cu in inoculated plants (Kong et al. 2017). Furthermore, co-inoculation with a consortium of rhizobacteria-containing *Rhizobium* sp. *CCNWSX0481*, *Rhizobium leguminosarum* bv. viciae, *Enterobacter cloacae*, and *Pseudomonas* sp. exerted a significant effect on reducing the accumulated Cu in roots compared to controlled plants, with a reduction rate of more than 35% (Fatnassi et al. 2015). Another study indicated that co-inoculation of lupines with a consortium of metal-resistant bacteria including *Bradyrhizobium* sp., *Pseudomonas* sp., and *Ochrobactrum cytisi* enhanced plant biomass production in addition to a very significant diminution of the accumulation of all studied metals, especially in roots. In fact, more than 50% reduction in the accumulation of Pb, Cd, and Zn in roots was observed. Furthermore, there was also a diminution of the amount of metal translocated to the shoot, between 30% and 60% depending on the metal, in plants inoculated with the consortium of three bacterial strains (*Bradyrhizobium* sp. *750*, *Ochrobactrum cytisi Azn6-2*, and *Pseudomonas* sp. *Az13*) (Dary et al. 2010; Kumar et al. 2016). Similar effects on plant growth and nutrient intake have been detected by co-inoculation with a consortium of Pb-resistant PGPRs including *A. tumefaciens*, *R. aquatilis*, and two *Pseudomonas* sp. strains. In addition, it was reported that co-inoculation by these bacteria in relatively low or moderately Pb-contaminated soil reduced Pb accumulation in roots and shoots by 22% and 80%, respectively. While in a heavily contaminated soil, a significant decrease in Pb level in shoots (66%) and an increase in roots (21%) were registered (Jebara et al. 2015). Furthermore, Kozdrój et al. (2007) found that the co-inoculation by ectomycorrhizal fungi associated with *Pseudomonas putida* or *Bacillus cereus* bacteria can positively affect pine seedlings growth in Cd stress conditions, thanks to the metal's absorption and immobilization promoting the ability of *P. putida* or *B. cereus*.

The positive effect of indigenous PGPRs on plant's growth and nodulation has been reported to enhance biomass and shoot nutrient uptake of co-inoculated maize by a consortium containing *Bacillus mycoides* and *Micrococcus roseus* strains, grown in heavy metal-contaminated soils (Malekzadeh et al. 2012).

Hence, co-inoculation with PGPRs, having great resistance to heavy metals, can form a symbiotic system useful for protecting plants from the toxic effects of heavy metals. This is done to improve the phytoremediation performances and enhance plant growth and essential nutrient uptake.

7.10 Transgenic Rhizobia for Improving Bioremediation

Recent advances based on the implementation of high-throughput technological methods for DNA sequencing and analysis of gene expression have invested most of the fields of current plant biology. This is done by harnessing genomic, transcriptomic, proteomic, and metabolomic means to modify the traits of "biological designers" to facilitate sustainable development and maximize phytoremediation efficiency (Abhilash et al. 2012).

The *Rhizobium*-legume interaction is considered as an important and appropriate tool for improving bioremediation of heavy metal-contaminated soils. Successful *Rhizobium*-legume symbiosis for more successful bioremediation under heavy metal stress requires selection of resistant and symbiotically effective rhizobia (Delgadillo et al. 2015; Verma et al. 2015). The selection of efficient rhizobial isolates can be harnessed for inoculation projects to promote plant growth or for genetic engineering (Lebrazi and Fikri-Benbrahim 2014). Bioengineering could be potentially exploited to manipulate the tolerance and the symbiotic interaction between plants and bacteria for potential degradation of pollutants. For example, the inoculation of *Astragalus sinicus* plants with a genetically modified strain of *Mesorhizobium huakii* subsp. *rengei* B3 able to express a resistance genes encoding the metal-binding protein, synthetic tetrameric metallothionein (MTL4), and a cDNA encoding the phytochelatin synthase from *Arabidopsis thaliana* (AtPCS) induces an increase in Cd accumulation in nodules (Teng et al. 2015). The copper resistance genes *copAB* from *Pseudomonas fluorescens* strain were expressed in *Sinorhizobium medicae* MA11 under the control of a nodulation-specific promoter *nifH*. The resulting *Sinorhizobium* strain was able to alleviate phytotoxicity generated by Cu on the inoculated *Medicago truncatula* at moderate concentrations (up to 300 μM). Moreover, the genetically modified strain did not show any effect on shoot and root dry weights, nodule number, N content, and photosynthetic rate in the presence of Cu. Furthermore, an increase in Cu accumulation in shoots and a decrease in roots were observed. Thus, engineered *S. medicae* MA11-copAB increased Cu rhizostabilization by *M. truncatula* and decreased translocation factor (Delgadillo et al. 2015).

A similar study was conducted by Pajuelo et al. (2016) to discuss the manipulation of both symbiotic partners. The expression of *mt4a* in composite plants increased tolerance toward Cu, reduced oxidative stress caused by this pollutant, and protected nodulation in the presence of Cu (60% decrease in nodulation); in addition, inoculation with the genetically modified Ensifer had a synergistic effect (only 20% decrease in nodulation in the presence of 200 μM Cu).

The effect of the symbiosis of common bean inoculated with engineered *Rhizobium etli* expressing the gene *vhb* and exposed to the herbicide paraquat which generates oxidative stress was studied. The expression of VHb in *R. etli* bacteroids showed higher nitrogenase activity and decreased sensitivity to oxidative stress (Ramírez et al. 2016). Positive effects of VHb expression in various nonsymbiotic bacterial species were well documented (Stark et al. 2015). Also, previous studies reported the positive effects of VHb expression in transgenic plants, i.e., *Arabidopsis*, tobacco, or cabbage, growing in control or in stress conditions (Li et al. 2005; Wang et al. 2009).

Some studies have demonstrated the advantage of simultaneous expression of several genes responsible for metal's uptake or degradation on improving plants' ability to tolerate high concentrations of metals which, in turn, makes it possible to improve the efficiency of pollutant removal (Ike et al. 2007; Meena et al. 2015d). In particular, the control of transfer efficiency and levels of gene expression in recipient cells appears to be critical; in fact, the effect of some individual genes may be limited to some impacts which may also limit its applications. The exploitation of genomic databases of these rhizobial strains and the identification of different gene functions are necessary to identify and select the most efficient genes to improve bioremediation strategies (Teng et al. 2015).

However, the use of genetically engineered bacteria may present some problems due to their poor survival in the contaminated sites, and this remains to be explored. The exploitation of new molecular approaches of modern recombinant DNA and of chemical industries and also the exploration and development of molecular plant-bacteria interactions can produce new strains capable of a broad hydrocarbon metabolism which could provide information on the sequence of functional modifications and adaptability to contaminated environments of the genetically engineered bacteria. Future studies should focus on combining strategy allowing the use of transgenic hyperaccumulating plants, with a high biomass-producing ability, and genetically engineered rhizobia which can present a powerful and effective approach for treating highly toxic chemicals in multiple heavy metal-contaminated conditions.

7.11 Conclusion

Rhizobia are soil bacteria which form a symbiotic relationship with legumes. This symbiosis can be used as an effective strategy to simultaneously improve soil N content and fertility and to significantly reduce chemical fertilizer use in agriculture. It may also play an important ecological and economical function since *Rhizobium*-legume symbiosis can be utilized as a model system to degrade and to remove environmental organic pollutants. As a result, this could reduce the risk associated with the presence of heavy metals in soil.

The toxicity of heavy metals to microbial life is particularly well known. These metals can cause dysfunction of rhizobia cell metabolism and consequently the number and survival of these bacteria in the rhizosphere. The presence of pollutants in the soil may also have a depressive effect on the parameters and processes involved in *Rhizobium*-legume symbiosis, which results in low bacteria-plant interactions and then poor N fixation. In addition, metals can also cause severe toxicity to various legume metabolic activities and may also have a detrimental effect on human health. Some rhizobial strains are characterized by their infectiveness and effectiveness in N fixation in heavy metal-contaminated soils. This feature makes them good candidates for inoculating legume. The co-inoculation of rhizobia and selected PGPR tolerant to heavy metals can improve not only the nodulation and the productivity of inoculated plants but can significantly reduce the toxicity of these metals and increase its bioremediation. Moreover, the exploitation of such technology could be considered as a cost-effective promising method and as a future approach to further facilitate bioremediation. The effect of heavy metals on *Rhizobium*-legume symbiosis and different strategies adopted by rhizobia and their host plants to withstand moderate concentrations of heavy metals and also the mechanisms used to revitalize contaminated soils are well established. Hence, rhizobia can produce polysaccharides or glutathione as an adaptation mechanism to heavy metals or can even develop active transport of HM by efflux systems such as HME-RND type or PIB- ATPase and multicopper oxidases. Currently, several bioengineering studies are looking for molecular and genetic method utilization to study the plant-*Rhizobium* symbiosis tolerance and the competitive pollutant's potential degradation for more successful bioremediation under field conditions even if the understanding of the factors regulating growth, metabolism, and functions of rhizobia degradation is not yet well known, and those researches had been mainly conducted under controlled laboratory conditions and not in fields.

References

Abd Alla M, Issa A, Ohyama T (2014) Impact of harsh environmental conditions on nodule formation and dinitrogen fixation of legumes. Adv Biol Ecol Nitrogen Fixat ISBN: 978–953

Abhilash PC, Powell JR, Singh HB, Singh BK (2012) Plant–microbe interactions: novel applications for exploitation in multipurpose remediation technologies. Trends Biotechnol 30:416–420

Abou-Shanab RAI, van Berkum P, Angle JS (2007) Heavy metal resistance and genotypic analysis of metal resistance genes in gram-positive and gram-negative bacteria present in Ni-rich serpentine soil and in the rhizosphere of Alyssum murale. Chemosphere 68:360–367

Ahemad M (2015) Phosphate-solubilizing bacteria-assisted phytoremediation of metalliferous soils: a review. 3 Biotech 5:111–121

Ahemad M, Kibret M (2014) Mechanisms and applications of plant growth promoting rhizobacteria: current perspective. J King Saud Univ - Sci 26:1–20

Ahmad E, Zaidi A, Khan M, Oves M (2012) Heavy metal toxicity to symbiotic nitrogen-fixing microorganism and host legumes. Toxic heavy Met to Legum bioremediation 29–44

Ahmadpour P, Ahmadpour F, Mahmud TM, Abdu A, Soleimani M, Tayefeh FH (2012) Phytoremediation of heavy metals: a green technology. Afr J Biotechnol 11:14036–14043

Aiking H, Kok K, Heerikhuizen HV, Van't Riet T (1982) Adaptation to cadmium by *Klebsiella aerogenes* growing in continuous culture proceeds mainly via formation of cadmium sulfide. Appl Environ Microbiol 44:938–944

Ali Q, Ahsan M, Khaliq I, Elahi M, Ali S, Ali F, Naees M (2011) Role of Rhizobacteria in phytoremediation of heavy metals: an overview. Int Res J Plant Sci 2:220–232

Ali H, Khan E, Sajad MA (2013) Phytoremediation of heavy metals—concepts and applications. Chemosphere 91:869–881

Angelovičová L, Fazekašová D (2014) Contamination of the soil and water environment by heavy metals in the former mining area of Rudňany (Slovakia). Soil Water Res 9.(2014:1

Arshad M, Saleem M, Hussain S (2007) Perspectives of bacterial ACC deaminase in phytoremediation. Trends Biotechnol 25:356–362

Ashoka P, Meena RS, Kumar S, Yadav GS, Layek J (2017) Green nanotechnology is a key for eco-friendly agriculture. J Clean Prod 142:4440–4441

Ashraf M, Maah M, Yusoff I (2012) Assessment of phytoextraction efficiency of naturally grown plant species at the former tin mining catchment. Fresenius Environ Bull 21:523–533

Azad MAK, Amin L, Sidik NM (2014) Genetically engineered organisms for bioremediation of pollutants in contaminated sites. Chin Sci Bull 59:703–714

Babu A, Shea P, Sudhakar D, Jung I, Oh B (2015) Potential use of *Pseudomonas koreensis* AGB-1 in association with *Miscanthus sinensis* to remediate heavy metal (loid)-contaminated mining site soil. J Environ Manag 151:160–166

Balestrasse K, Gardey L, Gallego S, Tomaro M (2001) Response of antioxidant defence system in soybean nodules and roots subjected to cadmium stress. Funct Plant 28:497–504

Bianucci E, Fabra A, Castro S (2012) Involvement of glutathione and enzymatic defense system against cadmium toxicity in *Bradyrhizobium sp.* strains (peanut symbionts). Biometals 25:23–32

Blasi B, Peca L, Vass I, Kos P (2012) Characterization of stress responses of heavy metal and metalloid inducible promoters in *Synechocystis* PCC6803. J Microbiol Biotechnol 22:166–169

Blindauer C, Harrison M, Robinson A, Parkinson JA, Bowness PW, Sadler PJ, Robinson NJ (2002) Multiple bacteria encode metallothioneins and SmtA-like zinc fingers. Mol Microbiol 45:1421–1432

Botsford J (1999) A simple method for determining the toxicity of chemicals using a bacterial indicator organism. Environ Toxicol 14:285–289

Buragohain S, Sharma B, Nath JD, Gogaoi N, Meena RS, Lal R (2017) Impact of ten years of bio-fertilizer use on soil quality and rice yield on an inceptisol in Assam, India. Soil Res. https://doi.org/10.1071/SR17001

Carrasco JA, Armario P, Pajuelo E, Burgos A, Caviedes MA, López R, Chamber MA, Palomares AJ (2005) Isolation and characterisation of symbiotically effective *Rhizobium* resistant to arsenic and heavy metals after the toxic spill at the Aznalcóllar pyrite mine. Soil Biol Biochem 37:1131–1140

Chaintreuil C, Rigault F, Moulin L, Jaffré T, Fardoux J, Giraud E, Dreyfus B, Bailly X (2007) Nickel resistance determinants in *Bradyrhizobium* strains from nodules of the endemic New Caledonia legume *Serianthes calycina*. Appl Environ Microbiol 73:8018–8022

Dadhich RK, Meena RS, Reager ML, Kansotia BC (2015) Response of bio-regulators to yield and quality of Indian mustard (*Brassica juncea* L. Czernj. and Cosson) under different irrigation environments. J Appl Nat Sci 7(1):52–57

Dal Corso G, Manara A, Furini A (2013) An overview of heavy metal challenge in plants: from roots to shoots. Metallomics 5:1117–1132

Dary M, Chamber-Pérez MA, Palomares AJ, Pajuelo E (2010) "In situ" phytostabilisation of heavy metal polluted soils using *Lupinus luteus* inoculated with metal resistant plant-growth promoting rhizobacteria. J Hazard Mater 177:323–330

Datta R, Baraniya D, Wang YF, Kelkar A, Moulick A, Meena RS, Yadav GS, Ceccherini MT, Formanek P (2017) Multi-function role as nutrient and scavenger off reeradical in soil. Sustain MDPI 9:402. https://doi.org/10.3390/su9081402

Delgadillo J, Lafuente A, Doukkali B, Redondo-Gómez S, Mateos-Naranjo E, Caviedes MA, Pajuelo E, Rodríguez-Llorente ID (2015) Improving legume nodulation and Cu rhizostabilization using a genetically modified rhizobia. Environ Technol 36:1237–1245

Dhakal Y, Meena RS, Kumar S (2016) Effect of INM on nodulation, yield, quality and available nutrient status in soil after harvest of green gram. Legume Res 39(4):590–594

Drewniak L, Matlakowska R, Sklodowska A (2008) Arsenite and arsenate metabolism of *Sinorhizobium sp.* M14 living in the extreme environment of the Zloty Stok gold mine. Geomicrobiol J 25:363–370

Duan J, Müller K, Charles T, Vesely S, Glick B (2009) 1-aminocyclopropane-1-carboxylate (ACC) deaminase genes in rhizobia from southern Saskatchewan. Microb Ecol 57:423–436

El Aafi N, Saidi N, Maltouf A, Perez-Palacios P, Dary M, Brhada F, Pajuelo E (2015) Prospecting metal-tolerant rhizobia for phytoremediation of mining soils from Morocco using *Anthyllis vulneraria* L. Environ Sci Pollut Res 22:4500–4512

EL Hilali I (2006) La Symbiose Rhizobium–Lupin: Biodiversité des Microsymbiotes et Mise en Evidence d'une Multi-Infection Nodulaire chez *Lupinus Luteus*. Dissertation, Rabat University Mohammed V

Ensley B (2000) Rational for use of phytoremediation. Phytoremediation of toxic metals: using plants to clean-up the environment. Wiley, New York, pp 3–12

Ezezika OC, Singer PA (2010) Genetically engineered oil-eating microbes for bioremediation: prospectus and regulatory challenges. Technol Soc 32:331–335

Ez-Zarhouny D, Hbaiz EM, Lebkiri M, Elanza S, Lebkiri A, Rifi EH, Habbadi N (2015) Evaluation of the accumulation of some heavy metals in the red beetroot (*Beta vulgaris* L.) grown on two different soils and irrigated by wastewater from the SETP Bouznika (Morocco). Int J Sci Eng Res 6:7p

Fan LM, Ma ZQ, Liang JQ, Li HF, Wang ET, Wei GH (2011) Characterization of a copper-resistant symbiotic bacterium isolated from *Medicago lupulina* growing in mine tailings. Bioresour Technol 102:703–709

Fantke P, Friedrich R, Jolliet O (2012) Health impact and damage cost assessment of pesticides in Europe. Environ Int 49:9–17

Fatnassi I, Chiboub M, Saadani O, Jebara M, Jebara S (2015) Phytostabilization of moderate copper contaminated soils using co-inoculation of *Vicia faba* with plant growth promoting bacteria. J Basic Microbiol 55:303–311

Figueira E, Gusmão Lima A, Pereira S (2005) Cadmium tolerance plasticity in *Rhizobium leguminosarum* bv. viciae: glutathione as a detoxifying agent. Can J Microbiol 51:7–14

Gamalero E, Glick B (2012) Plant growth-promoting bacteria and metals phytoremediation. Phytotechnol Remediat Environ Contam 361

Ghnaya T, Mnassri M, Ghabriche R, Wali M (2015) Nodulation by *Sinorhizobium meliloti* originated from a mining soil alleviates Cd toxicity and increases Cd-phytoextraction in *Medicago sativa* L. Front Plant Sci 6:863

Göhre V, Paszkowski U (2006) Contribution of the arbuscular mycorrhizal symbiosis to heavy metal phytoremediation. Planta 223:1115–1122

Gómez-Sagasti M, Marino D (2015) PGPRs and nitrogen-fixing legumes: a perfect team for efficient Cd phytoremediation? Front Plant Sci 6:81

González-Guerrero M, Argüello J (2008) Mechanism of Cu^+-transporting ATPases: soluble Cu+ chaperones directly transfer Cu^+ to transmembrane transport sites. Proc Natl Acad Sci 105:5992–5997

Gopalakrishnan S, Sathya A, Vijayabharathi R, Varshney R, Gowda C, Krishnamurthy L (2015) Plant growth promoting rhizobia: challenges and opportunities. 3 Biotech 5:355–377

Grass G, Fan B, Rosen B, Lemke K, Schlegel H, Rensing C (2001) Nre B from achromobacter xylosoxidans 31A is a nickel-induced transporter conferring nickel resistance. J Bacteriol 183:2803–2807

Grison C, Jackson S, Merlot S, Dobson A, Grison C (2015) *Rhizobium metallidurans* sp. nov., a symbiotic heavy metal resistant bacterium isolated from the *Anthyllis vulneraria* Zn-hyperaccumulator. Int J Syst Evol Microbiol 65:1525–1530

Guo J, Chi J (2014) Effect of Cd-tolerant plant growth-promoting rhizobium on plant growth and Cd uptake by *Lolium multiflorum* Lam. and *Glycine max* (L.) Merr. in Cd-contaminated soil. Plant Soil 375:205–214

Gupta R, Kumar T, Mittal A (2016) Isolation, identification and characterization of heavy metal resistant bacteria from soil of an iron industry, Haryana (India). Int J Pharm Sci Res 7:1308

Hadi F, Bano A (2010) Effect of diazotrophs (Rhizobium and Azatebactor) on growth of maize (*Zea mays L.*) and accumulation of lead (Pb) in different plant parts. Pak J Bot 42:4363–4370

Hao X, Lin Y, Johnstone L, Baltrus D, Miller S, Wei G, Rensing C (2012) Draft genome sequence of plant growth-promoting *rhizobium Mesorhizobium amorphae*, isolated from zinc-lead mine tailings. J Bacteriol 194:736–737

Hao X, Taghavi S, Xie P, Orbach MJ, Alwathnani HA, Rensing C, Wei G (2014) Phytoremediation of heavy and transition metals aided by legume-rhizobia symbiosis. Int J Phytoremediation 16:179–202

He Z, Yang X (2007) Role of soil rhizobacteria in phytoremediation of heavy metal contaminated soils. J Zhejiang Univ Sci B 8:192–207

Hengstler J, Bolm-Audorff U, Faldum A, Janssen K, Reifenrath M, Götte W et al (2003) Occupational exposure to heavy metals: DNA damage induction and DNA repair inhibition prove co-exposures to cadmium, cobalt and lead as more dangerous than hitherto expected. Carcinogenesis 24:63–73

Hider R, Kong X (2010) Chemistry and biology of siderophores. Nat Prod Rep 27:637–657

Huynh TMD (2009) Impact of heavy metals on plant / worm / soil microflora interaction "Impact des métaux lourds sur les interactions plante/ver de terre/microflore tellurique" (Doctoral dissertation, Université Paris-Est).

Ike A, Sriprang R, Ono H, Murooka Y, Yamashita M (2007) Bioremediation of cadmium contaminated soil using symbiosis between leguminous plant and recombinant rhizobia with the *MTL4* and the *PCS* genes. Chemosphere 66:1670–1676

Itoh H, Iwasaki M, Sawada N, Takachi R, Kasuga Y, Yokoyama S, Onuma H, Nishimura H, Kusama R, Yokoyama K, Tsugane S (2014) Dietary cadmium intake and breast cancer risk in Japanese women: a case–control study. Int J Hyg Environ Health 217:70–77

Jebara S, Saadani O, Fatnassi I, Chiboub M, Abdelkrim S, Jebara M (2015) Inoculation of *Lens culinaris* with Pb-resistant bacteria shows potential for phytostabilization. Environ Sci Pollut Res 22:2537–2545

Johnson M, Cooke J, Stevenson J (1994) Revegetation of metalliferous wastes and land after metal mining. Min Environ Impact R Soc Chem Camb:31–48
Joutey N, Bahafid W, Sayel H, Ghachtouli NE (2013) Biodegradation: involved microorganisms and genetically engineered microorganisms. Biodegrad Sci 72:289–320
Kalloniati C, Krompas P, Karalias G, Udvardi M, Rennenberg H, Herschbach C, Flemetakis E (2015) Nitrogen-fixing nodules are an important source of reduced sulfur, which triggers global changes in sulfur metabolism in *Lotus japonicus*. Plant Cell 27:2384–2400
Karthik C, Oves M, Sathya K, Sri Ramkumar V, Arulselvi P (2016) Isolation and characterization of multi-potential *Rhizobium* strain ND2 and its plant growth-promoting activities under Cr (VI) stress. Arch Agron Soil Sci:1–12
Khan AG (2005) Role of soil microbes in the rhizospheres of plants growing on trace metal contaminated soils in phytoremediation. J Trace Elem Med Biol 18:355–364
Khan M, Zaidi A, Wani P, Oves M (2009) Role of plant growth promoting rhizobacteria in the remediation of metal contaminated soils. Environ Chem Lett 7:1–19
Kinkema M, Scott P, Gresshoff P (2006) Legume nodulation: successful symbiosis through short- and long-distance signalling. Funct Plant Biol 33:707–721
Kong Z, Mohamad O, Deng Z, Liu X, Glick B, Wei G (2015) Rhizobial symbiosis effect on the growth, metal uptake, and antioxidant responses of *Medicago lupulina* under copper stress. Environ Sci Pollut Res 22:12479–12489
Kong Z, Deng Z, Glick B, Wei G, Chou M (2017) A nodule endophytic plant growth-promoting *Pseudomonas* and its effects on growth, nodulation and metal uptake in *Medicago lupulina* under copper stress. Ann Microbiol 67:49–58
Kozdrój J, Piotrowska-Seget Z, Krupa P (2007) Mycorrhizal fungi and ectomycorrhiza associated bacteria isolated from an industrial desert soil protect pine seedlings against Cd (II) impact. Ecotoxicology 16:449–456
Kuiper I, Bloemberg G, Lugtenberg B (2001) Selection of a plant-bacterium pair as a novel tool for rhizostimulation of polycyclic aromatic hydrocarbon-degrading bacteria. Mol Plant-Microbe Interact 14:1197–1205
Kumar S, Sheoran S, Kumar SK, Kumar P, Meena RS (2016) Drought: a challenge for Indian farmers in context to climate change and variability. Progress Res Int J 11:6243–6246
Kumar S, Meena RS, Yadav GS, Pandey A (2017) Response of sesame (*Sesamum indicum* L.) to sulphur and lime application under soil acidity. Int J Plant Soil Sci 14(4):1–9
Ladislas S, El-Mufleh A, Gérente C, Chazarenc F, Andrès Y, Béchet B (2012) Potential of aquatic macrophytes as bioindicators of heavy metal pollution in urban stormwater runoff. Water Air Soil 223:877–888
Lal R (2009) Agricultural sciences-volume 1. Encyclopedia of life support system, p 460
Larimer A, Bever J, Clay K (2012) Consequences of simultaneous interactions of fungal endophytes and arbuscular mycorrhizal fungi with a shared host grass. Oikos 121:2090–2096
Lebrazi S, Fikri-Benbrahim K (2014) Environmental stress conditions affecting the N_2 fixing Rhizobium-legume symbiosis and adaptation mechanisms. Afr J Microbiol Res 8:4053–4061
Li X, Peng R, Fan H, Xiong A, Yao Q, Cheng Z, Li Y (2005) *Vitreoscilla hemoglobin* overexpression increases submergence tolerance in cabbage. Plant Cell Rep 23:710–715
Li Z, Ma Z, Hao X, Wei G (2012) Draft genome sequence of *Sinorhizobium meliloti* CCNWSX0020, a nitrogen-fixing symbiont with copper tolerance capability isolated from lead-zinc mine tailings. J Bacteriol 194:1267–1268
Li Z, Ma Z, Hao X, Rensing C, Wei G (2014) Genes conferring copper resistance in *Sinorhizobium meliloti* CCNWSX0020 also promote the growth of *Medicago lupulina* in copper-contaminated soil. Appl Environ Microbiol 80:1961–1971
Liu X, Song Q, Tang Y, Li W, Xu J, Wu J, Wang F, Brookes PC (2013) Human health risk assessment of heavy metals in soil–vegetable system: a multi-medium analysis. Sci Total Environ 463:530–540
Lodewyckx C (2001) A potential role for bacterial endophytes in phytoremediation of heavy metal contaminated soils. PhD thesis, Limbg Univ (LUC), Belgium, 157pp
López M, Peralta-Videa J, Benitez T, Gardea-Torresdey J (2005) Enhancement of lead uptake by alfalfa (*Medicago sativa*) using EDTA and a plant growth promoter. Chemosphere 61:595–598

Lu M, Li Z, Liang J, Wei Y, Rensing C, Wei G (2016) Zinc Resistance Mechanisms of P 1B -type ATPases in *Sinorhizobium meliloti* CCNWSX0020. Sci Rep 6:12

Ma Y, Rajkumar M, Luo Y, Freitas H (2013) Phytoextraction of heavy metal polluted soils using *Sedum plumbizincicola* inoculated with metal mobilizing Phyllobacterium myrsinacearum RC6b. Chemosphere 93:1386–1392

Ma Y, Rajkumar M, Zhang C, Freitas H (2016) Beneficial role of bacterial endophytes in heavy metal phytoremediation. J Environ Manag 174:14–25

Macaskie LE, Bonthrone KM, Yong P, Goddard D (2000) Enzymically mediated bioprecipitation of uranium by a Citrobacter sp.: a concerted role for exocellular lipopolysaccharide and associated phosphatase in biomineral formation. Microbiology 146:1855–1867

Machehouri A, Ben Messaoud B, Nassiri L, Ibijbijen J (2017) Numeration of natural rhizobial populations from chickpea (Cicer arietinum) in different moroccan soils. "Dénombrement des populations naturelles de rhizobium du Pois chiche (Cicer arietinum) dans différents sols du Maroc". Eur Sci J 13(6):273–286

Mahanty T, Bhattacharjee S, Goswami M, Bhattacharyya P, Das B, Ghosh A, Tribedi P (2016) Biofertilizers: a potential approach for sustainable agriculture development. Environ Sci Pollut Res:1–21

Malekzadeh E, Alikhani H, Savaghebi-Firoozabadi G, Zarei M (2012) Bioremediation of cadmium-contaminated soil through cultivation of maize inoculated with plant growth–promoting rhizobacteria. Biorem J 16:204–211

Malusa E, Vassilev N (2014) A contribution to set a legal framework for biofertilisers. Appl Microbiol Biotechnol 98(15):6599–6607

Mandal S, Bhattacharyya R (2012) Rhizobium–legume symbiosis: a model system for the recovery of metal-contaminated agricultural land. In: Toxicity heavy met to legume bioremediation, pp 115–127

Mandal S, Pati B, Das A, Ghosh A (2008) Characterization of a symbiotically effective rhizobium resistant to arsenic: isolated from the root nodules of *Vigna mungo* (L.) Hepper grown in an arsenic-contaminated field. J Gen Appl Microbiol 54:93–99

Maynaud G, Brunel B, Yashiro E, Mergeay M, Cleyet-Marel J, Le Quéré A (2014) CadA of *Mesorhizobium metallidurans* isolated from a zinc-rich mining soil is a P IB-2-type ATPase involved in cadmium and zinc resistance. Res Microbiol 165:175–189

Meena H, Meena RS (2017) Assessment of sowing environments and bio-regulators as adaptation choice for clusterbean productivity in response to current climatic scenario. Bangladesh J Bot 46(1):241–244

Meena VS, Maurya BR, Meena RS, Meena SK, Singh NP, Malik VK (2014) Microbial dynamics as influenced by concentrate manure and inorganic fertilizer in alluvium soil of Varanasi, India. Afr J Microb Res 8(1):257–263

Meena RS, Yadav RS, Meena H, Kumar S, Meena YK, Singh A (2015a) Towards the current need to enhance legume productivity and soil sustainability worldwide: a book review. J Clean Prod 104:513–515

Meena RS, Meena VS, Meena SK, Verma JP (2015b) Towards the plant stress mitigate the agricultural productivity: a book review. J Clean Prod 102:552–553

Meena RS, Meena VS, Meena SK, Verma JP (2015c) The needs of healthy soils for a healthy world. J Clean Prod 102:560–561

Meena RS, Dhakal Y, Bohra JS, Singh SP, Singh MK, Sanodiya P (2015d) Influence of bioinorganic combinations on yield, quality and economics of mungbean. Am J Exp Agri 8(3):159–166

Meena HR, Meena RS, Lal R, Singh GS, Mitran T, Layek J, Patil SB, Kumar S, Verma T (2017a) Response of sowing dates and bio regulators on yield of clusterbean under current climate in alley cropping system in eastern U.P., Indian Leg Res, Accepted in press

Meena RS, Gogaoi N, Kumar S (2017b) Alarming issues on agricultural crop production and environmental stresses. J Clean Prod 142:3357–3359

Meena RS, Kumar V, Yadav GS, Mitran T (2017c) Response and interaction of *Bradyrhizobium japonicum* and *Arbuscular mycorrhizal* fungi in the soybean rhizosphere: a review. Plant Growth Reg, Accepted in press

Meena RS, Meena PD, Yadav GS, Yadav SS (2017d) Phosphate solubilizing microorganisms, principles and application of microphos technology. J Clean Prod 145:157–158
Mench M, Manceau A, Vangronsveld J, Clijsters H, Mocquot B (2000) Capacity of soil amendments in lowering the phytoavailability of sludge-borne zinc. Agronomie 20:383–397
Moreira H, Pereira SI, Marques APGC, Rangel AO, Castro PM (2016) Selection of metal resistant plant growth promoting rhizobacteria for the growth and metal accumulation of energy maize in a mine soil—effect of the inoculum size. Geoderma 278:1–11
Nies D (2003) Efflux-mediated heavy metal resistance in prokaryotes. FEMS Microbiol Rev 27:313–339
Nouairi I, Mrabet M, Rabhi M, Mhadhbi H, Zribi K (2015) Cu-tolerant *Sinorhizobium meliloti* strain is beneficial for growth, Cu accumulation, and mineral uptake of alfalfa plants grown in Cu excess. Arch Agron Soil Sci 61:1707–1718
Ögütçü H, Algur Ö, Elkoca E, Kantar F (2008) The determination of symbiotic effectiveness of *Rhizobium* strains isolated from wild chickpeas collected from high altitudes in Erzurum. Turkish J Agric For 7:241–248
Ögütçü H, Adiguzel A, Gulluce M, Karadayi M, Sahin F (2009) Molecular characterization of *Rhizobium* strains isolated from *wild Chickpeas* collected from high altitudes in Erzurum-Turkey. Rom Biotechnol Lett 14:4294–4300
Oves M, Khan M, Zaidi A, Ahmad E (2012) Soil contamination, nutritive value, and human health risk assessment of heavy metals: an overview. Springer, Vienna
Padmavathiamma P, Li L (2007) Phytoremediation technology: hyper-accumulation metals in plants. Water Air Soil Pollut 184:105–126
Pajuelo E, Rodríguez-Llorente ID, Dary M, Palomares AJ (2008) Toxic effects of arsenic on *Sinorhizobium–Medicago sativa* symbiotic interaction. Environ Pollut 154:203–211
Pajuelo E, Pérez-Palacios P, Romero-Aguilar A, Delgadillo, J., Doukkali, B., Rodríguez-Llorente I, Caviedes M (2016) Improving legume–rhizobium symbiosis for copper phytostabilization through genetic manipulation of both symbionts. Biol Nitrogen Fixat Benef Plant-Microbe Interact:183–193
Panhwar Q, Jusop S, Naher U, Othman R, Razi M (2013) Application of potential phosphate-solubilizing bacteria and organic acids on phosphate solubilization from phosphate rock in aerobic rice. Sci World J 2013:10
Park R, Bena J, Stayner L, Smith RJ, Gibb H, Lees P (2004) Hexavalent chromium and lung cancer in the chromate industry: a quantitative risk assessment. Risk Anal 24:1099–1108
Pastor J, Hernández AJ, Prieto N, Fernández-Pascual M (2003) Accumulating behaviour of *Lupinus albus* L. growing in a normal and a decalcified calcic luvisol polluted with Zn. J Plant Physiol 160:1457–1465
Paulsen I, Saier Jr M undefined (1997) A novel family of ubiquitous heavy metal ion transport proteins. J Membr Biol 156:99–103
Pazurkiewicz-Kocot K, Galas W, Kita A (2003) The effect of selenium on the accumulation of some metals in *Zea mays* L. plants treated with indole-3-acetic acid. Cell Mol Biol Lett 8:97–103
Pereira S, Castro P (2014) Diversity and characterization of culturable bacterial endophytes from Zea mays and their potential as plant growth-promoting agents in metal-degraded soils. Environ Sci Pollut Res 21:14110–14123
Pereira SIA, AIG L, EMDAP F (2006) Heavy metal toxicity in Rhizobium leguminosarum biovar viciae isolated from soils subjected to different sources of heavy-metal contamination: effects on protein expression. Appl Soil Ecol 33:286–293
Petruzzelli G, Pezzarossa B (2003) Ionic strength influence on heavy metal sorption processes by soil. J Phys IV 107:1061–1064
Petruzzelli G, Pedron F, Rosellini I, Barbafieri M (2015) The bioavailability processes as a key to evaluate phytoremediation efficiency. In: Phytoremediation. Springer, Cham, pp 31–43
Qureshi AS, Hussain MI, Ismail S, Khan QM (2016) Evaluating heavy metal accumulation and potential health risks in vegetables irrigated with treated wastewater. Chemosphere 163:54–61

Ramírez M, Íñiguez L, Guerrero G, Sparvoli F, Hernández G (2016) *Rhizobium etli* bacteroids engineered for Vitreoscilla hemoglobin expression alleviate oxidative stress in common bean nodules that reprogramme global gene expression. Plant Biotechnol Rep 10:463–474
Rangel W, Thijs S, Janssen J, Oliveira Longatti S, Bonaldi D, Ribeiro P, Jambo I, Eevers N, Weyens N, Vangronsveld J, Moreira F (2016) Native rhizobia from Zn-mining soil promotes the growth of *Leucaena leucocephala* on contaminated soil. Int J Phytoremediation 19:142–156
Reichman SM (2007) The potential use of the legume–rhizobium symbiosis for the remediation of arsenic contaminated sites. Soil Biol Biochem 39:2587–2593
Rittmann BE, Hausner M, Loffler F et al (2006) A vista for microbial ecology and environmental biotechnology. Environ Sci Technol 40:1096–1103
Roane T, Josephson K, Pepper I (2001) Dual-bioaugmentation strategy to enhance remediation of cocontaminated soil. Appl Environ Microbiol 67:3208–3215
Robinson B, Leblanc M, Petit D, Brooks R, Kirkman J, Gregg P (1998) The potential of *Thlaspi caerulescens* for phytoremediation of contaminated soils. Plant Soil 203:47–56
Scherer J, Nies D (2009) CzcP is a novel efflux system contributing to transition metal resistance in *Cupriavidus metallidurans* CH34. Mol Microbiol 73:601–621
Schurig-Briccio L, Gennis R (2012) Characterization of the PIB-Type ATPases present in *Thermus thermophilus*. J Bacteriol 194:4107–4113
Sepehri M, Rastin NS, Rahmani HAAH (2006) Effects of soil pollution by cadmium on nodulation and nitrogen fixation ability of native strains of *Sinorhizobium meliloti*. JWSS-Isfahan Univ Technol 10:153–163
Sethy SK, Ghosh S (2013) Effect of heavy metals on germination of seeds. J Nat Sci Biol Med 4(2):272–275
Sharma SS, Dietz KJ (2009) The relationship between metal toxicity and cellular redox imbalance. Trends Plant Sci 14:43–50
Singh A, Prasad SM (2015) Remediation of heavy metal contaminated ecosystem: an overview on technology advancement. Int J Environ Sci Technol 12(1):353–366
Singh OV, Labana S, Pandey G, Budhiraja R, Jain RK (2003) Mini-review phytoremediation: an overview of metallic ion decontamination from soil. Appl Microbiol Biotechnol 61:405–412
Singh JS, Abhilash PC, Singh HB et al (2011) Genetically engineered bacteria: an emerging tool for environmental remediation and future research perspectives. Gene 480:1–9
Singh S, Parihar P, Singh R, Singh VP, Prasad S (2016) Heavy metal tolerance in plants: role of transcriptomics, proteomics, metabolomics, and ionomics. Front Plant Sci 6:1143
Stark B, Pagilla K, Dikshit K (2015) Recent applications of *Vitreoscilla hemoglobin* technology in bioproduct synthesis and bioremediation. Appl Microbiol Biotechnol 99:1627–1636
Sun Y, Zhou Q, Xie X, Liu R (2010) Spatial, sources and risk assessment of heavy metal contamination of urban soils in typical regions of Shenyang, China. J Hazard Mater 174:455–462
Tangahu B, Sheikh Abdullah S, Basri H, Idris M, Anuar N, Mukhlisin M (2011) A review on heavy metals (As, Pb, and Hg) uptake by plants through phytoremediation. Int J Chem Eng 2011:31
Tekaya N, Saiapina O, Ouada H, Lagarde F, Namour P, Ouada H, Jaffrezic-Renault N (2014) Bi-Enzymatic conductometric biosensor for detection of heavy metal ions and pesticides in water samples based on enzymatic inhibition in arthrospira platensis. J Environ Prot (Irvine, Calif) 5:441–453
Teng Y, Wang X, Li L, Li Z, Luo Y (2015) Rhizobia and their bio-partners as novel drivers for functional remediation in contaminated soils. Front Plant Sci 6:32
Tian M, Qu J, Han X, Zhang M, Ding C, Ding J, Chen G, Yu S (2013) Microarray-based identification of differentially expressed genes in intracellular *Brucella abortus* within RAW264. 7 cells. PLoS One 8:e67014
US EPA 2000 Supplementary guidance for conducting health risk assessment of chemical mixtures, Risk Assessment Forum Technical Panel [EPA/630/R-00/002]
Vara Prasad M, de Oliveira FH (2003) Metal hyperaccumulation in plants: biodiversity prospecting for phytoremediation technology. Electron J Biotechnol 6:285–321
Verma JP, Jaiswal DK, Meena VS, Meena RS (2015) Current need of organic farming for enhancing sustainable agriculture. J Clean Prod 102:545–547

Vidal C, Chantreuil C, Berge O, Maure L, Escarre J, Bena G, Brunel B, Cleyet-Marel J (2009) Mesorhizobium metallidurans sp. nov., a metal-resistant symbiont of *Anthyllis vulneraria* growing on metallicolous soil in Languedoc, France. Int J Syst Evol Microbiol 59:850–855

von Rozycki T, Nies D (2009) *Cupriavidus metallidurans:* evolution of a metal-resistant bacterium. Antonie Van Leeuwenhoek 96:115–139

Wang Z, Xiao Y, Chen W, Tang K, Zhang L (2009) Functional expression of *Vitreoscilla hemoglobin* (VHb) in Arabidopsis relieves submergence, nitrosative, photo-oxidative stress and enhances antioxidants metabolism. Plant Sci 176:66–77

Wang X, Nie Z, He L, Wang Q, Sheng X (2017) Isolation of As-tolerant bacteria and their potentials of reducing As and Cd accumulation of edible tissues of vegetables in metal(loid)-contaminated soils. Sci Total Environ 579:179–189

Wani P, Khan M (2012) Bioremediaiton of lead by a plant growth promoting *Rhizobium* species RL9. Bacteriol J 2:66–78

Wani PA, Khan MS, Zaidi A (2007) Effect of metal tolerant plant growth promoting *Bradyrhizobium sp.* (vigna) on growth, symbiosis, seed yield and metal uptake by greengram plants. Chemosphere 70:36–45

Wani P, Khan M, Zaidi A (2008a) Effects of heavy metal toxicity on growth, symbiosis, seed yield and metal uptake in pea grown in metal amended soil. Bull Environ Contam Toxicol 81:152–158

Wani P, Khan M, Zaidi A (2008b) Chromium-reducing and plant growth-promoting *Mesorhizobium* improves chickpea growth in chromium-amended soil. Biotechnol Lett 30:159–163

Wani P, Khan M, Zaidi A (2008c) Impact of zinc-tolerant plant growth-promoting rhizobacteria on lentil grown in zinc-amended soil. Agron Sustain Dev 28:449–455

Wani P, Zaidi A, Khan M (2009) Chromium reducing and plant growth promoting potential of *Mesorhizobium* species under chromium stress. Biorem J 13:121–129

Wei G, Fan L, Zhu W, Fu Y, Yu J, Tang M (2009) Isolation and characterization of the heavy metal resistant bacteria CCNWRS33-2 isolated from root nodule of *Lespedeza cuneata* in gold mine tailings in China. J Hazard Mater 162:50–56

Williams PN, Lei M, Sun G, Huang Q, Lu Y, Deacon C, Meharg AA, Zhu Y-G (2009) Occurrence and partitioning of cadmium, arsenic and lead in mine impacted paddy rice: Hunan, China. Environ Sci Technol 43:637–642

Xie P, Hao X, Mohamad O, Liang J, Wei G (2013) Comparative study of chromium biosorption by *Mesorhizobium amorphae* strain CCNWGS0123 in single and binary mixtures. Appl Biochem Biotechnol 162:570–587

Yadav GS, Lal R, Meena RS, Datta M, Babu S, Das, Layek J, Saha P (2017a) Energy budgeting for designing sustainable and environmentally clean/safer cropping systems for rainfed rice fallow lands in India. J Clean Prod 158:29–37

Yadav GS, Lal R, Meena RS, Babu S, Das A, Bhomik SN, Datta M, Layak J, Saha P (2017b) Conservation tillage and nutrient management effects on productivity and soil carbon sequestration under double cropping of rice in North Eastern region of India. Ecol India. http://www.sciencedirect.com/science/article/pii/S1470160X17305617

Yu X, Li Y, Zhang C, Liu H, Liu J, Zheng W, Kang X, Leng X, Zhao K, Gu Y, Zhang X, Xiang Q, Chen Q (2014) Culturable heavy metal-resistant and plant growth promoting bacteria in V-Ti magnetite mine tailing soil from Panzhihua, China. PLoS One 9:e106618

Zaidi S, Usmani S, Singh B, Musarrat J (2006) Significance of Bacillus subtilis strain SJ-101 as a bioinoculant for concurrent plant growth promotion and nickel accumulation in *Brassica juncea*. Chemosphere 64:991–997

Zhang J, Min H (2010) Characterization of a multimetal resistant *Burkholderia fungorum* isolated from an e-waste recycling area for its potential in Cd sequestration. World J Microbiol Biotechnol 26:371–374

Zhengwei Z, Fang W, Lee H, Yang Z (2005) Responses of *Azorhizobium caulinodans* to cadmium stress. FEMS Microbiol Ecol 54:455–461

Harnessing Soil Rhizobacteria for Improving Drought Resilience in Legumes

8

Parvaze A. Sofi, Zahoor A. Baba, Burhan Hamid, and Ram Swaroop Meena

Contents

P. A. Sofi (✉)
Faculty of Agriculture, SKUAST-K, Wadura, Sopore, India

Z. A. Baba · B. Hamid
Biofertilser Research Laboratory, SKUAST-K, Wadura, Sopore, India

R. S. Meena
Department of Agronomy, Institute of Agricultural Sciences (BHU), Varanasi, UP, India

R. S. Meena et al. (eds.), *Legumes for Soil Health and Sustainable Management*,
https://doi.org/10.1007/978-981-13-0253-4_8

Abstract

There are multiple challenges like population growth, food security imperatives, hunger and malnutrition, climate change, economic access to food, decreasing factor in productivity. Such challenges necessitate an appropriate research and policy framework. The aim of this chapter is to focus on the sustainability of the soil and crop and commodity production support systems. Sustainability in agriculture is more important for input delivering resources rather than the output. It is also more important than all biological attributes, including the beneficial microbes, which is the key to it. The nexus of soil microbes-legumes-is an inseparable entity. This is the recent focus on soil biological health and legumes in achieving sustainability. Soil rhizobacteria are important especially in legume-based farming systems. This is because the resource constraints such as water and nutrients often limit the productivity of such systems. There are substantial evidences based on research findings to build a valid premise. This premise should deal with legume-rhizobial associations which can be optimally harnessed. The objective behind this is to not only enhance productivity under favourable systems but also to improve resilience to stresses such as drought. Soil rhizobacteria colonise the endo-rhizosphere/rhizosphere to drought tolerance by producing phytohormones, 1-aminocyclopropane-1-carboxylate (ACC), deaminase, volatile compounds and antioxidants, inducing accumulation of osmolytes. They also decrease the regulation of stress-responsive genes and alteration in root morphology during the acquisition of drought tolerance. The ability of soil microbes to transfer their intrinsic resilience to legume hosts opens up an amazing world of opportunities. These opportunities can be harnessed by identifying optimal legume-microbe associations. On the research front, there is a need to identify resilience mechanisms. Besides this, the underlying genetic factors and the mechanisms in host plants that optimise associations have to be identified. On the management front, there is need to create enabling conditions in soil systems that enhance the population and functionality of native and introduced microbial systems. These systems can favourably enhance output especially under resource-constrained conditions.

Keywords

Legumes · Drought stress · Rhizobacteria · Sustainability

Abbreviations

ACC	1-aminocyclopropane-1-carboxylate
BNF	Biological nitrogen fixation
CGIAR	Consultative Group on International Agricultural Research
EPS	Exopolysaccharides
PGPR	Plant growth-promoting bacteria
PSB	Phosphate-solubilising bacteria

8.1 Introduction

Global population is poised to reach ~8 billion by 2020, an increase of about 35% over the 1995 population and is projected to exceed 9.8 billion by 2050 and 11.2 billion by 2100 (UN 2017). Therefore, the major challenge would be ensuring food security while lowering the risk of climate change (Stagnari et al. 2017). As such, the demand for plant products that satiate the human requirements especially the food will rise by an even higher proportion. This is in the light of the fact that the average income is also growing, diets have diversified and greater urbanisation has occurred. A major challenge is that majority of the population living in developing countries will be at a serious risk of food and nutritional security. This situation is further complicated by looming threats of climate change. The sections of populations have less adaptive capacities for the climate change. Governments ensure sustained investments in this regard. However, apart from outputs from national and international agricultural research systems and a matching technology support, the food grain and livestock production, during the period 1995–2020, is expected to increase by only 1.5% and 2.7% per year, respectively. Even though the poverty situations may have improved, the malnutrition would continue as the greatest ever challenge before scientists and policymakers across the globe. Climate change is another major challenge to agriculture as it will dent one's efforts to ensure sufficient and economically accessible food to increasing global population (Vadez et al. 2011; Yadav et al. 2017a). With regard to food legumes, climate change implications will be manifested as increased frequencies and intensities of water stress. The change is visible especially in the legumes that are sensitive to water stress (Liu et al. 2006).

The food crops across diverse classes and all production systems are vulnerable to projected changes in climate. However, owing to certain inherent bottlenecks of smallholder subsistence farmers, such systems are more likely to be hit hard. This is because they lack adaptive capacities (Menike et al. 2015). Food legumes are important components of such farming systems, which are invariably resource constrained (Odendo et al. 2011). The food and nutritional security imperatives, coupled with population pressures and ecological costs of agriculture, are compelling situations. These situations demand a paradigm shift in research approach towards managing natural resources for better agricultural output. Food legumes are important components of farming systems that can help in transforming agriculture into a sustainable intensification mode from chemical input-based intensification mode. This will not only improve productivity and sustainability but also optimise resource utilisation and improve the resilience of the system (Mungai et al. 2016).

Chemical-based intensification of farming has, undeniably, transformed farming by contributing significantly to increases in food production worldwide. An appraisal of impacts reveals that there has been a disproportionate impact especially in marginal low-input systems (Bhattacharya and Majid 2013). Even in the high-input farming system, the productivity gains have come at the cost of deterioration of biophysical resource base of agriculture especially the soil. Chemical fertilisers may have favourably changed the physics and chemistry of the soil, but they have deteriorated the biology of the soil. The decline in soil health and microbial populations, as well as unfavourable changes in the structure of the soil, has been recognised as major offshoots of increased use of chemicals. This is in addition to the environmental as well as the energy cost of manufacturing chemical fertilisers (Gliessman 1998). Therefore, it is now increasingly being appreciated that the biology of the soil should be restored by harnessing the soil microbes and legume associations. The obvious benefits not only include a growth promotion but also an increase in the resilience of the systems (Prashar and Shah 2016; Meena et al. 2015a). This is all the more important, keeping in view that grain legumes are invariably grown in low-input marginal farming systems with minimal support and face resource constraints including water. It is imperative to improve the health to optimise the benefits of microorganisms because the grain legumes have evolutionary history of coexistence with microbes. This chapter will discuss the potential role of soil rhizobacteria in the context of improving sustainability and resilience of legume-based farming systems.

8.2 Agricultural Importance of Legumes

Food legume crops represent an indispensable component across all types of farming systems in both developing as well as developed countries. They are important determinants of food and nutritional and livelihood security. Nutritionally, food legumes are a cheap source of protein, especially in areas where animal protein is not affordable and more importantly in the light of increased carbon footprints of animal protein consumption and minerals, while agronomically they serve as important components of cereal-legume rotation. Besides this, they reduce soil-borne pests and supply nitrogen (N) to the companion crops through belowground complementation. Despite being valuable crops, the yields of legume crops have remained disappointingly low. This is due to interplay of various factors ranging from inherent physiological mechanisms, relegation to harsh environments and diseases and pests (Fig. 8.1). Even though food legumes are regarded as subsistence crops, they invariably fetch higher prices than most of cereals. Moreover, they supplement farmers' income for improving livelihood opportunities (Gowda et al. 2009). Currently around 78 million hectares are under various food legume crops globally with a production of about 85 million tonnes (FAO 2015). Consultative Group on International Agricultural Research (CGIAR) has identified food legumes as ideal target crops for achieving its developmental goals of tackling poverty and hunger and improving human health and nutrition. Above all, it is aimed at enhancing resilience of ecosystems. Under this umbrella programme, food legumes have found renewed research focus and policy support that has seen legume productivity increase in countries like India, where they are an indispensable component of farming and diets (CGIAR 2012).

The food legumes, soil and smallholder family farmers together form a favourable nexus in the developing countries. Food legumes or pulses are largely grown by smallholder family farmers accounting for a major proportion of food and nutritional

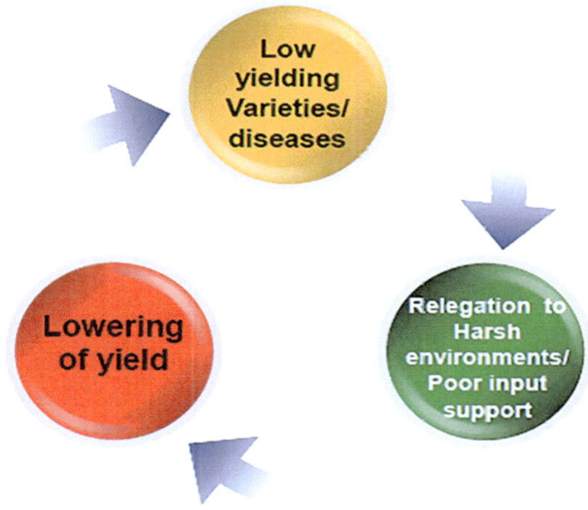

Fig. 8.1 Nexus of low yield in food legumes

and livelihood security. Soils are central to any agricultural development, and legumes now derive most of their nutrition from soil as they are grown under low-input farming systems. The United Nations (UN) General Assembly has in succession for 3 years highlighted the importance of these three components of the nexus by proclaiming 2014 as the International Year of Family Farmers, 2015 as International Year of Soils and 2016 as the International Year of Pulses (Belhassan 2017). The UN has during these 3 years made honest attempts to promote awareness about the importance of family farmers and also soils and pulses. This has been done through a series of programmes in collaboration with FAO, CGIAR, Crop Science Society of America, American Society of Agronomy and Soil Science Society of America as well as through National Agricultural Research systems (Anderson 2016). As part of this initiative, grain legumes were vigorously promoted for use as nearly perfect foods in terms of their nutritional benefits. This was for the protein source besides their environmental and economic benefits. Legumes form effective relationships with plant roots and in turn fix atmospheric N that not only increases the soil N supply to the legume crop (up to 70%) and in its companion crops in intercropping but also ensures an adequate N supply. This gives rise to better crop sequences when compared to sequences without legumes (Peoples et al. 2015; Verma et al. 2015a).

Therefore, food legumes can play a major role in ensuring global food security as well as sustainability and resilience of legume-based farming systems. However, despite their desirable features and a large number of species, only a handful of them have been harnessed to a level. Consequently, they define the productive capacities of farming systems, and most of them continue to enjoy the "orphan" or "underutilised" status (ICRISAT 1998). On one hand, the ecological benefits of incorporating food legumes to farming systems are fairly well recognised. On the other hand, the lack of recognition and evidence of the social and economic benefits precluded development and the utilisation of genetic and genomic resources of food legumes. Apart from this, they are invariably relegated to harsh environments with low-input support. As a result, they have become increasingly less competitive and cannot favourably compete with cereals. Even during the green revolution era, grain legumes have not witnessed similar research investment and output as that of cereals, which was partly driven by food security imperatives (Pingali 2012; Meena et al. 2015b).

8.2.1 Global Context: Rationale for Sustainable Intensification of Legumes in Cropping Systems

Increasing population will mean additional requirements of food legumes. The need is felt more so in resource-constrained countries where a large proportion of world's resource poor people live. Therefore, food legumes account for large requirement of protein. By the year 2050, the farm producers will need to create a supply, adequate enough, to feed an additional three billion people, most of whom will be added in developing regions of the world. Therefore, significant efforts would be needed to ensure the availability of adequate quantity of food legumes in the era of climate change and in a deteriorated biophysical resource base for agriculture. Together food and forage legumes are grown on some 180 million hectares that accounts for about 15% of the earth's arable area and contribute to 27% of the world's total primary crop production, with grain legumes alone accounting for 33% of the human dietary protein requirements (Vance et al. 2000). In order of rank, the food legumes that contribute to the dietary protein intake (Akibode and Maredia 2011).

The global demand for food legumes is expected to grow in the future, given the fact that incomes have increased, food consumption patterns are diversifying and consciousness towards healthy dieting is rather increasing (Tables 8.1 and 8.2). There is an increasing recognition of legumes, as nutraceuticals and functional foods. There are also greater health risks of consuming animal proteins, and the demand for legume-based products is expected to increase further. Most of the food legumes are rich sources of proteins (i.e. >20%) and soluble fibre and have a low glycaemic index and reduce the risk of cardiovascular diseases as well as certain forms of cancer (Duranti 2006). The global legume production has doubled from 150 million tons in the 1980s to around 300 million tons in the 2000s. Most of the

Table 8.1 Area, production and yield of major legume crops

Crop	Scientific name	Area (Mha)	Production (Mt)	Yield (Mg/ha)
Common bean	*Phaseolus vulgaris*	30.61	26.53	0.86
Chickpea	*Cicer arietinum*	13.98	13.73	0.98
Cowpea	*Vigna unguiculata*	12.61	5.59	0.44
Lentils	*Lens culinaris*	4.52	4.82	1.07
Pea	*Pisum sativum*	6.93	11.18	1.61
Pigeon pea	*Cajanus cajan*	7.03	4.89	0.69
Soybean	*Glycine max*	117.54	306.51	2.60
Groundnut	*Arachis hypogea*	26.54	43.91	1.65
Faba bean	*Vicia faba*	1.05	1.59	1.67

Data source: FAOSTAT (2014)

Table 8.2 Countries where pulses are major sources of protein intake (%)

Country	Percentage	Country	Percentage	Country	Percentage
Burundi	55	Rwanda	38	Uganda	20
Uganda	20	Kenya	20	Comoros	18
Eriteria	18	Comoros	18	Haiti	18
Cuba	16	Nicaragua	16	Malawi	15
Angola	15	Tanzania	14	India	13
Brazil	13	Mozambique	12	Korea	11
Mexico	10	Belize	10	Botswana	10

Data source: Maredia (2012)

production entails a nonfood legume like soybean, and the food legumes account for nearly 20% of total production during the same period (Gowda et al. 2009). However, with the projected increase of 40% in the population by 2050, food production needs to be increased by 70–100% to meet the increasing demands due to population growth (Price and Howitt 2014; Ashoka et al. 2017). This challenge is further compounded by the fact that legume crops will face severe competition for resources such as land and water from cereals, cash crops and biofuels. This will happen apart from developmental imperatives such as industrialisation and urbanisation. Such competition will further relegate legumes to marginal areas, where a lack of irrigation is a major constraint, limiting crop productivity. Moreover, the constraints of water availability and the season to season fluctuations in climatic parameters in such marginal areas tend to be relatively larger. This would significantly affect food security in such rain-fed systems (Postel 2000). Table 8.3 lists some of the major food legume crops that are components of different farming systems.

Food legumes are increasingly emerging as potential future crops in the light of certain desirable features such as shorter duration, low water requirements, deeper roots, ability to fix N and cheap source of proteins that make inherently the climate smart. However, given the current state of biophysical resources for agriculture, all the policymakers are equivocal in promoting a sustainable intensification (SI) of farming systems to address the concerns of food and nutritional security. This is done by increasing the production output without having an adverse effect on ecosystem services (Mungai et al. 2016). It involves increasing output per unit area, natural resources as well as social and human capital, with a concomitant increase in the flow of ecosystem services (Pretty et al. 2011). Food legumes are central to any sustainable intensification approach that seeks to transform the farming systems to become more productive, more resilient and more eco-friendly. Considering the increasing public concerns about the deleterious effects of chemical-based agriculture, more focus is now laid on optimising the biological balance, the microbial diversity and microbial dynamics in soil. SI is being proposed as a strategy for improved natural resource management with focus on reducing trade-offs between productivity, profitability and resilience (Kaczan et al. 2013; Pretty and Bharucha 2014; Meena et al. 2014). There are compelling situations encompassing social, economic and ecological justifications for promoting food legumes in the farming

Table 8.3 Major food legume crops

Crop	Scientific name
Common bean	*Phaseolus vulgaris*
Lima bean	*Phaseolus lunatus*
Scarlet runner bean	*Phaseolus coccineus*
Tepary bean	*Phaseolus acutifolius*
Adzuki bean	*Vigna angularis*
Mung bean	*Vigna radiate*
Rice bean	*Vigna umbellate*
Moth bean	*Vigna aconitifolia*
Bambara bean	*Vigna subterranean*
Faba bean (broad bean)	*Vicia faba*
Common vetch	*Vicia sativa*
Pea	*Pisum sativum*
Chickpea	*Cicer arietinum*
Cowpea	*Vigna unguiculata*
Pigeon pea	*Cajanus cajan*
Lentil	*Lens culinaris*
Hyacinth bean	*Lablab purpureus*
Sweet Jack bean	*Canavalia ensiformis*
Winged bean	*Psophocarpus tetragonolobus*
Guar bean (cluster bean)	*Cyamopsis tetragonoloba*
Velvet bean	*Mucuna pruriens*
African yam bean	*Sphenostylis stenocarpa*
Groundnut	*Arachis hypogea*
Soybean	*Glycine max*
Jack bean	*Canavalia ensiformis*
Sword bean	*C. gladiate*
Yam bean	*Pachyrhizus tuberosus*
Lablab bean	*Lablab purpureus*
Moth bean	*Vigna angularis*

Akibode and Maredia (2011); USDA National Nutrient Database (http://www.nal.usda.gov/fnic/foodcomp/search/)

systems. These farming systems help fulfil the food and nutritional security imperatives besides enhancing the ecosystem services associated with the reduced use of chemicals and lower water footprints and also promoting biodiversity on farms.

8.2.1.1 Social

8.2.1.1.1 Food Security

Food legume crops (consumed as grain, green pods, shelled beans and leaves) contribute to food security and dietary diversity goal besides improving overall nutrition and also preventing diseases due to their nutraceutical properties. Among the strategies that could help decrease the risk of hunger under looming threats of climate change, N-efficient crops are of paramount importance. This is followed by reduced tillage practices and integrated soil fertility management. They can also reduce hunger by 12%, 9% and 4%, respectively (FAO 2016a). Food legumes fulfil all these criteria and as such need to be promoted for ensuring food security. Compared to cereals, the food legumes contain twice the amount of protein and in most developing countries constitute the major source of protein. The nutritional qualities of legumes also make them extremely helpful in the fight against some non-communicable as well as some chronic diseases. Pulses also provide important vitamins and minerals such as iron, potassium, magnesium, zinc and B vitamins including folic acid, thiamin and niacin (FAO 2016b).

8.2.1.1.2 Farming System Compatibility

Most of the food legumes play diverse roles in the farming systems in most of the developing countries, where farming systems are characterised by marginality of scale, lower input support, fragility, inaccessibility, diversity and lower productivity. In such a situation, any sustainable intensification approach that seeks to increase the productivity of the system should seek to harness the advantages of food legumes integrated into the system. This will contribute to an overall social development in terms of adequate food, better nutrition and better livelihood opportunities (Dar et al. 2012).

8.2.1.1.3 Resilience to Harsh Environments

Most of food legumes can grow in harsh environments (drought prone) where the focus of production system is not on production per se but on the resilience of system as well. Intensive cereal-based farming cannot sustain profitably under such low-input farming systems. Simulation studies have suggested that such climate vulnerable sites may benefit from legume-based farming systems, although it still requires extensive on-farm validation (Smith et al. 2016).

8.2.1.1.4 Gender Specificities

Rural women are key agents for economic, environmental and social changes required for a sustainable development. On the other hand, limited access to education, resources and role in decision-making are among the many challenges they face (Sofi 2015). In fact, a recent FAO study indicates that, if women are equally empowered to make decisions in farming, the hunger will be reduced by about 15%. In smallholder situations and in low-income food-deficient nations, about 75% of women work in agriculture (Mungai et al. 2016). In developing countries, agriculture provides employment for about 38% of women with the proportion as high as 66% in Southeast Asia and 63% in Sub-Saharan Africa. In rural India, about 84% of women are dependent on agriculture for livelihood, about 33% formally participate in farming but only 10% own land and 9.7% are female-headed households (Kumar et al. 2016). In legume-based cropping systems, females are more involved and as such farming systems take care of gender specificities.

8.2.1.2 Economic

8.2.1.2.1 Livelihood

Food legumes are second to cereals in economic importance in agriculture with an annual production valued at around 31 billion US$. Legumes such as soybean, groundnut and common bean are an important cash crop, and they augment the income of smallholder farmers (ICRISAT 2013). Even crops like the common bean are an important cash crop especially in Central America where beans are among major income-generating field crops, while as in Myanmar, they contribute to about 10% of total export earnings. In fact, food legumes are only next to cereals in terms of contribution to food security (Akibode and Maredia 2011) and serve as a major source of subsistence and livelihood for more than 700 million smallholders in the developing countries valued at about US$ 31 billion annually. They mostly come from the soybean (83.8%), common bean (8.8%), groundnut (peanut) (4.9%) and chickpea (2.4%) (Abate et al. 2012).

8.2.1.2.2 N Economy

The major driving force for ensuring sustainability in agriculture is the efficient N management in the environment (Graham and Vance 2000). Since legume crops fix a sizeable proportion of N, they reduce the reliance on fossil fuels as nutrients such as N are produced by energy-intensive process of Haber-Bosch reaction. Nearly 40% of all population at present as well as in the future depend on the Haber-Bosch process for the synthesis of the key biomolecules (involving N) such as proteins, DNA and other N-containing molecules (Smil 1999). At present, more than 80 million tonnes of chemically fixed N are used in agriculture (Hawkesford 2014). The N fertiliser is extremely indispensable for the energy balance of the crop production. It is required in high quantities globally, for growing crops. The energy footprints of nitrogenous fertilisers are a little over 7.5 times larger than phosphatic and potash fertilisers. In countries like Canada, inorganic fertilisers (mostly N) account for nearly 70% of the total non-renewable energy used in crop production. Since energy is going to be a key limiting factor to determine the sustainability of fertiliser-intensive farming systems, the economic value attributable to food legumes is extremely enormous (Murrell 2016; Datta et al. 2017a).

8.2.1.2.3 Productivity of Farming Systems

Food legumes improve the productivity of rotations especially under marginal farming systems. There is substantial evidence that intercropping systems are better than sole crops in terms of productivity because they optimise the use of resources such as water and nutrients, both spatially and temporally. The observed yield advantages have been attributed to both above- and belowground complementation between component crops of intercropping sequences such as greater interception of sunlight, efficient conversion of the intercepted radiation, better root interactions as well as resource sharing (Rodrigo et al. 2001).

8.2.1.2.4 Lower Cost of Production

Most of the legumes are cultivated at a lower cost as they are grown under the conservation tillage system. In countries like Brazil, Australia and Turkey as well as in North America, agriculture is being implemented on a large scale using legume crops such as soybean, lentil, chickpea, pea and faba bean, involving reduced or no till (NT) systems that leads to a reduction in cost and makes significant positive impact on soil organic carbon (Alpmann et al. 2013; Christopher and Lal 2007).

8.2.1.3 Ecological

8.2.1.3.1 Biological Nitrogen Fixation

Sustaining optimum soil fertility is one of the major challenges in low-input marginal farming systems, and as such, the ability of grain legumes to fix N makes them important under such conditions. Food legume crops have a unique role in the global N cycle, as they fix atmospheric N in soils, thereby reducing chemical N requirements of the legume crop itself, as well as the succeeding crop. Yu et al. (2014) reported that the process of BNF favourably affects soil N availability and also furthers an enhancement of 9.7–20.5% in residual N content in the rice field; these were observed also in rice-bean and rice-vetch sequences. About 7–11% of total legume N is partitioned to roots and nodules with roughly around 11–14 kg N-fixed contributed towards per tonne of belowground dry matter, representing almost half of the total aboveground plant (Carranca et al. 2015). Annually, legume crops, put together, fix about 21.45 Tg of N, out of which a whopping 16.44 Tg is accounted for by soybean, whereas food legumes together account for 2.95 Tg of N-fixation mainly through chickpea followed by common bean, pea, faba bean, cowpea and lentil (Herridge et al. 2008).

An alarming situation has recently come up in a report by Fagodiya et al. (2016), wherein they have analysed the trends in sources of N use in agriculture from 1961 to 2010. As per the report, the N sourced from chemical fertilisers has increased from 15.47% to 51.38%, whereas the amount sourced from N-fixation has reduced from 29.33% to 12.31%. Similarly, N sourced from crop residues has reduced from 18.75% to 14.40% and that sourced from animal manure has reduced from 32.30% to 15.41%. The situation points to the decline in BNF possibly due to largely cereal-based farming systems under high-input agriculture and deterioration in soil health owing to less contribution of food legumes towards global agricultural productivity. The differential contribution of various legume crops in developing and developed world in terms of species diversity may also contribute to the drop in the share of N through BNF (Fig. 8.2). An appraisal of trends of changes in food legume area and production (Table 8.4) reveals that, during the period 1961–2012, the area under food legumes has increased by 27.86%, and the production has increased by 70.73%, much of which has possibly come from synthetic fertilisers. To add to the misery, it is estimated that, by 2015, more than half of the world population will rely on Haber-Bosch process increasing the N emissions. In countries like China and India, synthetic fertilisers account for about 60% of total N inputs (Ma et al. 2010).

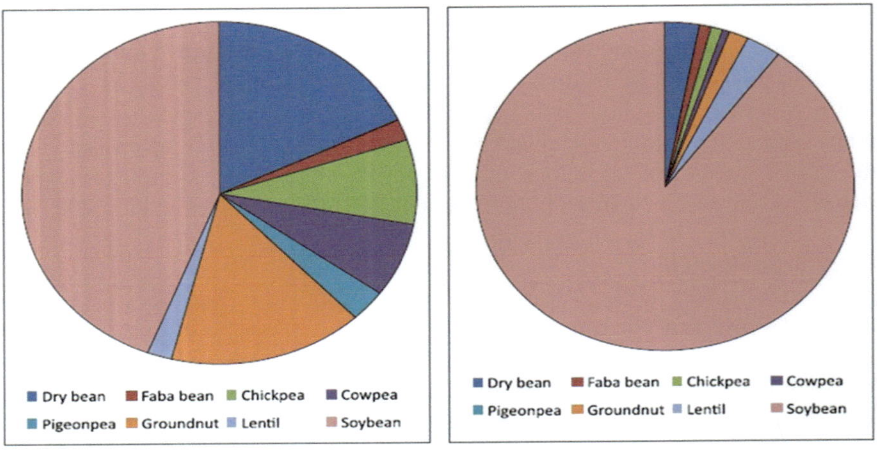

Fig. 8.2 Share of area under different legume crops in developing (left) and developed nations 2008–2010 (Nedumaran et al. 2015)

Table 8.4 World total pulse area, yield and production, 1961–2012

Area (m ha)			Production (M tones)			Yield (Kg/ha)		
1961	2012	Change (%)	1961	2012	Change (%)	1961	2012	Change (%)
61.0	78.0	+27.86	41.0	70.0	+70.73	637.0	908.0	+42.54

Data source: FAO STAT (2014)

8.2.1.3.2 Reduced Carbon Footprints

Legume crops are known as climate smart as they help in both adaptation and mitigation of climate change. In fact, an important climate change mitigation strategy would be the diversification of diets towards less animal-sourced foods especially for meeting protein requirements (FAO 2016a, b). The incorporation of food legumes into cropping system greatly reduces chemical fertiliser and energy footprints and consequently lowering the greenhouse gas emissions (Reckling et al. 2014; Yadav et al. 2017a). Food legumes offset the carbon footprints on account of reduced fertiliser use and consumption of fossil fuels on account of tillage and irrigation. Around one-fifth of all greenhouse gases are contributed by agriculture. This is a global responsibility and requires all economic sectors to shift to low emission intensity. The bulk of direct emissions of methane and nitrous oxide, two potent GHGs, is the result of fermentation in livestock, rice production, flooded fields and the application of N fertiliser and manure (FAO 2016a).

In Europe, the N economisation in legume-based rotations is around 277 kg ha^{-1} of CO_2 per year (1 kg $N = 3.15$ kg CO_2). Considering a baseline of 2.6–3.7 kg CO_2 generated per kg of N synthesised, approximately 300 Tg of CO_2 will be annually released into the atmosphere. All the CO_2 released during the process of N fertiliser synthesis is derived from fossil energy, thus indicating a net contribution of CO_2 to atmosphere (Jensen et al. 2012). On the contrary, the CO_2 released by root nodules wholly comes from the atmosphere through the photosynthetic activity. Legume-based cropping systems reduce nitrous oxide (N_2O) emissions by about 18–33%, compared to systems without legumes. N_2O accounts for about 5–6% of the total atmospheric greenhouse gases (GHGs), but are more damaging than CO_2 in terms of global warming potential (Crutzen et al. 2007). Agriculture contributes majority (60%) of total anthropogenic N_2O emissions (IPCC 2007). Most of GHG emissions result from the application of N fertilisers (Reay et al. 2012) evident from the fact about 1.0 kg of N_2O is released for every 100 kg of N fertiliser (Jensen et al. 2012) with denitrification as the major process of N_2O in most cropping systems (Soussana et al. 2010; Meena et al. 2017a). In the context of the quantum per unit area, legumes emit around five to seven times less GHGs compared to other crops (Jeuffroy et al. 2013)

8.2.1.3.3 Reduced Food Waste Footprints

Approximately, one-third of globally produced grains never reaches the plates as it is wasted along different stages of production value chain (amounting to a whopping US$ 750 billion), with most of the loss taking place during production and processing and lower during consumption (FAO 2011). Increasing primary food production is undoubtedly the major focus of farming research and policy. This is aimed at meeting the future increase in food demand. The economic accessibility of food to the poor is also important that can, to a large extent, be mitigated by reducing food losses. In view of a longer shelf life, the food waste footprints of legume crops are lower than cereals, vegetables and fruits. As per the FAO estimates, the oil crops and pulses together undergo a wastage of around 43.1 million tonnes as compared to 316.9 tonnes in cereals with around 70% occurring in low-income nations. These nations in turn have far greater food security concerns as compared to high-income nations (FAO 2011).

8.2.1.3.4 Sustainability

Legumes are an indispensable component of sustainable agriculture. The major advantages of legumes include the N-fixation to help both current and following crop as well as release of organic matter into the soil in terms of a balanced C/N ratio. Many of the legumes are characterised by deep root systems, which facilitate solubilisation of nutrients such as P, K and Zn by root exudates (Stagnari et al. 2017; Datta et al. 2017b). The food legumes used as rotation crop or intercrop in cereal-based farming systems reduce the soil pathogen inoculums and help achieve the N environmental sustainability goal. In case of the maize-legume system, the leguminous component contributes significantly to N requirement of the maize crop. As a result of such nutrient compensation, intercropped maize responds to a relatively lower fertiliser dosage as compared to sole maize. Similarly, in maize-cowpea intercropping system, it has also been observed that the N content of intercropped maize, especially under low N supply, is higher than that of sole maize (Francis 1986). This indicates some transfer of fixed N from cowpea to maize. In sorghum/black gram intercropping system, reduction of the N requirement of sorghum to the extent of 9 kg N/ha has been reported (Dusad and Morey 1979).

8.2.1.3.5 Increased Adaptive Capacities

There are ~475 million smallholder farmers worldwide (FAO 2016a, b) which are at a greater risk of climate change implications as they inherently lack adaptive capacities and resilience. The legume-based farming systems promote diversification and increased water use efficiency as compared to cereal crops, thereby increasing the adaptive capacity. The incorporation of legumes increases both structural diversity and genetic diversity that leads to pest suppression, nutrient recycling and land use optimisation and increases over all buffering capacity of farming systems (Hossain et al. 1984; Zhu et al. 2005).

8.2.1.3.6 Resource Recycling

Resource recycling is an important characteristic of multiple cropping systems that is accomplished through nutrient recycling and soil formation, especially in systems where legumes are component crops. Legumes have the ability to fix N and solubilise phosphorous and have the capacity to increase rhizosphere biodiversity that helps in such resource recycling. Additionally, the ability of legumes to promote higher rates of soil organic carbon accumulation compared to cereal-based systems can lead to enhanced sequestration of carbon (Bachinger et al. 2013).

8.2.1.3.7 Improving Soil Properties

Legume-based systems help improve soil fertility and soil organic carbon (SOC) as well as increase the availability of N and P. This is all the more important in the case of sandy soils where grain legumes can increase SOC content. The intercropping of soybean with maize has been reported to increase the SOC accumulation to up to 23.6 g C kg^{-1} as against 21.8 g C kg^{-1} under sole maize (Jensen et al. 2012). In fact, Bichel et al. (2016) reported that just by adding soybean residues to the soil increased the SOC by 38.5%. Hydrogen gas (H_2), which is as a by-product of BNF, favourably changes the composition microbial population of the soil, further enhancing the activities of rhizobacteria (Angus et al. 2015; Verma et al. 2015a).

8.2.1.3.8 Promoting Biodiversity

Modern agriculture is far more specialised and is mostly based on a monoculture involving very few crops that have reduced on-farm diversity. The dominance of cereals and other crops vis-à-vis legumes is largely due to the inherently low yielding ability of legume crops. However, with better input support and input-responsive varieties and better management, legumes could well become competitive and replace cereals. In fact, crop diversification is seen as a major driving force towards sustainable intensification. The intercropping systems are characterised by higher species diversity than monocropping. Greater species diversity increases the resource use efficiency that is invariably translated into higher yields and lowers the risk of crop failure due to increasing buffering of heterogeneous systems (FAO 2016a, b).

8.3 Implications of Drought Stress on Legumes

Abiotic stresses are major limiting factors that negatively affect crop growth. This consequently reduces crop productivity substantially, across all crops and farming systems, given that they are widespread. These factors often intensify in magnitude and occur almost every year (Wortmann et al. 1998; Mittler 2006). The crop growth rates are significantly influenced by the availability of water in the soil (Song et al. 2009). In fact, water availability is the most important abiotic factor that has shaped the plant evolution (Kijne 2006). The scenario of water availability is expected to deteriorate more owing to a decline in precipitation and unpredicted weather extremes that have created a proactive international policy and research interest in crop drought tolerance. In the United States alone, during the last 50 years, drought-related crop losses have been estimated at about 67% (Comas et al. 2013; Meena et al. 2015c). Drought stress is not a regional phenomenon, but a globally prevalent production constraint of all major crops including food legumes. The negative effects of prolonged water stress striking as early season, intermittent or terminal drought, are largely enhanced by high temperature (as drought and heat stress invariably come together) and low air relative moisture (vapour pressure deficit), and such situations are the most damaging for legume crops. Water stress especially during the flowering and grain filling periods causes precocity and greater reductions in the seed set, seed weight and seed yield in crops like dry bean (Singh 1995; Sofi et al. 2017).

Drought stress: Key concerns

- *Water is the most important resource for agriculture; any undesirable change in availability of water is going to severely dent our efforts to ensure food availability for the world population that is expected to cross 9 billion by 2050. In fact, agriculture accounts for the largest (70%) withdrawals of water (Siebert et al.* 2010*).*
- *The water utilisation between 1916 and 2016 has increased twofold in global population (UNO 2015).*
- *By 2025, there will be an upsurge in water withdrawals, and around 1.8 billion people will face acute shortage of water, and 66% people will be living under water-stressed conditions (UNO 2015).*
- *Alarmingly, 84% of economic impacts of drought will be on agriculture (www.fao.org).*
- *Under the predicted phenomenon of climate change, where temperatures are expected to increase by 1–2°, for each degree of increase in temperature will lead to 20% decrease in renewable water resources (www.fao.org).*
- *Increase in sea level will cause intrusion of saline water and cause salt stress.*

Legumes are considered as being inherently sensitive to water stress (Sprent 1972). For majority of grain legumes, the process of N_2 fixation seems especially more sensitive as it shows a decline early during the soil drying. Daryanto et al. (2015) has provided an excellent overview of the effect of drought stress on food legume production around the world. In dryland areas, legumes face up to 70% shortage of water, whereas under non-dryland systems, 60% reduction in water availability is reported causing a yield reduction of 30% and 40%, respectively. They reported that, across all food legume crops, early season water stress (at vegetative stage) reduces productivity by around 20%, water stress at either the early (flowering) or late (pod filling) reproductive stages caused yield reductions of up to 40%, whereas, water stress at both of these stages causes yield reduction in excess of 40%. Among crops, field pea, lentil, groundnut, soybean and pigeon pea suffer relatively lesser yield reduction under drought (20–30%); faba bean, chickpea, green gram and cowpea suffer mild reduction under drought (30–50%), whereas bambara bean, lablab bean, common bean and black gram suffer heavy reductions (60–80%) under drought (Verma et al. 2015b). The challenge of producing enough food for the increasing population is further complicated by the severe competition for already shrinking land and water from industry and urbanisation (Postel 2000) that further relegates legume-based farming to marginal areas, where water-limiting conditions often constrain crop productivity. The limitation posed by water availability in such areas is invariably persistent and season to season fluctuations tend to be large, significantly affecting food security in such rain-fed systems. Despite the fact that water stress adversely affects legume production, yet, about 70% of production occurs in regions such as sub-Saharan Africa, India and China that invariably experience water shortage (Gowda et al. 2009) resulting in lower yields. The variable rainfall patterns in these regions render legume cropping systems highly vulnerable to drought. Even in countries like Brazil where legumes like soybean are cultivated under sufficient precipitation, water deficiency may still surface resulting in significant yield reduction (Oya et al. 2004).

8.3.1 Implication of Water Stress on N-Fixation

The legume-*Rhizobium* symbiotic N-fixation is a biological phenomenon of paramount importance and also a major contributor towards improving soil fertility in legume-based farming systems. There is sufficient evidence to state that water and salt stress cause a substantial reduction in plant biomass accumulation (root and shoot), nodule development, and N activity as well as these strongly declined the yield (Egamberdieva et al. 2014). As a biological process, symbiotic N-fixation is highly sensitive to water stress, which results in decreased N accumulation and also in the yield of companion legume crops. With this premise, one can safely conclude that crops dependent on N-fixation in nodules, under drought stress, will experience N deficiency on account of reduced N_2 fixation. In fact, the increased sensitivity of N-fixation to water stress relative to leaf gas exchange is a major constraint on N_2

accumulation and yield of legumes grown under moisture stress (Serraj et al. 1999). The legume-rhizobacterial symbiotic system efficiency under stress is based on individual and interactive ability of various components such as plants, the N-fixing microbes and certain helper microbes to function optimally. Of the two component partners in the legume-rhizobium association, the *rhizobium* is more sensitive than the plant. This is evident from the observation that symbiotic association is more sensitive to environmental stress (especially drought) than the uninfected legume (Swaine et al. 2007; Meena et al. 2017b).

8.4 Association of Legumes with Soil Rhizobacteria

The interface of root and soil systems constitutes a dynamic associative as well as interactive system known as the *rhizosphere* where the roots, soil and microorganisms interact (Lynch 1990; Kennedy 1998). Hiltner (1904) was the first to describe the term "rhizosphere" as a zone of maximum microbial activity. Rhizosphere is a unique zone of soil near plant roots and formed under the influence of a plant root system (Berendsen et al. 2012). All the three components of soil environment, viz. physical, chemical and biological, of the rhizosphere are clearly different from the rest of the soil (outside the rhizosphere), where entirely different microbial diversity and activity have been reported (Kennedy and Smith 1995). Especially the biological component (microbes) of rhizosphere is clearly distinct from that of the bulk soil, both in abundance and in diversity, primarily due to the root exudates that create an enabling habitat by providing nutrition for microbial growth (German et al. 2000). The microorganisms may be present in the rhizosphere, root tissue and/or in a specialised root structure called a nodule. Very important and significant interactions have been reported among plant, soil and microorganisms present in the soil environment (Antoun and Prevost 2005). The rhizosphere comprises not only the wide array of microbial diversity (ranging from beneficial to harmful to neural microbes) but also a complex set of interactions (beneficial, harmful and/or neutral) between the roots and microbes (Ahmad et al. 2011; Bakker et al. 2013). The increased population of bacteria in the rhizosphere may be due to the root exudates that not only attract but also promote bacterial growth (Bais et al. 2006). The microorganisms colonising plant roots generally include bacteria, algae, fungi, protozoa and actinomycetes. Enhancement of plant growth and development by application of these microbial populations is well evident (Bhattacharyya and Jha 2012; Hayat et al. 2010; Meena et al. 2013). Of the different microbial populations present in the rhizosphere, bacteria are the most abundant of microorganisms (Kaymak 2010). Various genera of bacteria, *Pseudomonas*, *Enterobacter*, *Bacillus*, *Variovorax*, *Klebsiella*, *Burkholderia*, *Azospirillum*, *Serratia and Azotobacter*, cause a significant effect on overall plant growth and are referred to as plant growth-promoting rhizobacteria (PGPR). The PGPRs are reported to increase plant growth both under favourable and stressful conditions through various direct and indirect mechanisms (Nadeem et al. 2010) including biological N-fixation,

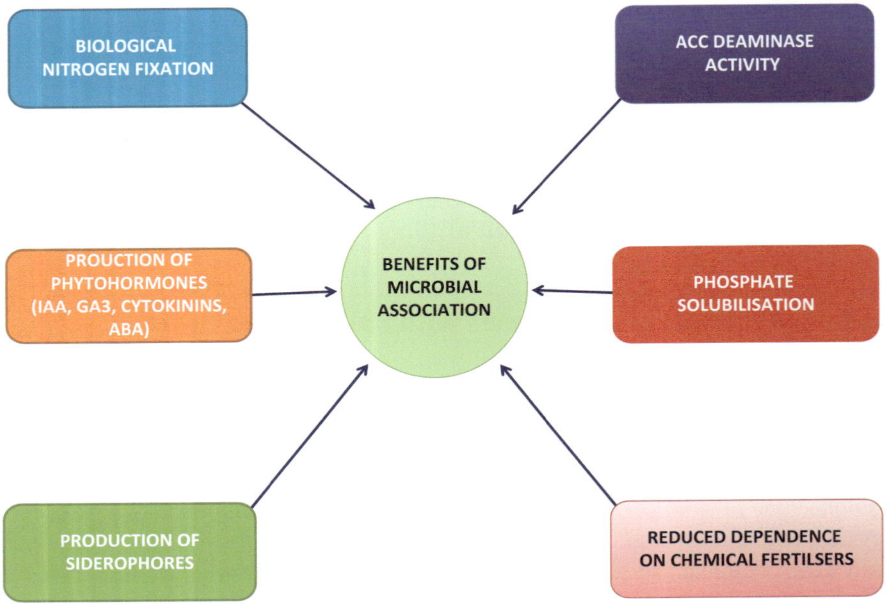

Fig. 8.3 Mechanisms of plant growth promotion by rhizobacterial association

phosphorus solubilisation, siderophore production, phytohormones and organic acids as well as enzymes such as ACC deaminase, chitinase and glucanase (Fig. 8.3) that have all protective roles under stress (Berg 2009; Glick et al. 2007).

The beneficial bacteria may either form a symbiotic relationship, involving the formation of nodules as in *rhizobia*, or may be free-living in the soil (Valdenegro et al. 2000). The beneficial free-living bacteria, commonly referred to as plant growth-promoting rhizosphere bacteria (PGPR), have been found in association with many different plant species (Majeed et al. 2015). The beneficial plant growth-promoting bacteria belong to a heterogeneous group of microorganisms inhabiting the rhizosphere, in diverse forms of associations, and thus enhancing the growth of plants as well as protecting them from various biotic and abiotic stresses (Dimkpa et al. 2009; Grover et al. 2011; Glick et al. 2007). In addition to the growth-promoting characteristics, some of these PGPRs have also been implicated in restricted pathogen proliferation through production of various biostatic compounds that stimulate the immune system (Berendsen et al. 2012).

Various types of plant-microbe (Fig. 8.4) interactions such as symbiotic, endophytic or associative are operative in the rhizosphere with distinct degrees of proximity with the roots and the rhizosphere. Endophytic rhizobacteria are invariably good candidates for use as inoculants on account of their ability to efficiently colonise roots. Therefore, they create a favourable environment for proliferation and effective function, while the non-symbiotic endophytes colonise the intercellular spaces of plant tissues, containing high levels of carbohydrates, amino acids and

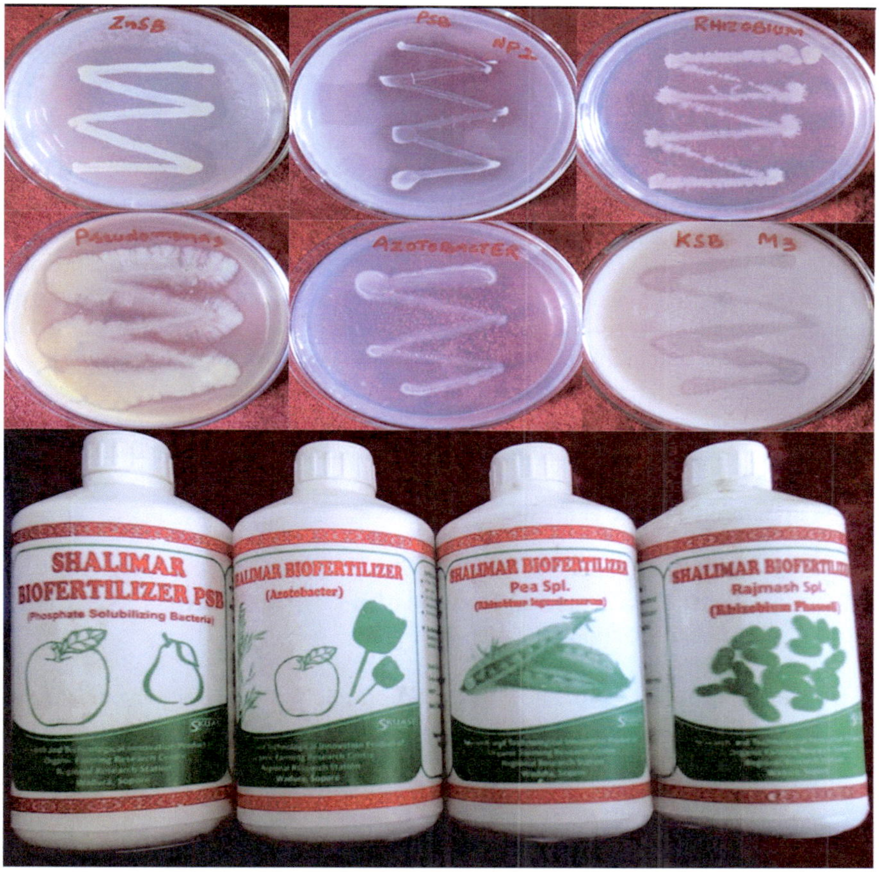

Fig. 8.4 Various types of rhizobacterial cultures and products for legume crops. (Source: Z A Baba)

inorganic nutrients (Bacon and Hinton 2006). The success and efficiency of the different types of PGPRs for agricultural crops is determined by factors such as:

- *Root-colonising ability of bacteria*: This is invariably influenced by the nature and extent of competition and survivability of the different microbes, as well as with the changes in gene expression (quorum sensing) in response to the population dynamics (Meneses et al. 2011).
- *Composition of root exudates*: Plant roots secrete a wide range of root exudates in response to the variations in the immediate environment. These root exudates influence the plant-microbe interaction that is an important consideration in the efficiency of the PGPR's use as inoculants (Carvalhais et al. 2013).
- *Soil health:* Soil health is collectively determined by physical, chemical and biological parameters of soil such as soil structure and texture caused, nutrient pool

and toxic metal concentrations, soil moisture, microbial diversity as well as management practices and affects the inoculation efficiency due to several characteristics.

Rhizosphere functionality significantly influences the overall fitness of plants and the soil health, especially under stress conditions as they can help the host plant to adapt to stress conditions, as well as ward of certain soil-borne pathogens (Bowen and Rovira 1991). Nearly all plants have a well-established association with a wide variety of soil microbes in the rhizosphere (Brundrett 2009). Such a system not only provides an enabling habitat for proliferation of microorganisms but also improves plant growth as well as soil health. A large number of such microorganisms establish an endophytic relationship with the root system of host plants (Stone et al. 2015). Climate change mainly increased CO_2 and is expected to change the plant physiology and consequently the root exudation, mainly by changes in carbon allocation to the roots. Such alterations may also include changes in the availability of attractants and/or signal molecules as well as changes in C/N ratio or the availability of certain nutrient (Haase et al. 2007; Meena et al. 2016).

8.5 Soil Rhizobacteria in Relation to Drought Stress Amelioration

There are ample evidences to show that the crop-microbial interactions in legumes enhance productivity, quality as well as resilience to various abiotic stresses through mechanisms outlined above (Fig. 8.3). Several limiting factors such as water, salt and high-temperature stress cause cell and tissue dehydration and irreversibly damage plant tissues. Symbiotic N-fixing bacteria such as *Rhizobium* can synthesise trehalose (Suárez et al. 2008), a sugar that accumulates in bacteroids as well as in nodules and helps retain water in cells. Water stress changes the phytohormonal balance in the plants resulting in higher abscisic acid (ABA) content in leaves, small decrease in indole acetic acid (IAA) and gibberellic acid (GA3) and a large decline in zeatin content in leaves (Figueiredo et al. 2008). For legume crops that depend on N-fixation, water stress causes a corresponding negative effect on nodulation, nodule functioning as well as N-fixation, biosynthesis of protein, malate and leghaemoglobin and changes in enzymatic activities, plant growth and metabolism (Aydi et al. 2004; Mhadhbi et al. 2004). Water stress may decrease persistence and the survival of *rhizobia* in the soil and root hair colonisation as well as in the colonisation and infection process (Gray and Smith 2005). Legume crops are colonised both by endocellular and intracellular microorganisms including bacteria and fungi that can enhance plant growth especially under stress conditions and improve yields (Dimkpa et al. 2009). PGPRs directly stimulate plant growth and development by providing fixed N, phytohormones, iron as well as phosphate (Hayat et al. 2010; Yadav et al. 2017b), plant physiology and growth under various abiotic stress conditions, and some examples are summarised in Table 8.5.

Table 8.5 Rhizobacterial species reported to alleviate various stresses in legume crops

Rhizobacterial species	Crop	Stress	References
Aeromonas hydrophila	Soybean	Heat stress	Zhang et al. (1997)
Azospirillum brasilense	Common bean	Water stress	German et al. (2000)
Azospirillum brasilense	Faba bean	Salt stress	Hamaoui et al. (2001)
Sinorhizobium meliloti	Medicago	Water stress	Vazquez et al. (2001)
Glomus intraradices	Soybean	Water stress	Porcel and Ruiz-Lozano (2004)
Bradyrhizobium	Soybean	Lead toxicity	Andrade et al. (2004)
Pseudomonas marginalis	Pea	Heavy metals	Safronova et al. (2006)
Pseudomonas brasiliense	Soybean	Mercury toxicity	Gupta et al. (2005)
Mesorhizobium	Common bean	Heat stress	Rodriguez et al. (2006)
Ochrobactrum	Mung bean	Chromium toxicity	Faisal and Hasnain (2006)
Mesorhizobium ciceri	Chickpea	Salt stress	Tejera et al. (2006)
Brevibacillus	Clover	Zinc toxicity	Vivas et al. (2006)
Mesorhizobium	Chickpea	Acidity	Rodriguez et al. (2006)
Rhizobium leguminosarum	Trifolium	Nickel toxicity	Vivas et al. (2006)
Glomus etunicatum	Soybean	Salt stress	Sharifi et al. (2007)
Ensifer meliloti	Common bean	Water stress	Mnasri et al. (2007)
Pseudomonas fluorescens	Groundnut	Salt stress	Saravanakumar and Samiyappan (2007)
Rhizobium tropici	Common bean	Water stress	Figueiredo et al. (2008)
Azospirillum brasilense	Pea	Salt stress	Dardanelli et al. (2008)
Rhizobium etli	Common bean	Water stress	Suárez et al. (2008)
Mesorhizobium mediterraneum	Chickpea	Water stress	Romdhane et al. (2009)
Variovorax paradoxus	Pea	Water stress	Belimov et al. (2009)
Rhizobium and Azotobacter	Faba bean	Water stress	Dashadi et al. (2011)
Glomus mosseae	Trifolium	Salt stress	Zou and Wu (2011)
Pseudomonas pseudoalcaligenes	Chickpea	Salt stress	Patel and Jain (2012)
Pseudomonas extremorientalis	Common bean	Salt stress	Egemberdieva (2011)
Glomus mosseae	Mung bean	Water stress	Habibzadeh et al. (2012)
Rhizophagus irregularis	Trigonella	Salt stress	Basrnawal et al. (2013)
Bradyrhizobium spp.	Common bean	Water stress	Uma et al. (2013)
Bradyrhizobium spp.	Mung bean	Water stress	Tittabutr et al. (2013)

(continued)

Table 8.5 (continued)

Rhizobacterial species	Crop	Stress	References
Glomus mosseae	Faba bean	Chromium toxicity	Ismail (2014)
Pseudomonas aeruginosa	Mung bean	Water stress	Sharma and Saikia (2014)
Bradyrhizobium japonicum	Soybean	Water stress	Prudent et al. (2015)
Pseudomonas putida	Faba bean	Salt stress	Metwali et al. (2015)
Bacillus thuringiensis	Soybean	Water stress	Prudent et al. (2015)
Rhizobium phaseoli	Common bean	Water stress	Sofi et al. (2017)
Rhizobium spp.	Chickpea	Water stress	Khadraji and Cherki (2007)
Mesorhizobium	Chickpea	Salt stress	Chaudhary and Sindhu (2017)

Table 8.6 Shifts in biomass partitioning under drought in common bean without rhizobial inoculation (Sofi et al. 2017)

Treatment	Root to total biomass ratio	Shoot to total biomass ratio	Root shoot ratio
Drought	0.465	0.535	0.930
Irrigated	0.263	0.737	0.381
% Increase/ decrease	+76.80	−27.40	+144.09

There are various strategies and mechanisms implicated in enhancement of plant drought stress tolerance mediated by rhizobacteria. However, the exact mechanisms of enhancement by rhizosphere bacteria have remained mostly speculative. In terms of gross morphology and biomass partitioning, rhizobium is reported to cause changes in allocation to help plant adapt to resource. This resource deficiency entail rooting depth, root biomass and root volume (Table 8.6). Rhizobium has been reported to increase height, leaf area, photosynthetic rate and dry matter production in plants under irrigated conditions (Thakur and Panwar 1995). Sofi et al. (2017) reported that among plant growth parameters, rhizobial inoculation caused the largest increase in shoot biomass (184.44%) followed by a root volume of (91.81%) and root biomass of (74.64%), whereas the smallest increase was recorded for rooting depth was (10.87%). Interestingly, the rhizobia caused a decrease in root/shoot ratio by −38.60%. This is interesting in view of the fact that drought stress, without any rhizobial treatment, invariably causes a shift in allocation towards roots to enhance resource acquisition. However, such shift always has penalties on the overall biomass production as well as the subsequent remobilisation of resources (Table 8.7).

The rhizobacteria-mediated alleviation of stressful conditions may be accomplished through either of the following possible mechanisms:

- Hormones like ABA, GA, cytokinins and auxin that promote growth and development and modulate plant response to stress by improving its soil resource acquisition. Phytohormones such as IAA produced by bacteria also stimulate the

Table 8.7 Effects of rhizobium on various root and shoot traits in common bean (*Phaseolus vulgaris* L.) (Sofi et al. 2017)

Treatment	Rooting depth (cm)	Root biomass (g)	Root volume (cm^3)	Plant height (cm)	Shoot biomass (g)	Root/shoot ratio
Without rhizobium	74.222	6.0037	6.527	46.527	10.283	0.583
With rhizobium	82.291	10.485	12.520	45.500	29.250	0.358
% increase or decrease	+10.87	+74.64	+91.81	−2.01	+184.45	−38.60

activity of the ACC deaminase (Glick 2005). Enhanced production of abscisic acid (ABA) plays a key role in closing stomata and regulating aquaporins either via their gene expression or via post-translational regulations. This helps plants to regulate water fluxes within as well as absorption of nutrients (Acharya and Assmann (2009).

- Enzymes such as ACC deaminase reduce ethylene level in the root of growing plants. Under stress conditions ACC deaminase reduces stress-induced ethylene production. The rhizosphere bacteria with ACC deaminase activity are reported to modify the sensitivity of the root and leaf growth to water stress through ethylene signalling. The reduced ethylene production upon inoculation improved recovery from water deficiency, although there was no effect on leaf relative water content (Mayak et al. 2004).
- Osmoprotective amino acids such as proline are enhanced under water stress. The *Medicago* plants infected by PGPR strains that produce higher quantity of phytohormones such as IAA were more tolerant to stress environmental conditions through accumulation of higher levels of proline (Verbruggen and Hermans 2008).
- Release of plant growth-promoting compounds in active form through hydrolysis of conjugated phytohormones and flavonoids in the root tissue as reported in case of *Azospirillum* inoculation (Dardanelli et al. 2008).
- Increased root growth, root biomass, enhanced lateral root formation as well as greater root hair proliferation that can result in higher tolerance to abiotic stress such as water stress as reported in case of *Rhizobia* (Sofi et al. 2017).
- PGPRs can also induce a reproductive delay leading to a better tolerance to water stress as in case of Arabidopsis (Bresson et al. 2013).
- Bacterially derived biofilms, i.e. extracellular matrix, trigger induced systemic resistance (Kim et al. 2013). Such extracellular matrix contains a wide variety of macromolecules, which are beneficial for plant growth and development. Biofilms also contain a number of sugars and polysaccharides that can play very essential roles in bacteria-plant interactions and in the improvement of their water retention capacity, thereby improving water availability in the root medium. Some of the polysaccharides have the capacity to retain water by severalfold of their mass (Timmusk and Nevo 2011). Even small polysaccharide such as

alginate in the biofilms can facilitate maintenance of hydrated microenvironment, by increasing the water retention capacity (Chang et al. 2007).

- Changes in the composition of cell envelope that may result in changes in proteins, periplasmic glucans and exo- and lipopolysaccharides. PGPRs such as *Pseudomonas* are known to survive under stress conditions due to the exopolysaccharide production that confers protection from water stress by increasing retention of water and regulating the diffusion of C sources in environment (Sandhya et al. 2009; Meena and Yadav 2015). Similarly, Klein et al. (1999) reported that, under osmotic stress, the composition of bacterial membrane is changed by changes in the length and branching of acyl chains as well as number of double bonds. Francius et al. (2011) reported the presence of loose, flexible surface appendage around the bacteria under low electrolyte concentration condition that acts as a protective barrier. Changes in phospholipid content in the cell membranes of cowpea have been observed upon inoculation with *Azospirillum* (Bashan et al. 1992).
- Certain osmolytes that increase the osmotic potential within cells are released into rhizosphere by root zone bacteria such as trehalose and glycine betaine (Farooq et al. 2009). Production of sugars like trehalose as reported in *R. elti* that help plants retain more water under stress conditions (Suárez et al. 2008).

The use of PGP microbes for stress alleviation can be used as a viable option for improving stress tolerance that is economically as well as ecologically sustainable. In this case, the native microbes could be more useful as they are relatively well adapted to the local environments on account of their competitive abilities (Mrabet et al. 2005). Rhizobacteria used as microbial inoculants have many direct and indirect growth-promoting properties including some tolerance mechanism against water stress, heavy metals and pesticides. Even though the response in legumes to various stresses is host plant centric reaction, this response can be favourably modulated by the rhizobia (Yang et al. 2009). Various accounts of role of *Rhizobium* and *Bradyrhizobium* in tolerance and nodulating capacity for alleviating problem soils, temperature and water stress conditions have been provided by Graham (1992) and Grover et al. (2010). The rhizobacteria have a remarkable ability to protect not only their own systems from stresses but also impart a fair amount of resilience to host plants, against abiotic stresses. Interaction of rhizobacteria with several crops in stress conditions is reported to reduce growth reduction and improve plant survival and performance in adverse conditions (Dimkpa et al. 2009).

8.6 Breeding Perspectives of Harnessing Soil Rhizobacteria

The crop-microbial interaction is a multi-partner association comprising plant roots, the rhizobacteria as well as certain helper microbes. These microbes facilitate recognition, colonisation as well as the functionality of the system. There are obvious genetic differences in both the partners of this association that determine the levels of interaction and effectiveness especially under stress conditions (Fig. 8.5). The

Fig. 8.5 Response of common bean under drought stress with and without rhizobium (top) irrigated (middle) drought with rhizobium (bottom) drought without rhizobium

different partners of the system are differentially susceptible to water stress, so they have to be considered as separate systems as well as on a whole system basis. This is done to find the combinations that effectively harness the potential of such combinations to alleviate water stress in food legume crops (Rengel 2002).

8.6.1 Selection for Differential Genotypic Response to Rhizobacterial Inoculation

The first step towards building effective crop microbial associations would be to understand natural variation in food legume crops, varieties, breeding lines and germplasm resources for their association capacities with different soil rhizobacteria. This would require extensive genotypic screening to identify the genotypes that have the ability to harbour effective microbial populations that can help improve overall growth and stress tolerance (Fig. 8.6). There are already ample evidences that indicate genotypic differences in legumes for rhizobacterial associations such as white clover (Ledgard 1989), faba bean (Caba et al. 2000), soybean (De Chueire and Hungria 1997), mung bean (Espiritu et al. 1993), common bean (Suárez et al. 2008), pea (Evans et al. 1995), groundnut (Ibrahim et al. 1995), lucerne (Hernandez

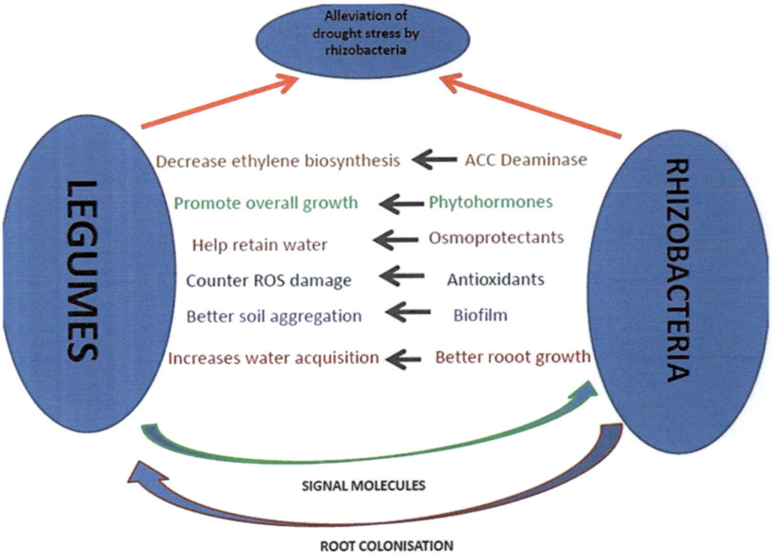

Fig. 8.6 Mechanistic representation of legume-rhizobacteria association for drought alleviation

et al. 1995) and chickpea (Sattar et al. 1995). The genotypic differences for glutamine synthetase and glutamate synthase activities in nodules have also been reported to be correlated to stress tolerance in faba bean (Caba et al. 2000). Therefore, increased activities of such enzymes can be used as selection criteria in the breeding programmes to increase stress tolerance of genotypes. Similarly, in chickpea, genotypic differences in the nodule number and weight have been reported (Dangaria et al. 1994). Genotypic differences for traits related to the nodule functionality such as N accumulation in shoots have also been reported in pea (Fesenko et al. 1995). The legume genotypes that are able to maintain a higher N-fixation under suboptimal levels of nitrate should also be selected (Blumenthal and Russelle 1996). This is especially relevant in the case of modern crop varieties that contain high N concentrations in the harvestable product by removing greater amounts of N on account of higher fertiliser use even in legume-based farming systems. Genotypic differences have also been reported in legumes in the level of tolerance of the N-fixation to nitrate. Such natural variation can be potentially harnessed for increasing the efficacy of symbiotic N-fixation by selecting the best combination of nitrate-tolerant plant genotype and rhizobia (Raffin and Roumet 1994).

Any breeding programme aimed at harnessing crop-microbial associations should tap this variation to identify the genotypes as well as elucidate the physiological and genetic factors underlying such a variation. However, a major problem in screening genotypes for differences in nodulation and N-fixing ability is the diversity of conditions that may occur in soils. This is especially the case of drought stress, where, in the screening process, different factors may confound the final results (Fig. 8.7). Under field conditions, a multitude of favourable and

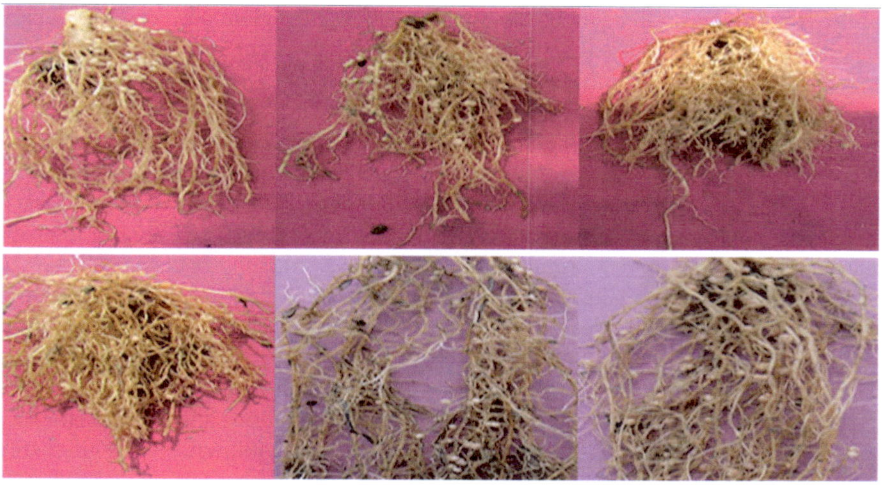

Fig. 8.7 Genotypic differences in nodulation in common bean. (Source: F. A. Sofi)

unfavourable effects may come into play, and delineation of actual genotypic differences becomes practically impossible. Therefore, the experiments should be, as far as possible, conducted in controlled conditions such as hydroponics and green houses. However, a major issue with such a setup is failure to replicate actual soil-like conditions, which represent the real farmer conditions. This problem may be further complicated by differences in functionality of rhizobacterial strains at different locations. Another major focus in identifying desirable genotypic variation for effective rhizobacterial associations is that the wild relatives of legume crops may provide valuable sources of variation/genes, since the wild relatives might have retained all the regulatory and structural components of effective associations as they have not really suffered changes under domestication such as fertilisers and chemical pesticides. Similarly, mutagenesis can also be used as an approach to generate desirable variation for this trait (Andriolo et al. 1994).

8.6.2 Selection for Competitive Rhizobacterial Strains

Corresponding to the natural variation in the legume crop species, the rhizobacterial strains that have come under evolutionary forces also developed a variation in cross compatibility relationships for colonising, nodulating, and effective functioning. This process was in association with the appropriate legume species and consequently the rate of change of such traits that define the ability of the rhizobacteria to enter into symbiotic relationship with legume crops might be higher than host plants. This is due to their small size, greater generation turnover and ability to generate enormous variability. From a functionality point of view, the potential of association and colonisation is important, apart from the efficiency of N-fixation and the ability to transfer some degree of tolerance to stresses to plant partner. Since rhizobacteria have been reported to differ in their N-fixation capacity, it is possible to select the efficient strains that fix a higher quantity of N even under stress (Hungria et al. 2013). Since substantial natural variation for nodulation and the N_2-fixation is widely reported in different rhizobia-host combinations, it should be possible to identify the most efficient rhizobial strains as well as the underlying genetic factors that regulate the efficiency of the symbiosis. Once the genes are characterised, it would be possible to transfer them into commercial strains of rhizobia to improve the efficiency of their N-fixation. Even within the nodulating bacteria, substantial variability has been found in tolerance of the N-fixation process to nitrate (Nour et al. 1994; Varma and Meena 2016), an issue of paramount significance, while breeding for better symbiosis (Rengel 2002).

Superior rhizobia have better N-fixation. However, an increased N-fixing ability should not be the only index of selection. More than that, one should also consider differential competitive abilities vis-à-vis the native rhizobia which are invariably ineffective in N-fixation, yet competitive due to the adaptive advantage. Superior N-fixing strains invariably have the ability to outcompete the native rhizobial strains and occupy a greater proportion of the nodules. In order to increase the selection efficiency, the selection for competitive ability and functional efficiency should be done under natural conditions to identify superior ones. The effective rhizobia are characterized by better establishment in the soil and the rhizosphere (saprophytic competence), causing better nodulation, occupying larger proportion of nodules, and having better nitrogenase activity. Even though functionally efficient rhizobial strains can be selected from native populations (Howieson et al. 1995), achieving a larger nodule occupancy by rhizobia is also an important practical constraint as mass inoculation does not always ensure improved nodule occupancy (Kuykendall 1989). In fact, the quantum of required inoculum to outcompete the native rhizobia is invariably uneconomical (Vlassak and Vanderleyden 1997). Moreover, the inoculums load as well as number of bacteria is not the single deciding parameter for competitiveness. Other factors such as mobility of rhizobial may also equally be important in determining the capacity of rhizobial strains to effectively nodulate the crown and lateral roots. However, the competitive ability of introduced rhizobia strains can be increased through genetic engineering to produce compounds that inhibit the nod gene expression in native rhizobia. In fact, the nod gene repressor (NolA) in strain USDA110 upon transfer to *B. japonicum* caused a decline in Nod factor production. The native rhizobacterial strains of a particular geographical area hold greater promise on account of their adaptability to environmental conditions and long evolutionary history of coexistence with the local legume crops. The rhizobacterial cultures brought into the system by way of ruthless import of microbial formulations which may have more negative implications rather than potential benefits. This is because they may lack adaptability to the agroecological conditions, may not colonise the host plant optimally, may be invasive and may erode local microbial biodiversity. Besides, they may sometimes negatively influence plant growth parameters (Vlassak and Vanderleyden 1997).

8.6.3 Identification of Adaptive Crop-Microbial Associations

In order to identify stress adaptive crop-microbe associations, one needs to test one's functionality under appropriate stress environments. The host legume crop grown under stressful conditions have exhibited substantial natural variation in nod gene inducers found in their root exudates (Raghuwanshi et al. 1994) such as under low P (Mullen et al. 1988), low Ca (Munns 1970) and the soil acidity. Rhizobial strains with relatively higher nodulation capacity under low P or in acidic soils have been isolated (Howieson et al. 1995) and found to possess better symbiotic abilities under such stress conditions especially in acidic soils in tropical areas where subsistence farmers do not have sustainable options to alleviate soil acidity. In terms of the root phenes, the phenomic level potential root traits to be considered are length, angle, biomass and branching, while as at the level of transcriptome, proteome and metabolome, one needs to consider osmoprotectants, auxin-responsive factors, transcription factors, methionine and coumestrol. Similarly, in nodule partner, the traits at the phenomic level include number, density and biomass, while as traits at the level of transcriptome, proteome and metabolome include C/N/S metabolism, protein turnover and lipoxygenase Kunert et al. 2016; Dhakal et al. 2016).

8.6.4 Genetic Modification of Rhizobacteria

Use of genetically engineered microbes that can suitably alter plant response to stresses by overexpression of certain osmolytes is a viable option. There are already reports about the use of engineered *R. elti* over expressing trehalose conferring drought stress tolerance in common bean (Suárez et al. 2008). The strategy of using genetic transformation in rhizobacteria as against developing transgenic plants for improved plant performance under stresses has many practical advantages (Carman and Defez 2011; Hays et al. 2015):

- They are more robust in that they possess diverse mechanisms to environmental disturbances and in association can transfer a fair amount of tolerance to the host plant.
- It is comparatively much easier to perform genetic modification in bacteria as compared to complex higher plants.
- Many traits that promote plant growth can be simultaneously combined in a single organism thereby overcoming the need to engineering diverse crops, especially in the case of *Azospirillum* which is nonspecific.

8.7 Conclusion

Ensuring food and nutritional securities will be central to all research efforts and policy support systems and will assume much greater significance in the years to come due to predicted projections of climate change implications. Legumes will for sure be a key to any such endeavour owing to their unique features that in addition to increasing farm productivity also lend sustainability to the system at different levels. The sustainability would encompass the food system level where they ensure food and nutritional security to both human and animals; production system level where they ensure reduced dependence on chemical fertilisers especially for low-input farming systems, as well as their role in reducing carbon foot prints, improving soil health and mitigating greenhouse gases emissions; and cropping system levels where their role in diversification of agroecosystems helps enhance farm biodiversity, in reducing pest and disease vulnerability.

8.8 Future Prospective

Plant breeding has undeniably contributed heavily to enhancing productivity across a broad range of growing environments. New crop varieties, with better yields as well as inbuilt resilience to various stresses such as drought, have been continuously developed and released using both the conventional plant breeding and molecular breeding using MAS and genetic engineering through transgenic varieties (Atkinson and Urwin 2012). Depending upon the trait complexity, conventional plant breeding does suffer from being time consuming, as well as laborious and cost intensive (Ashraf 2011). The techniques of molecular biology and biotechnology have largely helped to reduce the time and labour costs of conventional breeding, as well as increase precision. Transgenics have been developed in crops like cotton, maize and soybean carrying genes for economically important traits, more often imparting resilience to biotic and abiotic stresses. Recently the focus has also shifted to traits such as growth regulators, antioxidants, osmolytes and other factors that have been implicated in improved tolerance to stresses (Eisenstein 2013).

Both, the conventional and molecular breeding, work on the premise that the plants function as autonomous organisms and are regulated solely by their gene composition (Barrow et al. 2008). However, there is substantial evidence to safely assume that plant-microbe interactions cannot be ruled out as significant modulators of crop response to abiotic stresses. Despite that, the field trials of newly released stress-tolerant cultivars barely address microbial influence on improved performance (Cooper et al. 2014). Even the greenhouse trials are invariably conducted with sterilised soil and soil amendments (Witt et al. 2012) to create a microbe-free growth environment, a situation hardly found in actual field conditions (Friesen et al. 2011). In doing so, the important determinants of phenotypic output are neglected, which often lead to overestimation of the effect of host genotype on plant phenotype and make the basis for our hypothesised view of plants as individual autonomous systems (Barrow et al. 2008).

Coleman-Derr and Tringe (2014) outlined the comparative advantages of using rhizobacteria approaches to improving stress tolerance as compared to seeking plant improvement per se for stress tolerance. They pointed out the following advantages of harnessing crop-rhizobacterial associations and for improving drought tolerance:

- Microbial systems are capable of conferring stress tolerance to a wide variety of diverse plants, across different genera and species (Zhang et al. 2008). One's ability to harness the stress-resistant substances across crop species through microbial inoculation can potentially save one several years of plant breeding effort. It has been experimentally shown that microbes isolated from the rhizosphere of a desert crop can equally improve the growth of a different host species when grown under water-limited conditions (Marasco et al. 2013).
- Rhizobacteria invariably confer tolerance through a diversity of tolerance mechanisms (Rodriguez et al. 2008). This is all the more important in view of the fact that crops grown in areas where water stress and high temperatures are more prevalent (arid and semiarid) typically suffer from multiple stress. Rhizobacteria, possibly modulate plant response to stress through manipulation of plant hormone pathways that involve substantial crosstalk (Glick et al. 2007; Atkinson and Urwin 2012).
- The plant-microbe system represents a highly flexible co-evolved system that can favourably add genetic flexibility to the stress adaptation of plants (Barrow et al. 2008). In fact, the concept of "habitat specific symbiosis" is one of the most intriguing scientific discoveries that defines the contribution of soil microbes to stress tolerance (Rodriguez et al. 2008).
- It is now fairly easy to characterise vast diversity of rhizobacterial species than ever before. A substantial proportion of microbes isolated from crops have exhibited significant effects on overall fitness of hosts (Friesen et al. 2011). Even for the strains that are otherwise difficult to culture, metagenomics has evolved as a powerful approach to asses such vast diversity spectrum of microbes in rhizosphere (Berg et al. 2016).

References

Abate T, Alene AD, Bergvinson D, Shiferaw B, Silim S, Orr A (2012) Tropical grain legumes in Africa and South Asia: knowledge and opportunities. Kenya International Crops Research Institute for the Semi-Arid, Nairobi

Acharya B, Assmann S (2009) Hormone interactions in stomatal function. Plant Mol Biol 69:451–462. https://doi.org/10.1007/s11103-008-9427-0

Ahmad F, Husain FM, Ahmad I (2011) Rhizosphere and root colonization by bacterial inoculants and their monitoring methods: a critical area in PDPR research. In: Ahmad I et al (eds) Microbes and microbial technology: agricultural and environmental applications. Springer Science LLC, New York

Akibode S, Maredia M (2011) Global and regional trends in production, trade and consumption of food legume crops. Department of Agricultural, Food and Resource Economics: Michigan State University, East Lansing

Alpmann D, Braun J, Schäfer BC (2013) Analyse einer Befragung unter erfol- greichen Körnerleguminosen anbauern im konventionellen Landbau. Erste Ergebnisse aus dem Forschungsprojekt LeguAN In: Wintertagung DLG, Im Fokus: Heimische Körnerleguminosen vom Anbau bis zur Nutzung. Berlin

Anderson RM (2016) International year of pulses 2016 in review. Pulse India 2(3):9–14

Andriolo J, Pereira PAA, Henson RA (1994) Variability among wild Phaseolus vulgaris lines for traits related to biological N2 fixation. Pesq Agrop Brasileira 29:831–837
Andrade SAL, Abreu CA, de Abreu MF, Silveira APD (2004) Influence of lead additions on arbuscular mycorrhiza and *Rhizobium* symbioses under soybean plants. Appl Soil Ecol 26:123–111
Angus JF, Kirkegaard JA, Hunt JR, Ryan MH, Ohlander L, Peoples MB (2015) Break crops and rotations for wheat. Crop Pasture Sci 66:523–552
Antoun H, Prevost D (2005) Ecology of plant growth promoting Rhizobacteria In: Siddiqui ZA (ed) PGPR: Biocontrol and Biofertilization, pp 1–38
Ashoka P, Meena RS, Kumar S, Yadav GS, Layek J (2017) Green nanotechnology is a key for eco-friendly agriculture. J Clean Prod 142:4440–4441
Ashraf M (2011) Inducing drought tolerance in plants: recent advances. Biotechnol Adv 28:169–183. https://doi.org/10.1016/j.biotechadv.2009.11.005
Atkinson NJ, Urwin PE (2012) The interaction of plant biotic and abiotic stresses: from genes to the field. J Exp Bot 63:3523–3543. https://doi.org/10.1093/jxb/ers100
Aydi S, Drevon JJ, Abdelly C (2004) Effect of salinity on root-nodule conductance to the oxygen diffusion in the *Medicago truncatula– Sinorhizobium meliloti* symbiosis. Plant Physiol Biochem 42:833–840. https://doi.org/10.1016/j.plaphy.2004.10.003
Bacon CW, Hinton DM (2006) Bacterial endophytes: The endophytic niche, its occupants, and its utility. In: Gnanamanickam SS (ed) Plant-associated bacteria. Springer, Dordrecht, pp 155–194
Bachinger K, Reckling M, Hufnagel J et al (2013) Ecological recycling agriculture. Guidelines for farmers and advisors, vol IV, 71 pp
Bais HP, Weir TL, Perry LG et al (2006) The role of root exudates in rhizosphere interactions with plants and other organisms. Annu Rev Plant Biol 57(1):233–266. https://doi.org/10.1146/annurev.arplant.57.032905.105159
Bakker PA, Berendsen RL, Doornbos RF et al (2013) The rhizosphere revisited: root microbiomics. Front Plant Sci 4:165
Barrow JR, Lucero ME, Reyes-Vera I, Havstad KM (2008) Do symbiotic microbes have a role in plant evolution, performance and response to stress? Commun Integr Biol 1:69–73. https://doi.org/10.4161/cib.1.1.6238
Bashan Y, Alcaraz-Melendez L, Toledo G (1992) Responses of soybean and cowpea root membranes to inoculation with Azospirillum brasilense. Symbiosis 13:217–228
Basrnawal D, Maji D, Bharti N (2013) ACC deaminase-containing *Bacillus subtilis* reduces stress ethylene-induced damage and improves mycorrhizal colonization and rhizobial nodulation in *Trigonella foenum-graecum* under drought stress. J Plant Growth Regul 32:809. https://doi.org/10.1007/s00344-013-9347-3
Belhassan BB (2017) The International year of pulses: keeping the momentum beyond 2016. In: The pulse Pod Magazine (Global Pulse Confederation), p 4
Belimov AA, Dodd IC, Hontzeas N, Theobald JC, Safronova VI, Davies WJ (2009) Rhizosphere bacteria containing 1-aminocyclopropane-1-carboxylate deaminase increase yield of plants grown in drying soil via both local and systemic hormone signalling. New Phytol 181:413–423
Berendsen RL, Pieterse CM, Bakker PA (2012) The rhizosphere microbiome and plant health. Trends Plant Sci 17(8):478–486
Berg G (2009) Plant-microbe interactions promoting plant growth and health: perspectives for controlled use of microorganisms in agriculture. Appl Microbiol Biotechnol 84(1):11–18
Berg G, Rybakova D, Grube M, Köberl M (2016) The plant microbiome explored: implications for experimental botany. J Exp Bot 67:995–1002. https://doi.org/10.1093/jxb/erv466
Bhattacharyya PN, Jha DK (2012) Plant growth-promoting rhizobacteria (PGPR): emergence in agriculture. World J Microbiol Biotechnol 28:1327. https://doi.org/10.1007/s11274-011-0979-9
Bhattacharya P, Majid A (2013) Impact of green revolution on output, cost and income of small and big farmers, pp A147–A150. JSTOR. http://www.jstor.org/stable/4365209&gt
Bichel A, Oelbermann M, Voroney P, Echarte L (2016) Sequestration of native soil organic carbon and residue carbon in complex agroecosystems. Carbon Manag 7:1–10

Blumenthal JM, Russelle MP (1996) Subsoil nitrate uptake and symbiotic dinitrogen fixation by alfalfa. Agron J 88:909–915

Bowen GD, Rovira AD (1991) The rhizosphere, the hidden half of the hidden half. In: Waisel Y, Eshel A, Kafkafi U (eds) The plant roots, the hidden half. Marcel Dekker, New York, pp 641–669

Bresson J, Varoquaux F, Bontpart T, Touraine B, Vile D (2013) The PGPR strain *Phyllobacterium brassicacearum* STM196 induces a reproductive delay and physiological changes that result in im- proved drought tolerance in Arabidopsis. New Phytol 200:558–569. https://doi.org/10.1111/nph.12383

Brundrett MC (2009) Mycorrhizal associations and other means of nutrition of vascular plants: understanding the global diversity of host plants by resolving conflicting information and developing reliable means of diagnosis. Plant Soil 320:37. https://doi.org/10.1007/s11104-008-9877-9

Caba JM, Centeno ML, Fernandez B, Gresshoff PM, Ligero F (2000) Inoculation and nitrate alter phytohormone levels in soybean roots: differences between a super nodulating mutant and the wild type. Planta 211:98–104

Carman B, Defez R (2011) Soil bacteria protect plants against stresses. In Shankar A, Venkateshwarlu B (eds) Abiotic stress in plants-adaptations and mechanisms. InTech Open Access

Carranca C, Torres MO, Madeira M (2015) Underestimated role of legume roots for soil N fertility. Agron Sustain Dev 35:1095–1102

Carvalhais LC, Dennis PG, Fan B, Fedoseyenko D, Kierul K, Becker A, Von Wiren N, Borriss R (2013) Linking plant nutritional status to plant-microbe interactions. PLoS One 8:e68555

CGIAR (2012) Leveraging legumes to combat poverty, hunger, malnutrition and environmental degradation. Avaialble at https//:www.library.cgiar.org

Chang WS, van de Mortel M, Nielsen L, Nino de Guzman G, Li X (2007) Alginate production by Pseudomonas putida creates a hydrated microenvironment and contributes to biofilm architecture and stress tolerance under waterlimiting conditions. J Bacteriol 189:8290–8299

Chaudhary D, Sindhu SS (2017) Amelioration of salt stress in chickpea by coinoculation ACC deaminase containing rhizosphere bazteria with Mesorhizobium strains. Legum Res 40:80–86

Christopher SF, Lal R (2007) Nitrogen management affects carbon sequestration in North American cropland soils. Crit Rev Plant Sci 26:45–64

Chueire LM, Hungria M (1997) N_2 fixation ability of Brazilian soybean cultivars with *Sinorhizobium fredii* and *Sinorhizobium xinjiangensis*. Plant Soil 196:1–5

Coleman-Derr D, Tringe SG (2014) Building the crops of tomorrow: advantages of symbiont-based approaches to improving abiotic stress tolerance. Front Microbiol 5:283–288

Comas L, Becker S, Cruz V, Byrne P, Dierig D (2013) Root traits contributing to plant productivity under drought. Front Plant Sci 4:442–447

Cooper M, Gho C, Leafgren R, Tang T, Messina C (2014) Breeding drought-tolerant maize hybrids for the US corn-belt: discovery to product. J Exp Bot 65:6191–6204. https://doi.org/10.1093/jxb/eru064[Epubaheadofprint

Crutzen PJ, Mosier AR, Smith KA, Tang T, Messina C (2007) N_2O release from agrobiofuel production negates global warming reduction by replacing fossil fuels. Atmos Chem Phys Discuss 7:11191–11205

Dangaria CJ, Parameshwarappa R, Salimath PM, Annigeri BS (1994) Genetic divergence for nodulating characters in chickpea. Legum Res 17:32–36

Dar WD, Echeverrea RG, Solh M, Sanginga N (2012) CGIAR Research Programme on Grain Legumes: leveraging legumes to combat poverty, hunger, malnutrition and environmental degradation. ICRISAT, 210 pp

Dardanelli MS, de Cordoba FJF, Espuny MR, Carvajal MAR, Diaz MES, Serrano AMG (2008) Effect of Azospirillum brasilense coinoculated with rhizobium on *Phaseolus vulgaris* flavonoids and Nod factor production under salt stress. Soil Biol Biochem 40(11):2713–2721. https://doi.org/10.1016/j.soilbio.2008.06.016

Daryanto S, Wang L, Jacinthe PA (2015) Global synthesis of drought effects on food legume production. PLoS One 10(6):e0127401. https://doi.org/10.1371/journal.pone.0127401
Dashadi M, Khosravi H, Moezzi A, Nadian H, Heidari M, Radjabi R (2011) Co-inoculation of Rhizobium and Azotobacter on growth indices of faba bean under water stress in the green house condition. Adv Stud Biol 3:373–385
Datta R, Baraniya D, Wang YF, Kelkar A, Moulick A, Meena RS, Yadav GS, Ceccherini MT, Formanek P (2017a) Multi-function role as nutrient and scavenger of free radical in soil. Sustain MDPI 9:402. https://doi.org/10.3390/su9081402
Datta R, Kelkar A, Baraniya D, Molaei A, Moulick A, Meena RS, Formanek P (2017b) Enzymatic degradation of lignin in soil: a review. Sustain MDPI 9:1163. https://doi.org/10.3390/su9071163. 1–18
Dhakal Y, Meena RS, Kumar S (2016) Effect of INM on nodulation, yield, quality and available nutrient status in soil after harvest of green gram. Legum Res 39(4):590–594
Dimkpa C, Weinand T, Asch F (2009) Plant-rhizobacteria interactions alleviate abiotic stress conditions. Plant Cell Environ 32:1682–1694. https://doi.org/10.1111/j.1365-3040.2009.02028.x
Duranti M (2006) Grain legume proteins and nutraceutical properties. Fitoterapia 77:67–82. PMID: 16406359
Dusad LR, Morey DK (1979) Effect of intercropping sorghum with legumes on the yield, economics and nitrogen economy. J Maharashtra Agric Univ 4:314–317
Egemberdieva D (2011) Survival of *Pseudomonas extremorientalis* TSAU20 and P. chlororaphis TSAU13 in the rhizosphere of common bean (*Phaseolus vulgaris*) under saline conditions. Plant Soil Environ 57:122–127
Egamberdieva D, Shurigin V, Gopalakrishnan S, Sharma R (2014) Growth and symbiotic performance of chickpea (*Cicer arietinum*) cultivars under saline soil conditions. Journal of Biological and Chemical Research 31(1):333–341
Eisenstein M (2013) Discovery in a dry spell. Nature 501:S7–S9. https://doi.org/10.1038/501S7a
Espiritu BM, Lales EH, Palacpac NQ (1993) Interaction ef- fects of Bradyrhizobium strain and cultivar in mungbean (Vigna radiata (L.) Wilczek). Philippine J Biotechnol 4:61–68
Evans J, Chalk PM, O'Connor GE (1995) Potential for increasing N2 fixation of field pea through soil management and genotype. Biol Agric Hortic 12:97–112
Fagodiya RK, Pathak H, Kumar A, Bhatia A, Jain N (2016) Global temperature change potential of nitrogen use in agriculture: A 50-year assessment. Sci Rep 7:44928. https://doi.org/10.1038/srep44928
FAO (2011) The state of the world's land and water resources for food and agriculture (SOLAW) – managing systems at risk. Food and Agriculture Organization of the United Nations, Rome and Earthscan, London
FAO (2014) FAOSTAT database. Available at: http://faostat3.fao.org/faostat gateway/ go/to/home/
FAO (2015) FAOSTAT database. Available at: http://faostat3.fao.org/faostat gateway/ go/to/home/
FAO (2016a) Pulses contribute to food security. http://www.fao.org/3/a-c0063e.pdf
FAO (2016b) Nutritional benefits of pulses. http://www.fao.org/3/a-c0049e.pdf
Farooq M, Wahid A, Kobayashi N, Fujita D, Basra SMA (2009) Plant drought stress: effects, mechanisms and management. Agron Sustain Dev 29(1):185–212
Fesenko AN, Provorov NA, Orlova IF, Orlov VP, Simarov BV (1995) Selection of *Rhizobium leguminosarum* bv. *viceae* strains for inoculation of *Pisum sativum* L. cultivars: analysis of symbiotic efficiency and nodulation competitiveness. Plant Soil 172:189–198
Figueiredo MVB, Burity HA, Martinez CR, Chanway CP (2008) Alleviation of drought stress in common bean (Phaseolus vulgaris L.) by co-inoculation with Paenibacillus polymyxa and Rhizobium tropici. Appl Soil Ecol 40:182–188
Francis CA (1986) Multiple cropping systems. Macmillan, New York
Francius G, Polyakov P, Merlin J, Abe Y, Ghigo JM, Merlin C, Beloin C, Duval JF (2011) Bacterial surface appendages strongly impact nanomechanical and electrokinetic properties of Escherichia coli cells subjected to osmotic stress. PLoS One 6(5):e20066. https://doi.org/10.1371/journal.pone.0020066

Friesen ML, Porter SS, Stark SC, von Wettberg EJ, Sachs JL, Martinez-Romero E (2011) Microbially mediated plant functional traits. Annu RevEcolEvolSyst 42:23–46. https://doi.org/10.1146/annurev-ecolsys-102710–145039

German MA, Burdman S, Okon Y et al (2000) Effects of *Azospirillum brasilense* on root morphology of common bean (*Phaseolus vulgaris* L.) under different water regimes. Biol Fertil Soils 32:259–264

Gllick BR. (2012). Plant growth-promoting bacteria: mechanisms and applications: 963401. https://doi.org/10.6064/2012/963401

Glick BR, Todorovic B, Czarny J, Cheng Z, Duan J, McConkey B (2007) Promotion of plant growth by bacterial ACC-deaminase. Crit Rev Plant Sci 26:227–242

Gliessman SR (1998) Agroecology: ecological processes in sustainable agriculture. Ann Arbor Press, Chelsea

Gowda CLL, Parthasarathy Rao P, Bhagavatula S (2009). Global trends in production and trade of major grain legumes. International conference on grain legumes: quality improvement, value addition and trade, February 14–16, Indian Institute of Pulses Research, Kanpur. Indian Society of Pulses Research and Development, India, pp 282–301

Graham PH (1992) Stress tolerance in Rhizobium and Bradyrhizobium and nodulation under adverse soil conditions. Can J Microbiol 38:475–484

Graham PH, Vance CP (2000) Nitrogen fixation in perspective: an overview of research and extension needs. Field Crop Res 65:93–106

Gray EJ, Smith DL (2005) Intracellular and extracellular PGPR: commonalities and distinctions in the plant–bacterium signaling processes. Soil Biol Biochem 37:395–412

Grover M, Ali SZ, Sandhya V, Rasul A, Venkateswarlu B (2010) Role of microorganisms in adaptation of agriculture crops to abiotic stresses. World J Microbiol Biotechnol 27:1231–1240. https://doi.org/10.1007/s11274-010-0572–577

Grover M, Ali SKZ, Sandhya V, Rasul A, Venkateswarlu B (2011) Role of microorganisms in adaptation of agriculture crops to abiotic stresses. World J Microbiol Biotechnol 27:1231–1240

Gupta A, Rai V, Bagdwal N, Goel R (2005) In situ characterization of mercury-resistant growth-promoting fluorescent pseudomonads. Microbiol Res 160:385–388

Haase S, Neumann G, Kania A, Kuzyakov Y, Romheld V, Kandeler E (2007) Elevation of atmospheric CO_2 and N nutritional status modify nodulation, nodule-carbon supply, and root exudation of *Phaseolus vulgaris* L. Soil Biol Biochem 39:2208–2221

Habibzadeh Y, Pirzad A, Zardashti MR, Jalilian J, Eini O (2012) Effects of Arbuscular Mycorrhizal fungi on seed and protein yield under water-deficit stress in Mung Bean. Agron J 105:79–84

Hamaoui B, Abbadi JM, Burdman S, Rashid A, Sarig S, Okon Y (2001) Effects of inoculation with *Azospirillum brasilense* on chickpeas (*Cicer arietinum*) and faba beans (*Vicia faba*) under different growth conditions. Agronomie 21(6–7):553–560. https://doi.org/10.1051/agro

Hawkesford MJ (2014) Reducing the reliance on nitrogen fertilizer for wheat production. J Cereal Sci 59(2014):276e283

Hayat R, Ali S, Amara U, Khalid R, Ahmed I (2010) Soil beneficial bacteria and their role in plant growth promotion: a review. Ann Microbiol 60(4):579–598. https://doi.org/10.1007/s13213-010-0117-1

Hays SG, Patrick WG, Ziesack M (2015) Better together: engineering and application of microbial symbioses. Curr Opin Biotechnol 36:40–49

Hernandez G, Ramirez M, Suarez R, Fuentes SI (1995) Root exuded nod-gene inducing signals limit the nodulation capacity of different alfalfa varieties with *Rhizobium meliloti*. Plant Cell Rep 14:626–629

Herridge D, Peoples M, Boddey R (2008) Global inputs on biological nitrogen fixation in agricultural systems. Plant Soil 311:1–18

Hiltner L (1904) Recent experiments and new problems in soil bacteriology, with especial reference to green manuring and fallows. Arbeiten Deutscher Landwirtschafts Gesellschaft 98:59–78

Hossain MA, Nakano Y, Asada K (1984) Monodehydroascorbate reductase in spinach chloroplasts and its participation in regeneration of ascorbate for scavenging hydrogen peroxide. Plant Cell Physiol 25(3):385–395

Howieson JG, Loi A, Carr SJ (1995) *Biserrula pelecinus* L. -a legume pasture species with potential for acid, duplex soils which is nodulated by unique root-nodule bacteria. Aust J Agric Res 46:997–1009

Hungria M, Nogueira MA, Araujo RS (2013) Co-inoculation of soybeans and common beans with rhizobia and azospirilla: strategies to improve sustainability. Biol Fertil Soils 49:791–801

Ibrahim KK, Arunachalam V, Kesava PS, Rao K, Tilak VBR (1995) Seasonal response of groundnut genotypes to *arbuscular mycorrhiza Bradyrhizobium* inoculation. Microbiol Res 150:218–224

ICRISAT (1998) From orphan crop to pacesetter: pigeonpea improvement at ICRISAT: International Crops Research Institute for the Semi-Arid Tropics, 24 pp. ISBN 92-9066-401-0

ICRISAT (2013) Bulletin of tropical legumes 21. Available at http://www.icrisat.org/TropicallegumesII/Bulletin of Tropical-Legumes.htm

IPCC (2007) Climate change: synthesis report. Summary for policymakers. Intergovernmental panel on climate change (IPCC)

Ismail MA (2014) Genotoxicity of Methyl Tert Butyl Ether (MTBE) to *Vicia faba* L. plants. Annual Research & Review in Biology 4(11):1867–1878

Jensen ES, Peoples MB, Boddey RM, Gresshoff PM, Hauggaard-Nielsen H, Alves BJ, Morrison MJ (2012) Legumes for mitigation of climate change and the provision of feedstock for biofuels and biorefineries. Rev Agron Sustain Dev 32:329–364

Jeuffroy MH, Baranger E, Carrouee B, Chezelles ED, Gosme M, Henault C (2013) Nitrous oxide emissions from crop rotations including wheat, oilseed rape and dry peas. Biogeosciences 10:1787–1797

Kaczan D, Swallow BM, Adamowicz WL (2013) Designing a payments for ecosystem services (PES) program to reduce deforestation in Tanzania: an assessment of payment approaches. Ecol Econ 95:20–30

Kaymak HC (2010) Potential of PGPR in agricultural innovations. In: Maheshwari D (ed) Plant growth and health promoting bacteria. Microbiology monographs, vol 18. Springer, Berlin/Heidelberg

Kennedy AC (1998) The rhizosphere and spermosphere. In: Sylvia DM, Fuhrmann JJ, Hartel PG, Zuberer DA (eds) Principles and applications of soil microbiology. Prentice Hall, Upper Saddle River, pp 389–407

Kennedy AC, Smith KL (1995) Soil microbial diversity and the sustainability of agricultural soils. Plant Soil 170:75–86

Khadraji A, Cherik G (2007) Effect of drought on growth, physiological and biochemical processes of chickpea-rhizobia symbiosis. Legum Res 40:921–926

Kim YC, Glick B, Bashan Y, Ryu CM (2013) Enhancement of plant drought tolerance by microbes. In: Aroca R (ed) Plant responses to drought stress. Springer Verlag, Berlin

Klein W, Weber MHW, Marahiel MA (1999) Cold shock response of *Bacillus subtilis*: isoleucine-dependent switch in the fatty acid branching pattern for membrane adaptation to low temperatures. J Bacteriol 181:5341–5349

Kumar S, Sheoran S, Kumar SK, Kumar P, Meena RS (2016) Drought: a challenge for Indian farmers in context to climate change and variability. Progress Res Int J 11:6243–6246

Kunert KJ, Vorster BJ, Fenta BA, Kibido T, Dionisio G, Foyer CH (2016) Drought stress responses in soybean roots and nodules. Front Plant Sci 7:1015. https://doi.org/10.3389/fpls.2016.01015

Kuykendall LD (1989) Influence of *Glycine max* nodulation on the persistence in soil of a genetically marked *Bradyrhizobium japonicum* strain. Plant Soil 116:275–277

Ledgard SF (1989) Nitrogen fixation and transfer to associated grasses by white clover cultivars under dairy cow grazing. In: Proceedings of the XVI international Grassland Congress, Nice, France, 4–11 October 1989, pp 169–170

Liu XM, Feng ZB, Zhang FD, Zhang SQ, He XS (2006) Preparation and testing of cementing and coating nano-subnanocomposites of slow/controlled-release fertilizer. Agric Sci China 5:700–706

Lynch JM (1990) The rhizosphere. Wiley-Interscience, Chichester. 458 p

Ma L, Ma WQ, Velthof GI et al (2010) Modeling nutrient flow in food chains of China. J Environ Quality 39:1279–1289

Majeed A, Abbasi MK, Hameed S, Imran A, Rahim N (2015) Isolation and characterization of plant growth-promoting rhizobacteria from wheat rhizosphere and their effect on plant growth promotion. Front Microbiol 6

Marasco R, Rolli E, Vigani G, Borin S, Sorlini C, Ouzari H, Zocchi G, Daffonchio D (2013) Are drought-resistance promoting bacteria cross-compatible with different plant models? Plant Signal Behav 8:10. https://doi.org/10.4161/psb.26741

Maredia M (2012) Global pulse production and consumption trends: the potential of pulses to achieve 'feed the future' food and nutritional security goals. Global Pulse Researchers Meeting, Rwanda, February 13–19, 2012

Mayak S, Tirosh T, Glick BR (2004) Plant growth-promoting bacteria that confe resistance to water stress in tomatoes and peppers. Plant Sci 166:525–530. https://doi.org/10.1016/j.plantsci.2003.10.025

Meena RS, Yadav RS (2015) Yield and profitability of groundnut (*Arachis hypogaea* L) as influenced by sowing dates and nutrient levels with different varieties. Legum Res 38(6):791–797

Meena VS, Maurya BR, Verma R, Meena RS, Jatav GK, Meena SK, Meena R, Meena SK (2013) Soil microbial population and selected enzyme activities as influenced by concentrate manure and inorganic fertilizer in alluvium soil of Varanasi. Bioscan 8(3):931–935

Meena VS, Maurya BR, Meena RS, Meena SK, Singh NP, Malik VK (2014) Microbial dynamics as influenced by concentrate manure and inorganic fertilizer in alluvium soil of Varanasi, India. Afr J Microb Res 8(1):257–263

Meena RS, Meena VS, Meena SK, Verma JP (2015a) The needs of healthy soils for a healthy world. J Clean Prod 102:560–561

Meena RS, Yadav RS, Meena H, Kumar S, Meena YK, Singh A (2015b) Towards the current need to enhance legume productivity and soil sustainability worldwide: a book review. J Clean Prod 104:513–515

Meena RS, Meena VS, Meena SK, Verma JP (2015c) Towards the plant stress mitigate the agricultural productivity: a book review. J Clean Prod 102:552–553

Meena RS, Bohra JS, Singh SP, Meena VS, Verma JP, Verma SK, Shiiag SK (2016) Towards the prime response of manure to enhance nutrient use efficiency and soil sustainability a current need: a book review. J Clean Prod 112:1258–1260

Meena RS, Meena PD, Yadav GS, Yadav SS (2017a) Phosphate solubilizing microorganisms, principles and application of microphos technology. J Clean Prod 145:157–158

Meena RS, Gogaoi N, Kumar S (2017b) Alarming issues on agricultural crop production and environmental stresses. J Clean Prod 142:3357–3359

Meneses CH, Rouws LF, Simões-Araújo JL, Vidal MS, Baldani JI (2011) Exopolysaccharide production is required for biofilm formation and plant colonization by the nitrogen- fixing endophyte Gluconacetobacter diazotrophicus. Mol Plant-Microbe Interact 24:1448–1458

Menike LCM, Dunusinghe PM, Ranasinghe A (2015) Macroeconomic and firm specific determinants of stock returns: a comparative analysis of stock markets in Sri Lanka and in the United Kingdom. J Fin Accoun 3(4):86–96. https://doi.org/10.11648/j.jfa.20150304.14

Metwali EMR, Abdelmoneim TS, Bakheit MA, Kadasa NMS (2015) Alleviation of salinity stress in faba bean ('Vicia faba' L.) plants by inoculation with plant growth promoting rhizobacteria (PGPR). Plant Omics 8:449–460

Mhadhbi H, Jebara M, Limam F, Aouani ME (2004) Rhizobial strain involvement in plant growth, nodule protein composition and antioxidant enzyme activities of chickpea–rhizobia symbioses: modulation by salt stress. Plant Physiol Biochem 42:717–722

Mittler R (2006) Abiotic stress, the field environment and stress combination. Trends Plant Sci 11(1):15–19

Mnasri B, Tajini F, Trabelsi M, Aouani ME, Mhamdi R (2007) *Rhizobium gallicum* as an efficient symbiont for bean inocula- tion. Agron Sustain Dev 27:331–336. https://doi.org/10.1051/agro:2007024

Mrabet M, Mhamdi R, Tajini F, Tiwari R, Trabelsi M, Aouani ME (2005) Competitiveness and symbiotic effectiveness of a *R. gallicum* strain isolated from root nodules of *Phaseolus vulgaris*. Eur J Agron 22:209–216

Mullen MD, Israel DW, Wollum AG (1988) Effects of Bradyrhizobium japonicum and soybean (Glycine max (L.) Merr.) phosphorus nutrition on nodulation and dinitrogen fixation. Appl Environ Microbiol 54:2387–2392

Mungai LM, Snapp S, Messina J, Chikova R, Smith A, Anders E, Richardson R, Li G (2016, 1720) Smallholder farms and the potential for sustainable intensification. Front Plant Sci:7. https://doi.org/10.3389/fpls.2016.01720

Munns ND (1970) Nodulation of *Medicago sativa* in solution culture. V. Calcium and pH requirements during infection. Plant Soil 32:90–102

Murrell D (2016) Global research and funding survey on pulse productivity and sustainability. Global Pulse Confederation Report, 71 pp

Nadeem SM, Zahir ZA, Naveed M, Asghar HN, Arshad M (2010) Rhizobacteria capable of producing ACC-deaminase may mitigate salt stress in wheat. Soil Sci Soc Am J 74(2):533–542. https://doi.org/10.2136/sssaj2008.0240

Nedumaran S, Abinaya P, Jyosthnaa P, Shraavya B, Rao P, Bantilan C (2015) Grain legumes production, consumption and trade trends in developing countries. Working paper series no 60. ICRISAT research program, markets, institutions and policies. International Crops Research Institute for the Semi-Arid Tropics, Patancheru, 64 pp

Nour SM, Cleyet-Marel JC, Beck D, Effosse A, Fernandes MP (1994a) Genotypic and phenotypic diversity of Rhizobium isolated from chickpea (Cicer arietinum L.). Can J Microbiol 40:345–354

Odendo M, Batino A, Kimani S (2011) Socio-economic contribution of legumes to livelihoods in sub-Saharan Africa. In: Bationo A, Waswa B, Okeyo JM et al (eds) Fighting poverty in Sub-Saharan Africa: The multiple roles of legumes in integrated soil fertility management. Springer. ISBN: 978-94-007-1535-6, New York, pp 27–46

Oya T, Nepomucemo AL, Neumaier N, Farias JRB, Tobita S, Ito O (2004) Drought tolerance characteristics of Brazilian soybean cultivars: evaluation and characterization of drought tolerance of various Brazilian soybean cultivars in the field. Plant Production Science 7(2):129–137

Patel RK, Jain M (2012) NGS QC toolkit: a toolkit for quality control of next generation sequencing data. PLoS One 7(2):e30619. https://doi.org/10.1371/journal.pone.0030619

Peoples M, Swan T, Goward J, Hunt J, Li G, Harris R, Ferrier D,Browne C, Craig S, van Rees H, Mwendwa J, Pratt T, Turner F, Potter T, Glover A, Midwood J (2015). Legume effects on soil N dynamics comparisons of crop response to legume and fertiliser N. Grains Research and Development Corporation, Government of Australia. Available at: http://grdc.com.au/Research-and-Development/GRDC-UpdatePapers/2015/02/Legume

Pingali P (2012) Green revolution: impacts, limits, and the path ahead. PNAS 109:12302–12308. https://doi.org/10.1073/pnas.0912953109

Porcel R, Ruiz-Lozano JM (2004) Arbuscular mycorrhizal influence on leaf water potential, solute accumulation, and oxidative stress in soybean plants subjected to drought stress. J Exp Bot 55:1743–1750. https://doi.org/10.1093/jxb/erh188

Postel SL (2000) Entering an era of water scarcity: the challenges ahead. Ecol Appl 10(4):941–948

Prashar P, Shah S (2016) Impact of Fertilizers and Pesticides on Soil Microfl ora in Agriculture. In: Lichtfouse E (ed) Sustainable agriculture reviews. Sustainable agriculture reviews, vol 19. https://doi.org/10.1007/978-3-319-26777-7_8

Pretty J, Bharucha ZP (2014) Sustainable intensification in agricultural systems. Ann Bot 14:1571–1596

Pretty J, Toulmin C, Williams S (2011) Sustainable intensification in African agriculture. Int J Agric Sustain 9(1):5–24

Price GD, Howitt SM (2014) Plant science: towards turbocharged photosynthesis. Nature. https://doi.org/10.1038/nature13749

Prudent M, Salon C, Soulemonov A, Emery RJ, Smith DL (2015) Soybean is less impacted by water stress using *Bradirhizobium japonica* and Thuricin-17 from *B. thuringiencis*. Agron Sustain Dev 35:749–757

Raffin A, Roumet P (1994) Shoot-root control of nitrate tolerance of N2 fixation in spontaneously tolerant soybean lines: reciprocal grafting experiments. Agronomie 14:473–480

Raghuwanshi A, Dudeja SS, Khurana AL (1994) Effect of temperature on flavonoid production in pigeonpea (Cajanus cajan (L.) Millsp) in relation to nodulation. Biol Fertil Soils 17:314–316

Reay DS, Davidson EA, Smith KA, Smith P, Melillo JM (2012) Global agriculture and nitrous oxide emissions. Nat Clim Chang 2:410–416

Reckling M, Preissel S, Zander P et al (2014) Effects of legume cropping on farming and food systems. Legume Futures Report 1:6. Available from www.legumefutures.de

Rengel Z (2002) Breeding for better symbiosis. Plant Soil 245:147–162

Rodrigo VH, Stirling CM, Teklehaimanot Z, Nugawela A (2001) Intercropping with banana to improve fractional interception and radiation-use efficiency of immature rubber plantations. Field Crop Res 69:237–249

Rodriguez U, Mary A, O'Connell O (2006) A root-specific bZIP transcription factor is responsive to water deficit stress in tepary bean (*Phaseolus acutifolius*) and common bean (*P. vulgaris*). J Exp Bot 57:1391–1398. https://doi.org/10.1093/jxb/erj118

Rodriguez RJ, Henson J, Van Volkenburgh E, Hoy M, Wright L, Beckwith F (2008) Stress tolerance in plants via habitat-adapted symbiosis. ISME J 2:404–416. https://doi.org/10.1038/ismej.2007.106

Romdhane SB, Mohamed MT, Aouani E, Lajudie P, Mhamdi R (2009) The diversity of rhizobia nodulating chickpea (*Cicer arietinum*) under water deficiency as a source of more efficient inoculants. Soil Biol Biochem 41:2568–2572

Safronova VI, Stepanok VV, Engqvist GL (2006) Root-associated bacteria containing 1-aminocyclopropane-1-carboxylate deaminase improve growth and nutrient uptake by pea genotypes cultivated in cadmium supplemented soil. Biol Fertil Soils 42:267. https://doi.org/10.1007/s00374-005-0024-y

Sandhya V, Ali SKZ, Grover M (2009) Alleviation of drought stress effects in sunflower seedlings by the exopolysaccharides producing *Pseudomonas putida* strain GAP-P45. Biol Fertil Soils 46:17. https://doi.org/10.1007/s00374-009-0401-z

Saravanakumar D, Samiyappan R (2007) ACC deaminase from *Pseudomonas fluorescens* mediated saline resistance in groundnut (*Arachis hypogea*) plants. J Appl Microbiol 102:1283–1292. https://doi.org/10.1111/j.1365-2672.2006.03179.x

Sattar MA, Quader MA, Danso SKA (1995) Nodulation, N_2 fixation and yield of chickpea as influenced by host cultivar and *Bradyrhizobium* strain differences. Soil Biol Biochem 27:725–727

Serraj R, Sinclair TR, Purcell LC (1999) Symbiotic N2 fixation response to drought. J Exp Bot 50:143–155

Sharifi M, Ghorbanli M, Ebrahimzadeh H (2007) Improved growth of salinity-stressed soybean after inoculation with salt pre-treated mycorrhizal fungi. J Plant Physiol 164:1143–1151

Sharma RK, Saikia R (2014) Alleviation of drought stress in mung bean by strain *Pseudomonas aeruginosa* GGRJ2. Plant Soil 377:111–126

Siebert S, Burke J, Frenken K, Hoogeven J, Doll P, Portmann FT (2010) Ground water use for irrigation- global inventory. Hydrol Earth Syst Sci 14:1863–1880

Singh SP (1995) Selection for water stress tolerance in interracial populations of common bean. Crop Sci 35:118–124

Smil V (1999) Nitrogen in crop production: an account of global flows. Glob Biogeochem Cycles 13:647–662

Smith A, Snapp S, Dimes J, Gwenambira C, Chikowo R (2016) Doubled-up legume rotations improve soil fertility and maintain productivity under variable conditions in maize-based cropping systems in Malawi. Agric Syst 145:139–149

Sofi PA (2015) Women empowerment as a component of social responsibility in participatory varietal selection. J Krishi Vigyan 3:113–118

Sofi PA, Saba I, Amin Z (2017) Root architecture and rhizobial inoculation in relation to drought stress response in common bean (*Phaseolus vulgaris* L.). J Appl Nat Sci 9(1):502–507

Song L, Li FM, Fan XW, Xiong YC, Wang WQ, Wu XB, Turner NC (2009) Soil water availability and plant competition affect the yield of spring wheat. Eur J Agron 31(1):51–60

Soussana JF, Tallec T, Blanfort V (2010) Mitigating the greenhouse gas balanceof ruminant production systems through carbon sequestration in grasslands. Animal 4:334–350

Sprent JJ (1972) Nitrogen fixation. In: Paleg LC, Aspinall D (eds) Physiology and biochemistry of drought resistance in plants, pp 131–143. Academic Press. 1981

Stagnari F, Maggio A, Galieni A, Pisante M (2017) Multiple benefits of legumes for agriculture sustainability: an overview. Chem Biol Technol Agric 4:2

Stone V, Brown DM, Watt N, Wilson M, Donaldson K, Ritchie H, MacNee W (2015) Ultrafine particle-mediated activation of macrophages: intracellular calcium signaling and oxidative stress. Inhal Toxicol 12:345–351. https://doi.org/10.1080/08958378.2000.11463244

Suárez R, Wong A, Ramírez M, Barraza A, Carmen Orozco M, Cevallos M, Lara M, Hernández G, Iturriaga G (2008) Improvement of drought tolerance and grain yield in common bean by overexpressing trehalose-6-phosphate synthase in rhizobia. Mol Plant Microb Interact 21:958–966

Swaine EK, Swaine MD, Killham K (2007) Effects of rought on isolates of Bradyrhizobium elkanii cultured from Albizia adianthifolia seedlings of different provenances. Agrofor Syst 69:135. https://doi.org/10.1007/s10457-006-9025-6

Tejera NA, Soussi M, Lluch C (2006) Physiological and nutritional indicators of tolerance to salinity in chickpea plants growing under symbiotic conditions. Environ Exp Bot 58:17–24

Thakur AK, Panwar JDS (1995) Effect of rhizobium-VAM interactions on growth and yield in mungbean (*Vigna radiata* (L) Wilczek) under field conditions. Indian J Plant Pathol 38:62–65

Timmusk S, Nevo E (2011) Plant root associated biofilms. In: Maheshwari DK (ed) Bacteria in agrobiology: plant nutrient management, vol 3. Springer, Berlin, pp 285–300

Tittabutr P, Piromyou P, Longtonglang A, Noisa-Ngiam R, Boonkerd N, Teaumroong N (2013) Alleviation of the effect of environmental stresses using co-inoculation of mungbean by *Bradyrhizobium* and rhizobacteria containing stress-induced ACC deaminase enzyme. Soil Sci Plant Nutr 59(4):559–571. https://doi.org/10.1080/00380768.2013.804391

Uma C, Sivagurunathan P, Sangeetha D (2013) Performance of *bradyrhizobial* isolates under drought conditions. Int J Curr Microbiol App Sci 2:228–232

United Nations Organisation (UNO) (2015) The united nations world water development report 2015: water for a sustainable world. UNESCO, Paris

UNO (United Nations Organisation) 2017. World population prospects 2017. www.esa.un.org/unpd/wpp/Download/Standard/Population/

Vadez V, Berger JD, Warkentin T, Asseng S, Ratnakumar P, Rao KPC, Gaur PM, Munier-Jolain N, Larmure A, Voisin AS, Sharma HC, Pande S, Sharma M, Krishnamurthy L, Zaman MA (2011) Adaptation of grain legumes to climate change: a review. Agron Sustain Dev 32:31–44

Valdenegro M, Barea JM, Azcón R (2000) Influence of arbuscular-mycorrhizal fungi, *Rhizobium meliloti* strains and PGPR inoculation on the growth of *Medicago arborea* used as model legume for re-vegetation and biological reactivation in a semi-arid mediterranean area. Plant Growth Regul 34:233–240

Vance CP, Graham PH, Allan DL (2000) Biological nitrogen fixation. Phosphorus: a critical future need. In: Pedrosa FO, Hungria M, Yates MG, Newton WE (eds) Nitrogen fixation: from molecules to crop productivity. Kluwer Academic Publishers, Dordrecht, pp 506–514

Varma D, Meena RS (2016) Mungbean yield and nutrient uptake performance in response of NPK and lime levels under acid soil in Vindhyan region, India. J App Nat Sci 8(2):860–863

Vazquez M, Barea J, Azcon R (2001) Impact of soil nitrogen concentration on Glomus spp.-Sinorhizobium interactions as affecting growth, nitrate reductase activity and protein content of Medicago sativa. Biol Fertil Soils 34:57. https://doi.org/10.1007/s003740100373

Verma JP, Meena VS, Kumar A, Meena RS (2015a) Issues and challenges about sustainable agriculture production for management of natural resources to sustain soil fertility and health: a book review. J Clean Prod 107:793–794

Verma JP, Jaiswal DK, Meena VS, Meena RS (2015b) Current need of organic farming for enhancing sustainable agriculture. J Clean Prod 102:545–547

Vivas A, Biró B, Ruíz-Lozano JM, Barea JM, Azcón J (2006) Two bacterial strains isolated from a Zn-polluted soil enhance plant growth and mycorrhizal efficiency under Zn-toxicity. Chemosphere 62:123–133

Vlassak KM, Vanderleyden J (1997) Factors influencing nodule occupancy by inoculant rhizobia. Crit Rev Plant Sci 16:163–229

Witt S, Galicia L, Lisec J, Cairns J, Tiessen A, Araus JL (2012) Metabolic and phenotypic responses of greenhouse-grown maize hybrids to experimentally controlled drought stress. MolPlant 5:401–417. https://doi.org/10.1093/mp/ssr102

Wortmann CS, Kirkby RA, Eledu CA et al (1998) Atlas of common bean (*Phaseolus vulgaris* L.) production in Africa. CIAT publication no. 297. CIAT, Cali, Colombia, 131 pp

Yadav GS, Lal R, Meena RS, Datta M, Babu S, Das LJ, Saha P (2017a) Energy budgeting for designing sustainable and environmentally clean/safer cropping systems for rainfed rice fallow lands in India. J Clean Prod 158:29–37

Yadav GS, Lal R, Meena RS, Babu S, Das A, Bhomik SN, Datta M, Layak J, Saha P (2017b) Conservation tillage and nutrient management effects on productivity and soil carbon sequestration under double cropping of rice in North Eastern Region of India. Ecol Indi. http://www.sciencedirect.com/science/article/pii/S1470160X17305617

Yang J, Kloepper JW, Ryu CM (2009) Rhizosphere bacteria help plants tolerate abiotic stress. Trends Plant Sci 14:1–4

Yu Y, Xue L, Yang L (2014) Winter legumes in rice crop rotations reduces nitrogen loss, and improves rice yield and soil nitrogen supply. Agron Sustain Dev 34:633–640

Zhang DS, Dong W, Hui DW, Chen SY, Zhuang BC (1997) Construction of a soybean linkage map using an F2 hybrid population from a cultivated variety and a semi-wild soybean. Chin Sci Bull 42:1326–1330

Zhang H, Kim MS, Sun Y, Dowd SE, Shi H, Paré PW (2008) Soil bacteria confer plant salt tolerance by tissue-specific regulation of the sodium transporter HKT1. Mol Plant-Microbe Interact 21:737–744. https://doi.org/10.1094/MPMI-21-6-0737

Zhu H, Choi HK, Cook DR, Shoemaker RC (2005) Bridging model and crop legumes through comparative genomics. Plant Physiol 137:1189–1196

Zou YN, Wu QS (2011) Efficiencies of Five Arbuscular Mycorrhizal Fungi in Alleviating Salt Stress of Trifoliate Orange. Int J Agric Biol 13:991–995

Nitrogen and Legumes: A Meta-analysis

9

S. K. Kakraliya, Ummed Singh, Abhishek Bohra,
K. K. Choudhary, Sandeep Kumar, Ram Swaroop Meena,
and M. L. Jat

Contents

S. K. Kakraliya (✉) · S. Kumar
Department of Agronomy, CCS Haryana Agricultural University, Hisar, Haryana, India

U. Singh · A. Bohra
ICAR – Indian Institute of Pulses Research, Kanpur, Uttar Pradesh, India

K. K. Choudhary
ICAR – Central Arid Zone Research Institute, RRS Pali, Rajasthan, India

R. S. Meena
Department of Agronomy, Institute of Agricultural Sciences (BHU), Varanasi, UP, India

M. L. Jat
International Maize and Wheat Improvement Centre (CIMMYT)- NASC Complex,
New Delhi, India

R. S. Meena et al. (eds.), *Legumes for Soil Health and Sustainable Management*,
https://doi.org/10.1007/978-981-13-0253-4_9

Abstract

The current progress in agricultural production does not really cater to the demand of the burgeoning human population. Consequently, this puts global food and nutritional security at a great risk. This challenge calls for concerted efforts of all stakeholders to produce required quantity and quality of assured foods for ensuring food security. In the past, the principal driving force was to increase the yield potential of food crops and to maximize productivity. Today, the drive for productivity is increasingly combined with a desire for sustainability. For farming systems to remain productive and to be sustainable in the long term, it will be necessary to replenish the reserves of nutrients which are removed or lost from the soil. The nitrogen (N) inputs derived from atmospheric N via biological N fixation (BNF). Therefore, current farming systems need sustainable intensification through the inclusion of legume crops. This facilitates the precise use of nitrogen (N) by reducing their losses into the environment and ensures self-sufficiency in protein. The relevance of legumes in this context is enhanced as these crops offer numerous amenities that remain in line with prevalent sustainability principles. Legume crops provide protein-rich food, oil and fibre while supplying the 195 Tg N year^{-1} (also includes actinorhizal species) to the agroecosystem through the process of biological nitrogen fixation (BNF). Besides serving as the fundamental global source of good-quality food and feed, legume crops contribute to 15% of the N in an intercropped cereal and mitigate the emission of greenhouse gases (GHGs) by reducing the application demand of synthetic nitrogenous fertilizers. Legume cultivation releases up to seven times less GHGs per unit area than non-legume crops. Legumes allow the sequestration of carbon (1.42 Mg C ha^{-1} year^{-1}) in soils and induce the conservation of fossil energy inputs in the system. The other benefits of legume crops include their significant positive impacts on biodiversity and soil health. Rotating legume crops with non-legume crops has the dual advantage of cultivating the legumes with slight or no extra N fertilizer. Care should be taken to ensure the availability of adequate N for the succeeding non-legume crops. The legume crops respond very well to conservation of agricultural practices. Overall, these characteristics are crucial to agriculture both in developing and developed countries apart from the conventional farming systems. Legumes in rotation promote exploration of nutrients by crops from different soil layers. They also help in reducing pressure on soil created by monocropping. Thus, crop rotation acts like a biological pump to recycle the nutrients. Hence, inclusion of legumes in the cropping system is inevitable to advance soil sustainability and food and nutritional security without compromising on the long-term soil fertility potential.

Keywords
Legumes' effects on succeeding crops · Legumes mitigate GHGs · Nitrate leaching · N fixation · Residual N in soil

Abbreviations

ADP	Adenosine diphosphate
ATP	Adenosine-5′-triphosphate
BNF	Biological nitrogen fixation
GHGs	Greenhouse gases
GWP	Global warming potential
N	Nitrogen
NUE	Nitrogen use efficiency
SOC	Soil organic carbon
SPA	Soil-plant-atmosphere

9.1 Introduction

Legumes belong to the Leguminosae or Fabaceae family and rank third in global production after cereal and oilseed. These hold immense agricultural significance worldwide contributing an area ~14% of total land under cultivation (Sulieman and Tran 2015). They largely contribute to global food and nutritional security, besides soil health. Moreover, they generate income for millions of smallholder farmers at the regional and global level, and their role in environmental safety measures is well documented (Peoples et al. 2009; Yadav et al. 2015; Guardia et al. 2016). Legumes fix the atmospheric N through symbiotic associations. They are also important sources for proteins, minerals and micronutrients suitable for human and animal consumption besides being sources of fibre and oils (Voisin et al. 2014; Stagnari et al. 2017). Half of the entire N used in agriculture production system is delivered by the legume crops (Graham and Vance 2003). So, the biologically fixed N remains adequate to cater to the requirements of the plant. This is apart from leaving some N (as residual N) in the soil for the succeeding non-legume crops (Mayer et al. 2003; Peoples et al. 2009; Dhakal et al. 2016). This underlines the great potential of legume crops for use in soil restoration and stabilization. The scope of legume crops to agricultural systems could be further enhanced manifold. This could be done by attending to soil constraints such as soil acidity, salinity and drought and through undertaking modern plant breeding programmes (Graham and Vance 2003).

The use of legume crops as green manure in non-legume-based cropping system was prevalent since agriculture began to be developed. However, a drastic decline in the practice was seen with the increasing availability of industrially produced fertilizers (especially the N fertilizer). Green manure adds N to the soil and improves quality by increasing the soil organic carbon (SOC), macro- and micronutrients and humus content (Graham and Vance 2003; Jensen et al. 2012; Hajduk et al. 2015). Legumes enrich the soil with N and thus facilitate a better environment to subsequent crops for better growth and productivity (Meena et al. 2015a). Legumes can fix substantial amounts of free atmospheric N, which allows them to be grown in N-stored soils without using synthetic N fertilizers. The BNF by legumes and actinorhizal species is estimated for about 195 Tg year^{-1} (Vitousek et al. 2013). According to Frame (2005), with an estimated capacity to fix 72–350 kg N ha^{-1} year^{-1}, the legumes facilitate the transformation of environmental N into various nitrogenous compounds including amino acids and proteins. These amino acids and proteins, which are being used by the growing plants, also contribute to improved soil fertility (Nulik et al. 2013). Peoples et al. (2009) reported that nearly 30–40 kg of N is fixed on a whole plant basis for each ton of dry matter produced by legume crops. According to Mayer et al. (2003), total N uptake by the following non-legume crops is strongly influenced by the preceding legume crop. It relies upon the residual N input and the N fixation capacity of the different legume crops. Up to 12% of the residual N is recovered from the succeeding crops at maturity. Berg (1997) highlighted that wheat (*Triticum aestivum*) yield of 3070 kg ha^{-1} year^{-1} over 5 years following alfalfa, 2580 kg ha^{-1} year^{-1} following milk vetch and 950 kg ha^{-1} year^{-1} following grass with N uptake was attributed to the residual effect from legumes averaged 34 kg N ha^{-1} year^{-1} from alfalfa and 25 kg ha^{-1} year^{-1} from milk vetch. Mineral N in root-zone soil is often 30–60 kg N ha^{-1} higher when crops are preceded by legumes as compared to cereal crops (Dalal et al. 1998; Meena et al. 2015d). This enhancement is accounted to both nitrate sparing by the legume species and mineralization of the N-rich residues (Evans et al. 1991). The increasing cost of industrial fertilizers and the availability of fertilizers at the proper time raise serious concerns among farmers. This is particularly felt by the farming community with resources of a poor and marginal nature (Luce et al. 2015). These in turn incorporate legumes in the cropping systems as alternate ways to minimize the use of synthetic fertilizers (Yadav et al. 2000; Ram and Meena 2014). In this context, incorporating legumes in the cropping and intercropping system can contribute substantially to improved soil fertility and better plant growth. This in turn could improve the productivity of succeeding non-legume crops (Banyong et al. 2000; Yusuf et al. 2009; Bonilla et al. 2017).

Furthermore, Jeuffroy et al. (2013) observed that the legume crops release approximately five to seven times lower greenhouse gases (GHGs) to the atmosphere per unit area compared to non-legume crops. This process highlights their growing significance in the face of the global climate change. Further, it is reported that the peas (*Pisum sativum* L.) released 69 kg N_2O ha^{-1}, which is far less when compared to rape (*Brassica napus*) (534 kg N_2O ha^{-1}) and wheat (*Triticum aestivum*) (368 kg N_2O ha^{-1}) (Stagnari et al. 2017). Similarly, Clune et al. (2017) further reinforced the relevance of legume crops in the climate change scenario (Meena et al. 2017a). This was done by stating their extremely low global warming potential (GWP) values (0.50–0.51 kg CO_2 eq. kg^{-1} produce). Schwenke et al. (2015) showed that the emissions of N_2O from mineral N-fertilized canola (385 N_2O ha^{-1}) largely exceeded those from the faba bean (*Vicia faba* L.) (166 N_2O ha^{-1}), chickpea (*Cicer arietinum* L.) (166 N_2O ha^{-1}) and field pea (135 N_2O ha^{-1}). The authors also reported that the N fixed by legumes represented a less-emissive form of N input to the soil when compared to the nitrogenous fertilizers of industrial origin.

Given the above description, the objective of this chapter is to offer an overview of the mechanism of N fixation by legume crops. This is followed by a brief description of the impact of legumes on soil fertility and productivity of succeeding crops. The role of legume crops in mitigating environmental N emission is also highlighted. Hence, the food and nutritional security can achieve great heights under intense agriculture and thereby improve long-term soil fertility potential.

9.2 Legumes and Nitrogen Cycle

9.2.1 Mechanics of the Nitrogen Cycle

The predominant form of N, the seventh most abundant element on the earth, is the N gas (that constitutes ~78% of the earth's atmosphere). This form of N is relatively inert and therefore warrants its conversion to available form for its subsequent use by different organisms and plants (Carroll and Salt 2004; Sergei 2012) (Table 9.1). As illustrated in Fig. 9.1, this occurs through a process called the "N cycle" that converts N into compounds to be used by living organisms (plants and animals). In the atmosphere, N is the most important limiting nutrient for plant growth and development. N is generally taken in two forms, i.e. ammonium (NH_4^+) and nitrate (NO_3^-) (Sergei 2012; James 2013). Also, N is a constituent of organic molecules like proteins, amino acid and nucleic acids. In the N cycle, the important processes include N fixation, ammonification, nitrification, assimilation of N and denitrification (James 2013; Varma et al. 2017).

Table 9.1 N inputs, outputs and cycling in the soil-plant-atmosphere (SPA) system

N inputs (gain)	N output (loss)	No gain or loss in net N (cycling)	References
BNF	N uptakes by plants	Mineralization	Gonzalez et al. (2005)
Physical N_2 fixation	Denitrification (N_2O, NO, N_2)		Groffman (2012)
1. Industrial (Haber-Bosch process)			
2. Electrical (lighting)			
3. Combustion (fossil fuel)			
Animal manures	Volatilization (NH_3)	Immobilization	Rochette et al. (2009)
Crop residues	Leaching (NO_3^-)	Nitrification	Weil and Brady (2017)
	NH_4^+ fixation		Weil and Brady (2017) and Dhakal et al. (2015)

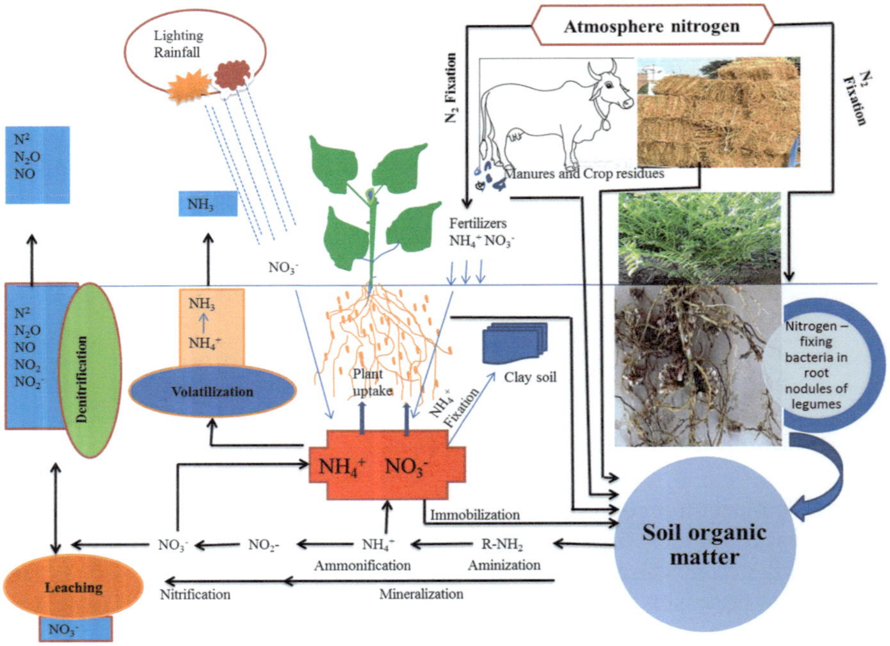

Fig. 9.1 The N cycle; most of the nitrogen conversions are facilitated by various microorganisms and cycle through the organic fraction in the soil (The complete N cycle is also described in brief as steps A–F)

Step A: N is found in manures and plant residues and is derived from the atmosphere through the process of physical electrical discharges/lighting, industrial processes and biological N fixation through living organisms, i.e. legumes. Consequently upon the completion of the N fixation, N is added to the soil (Weil and Brady 2017).

Step B: Organic N is mineralized to ammonium (NH_4^+) by certain organisms in the soil. Abundant ammonia is then transformed to nitrate (NO_3^-) by bacteria through the process called "nitrification" (Weil and Brady 2017).

Step C: Nitrate and ammonium ions are taken up by active roots of plants in soil solution (Weil and Brady 2017).

Step E: Soil solution ammonium and nitrate are converted back to N compounds through immobilization. The solution of NO_3^- can be lost by leaching to groundwater drainage system as a result of the vertical movement beneath the root zone in percolation water (Weil and Brady 2017).

Step D: Ammonium is absorbed and fixed by clay colloids (Weil and Brady 2017).

Step E: Ammonium can be volatilized into the gaseous NH_3 (Weil and Brady 2017).

Step F: Nitrate (NO_3^-) derived from nitrification, fertilization or rainfall can be converted by denitrifying bacteria to N_2, N_2O and NO gases which are emitted into the atmosphere (denitrification) (Weil and Brady 2017; Meena et al. 2014a, b).

9.2.1.1 Soil Nitrogen Forms

N is a nutrient that is usually deficient in most of the crop rotations involving non-leguminous crops (Ladha and Peoples 1995). A variety of sources of N including organic and inorganic types are supplied to non-legume crops (Shaha et al. 2003; Meena et al. 2015b). The quantity of N fixed by legumes is usually adequate to allow their growth and development. The N in soils exists in two forms: (i) organic and (ii) inorganic nitrogen (Weil and Brady 2017). Nearly all N is present in the organic form in contrast to the inorganic form which constitutes only around 2%. According to Weil and Brady (2017), clay form is represented by 8% and 40% of the total N in surface and sub-surface soils, respectively.

Organic Nitrogen

The organic form of soil N is represented by compounds such as amino acids, amino sugars, proteins and more resistant N compounds such as humus. This organic N in soil (mostly in hydrolysable form) is gradually mineralized and converted into mineral N through the process of aminization, ammonification and nitrification, thus ultimately rendering N available to plants (Gonzalez et al. 2005). The organic soil N is found in manure, compost, crop residues, green manure, bio-fertilizer and several waste materials (Ladd et al. 1983). Amino acids, proteins and polypeptides are the most common organic constituents of living organisms including plants (Gonzalez et al. 2005).

Inorganic Nitrogen

The inorganic forms of N are represented by ammonium (NH_4^+), ammonia (NH_3), nitrate (NO_3^-) and nitrite (NO_2^-) (James 2013), which can be utilized by plants as plant roots absorb N from the soil in the form NO_3^- and NH_4^+. Also, NH_4^+, NO_2^-, NO_3^-, N_2O, NO and elemental N are important sources of nutrition for N-fixing microorganisms (Havlin et al. 2014; Meena et al. 2014a, b). N in its inorganic forms remains "available" to plants and microorganisms or could move downward in the soil along with the movement of water. By contrast, the majority of the N in the soil remains unavailable to plants due to its organic form (Havlin et al. 2014). The N that is absorbed by plants or any other living organism is incorporated into soil organic matter after the death and subsequent decomposition of the organisms. Nitrate is the dominant form of N in aerobic soil, while N remains predominantly as ammonium in case of anaerobic soils (Sqrensen and Sessitsch 2007).

The N Cycle Involves the Following Processes

1. Nitrogen fixation
2. Ammonification
3. Nitrification
4. Denitrification
5. Volatilization
6. Leaching of nitrate

The conversion of N can be accomplished through both biological and physical processes.

Fixation of Nitrogen

N fixation yields ammonia (NH_3) and then N-containing organic compounds as a result of the transformation of earth's atmospheric N. This is a process that makes N accessible to the entire living organisms (Postgate 1998). In nature, the process of N fixation is mediated by certain N-fixing rhizobia bacteria (*Rhizobiaceae*, *α-Proteobacteria*) (Sqrensen and Sessitsch 2007; Buragohain et al. 2017). Alternatively, the N fixation can be accomplished by natural means like lightning and/or processes including the Haber-Bosch that is used to produce fertilizers such as urea and other chemical fertilizers (Havlin et al. 2014). Among all N fixation processes, BNF is the most common one in plants. Due to their property to fix atmospheric N and accumulate a great quantity of N in their organs, legumes serve as bio-fertilizers in crop production systems (Peter et al. 2002). Accordingly, a leguminous crop when applied as green manure in the soil confers the subsequent non-legume crops with a huge quantity of sources comprising N and C (Stagnari et al. 2017).

Ammonification/Mineralization of Nitrogen

The N mineralization generates inorganic N (NH_4^+) from organic N involving two major processes, viz. ammonization and ammonification.

(i) *Aminization*: An enzymatic reaction mediated by soil microorganisms (aerobic and anaerobic bacteria, fungi and actinomycetes) converts proteinous and protein compounds into amino acid and amines (James 2013; Stagnari et al. 2017).

$$\text{Proteins} \underset{\text{Soil microbes}}{\overset{\text{Enzyme}}{\rightarrow}} \text{polypeptide} \underset{H_2O}{\overset{\text{Enzyme}}{\rightarrow}} \underbrace{R-\overset{\displaystyle NH_2 \atop |}{\underset{| \atop \displaystyle H}{C}}-COOH}_{\text{Amino acid}} + \underbrace{R-NH_2}_{\text{Amine}} + CO_2 + \text{Energy}$$

9.2.1.2 Ammonification

Ammonification driven by certain soil microorganism enables organic N compounds to be transformed into ammonia (NH_3) or ammonium (NH_4^+). The NH_4^+ ions are produced as a waste of animal, organic matter, crop residues and manure by bacteria (aerobic and anaerobic), fungi and actinomycetes (Sergei 2012; Meena et al. 2014a, b). The process of ammonification takes place in aerobic environments with the liberation of NH_3 or NH_4^+ ions, which are either released into the atmosphere or used by selective plants (e.g. rice) and microorganisms. Also, the ions may be oxidized to nitrites and finally to nitrates under favourable soil conditions (Havlin et al. 2014).

The process of ammonification as shown below is commonly mediated by *Clostridium* spp., *Micrococcus* sp., *Proteus* spp., etc. (Groffman 2012).

$$R-NH_2+H_2O \underset{\text{Hydrolysis}}{\overset{\text{Enzyme}}{\rightarrow}} NH_3 \overset{H_2O}{\rightarrow} NH_4^+ + OH^- + R-OH + \text{Energy}$$

Nitrification

Enzymatic oxidation of NH_4^+ to NO_2^- and ultimately to nitrate (NO_3^-) by certain soil microorganism is termed as nitrification. Two groups of bacteria participate in the process of nitrification: the one that causes oxidation of ammonia to produce nitrite (NO_2^-) and the other that further oxidizes nitrite to NO_3^- (Sergei 2012). These bacteria obtained energy from N compound (proteins, polypeptides and amino acids) and carbon from CO_2. Nitrate (NO_3^-), the end product of nitrification, is extremely important for plant growth (Bundy 1998).

As mentioned above, the oxidation process is completed in two steps, and each step is performed by different groups of bacteria as follows:

Step I: The process referred to as "nitrification" leads to the generation of nitrite and is mediated by ammonia-oxidizing bacteria (*Nitrosomonas*, *Micrococcus*, *Europaea*, *Nitrosococcus*, *Nitrosospira*, *Briensis*, *Nitrosovibrio* and *Nitrocystis*). The chemical reaction underlying the process is denoted as follows (James 2013; Meena et al. 2014a, b).

$$\underset{\text{Ammonia}}{2NH_3} + \tfrac{1}{2}O_2 \rightarrow \underset{\text{Nitrite}}{NO_2} + 2H + H_2O\,(\text{Nitrosomonas bacteria})$$

Step II: In the second step, nitrite is oxidized to nitrate by nitrite-oxidizing bacteria (*Nitrobacter winogradskyi*, *Nitrosococcus mobilis*, *Nitrocystis*, *Nitrospina gracilis*, etc.) and some fungi (e.g. *Penicillium*, *Aspergillus*) and actinomycetes (e.g. *Streptomyces*, *Nocardia*).

$$\underset{\text{Nitrite ions}}{NO_2^-} + \tfrac{1}{2}O_2 \rightarrow \underset{\text{Nitrate ions}}{NO_3^-}\ (\text{Nitrobacter})$$

Nitrification Is Affected by Several Factors, Which Include

(i) Supply of ammonium ions
(ii) Soil moisture
(iii) Soil temperature
(iv) Soil pH
(v) Soil aeration
(vi) C/N ratio

9.2.1.3 N Losses from Soil-Plant System

Denitrification

As a reverse process to nitrification, denitrification causes reduction of NO_3^- and NO_2^- by anaerobic bacterial, thus resulting in the release of nitric oxide (NO), nitrous oxide (N_2O) and N_2 that eventually are lost to the atmosphere (Seitzinger et al. 2006; Groffman 2012). Consequently, the plant-available N (an inorganic form of N) in the soil is lost to the atmosphere; this process is also called dissimilarity nitrate reduction, and NO_3^- is also reduced to NH_4^+, and this is assimilated to the protein through the formation of amino acids. This process is called assimilatory nitrate reduction.

The process of denitrification is as follows (Seitzinger et al. 2006; Datta et al. 2017):

Denitrification reaction series

$$\underset{\text{Nitrate}}{NO_3^-} \rightarrow \underset{\text{Nitrite}}{NO_2} \rightarrow \underset{\text{Nitric oxide}}{NO} \rightarrow \underset{\text{Nitrous oxide}}{N_2O} \longrightarrow \underset{\text{N}}{N_2} -$$

When oxygen is depleted in the soil, some of the NO_3^- can change to N_2O, and N_2 gaseous forms are lost to the atmosphere. The sequence of intermediate products of denitrification is as follows. Some of the organisms (*Thiobacillus thioparus*, *Thiobacillus denitrificans*, *Pseudomonas*, *Micrococcus*, *Bacillus* and *Achromobacter*) are involved in this process (Groffman 2012).

A range of factors influence the process of denitrification such as:

(i) Supply of nitrate substrate
(ii) Soil texture
(iii) Aeration and water status
(iv) Soil pH
(v) Available soil organic carbon
(vi) Temperature

Ammonium Volatilization

Ammonia gas (NH_3) is produced in the soil-plant system from the mineralization of crop residues, organic matter, farmyard manure (FYM), compost and industrial chemical fertilizer (like anhydrous ammonia and urea) (Rochette et al. 2009). This process reflects a reversible reaction as follows:

Reversible Process

$$\underset{\text{Dissolved ions}}{NH_4^+} + OH \leftrightarrow H_2O + \underset{\text{Ammonia gas}}{NH_3} -$$

Volatilization of NH_3 depends on the concentration of ammonium and ammonia ions in the soil solution and the soil pH. At pH 9.5, the ammonium and ammonia are

Table 9.2 Global N fixation from different sources

Source of N fixation	N fixation rate kg ha^{-1}	Nitrogen fixed (10^6 tons $year^{-1}$)
Legume crops	140	35
Non-legume crops	8	9
Meadows and grassland	15	45
Forest and woodland	10	40
Other vegetated lands	2	10
Ice-covered land	0	0
Total land		139
Sea	1	36
Total biological fixation		175
Lightning		8
Fertilizer industry		77
Total non-biological fixation		85
Grand total		260

Source of data: The Nature and Properties of Soils (2002): Weil and Brady (2017)

of equal concentration (50% each) with ammonia increasing constantly with increasing soil pH (Havlin et al. 2014; Varma and Meena 2016) A large proportion of mineral N fertilizer applied can be lost through the process of volatilization, if not properly managed, and in this way, it is not incorporated into the soil. When soil pH reaches above 7.5, a percentage of the NH_4^+ can be converted into ammonia gas (NH_3) and thus released to the atmosphere (Havlin et al. 2014).

In addition to soil pH, higher soil moisture and temperature contribute to NH_3 loss. Sometimes, NH_3 volatilization can occur under neutral and acidic soils.

Nitrogen Fixation in Soil

As mentioned earlier, notwithstanding the existence of nearly 80% of N in the atmosphere, the maximum abundant forms cannot be accessed directly by plants (Ladha and Peoples 1995). This makes the N often a limiting factor in the agricultural production system, especially for non-legume crops that show a greater demand for a high amount of N. The free atmospheric N gas can become available to plant through N_2 fixation.

Biological N fixation, the most common N fixation process, facilitates fixation of an estimated 175 × 10^6 tons of N each year worldwide (Table 9.2).

9.2.2 Types and Process of Soil Nitrogen Fixation

Atmospheric N is fixed by two major means, viz. (i) physicochemical and (ii) biological processes, which enable nearly 10 and 90% of natural N fixation, respectively.

9.2.2.1 The Low Type of Soil Nitrogen Fixation

Physical Nitrogen Fixation

Natural Nitrogen Fixation

With the influence of lightning strikes (i.e. electric discharge in the clouds) and thunder, N and oxygen (O_2) of the air react to form nitric oxide (NO). The nitric oxides are again oxidized with oxygen to form N peroxide (NO_2) (Sergei 2012; Havlin et al. 2014).

$$N_2 + O_2 \text{ Lightning} \rightarrow \text{Thunder } 2NO(\text{Nitric Oxide});$$
$$2NO + O_2 \rightarrow 2NO_2 \text{Oxidation} (\text{N peroxide})$$

During the rains, NO_2 combines with rainwater to form nitrous acid (HNO_2) and nitric acid (HNO_3). The acids fall on the soil with rainfall and react with the alkaline radicals to form water-soluble nitrates (NO_3^-) and nitrites (NO_2^-) (Sergei 2012; Havlin et al. 2014).

$$2NO_2 + H_2O \rightarrow HNO_2 + HNO_3$$
$$HNO_3 + \text{Ca or K salts} \rightarrow \text{Ca or K nitrates}$$

The nitrates are soluble in water and are directly absorbed by the roots of the plants (Vitousek et al. 2013).

Industrial N Fixation

Ammonia is produced industrially by direct combination of atmospheric N with hydrogen (obtained from water) at high temperature (400–500 °C) and pressure (15–25 MPa). Further, it is converted into different types of fertilizers, such as urea, etc.

Biological Nitrogen Fixation

Mechanism and Process of BNF

N is an essential element for plant growth and development (Sergei 2012). Plants instead depend upon combined or fixed forms of N, such as ammonium and nitrate. A considerable proportion of this N is supplied to the cropping systems in the form of industrially produced N fertilizers. The prime sources of N include N available in the soil, the BNF and synthetic fertilizers. Soil organic N in natural or human-made ecosystems is constantly lost through plant exclusion and further losses through leaching, denitrification and NH_3 volatilization. BNF is known to be a key to sustain agriculture production and to increase soil fertility (Vitousek et al. 2013; Meena et al. 2017b).

Research on microorganisms and plants capable of fixing atmospheric N contribute fundamentally to bio-fertilizer production. Thus, it is important to ensure that BNF research and development will take into account the needs of smallholder farmers in the developing countries (Bhat et al. 2015). BNF refers to the conversion of atmospheric N_2 into NH_3 and then to N-containing organic compounds that can become available to form life through the N cycle (Herridge et al. 2008).

Globally, a huge amount of N is fixed biologically each year and ranges from 130 to 180 × 10^6 tons, with 50% fixed by *Rhizobium* (Havlin et al. 2014). In contrast,

Table 9.3 A short list of *Rhizobium* species and their corresponding hosts

Nodulating bacteria	Host crop/plant	References
Rhizobium phaseoli, R. leguminosarum biovar *phaseoli* and *R. tropics*	*Phaseolus vulgaris* (common bean)	Kahindi et al. (2009)
R. leguminosarum bv. *viciae*	*Lens* (lentils), *Vicia* (vetch), *Pisum sativum* L. (peas)	
R. leguminosarum bv. *trifolii*	*Trifolium* sp. (clovers)	
Mesorhizobium loti	*Lotus*	
R. leguminosarum bv. *phaseoli*	*Vicia faba* (broad bean)	
Bradyrhizobium japonicum, B. elkanii, R. fredii	*Glycine max* (soybean)	Kahindi et al. (2009) and Meena et al. (2017b)
Azorhizobium caulinodans	*Sesbania* sp., *Sesbania rostrata* (stem nodulating)	
R. meliloti, Sinorhizobium meliloti	*Medicago sativa* (alfalfa), *Trigonella* (fenugreek)	
R. loti	*Lotus* (trefoils), *Lupinus* (lupin), *Cicer* (chickpea), *Leucaena*	
"Cowpea rhizobia" group or *Rhizobium* sp.	*Vigna unguiculata* (cowpea)	Kahindi et al. (2009)
Bradyrhizobium sp.	*Arachis hypogaea* (peanut), *Cajanus* (pigeon pea) and *Crotalaria* (crotalaria)	

global fertilizer N use was about 109 × 10^6 tons in 2014 (FAOSTAT 2014), about twice as much as is industrially fixed in the manufacture of N fertilizers.

BNF was discovered by the German agronomist Hermann Hellriegel and Dutch microbiologist Martinus Beijerinck. In the BNF equation, 2 moles of NH_3 ions are made by a single mole of N_2 gas, at the expense of 16 moles of ATP and a supply of electrons and protons (hydrogen ions). The nitrogenase enzyme is the key to biological N fixation, which catalyses the reduction of N gas to ammonia.

$$N_2 + 8H^+ + 8e^- + 16\,ATP = 2NH_3 + H_2 + 16ADP + 16\,Pi$$

The NH_3, in turn, is combined with organic acids to amino acid and, finally, protein:

$$NH_3 + \text{organic acids} \rightarrow \text{amino acids} \rightarrow \text{protein}$$

The BNF occurs through a number of microorganisms in the system, with or without direct association with higher plants (Tables 9.3 and 9.4); while the legume-bacteria symbiotic system has received the most attention, recent findings suggest that the other system involves many more families of plants worldwide and may even rival the legume-associated system as supplier of biological N to the soil. Each major system will be discussed below briefly (Timothy 1999; Herridge et al. 2008). As mentioned already, large quantities of N can be fixed through the process of symbiosis of microorganisms and legumes (Meena et al. 2014a, b). In this process, the plants produce the energy through the process of photosynthesis, and the

Table 9.4 Range in quantity of N_2 fixed by selected legumes

Sr. no.	Legumes	Botanical/ scientific name	Associated organism	N fixing (kg ha^{-1} $year^{-1}$)	References
1.	Soybean	*Glycine max* L.	Bacteria (*Bradyrhizobium*)	100–150	Ahlawat and Gangaiah (2004), Mugwe et al. (2011) and Meena et al. (2017b)
2.	Chickpea	*Cicer arietinum* L.	Bacteria (*Rhizobium*)	40–50	Ahlawat and Gangaiah (2004) and Seymour et al. (2015)
3.	Lentil	*Lens esculents* Medik.	Bacteria (*Rhizobium*)	40–68	Shaha et al. (2003) and Mugwe et al. (2011)
4.	Groundnut	*Arachis hypogaea* L.	Bacteria (*Bradyrhizobium*)	150	Mugwe et al. (2011) and Seymour et al. (2015)
5.	Field pea	*Pisum sativum* L.	Bacteria (*Rhizobium*)	65–100	Peoples et al. (2009)
6.	Pigeon pea	*Cajanus cajan* L.	Bacteria (*Bradyrhizobium*)	100,200	Mugwe et al. (2011)
7.	Mung bean	*Vigna radiata* L.	Bacteria (*Rhizobium*)	60,112	Shaha et al. (2003) and Seymour et al. (2015)
8.	Urdbean	*Vigna sinensis* L.	Bacteria (*Rhizobium*)	30	Ahlawat and Gangaiah (2004)
9.	Cowpea	*Vigna unguiculata* L.	Bacteria (*Bradyrhizobium*)	90	Mugwe et al. (2011)
11.	Lupins	*Lupinus* sp. L.	Bacteria (*Rhizobium*)	60–100	Havlin et al. (2014)
13.	Beans	*Phaseolus vulgaris* L.	Bacteria (*Rhizobium*)	20–80	Mugwe et al. (2011) and Havlin et al. 2014
14.	Alfalfa	*Medicago sativa* L.	Bacteria (*Rhizobium*)		Carlsson and Huss-Danell (2003), Aranjuelo et al. (2009) and Mugwe et al. (2011)
15.	Cluster bean	*Cyamopsis tetragonoloba* L.	Bacteria (*Rhizobium*)	60–150	Mugwe et al. (2011) and Meena et al. (2017b)
17.	Fenugreek		Bacteria (*Rhizobium*)	45	Mugwe et al. (2011)
18.	Black gram	*Vigna mungo* L.	Bacteria (*Rhizobium*)	100	Mugwe et al. (2011)

(continued)

Table 9.4 (continued)

Sr. no.	Legumes	Botanical/ scientific name	Associated organism	N fixing (kg ha^{-1} $year^{-1}$)	References
20.	Faba bean	*Vicia faba* L.	Bacteria (*Rhizobium*)	130	Peoples et al. (2009) and Seymour et al. (2015)
21.	Clover	*Trifolium pratense* L.	Bacteria (*Rhizobium*)	100–150	Aranjuelo et al. (2009)
23	Red clover	*Trifolium pratense* L.	Bacteria (*Rhizobium*)		Carlsson and Huss-Danell (2003)
24	White clover	*Trifolium repens* L.	Bacteria (*Rhizobium*)		Carlsson and Huss-Danell (2003)
Non-legumes (nodulated)					
25	Species of *Gunnera*		Cyanobacteria[a] (*Nostoc*)	10–20	Weil and Brady (2017)
26	Alders	*Alnus* sp.	Actinomycetes (*Frankia*)	50–150	Weil and Brady (2017)
Non-legumes (non-nodulated)					
27	Pangola grass	*Digitaria decumbens*	Bacteria (*Azospirillum*)	5–30	Weil and Brady (2017)
28	Bahia grass	*Paspalum notatum*	Bacteria (*Azotobacter*)	5–30	Weil and Brady (2017)
29	Azolla		Cyanobacteria[a] (*Anabaena*)	150–300	Weil and Brady (2017)
	Non-symbiotic		Bacteria (*Azotobacter*, *Clostridium*)	5–20	Weil and Brady (2017)
	Non-symbiotic		Cyanobacteria[a] (various)	10–50	Weil and Brady (2017)

[a]Sometimes referred to as blue-green algae

microorganisms utilize this energy to fix N. The process of BNF is carried out by a different group of bacteria that are either free living or in symbiotic associations with plants (*Rhizobium* and actinomycetes). Legume-bacteria symbiosis is the major form of N fixation that delivers N to crops.

The BNF can take many forms in nature, including blue-green algae (a bacterium), lichens and free-living soil bacteria. These kinds of N fixation contribute significant quantities of ammonia (NH_3) to ecosystems but not to most cropping systems, except paddy rice. Their contributions are less than 6 kg N ha^{-1} $year^{-1}$ (Fig. 9.2). However, N fixation by legumes can be in the range of 25–190 kg N ha^{-1} $year^{-1}$ in a natural ecosystem and several hundred kilogrammes in a cropping system (Frankow and Dahlin 2013). BNF of the atmospheric N can be estimated at around 175 million metric tons $year^{-1}$ or nearly 70% of all N fixed on the soil in each year; the remaining is by some microorganisms, autotrophs or heterotroph "free" fixers (Peter et al. 2002).

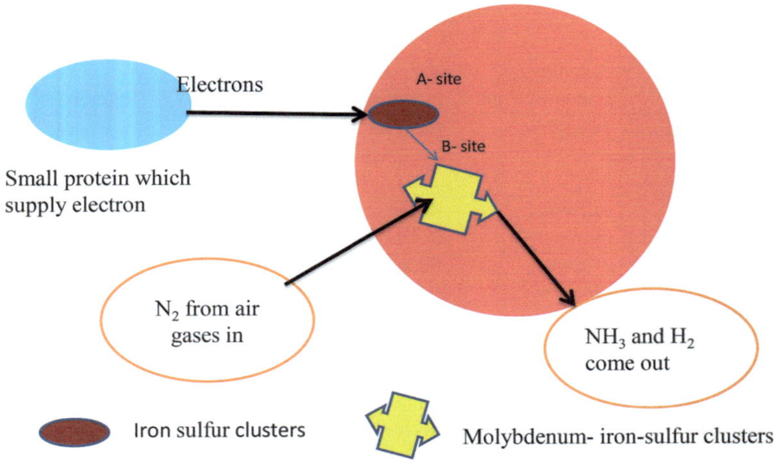

Fig. 9.2 Nitrogenase enzyme consists of two proteins. The bigger protein converts atmospheric N into ammonia using electrons which are provided by the smaller protein. The B sites on the bigger protein arrest N_2 from the air; however, the A site gets the electrons from the small protein, so finally N_2 can be reduced to NH_3

The conversion or fixation of N is from the unavailable gaseous form in the atmosphere to forms that plants and other living organisms can take (either ammonia or nitrate) and is mediated by (i) bacteria in symbiotic relationships with vascular plants, (ii) symbioses between cyanobacteria and fungi (lichens) or plants, (iii) free-living autotrophic or heterotrophic bacteria that are naturally associated with soil or litter and (iv) abiotic reactions that occur with lightning in the atmosphere (Timothy 1999; Meena et al. 2017b).

The Following Are Three Types of N-fixing Microorganisms of Symbiotic and Asymbiotic Nature

1. Symbiotic N fixation: fixes N_2 only by the formation of nodules in legume, e.g. *Rhizobium*, *Bradyrhizobium* and *Sinorhizobium*, and in some selective non-legumes, e.g. *Anabaena*, *Azolla* and *Frankia*
2. Associative N fixation: requires oxygen for growth and fixes N in the existence of oxygen (*Azospirillum*)
3. Free-living N fixation: fixes N both in aerobic and anaerobic (*Azotobacter*, *Thiobacillus*, *Bacillus* and *Clostridium* and *Klebsiella*)

Symbiotic Nitrogen Fixation

Many microorganisms fix N_2 symbiotically by partnering with a host plant. The plant provides food (sugars) from photosynthesis that is used by the N-fixing microorganism for the energy it needs for N_2 fixation. In exchange for these carbon sources, the microbe provides fixed N_2 to the host plant for its growth (Peter et al. 2002).

The symbiotic microorganisms are not only bacteria but also involve fungi, actinomycetes (e.g. *Frankia*) and cyanobacteria (e.g. *Anabaena*). These microorganisms create multiple kinds of relationships with different parts of plants and develop a special structure. Species from angiosperm family Leguminosae such as *Pisum sativum*, *Cajanus cajan*, *Glycine soja* and *Cicer arietinum* build a symbiotic relationship with *Bradyrhizobium*, *Azorhizobium*, *Allorhizobium*, *Photorhizobium*, *Sinorhizobium*, *Mesorhizobium* and *Rhizobium* (Herridge et al. 2008: Havlin et al. 2014; Meena et al. 2017b).

For the first time, rhizobia were isolated from root nodules by M. Beijerinck and were shown to have the capacity to reinfect their legume hosts and to fix N in symbiosis. N-fixing bacteria (rhizobia) are known for their ability to establish symbiotic interactions with leguminous plants through the development and colonization of root nodules, where the bacteria fix N to ammonia and make it available for the plant. The bacteria are mostly rhizospheric microorganisms, in spite of their ability to live in the soil for a long period (Gonzalez et al. 2005).

Rhizobium is a free-living, gram-negative, aerobic, non-sporulating and rod-shaped (0.5–0.9 mm and 1.2–3 mm) bacteria, which produce nodules in the leguminous plant. It is a fast-growing bacterium; however, *Bradyrhizobium* is a slow-growing strain which possesses subpolar flagella (Vieira et al. 2010).

Acetobacter diazotrophicus colonizes the stem apoplast in maize (*Zea mays*). The N_2-fixing microorganism forms a symbiotic association with the grasses without nodule formation, and such association is called associative N fixation. The *Azotobacter paspali* remains alive in the rhizospheric zone of *Paspalum notatum*, a tropical grass (Yusuf et al. 2009; Frankow and Dahlin 2013).

The *Beijerinckia* living in the rhizosphere of sugarcane (*Saccharum officinarum*) and *Klebsiella* in leaf nodules of *Psychotria*, *Casuarina equisetifolia*, *Alnus*, *Myrica* and *Parasponia* do not form nodules but fix N_2 through harbouring *Frankia* and *Rhizobium*. They show host specificity. The two partners in the N_2 fixation recognize each other with the help of the chemical substance lectins which are phytohaemagglutinins (carbohydrates having plant proteins). All the 42 bacterial isolates (grouped in the genera *Sinorhizobium* (27), *Rhizobium* (13) and *Agrobacterium* (2)), *Sinorhizobium* sp. strains STM 4036, STM 4034 and STM 4039 forming the most effective symbiosis are potential candidates for inoculants in revitalisation programmes (Mahdhi et al. 2008).

Nitrogen-Fixing Mechanism in Legumes

When legume root growth is initiated, N-fixing bacteria in soil enter root hair and multiply (Figs. 9.3 and 9.4). Legume root responds by developing tumour-like structure called nodules on the plant root surface (Kahindi et al. 2009). The particular bacteria called rhizobia inside the nodule absorb N_2 from soil air and convert it to ammonia. Rhizobia use the enzyme nitrogenase and energy from the transformation of adenosine-5′-triphosphate (ATP) to adenosine diphosphate (ADP) to break the strong triple bond in nitrogen (Peix et al. 2010). The symbiotic connection between nodule bacteria and the legume host plant is mutually beneficial. Growth-stimulating substances like biotin, thiamine, amino acids, etc. are secreted by the root of the symbiotic bacteria which enhances the growth of rhizobia and other microbes

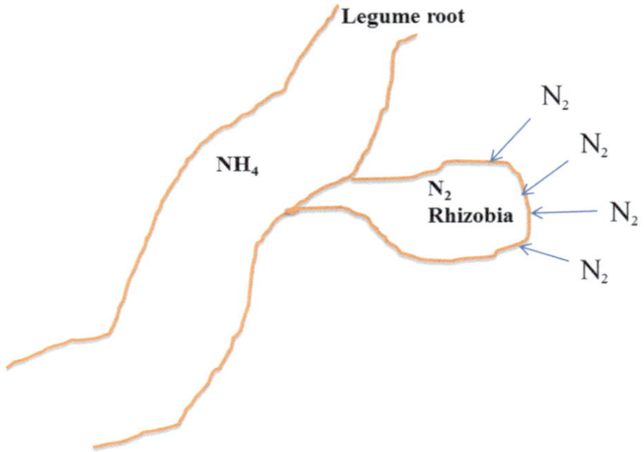

Fig. 9.3 Conversion of N_2 to NH_4 by rhizobia inside legume root nodules

Fig. 9.4 Example of nodules on lucerne (left) and clover (right) showing difference in nodulation

(Skorupska et al. 2010; Meena et al. 2014a, b). The reactions between polysaccharide (callose) present on the surface of the rhizobial cell and the lectin secreted by the plant root hairs help in the recognition of the correct host plant by the specific *Rhizobium* (Skorupska et al. 2010). Some other compounds (specific flavonoids) are also secreted by plant root hairs, e.g. alfalfa secretes luteolin which activates the "*nod*" genes in the bacteria, which results in nodule formation. Except for the two *Agrobacterium* isolates, all strains induced nodulation on *Argyrolobium* uniformly, but the number of nodules and N fixation efficiency varied among them (Mahdhi et al. 2008). The host plant delivers energy (carbohydrates, sugar and ATP) for rhizobia to fix N, and rhizobia, in turn, provide ammonium for production of protein by the host plant, and most of the fixed N is utilized by the host plant. However, some may be excreted from the nodule into the soil and used by nearby plants or resealed as nodules which decompose after the plants die (Sindhu et al. 2010).

Many rhizobium species exist in soil, each requiring a specific host legume plant. For example, *Rhizobium leguminosarum* biovar. *trifolii* will only nodulate clover (*Trifolium*), while *Rhizobium meliloti* will only nodulate alfalfa. This host specificity is referred to as cross inoculation group cell signalling between the legume host and the bacteria. The above-mentioned Nod factors have been identified as lipochition oligosaccharides. Dissimilarities in the structures of these oligosaccharides determine the host specificity for the bacterium. The presence of nodules on legume root does not necessarily indicate N_2 fixation by active rhizobia. Mature, effective chickpea nodules tend to be elongated and clustered on the primary roots and have pink to red centres. This red colour is due to leghaemoglobin and indicates that the rhizobia are actively fixing N. The main cross inoculation groups of bacteria are presented in Table 9.4.

Curling of Root Hairs

Certain soil bacteria release Nod factors; this results in curling of root hairs, which is accompanied by the formation of the infection thread by the hair tip (Kahindi et al. 2009). There is a continuation of the wall of the infection thread with the cell wall of the root hair. The branching of infection thread then occurs, and bacteria continue to produce nod factors which result into the stimulation of root cells to proliferate, resulting in nodule formation (Peix et al. 2010; Sindhu et al. 2010). Thousands of N-fixing bacteria (rhizobia) inhabit the root nodule and form the bacteroids (the bacterial cells become dormant and are called bacteroids). The membrane that is formed by the plant cell, which surrounds the bacteroid, is called symbiosome or peribacteroid (Kahindi et al. 2009).

Nodule Formation and Leghaemoglobin

The mucopolysaccharide released by N-fixing bacteria reacts with a component of root hair cell to form a compound which induces the production of polygalacturonase. When all these processes occur, rhizobia enter into the cell (Kahindi et al. 2009). The inner cortical tissue stimulates by bacteria divide and forms an organized mass of infected plant tissue which is protruded out and appears as the nodule. Rhizobia are released from the infection tube and occupy the central position in the nodule. The central nodule is tetraploid which is a peculiar characteristic (Peix et al. 2010; Dhakal et al. 2016). The available space in the host cell is totally filled. The free-living microorganisms develop mechanisms to protect the enzyme of nitrogenase from oxygen such as high rates of metabolism, physical barriers, etc.; the level of oxygen in nodules is controlled by leghaemoglobin. Nodules have an

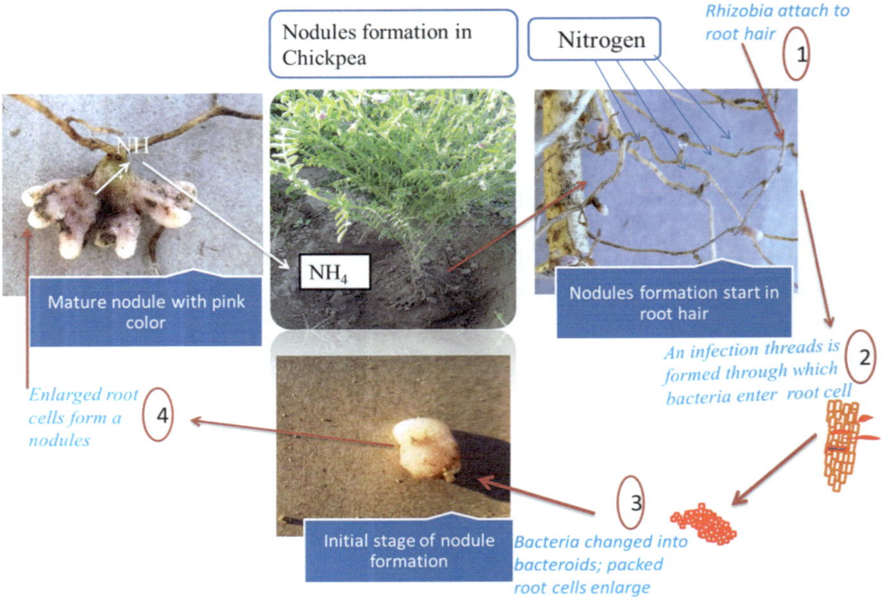

Fig. 9.5 Schematic view of N_2-fixing bacterial association with leguminous plant and development of root nodules by *Rhizobium* sp.

oxygen-binding heme protein, i.e. leghaemoglobin, and hence the colour of nodules appears pink (Fig. 9.5). When the nodule is fully mature after that it dies, bacteria are released into the soil. The bacteroids are the main sites of the N fixation (Kahindi et al. 2009; Peix et al. 2010). Bacteroids may be swollen, irregular, star shaped, branched, etc. Leghaemoglobin has about ten times higher affinity for oxygen than human haemoglobin. The prosthetic group protohaem synthesizes the bacteroids, while the synthesis of the protein part involves the plant cell (Peix et al. 2010). It supplies O_2 to the respiring symbiotic bacterial cells. It enhances the transport of oxygen at low partial pressure and also provides protection to nitrogenase against oxygen and stimulates ATP production needed for N_2 fixation (Peix et al. 2010; Skorupska et al. 2010).

Nitrogen Release to the Soil and Other Crops
Almost the entire N fixed is taken directly by the plant; the minute leaks into the soil for a neighbouring non-legume plant (Herridge et al. 2008). Nevertheless, N finally returns to the soil for a neighbouring plant when vegetation (roots, stem, leaves and fruit) of the legume dies and decomposes. The yield of non-legume crops is often increased when grown following legumes. For example, when maize is grown after soybean, the N requirement is far less than that required for maize after maize (Havlin et al. 2014; Ram and Meena 2014). Similarly, less N is required to improve wheat yield following legumes (Stagnari et al. 2017).

Reduced N rates with the non-legume crop follow legumes and then continuous non-legume rotations. This is mainly due to:

- Readily decomposition of legumes residue providing plant-available N (Stagnari et al. 2017).
- Greatly reduced N immobilization drives continuous legume rotation (Shaha et al. 2003; Peoples et al. 2009).
- Improved soil microbial activity results in increased N mineralization in legume rotation optimum (Stagnari et al. 2017).

Asymbiotic N_2 Fixation

Certain free-living microorganisms present in soil and water can fix atmospheric N_2 because these organisms are not directly associated with plants; the conversion is called non-symbiotic or free-living microorganism. The N fixation by tropical grasses including some cereals by a non-symbiotic process was first time recognized by a Brazilian scientist Johanna Döbereiner in the 1960s. She found out the considerably greater population of *Beijerinckia* and *Azotobacter* in the rhizospheric zone of batatais grass (*Paspalum notatum*) under acidic soil environment (Ruschel and Döbereiner 1965). Some of the bacteria and majority of the cyanobacteria involve this class of microorganisms. These microorganisms are generally called free-living diazotrophs. Among the cyanobacteria, unicellular, filamentous non-heterocystous and filamentous heterocystous fix N independently. Both aerobic and anaerobic bacteria are free-living diazotrophs (Table 9.5). Water, nutrients and oxygen are required in an appropriate amount, so that the microorganism can grow. Cyanobacteria grow commonly in the crop fields. The site of N fixation in the cyanobacteria is the heterocyst because of the nitrogenase enzyme required for N fixation which acts under an anaerobic situation (Kumari and Rajeshwari 2011). Asymbiotic N_2-fixation process is complete through two major groups of bacteria which are as follows:

Asymbiotic N_2 Fixation by Heterotrophs

The major fixation is brought about by species of two genera of heterotrophic aerobic bacteria, *Azotobacter* and *Beijerinckia*, which belong to temperate zones and

Table 9.5 Bacterial types fixing N symbiotically

Aerobic bacteria	Anaerobic bacteria	Facultative bacteria	Photosynthetic bacteria
Azomonas	*Clostridium*	*Bacillus*	*Chlorobium*
Azotobacter	*Desulfovibrio*	*Enterobacter*	*Chromatium*
Beijerinckia		*Klebsiella*	*Rhodomicrobium*
Derxia			*Rhodopseudomonas*
Methylomonas			*Rhodospirillum*
Mycobacterium			

Source: Kumari and Rajeshwari (2011) and Havlin et al. (2014)

tropical soils, respectively. Other aerobic bacteria of the genus *Clostridium* are also able to fix N_2. Because pockets of low O_2 supplies exist in the soil, despite good tilth, the anaerobic bacteria may work side by side in many well-drained soils. The amount of N_2 fixed by these heterotrophs varies greatly with soil properties such as pH, soil N level and sources of organic matter available. Because of their limited energy supply, under normal agriculture conditions, the rate of N fixation by these organisms is brought to be in the range of 5–20 kg N ha^{-1} $year^{-1}$ only a small fraction of the needed N by crops.

Asymbiotic N_2 Fixation by Autotrophs

Among the autotrophs able to fix N are certain photosynthetic bacteria and cyanobacteria. With the presence of light, these organisms can fix CO_2 and N_2 simultaneously. The contribution of the photosynthetic bacterias are uncertain, if bacteria are thought to be of some significance, particularly in wetland areas and in a rice field. In some case, the algae have been found to fix sufficient N_2 for reasonable rice yields, but the usual level may be no more than 20–30 kg N ha^{-1} $year^{-1}$. Cyanobacteria also fix N_2 in the upland soil, but the level is much lower than that which is found under wetland conditions.

9.2.3 How to Increase BNF and N_2-Fixing Ability

Biological N_2 fixed represents N advantage and determines mineral N fertilizer savings in cropping systems. Legumes can fix more than 40–250 kg N ha^{-1}. However, the amounts of N_2 fixed can differ considerably in time and space.

Four common approaches to enhance biological N fixation are:

- Inoculation with proven strains (covered above)
- Microbial screening for improved strains
- Host plant screening and breeding and adoption of cropping systems and cultural practices

9.2.4 Factors Affecting N_2 Fixation

The following factors will affect the rate of N_2 fixation by legume-bacteria symbiosis: the amount of applied manure or fertilizer N; because N_2 fixation requires higher energy, the amount of N fixed through BNF will be much less when the soil contains large inorganic N from other sources (Meena et al. 2013). The N_2 fixation increases with the decrease of N availability in soil. Excess NO_3^- availability reduces nitrogenase activity, which resulted in reduction of N_2 fixation by competition for photosynthate between NO_3^- reduction and N_2^- fixation reactions (Havlin et al. 2014).

Soil fertility: the deficiencies of some nutrients in soil such as Mo, Fe, P, Mg and S will result in a reduced N_2 fixation as these elements are part of the nitrogenase complex which permits N_2 fixation to take place. N fixation requires more Mo than the host plant; because of this, Mo is a main component of the nitrogenase (Hungria and Vargas 2000; Verma et al. 2015).

Soil pH: N fixation reduces when soil pH is lower than 6.7 (Hungria and Vargas 2000).

Soil temperature: soil temperature from 24 to 30 °C is optimum for N_2-fixing bacteria (rhizobia). Effective N_2 fixation will be inhibited below 10 °C (Havlin et al. 2014).

9.3 Leguminous Effect on Succeeding Crops

Legume crops that generate cash or economic income will fit better in the production systems practised by small farmers with limited resources (e.g. maize-mung bean-wheat and rice-mung bean-wheat). Legumes can be successfully accommodated into cereal-based cropping systems through the following means:

1. Crop rotation
2. Green manure
3. Intercropping

9.3.1 Crop Rotation

Crop rotation is the system of growing a sequence of different crops on the same ground so as to maintain or increase soil fertility and crop productivity. In crop rotation, legumes contribute to the diversification of cropping systems and act as free atmosphere N_2-fixing plants; it can reduce the synthetic N fertilizer demand. In the rotation of crops, leguminous crops like pulses, chickpea, beans, peas, groundnut, soybean, lentil, Bengal gram and cluster bean are sown in between the seasons of cereal crops like rice, wheat, maize, sorghum and pearl millet and cash crop like cotton, sugarcane, etc. (Shaha et al. 2003; Mayer et al. 2003; Luce et al. 2015).

Presently, several groups are intensively researched on sustainable reintroduction of grain legumes into non-legume crop rotations, based on their economic advantage on crop yield and quality characteristics on succeeding non-legume crops (Kirkegaard et al. 2008; Luce et al. 2015; Yadav et al. 2017). Legumes could be competitive crops regarding ecological and socioeconomic benefits with the potential to be introduced in modern cropping systems, which are characterized by decreasing crop diversity (FAO 2011). Legumes enrich the soil with N which provides a more favourable environment to succeeding cereals or non-legume crops for better growth and grain yield (Fig. 9.6). However, their ability to fix atmospheric N_2 did not make them independent of other sources of N, even when symbiosis was

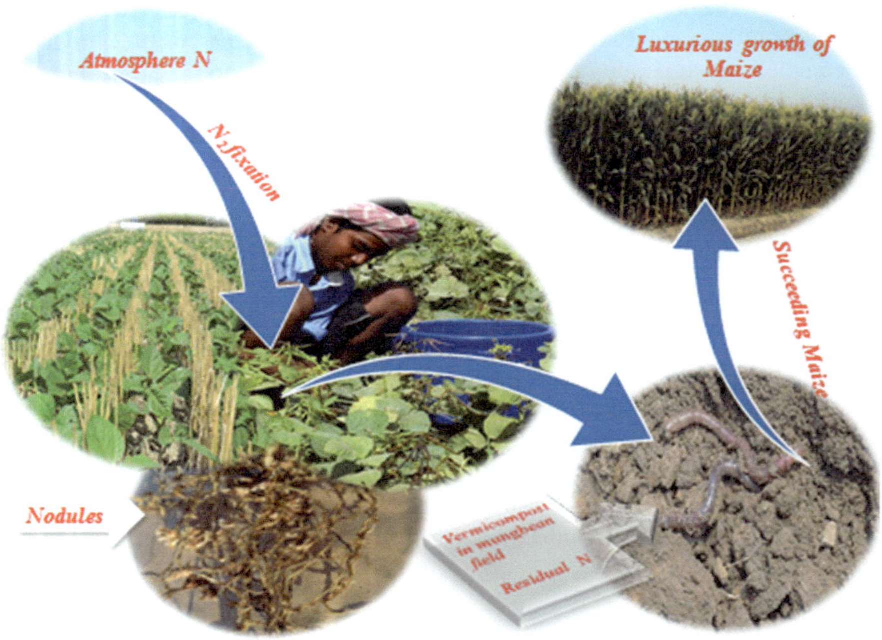

Fig. 9.6 Beneficial effects of legumes on succeeding crop in maize/wheat cropping systems

fully effective (Gibson 1976). Legumes cause significant, positive effect on growth and yield on subsequent non-legumes when compared with rotations with non-legumes (Chalk 1998; Adeleke and Haruna 2012; Dhakal et al. 2016). In addition to its beneficial factors, such as improving soil biodiversity, breaking pest and disease cycles and the phytotoxic and allelopathic effects of crop residues, N is a key factor in the positive response of cereals following legumes (Chalk 1998).

Several studies have reviewed the yield advantage of legumes for the following cereal and cash crops; the example is described as follows:

9.3.1.1 Maize-Based Cropping Systems

Banyong et al. (2000) while examining the amount of N fixed by preceding legume crops observed that the amount of N fixed varied from 20 to 104 kg N ha^{-1} and the net N benefit to the subsequent crops was up to 51 kg N ha^{-1}. Concerning the grain yield advantage in succeeding maize, an increase of up to 34% was obtained using legume crops when compared to the non-legume treatment (Yusuf et al. 2009).

Bonilla et al. (2017) concluded that incorporation of legumes and cover crops causes reduced requirements of N-based chemical fertilizers (by 13–30% for wheat and 49–61% at the rotation level) without experiencing any deterioration in wheat yield and quality. Similarly, the use of green manure to rice was reported to increase the yield of the subsequent wheat crop due to residual effect (Yadav et al. 2000;

Meena et al. 2015d). A residual benefit of legume incorporation is commonly assessed on increased grain and dry matter yields of succeeding crops (Mubarak et al. 2002).

9.3.1.2 Rice-Based Cropping Systems

The practice of using crop rotations with green/brown manure crops and farmyard manure (FYM) is critical to the sustenance of soil fertility and enhanced soil microbial productivity. The addition of leguminous crops into the cereal cropping system is vital for their long-term sustainability, primarily for the legume-led fixing of atmospheric N_2 (Chalk 1998). Further, legumes in rotation with crops increase the organic matter content of the soil and also improve the soil fertility status (Schulz et al. 1999). Biological N-fixing systems offer an economically attractive and ecologically sound means of reducing internal inputs of industrial N fertilizers and in saving internal resources (Ladha and Peoples 1995; Meena et al. 2017b).

Crop rotation with legumes improves soil physical properties and the microbial population (Yusuf et al. 2009) and might, therefore, reduce mineral fertilizer requirements of succeeding leguminous crops. In an experiment involving sequential cropping, a significantly higher N_2 fixation and residual N effect on the succeeding non-legume rice crop were noted for groundnut (*Arachis hypogaea*) than black gram, mung bean and pigeon pea (Ahmad et al. 2001). The growth and N yield of the rice crop were positively correlated with the quantity of N_2 fixed by the preceding legume crop, which led rice yield to become 0.6–1.1 t ha^{-1} higher in the legume-cereal rotation than in the cereal-cereal sequence (Ahmad et al. 2001).

9.3.1.3 Wheat-Based Cropping Systems

Higher yield (by 30%) of wheat was recorded after legumes (field peas, lupins, faba beans, chickpeas and lentils) compared to wheat monocropping (wheat-wheat yield of 4.0 t ha^{-1}) (Angus et al. 2015; Meena et al. 2015d). In temperate environments, cereal yield is on an average 17 and 21% higher in legume-based systems than the wheat-wheat-based system, under standard and moderate fertilization levels, respectively (Jensen et al. 2004).

9.3.1.4 Cotton-Based Cropping Systems

A 3-year cotton-corn-soybean rotation with 134 kg N ha^{-1} $year^{-1}$ had higher soil organic matter (SOM) and crop yield compared to cotton grown every year without a legume crop (Entry et al. 1996). However, the cotton crop cultivation after legume produced with higher oil content (22.87%), seed cotton yields (2428 kg ha^{-1}) and N intake increased up to 91.17 kg ha^{-1}. For improvement in crop productivity, the inclusion of the leguminous crop at least once in a 2-year cropping rotation was suggested, because leguminous crops enrich soil fertility (Kumbhar et al. 2008).

9.3.1.5 Sugarcane-Based Cropping Systems

Leguminous plants can accumulate 5 t ha^{-1} of dry mass in a short period during the summer season and subsequently accumulate more amounts of N and potassium. Most of this N comes from the association of legumes with N-fixing bacteria rhizobia. In this context, crop rotation with legume crops can replace partially or fully the N mineral fertilization of sugarcane, at least for the first ratoon (Ambrosano et al. 2005).

9.3.2 Green/Brown Manures

The addition of crop residue into soils, with the objective of sustaining or improving productivity and soil fertility for the succeeding non-legumes, is known as green manuring. The introduction of green manure's biomass in crop rotation improves soil quality and their beneficial N effects (Jannink et al. 1996). Incorporation of legumes residue using ^{15}N label highlights that 10–34% of the legume N can be recovered in the succeeding rye or wheat crop, 42% in rice, 24% recovery from velvet bean by corn crop, around 15% of N recovery from sunn hemp by corn plants in no-till system, 30% by maize (Ambrosano et al. 2009) and 5% of N recovery from sunn hemp by sugarcane (Ambrosano et al. 2005) and ranged from 19% to 21% when the recovery was observed from sunn hemp by two sugarcane harvests (Ambrosano et al. 2011). Legumes develop deep root systems which enable the acquisition of nutrients from deeper soil layers, and symbiotic N_2-fixing bacteria convert the environmental N into a form. This form is directly available for plant intake.

9.3.3 Intercropping Systems

Intercropping systems consist of synchronized growth of two or more crop species in the same area and at the same time (Brooker et al. 2015). Legumes can contribute up to 15% of the N in an intercropped cereal (Li et al. 2009), thus increasing biomass production (Pappa et al. 2012; Ram and Meena 2014) and reducing synthetic mineral N fertilizer use and mitigating N_2O fluxes. Osman et al. (2011) reported that intercropping with two rows of cowpea and one row of millet gave significantly higher economic benefit than a mixture with one row of each of the crops. Nair et al. (1979) revealed that legumes like cowpea, soybean, pigeon pea and groundnut when grown as intercrops with corn had a beneficial residual effect on the grain yield of the succeeding wheat crop. Similarly, intercropping of sorghum with groundnut, green gram and cowpea reduced by 61, 83 and 38 kg ha^{-1}, respectively, the mineral N fertilizer requirements of wheat for a target yield of 4.0 tons ha^{-1}.

9.4 Leguminous Residual N in Field

The benefits of legumes are usually associated with their N contribution to succeeding crops. Fixation of atmosphere N_2 by legumes in symbiosis with *Rhizobium* bacteria contributes to subsequent non-fixing crops upon decomposition of legume shoot (above portion of the ground) and root material (Bruulsema and Christie 1987; Meena et al. 2015d). Maize grown without mineral N fertilizer following crimson clover produced a higher yield of maize as maize grown following rye with 44 kg N acre^{-1} (Mitchell and Teel 2007). Bruulsema and Christie (1987) reported 56 kg N acre^{-1} contribution from alfalfa residues which resulted in a 2717 kg acre^{-1} maize yield and 56 kg N acre^{-1}contribution from red clover residues that resulted in a 2870 kg acre^{-1} maize yield. Hestermann et al. (1986) observed that crimson clover could replace 48 kg acre^{-1} of N fertilizer. Peanut residues were reported to release 17 kg N acre^{-1} to a succeeding maize crop (Mubarak et al. 2002).

Yano et al. (1994) reported that peanut residue contributed 11.2% N for succeeding wheat (*Triticum aestivum* L.) crop upon decomposition. This was comparable with the application of 30 kg N acre^{-1} as fertilizer. A small amount of N (5–15 kg) is recommended for legume at the initial stage of plant growth. This N gets the host plants off to a vigorous start, allowing rapid development of nodules and subsequent N fixation. However, studies have shown that a large amount of residual N in the soil, either from carry-over or added N, reduces N fixation (Havlin et al. 2014). In general, the host plant expends less energy by utilizing residual soil N than by fixing N through the rhizobia.

Mayer et al. (2003) reported that total N intake of the subsequent crop influenced by the legume used as preceding crop determines the residual N input and the N_2 fixation capacity of the legumes. The succeeding crops recovered 8.6–12.1% of the residual N at harvesting. Similar patterns were found for the microbial biomass, which recovered 8.2–10.6% of the residual N. Berg (1997) highlighted that wheat hay yields averaged 3.1 t ha^{-1} year^{-1} over 5 years following alfalfa, 2.6 t ha^{-1} year^{-1} following milk vetch and 0.95 t ha^{-1} year^{-1} following grass with N intake attributed to the residual effect from legumes averaged 34 kg N ha^{-1} year^{-1} from alfalfa and 25 kg ha^{-1} year^{-1} from milk vetch. Mineral N in root-zone soil following legumes is often 30–60 kg N ha^{-1} higher than after cereal crops in the same environment (Dalal et al. 1998).

9.5 Leguminous Residual Nitrogen in Field

Crop residue of legumes as a source of carbon and N for subsequent non-legume crops was found in low-input agriculture production systems. Several studies have reported that the increase of crop yield with legume residue inclusion in the field (Paré et al. 1992) and enhanced soil fertility by the providing of BNF (Ladd et al. 1983; Dhakal et al. 2016). Crop residues provide SOC and N to soil organism and physically protect soil from erosion. Legume crop residues are robust; they protect soil from erosion (wind and water) and help in improved soil physical properties and fertility. The ecosystem's nutrient retention, conserve soil moisture, help in carbon sequestration, reduce weed, help in hydraulic conductivity, help in water holding capacity of the soil and water infiltration and can contribute to climate change adaptation and mitigation. On the whole, they help ensure food, soil health and water security over a long term (Kabir and Koide 2002).

Legume crop residues decompose due to the presence of microorganism. This is done through the process of mineralization or immobilization and the release of plant nutrients into soil solution. They are easily available to subsequent non-legume crops. If properly incorporated, these crops do not require the application of N fertilizer. Legume residues contain huge amounts of N and have a relatively low C/N residue, leading to the more rapid release of N than lower N-containing cereal residues. Work by Sawatsky and Soper (1991) at the University of Manitoba reported that up to 44% of N fixed by legumes remained in the soil. This fixed N continuing in the soil would become available for succeeding non-legume crops. Enhanced N (N) availability to crops following legumes may also be due to reduced immobilization, as legume crops commonly produce lower amounts of residues along with higher N concentration than do cereal crops.

9.6 N Leaching

An appropriate cropping system and best management practices can help minimize the leaching risk besides improving N use efficiency (NUE). Legume intercropping in cereal-based system grown in wider crop rows can reduce the nitrate leaching risk (Weil and Brady 2017). Parallel multiple cropping (a system of growing two crops with dissimilar growth habits with minimum competition) of sugarcane and black gram or pigeon pea and maize resulted in low nitrate content in the soil profile when compared to sole cropping (Yadav et al. 2000). Soybean (*Glycine max*) seems to reduce the nitrate concentration in the soil profile more than maize.

9.7 Legumes and Soil Properties

Legume-based cropping systems improve several aspects of soil fertility, such as SOC, and major and micronutrient availability (Jensen et al. 2012). With respect to SOC, grain legumes can increase it in several ways, by supplying biomass, organic C and N (Lemke et al. 2007; Garrigues et al. 2012), as well as releasing the H_2 gas as by-product of BNF, which promotes bacterial legume nodules' development in the rhizosphere (La Favre and Focht 1983; Ram and Meena 2014). Although there is a general agreement on the influence of grain legumes on rhizosphere properties in terms of N supply, SOC and P availability, the magnitude of the impact varied across legume species, soil properties and climatic conditions. Among these, the soil type represents the major factor determining plant growth, rhizosphere nutrient dynamics and microbial community structure (Stagnari et al. 2017). The pattern of depletion and accumulation of some macro- and micronutrients differed also between cropping systems (i.e. monoculture, mixed culture, narrow crop rotations) as well as among soil management strategies (i.e. tillage, no tillage) (Shaha et al. 2003). Legume-based cropping system increased the soil organic matter. The average rate of sequestration or addition is ~1.42 Mg C ha^{-1} year^{-1} in the soil profile of metre depth in soil with legume-based cropping system (Ahmad et al. 2001)

9.8 Legumes Mitigate Environmental N Emission

The nitrogen fertilization accounts for 60% of N_2O anthropogenic emissions through agricultural practices. Agriculture is also contributing to other greenhouse gas (GHG) emissions, such as CH_4 and CO_2. In this context, Rees et al. (2013) considered that potential strategies for reducing GHG emissions in a cropping system could be developed by making changes in the variables/interventions (Meena et al. 2015c). These interventions influence the biochemical processes that trigger GHG emissions from soils, as a result of agricultural operations (e.g. tillage, fertilization, irrigation and crop rotation). Total emissions of CO_2 and N_2O from legumes are less than those from N-fertilized crops. Legumes contribute to the mitigation of climate change.

9.8.1 Legumes and Mitigation Potential of GHGs

Nitrous oxide is an important anthropogenic GHG which contributes around 5–6% of the total atmospheric GHGs, but it is abundantly more active than CO_2 (Crutzen et al. 2007). It plays a major role in ozone depletion (Ravishankara et al. 2009). Agriculture is considered as the largest N_2O source (Robertson et al. 2004; Takle et al. 2008). Total global N_2O emissions from agricultural soils are estimated to be 2.1 million tons N year^{-1} (Jensen and Hauggaard 2003) and will continue to increase annually at a rate of 0.25% (Kaiser et al. 1998). In general, N_2O is produced from the soil by microbial conversions, that is, ammonification, nitrification and denitrification, especially when N availability exceeds plant requirements (Smith and Conen 2004).

Globally, it is estimated that 150–200 million tons of mineral N is required annually by the plants, out of which nearly 100 million tons of N is fixed through the industrial Haber-Bosch process (Unkovich et al. 2008) and 175 million tons of N fixed through biological N fixation of atmospheric N_2 yearly (Chafi and Bensoltane 2003). Undoubtedly, the N_2-fixing ability of the legumes minimizes synthetic N input in soil and does minimization of negative environmental impact (Lupwayi et al. 2010; Kumar et al. 2016).

9.8.2 The Role of Cropping Systems and Inclusion of Legumes in Mitigation of GHGs

Appropriate management of cropping systems allows greater carbon sequestration; intensification of the cropping system is one of the strategies used to mitigate climate change. It may include incorporation of pulses such as field pea, common bean, soybean, faba bean and lentil; rotation of forages such as alfalfa, meadow brome, timothy grass and cocksfoot; and use of cover crops such as annual clover, red clover, hairy vetch, ryegrass and yellow sweet clover. Cropping systems provide

Table 9.6 Total N_2O emissions from field-grown legumes, N-fertilized grass pastures and crops or unfertilized soils

S. no.	Crop/species	Total N_2O emission per growing season or year (kg N_2 ha^{-1}) Range	Mean	References
N-fertilized crops	Rice	0.2–5.0 (3)[a]	2.21	Gupta et al. (2016)
	Wheat	0.09–8.57 (18)	2.73	Jensen et al. (2012)
	Maize	0.16–12.67 (22)	2.72	Jensen et al. (2012)
	Canola	0.13–8.60 (8)	2.65	Jensen et al. (2012)
N-fertilized pasture[a]	Grass	0.3–18.16 (19)	4.49	Jensen et al. (2012)
Mean of fertilized systems				
Pure legume stands[b]	Alfalfa	0.67–4.57 (14)	1.99	Jensen et al. (2012)
	White clover	0.50–0.90 (3)	0.79	Jensen et al. (2012)
Mixed pasture sward[b]	Grass-clover	0.10–1.30 (8)	0.54	Jensen et al. (2012)
Legume crops[b]	Faba bean	(1)	0.41	Jensen et al. (2012)
	Chickpea	(5)	0.05	Jensen et al. (2012)
	Lupin	0.03–0.16 0 (1)	0.06	Jensen et al. (2012)
	Field pea	0.38–1.73 (6)	0.65	Jensen et al. (2012)
	Soybean	0.29–7.09 (33)	1.58	Jensen et al. (2012)
Mean of all legumes			1.29	
Soil	No N fertilizer or legume	0.03–4.80 (33)	1.20	Jensen et al. (2012)

[a]Parenthesis values are highlighting the total number of experimentation sites
[b]Data come from systems where either zero N fertilizer was used or legume crops were provided with just 5 kg mineral N fertilizer ha^{-1} as starter N at the time of sowing. This is barring two experiments with grass-clover pastures and three soybean studies where 35–44 kg N fertilizer ha^{-1} had been applied

opportunities in reducing N_2O while improving energy use efficiency and in increasing C sequestration (Table 9.6). The introduction of legume crops could have an impact on soil microorganisms, including symbiotic and asymbiotic N_2-fixing bacteria, mycorrhiza and soil fauna. Legume crops contribute to mitigation of GHGs by replacing the N requirement of the cropping system than without the inclusion of pulses (Lupwayi et al. 2010; Ram and Meena 2014; Gregorich et al. 2005) found that emissions of N_2O from soils increased linearly with the quantity of mineral N fertilizer applied and because systems containing legumes produce lower annual N_2O emissions. Legumes emit around five to seven times less GHGs per unit area when compared to non-legume crops. Measuring N_2O fluxes, it was shown that peas emitted 69 kg N_2O ha^{-1}, much less than wheat (368 kg N_2O ha^{-1}) and rape (534 kg N_2O ha^{-1}) (Jeuffroy et al. 2013).

Guardia et al. (2016) reported that the emission of N_2O was higher for barley (non-legume) compared to vetch and lentil (legumes); moreover, the N_2O fluxes derived from the chemical fertilizers added to the crops were 2.5 times higher in barley than vetch. Schwenke et al. (2015) demonstrated that the cumulative N_2O

emissions from N-fertilized canola (385 g N_2O ha^{-1}) greatly exceeded those from chickpea (166 g N_2O ha^{-1}), faba bean (166 g N_2O ha^{-1}) and field pea (135 g N_2O ha^{-1}). When faba bean (441 g N_2O ha^{-1}) was grown through monocropping, it led to three times higher cumulative N_2O emissions than that of unfertilized wheat (152 g N_2O ha^{-1}); conversely, when faba bean was mixed with wheat (intercropping system), cumulative N_2O emission fluxes were 31% lower than that of N-fertilized wheat (Senbayram et al. 2016). The mitigation of GHG emissions is also achieved by adopting sustainable agricultural systems, such as conservation tillage and conservation agriculture systems, which are suitable for the cultivation of both grain and green manure legumes. Emissions of N_2O tend to be lower under legumes when compared to N-fertilized crops and pastures, mainly when commercially relevant rates of N fertilizer are applied.

9.9 Conclusion

Restoring the essential plant nutrients that are removed or lost from the soil remains central to long-term productivity, profitability and sustainability of any farming system. The emerging role of legume crops becomes evident in enhancing crop productivity along with retaining soil fertility and environmental quality. Besides serving as high-quality food and feed worldwide and allowing BNF, legume crops offer a range of other benefits. These benefits include considerable positive impact on biodiversity and soil health. Introducing legumes to modern cropping systems will not only improve crop diversity but also contribute to reduced use of imported inputs (especially N fertilizer). This would be in conjunction with enhanced yields of succeeding non-legumes. This also demands great efforts to broaden the scope of legume crops and their numerous positive advantages. These advantages are aimed towards sustainable intensification of agriculture besides the livelihoods of millions of farmers across the world.

Future Prospective

Pulses are always praised for dual benefits of soil health and human health. The dual benefits are directly related to ecological and economical upliftment of society. In a bid to heighten the public awareness of the environmental and nutritional benefits of pulses, as part of sustainable food production for achieving food, nutrition and environmental security, the 68th United Nations General Assembly declared 2016 as the International Year of Pulses. To envisage the importance of pulses which are equally good for people and soil, their sustainability should be given priority. It is also well understood and proved by the researchers globally that pulses have the ability to fix hefty quantities of nitrogen in the soil, by boosting soil fertility and reducing dependence on external nutrients (millions of tons globally). It is likely, of the other crops, pulses use half the non-renewable energy inputs including nitrogen. This results in remarkably small carbon footprints. The key message “Soil and Pulses: Symbiosis for Life” is the prelude to the sustainability, which will remain unfulfilled without pulses. Additionally, the importance of pulses in crop rotation, cropping system, intensification and diversification has also been explained in the text.

The vagaries of unforeseen negative factors of productivity need to be tackled smartly and cleverly. There are a few lines of research milestones that can help in augmenting soil and environmental benefits via pulses. Some of them are:

1. Developing short-duration high-yield yielding pulse varieties: Having the pulses and their varieties matching crop maturity duration to existing cropping window could be beneficial to cereals and vice versa. This way crop intensification, diversification and mutual sharing of critical natural resources especially water, inherent soil nutrients, sunlight, etc. may be utilized efficiently.
2. Pulses in new niches: Popularizing pulses in unexploited or untapped area, viz. rice fallows, which have approximately 14.3 million hectare area in Indo-Gangetic Plains (IGP), especially in Eastern India, spread over four Asian countries—Pakistan, India, Nepal and Bangladesh. These rice fallows offer a huge potential and scope to expand pulses, thereby improving soil, environmental and nutritional health. Adding more area under pulses directly helps in adding more SOC and releasing external nutrient input dependence. Further, development of short-duration/extra-early/super-early pigeon pea varieties for different agro-ecologies might enhance cropping intensity through pulses.
3. Nodulation engineering: As nodulation degeneration starts in pulses after the flowering initiation stage, this leads to declining or stoppage of nitrogen fixation. If the pulse roots are so engineered and nodulation period is prolonged, then it may be a marvellous nitrogen economy for the pulses as well as the succeeding crops.
4. Smart microbial strains: Due to non-availability and production of effective and efficient strains of *Rhizobium*, the full potential on nitrogen fixation by the legumes is tingling. Additionally, developing efficient strains for other nutrients (P, K, S, Zn, Fe, Mo) might have positive interaction with N, and thereby it may save millions of tons of respective nutrient fertilizers too.
5. Smart pulses: The development of smart pulse plant types could tackle the future problems of climate change and speedy depletion of natural resource base. Further, developing multi-nutrient-efficient genotypes/cultivars of pulses through the use of high-end technologies, viz. transgenic, next-generation genomic tools, coupling with classical breeding might become a boon to avoid external nutrient use.

Summing up, this chapter points out that pulses are not only N factory but also could be futuristic nutrient factory.

References

Adeleke MA, Haruna IM (2012) Residual nitrogen contributions from grain legumes to the growth and development of succeeding maize crop. ISRN Agron. https://doi.org/10.5402/2012/213729

Ahlawat IPS, Gangaiah B (2004) Grain legumes farmers own nitrogen fertilizer units role revisited. In: Souvenir: national symposium on resource conservation and agricultural productivity, 22–25 November, Ludhiana

Ahmad T, Hafeez FY, Mahmood T, Malik K (2001) Residual effect of nitrogen fixed by mungbean (*Vigna radiata*) and blackgram (*Vigna mungo*) on subsequent rice and wheat crops. Aus J Exp Agric 41:245–248

Ambrosano EJ, Trivelin PCO, Cantarella H, Ambrosano GMB, Schammass EA, Guirado N, Rossi F, Mendes PCD, Muraoka T (2005) Utilization of nitrogen from green manure and mineral fertilizer by sugarcane. Sci Agric 62:534–542

Ambrosano EJ, Trivelin PCO, Cantarella H, Ambrosano GMB, Schammass EA, Muraoka T, Guirado N, Rossi F (2009) Nitrogen supply to maize from sunn hemp and velvet bean green manures. Sci Agric 66:386–394

Ambrosano EJ, Ambrosano GMB, Azcón R, Cantarella H, Dias FLF, Muraoka T, Trivelin PCO, Rossi F, Schammass EA, Sachs RCC (2011) Productivity of sugarcane after previous legumes crop. Bragantia 70(4):1–9. https://doi.org/10.5772/53825

Angus JF, Kirkegaard JA, Hunt JR, Ryan MH, Ohlander L, Peoples MB (2015) Break crops and rotations for wheat. Crop Pasture Sci 66:523–552

Aranjuelo I, Irigoyen JJ, Nogués S, Sánchez DM (2009) Elevated CO_2 and water availability effect on gas exchange and nodule development in N_2-fixing alfalfa plants. Environ Exp Bot 65:18–26

Banyong T, Viriya L, Sanun J, Aran P, Prabhakar P, Suhas P W, Sahrawat KL (2000) Role of legumes in improving soil fertility and increasing crop productivity in Northeast Thailand report

Berg WA (1997) Residual nitrogen effects on wheat following legumes in the Southern Plains. J Plant Nutr 20:247–254

Bhat TA, Ahmad L, Ganai MA, Ul-Haq S, Khan OA (2015) Nitrogen fixing biofertilizers; mechanism and growth promotion: a review. J Pure Appl Microbiol 9(2):1675–1690

Bonilla DP, Nolot JM, Raffaillac D, Justes E (2017) Innovative cropping systems to reduce N inputs and maintain wheat yields by inserting grain legumes and cover crops in southwestern France. Eur J Agron 82:331–341

Brooker RW, Bennett AE, Cong WF, Daniell TJ, George TS, Hallett PD (2015) Improving intercropping: a synthesis of research in agronomy, plant physiology and ecology. New Phytol 206:107–117

Bruulsema TW, Christie BR (1987) Nitrogen contribution to succeeding corn from alfalfa and red clover. Agron J 79:96–100

Bundy LG (1998) Soil and applied, University of Wisconsin System Board of Regents and University of Wisconsin-Extension. A2519 Soil and applied nitrogen, R-07-98-IM-75 cecommerce.uwex.edu

Buragohain S, Sharma B, Nath JD, Gogaoi N, Meena RS, Lal R (2017) Impact of ten years of bio-fertilizer use on soil quality and rice yield on an inceptisol in Assam, India. Soil Res. https://doi.org/10.1071/SR17001

Carlsson G, Huss-Danell K (2003) Nitrogen fixation in perennial forage legumes in the field. Plant Soil 253:353–372

Carroll SB, Salt SD (2004) Ecology for gardeners. Timber Press, Portland, p 93. ISBN:978-0-88192-611-8

Chafi MH, Bensoltane A (2003) *Vicia faba* (L), a source of organic and biological manures for the Algerian Arid regions. World J Agric Sci 5:698–706

Chalk PM (1998) Dynamics of biologically fixed N in legume-cereal rotation: a review. Aust J Agric Res 49:303–316

Clune S, Crossin E, Verghese K (2017) Systematic review of greenhouse gas emissions for different fresh food categories. J Clean Prod 140:766–783

Crutzen PJ, Mosier AR, Smith KA, Winiwarter W (2007) N_2O release from agrobiofuel production negates global warming reduction by replacing fossil fuels. Atmos Chem Phys Discuss 7:11191–11205

Dadhich RK, Meena RS, Reager ML, Kansotia BC (2015) Response of bio-regulators to yield and quality of Indian mustard (*Brassica juncea* L. Czernj. and Cosson) under different irrigation environments. J App and Nat Sci 7(1):52–57

Dalal RC, Doughton JA, Strong WM, Weston EJ, Cooper JE, Wildermuth GB, Lehane KJ, King AJ, Holmes CJ (1998) Sustaining productivity of a vertisol at Warra, Queensland, with fertilisers, no-tillage or legumes. Wheat yields, nitrogen benefits and water-use efficiency of chickpea-wheat rotation. Aust J Exp Agric 38:489–501

Datta R, Kelkar A, Baraniya D, Molaei A, Moulick A, Meena RS, Formanek P (2017) Enzymatic degradation of lignin in soil: a review. Sustainability MDPI 9:1163. https://doi.org/10.3390/su9071163. 1–18

Dhakal Y, Meena RS, Kumar S (2016) Effect of INM on nodulation, yield, quality and available nutrient status in soil after harvest of green gram. Legum Res 39(4):590–594

Entry JA, Mitchell CC, Backman CB (1996) Influence of management practices on soil organic matter, microbial biomass and cotton yield in Alabamas Old Rotation. Biol Fertil Soils 23(4):353–358

Evans J, Fettell NA, Coventry DR, Connor GE, Walsgott DN, Mahoney J, Armstrong EL (1991) Wheat responses after temperate crop legumes in South-Eastern Australia. Aust J Agric Res 42:31–43

FAO (2011) The state of the world's land and water resources for food and agriculture (SOLAW) managing systems at risk. Food and Agriculture Organization of the United Nations/Earthscan, Rome/London

FAOSTAT (2014) http://www.fao.org/faostat/en/#data/RF. Update on 9/03/2017

Frame J (2005) Forage legumes for temperate grasslands. Science Publishers, Enfield USA and Food and Agriculture Organization of the United Nations, Rome

Frankow BE, Dahlin AS (2013) N2-fixation, N transfer, and yield in grassland communities including a deep-rooted legume or non-legume species. Plant Soil 370:567–581

Garrigues E, Corson MS, Walter C, Angers DA, Van der Werf H (2012) Soil quality indicators in LCA: method presentation with a case study. In: Corson MS, van der Werf HMG (eds) Proceedings of the 8th international conference on life cycle assessment in the agri-food sector, 1–4 October 2012. INRA, Saint Malo, pp 163–168

Gibson AH (1976) In: Nutman PS (ed) Symbiotic nitrogen fixation in plants, In Biological Programme No. 7. Cambridge University Press, Cambridge, pp 385–403

Gonzalez LJ, Rodelas C, Pozo V, Salmeron MV, Salmeron V (2005) Liberation of amino acids by heterotrophic nitrogen fixing bacteria. Amino Acids 28:363–367

Graham PH, Vance CP (2003) Legumes: importance and constraints to greater use. Plant Physiol 131:872–877

Gregorich EG, Rochette PV, Bygaart AJ, Angers D (2005) Greenhouse gas contributions of agricultural soils and potential mitigation practices in Eastern Canada. Soil Tillage Res 81:53–72

Groffman P (2012) Terrestrial denitrification: challenges and opportunities. Ecol Process 1:112. https://doi.org/10.1186/2192:1709-1-11

Guardia G, Tellez-Rio A, García-Marco S, Martin-Lammerding D, Tenorio JL, Ibáñez MÁ, Vallejo A (2016) Effect of tillage and crop (cereal versus legume) on greenhouse gas emissions and Global Warming Potential in a non-irrigated Mediterranean field. Agric Ecosyst Environ 221:187–197

Gupta DK, Bhatiaa A, Kumar A, Das TK, Jain N, Tomar R, Sandeep KM, Fagodiya RK, Dubey R, Pathak H (2016) Mitigation of greenhouse gas emission from rice-wheat system of the Indo-Gangetic plains: through tillage, irrigation and fertilizer management. Agric Ecosyst Environ 230:1–9

Hajduk E, Właśniewski S, Szpunar KE (2015) Influence of legume crops on content of organic carbon in sandy soil. Soil Sci Annu 66:52–56

Havlin JL, Tisdale SL, Nelson WL, Beaton JD (2014) Soil fertility and fertilizers; An introduction to nutrient management, 8th edn. ISBN:978-81-203-4868-4

Herridge DF, Peoples MB, Boddey RM (2008) Global inputs of biological nitrogen fixation in agricultural systems. Plant Soil 311:1–18

Hestermann OB, Sheaffer CC, Barnes DK, Lueschen WE, Ford JH (1986) Alfalfa dry matter and nitrogen production, and fertilizer nitrogen response in legume corn rotations. Agron J 78:19–23

Hungria M, Vargas MAT (2000) Environmental factors affecting N_2 fixation in grain legumes in the tropics, with an emphasis on Brazil. Field Crop Res 65:151–164

James W (2013) Nitrogen in soil and the environment. College of Agriculture and Life Science, Tucson

Jannink JL, Liebman M, Merrick LC (1996) Biomass production and nitrogen accumulation in pea, oat, and vetch green manure mixtures. Agron J 88(2):231–240

Jensen ES, Hauggaard NH (2003) How can increased use of biological N_2 fixation in agriculture benefit the environment? Plant Soil 252:177–186

Jensen CR, Joernsgaard B, Andersen MN, Christiansen JL, Mogensen VO, Friis P, Petersen CT (2004) The effect of lupins as compared with peas and oats on the yield of the subsequent winter barley crop. Eur J Agron 20:405–418

Jensen ES, Peoples MB, Boddey RM, Gresshoff PM, Hauggaard-Nielsen H, Alves BJ, Morrison MJ (2012) Legumes for mitigation of climate change and the provision of feedstock for biofuels and bio refineries. A review. Agron Sustain Dev 32:329–364

Jeuffroy MH, Baranger E, Carrouée B, Chezelles ED, Gosme M, Hénault C (2013) Nitrous oxide emissions from crop rotations including wheat, oilseed rape and dry peas. Biogeosciences 10:1787–1797

Kabir Z, Koide RT (2002) Effect of autumn and winter mycorrhizal cover crops on soil properties, nutrient uptake and yield of sweet corn in Pennsylvania, USA. Plant Soil 238(2):205–215

Kahindi J, Karanja N, Gueye M (2009) In: Doelle HW, Raken S, Berovic M (eds) Biological nitrogen fixation biotechnology, vol XV. Fundam Biotechnol. ISSBN:978-1-184826-269-0

Kaiser EA, Kohrs K, Kücke M, Schnug E, Munch JC, Heinemeyer O (1998) Nitrous oxide (N_2O) release from arable soil: importance of perennial forage crops. Biol Fertil Soils 28(1):36–43

Kirkegaard JA, Christen O, Krupinsky J, Layzell DB (2008) Break crop benefits in temperate wheat production. Field Crop Res 107:185–195

Kumar S, Sheoran S, Kumar SK, Kumar P, Meena RS (2016) Drought: a challenge for Indian farmers in context to climate change and variability. Prog Res, An Int J 11:6243–6246

Kumari S, Rajeshwari PS (2011) Symbiotic and asymbiotic nitrogen fixation. https://www.researchgate.net/publication/279534008

Kumbhar AM, Buriro UA, Junejo S, Oad FC, Jamro GH, Kumbhar BA, Kumbhar SA (2008) Impact of different nitrogen levels on cotton growth, yield and N-uptake planted in legume rotation. Pak J Bot 40(2):767–778

La Favre JS, Focht DD (1983) Conservation in soil of H_2 liberated from N_2 fixation by H up-nodules. Appl Environ Microbiol 46:304–311

Ladd JN, Amato M, Jackson RB, Butler JHA (1983) Utilization by wheat crops of nitrogen from legume residues decomposing in soils in the field. Soil Biol Biochem 15:231–238

Ladha JK, Peoples MB (1995) Management of biological nitrogen fixation for the development of more productive and sustainable agricultural systems. Kluwer Academic Publishers, Dordrecht, p 287

Lemke RL, Zhong Z, Campbell CA, Zentner RP (2007) Can pulse crops play a role in mitigating greenhouse gases from North American agriculture? Agron J 99:1719–1725

Li YF, Ran W, Zhang RP, Sun SB, Xu GH (2009) Facilitated legume nodulation, phosphate uptake and nitrogen transfer by arbuscular inoculation in an upland rice and mung bean intercropping system. Plant Soil 315:285–296

Luce SM, Grant CA, Zebarth BJ, Ziadi N, O'Donovan JT, Blackshaw RE (2015) Legumes can reduce economic optimum nitrogen rates and increase yields in a wheat-canola cropping sequence in western Canada. Field Crop Res 179:12–25

Lupwayi NZ, Kennedy AC, Chirwa RM (2010) Grain legume impacts on soil biological processes in Sub-Saharan Africa. African J Plant Sci 5:1–7

Mahdhi M, Lajudie P, Mars M (2008) Phylogenetic and symbiotic characterization of rhizobial bacteria nodulating Argyrolobium uniflorum in Tunisian arid soils. Can J Microbiol 54:209–217

Mayer J, Buegger FL, Jensen SE, Schloter M, Heb J (2003) Residual nitrogen contribution from grain legumes to succeeding wheat and rape and related microbial process. Plant Soil 255:541–554

Meena VS, Maurya BR, Verma R, Meena RS, Jatav GK, Meena SK, Meena R, Meena SK (2013) Soil microbial population and selected enzyme activities as influenced by concentrate manure and inorganic fertilizer in alluvium soil of Varanasi. The Biosc 8(3):931–935

Meena RS, Yadav RS, Meena VS (2014a) Response of groundnut (*Arachis hypogaea* L.) varieties to sowing dates and NP fertilizers under western dry zone of India. Bangladesh J Bot 43(2):169–173

Meena VS, Maurya BR, Meena RS, Meena SK, Singh NP, Malik VK (2014b) Microbial dynamics as influenced by concentrate manure and inorganic fertilizer in alluvium soil of Varanasi, India. African J Microb Res 8(1):257–263

Meena RS, Yadav RS, Meena H, Kumar S, Meena YK, Singh A (2015a) Towards the current need to enhance legume productivity and soil sustainability worldwide: a book review. J Clean Prod 104:513–515

Meena RS, Dhakal Y, Bohra JS, Singh SP, Singh MK, Sanodiya P (2015b) Influence of bioinorganic combinations on yield, quality and economics of mungbean. Am J Exp Agric 8(3):159–166

Meena RS, Meena VS, Meena SK, Verma JP (2015c) Towards the plant stress mitigate the agricultural productivity: a book review. J Clean Prod 102:552–553

Meena VS, Maurya BR, Meena RS (2015d) Residual impact of well-grow formulation and NPK on growth and yield of wheat (*Triticum aestivum* L.). Bangladesh J Bot 44(1):143–146

Meena RS, Gogaoi N, Kumar S (2017a) Alarming issues on agricultural crop production and environmental stresses. J Clean Prod 142:3357–3359

Meena RS, Kumar V, Yadav GS, Mitran T (2017b) Response and interaction of *Bradyrhizobium japonicum* and *Arbuscular mycorrhizal* fungi in the soybean rhizosphere: a review. Plant Growth Reg. Accepted in press

Mitchell WH, Teel MR (2007) Winter annual cover crops for no tillage corn production. Agron J 69:569–573

Mubarak AR, Rosenani AB, Anuar AR, Zauyah S (2002) Decomposition and nutrient release of maize stover and groundnut haulm under tropical field conditions of Malaysia. Commun Soil Sci Plant Anal 33(3&4):609–622

Mugwe J, Mugendi DN, Mucheru-Muna M, Kungu J (2011) Soil inorganic N and N uptake by maize following application of legume biomass, tithonia, manure and mineral fertilizer in central Kenya. In: Bationo A (ed) Innovations as key to the Green Revolution in Africa. Springer, pp 605–612

Nair KPP, Patel UK, Singh RP, Kaushik MK (1979) Evaluation of legume intercropping in conservation of fertilizer nitrogen in maize culture. J Agric Sci 93:189–194

Nulik J, Dalgliesh N, Cox K, Gabb S (2013) Integrating herbaceous legumes into crop and livestock systems in eastern Indonesia. Australian Centre for International Agricultural Research (ACIAR), Canberra

Osman AN, Ræbild A, Christiansen JL, Bayala J (2011) Performance of cowpea (*Vigna unguiculata*) and pearl millet (*Pennisetum glaucum*) intercropped under Parkia biglobosa in an agroforestry system in Burkina Faso. Afr J Agric Res 6(4):882–891

Pappa VA, Rees RM, Walker RL, Baddeley JA, Watson CA (2012) Legumes intercropped with spring barley contributes to increased biomass production and carry-over effects. J Agric Sci 150:584–594

Paré T, Chalifour FP, Bourassa J, Antoun H (1992) Forage-corn dry-matter yields and N uptake as affected by previous legumes and N fertilizer. Can J Plant Sci 72:699–712

Peix A, Velazquez E, Silva LR, Mateos PF (2010) Key molecules involved in beneficial infection process in rhizobia-legume symbiosis. In: Microbes for legume improvement. https://doi.org/10.1007/978-3-211-99753-6

Peoples MB, Brockwell J, Herridge DF, Rochester IJ, Alves BJR, Urquiaga S, Boddey RM, Dakora FD, Bhattarai S, Maskey SL, Sampet C, Rerkasem B, Khan DF, Hauggaard NH, Jensen ES (2009) The contributions of nitrogen-fixing crop legumes to the productivity of agricultural systems. Symbiosis 48:1–17

Peter VM, Cassman K, Cleveland C, Crews T, Christopher BF, Grimm BN, Howarth WR, Marinov R, Martinelli L, Rastetter B, Sprent IJ (2002) Towards an ecological understanding of biological nitrogen fixation. Biogeochemistry 57:1–45

Postgate J (1998) Nitrogen fixation, 3rd edn. Cambridge University Press, Cambridge

Ram K, Meena RS (2014) Evaluation of pearl millet and mungbean intercropping systems in Arid Region of Rajasthan (India). Bangladesh J Bot 43(3):367–370

Ravishankara R, Daniel JS, Portmann RW (2009) Nitrous oxide (N_2O): the dominant ozone-depleting substance emitted in the 21st century. Science 5949:123–125

Rees RM, Augustin J, Alberti GB, Ball C et al (2013) Nitrous oxide emissions from European agriculture an analysis of variability and drivers of emissions from field experiments. Biogeosciences 10:2671–2682

Robertson GP, Paul EA, Harwood RR (2004) Greenhouse gases (GHGs) in intensive agriculture: contributions of individual gases to the radiative forcing of the atmosphere. Science 5486:1922–1925

Rochette P, Angers DA, Chantigny MH, MacDonald JD, Bissonnette N, Bertrand N (2009) Ammonia volatilization following surface application of urea to tilled and no-till soils: a laboratory comparison. Soil Tillage Res 103:310–315

Ruschel AP, Döbereiner J (1965) Bactérias assimbióticas fixadoras de N na rizosfera de gramíneas forrageiras. In: Congresso Internacional de Pastagens, vol 9. São Paulo, pp 1103–1107

Sawatsky N, Soper RJ (1991) A quantitative measurement of the nitrogen loss from the root system of field peas (*Pisum avense* L.) grown in the field. Soil Biol Biochem 23:255–259

Schulz S, Keatinge JDH, Wells GJ (1999) Productivity and residual effects of legumes in rice based cropping systems in a warm temperature environment. Legume biomass production and N- fixation. Field Crop Res 61:23–35

Schwenke GD, Herridge DF, Scheer C, Rowlings DW, Haigh BM, McMullen KG (2015) Soil N2O emissions under N2-fixing legumes and nitrogen-fertilised canola: a reappraisal of emissions factor calculations. Agric Ecosyst Environ 202:232–242

Seitzinger S, Harrison JA, Bohlke JK, Bouwman AF, Lowrance R, Peterson B, Tobias C, Van Drecht G (2006) Denitrification across landscapes and waterscapes: a synthesis. Ecol Appl 16:2064–2090

Senbayram M, Wenthe C, Lingner A, Isselstein J, Steinmann H, Kaya C, Köbke S (2016) Legume-based mixed intercropping systems may lower agricultural born N_2O emissions. Energy Sustain Soc 6:2

Sergei AM (2012) Nitrogen cycle. In: Spradley J, Elliott KD, Dutch SI, Boorstein M (eds) Earth's weather, water, and atmosphere. EBSCO. Earth Sci, pp 347–350

Seymour N, McKenzie K, Krosch S (2015) Fixing more nitrogen in pulse crops. In Rachaputi RCN (ed) Nindigully GRDC grains research update. Department of Agriculture and Fisheries

Shaha Z, Shahb SH, Peoplesc MB, Schwenked GD, Herridge DF (2003) Crop residue and fertiliser N effects on nitrogen fixation and yields of legume-cereal rotations and soil organic fertility. Field Crop Res 83:1–11

Sindhu SS, Dua S, Verma MK, Khandelwal A (2010) Growth promotion of legumes by inoculation of rhizosphere bacteria, Microbes for legume improvement, vol 9. Springer, Wien, pp 195–237. https://doi.org/10.1007/978-3-211-99753-6_10

Skorupska A, Wielbo J, Kidaj D, Kozaczuk MM (2010) Enhancing rhizobium-legume symbiosis using signaling factors. In: Khan MS et al (eds) Microbes for legume improvement, vol 2. Springer, Wien, pp 27–54. https://doi.org/10.1007/978-3-211-99753-6_2

Smith KA, Conen F (2004) Impacts of land management on fluxes of trace greenhouse gases. Soil Use Manag 20:255–263

Sqrensen J, Sessitsch A (2007) Plant-associated bacteria lifestyle and molecular interactions. In: van Elsas JD, Jansson JK, Trevors JT (eds) Modern soil microbiology, 2nd edn. CRC Press, Taylor and Francis Group, Boca Raton, pp 211–236

Stagnari F, Maggio A, Galieni A, Pisante M (2017) Multiple benefits of legumes for agriculture sustainability. Chem Biol Technol Agric 4:2. https://doi.org/10.1186/s40538-016-0085-1

Sulieman S, Tran LP (2015) Legume nitrogen fixation in a changing environment, achievements and challenges. www.springer.com. https://doi.org/10.1007/978-3-319-06212-9-1

Takle E (2008) Global warming: agriculture's impact on greenhouse gas emission, Ag DM Newsletter Article
Timothy CE (1999) The presence of nitrogen fixing legumes in terrestrial communities: evolutionary vs ecological considerations. Biogeochemistry 46:233–246
Unkovich M, Herridge D, Peoples M (2008) Measuring plant-associated nitrogen fixation in agricultural systems, ACIAR Monograph No. 136
Varma D, Meena RS (2016) Mungbean yield and nutrient uptake performance in response of NPK and lime levels under acid soil in Vindhyan region, India. J App Nat Sci 8(2):860–863
Varma D, Meena RS, Kumar S, Kumar E (2017) Response of mungbean to NPK and lime under the conditions of Vindhyan Region of Uttar Pradesh. Legum Res 40(3):542–545
Verma JP, Jaiswal DK, Meena VS, Meena RS (2015) Current need of organic farming for enhancing sustainable agriculture. J Clean Prod 102:545–547
Vieira RF, Mendes IC, Junior FBR, Hungria M (2010) Symbiotic nitrogen fixation in tropical food grain legumes: current status. Microbes for legume improvement Chapter 8. In: Khan MS et al (eds) Microbes for legume improvement. https://doi.org/10.1007/978-3-211-99753-6_18
Vitousek PM, Menge DNL, Reed SC, Cleveland CC (2013) Biological nitrogen fixation: rates, patterns and ecological controls in terrestrial ecosystems. Philos Trans R Soc B 368:20130119. https://doi.org/10.1098/rstb.2013.0119
Voisin AS, Guéguen J, Huyghe C, Jeuffroy MH, Magrini MB, Meynard JM (2014) Legumes for feed, food, biomaterials and bioenergy in Europe: a review. Agron Sustain Dev 34:361–380
Weil RR, Brady NC (2017) The nature and properties of soil, 17th edn. Pearson and Prentice Hall, Upper Saddle River. ISBN:978-0133254488
Yadav RL, Dwivedi BS, Pandey PS (2000) Rice-wheat cropping system: assessment of sustainability under green manuring and chemical fertilizer inputs. Field Crop Res 65:15–30
Yadav SS, Hunter D, Redden B, Nang M, Yadava DK, Habibi AB (2015) Impact of climate change on agriculture production, food, and nutritional security. In: Redden R, Yadav SS, Maxted N, Dulloo MS, Guarino L, Smith P (eds) Crop wild relatives and climate change. Wiley, Hoboken, pp 1–23
Yadav GS, Lal R, Meena RS, Datta M, Babu S, Das, Layek J, Saha P (2017) Energy budgeting for designing sustainable and environmentally clean/safer cropping systems for rainfed rice fallow lands in India. J Clean Prod 158:29–37
Yano KH, Daimon H, Mimoto (1994) Effect of Sunhemp and peanut incorporated as green manure on growth and nitrogen uptake of the succeeding wheat. Jpn J Crop Sci 63(1):137–143
Yusuf AA, Abaidoo RC, Iwuafor ENO, Olufajo OO, Sanginga N (2009) Rotation effects of grain legumes and fallow on maize yield, microbial biomass and chemical properties of an Alfisol in the Nigerian savanna. Agric Ecosyst Environ 12:325–331

Leguminous Trees an Innovative Tool for Soil Sustainability

10

M. K. Jhariya, A. Banerjee, D. K. Yadav, and A. Raj

Contents

Abstract

World food production is to some extent dependent upon biological nitrogen (N) fixation (about 100 million tons per year globally) in agroecosystem. Legumes reflect multidimensional activity towards developing soil nutrient pool and improving soil fertility. Increased level of CO_2 (0.04%) associated with addition of N in a system is dependent upon various abiotic (temperature, humidity, soil) and biotic (species interaction, resource partitioning, biotic interference) factors.

M. K. Jhariya (✉) · D. K. Yadav (✉)
Department of Farm Forestry, Sarguja University, Ambikapur, Chhattisgarh, India

A. Banerjee
Department of Environmental Science, Sarguja University, Ambikapur, Chhattisgarh, India

A. Raj
Department of Forestry, College of Agriculture, I.G.K.V., Raipur, Chhattisgarh, India

R. S. Meena et al. (eds.), *Legumes for Soil Health and Sustainable Management*,
https://doi.org/10.1007/978-981-13-0253-4_10

As a consequence there may be a significant level of variation in the N cycle in different ecosystems. In comparison with cropland soils of Europe and North America, soils of India are strongly depleted of their N reserves. Such deficiency can be mitigated through the inherent N-fixing ability and improvement of soil condition by leguminous tree species. Such approaches also promote proper enhancement of forest floor biodiversity in terms of various living communities. Leguminous trees are often found to be a key instrument towards combating climate change due to their higher C sequestration potential and wide ecological amplitude at various conditions. Such potentiality often hampers the flourishment of legume trees in nature due to over exploitation and improper regeneration. Community-based natural resource management practices are the suitable solution for these problems. Exploration of areas with higher density of legumes and management of legumes in captivity and under natural condition needs to be prioritized. In this context appropriate research work should be aimed towards proper exploration of potentiality among leguminous vegetation in fixing atmospheric N. Wider application of such species has become a thrust area of research in modern science perspectives. All these issues are periodically reviewed with research-oriented database for the benefits of soil sustainability. The present chapter deals with the beneficial and multipurpose role of leguminous tree species towards soil sustainability and plant growth.

Keywords
C and N sequestration · N fixation · Nutrient pool · Tree species

Abbreviations

AMF	Arbuscular mycorrhizae fungi
BNF	Biological nitrogen fixation
C	Carbon
CO_2	Carbon dioxide
FACE	Free-air CO_2 enrichment
FAO	Food and Agricultural Organization
GHG	Greenhouse gases
INM	Integrated nutrient management
N	Nitrogen
NFP	Nitrogen-fixing potential
NFT	Nitrogen-fixing trees
OM	Organic matter
R&D	Research and development
SCP	Soil carbon pool
SNF	Symbiotic nitrogen fixation
SNP	Soil nitrogen pool
SOCP	Soil organic carbon pool
SOM	Soil organic matter

10.1 Introduction

Legumes show considerable promise for sustained supply of N into the soil systems for ready uptake of crop species along with checking soil health problems. Flowering plants are to some extent dominated by legumes. The Gramineae family which includes cereals and grasses while family Leguminosae (Fabaceae) includes legumes or the bean family has a wider contribution to the soil health and nutritional security in world agriculture. The Leguminosae family is represented by 750 genera and 19,000 species (Stevens 2001; Bargali 2016; Dhakal et al. 2016) and is divided as the Caesalpinioideae, Papilionoideae and Mimosoideae.

Among the flowering plants, members of Leguminosae family are providing a variety of products like food, gums, fodder, timber, etc. in various climatic regimes of the world (Rao and Husain 1993; Bargali 2016). In India, legumes are also widely adapted plant species under various environmental conditions and are represented by 1152 species under 179 genera (Husain and Kapoor 1990; Sanjappa 1991). The family comprises diverse growth form in terms of height, growth, life cycle and various life forms (Rao and Husain 1993; Bargali 2016). Symbiotic association of microbes with root system of legumes is often designated as nodular growth of root (Allen and Allen 1981).

Legumes provide good quality foods as well as produce lesser GHGs (five to seven times) when compared to other crops. They also sequester 7.21 g kg^{-1} DM, 23.6 versus 21.8 g C kg^{-1} year. In agricultural system, legumes can be widely grown in conservation system, low-input farming system as well as intercropping system (Stagnari et al. 2017; Varma et al. 2017a). Nitrogen fixer (mostly legumes) maintains harmony between productivity and sustainability (Rao et al. 2007). Legumes perform various ecological functions like improvement in the soil quality, reduction in N requirement for plant species for growth purpose and enrichment of wildlife habitat, improving land capability which stops further land degradation (Bargali and Bargali 2009). They can be successfully incorporated through practices such as crop rotation/intercropping to improve soil health with minimum amount of fertilizer application. Legume-rhizobium association stands to be the most promising N fixation system providing economic benefits in terms of lesser fertilizer application and soil sustainability (Crews and Peoples 2004; Kumar et al. 2013, 2014; Bhagat et al. 2014). Such potentialities promote ecological restoration of degraded land habitat. The biological N fixation (BNF) process is often dependent upon various abiotic factors as well as soil nutrient status which influence the rate of N fixation of legumes at molecular and functional level which regulates the N-fixing potential at a certain time interval in a certain area (Bommarco et al. 2013). Symbiotic process includes N fixation as a natural process that helps to maintain soil fertility along with crop productivity under semiarid tropical condition. From sustainable agriculture perspective, symbiotic N fixation (SNF) is a suitable strategy with growing dimension for future to boost up agricultural productivity. Qualitative assessment on SNF and its impact on crop and leguminous species indicated that leguminous SNF is highly susceptible towards environmental changes (Galiana et al. 2004), and therefore N fixation potential (NFP) can be hindered. Leguminous N-fixing trees

(NFT) often have multipurpose uses, e.g. *Leucaena* can be effectively utilized for fodder, fuel and fibre. Other species such as wattles (*Acacia* spp.) has been reported to produce gums and resins. Lesser nodulation was observed in Caesalpinioideae than the other subfamilies. The present chapter deals with the potential role of legumes in different directions towards environmental sustainability. Further, it addresses the ecofriendly roles of legumes towards soil sustainability under tropical condition of India.

10.2 Leguminous Trees and Their Role in Soil Sustainability

Nowadays, soil sustainability is a major challenge under the complex influence of various phenomena such as nutrient loss, soil erosion as well as agricultural pollution (Zentner et al. 2004). Legumes perform multifaceted activities in various spheres of agricultural sector as well as help fix atmospheric carbon and provide daily basic needs (Fig. 10.1). Legumes also promote soil carbon sequestration. They also serve as high-protein feeds and promote biodiversity and soil quality. Multifaceted role of legumes and their agro-productivity along with socio-economic upliftment of the farming community in different agroecological zones of the world should be emphasized in research and development activities.

Legume has the potential to stop the splashing action of rain drops to check soil erosion. *Rhizobium* colonizes the roots of legume plant in the form of symbiotic relationship (Bilyaminu and Wani 2016). Root nodule formation by *Rhizobium* species has a positive influence on rates of BNF under nutrient-deficient condition. *Rhizobium* bacterium performs the vital ecological role in solubilization of nutrient with vesicular-arbuscular mycorrhizae (VAM) association. Such type of association positively influences the root system to explore and gain more nutrients from the soil nutrient pool such as phosphorous under stress condition. As per earlier reports, it was observed that plant species reflects a significant level of variation over soil structure. Legumes are found to be much more promising in this context (Drury et al. 1991) because legume promotes higher mobilization of nitrate in the plant-soil ecosystem (Holtham et al. 2007; Meena et al. 2015a).

Presently throughout the world, half of the cultivated land is under the threat of degradation and future prediction reveals that present rate of land degradation would lead to huge area of agricultural area to become less productive by near future. Under tropical condition building up of SOM pool is an essential requirement under acidic condition by the elevated saturation of Al (aluminium) and lesser availability of P (phosphorus). OM condition is a key factor towards proper nutrient utilization by the crop plants on such type of surfaces. N and biotic constituents represent better relationship towards rate of mineralization which indicates quality of OM (Fox et al. 1990; Thomas and Asakawa 1993).

Legumes further promote checking soil erosion by stabilizing ravines and gullies. Legumes have the potential to reduce N pollution through chemical fertilizer which subsequently reduces fossil fuel consumption (Zentner et al. 2001, 2004). Without crop rotation, productivity of agroecosystem decreases under various biotic

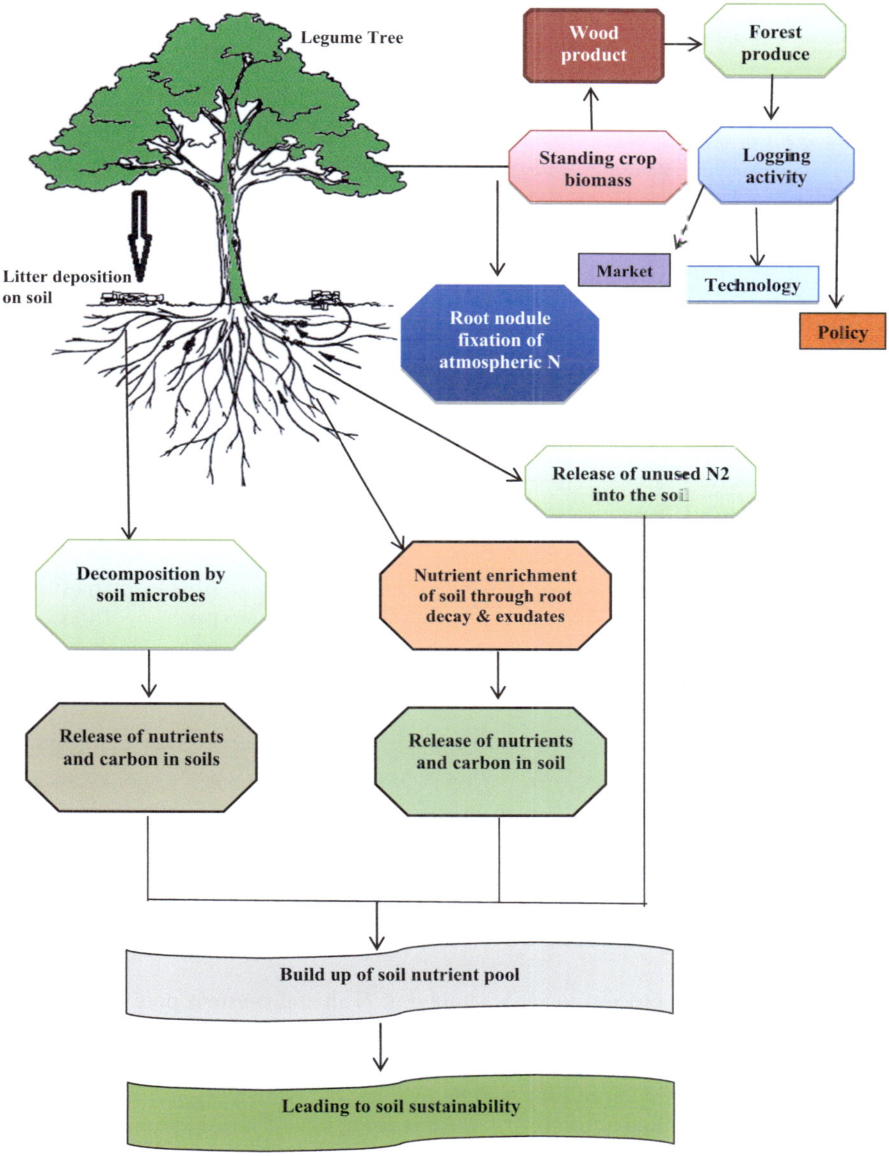

Fig. 10.1 Role of tree towards soil sustainability and meeting human needs

and abiotic factors. Non-judicious crop rotation practices lead to decline in productivity, the yield of crops with subsequent degradation of the soil quality with the gradual invasion of various biotic factors (Dumanski et al. 1998; Jhariya and Yadav 2017). Studies on mixed cropping on long-term experimental tenure in specific cropping sequence reported higher productivity and higher biomass turnover in

Table 10.1 Legume biomass (t/ha) in tropics of Chhattisgarh, India (Jhariya et al. 2014)

Species	Bole	Branch	Leaf	Root	Total
Tree stage					
Cassia fistula Linn.	0.59–0.98	0.64–1.18	0.06–0.10	0.22–0.37	1.52–2.63
Dalbergia paniculata Roxb.	0.64	0.62	0.07	0.24	1.57
Ougeinia oojeinensis (Roxb.) Hochr.	3.59–5.49	3.90–7.94	0.40–054	1.34–2.06	9.23–16.03
Sapling stage					
Butea monosperma (Lamk) Taub.	0.016–0.03	0.007–0.01	0.003–0.005	0.006–0.01	0.03–0.07
Cassia fistula Linn.	0.03	0.01	0.05	0.01	0.07
Dalbergia paniculata Roxb.	0.16	0.09	0.02	0.06	0.32
Ougeinia oojeinensis (Roxb.) Hochr.	0.11–0.36	0.066–0.21	0.01–0.05	0.04–0.14	0.23–0.76
Seedling stage					
Cassia fistula Linn.	0.24	0.05	0.047	0.087	0.42
Dalbergia paniculata Roxb.	0.06	0.01	0.01	0.02	0.11
Ougeinia oojeinensis (Roxb.) Hochr.	0.28	0.06	0.056	0.10	0.50
Shrubs					
Bauhinia racemosa Lam.	0.02	0.01	0.10	0.010	0.15
Bauhinia vahlii (W.) A.	0.03	0.02	0.17	0.02	0.24
Butea superba Roxb. ex Willd.	0.53	0.28	1.81	0.09	2.71
Spatholobus roxburghii Benth.	0.007	0.005	0.034	0.003	0.05

annual rotation system instead of monoculture (Drinkwater et al. 1998). This, therefore, revealed lower C/N ratio in the USA, which has increased C pool that promotes a certain amount (1–2%) of emission reduction of carbon in atmosphere through burning of fossil fuel (Marland and Boden 1997; Yadav et al. 2017a).

10.3 Leguminous Trees and Soil Biomass

BNF is an effective process towards supplying N in soil nutrient pool due to their capability to fix atmospheric N into the soil. Legume can be effectively utilized in this case due to their fast-growing nature, drought resistancy as well as the ability to fix atmospheric N. *Acacia* species has a high potential for N fixation, which grows very fast in the wasteland and agroforestry systems throughout India. Legumes such as *A. nilotica* have a significant stimulatory impact over paddy cultivation regarding N-fixing potential as well as higher organic matter (OM) accumulation which promotes its development in rice-cultivated area (Jhariya et al. 2015). The species have been found to have higher carbon-sequestering potential when compared to other plants which are nonleguminous in nature. These species have substantial potential to produce higher biomass and carbon storage (Tables 10.1 and 10.2) supports to build high soil organic biomass depending upon the age, site quality and successive stage. *A. nilotica* approximately fixes 228.45 kg per tree biomass carbon which is added to

Table 10.2 Legume C storage (t/ha) potential in tropics of Chhattisgarh, India (Jhariya et al. 2014)

Species	Bole	Branch	Leaf	Root	Total
Tree stage					
Cassia fistula Linn.	0.49	0.59	0.05	0.18	1.32
Dalbergia paniculata Roxb.	0.32	0.31	0.04	0.12	0.79
Ougeinia oojeinensis (Roxb.) Hochr.	2.74	3.97	0.27	1.03	8.01
Sapling stage					
Butea monosperma (Lamk) Taub.	0.02	0.01	0.003	0.01	0.03
Cassia fistula Linn.	0.02	0.01	0.003	0.01	0.03
Dalbergia paniculata Roxb.	0.08	0.04	0.01	0.03	0.16
Ougeinia oojeinensis (Roxb.) Hochr.	0.18	0.11	0.03	0.07	0.38
Seedling stage					
Cassia fistula Linn.	0.118	0.026	0.024	0.044	0.21
Dalbergia paniculata Roxb.	0.032	0.007	0.006	0.012	0.06
Ougeinia oojeinensis (Roxb.) Hochr.	0.139	0.030	0.028	0.052	0.25
Shrubs					
Bauhinia racemosa Lam.	0.01	0.01	0.05	0.005	0.07
Bauhinia vahlii (W.) A.	0.02	0.01	0.08	0.008	0.12
Butea superba Roxb. ex Willd.	0.26	0.14	0.90	0.044	1.35
Spatholobus roxburghii Benth.	0.01	0.005	0.03	0.003	0.05

the soil organic carbon (SOC) stock rendering higher fertility of the soil. Carbon content sequence appears to be *Eucalyptus tereticornis* = *Azadirachta indica* = *Acacia nilotica* = *Butea monosperma* > *Albizia procera* = *Dalbergia sissoo* > *Emblica officinalis* = *Anogeissus pendula*. *Albizia procera* were most efficient in CO_2 fixation, and *Anogeissus pendula* were least in these aspects (Rajendra Prasad et al. 2010).

10.4 Leguminous Trees' Role in Carbon Sequestration

Leaves of leguminous trees add a considerable level of SOC due to decomposition by soil microorganisms (Fig. 10.2). Trees have the potentiality to entrap atmospheric carbon which leads to carbon sequestration in forests (Tables 10.3 and 10.4). Legume tree like *Acacia nilotica* can fix substantial level of C which leads to enhancement of SOC pool (SOCP) as well as soil fertility. Different plant parts of aged trees reported the variable level of carbon concentration. A sum of stored carbon in different plant parts varied from 1.36 to 3.08 t/ha (Dhruw et al. 2009; Datta et al. 2017). Maximum amount (70%) of stored carbon were represented in the above ground parts, and lesser amount (30%) were obtained from below ground.

A higher level of carbon sequestration in soil leads to increased biomass which further promotes soil fertility. Increasing level of soil fertility adds more C to the soil. *Leucaena leucocephala* recorded fixation of carbon up to 0.575 t/ha annually and a higher level of C storage in legume-based tropical pastures in comparison to grassland. Some earlier works (Cadisch et al. 1998) reported major contribution of N-fixing species towards gradual building up of SOM in tropical soil.

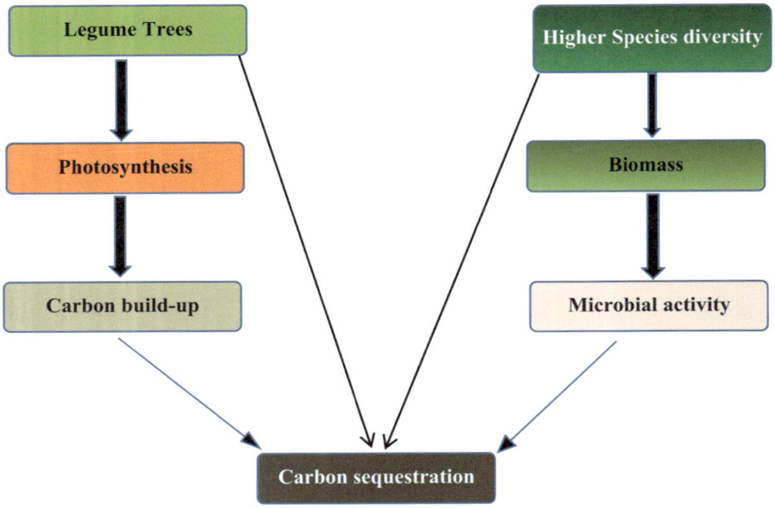

Fig. 10.2 Role of legumes trees in soil carbon sequestration

Table 10.3 Carbon storage potential of some legume species

Tree species	Carbon storage	Author
Dalbergia sissoo	151.84 t/tree	Bilyaminu and Wani (2016)
Acacia nilotica	86.52 t/tree	
Leucaena leucocephala	71.27 t/tree	
Albizia lebbeck	158.20 t/tree	
Bauhinia variegata	19.54 t/tree	
Leucaena leucocephala	13.660 kg/tree	Deka et al. (2016)
Bauhinia variegata	26.020 kg/tree	
Mixed plantation of *D. sissoo* and *L. leucocephala*	93.47 ± 0.67 t/ha	Sheikh et al. (2015)
Dalbergia sissoo	74.54 ± 0.53 t/ha	
L. leucocephala	53.98 ± 1.21 t/ha	

L. leucocephala along with *D. sissoo* promoted higher carbon sequestration potential instead of the separate plantation of each. Such strategies may lead to greater reduction of the atmospheric CO_2 level and thus will be helpful to combat climate change (Sheikh et al. 2015; Kumar et al. 2016). Inclusion of leguminous trees within *Eucalyptus* plantations promoted carbon sequestration in the vegetation stand (Kaye et al. 2000). The presence of NFT within forests accelerates the higher level of carbon sequestration in soils (Resh et al. 2002) and therefore improves the level of SOM and carbon. As per earlier reported works, 1 g N is associated with fixation of 12–15 g C (Binkley and Menyailo 2005). To promote higher absorption of light and CO_2, mobilization of carbon in the upper canopy part of a plant body from roots and the mycorrhizal association is connected with nutrient availability (McConnaughay and Coleman 1999). The amount of carbon and N level were proportional to the carbon content in lithospheric zone of forested area (Macedo et al. 2008).

Table 10.4 Carbon sequestration potential of some legume species

Tree species	Carbon sequestration	Author
Mixed plantation of *D. sissoo* and *L. leucocephala*	34.30 ± 0.24 t/ha/year	Sheikh et al. (2015)
Dalbergia sissoo	27.35 ± 0.19 t/ha/year	
L. leucocephala	19.81 ± 0.44 t/ha/year	

Table 10.5 Various sources of N fixation (Dashora 2011)

N fixation source	N fixed (10^6 tons/year)
Land	155
Legume	40
Nonlegume	10
Others	105
Sea	40
Total biological	195
Lightning	10
Industry	85
Total non-biological	95

10.5 Leguminous Trees' Role in N Sequestration

Several works on different legume species have emphasized the importance of mixed culture system in comparison to monoculture plantation of legumes for BNF. N_2-fixing potential (NFP) refers to the relative capability of species to fix atmospheric N_2 in the absence of limiting factor (Table 10.5). The presence of limiting factor influences N_2 fixation negatively, and, therefore, measurement of actual fixation of N_2 was measured (Galiana et al. 2004). Hereditary characteristics of both the symbionts (host plant and associated bacterial strain) influence NFP of legumes. Species such as *Leucaena leucocephala*, *Calliandra* spp., *Acacia mangium*, *Acacia auriculiformis*, *Acacia crassicarpa*, *Acacia mearnsii*, *Gliricidia sepium*, *Sesbania spp*., *Casuarina equisetifolia* and *Casuarina cunninghamiana* have high NFP (60–100 kg/ha annually); *Prosopis juliflora* and *Acacia saligna* (syn.: *A. cyanophylla*) are reported to have medium NFP, and *Acacia raddiana*, *Acacia senegal*, *Acacia cyclops* and *Faidherbia albida* have low NFP (Ganry and Dommergues 1995).

BNF is an effective mechanism to improve soil N status through the cultivation of leguminous crops under N-deficient soil condition. These species have the inherent capability to fix atmospheric N to be utilized by both plant and soil system (Fig. 10.3). During summer leguminous species like *Cajanus*, *Crotalaria*, *Gliricidia*, *Sesbania* and *Tephrosia* have been used as intercrop which promoted 100–200 kg N/ha biannually in African subcontinent (Rao et al. 2007).

Leguminous plants may fix 15–200 kg N/ha annually (Dakora and Keya 1997; Unkovich and Pate 2000), which renders them to be utilized as an intercrop or as cover crops (Table 10.6). As per earlier reports, it can be formulated that higher ambient concentration of CO_2 will promote addition of more N into the soil through

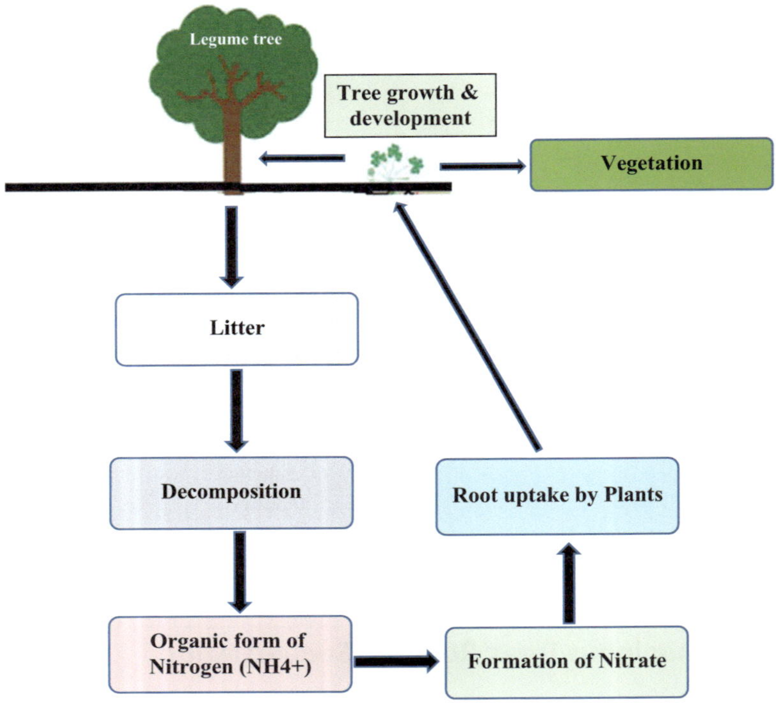

Fig. 10.3 Nitrogen fixation cycle in trees

BNF mechanism. Within agricultural ecosystems, biologically fixed N is mainly contributed towards different vegetative parts, but the fewer amount of N is emitted to the soil (Hardarson et al. 1987). Various scientific reports revealed that climate change has its influence upon BNF species, but lacunae in scientific knowledge still prevail about the increased level of CO_2 upon ecosystem influencing nutrient cycle. Recent development regarding free-air CO_2 enrichment (FACE) approach (Miglietta et al. 2001; Okada et al. 2001) reveals the elevated level of CO_2 without altering other environmental, climatic elements and other living biological organisms. FACE experiments are now exploring the role of various climate change segments altering addition of N in ecological systems through BNF technology. A comparative account was prepared under various treatment systems which includes CO_2 and lower atmospheric O_3 are applied in a specific crop rotation system, and addition of CO_2 and fertilizer was done as treatments of N on pasture land comprising of clover and ryegrass (Miglietta et al. 2001; Okada et al. 2001; Meena et al. 2013a, b). Swiss FACE experiment appears to be the primary source of information about the role of elevated CO_2 on N dynamics in the ecosystem. Proper extrapolation needs to be done for Swiss FACE results in comparison to other agricultural systems due to various factors such as higher density of these plant species are pasture grown, mixed vegetation with clumpy planting and without crop rotation. Zak et al. (2000)

Table 10.6 Biological N capture in different ecological systems

Common name	Scientific name	Amount of N fixed (kg/ha/year)	References
Agri-pastoral component			
Soybean	*Glycine max*	15–140	Unkovich and Pate (2000)
Common bean	*Phaseolus vulgaris*	17–85	
Peanut	*Arachis hypogaea*	30–175	
Chickpea	*Cicer arietinum*	60	
Field pea	*Pisum sativum*	105–200	
Lentil	*Lens culinaris*	80	
Faba bean	*Vicia faba*	90	
Narrow-leaf lupin	*Lupinus angustifolius*	230	
Cowpea	*Vigna unguiculata*	24–200	Dakora and Keya (1997)
Bambara groundnut	*Vigna subterranean*	40–65	
Agriculture with forestry component			
Leucaena	*Leucaena leucocephala*	110–550[a, b]	Danso et al. (1992) and Dakora and Keya (1997)
Australian pine	*Casuarina equisetifolia*	43–60[a, b]	Danso et al. (1992)
Sesbania	*Sesbania rostrata*	505–601[a, b]	
Egyptian riverhemp	*Sesbania sesban*	45–100[a, b]	
Siris	*Albizia lebbeck*	94[a, b]	
Soapbush wattle or strap wattle	*Acacia holosericea*	36–110[a, b]	Dakora and Keya (1997)
Gliricidia	*Gliricidia sepium*	108[a, b]	Danso et al. (1992)
Naturally occurring components			
Hawaii ashflow	*Myrica faya*	20	Vitousek et al. (1987)
SE coastal plain	*Myrica cerifera*	<2–10	Permar and Fisher (1983)
Appalachian oak forest	*Robinia pseudoacacia*	30–75	Boring and Swank (1984a, b)
Sonoran desert	*Prosopis glandulosa*	25–35	Rundel et al. (1982)
Massachusetts peatland	*Myrica gale*	35	Schwintzer (1983)
Pacific northwest	*Ceanothus velutinus*	0–100	Tarrant (1983)
	Alnus rubra	40–160	Luken and Fonda (1983)
Alaskan boreal forest	*Alnus incana*	155–360	Van Cleve et al. (1971)

[a]Above ground only
[b]Using total N-difference methods

reported the influence of CO_2 enrichment has got significant influence over N cycle and mineralization as well as over dry wet of microbes.

Pueraria phaseoloides or *Centrosema pubescens* are frequently used as leguminous cover crops. These species can fix up to 150 kg/ha/year of ambient N (Giller

Table 10.7 N capture by woody leguminous species in various regions

Scientific name (English name)	Region	N fixed (kg/ha/year)	References
Leucaena leucocephala (Leucaena)	Tanzania	110	Hogberg and Kvarnstrom (1982)
	Nigeria	305	Danso et al. (1992)
Sesbania sesban (Egyptian riverhemp)	Senegal	43–100	Ndoye and Dreyfus (1988)
	Kenya	52	Gathumbi et al. (2002)
Gliricidia sepium (Gliricidia)	Nigeria	110	Danso et al. (1992)
	Brazil	110	Apolinário et al. (2016)
Cajanus cajan (pigeon pea)	Kenya	90	Gathumbi et al. (2002)
Calliandra calothyrsus (Calliandra)		25	
Mimosa caesalpiniifolia (Sabia)	Brazil	160	Apolinário et al. (2016)

and Wilson 1991), act as a weed suppressant and stimulate the activity of soil dwelling forms (Agamuthu and Broughton 1985; Sanginga et al. 1992). Plant species used as cover crops often requires labour inputs in terms of initiation and progression to facilitate protection to woody species from aerial legumes. During summer season considerable competitions were observed among various cover crops with the woody crops for physiological requirements (Lehmann et al. 2000).

N has its origin in the ecosystem by means of BNF includes biomass decomposition in terms of litter and root turnover. Prediction of effect on climate change on ecosystems reveals a reduction in litter quality and its decomposition rate promoting higher mobility of nutrients from plant parts to litter in terms of high ratio of C and N (Strain and Bazzaz 1983). N-fixing and non-N-fixing plant species have not supported the findings above with sufficient consistency (Norby et al. 2001). As per Ross et al. (1996), in New Zealand within pasture land, ecosystem has higher CO_2 due to higher decomposition rates promoted by the production of CO_2 in soils, with least effect on N content in soil. Treatments of CO_2 and fertilizer to promote N input in soil revealed nutrient content in dried plant parts reflected unchanged in pasture under a higher level of CO_2 as a result of higher N fixation by plant community (Hartwig et al. 2000).

There is a significant level of variation in N_2 fixation (Table 10.7) under the influence of biotic and abiotic factors. BNF by tree legumes has got higher potentiality for making economic gains in the agricultural sector regarding lesser fertilizer inputs as well as minimizing C footprint of the agroecosystem. Synthetic N fertilizer contributes mostly C to the agroecosystem. Natural gas is the driving energy source for synthetic N fertilizer production. As per earlier report (Lal 2004) C emission for manufacturing, allocation of synthetic nitrogenous fertilizer appears to be approximately as high as up to 2 kg of C per kg N which is proportionate to amount of N fixed. From sustainability perspective, BNF seems to be promising in comparison to synthetic N fertilizers which creates multi-facets of environmental hazards (Crews and Peoples 2004; Yadav et al. 2017b).

Legumes often undergo a close association with soil bacteria and utilize the biologically fixed N for their growth. Peoples et al. (2012) reported that in leguminous system when grass was added, the level of symbiotic association changes from legume bacteria to other understorey vegetation. From productivity perspectives, combination of grass and legume is much more productive than legume alone due to bimodal N uptake in legume-grass association (Nyfeler et al. 2011). N transfer in ecosystem occurs in various ways. Recycling of N to soil takes place through cattle excreta with a minute amount added as animal products (Dubeux Jr et al. 2007). As per Cantarutti et al. (2002) litter deposition aids N transfer in sward, transfer within soil compartment through exudates of roots, various anatomizing network systems which include common fungal networks along with turnover from root system (Sierra et al. 2007; Verma et al. 2015a).

Few reports are available regarding mobilization of N from N fixer to non-N fixer species. It appears to be difficult approach for measuring the indirect nutrient transfer process through litter recycling under field conditions due to the utilization of N for this purpose for N-fixing species. As per Dommergues et al. (1999), in situ mobilization of N by mycorrhizal involvement is yet to be explored. A quantitative estimation of fixed N from a N fixer to a non-N fixer were done through radio isotopic study of natural sources of isotopic N (Van Kessel et al. 1995). 15N radio isotope in *Leucaena leucocephala* and its associated vegetation reflected a declining pattern of fixation of N on a long time basis (1–6 year) due to litter deposition by the leguminous tree species. Recycling of N from leaf litter and roots of trees is mediated by soil microorganisms. Mobilization of nutrients within different plant parts needs to incorporate N. As per some earlier works, recycled N level in N fixer is less in amount in comparison to non-N fixer (Wheeler 1991). Trees usually excavate N in the form of nutrients from the deeper part of the soil and uplifted in the surface soil (Dupuy and Dreyfus 1992).

From Indian perspective, Garg and Jam (1992) reported the increment of the level of soil N in case of *Acacia nilotica* and *Prosopis juliflora* plantation. In Senegal, soil N content was found to be 309 kg ha^{-1} under *Casuarina equisetifolia*. Areas without the establishment of legume trees have much lower soil N content (Dommergues 1963). Different conditions soil and other abiotic factors reflect significant level of variation in productivity along with N balance in plantation schemes. Macedo et al. (2008) reported the increment in the level of N due to BNF is associated with carbon assimilation in the sites. NFP of legumes has promoted them to be cultivated in the various ecosystems such as pastures (Tarre et al. 2001), zero tillage land (Boddey et al. 2010), woody plantations (Balieiro et al. 2008) and agriculture with forestry (Handayanto et al. 1995) which have increased soil N and SOCP. SOM maintains soil fertility and quality under tropical conditions (Six et al. 2002). BNF activity of N-fixing legumes positively promotes carbon stock in degraded areas (Boddey et al. 2009; Meena et al. 2015b). As per Banning et al. (2008), litter of leguminous species is very much beneficial for stimulating humification process along with biogeochemical cycling (Costa et al. 2004).

10.6 Leguminous Trees' Variable Traits for Soil Improvement

Legumes have the phenomenal trait of developing root nodules in the presence of soil *rhizobacteria*. Such type of symbiotic association often may act as limiting factor for their potential to be used as biological N fixers. Root nodule formation takes place between host and its symbiont species based on proper cell signalling mechanism. The mechanism involves induction of transcription by *nif* genes present in *rhizobacteria* through stimulation obtained from the root exudates of legume species. Induction of *nif* genes leads to the formation of lipochitooligo-saccharide molecules which promotes nodule formation in host plants (Long 1996). The event of nodulation involves hundreds of genes present in legume species and its rhizobial counterpart (Vance 2002).

Woody plants capable of fixing atmospheric N are utilized in terms of cover crops for other agricultural plantation crops (Beer et al. 1998). Earlier research reports suggest multifaceted uses of legume species in relation to fruits and firewood (*Inga* spp.); 'service functions', such as BNF; and production of nutrient-rich litter and shade (*Gliricidia sepium* and *Erythrina* spp.). Beer (1988) reported the N-fixing ability of legume shade trees as high as 60 kg N/ha/year in smallholder plantations with deficiency of N fertilization. According to a research report in Latin American country (Brazil), enhanced level of N pool were recorded with respect to certain distance from specific tree species (*Erythrina glauca* and *E. poeppigiana*) which grow under shade condition when compared to adjacent crop species. As per Santana and Cabala-Rosand (1982), lower level of total N in soil under cacao plantation in comparison to *Erythrina* inhabitates habitat. In the coastal sandy region of Cote d'Ivoire, regeneration of coconut plantation was promoted by N_2-fixing trees (Zakra et al. 1996). Shoot pruning is an effective measure to increase N capture process among N fixers along with N mobilization within associated crop species (Beer et al. 1998; Schroth et al. 2000; Dhakal et al. 2015).

N_2 fixation by legume is a subsidy for cropping plants under N scarce condition. Research reports reveal unscientific approach of a higher rate of application of N fertilizer (e.g. up to 270 kg N/ha/year in specific agricultural plantation crop species in Costa Rica) leads towards lower N_2 fixation due to wrong screening of low-economic potential plants growing under shade condition (Beer 1988). According to Giller and Wilson (1991), the higher rate of application of N fertilizer increases nitrate levels in the soil which hinders the BNF activity (Giller and Wilson 1991). Further, plantation of legume under shade conditions is an effective strategy due to their ease of pruning and rapid regrowth as well as higher biomass production that leads to maintaining adequate SOM levels which are good for soil health and fertility (Beer 1988).

N distribution in different plant parts of different crops reflects significant contributions in terms of N assimilation and accumulation in the residual parts of legume crops which may be used by the next successive cropping sequence as well as applicable for nonlegume in intercropping systems. BNF has its two principle sources which include decomposed plant parts of legume species as well as from excreta of domestic animal. The root-mediated deposition process adds more N to the soil through root system (Herridge et al. 2008; Fustec et al. 2010). Therefore, N fixers are effective machinery towards sustainable agricultural practices and promote lesser use of chemical fertilizer.

Legume species associated with mycorrhizal species along with nodulation proved to be most effective to colonize substrate without OM (Franco and Faria 1997). Franco and Faria (1997) reported the effectiveness of rhizobial species with legumes to utilize for land reclamation purposes which include various *Acacia* species, *Albizia* and other species.

Legumes' association with arbuscular mycorrhizae fungi (AMF) reflect their potential to be utilized in the land reclamation process. The fungal increases the root anchorage potential for better absorption of nutrients and water (Siqueira 1996). A significant level of variation was observed in case of soil rhizobial or arbuscular fungi activity in relation to nutrient mobilization in plant species and reflects a combined effect of both. In the case of *Leucaena leucocephala* grown in greenhouse under latossol, the mycorrhizal activity was greater in comparison to rhizobia, reflecting phosphorous as a growth-limiting factor instead of N.

Plant colonization is a natural process which is regulated by the availability of nutrients in soil regarding N, but SOM may impose its impact over the invasion of colonizing plants. In this context, legumes have the advantage of depositing more OM regarding biomass in comparison to other nonlegumes. As per reports, tissues of leguminous vegetation have higher N content in comparison to nonleguminous species which adds more N into soil N pool (SNP). Such available N can be effectively used by non-N fixer. Indigenous and inherent SNP often acts as a major influencing issue, promoting optimum crop proliferation. Therefore, utilization of N-fixing species within the cropping pattern in a cultivation system can be an effective strategy to overcome such limitations.

In northwestern part of India, soil with higher pH value (10.2) and trees with fodder crop such as *Acacia nilotica*, *Dalbergia sissoo* and *Prosopis juliflora* along with herbs and grasses reflected improvement of soil conditions (Kaur et al. 2002). Positive influence of woody vegetation including *Acacia* upon soil conductivity as well as good rhizospheric development for next crop has been reported (Yunusa et al. 2002). Under dry or saline condition, *Acacia* and *Prosopis* have been found to enhance soil nutrient status (Zuzana and Ward 2002; Meena 2013).

10.7 Leguminous Trees Improve Forest Flora

Biodiversity of an area can be promoted through legume plantation on wider dimension. Woody leguminous species reflects better ecological adaptability in terms of species interaction for fulfilling basic needs for survival with respect to other species. Benefits received from the woody leguminous species under various climatic zones are yet to be scientifically explored. As per Franco and Faria (1997), NFT legumes with mycorrhizal association promote phosphorous in organic form in the soil and revegetation of altered pedon with lesser agrochemical addition.

Legume root, shoot and leaf biomass acts as a nutrient source for above- and belowground fauna. Surface litter and incorporated SOM serve as a nutrient source for the decomposer community and soil microorganisms after senescence and recombination (Mattson 1980).

Species interactions at various trophic levels of a food chain influence the distribution of N resources under the presence of legume trees. Other forest flora such as insect-feeding organisms, herbivores and other associated organisms are influenced by N-fixing legumes. The diverse functions include vegetation diversity-productivity gradients, vegetation community invisibility, natural enemy dynamics, soil structure and functioning, ecology and conservation of avifauna, emissions of GHG from agriculture and soil C sequestration (Drinkwater et al. 1998; Bullock et al. 2001; Birkhofer et al. 2011). Legume ecology reflects significant promise to mitigate the problems of biodiversity crisis and change in global climate.

Legumes can promote diversity in some plant community (Tilman et al. 1997; Fargione et al. 2007). Legumes have a positive impact on the growth of nonleguminous plant species, reduce the competition within vegetation communities, increase ecological invasion and maintain vegetational diversity (Smith and Gross 2007). Newer methods are being developed to assess the functional role of ecosystems (Mace et al. 2012; Bommarco et al. 2013; Meena et al. 2013).

Legume often competes for resources and food with non-crop plant species. Legume canopy with broad leaves reduces the infiltration of sunlight to understorey stratum which promotes ground vegetation with higher leaf area index (Bilalis et al. 2010). N can be effectively mobilized by legume crops through reducing N uptake from soil supplemented by their own biologically fixed N as well as direct supply to other neighbouring plants. Resource partitioning in such way leads to higher diversity over a geographical area within a specified time (Tilman et al. 1997). Legumes have wide dimensions of advantages regarding having a positive influence over vegetation communities through the betterment of soil structure along with high seed bank (Albrecht 2003). Legumes have a significant impact on weed community and bring significant level of changes.

Leguminous tree species has significant influence over pollination of flowering plants which proliferates the diversity of forest flora (Ghazoul 2006). In legume-supported system, it has been observed that legumes have significant influence over the associated vegetation which is dependent upon the interaction between them.

Legumes may promote herbaceous vegetation of the forest through proper transportation of water from deeper soil layers. Such type of relationship may not be visible for legumes and its associated species which might be attributed towards moisture regimes of the soil (Gea-Izquierdo et al. 2009). Among the various plant families, Fabaceae is the dominant representative in the forest of Neotropics and Africa (including Madagascar). As per Ter Steege et al. (2006), prevalence was reflected by legumes at Amazon rain forest, neotropical dry forests (Pennington et al. 2006; Sarkinen et al. 2011) as well as savannas (Ratter et al. 2006). Higher density of legumes is an important factor towards carbon and N sequestration in forest ecosystem (Knops and Tilman 2000). Sequestering higher N in the soil of the forest legume promotes canopy structure and differentiation which leads to higher biomass accumulation (Spehn et al. 2002). Further N fixer such as legumes (Sprent 2009) is the common habitant on several continents (Lewis et al. 2009) and oceanic islands (Caetano et al. 2012), including savanna and grassland ecosystems (Chaneton et al. 2004). Slow-growing trees such as *Dalbergia* can sequester carbon storage up to significant level due to low decomposition rate (Weedon et al. 2009; Varma et al. 2017b).

10.8 Leguminous Trees' Key Players to Mitigate Climate Change

Higher ambient CO_2 level along with other factors contributing climate change significantly influences the N-fixing process, its associated species and amount of fixed N along with addition into soil. Various anthropogenic activities and changes in the land-use pattern have raised the CO_2 level from the time period before industrialization to recent time frame, which may further double in the upcoming century. As per current updates, human-mediated emission of CO_2 reached 32 (±2.7) $GtCO_2$/year in 2010 and progressed further by about 3% between 2010 and 2011 and by about 1–2% between 2011 and 2012 (IPCC 2014). CO_2 happens to be a main driving force for global warming phenomenon in comparison to other GHGs. CO_2 may promote a considerable level of increase in global-mean temperature (IPCC 2014), which may influence the regional rainfall pattern throughout the world (Rind et al. 1990). Three of the important factors such as CO_2, temperature and water may cause alterations in plant growth and development on a spatial basis. Carbon sequestration by plants can affect the factors influencing the climate change. From future perspectives, exact quantification of CO_2 inputs in various ecosystems along with the release from the biosphere needs to be explored.

Tropospheric ozone imposes negative influences over legume species (Morgan et al. 2003). A higher level of N in plant tissues and the soil often inhibits N capture (Hunt and Layzell 1993). NFP is not very much compatible with soil N uptake in comparison to non-fixing plants. Several natural and human-made issues influence the output of symbiotic association of N fixer in context of climate change. Root nodule formation and enzymatic activity related to BNF reflect significant level of variation under environmental factors. Supply of required resources for N fixation is regulated by carrying capacity of environment. As per source control hypothesis, availability of light influences photosynthetic rate which in turn regulates the level of carbohydrate required for N_2 fixation in nodules, thereby hindering nodulation and nitrogenase activity.

A higher level of CO_2 promotes higher nutrient uptake from soil (Berntson 1994). Earlier works reveal that elevated CO_2 level may change the structural configuration of the root (Berntson and Bazzaz 1996) and nutrient uptake ability of fine roots (Jackson and Reynolds 1996). Symbiotic associations of root with symbiont species positively influence nutrient uptake through CO_2 enrichment (Thomas et al. 2000). The proliferation of N fixers in a higher level of CO_2 reveals unencing availability of N through BNF (Lee et al. 2003). Under the condition of limited availability of N in the soil, the N fixers are much more promising in comparison to non-fixing plants (Poorter and Navas 2003; Meena et al. 2016).

The beneficial effects of CO_2 enrichment in the ambient environment contribute towards a higher level of N content in the whole plant body. Thomas et al. (2000) reported the relationship between higher CO_2 level and enriched N content in plant body might be attributed towards increased carbohydrate content in the nodule as well as in the roots (Cabrerizo et al. 2001). By using radioactive isotopes of carbon, Tissue et al. (1997) reported a higher rate of photosynthesis with higher carbon assimilation and transport within the plant body in commonly used agroforestry tree species such as *Gliricidia sepium*. Elevated CO_2 also promotes higher microbial growth in soil rhizosphere zone of N_2-fixing plants (Marilley et al. 1999) and found to promote the soil rhizosphere microbial population in aquatic condition (Dakora and Keya 1997). Higher level of CO_2 promotes lesser availability of soluble N compounds, which may otherwise inhibit N capture process (Serraj and Sinclair 2003).

Dixon and Wheeler (1983) reported that species with BNF activity includes N-fixing legumes along N_2-fixing microbes display a wide variability in activity under variable environmental conditions. Research reports revealed that in the case of *Alnus incana* and other N_2-fixing species show their growth in Arctic range and on the other hand *Casuarina* spp. occur above 30 °C or more and found in developed regions. Mulder et al. (1977) reported that wider thermal regime prevails from a development perspective in comparison to optimal N_2 capture. Present scenario such as climate change would have the least impact on tropical legumes as reflected through prevailing rates of BNF.

Enzyme activity for N fixation reflects wide thermal spectrum with an optimum temperature range between 20 and 30 °C. As per Waughman (1977) within this wide temperature spectrum, temperate legumes prefer the lowest range, and species of tropics prefer highest range. Furthermore, Ryle et al. (1989) reported that *Trifolium repens* reflected elevated Nase activity linearly within 25 °C. Further increase in temperature leads to have no effect on enzyme activity. As per earlier reports, elevated Nase activity within temperature range above 18 °C and below 28 °C was reported for specific species (Crush 1993).

Fungal species inhabiting rhizosphere have shown variability in Nase activity due to variation in temperature. Hensley and Carpenter (1979) reported various temperate zone species reflect optimum temperature range between 20 and 25 °C. Species of subtropics such as *Casuarina* species, optimum temperature was reported to be above 30 °C or more (Bond and Mackintosh 1975). Enzyme activity for N fixation reflects variability under various temperature ranges along with species (Waughman 1977) and temperature conditions used for growing plants (Gibson 1976).

Research reports reveal that increment of the CO_2 level associated with water stress condition helps to combat such abiotic form of stress through increasing photosynthetic activity as well as efficient water use during photosynthetic. Very few reports are there regarding the interrelationship between elevated CO_2, water stress and N fixation. The increment in the level of CO_2 has partially mitigated the problem of drought-mediated reduction in N fixation through stimulation of mass nodule along with nodular activity (Serraj et al. 1999). The proposed mechanism behind such type of process includes biosynthesis of some nonstructural carbohydrates promoting a reduction in the level of soluble N compounds thereby reducing N fixation process (Serraj and Sinclair 2003; Meena et al. 2015c).

Ecosystem services refer to the processes which are beneficial to mankind. These includes services from various angle dimensions such as resource-based functions such as fulfilling basic needs, ecological and environmental services, social and spiritual functions which include multi-facets of benefits to mankind and other sustainable ecological functioning (Millennium Ecosystem Assessment 2005). N fixer facilitates such services on a sustainable basis from environmental perspectives. Legumes perform mitigating changes in global climate which includes a reduction of GHG emissions in comparison to N fertilizer-based cropping system, reduction in fossil fuel energy, sequestering carbon in soil and providing biomass as biofuel (Jensen et al. 2012). Several research reports reveal cycling of carbon is associated with N cycle, C to N ratios in various pedons. Therefore, woody BNF can promote sequestering of C in the soil through BNF (Nair et al. 2010; Kirkby et al. 2011).

10.9 *Acacia nilotica*: A New Promise of Legume-Based Agroforestry

Babool (*Acacia nilotica*) is a BNF species having its wider distribution throughout Asia, Africa as well as in southern America, Australia and Mexico. *A. nilotica* comprises of nine varieties/subspecies, of which majority (six) of species are native to tropical Africa and remaining (three) species are native to the Indian subcontinent.

Genus *Acacia* is represented by numerous species in the tropical part of the world. One-fourth of the area of the world under subtropical and semiarid condition was found to be represented dominantly by genera *Acacia*. Mixed interpretations have been suggested by several workers towards the management of arid ecosystems. On one hand, woody legumes should be eliminated to promote herbaceous growth for livestock feeding (Fisher 1977) and on the other hand higher possibility of output regarding arid forestry (Felker 1998). A comparative assessment has been shown by several workers between N fixer and non-N fixer tree leading to the higher availability of N in top soil from deeper layers (Barth and Klemmedson 1982). A significant level of benefits can be harvested from *Acacia* in terms of environmental, social and economic benefits associated with C and N sequestration under tropical condition, and therefore approaches need to be designed for sustainable management of these species. Besides improving soil condition, *Acacia* also provides diverse nature of NTFPs like gum, fodder and small timber. Although the fact that trees promote higher N fixation and thus help to maintain the productivity under tropical system is widely known, very few research approaches have been oriented towards this direction.

Agroforestry is a time-oriented concept of sustainable agriculture which promotes production in various spheres aiming towards optimum utilization of resources in a sustainable manner. Agroforestry appears to be a holistic approach involving multistrata plantation along prevailing cultivation system and livestock promoting agricultural productivity. Moreover, agroforestry can enhance the population of beneficial soil microorganism by increasing the SOM through litter production (Raj et al. 2016; Singh and Jhariya 2016; Varma and Meena 2016). Chhattisgarh plain within the central part of India is characterized by traditional agroforestry systems which are a widely practiced land-use system. The farming system needs to be integrated with the plantation scheme in judicious manner to boost up the agricultural productivity. *A. nilotica* represents mostly in agroforestry systems of Chhattisgarh plains due to its hardy nature and adjusting with diverse environmental condition giving agro-products of diverse nature. Therefore, *A. nilotica* with rice cultivation system is prevalent in Chhattisgarh (Jhariya et al. 2015).

Soil productivity in the modern perspective is a major challenge due to nutrient loss, soil erosion as well as non-judicious approach in the field of agricultural productivity. N-deficient soil condition can be effectively managed using plantation or integration of leguminous crop plants through BNF activity. Leguminous species have their inherent capability to fix atmospheric N to be used in soil and plant systems. In such conditions *A. nilotica* shows significant promise due to higher growth rate, stress tolerant along BNF ability. It has higher growth rate in degraded lands and important components of agroforestry practices for its multipurpose uses in India along with BNF activity. *A. nilotica* has a positive impact over paddy cultivation regarding addition of N and OM in soil under paddy cultivation.

A. nilotica reflects the positive influence on soil physical attributes by reducing the splashing action of raindrops over top soil particles. Roots of *A. nilotica* are very much colonized by *Rhizobium* species. Higher number of root nodule formation by *Rhizobium* species significantly promotes a higher rate of N fixation to nutrient-deficient soils. *Rhizobium* bacterium plays the secondary role of nutrient solubilization in association with vesicular-arbuscular mycorrhizae. Such relationship helps higher adherence and accessibility of roots towards soil nutrients such as phosphorous and also mobilizes unavailable form of nutrient into available form.

This legume tree is a boon for the resource-poor farming community in the context of Chhattisgarh who have age old practice of cultivating different tree species along with agricultural crops. *A. nilotica* contributes gum (Das et al. 2014; Raj 2015a) which has its diverse use for the pharmaceutical purpose, calico-printing, sizing paper, cloth and textiles, encapsulation, etc., and tapping of gum would promote the socio-economic upliftment of the local community stakeholders as well as implementation of conservative measures towards environment and bio-resources (Raj 2015b; Meena et al. 2015d). Moreover, the species has a higher potential for fuelwood production in tropical climatic conditions.

It has been reported that the species have higher carbon-sequestering potential in comparison to other nonlegume plant species. Some earlier works reported that *A. nilotica* is capable of fixing 228.42 kg/tree biomass carbon which may be later on added to the soil to improve the SOCP leading to improvement in soil fertility. *A. nilotica* will help in stabilizing ravines and gullies and checking their spread.

10.10 Threats to Legumes

Due to multi-facets of utilization potentiality, abuse of legume species is often over-exploited. In India, different types of human-made perturbances have created a significant loss in the legume species. Legumes can be considered as a reservoir of various valuable resources having diverse distribution pattern in different regions. Mostly they are used by the local community stakeholders for maintaining their daily livelihood as well as to economic gains. Improper way of collection due to lack of knowledge, technical expertise also leads to the destruction of the legume species without appropriate regeneration which renders the species under severe threat of extinction.

Wild relatives of legumes cultivated in tribal area are being replaced by high-yielding legume varieties which may lead to degeneration of the genetic base of the wild relatives. Habitat degradation of wild relatives of legume species leads to the destruction of natural stock of legume species under high population pressure accompanied by climate change (Rao and Husain 1993). The problem of loss of genetic resource base of wild relatives of legume is further aggravated with limited distribution along with the rapid rate of urbanization and unprecedented growth of grazing activity. As per Lane and Jarvis (2007), such activities have generated their possibilities of extinction within the next upcoming future.

10.11 Legumes and Their Conservation Perspectives

Equitable and sustainable use of biodiversity is a new challenge under modern era (Reid 1992). The West Himalayan region is enriched with endemic plants which include endemic legumes which required conservation priority (Rao and Husain 1993).

Conservation includes appropriate strategies which include identifying species under potential threat of extinction and their subsequent conservation along with their habitat. Community-based natural resources management practices along with the optimum use of legumes need to be prioritized. Areas with the higher abundance of leguminous species need to be explored, and proper care should be taken for their protection and conservation. Traditional knowledge-based plant-resource management should be emphasized by the scientific community. Local community stakeholders should be promoted for plant collection in various regions for efficient management of plant resources. The barren land area is commonly associated with low fertility. Leguminous species can be effectively planted to build up the SNP through BNF for their growth and development (Bargali 2011; Verma et al. 2015b). Utilization of BNF species for conservation of soil nutrient status as a part of INM (integrated nutrient management) as well as eco-restoration of degraded land is an effective strategy towards maintaining the sustainability of the ecosystem.

Under semiarid and tropical ecosystems production of cattle, fuel wood and small timber, wood often hampers N balance in the ecosystem which could be ameliorated through a plantation of tree legume species. Suitable approaches need to be focussed to assess the carbon sequestration potential of legume species. Land degradation is a major environmental problem in present days which hampers the soil fertility to a great extent. Eco-restoration is a serious challenge under the arena of land degradation. Rejuvenation of soil fertility through legume plantation is a major step as they can supply extra N through BNF process apart from SNP.

10.12 Future Prospectus of Leguminous Trees

Despite the advantages reflected by BNF technology, it has got some inherent problems such as transfer of technology from lab to field conditions. In developing nation farming community is very much reluctant to cope with such type of technology considering the economical output of such technologies. Besides the multifaceted sustainable role of legume tree species, they have always been treated as minor crops. In this context the quantification of N inflow and outflow from the system is an essential prerequisite to harmonize BNF potential of legume trees. Several advanced methods are available to quantify these dimensions of measurement of N input. To measure the N input in any ecosystem, various advanced methods are available nowadays such as isotopic methods (isotopic dilution, natural abundance in '15N, A-value method), nodulation observation, ureides and amides in xylem sap and acetylene reduction method (Danso et al. 1992). The combination of these various methods can determine the N input into the ecosystem much more accurately. Isotope study appears to be the most appropriate methods for quantification of BNF activity. Characteristics of tree species as well as other sources of variation can significantly influence the evaluation of BNF activity (Boddey et al. 2000; Meena et al. 2017). The amount of N fixed significantly varies upon variable climatic and soil conditions.

Maintaining N balance in a natural ecosystem is an essential prerequisite (Ganry et al. 2001). N balance can be maintained through optimum N fertilizer dose application as well as BNF activity. In the present perspective, very few works have been reported regarding utilization of agroforestry system to maintain soil N balance. However, biotechnology shows significant promise to improve the potential of BNF species fixed atmospheric N on long-term basis. In this context use of BNF inoculation in the nursery, application of clones of N fixer having higher NFP is a stimulatory approach under lab condition.

Acceptability of BNF technology in the farming community is the major challenge for Third World countries. Low economy, political influences as well as the lack of adequate scientific knowledge often hinder this process. Unknown facts such as identification of host legume species, screening of efficient rhizobial strains as well as their interactions often influence the applicability of the process. BNF appears to be a key element of INM system which produces products on a sustainable basis. Huge potential of N fixation may increase the importance of legumes on biosphere perspective. To combat soil pollution as well as depletion of land resource plantation of legumes is a suitable ecofriendly approach. Such process due to its lower cost has got wider acceptability throughout the world which sustains the food productivity and security. Under tropical conditions, maintenance of food production level and quality is a big challenge due to ever-increasing population growth. Good quality foods with optimum nutrient supplements are also a bigger challenge for poor people of the Third World nation. Subsequently, the impact of climate change would further aggravate the problem in areas such as South Asia and sub-Saharan Africa, where optimum conditions for cultivation are not available. Under these circumstances BNF technology and legume trees seem to be a suitable tool in the hand of mankind to combat various dimensions of environmental perturbances.

10.13 Conclusion

Plantation of leguminous tree species under agroforestry systems would promote optimum fixation of nitrogen into the soil environment. Leguminous plants also have the potentiality to fix carbon at higher rate which contribute towards gradual build-up of SOC pool through biomass accumulation. Legumes have been reported to undergo close association with vesicular-arbuscular mycorrhizae and other soil bacteria which promote improvement in the fertility status of the soil. Therefore, legumes can be effective in wasteland reclamation process. Legumes have also been reported to promote forest floor biodiversity through nutrient build-up mechanism. Higher photosynthetic rate of leguminous species under elevated CO_2 level helps to combat climate change. Due to its multifaceted use, there is a progressive threat of abuse and overexploitation of legumes. From this perspective effective conservative measures in terms of exploring areas with high density of legume along with community-based conservation system need to be properly implemented. Overall the BNF potential of legume species needs to be fully explored for better future into a sustainable world.

References

Agamuthu P, Broughton WJ (1985) Nutrient cycling within the developing oil palm-legume ecosystem. Agric Ecosyst Environ 13:111–123

Albrecht H (2003) Suitability of arable weeds as indicator organisms to evaluate species conservation effects of management in agricultural ecosystems. Agric Ecosyst Environ 98:201–211

Allen ON, Allen EK (1981) The Leguminosae, a source book of characteristics, use and nodulation. The University of Wisconsin Press, Madison

Apolinário VXO, Dubeux JCB, Mello ACL Jr, Vendramini JMB, Lira MA, Santos MVF, Muir JP (2016) Decomposition of arboreal legume fractions in a silvipastoral system. Crop Sci 56:1356–1363

Balieiro FC, Pereira MG, Alves BJR, Resende AS, Franco AA (2008) Soil carbon and nitrogen in pasture soil reforested with eucalyptus and guachapele. Rev Bras Ci Solo 32:1253–1260

Banning NC, Grant CD, Jones DL, Murphy DV (2008) Recovery of soil organic matter, organic matter turnover and nitrogen cycling a post-mining rehabilitation chronosequence. Soil Biol Biochem 40:2021–2031

Bargali K (2011) Screening of leguminous plants for VAM association and their role in restoration of degraded lands. J Am Sci 7(1):7–11

Bargali K (2016) Leguminous plants of Kumaun Himalaya: diversity, distribution, threats and management. In: Chand R, Leimgruber W (eds) Globalization and marginalization in mountain regions, perspectives on geographical marginality 1, pp 199–204. https://doi.org/10.1007/978-3-319-32649-8-15

Bargali K, Bargali SS (2009) *Acacia nilotica*: a multipurpose leguminous plant. Nat Sci 7(4):11–19

Barth RC, Klemmedson JO (1982) Amount and distribution of dry matter, nitrogen and organic carbon in soil-plant systems of mesquite and Palo Verde. J Range Manag 35:412–418

Beer J (1988) Litter production and nutrient cycling in coffee (*Coffea arabica*) or cacao (*Theobroma cacao*) plantations with shade trees. Agrofor Syst 7:103–114

Beer J, Muschler R, Kass D, Somarriba E (1998) Shade management in coffee and cacao plantations. Agrofor Syst 38:139–164

Berntson GM (1994) Modeling root architecture-are there tradeoffs between efficiency and potential of resource acquisition. New Phytol 127:483–493

Berntson GM, Bazzaz FA (1996) The allometry of root production and loss in seedlings of *Acer rubrum* (Aceraceae) and *Betula papyrifera* (Betulaceae): implications for root dynamics in elevated CO2. Am J Bot 83:608–616

Bhagat PK, Dash D, Raj A, Jhariya MK (2014) Effect of rhizobium inoculation on growth and biomass accumulation in *Leucaena leucocephala*. Ecoscan (Spec Issues Published Elsevier) V:65–74

Bilalis D, Papastylianou P, Konstantas A, Patsiali S, Karkanis A, Efthimiadou A (2010) Weed-suppressive effects of maize-legume intercropping inorganic farming. Int J Pest Manage 56:173–181

Bilyaminu H, Wani AM (2016) Carbon sequestration potential of different tree species in Allahabad, Uttar Pradesh. Int J Farm Sci 6(2):153–158

Binkley D, Menyailo O (eds) (2005) Tree species effects on soils: implication for global change. NATO Sci IV: Earth Environ Sci 55:155–164

Birkhofer K, Diekoetter T, Boch S, Fischer M, Mueller J, Socher S, Wolters V (2011) Soil fauna feeding activity in temperate grassland soils increases with legume and grass species richness. Soil Biol Biochem 43:2200–2207

Boddey RM, Peoples MB, Palmer B, Dart PJ (2000) Use of the 15N natural abundance method to quantify biological nitrogen fixation in woody perennials. Nutr Cycl Agroecosyst 57:235–270

Boddey RM, Alves BJR, Soares LHDB, Jantalia C, Urquiaga S (2009) Biological nitrogen fixation and the mitigation of greenhouse gas emissions. In: Merich DW, Krishnan HB (eds) Nitrogen fixation in crop production. ASA-CSSA-SSSA, Madison, pp 387–413

Boddey RM, Jantalia CP, Conceicao PC, Zanatta JA, Bayer C, Mielnizuk J, Dieckow J, Santos HP, Denardin JE, Aita C, Giacomini SJ, Alves BJR, Urquiaga S (2010) Carbon accumulation at depth in Ferralsols under zero-till subtropical agriculture in southern Brazil. Glob Chang Biol 16:784–795

Bommarco R, Kleijn D, Potts SG (2013) Ecological intensification: harnessing ecosystem services for food security. Trends Ecol Evol 28:230–238

Bond G, Mackintosh AH (1975) Diurnal changes in nitrogen-fixation in root nodules of *Casuarina*. Proc R Soc Lond B-Biol Sci 192:1–12

Boring LR, Swank WT (1984a) Symbiotic nitrogen-fixation in regenerating black locust (*Robinia pseudoacacia* L.) stands. For Sci 30:528–537

Boring LR, Swank WT (1984b) The role of black locust (*Robinia pseudoacacia*) in forest succession. J Ecol 72:749–766

Bullock JM, Pywell RF, Burke MJ, Walker KJ (2001) Restoration of biodiversity enhances agricultural production. Ecol Lett 4:185–189

Cabrerizo PM, Gonzalez EM, Aparicio-Tejo PM, Arrese-Igor C (2001) Continuous CO2 enrichment leads to increased nodule biomass, carbon availability to nodules and activity of carbon-metabolising enzymes but does not enhance specific nitrogen fixation in pea. Physiol Plant 113:33–40

Cadisch G, Oliveira OC, de Cantarutti R, Carvalho E, Urquiaga S (1998) The role of legume quality in soil carbon dynamics in savannah ecosystems. In: Bergstrom I, Kirchmann H (eds) Carbon and nutrient dynamics in natural and agricultural tropical ecosystems. CAB International, Wallingford

Caetano S, Currat M, Pennington RT, Prado DE, Excoffier L, Naciri Y (2012) Recent colonization of the Galapagos by the tree *Geoffroea spinosa* Jacq. (Leguminosae). Mol Ecol 21:2743–2760

Cantarutti RB, Tarre R, Macedo R, Cadisch G, de Rezende CP, Pereira JM, Braga JM, Gomide JA, Ferreira E, BJR A, Urquiaga S, Boddey RM (2002) The effect of grazing intensity and the presence of a forage legume on nitrogen dynamics in *Brachiaria* pastures in the Atlantic forest region of the south of Bahia, Brazil. Nutr Cycl Agroecosyst 64:257–271

Chaneton EJ, Mazia CN, Machera M, Uchitel A, Ghersa CM (2004) Establishment of Honey Locust (*Gleditsia triacanthos*) in burned Pampean grasslands. Weed Technol 18:1325–1329

Costa GS, Franco AA, Damasceno RN, de Faria SM (2004) Aporte de nutrientes pela serapilheira em uma área degradada e revegetada com leguminosas arbóreas. Rev Bras Ci Solo 28:919–927

Crews TE, Peoples MB (2004) Legume versus fertilizer sources of nitrogen: ecological tradeoffs and human needs. Agric Ecosyst Environ 102:279–297

Crush JR (1993) Hydrogen evolution from root-nodules of *Trifolium repens* and *Medicago sativa* plants grown under elevated atmospheric CO2. N Z J Agric Res 36:177–183

Dakora FD, Keya SO (1997) Contribution of legume nitrogen fixation to sustainable agriculture in Sub-Saharan Africa. J Soil Biol Biochem 29:809–817

Danso SKA, Bowen GD, Sanginga N (1992) Biological nitrogen-fixation in trees in agroecosystems. Plant Soil 141:177–196

Das I, Katiyar P, Raj A (2014) Effects of temperature and relative humidity on ethephon induced gum exudation in *Acacia nilotica*. Asian J Multidiscip Stud 2(10):114–116

Dashora K (2011) Nitrogen yielding plants: the pioneers of agriculture with a multipurpose. Am-Eurasian J Agron 4(2):34–37

Datta R, Baraniya D, Wang YF, Kelkar A, Moulick A, Meena RS, Yadav GS, Ceccherini MT, Formanek P (2017) Multi-function role as nutrient and scavenger of free radical in soil. Sustain MDPI 9:402. https://doi.org/10.3390/su9081402

Deka M, Wani AM, Hussain M (2016) Assessment of carbon sequestration of different trees species grown under agroforestry system. J Adv Environ Sci 1(4):149–153

Dhakal Y, Meena RS, De N, Verma SK, Singh A (2015) Growth, yield and nutrient content of mungbean (*Vigna radiata* L.) in response to INM in eastern Uttar Pradesh, India. Bangladesh J Bot 44(3):479–482

Dhakal Y, Meena RS, Kumar S (2016) Effect of INM on nodulation, yield, quality and available nutrient status in soil after harvest of green gram. Leg Res 39(4):590–594

Dhruw SK, Singh L, Singh AK (2009) Storage and sequestration of carbon by leguminous and non-leguminous trees on red lateritic soil of Chhattisgarh. Indian For 135(4):531–538

Dixon ROD, Wheeler CT (1983) Biochemical, physiological and environmental aspects of symbiotic nitrogenN fixation. In: Gordon JC, Wheeler CT (eds) Biological nitrogen fixation in forest ecosystems: foundations and applications. Nijhoff/Junk Publishers, The Hague, pp 107–171

Dommergues YR (1963) Evaluation du taux de fixation de l'azote dans un sol dunaire reboisé en filao (Casuarina equisetifolia). Agrochimica 7:335–340

Dommergues YR, Duhoux E, Diem HG (1999) In: Ganry F (ed) Les arbres fixateurs d'azote. Caractéristiques fondamentales et role dans l'aménagement des méditerranéens et tropicaux. CIRAD/Editions Espaces 34/FAO/IRD, Montpellier, p 499

Drinkwater LE, Wagoner P, Sarrantonlo M (1998) Legume-based cropping systems have reduced carbon and nitrogen losses. Nature 396(6708):262–265

Drury CF, Stone JA, Findlay WI (1991) Microbial biomass and soil structure associated with corn, grasses and legumes. Soil Sci Soc Am J 55:805–811

Dubeux JCB Jr, Sollenberger LE, Mathews BW, Scholberg JM, Santos HQ (2007) Nutrient cycling in warm-climate grasslands. Crop Sci 47:915–928

Dumanski J, Desjardins RL, Tarnocai C, Moreal C, Gregorich EG, Kirkwood V, Campbell CA (1998) Possibilities for future carbon sequestration in Canadian agriculture in relation to land use changes. J Clim Res 40(1):81–103

Dupuy N, Dreyfus BL (1992) Bradyrhizobium populations occur in deep soil under the leguminous tree *Acacia albida*. Appl Environ Microbiol 58:2415–2419

Fargione J, Tilman D, Dybzinski R, Lambers JHR, Clark C, Harpole WS, Knops JMH, Reich PB, Loreau M (2007) From selection to complementarity: shifts in the causes of biodiversity–productivity relationships in a long-term biodiversity experiment. Proc R Soc B Biol Sci 274:871–876

Felker P (1998) The value of mesquite for the rural southwest: fine lumber and soil improvement. J For 96:16–21

Fisher CE (1977) Mesquite and modern man in southwestern North America. In: Simpson BB (ed) Mesquite: its biology in two desert ecosystems. US/IBP Synthesis Series 4. Dowdon, Hutchison and Ross, Stroudsburg, p 250

Fox RH, Myers RJK, e Vallis I (1990) The nitrogen mineralization rate of legumes residues in soil as influenced by polyphenol, lignin and nitrogen contents. Plant Soil 129:251–259

Franco AA, Faria SM (1997) The contribution of N2-fixing tree legumes to land reclamation and sustainability in the tropics. Soil Biol Biochem 29(5/6):897–903

Fustec J, Lesuffleur F, Mahieu S, Cliquet JB (2010) Nitrogen rhizodeposition of legumes. A review. Agron Sustain Dev 30:57–66

Galiana A, Bouillet JP, Ganry F (2004) The importance of biological nitrogen fixation by trees in agroforestry. In: Serraj R (ed) Symbiotic nitrogen fixation: prospects for enhanced application in tropical agriculture. Oxford & IBH Publishing Co. Pvt. Ltd., New Delhi, pp 185–199

Ganry F, Dommergues YR (1995) Arbres fixateurs d'azote: champ ouvert pour la recherche. Agric Dev 7:38–55

Ganry F, Feller C, Harmand JM, Guibert H (2001) Management of soil organic matter in semiarid Africa for annual cropping systems. Nutr Cycl Agroecosyst 61:105–118

Garg VK, Jam RK (1992) Influence of fuelwood trees on sodic soils. Can J For Res 22:729–735

Gathumbi SM, Cadisch G, Giller KE (2002) 15N natural abundance as a tool for assessing N2-fixation of herbaceous, shrub and tree legumes in improved fallows. Soil Biol Biochem 34:1059–1071

Gea-Izquierdo G, Montero G, Canellas I (2009) Changes in limiting resources determine spatio-temporal variability in tree-grass interactions. Agrofor Syst 76:375–387

Ghazoul J (2006) Floral diversity and the facilitation of pollination. J Ecol 94:295–304

Gibson AH (1976) Recovery and compensation by nodulated legumes to environmental stress. In: Nutman PS (ed) Symbiotic nitrogen fixation in plants. Cambridge University Press, London

Giller KE, Wilson KJ (1991) Nitrogen fixation in tropical cropping systems. CAB International, Wallingford, p 313

Handayanto E, Cadisch G, Giller KE (1995) Manipulation of quality and mineralization of tropical legume tree prunings by varying nitrogen supply. Plant Soil 176:149–160

Hardarson G, Danso SKA, Zapata F (1987) Biological nitrogen fixation in field crops. In: Christie BR (ed) Handbook of plant science in agriculture. CRC Press, Boca Raton, pp 165–192

Hartwig UA, Luscher A, Daepp M, Blum H, Soussana JF, Nosberger J (2000) Due to symbiotic N2 fixation, five years of elevated atmospheric pCO2 had no effect on the N concentration of plant litter in fertile, mixed grassland. Plant Soil 224:43–50

Hensley DL, Carpenter PL (1979) Effect of temperature on N2 fixation (C2H2 reduction) by nodules of legume and actinomycete-nodulated woody species. Bot Gaz 140:S58–S64

Herridge DF, Peoples MB, Boddey RM (2008) Global inputs of biological nitrogen fixation in agricultural systems. Plant Soil 311:1–18

Hogberg P, Kvarnstrom M (1982) Nitrogen fixation by the woody legume Leucaena leucocephala in Tanzania. Plant Soil 66:21–28

Holtham DAL, Matthews GP, Scholefield D (2007) Measurement and simulation of void structure and hydraulic changes caused by root-induced soil structuring under white clover compared to ryegrass. Geoderma 142:142–151

Hunt S, Layzell DB (1993) Gas-exchange of legume nodules and the regulation of nitrogenase activity. Annu Rev Plant Physiol Plant Mol Biol 44:483–511

Husain T, Kapoor SL (1990) Enumeration of legumes in India (indigenous and introduced). National Botanical Research Institute, Lucknow

IPCC (2014) Climate change 2014, mitigation of climate change. Cambridge University Press, Cambridge

Jackson RB, Reynolds HL (1996) Nitrate and ammonium uptake for single- and mixed species communities grown at elevated CO2. Oecologia 105:74–80

Jensen ES, Peoples MB, Boddey RM, Gresshoff PM, Hauggaard-Nielsen H, Alves BJR, Morrison MJ (2012) Legumes for mitigation of climate change and the provision of feedstock for biofuels and biorefineries. Rev Agron Sustain Dev 32:329–364

Jhariya MK, Yadav DK (2017) Invasive alien species: challenges, threats and management. In: Rawat SK, Narain S (eds) Agriculture technology for sustaining rural growth. Biotech Books, New Delhi, pp 263–285. ISBN:978-81-7622-381-2

Jhariya MK, Bargali SS, Swamy SL, Kittur B, Bargali K, Pawar GV (2014) Impact of forest fire on biomass and carbon storage pattern of tropical deciduous forests in Bhoramdeo Wildlife Sanctuary, Chhattisgarh. Int J Ecol Environ Sci 40(1):57–74

Jhariya MK, Bargali SS, Raj A (2015) Possibilities and perspectives of agroforestry in Chhattisgarh. In: Zlatic M (ed) Precious forests-precious earth. InTech, Croatia, pp 237–257. https://doi.org/10.5772/60841. ISBN:978-953-51-2175-6, 286 pages

Kaur B, Gupta SR, Singh G (2002) Carbon storage and nitrogen cycling in silvopastoral systems on a sodic soil in northwestern India. Agrofor Syst 54:21–29

Kaye JP, Resh SC, Kaye MW, Chimner RA (2000) Nutrient and carbon dynamics in a replacement series of *Eucalyptus* and *Albizia* trees. Ecology 81(12):3267–3273

Kirkby CA, Kirkegaard JA, Richardson AE, Wade LJ, Blanchard C, Batten G (2011) Stable soil organic matter: a comparison of CNPS ratios in Australian and international soils. Geoderma 163:197–208

Knops JMH, Tilman D (2000) Dynamics of soil carbon and nitrogen accumulation for 61 years after agricultural abandonment. Ecology 81:88–98

Kumar A, Dash D, Jhariya MK (2013) Impact of *Rhizobium* on growth, biomass accumulation and nodulation in *Dalbergia sissoo* seedlings. Bioscan 8(2):553–560

Kumar A, Dash D, Jhariya MK (2014) Influence *of Rhizobium* inoculation on N, P and K content *in Dalbergia sissoo* Rox*b*. Ecol Environ Conserv 20(3):1059–1065

Kumar S, Sheoran S, Kumar SK, Kumar P, Meena RS (2016) Drought: a challenge for Indian farmers in context to climate change and variability. Progress Res, An Int J 11:6243–6246

Lal R (2004) Carbon emission from farm operations. Environ Int 30:981–990

Lane A, Jarvis A (2007) Change in climate will modify the geography of crop suitability: agricultural biodiversity can help with adaptation. Paper presented at ICRISAT/CGIAR 35th anniversary symposium, climate-proofing innovation for poverty reduction and food security. 22–24 Nov. CCRISAT, Patancheru

Lee TD, Tjoelker MG, Reich PB, Russelle MP (2003) Contrasting growth response of an N2-fixing and non-fixing forb to elevated CO2: dependence on soil N supply. Plant Soil 255:475–486

Lehmann J, da Silva Jr JP, Trujillo L, Uguen K (2000) Legume cover crops and nutrient cycling in tropical fruit tree production. Acta Hortic 531:65–72

Lewis JP, Noetinger S, Prado DE, Barberis IM (2009) Woody vegetation structure and composition of the last relicts of Espinal vegetation in subtropical Argentina. Biodivers Conserv 18:3615–3628

Long SR (1996) *Rhizobium* symbiosis: nod factors in perspective. Plant Cell 8:1885–1898

Luken JO, Fonda RW (1983) Nitrogen accumulation in a chronosequence of red alder communities along the Hoh River, Olympic National Park, Washington. Can J For Res Rev Can De Rech Forestiere 13:1228–1237

Mace GM, Norris K, Fitter AH (2012) Biodiversity and ecosystem services: a multilayered relationship. Trends Ecol Evol 27:19–26

Macedo MO, Resende AS, Garcia PC, Boddey RM, Jantalia CP, Urquiaga S, Campello EFC, Franco AA (2008) Changes in soil C and N stocks and nutrient dynamics 13 years after recovery of degraded land using leguminous nitrogen-fixing trees. For Ecol Manag 255:1516–1524

Marilley L, Hartwig UA, Aragno M (1999) Influence of an elevated atmospheric CO2 content on soil and rhizosphere bacterial communities beneath *Lolium perenne* and *Trifolium repens* under field conditions. Microb Ecol 38:39–49
Marland G, Boden TA (1997) Trends: a compendium of data on global change. Carbon dioxide information analysis center. Oak Ridge National Laboratory, Oak Ridge
Mattson WJ (1980) Herbivory in relation to plant nitrogen content. Annu Rev Ecol Syst 11:119–161
McConnaughay KDM, Coleman JS (1999) Biomass allocation in plants: ontogeny or optimality? A test along three resource gradients. Ecology 80:2581–2593
Meena RS (2013) Response to different nutrient sources on green gram (*Vigna radiata* L.) productivity. Indian J Ecol 40(2):353–355
Meena KK, Meena RS, Kumawat MS (2013a) Effect of sulphur and iron fertilization on yield attribute, yield, nutrient uptake of mungbean (*Vigna radiata* L.). Indian J Agri Sci 83(4):108–112
Meena VS, Maurya BR, Verma R, Meena RS, Jatav GK, Meena SK, Meena R, Meena SK (2013b) Soil microbial population and selected enzyme activities as influenced by concentrate manure and inorganic fertilizer in alluvium soil of Varanasi. Bioscan 8(3):931–935
Meena RS, Yadav RS, Meena H, Kumar S, Meena YK, Singh A (2015a) Towards the current need to enhance legume productivity and soil sustainability worldwide: a book review. J Clean Prod 104:513–515
Meena RS, Dhakal Y, Bohra JS, Singh SP, Singh MK, Sanodiya P (2015b) Influence of bioinorganic combinations on yield, quality and economics of mungbean. Am J Exp Agric 8(3):159–166
Meena RS, Meena VS, Meena SK, Verma JP (2015c) Towards the plant stress mitigate the agricultural productivity: a book review. J Clean Prod 102:552–553
Meena RS, Meena VS, Meena SK, Verma JP (2015d) The needs of healthy soils for a healthy world. J Clean Prod 102:560–561
Meena RS, Bohra JS, Singh SP, Meena VS, Verma JP, Verma SK, Shiiag SK (2016) Towards the prime response of manure to enhance nutrient use efficiency and soil sustainability a current need: a book review. J Clean Prod 112:1258–1260
Meena HR, Meena RS, Lal R, Singh GS, Mitran T, Layek J, Patil SB, Kumar S, Verma T (2017) Response of sowing dates and bio regulators on yield of clusterbean under current climate in alley cropping system in eastern U.P., Indian Leg Res, Accepted in press
Miglietta F, Peressotti A, Vaccari FP, Zaldei A, deAngelis P, Scarascia-Mugnozza G (2001) Free-air CO2 enrichment (FACE) of a poplar plantation: the POPFACE fumigation system. New Phytol 150:465–476
Millennium Ecosystem Assessment (2005) Ecosystems and human well-being: synthesis. Island Press, Washington, DC
Morgan PB, Ainsworth EA, Long SP (2003) How does elevated ozone impact soybean? A meta-analysis of photosynthesis, growth, and yield. Plant Cell Environ 26:1317–1328
Mulder EG, Lie TA, Houwers A (1977) The importance of legumes under temperate conditions. In: Hardy RWF, Gibson AH (eds) A treatise on dinitrogen fixation. IV. Agronomy and ecology. Wiley, New York, pp 221–242
Nair PKR, Nair VD, Kumar BM, Showalter JM (2010) Carbon sequestration in agroforestry systems. Adv Agron 108:237–307
Ndoye I, Dreyfus B (1988) N2 fixation by Sesbania rostrate and Sesbania sesban estimated using 15N and total N difference methods. Soil Biol Biochem 20:209–213
Norby RJ, Cotrufo MF, Ineson P, O'Neill EG, Canadell JG (2001) Elevated CO2, litter chemistry, and decomposition: a synthesis. Oecologia 127:153–165
Nyfeler D, Huguenin-Elie O, Suter M, Frossard E, Luscher A (2011) Grass-legume mixtures can yield more nitrogen than legume pure stands due to mutual stimulation of nitrogen uptake from symbiotic and non-symbiotic sources. Agric Ecosyst Environ 140:155–163
Okada M, Lieffering M, Nakamura H, Yoshimoto M, Kim HY, Kobayashi K (2001) Free-air CO2 enrichment (FACE) using pure CO2 injection: system description. New Phytol 150:251–260
Pennington RT, Ratter JA, Lewis GP (2006) An overview of the plant diversity, biogeography and conservation of Neotropical savannas and seasonally dry forests. In: Pennington RT, Lewis GP,

Ratter JA (eds) Neotropical savannas and seasonally dry forests: plant diversity, biogeography and conservation. CRC Press, Florida, pp 1–29. https://doi.org/10.1201/9781420004496.ch1
Peoples MB, Brockwell J, Hunt JR, Swan AD, Watson L, Hayes RC, Li GD, Hackney B, Nuttall JG, Davies SL, Fillery IRP (2012) Factors affecting the potential contributions of N2 fixation by legumes in Australian pasture systems. Crop Pasture Sci 63:759–786
Permar TA, Fisher RF (1983) Nitrogen-fixation and accretion by wax myrtle (*Myrica cerifera*) in slash pine (*Pinus elliottii*) plantations. For Ecol Manag 5:39–46
Poorter H, Navas ML (2003) Plant growth and competition at elevated CO2 on winners, losers and functional groups. New Phytol 157:175–198
Prasad R, Saroj NK, Newaj R, Venkatesh A, Dhyani SK, Dhanai CS (2010) Atmospheric carbon capturing potential of some agroforestry trees for mitigation of warming effect and climate change. Indian J Agroforest 12(2):37–41
Raj A (2015a) Evaluation of gummosis potential using various concentration of ethephon. M.Sc. thesis, I.G.K.V., Raipur (C.G.), pp 89
Raj A (2015b) Gum exudation in *Acacia nilotica*: effects of temperature and relative humidity. In Proceedings of the national expo on assemblage of innovative ideas/work of post graduate agricultural research scholars, Agricultural College and Research Institute. Madurai (Tamil Nadu), pp 151
Raj A, Jhariya MK, Bargali SS (2016) Bund based agroforestry using eucalyptus species: a review. Curr Agric Res J 4(2):148–158
Rao RR, Husain T (1993) Himalayan legumes: diversity and conservation. In: Dhar U (ed) Himalayan biodiversity: conservation strategies. Gyanodaya Prakashan, Nainital, pp 253–266
Rao KPC, Verchot LV, Laarman J (2007) Adaptation to climate change through sustainable management and development of agroforestry systems. SAT eJournal. An Open Access Journal published by ICRISAT 4(1):1–30
Ratter JA, Bridgewater S, Ribeiro JF (2006) Biodiversity patterns of the woody vegetation of the Brazilian cerrado. In: Pennington RT, Lewis GP, Rartter JA (eds) Neotropical savannas and seasonally dry forests: plant diversity, biogeography and conservation. CRC Press, Florida, pp 31–66
Reid WV (1992) Conserving life's diversity. Environ Sci Technol 26:1090–1095
Resh SC, Binkley D, Parrotta JA (2002) Greater soil carbon sequestration under nitrogen-fixing trees compared with Eucalyptus species. Ecosystems 5:217–231
Rind D, Suozzo R, Balachandran NK, Prather MJ (1990) Climate change and the middle atmosphere. 1. The doubled CO2 climate. J Atmos Sci 47:475–494
Ross DJ, Saggar S, Tate KR, Feltham CW, Newton PCD (1996) Elevated CO_2 effects on carbon and nitrogen cycling in grass/clover turves of a Psammaquent soil. Plant Soil 182:185–198
Rundel PW, Nilsen ET, Sharifi MR, Virginia RA, Jarrell WM, Kohl DH, Shearer GB (1982) Seasonal dynamics of nitrogen cycling for a prosopis woodland in the Sonoran desert. Plant Soil 67:343–353
Ryle GJA, Powell CE, Timbrell MK, Gordon AJ (1989) Effect of temperature on nitrogenase activity in white clover. J Exp Bot 40:733–739
Sanginga N, Mulongoy K, Swift MJ (1992) Contribution of soil organisms to the sustainability and productivity of cropping systems in the tropics. Agric Ecosyst Environ 41:135–152
Sanjappa M (1991) Legumes of India. Bishen Singh Mahandra Pal Singh, Dehradun
Santana MBM, Cabala-Rosand P (1982) Dynamics of nitrogen in a shaded cacao plantation. Plant Soil 67:271–281
Sarkinen T, Iganci JR, Linares-Palomino R, Simon MF, Prado D (2011) Forgotten forests-issues and prospects in biome mapping using Seasonally Dry Tropical Forests as a case study. BMC Ecol 11:27. https://doi.org/10.1186/1472-6785-11-27
Schroth G, Krauss U, Gasparotto L, Duarte Aguilar JA, Vohland K (2000) Pests and diseases in agroforestry systems of the humid tropics. Agrofor Syst 50:199–241
Schwintzer CR (1983) Non-symbiotic and symbiotic nitrogen-fixation in a weakly minerotrophic peatland. Am J Bot 70:1071–1078

Serraj R, Sinclair TR (2003) Evidence that carbon dioxide enrichment alleviates ureide induced decline of nodule nitrogenase activity. Ann Bot 91:85–89

Serraj R, Allen LH, Sinclair TR (1999) Soybean leaf growth and gas exchange response to drought under carbon dioxide enrichment. Glob Chang Biol 5:283–291

Sheikh MA, Kumar M, Todaria NP (2015) Carbon sequestration potential of nitrogen fixing tree stands. For Stud Metsanduslikud Uurimused 62:39–47

Sierra J, Daudin D, Domenach AM, Nygren P, Desfontaines L (2007) Nitrogen transfer from a legume tree to the associated grass estimated by the isotopic signature of tree root exudates: a comparison of the 15N leaf feeding and natural 15N abundance methods. Eur J Agron 27:178–186

Singh NR, Jhariya MK (2016) Agroforestry and agrihorticulture for higher income and resource conservation. In: Narain S, Rawat SK (eds) Innovative technology for sustainable agriculture development. Biotech Books, New Delhi, pp 125–145. ISBN:978-81-7622-375-1

Siqueira JO (ed) (1996) Avanços em fundamentos e aplicação de micorrizas. Univ. Federtal de Lavras /DCS e DCF, Lavras, p 290

Six J, Conant RT, Paul EA, Paustian K (2002) Stabilization mechanisms of soil organic matter: implications for C-saturation of soils. Plant Soil 241:155–176

Smith RG, Gross KL (2007) Assembly of weed communities along a crop diversity gradient. J Appl Ecol 44:1046–1056

Spehn EM, Scherer-Lorenzen M, Schmid B, Hector A, Caldeira MC, Dimitrakopoulos PG, Finn JA, Jumpponen A, O'Donnovan G, Pereira JS, Schulze ED, Troumbis AY, Korner C (2002) The role of legumes as a component of biodiversity in a cross-European study of grassland biomass nitrogen. Oikos 98:205–218

Sprent JI (2009) Legume nodulation: a global perspective. Wiley-Blackwell, Chichester. https://doi.org/10.1002/9781444316384

Stagnari F, Maggio A, Galieni A, Pisante M (2017) Multiple benefits of legumes for agriculture sustainability: an overview. Chem Biol Technol Agric 4(2):1–13. https://doi.org/10.1186/s40538-016-0085-1

Stevens PF (2001) Angiosperm phylogeny, Website version 9, June 2008. Assessed on 15 Mar 2012

Strain BR, Bazzaz FA (1983) Terrestrial plant communities. In: Lemon ER (ed) CO2 and plants. Westview, Boulder, pp 177–222

Tarrant RF (1983) Nitrogen fixation in American forestry: research and application. In: Gordon JC, Wheeler CT (eds) Biological nitrogen fixation in Forest ecosystems: foundations and applications. M. Nijhoff/W. Junk, The Hague, pp 261–278

Tarre RM, Macedo R, Cantarutti RB, Resende CP, Pereira JM, Ferreira E, Alves BJR, Urquiaga S, Boddey RM (2001) The effect of the presence of a forage legume on nitrogen and carbon levels in soils under Brachiaria pastures in the Atlantic Forest region of the South of Bahia, Brazil. Plant Soil 234:15–26

Ter Steege H, Pitman NCA, Phillips OL, Chave J, Sabatier D, Duque A, Molino JF, Prevost MF, Spichiger R, Castellanos H, von Hildebrand P, Vasquez R (2006) Continental- scale patterns of canopy tree composition and function across Amazonia. Nature 443:444–447

Thomas RJ, Asakawa NM (1993) Decomposition of leaf litter from tropical forage grasses and legumes. Soil Biol Biochem 25:1351–1361

Thomas RB, Bashkin MA, Richter DD (2000) Nitrogen inhibition of nodulation and N2 fixation of a tropical N2-fixing tree (*Gliricidia sepium*) grown in elevated atmospheric CO2. New Phytol 145:233–243

Tilman D, Knops J, Wedin D, Reich P, Ritchie M, Siemann E (1997) The influence of functional diversity and composition on ecosystem processes. Science 277:1300–1302

Tissue DT, Megonigal JP, Thomas RB (1997) Nitrogenase activity and N2 fixation are stimulated by elevated CO2 in a tropical N2-fixing tree. Oecologia 109:28–33

Unkovich MJ, Pate JS (2000) An appraisal of recent field measurements of symbiotic N2 fixation by annual legumes. Field Crop Res 65:211–228

Van Cleve K, Viereck LA, Schlentor RL (1971) Accumulation of N in alder (*Alnus*) ecosystems near Fairbanks, Alaska. Arct Alp Res 3:101–114

Van Kessel C, Farrell RE, Roskoski JP, Keane KM (1995) Recycling of the naturally 15N occuring in an established stand of Leocaena leucocephala. Soil Biol Biochem 25:757–762

Vance CP (2002) Root-bacteria interactions: symbiotic nitrogen fixation. In: Waisel Y, Eshel A, Kafkati U (eds) Plant roots: the hidden half, 3rd edn. Marcel Dekker Publishers, New York, pp 839–867

Varma D, Meena RS (2016) Mungbean yield and nutrient uptake performance in response of NPK and lime levels under acid soil in Vindhyan region, India. J App and Nat Sci 8(2):860–863

Varma D, Meena RS, Kumar S, Kumar E (2017a) Response of mungbean to NPK and lime under the conditions of Vindhyan Region of Uttar Pradesh. Leg Res 40(3):542–545

Varma D, Meena RS, Kumar S (2017b) Response of mungbean to fertility and lime levels under soil acidity in an alley cropping system in Vindhyan Region, India. Int J Chem Stu 5(2):384–389

Verma JP, Meena VS, Kumar A, Meena RS (2015a) Issues and challenges about sustainable agriculture production for management of natural resources to sustain soil fertility and health: a book review. J Clean Prod 107:793–794

Verma JP, Jaiswal DK, Meena VS, Meena RS (2015b) Current need of organic farming for enhancing sustainable agriculture. J Clean Prod 102:545–547

Vitousek PM, Walker LR, Whiteaker LD, Mueller-Dombois D, Matson PA (1987) Biological invasion by Myrica faya alters ecosystem development in Hawaii. Science 238:802–804

Waughman GJ (1977) Effect of temperature on nitrogenase activity. J Exp Bot 28:949–960

Weedon JT, Cornwell WK, Cornelissen JHC, Zanne AE, Wirth C, Coomes DA (2009) Global meta-analysis of wood decomposition rates: a role for trait variation among tree species? Ecol Lett 12:45–56

Wheeler CT (1991) Symbiotic nitrogen fixation. In: Raghavendra AS (ed) Proceedings of physiology of trees. Wiley, London, pp 111–135

Yadav GS, Lal R, Meena RS, Babu S, Das A, Bhomik SN, Datta M, Layak J, Saha P (2017a) Conservation tillage and nutrient management effects on productivity and soil carbon sequestration under double cropping of rice in north eastern region of India. Ecol Indic. http://www.sciencedirect.com/science/article/pii/S1470160X17305617

Yadav GS, Lal R, Meena RS, Datta M, Babu S, Das LJ, Saha P (2017b) Energy budgeting for designing sustainable and environmentally clean/safer cropping systems for rainfed rice fallow lands in India. J Clean Prod 158:29–37

Yunusa IAM, Mele PM, Rab MA, Schefe CR, Beverly CR (2002) Priming of soil structural and hydrological properties by native woody species, annual crops, and a permanent pasture. Austral J Soil Res 40:207–219

Zak DR, Pregitzer KS, Curtis PS, Vogel CS, Holmes WE, Lussenhop J (2000) Atmospheric CO2, soil-N availability, and allocation of biomass and nitrogen by *Populus tremuloides*. Ecol Appl 10:34–46

Zakra N, Domenach AM, Sangare A (1996) Bilan positif de l'association cocotiers/acacias pour la restitution de l'azote, de la potasse et du magnésium. Plant Rech Dveloppement 3:39–48

Zentner RP, Campbell CA, Biederebeck VO, Miller PR, Selles F, Fernandez MR (2001) In search of a suitable cropping system for the semi-arid Canadian prairies. J Sustain Agric 18(2–3):117–136

Zentner RP, Campbell CA, Biederebeck VO, Selles F, Lemke R, Jefferson PG, Gan Y (2004) Long-term assessment of management of an annual legume green manure crop for fallow replacement in the brown soil zone. Can J Plant Sci 83:475–482

Zuzana M, Ward D (2002) *Acacia* trees as keystone species in Negev desert ecosystems. J Veg Sci 13:227–236

Cereal+Legume Intercropping: An Option for Improving Productivity and Sustaining Soil Health

11

Jayanta Layek, Anup Das, Tarik Mitran, Chaitanyaprasad Nath, Ram Swaroop Meena, Gulab Singh Yadav, B. G. Shivakumar, Sandeep Kumar, and Rattan Lal

Contents

J. Layek (✉)
Division of Crop Production, ICAR Research Complex for NEH Region, Umiam, Meghalaya, India

Carbon Management and Sequestration Center, Ohio State University, Columbus, USA

A. Das
Division of Crop Production, ICAR Research Complex for NEH Region, Umiam, India

T. Mitran
Carbon Management and Sequestration Center, Ohio State University, Columbus, USA

C. Nath
ICAR-Indian Institute of Pulse Research, Kanpur, India

R. S. Meena
Department of Agronomy, Institute of Agricultural Sciences (BHU), Varanasi, UP, India

G. S. Yadav
Division of Crop Production, ICAR Research Complex for NEH Region, Lembucherra, India

B. G. Shivakumar
Southern Regional Research Station, Indian Grassland and Fodder Research, Dharwad, Karnataka, India

S. Kumar
Department of Agronomy, CCS Haryana Agricultural University, Hisar, Haryana, India

R. Lal
Carbon Management and Sequestration Center, SENR/FAES, The Ohio State University, Columbus, OH, USA

R. S. Meena et al. (eds.), *Legumes for Soil Health and Sustainable Management*,
https://doi.org/10.1007/978-981-13-0253-4_11

Abstract

Intercropping is the system of simultaneously growing two or more crops on the same land area with a definite row arrangement. However, it is very much important to ensure that component crops do not compete with each other for space, moisture, nutrients, and solar radiation. Cereal+legume intercropping provides a greater scope for minimizing the adverse impact of moisture and nutrient stress in addition to improving system productivity and soil health. Researchers reported significant enhancement of system productivity in intercropping besides supplying diversified food. By improving chemical, biological, and physical environment in the soil, legumes can check the declining productivity of cereal-based cropping system. It is important to identify the best intercrops and to assess appropriate doses of nitrogen (N) for the cereal components in intercropping systems considering the sparing effect of biological nitrogen fixation (BNF) from the leguminous component. A number of indices such as land equivalent ratio, crop equivalent yield, relative crowding coefficient, competition ratio, aggressivity, actual yield loss, etc. have been suggested to evaluate the competition between cereal and legume intercrops and advantage of intercropping compared to sole cropping. Although there are some constraints for practicing cereal+legume intercropping systems in large scales like limited availability of good-quality seeds, biofertilizers, and technical and proper scientific knowledge for the complex intercropping system, there is a very good potential to increase the productivity and profitability from resource-poor agricultural systems by adopting this system besides reducing farmers' risks and improving the soil quality in the long term.

Keywords
Biological nitrogen fixation · Cereal · Legume · Nitrogen · Soil quality · Yield

Abbreviations

A	Aggressivity
AYL	Actual yield loss
BNF	Biological nitrogen fixation
CEY	Crop equivalent yield
CR	Competition ratio
GHG	Greenhouse gas
IA	Intercropping advantage
LAI	Leaf area index
LER	Land equivalent ratio
MAI	Monetary advantage index
N	Nitrogen
NER	Northeast region
RCC	Relative crowding coefficient
RDF	Recommended dose of fertilizer
RDN	Recommended dose of nitrogen
SOM	Soil organic matter
WUE	Water use efficiency

11.1 Introduction

The global human population is projected to increase beyond 9.8 billion by the end of the year 2050 (UN 2017). Agriculture must use the scientific technologies and inputs developed over the decades to meet this challenge to feed this burgeoning population (Dietrich et al. 2014). However, this should not be a cause for excessive consumption of fossil fuel leading to greenhouse gas (GHG) emission, loss of biodiversity, and environmental pollution (Milder et al. 2011, Sainju et al. 2012). Thus, ensuring food security while sustaining the soil and environment is a major challenge to the agricultural planners and researchers (Lal et al. 2003; Drechsel et al. 2015). The productivity level of crops should be increased further without deteriorating the soil fertility, environment, and food quality (Lal et al. 2003; Bedoussac et al. 2015; Meena et al. 2015a). Achieving higher plant diversity within an agricultural ecosystem is necessary for sustainable development (IAASTD 2009; Davies et al. 2009). Synthetic N fertilizers are used in a heavy amount to increase the crop productivity in the short term (Bedoussac et al. 2015). Indiscriminate and imbalanced use of synthetic fertilizers, however, deteriorates the soil health in the long term (Savci 2012; Meena et al. 2015b). The new sustainable crop production systems will heavily rely on symbiotic N fixation by legumes (Ladha et al. 2013). The situation was in some European countries in the 1950s, where about half of all available N may have come from leguminous food, fodder, and green manure crops by symbiotically N fixation (Peoples et al. 2009). The land used to grow legume either in rotations or in intercropped with other component crops sequestered carbon (C), stored N, and enriched the biodiversity (Peoples et al. 2009). Hence, exploitation of the leguminous biological nitrogen fixation (BNF) is the need of the hour to reduce the dependency on synthetic N fertilizer (Nieder and Benbi 2008). It will also lower down the C footprints of agricultural products (Gan et al. 2011).

The legume-based intercropping aims to produce higher yield from a unit area by making optimal use of all available resources that could not be utilized by a single crop (Zhang et al. 2011; Ram and Meena 2014). It is important to ensure that component crops do not compete with each other for space, solar radiation, and nutrients (Lithourgidis et al. 2011). In an ideal intercropping system, most of the available natural resources are efficiently utilized to enhance productivity from a unit area of land in unit time and minimize the risk of crop failure (Seran and Brintha 2010). Biological efficiency of intercropping is generally higher as compared to sole cropping as it (intercropping) explored the relatively larger amount of soil mass than that of sole cropping (Gao et al. 2010). This advanced agro-technique has been practiced since time immemorial and greatly contributed to achieve the goal of sustainable agriculture (Wezel et al. 2014; Dwivedi et al. 2015; Yadav et al. 2017a). Intercropping of suitable component crops has several socioeconomic (Ofori and Stern 1987; Ghosh 2004), biological (Kremen and Miles 2012; Bedoussac et al. 2015), and ecological (Seran and Brintha 2010; Brooker et al. 2015) advantages over monocropping. Intercropping also enhances ecosystem biodiversity (Tscharntke et al. 2005) as the component crops provide suitable habitat for numbers of insects and soil organisms which otherwise not present in a monocrop situation (Cai et al. 2010). Natural enemies like spiders, parasitic wasps, etc. help to control outbreaks of crop pests by controlling their population (Altieri 1994; Gianoli et al. 2006; Veres et al. 2013). As the legumes are known for BNF, they should be included in arable cropping systems as intercrops or sequential crops (Liu et al. 2011; Bedoussac et al. 2015). Legumes help in improving the soil fertility via BNF and reduce the competition for available N in soil due to the more competitive character of the cereal (Layek et al. 2014a) and thus contribute to the complementary and efficient use of available N (Bedoussac and Justes 2010).

11.2 Concept and Goal of Intercropping

Intercropping is the practical application of basic ecological principles, viz., diversity, competition, and facilitation, for crop production (Gomes and Gomez 1983; Lithourgidis et al. 2011; Bedoussac et al. 2015). In general, the productivity of intercropping is greater than that of the sole cropping systems (Lithourgidis et al. 2006; Layek et al. 2014a;). As the limiting resources like water, light, and nutrients are efficiently utilized in intercropping systems as against their respective sole cropping, it leads to higher yield (Li et al. 2006; Lithourgidis et al. 2011; Bedoussac et al. 2015). As cereal and legumes widely differ for their rooting patterns, intercropping of them increases the water uptake as well as transpiration and reduces the water loss from soil either through evaporation or deep percolation (Carlson 2008). With increased leaf cover in the intercropping system, transpiration makes the microclimate cooler (Innis 1997), which helps to reduce the soil temperature and associated evaporation (Chai et al. 2011; Miao et al. 2016). It is highly significant when moisture content in soil is limited as a higher amount of available water is being used in intercropping as against sole cropping (Mao et al. 2012). The intercropping system also explored the soil more efficiently as against growing of the single crop (Hinsinger et al. 2011). A mixture of two or more crops will often give a better coverage of the soil and reduce the growth of weeds, runoff, and loss of soil and nutrients (Banik et al. 2006). As component crops in an intercropping system differed in their competitive ability, they can use the available resources of water, nutrient, and solar radiation more efficiently (Hauggaard-Nielsen et al. 2008) and convert them regarding crop biomass or productivity (Lithourgidis et al. 2011). There are several factors, viz., selection of suitable cultivars, seeding ratio, competition between component crops, etc. which can affect the performance of intercropping systems (Caballero et al. 1995; Pandita et al. 2000; Lithourgidis et al. 2011). In general, the competition in intercropping is lower for component crops of different species, and productivity is higher as against mixture of the same species (Vandermeer 1989; Zhang and Li 2003).

11.3 Cereal+Legume Intercropping System

Although the cereal+legume intercropping system being popularized as an insurance against crop failure for monocropping under rainfed conditions, the chief goal of intercropping is to ensure improved and sustainable production (Seran and Brintha 2010; Ali et al. 2012). The intercropping system utilizes resources like water, soil nutrients, light, etc. efficiently, and their productivity is increased (Ghanbari et al. 2010; Zhang et al. 2011). Generally, cereals are nutrient-exhaustive crops and absorb nutrients from upper soil layers (Ali et al. 2012; Layek et al. 2014a). Legume, being able to fix atmospheric N in soil, improves the soil fertility and reduces the completion of limited soil nutrients within the soil (Fujita and Ofosu-Budu 1996; Meena et al. 2015a). Further, legumes help in absorbing nutrients from deeper soil layers due to their robust tap root system (Jat et al. 2012). Cereal+legume intercropping has a great role for subsistence agriculture and provides a diversified food crops in both developed and developing countries particularly in areas having limited irrigation facilities (Tsubo et al. 2005). Cereal+legume intercropping has been reported to conserve soil and water within particular land forms (Anil et al. 1998), provide stable yield (Lithourgidis et al. 2006), prevent lodging (Anil et al. 1998), help in weed management (Banik et al. 2006), and improve the quality, supply period, as well as preservation of fodder to animals (Qamar et al. 1999; Papastylianou 2004).

Soybean+maize intercropping (2:1)

Soybean+sorghum intercropping (2:1)

Soybean+pearl millet intercropping (2:1)

Fig. 11.1 Soybean+cereal intercropping (2:1 ratio) in Delhi, India

The intercropping of legumes like soybean (*Glycine max* L.), groundnut (*Arachis hypogaea* L.), etc. with cereals like rice (*Oryza sativa* L.), maize (*Zea mays* L.), sorghum (*Sorghum bicolor* L.), pearl millet (*Pennisetum typhoides*), etc. (Fig. 11.1) provide great scope for minimizing the adverse impact of moisture stress in lean rainfall years as well as excess moisture during high rainfall years (Singh et al. 2008; Layek et al. 2014a). Besides, these crops with their varied morphology can exploit the edaphic and climatic conditions more efficiently as compared to their cultivation as sole crops (Singh et al. 2008; Layek et al. 2012; Ram and Meena 2014).

The diverse rooting pattern, growth pattern, differences in nutrient requirement, crop duration, etc. tend to impart them a certain degree of ability to come up well even under stressed conditions. The soybean being legume and maize, sorghum, and pearl millet being cereals, their intercropping systems minimize the competition for common growth factors (Layek et al. 2014a). However, the legumes like soybean and groundnut being small-statured crop as compared to cereals in the intercropping system are likely to face a shortage of solar radiation, competition for nutrients, and also competition for moisture under moisture scarcity conditions (Jeyabal and Kuppuswamy 2001; Maingi et al. 2001; Layek et al. 2014b). Under such conditions, it is possible that growth of intercropped soybean, groundnut, etc. may be hampered leading to reduced productivity and impaired quality. Thus, for successful cereal+legume intercropping, the following principles should be followed:

1. The time of peak nutrient demands of component crops should not be overlapped. For example, in maize+green gram [*Vigna radiata* (L.)] intercropping, peak demand for nutrients for green gram comes about 35 days after sowing (DAS), while for maize it comes after 50 DAS (Ofori and Stern 1987).
2. There should be minimum competition for light among the component crops. As all the plants use the same resources as light, water, nutrients, etc., there is competition within and between the species for these resources. When the non-legume or cereal component crop in an intercropping system is relatively competitive in nature for using soil inorganic N, the legume relied mostly on the biologically fixed N. It suits best, if the cereal crops (taller) are sun-loving plants and the intercropped legumes (shorter height) are shade-tolerant plants (Ofori and Stern 1987; Meena et al. 2017a, b).
3. Complementarity should exist between the component crops for the use of growth resources in both time and space. The most important relation between cereals and legumes is the N-use complementarity (Ofori and Stern 1987; Jensen 1996). For example, in cereal+legume intercropping, the legumes have the ability to use atmospheric N, while the associated cereal crops mostly depend on their N from the soil sources. Some parts of the fixed N by the legumes are also being shared to the cereals; hence there is more of complementarity and less of competition for N for the associated crops (Herridge et al. 1995).
4. There should be difference in maturity for component crops by at least 30 days for least competition in intercropping.

When cereals and legumes are grown as intercrops in an intercropping system, each of them should have sufficient space to maximize use of available resources, so that there will be minimum competition between them. For making cereal+legume intercropping system a successful one, spatial arrangements, plant density, and maturity of component crops should be taken care of.

11.3.1 Spatial Arrangement

Growth, yield attributing characters and yield of any intercropping system depends on spatial arrangement of component crops (Musa et al. 2010). Spatial arrangement of intercrops is a management practice that can improve radiation interception by leaves (Undie et al. 2012).

The following types of spatial arrangements are used in intercropping:

Row intercropping – growing two or more crops simultaneously in the field with at least one crop planted in rows. Most of the improved agricultural practices throughout the world use this concept to optimize productivity and resource use efficiency (Varma et al. 2017).

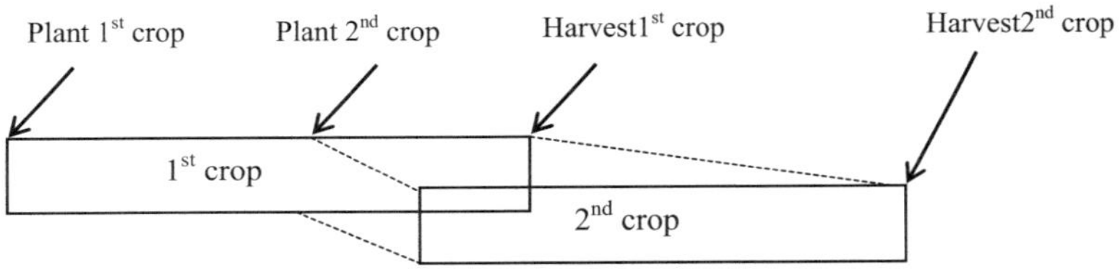

Fig. 11.2 Schematic diagram of relay cropping

Strip intercropping – growing two or more crops together in strips. The stripes should be wide enough to follow individual crop production principles using implements but close enough for the crops to interact with each other (Wang et al. 2015). Strip intercropping is becoming popular across different parts of the world due to its higher radiation use efficiency (Yang et al. 2015).

Mixed intercropping – growing two or more crops together without any distinct row arrangement. Seeds of different crops are mixed before planting and generally sown by randomly broadcasting them in the field. The mixed intercropping is generally used in agriculturally less developed countries to meet the diverse need of food products from the limited area of land (Undie et al. 2012). It is also being used in the pasture to provide diversified and quality products to grazing animals (Malezieux 2012). Cultivation in slash-and-burn agriculture (shifting cultivation), where farmers used to grow 10–12 crops in a particular land is the best example for mixed cropping (Ramakrishnan 2007).

Relay intercropping – planting a second crop before harvesting of the first crop. The second crop is generally sown/planted into a standing crop in its reproductive stage. In areas where the cropping season is limited to grow two individual sequential crops in succession, relay intercropping permits to grow two separate crops successfully in the same area within a year (Balde et al. 2011). By sowing the succeeding second crop before harvesting the proceeding crop (standing situation), the second crop gets a good start on its growing season and able to mature in due time (Fig. 11.2). In other areas, where generally double cropping is practiced, aberrant weather situation or labor or machinery shortage delayed the harvesting of the first crop and subsequent sowing of the second one. In this condition, this relay intercropping can play a great role to obtain two crops from the same field within a calendar year (Thiessen Martens et al. 2005).

11.3.2 Plant Density

Plant density of component crops in legume+cereal intercropping is important to achieve the highest potential (Hauggaard-Nielsen et al. 2008). To get the optimum results, a seeding rate of component crops in intercropping is reduced than its full seed rate in sole cropping (Ren et al. 2016). Crop density greatly influences the competition between cereal and legume in an intercropping system, and reduced competition among intercrops has a positive role on production efficiency and yield

(Ijoyah and Fanen 2002; Ram and Meena 2014). If both the component crops are grown by their full seeding rate, none of the crops can grow optimally as a result of overcrowding and competition (Caballero et al. 1995). By reducing the seeding rates of intercrops, component crops have a greater chance to perform optimally within the mixture. The question is that what will be the appropriate seeding rate for the mixture. For example, one row of maize intercropped with two rows of soybean recorded the highest system productivity in moisture- and nutrient-limited condition (Layek et al. 2014b).

11.3.3 Maturity Dates

If the maturity dates of intercropped legumes differ from their associated cereal crops, the competition between them for water, nutrient, and space reduced significantly and helped in performing their full potential (Ullah et al. 2007). As maturity of component crops differ, peak demand for resources like solar radiation, water, and nutrients at their critical growth stages also differs. If any one of the component crops matures earlier than the other, the competition among them reduces significantly especially at the reproductive stage of the second crop. An aggressive cereal crop pearl millet sometimes gave tough competition to associated intercrop soybean and reduced the yield significantly (Layek et al. 2012). In that condition, if the sowing time of pearl millet is advanced by fewer days as compared to the sowing of associated intercrop soybean, the competition as well as maturity differs and thus reduces the negative impact on soybean. In the southern USA, farmers generally grow cowpeas (*Vigna unguiculata* L.) within standing maize at last maize cultivation. Maize is sown on wide 40-in. rows with lower plant population to allow sufficient sunlight for the intercropped beans or cowpeas. When the maize crop matures, the intercrops use the maize stalk for support and climbing. Thus, two or more crop can be grown simultaneously. After harvesting the economic products, cattle are allowed to feed on the remaining crop residues. In India, in sorghum+pigeon pea (*Cajanus cajan* L.) intercropping system, the sorghum growth is very high at the initial phase of intercropping but matures within a very short span of time (about 4 months). The flowering of pigeon pea starts after harvesting of the sorghum and ripens subsequently. So, there is very less competition between sorghum and associated intercrop pigeon pea for most of the resource-limiting factors.

However, in intensive intercropping systems like additive series where the base crop population is maintained at 100% and intercrops are introduced though modified crop geometry, there is the likelihood of some influence of base crops on the performance of intercrops. This may affect the normal growth and physiological processes of the intercrops leading to below par performance. Further in intercropping systems involving legumes as base crop and cereals as intercrops, one needs to tailor optimum dose of N keeping in view the contribution of BNF from legume component to the cereal companion (Subba Rao et al. 2001; Sharma and Behera 2009; Dhakal et al. 2016). Here it is necessary to understand the actual N requirement for cereals in intercropping vis-a-vis the N in sole crops of cereals (Lupwayi and Kennedy

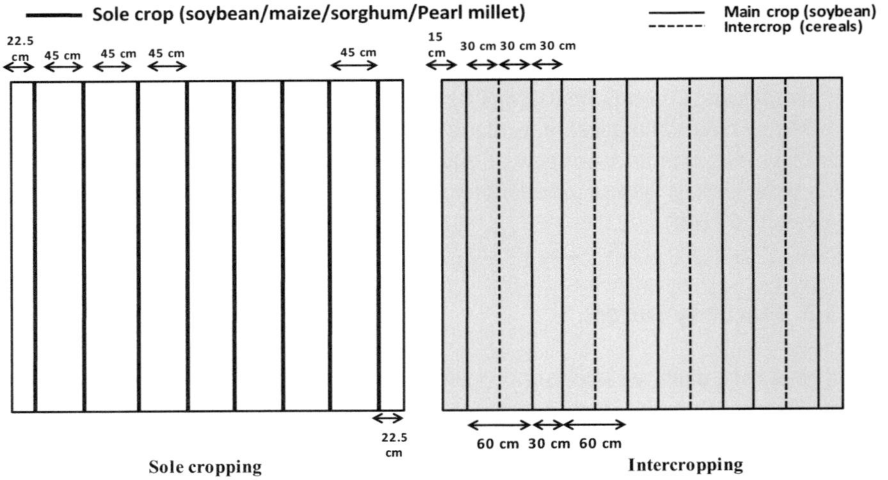

Fig. 11.3 Planting geometry of sole crops and soybean+cereal intercropping (2:1) in additional series

2007). Often due to better N supply and also the reduced intracrop competition, the cereals are likely to exhibit better growth and physiological parameters as compared to sole crops with recommended dose of N (Seran and Brintha 2010; Layek et al. 2014a). In addition to realizing their full potential, the cropping system should also be economically viable in terms of higher monetary returns.

11.4 Types of Intercropping and Crop Geometry

The spatial arrangement of component crops in an intercropping system determines whether there is any advantage of intercropping in comparison to their sole cropping (Yang et al. 2015). Based on the percentage of plant population of intercrops used in an intercropping system about their pure stand, it is categorized in two series, viz., additive series and replacement series.

Additive Series This type of intercropping is mostly popular in developing countries like India where one component crop is grown with its 100% recommended population (same as a sole crop or pure stand), known as the base crop. The associated intercrop grew within the base crop by changing the crop geometry of the base crop. Here, the plant density of the intercrops is lower than its pure stand (Fig. 11.3). This series is highly efficient regarding total system productivity and utilization of available farm resources. There is always an advantage of the land equivalent ratio (LER) as compared to their pure stand.

Figure 11.3 depicts the actual situation in additional series of intercropping. All the sole crops (soybean as well as maize, sorghum, and pearl millet) are sown in

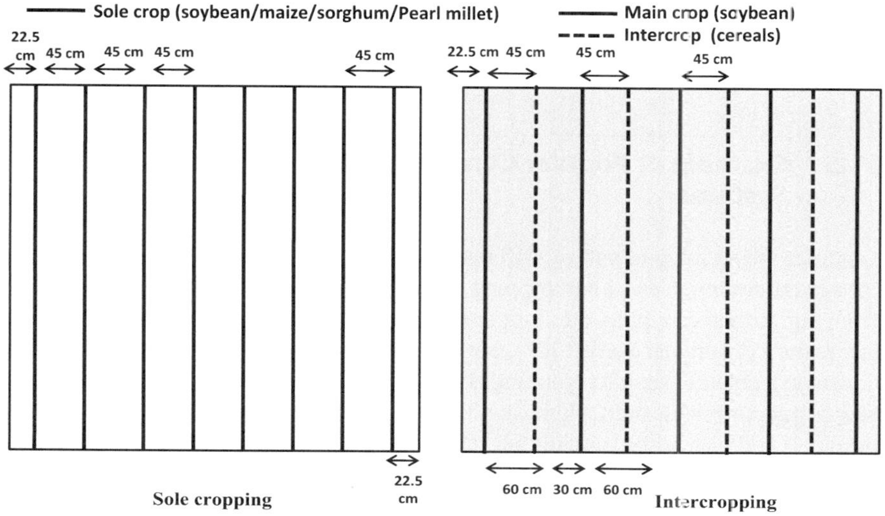

Fig. 11.4 Planting geometry of sole crops and soybean+cereal intercropping (1:1) in replacement series

lines with a spacing of 45 cm between rows. For example, the inter-row spacing between two soybean lines (45 cm spacing for sole soybean) is reduced to 30 cm to accommodate one row of intercropped cereal after every two rows of soybean (Fig. 11.1). Thus, in additional series of intercropping apart from maintaining 100% soybean population, cereal crops are introduced within soybean (base crop) with 50% population of their pure stand (sole crop) without affecting the soybean population (Layek et al. 2014a; Meena et al. 2017a, b).

Replacement Series In replacement series of intercropping, both crops are called component crops or intercrops. None of the component crops are raised with their 100% recommended population in pure stand. By reducing some percentage of the population of one intercrop, another intercrop is introduced (Fig. 11.2). It is mostly done for component crops of similar phenology. The gaining in yield from such intercropping is from a simple response due to "reduced" population and the complementary use of either space or time, or both. However, in many intercropping situations, yield advantages are maximized by increasing population density in excess than their recommended population in the pure stand (Baker and Blamey 1985). The replacement series of intercropping is practiced mostly in the western countries. Here, the completion is relatively lesser in between component crops as compared to additional series. However, there was not any advantage in land equivalent ratio (LER) here. It is mostly done to produce diversified products as well as to improve the forage quality.

In Fig. 11.4, one component crop soybean is sown with it 50% population of the pure stand (sole crop). The 50% population of soybean is sacrificed to accommodate the intercrop cereals (maize/sorghum/pearl millet) in the system.

11.5 Example of Popular Cereal+Legume Intercropping Systems

Legumes are an integral part of different cropping systems as they fit well in different mixed cropping and intercropping, catch cropping, sequential cropping, relay cropping, ratoon cropping, etc. In intercropping, the crops are organized in distinct row pattern (Ofori and Stern 1987). Sowing of both crops might be done at the same time or in a phase wise. Intercropping is an enhanced arrangement of mixed cropping which guarantees desired plant stand; ease in intercultural operation, chemical spraying, and harvesting; and higher returns. The important determining factors for intercropping are the differentiating development, growth pattern, and stature and rooting behavior so that this supplements each other instead of going after the resources and making preparations for climate difficulties (Zhang et al. 2011). Developing of crops in intercropping frameworks is more beneficial especially under rainfed conditions (Singh et al. 2009; Yadav et al. 2017a). Intercropping of pulses with cereals is common in rainfed areas of the world. Here, few dominating pulse-inclusive intercropping systems with cereals are mentioned.

In India, the intercropping systems comprising cereals and legumes are very common. Several pulse-inclusive cropping systems are prevailing across the country. For instance, maize (*Zea mays* L.) is generally grown both in rainy (summer) season and dry (winter) season in India. Pulses like mung bean (*Vigna radiata* L.) and urd bean (*Vigna mungo* L.) are sown between the rows of the maize crop. This system is practiced in North India and its adjoining hilly areas. It has been seen that one row of maize sown after two to four rows of urd bean and mung bean has been found appropriate. The growth of maize in rainy season is faster; hence, sowing of maize after four rows of mung bean/urd bean gives higher equivalent yield than closer spacing. But in winter sown maize, the maize crop sown after every row of vegetable pea (*Pisum sativum* L.) gave high maize equivalent yield over maize+lentil (*Lens culinaris* Medikus) or sole maize. Further, to minimize the shading effects of maize to legumes, north-south direction sowing was found to be the best (Singh et al. 2009). Given the advantage of higher economic return, soil health maintenance, weed control, and less risky intercropping are very popular in the areas of smallholding farmers (Ghosh et al. 2007; Kumar et al. 2016). Some of the potential pulse-inclusive intercropping systems are pigeon pea+sorghum, pigeon pea+maize, etc. (Ghosh et al. 2006). In dry lands of subtropical India, pearl millet is grown extensively, and different grain legumes, viz., green gram, black gram, cowpea, and groundnut, can be accommodated as intercrops (Ghosh et al. 2007).

Traditionally in India, chickpea (*Cicer arietinum* L.) was commonly grown with wheat (*Triticum aestivum* L.) and barley (*Hordeum vulgare* L.) under rainfed conditions. Under limited availability of water, wheat+chickpea is more remunerative than that of wheat+mustard (*Brassica* spp.). But under irrigated conditions, wheat+mustard intercropping proved more profitable over wheat+chickpea intercropping. For optimum profit, proper row ratio is important besides the selection of appropriate varieties. But with the increasing demands and advent of high yielding dwarf varieties of wheat and barley, the area under such cropping is decreasing progressively. But intercropping of pulses with wheat is not always profitable due to its closer spacing (Singh et al. 2009; Meena et al. 2015).

Intercropping of cereals with legumes is very common for rainfed areas of Mediterranean countries. In the Mediterranean nations, common vetch (*Vicia sativa* L.), a yearly legume with a climbing habit and high protein, is very popular to grow with cereals in intercropping (Anil et al. 1998). Various distinctive cereal crops like wheat, oat (*Avena sativa* L.), and barley are tried to fit in intercropping with common vetch (Lithourgidis et al. 2006). Intercropping of cereal+legume enhances soil preservation, smothers weed, gives anchorage to crops, yields strength, and feeds curing and may expand crude protein rate, protein yield, and length of ideal harvest period of hay over grasses (Banik et al. 2006).

In Europe, field pea and spring barley are grown in a typical intercropping framework. In an analysis, field pea and spring barley were intercropped to look at the impacts of harvest assorted qualities on profitability and utilization of N sources from a weed-infested crop land. Pea+barley intercropping proved to be promising with regard to protein enhancement in intercropping systems severely infested with weed and soils with low N availability (Hauggaard-Nielsen et al. 2001). Field trials were conducted on a sandy soil in Denmark for more than three continuous seasons including dual-purpose legumes [pea and faba bean (*Vicia faba* L.)]+barley intercropping and compared with the separate sole crops. The yield strength of intercrops was not more prominent than grain legumes grown as sole crops, except for the intercropping of crops having faba bean. The intercropped faba bean had brought down yield steadiness of barley as compared to pea and barley (Hauggaard-Nielsen et al. 2008).

In sub-Saharan Africa, intercropping of cereals like rice, maize, sorghum, millets, etc. is very common to grow with legumes like cowpea, soybean, groundnut, and beans (Beets 1982). However, the components of an intercropping generally vary with soil fertility, soil pH, agroclimatic condition, economic status, and food preferences of the farming community (Steiner 1982). While, in Eastern Africa, maize+bean intercropping systems are very popular, maize+cowpea/groundnut is popular in Southern Africa (Odendo et al. 2011). Cereal+legume intercropping is very much important to sustain the fertility and production in the sub-Saharan African region, as the soils are deficient in available N and the legume can supply a constant source of N to the soil by BNF. Maize+cowpea intercropping is reported to increase the amount of soil N, phosphorus (P), and potassium (K) contents in this region as compared to monocrops of maize (Mugwe et al. 2011). The maize+soybean intercropping also improved the soil-available nutrients besides increasing the N uptake through grain (Chalka and Nepalia 2006). The introduction of legume component either through intercropping or in rotation with cereal crops with the minimum use of external inputs is the important way to improve the soil fertility and crop productivity in the region, and recent efforts are going on in this direction (Bedoussac et al. 2015; Meena et al. 2015a).

11.6 Role of Cereal+Legume Intercropping on Soil Quality

Pulses are known for their soil fertility restoration ability (Ghosh 2004: Bedoussac et al. 2015; Dhakal et al. 2016). Deep rooting, nitrogen fixation, leaf shedding ability, and mobilization of insoluble soil nutrients are some of the unique characteristics of pulses (Ofori and Stern 1987). By improving chemical, biological, and physical environment in the soil, pulses can check the declining productivity trend of the continuous cereal-cereal system (Savci 2012). The inclusion of pulses in intensive cereal-based system itself is a component of integrated plant nutrient supply (IPNS) system. Therefore, pulses have become a viable alternative to improve the soil health and conserve the natural resources and agricultural sustainability. Here in a nutshell, the effect of pulses as an intercrop with cereal on soil physicochemical and biological properties is discussed.

11.6.1 Soil Physical Properties

The cereal+legume intercropping is common among smallholder farmers because of the capacity of the legume to adapt to soil disintegration and with declining soil health. The essential purposes behind smallholder agriculturists to intercrop are adaptability, benefit amplification, the chance of risk minimization against complete crop loss, soil preservation, weed control, and integrated nutrient management (Shetty et al. 1995). The accessibility of water is a standout among essential components deciding profitability in grain legume intercropping frameworks. Enhancing water use efficiency (WUE) in intercropping frameworks expands the employments of different assets, and it has been recognized to save water to a great extent due to early foliage cover and greater leaf area (Ogindo and Walker 2005). The pearl millet and legume framework was the most effective as far as productivity and WUE are concerned. Whereas, under restricted water supply, WUE in the intercrop contrasted with sole grain can be higher bringing about decreased yield (Hulugalle and Lal 1986).

Intercropping of diversified crops controls soil disintegration by checking precipitation drops from directly hitting the soil surface and possible sealing of surface pores, increases the water infiltration, and reduces the runoff volume (Seran and Brintha 2010). Kariaga (2004) reported that in the maize-cowpea intercropping framework, cowpea was used as best cover crop and decreased soil disintegration than a maize-bean framework. The taller crops go about as wind hindrance for short-statured crops, in mixed intercropping of taller and shorter crops. Also, sorghum+cowpea intercropping decreased surface runoff to the extent of 20–30%

against sole cropping of sorghum and by 45–55% contrasted to sole cowpea. Additionally, soil loss reduced by 50% due to intercropping of sorghum and cowpea against growing them separately (Zougmore et al. 2000).

Grain legumes are true component crops in cereal dominating cropping systems of South Asia for enhancing soil physicochemical and biological properties. Legume crops enhance soil organic matter (SOM) which balances out soil aggregates, makes soil easily cultivable, and increases air circulation, soil water holding, and buffering limits. Further, SOM breakdown delivers accessible nutrients to plants. Lado and Ben-Hui (2004) and Yadav et al. (2017b) reported that SOM physically and chemically chelates the soil forming the better soil aggregation, thereby stabilizing the soil and resisting soil from disintegration. The narrow C/N ratios of grain legume residues fasten their decomposition and improve SOM, thereby impacting soil aggregations and lessening soil bulk density. Ganeshamurthy et al. (2006) reported that incorporation of mung bean stover in rice-wheat-mung bean sequence resulted in lower bulk density and hydraulic conductivity.

11.6.2 Soil Chemical Properties

Expanded nutrient take-up in intercropping systems can happen over space and time. Spatial nutrient take-up can be expanded through the expanding root mass, while the temporal benefit of enhanced nutrient uptake happens when there is no synchronization in nutrient demand by component crops in an intercropping framework. Additionally, if the species have diverse establishing and take-up behaviors, for example, cereal+legume intercropping framework, more productive utilization of accessible supplements may happen, and higher N take-up in the intercrop has been accounted for, over monocrops (Fujita and Ofosu-Budu 1996). While, when just a single crop is grown, all roots have a tendency to rival each other since they are all comparative in their architecture and underneath surface profundity. A few reviews have demonstrated the relative advantage of intercrops to monocrops. For example, Vesterager et al. (2008) found that maize+cowpea intercropping is very much suitable for N-deficient soil and it improved the available N, phosphorus, and potassium content in the soil as compared with monocropping of maize. Legume crops can modify the pH of the soil in the rhizosphere and make small-scale condition good for nutrient accessibility to cereals (Yan et al. 1996). Since pulses acquire a greater part of their N requirement from the air as diatomic N rather than from the soil as NO_3, their net effect is to lower the pH of the soil. Among grain legumes, chickpea reduces the pH most followed by pea and pigeon pea (Singh et al. 2009).

Although legume residues cause partial immobilization of N, they benefit soil nutrient status by expanding SOM when utilized as a part of intercropping in the long run (Fox et al. 1990). SOM enhances soil physicochemical and biological properties that may diminish soil disintegration and increment water and nutrient availability (Sharma et al. 2005; Dhakal et al. 2016). Nitrogen "sparing" is a path in which legumes supply N to intercrop. Since, some portion of their N prerequisite is fulfilled by N fixation, legumes use less of the accessible soil N than cereals, subsequently “saving” or “conserving” inorganic N for the intercrop (Herridge et al. 1995). Meanwhile, N saving does not always hold true since, legumes in some instances may take up significantly more soil inorganic N than equivalent non-legume crops. Some grain legumes, for example, pigeon pea, have solid profound roots. This crop mine plant supplements from profound layers of the soil that cereal crops can’t reach. The supplements are stored on the soil surface because of residue decay, adding to nutrient cycling. Likewise, the *Rhizobia* act as plant growth-promoting rhizobacteria (PGPR) (Antoun et al. 1998) and are known to increase the rooting depth and density (Baudoin et al. 2009), which ultimately helps plants to absorb higher amount of primary, secondary, and micronutrients from a larger volume of soil (Elkoca et al. 2010).

11.6.3 Soil Biological Properties

The legumes are known to enrich the soil by supplying N in the soil through N fixing from the atmosphere by the process of BNF especially when N fertilizer is restricted (Fujita and Ofosu-Budu 1996). The measure of N settled by the legume part in legume intercropping frameworks relies upon a few components, for example, legume types, the morphology of the crop, plant density of intercrops, sort of cultivation practices, and aggressiveness of component crops. In maize+cowpea intercropping system, it was estimated that around 41 kg N ha^{-1} is added to the soil by cowpea through BNF (Eaglesham et al. 1981). Again, Fujita et al. (1992) recorded 20% higher yield of sorghum in soils with high available N in the sorghum+soybean intercropping system. While the sorghum relied mostly on the available N in soil and added fertilizer, the soybean depended most on BNF. Studies reported that related non-legumes might benefit through N exchange from legumes (Fujita et al. 1992; Meena et al. 2017a, b). This N exchange is considered to happen through root discharge, N drained from leaves, leaf fall, and faunal excreta if present in the intercropping framework. The reviews proposed that N settled by a leguminous part might be accessible to the cereal intercrop in the growing season, known as immediate N exchange (Eaglesham et al. 1981). Additionally, Fujita et al. (1992) indicated that advantages to the related non-leguminous product in intercropping frameworks are impacted by crop densities, which decide the compactness of legume and non-legume crops and legume development stages. Pulses are also known to improve the microbial environment in the soils (Kumar and Goh 2000; Meena et al. 2014). They are known to release a part of unused nitrate fixed through symbiotic nitrogen fixation to the soil (Herridge et al. 1995). Also, low molecular weight organic compounds are released to the soil as exudates. This serves as a substrate to soil microorganisms resulting in the buildup of the population of soil microbes. Grain legumes may give great natural conditions for the growth and development of soil microorganisms. The soil microbial biomass is the active portion of the soil that includes viable microorganisms and parasites, including soil microfauna and green growth (Kumar and Goh 2000). The enzymatic activity in the soil is generally the product of the magnitude of the microbial population in soil, being gotten from intracellular, cell-related, or free mixes. The grain legume crops boost the dehydrogenase, urease, protease, phosphatase, and β-glucosidase reactions in the soil (Roldan et al. 2003).

11.7 Nitrogen Management in Cereal+Legume Intercropping

Among the major nutrients, N is found to be very important for the growth of cereals. Nitrogen is an essential constituent of chlorophyll and imparts a green color to the leaves. The higher availability of nutrients especially N, in the initial stage, helps in acquiring a definite advantage over other treatments in respect to growth. Nitrogen application also helps the plant in making quick growth and vigorous development regarding root and shoot dry weight. Supply of N in adequate amount and available form has a high degree of positive correlation with crop productivity (Mashhadi and Zand 2004; Dhakal et al. 2015). Variation in the leaf carbon dioxide exchange rate is strongly associated with leaf N content. Over the years, N deficiency has become a major constraint for the production of cereals due to low N status of Indian soils and high N requirement of cereal crops. When a cereal crop is grown in intercropping with a legume, the cereal may be benefitted by direct N transfer from the legume through BNF (Giller and Wilson 1991). As legumes have the capacity to fit in different cropping patterns and fix N from the atmosphere in the soil, they may offer opportunities to increase the productivity of the intercropping system (Jeyabal and Kuppuswamy 2001). This may be due to soil fertility enhancement either through the supply of biologically fixed N or root excretion from the associated legume crop (Ghosh 2004).

Soybean being a leguminous crop can fix biological N_2 through nodules. As the legumes are often overlooked owing to their lower yield ability and susceptibility to pests and diseases, the stability in productivity and profitability through intercropping may help in making the cultivation of legumes more attractive. The location-specific intercropping systems with suitable cereal crops will again strengthen its cultivation under different agro-environments faced with the challenge of declining partial factor productivity in major cropping systems already, as is the case with rice-wheat cropping system in the Indo-Gangetic Plains of India. The increased overall productivity, stability in production over a period, improvement in soil conditions, breakdown in the perpetuation of the pests and disease cycles, weed problems, etc. may be effectively achieved through the new innovative intercropping systems involving soybean and cereals. The complementarities with such systems further help in addressing the nutrient management problems common in cereal-cereal or legume-legume cropping systems.

As the *Rhizobium* bacteria that infect legume roots normally supply adequate N to the host plant, well-nodulated legumes rarely respond to additions of N fertilizer. Occasionally, however, soybean may respond to applications of N late in the season, presumably because of a significant decline in the activity of N fixation in the nodules. Such responses are quite erratic, and late-season applications of N to soybean are not routinely recommended. The amount of atmospheric N_2 fixed by nonsymbiotic soil organisms varies with soil types, organic matter present, and soil pH. The N plays a very significant role in crop production, and the appropriate rate of N promotes vigorous growth of plants and higher yields. Variuos researchers like Shah et al. (2003), Marchiol et al. (1992), and Meena et al. (2017a, b) reported that intercropping increased soybean plant height but decreased vegetative biomass and pod production compared with the pure stand. As the density of component crop soybean increased, the biomass of intercropped maize decreased proportionately. However, the effect of N application in the system had very limited effects on dry matter production of the intercrops. While the dry matter yield of the intercrop soybean was about 89% of the pure stand, the dry matter yield of maize was 104% of its pure stand. The LER of the maize+soybean intercropping system was recorded to be about 1.30, which means that there is 30% advantage in intercropping against sole cropping of maize and soybean separately (Marchiol et al. 1992).

The row/population ratio by which intercrops in a cereal+legume intercropping are grown also influences the growth and productivity of the system. Behairy (1994) reported that soybean plants intercropped with maize in 4:2 ratio received more radiation than those intercropped in a 2:2 ratio. Leaf chlorophyll content of soybean leaves was greatly decreased under the 2:2 ratio and to a lesser extent under the 4:2 ratio compared to soybean grown alone. Although the plant height of soybean was recorded to be increased at flowering stages in soybean+maize intercropping, the weight of stems, leaves, and pods, as well as leaf area per plant, was decreased in the intercropping system as compared with soybean grown alone. Plant height was negatively correlated with light intensity in the middle or above the soybean canopy. Light intensity in the middle or below the soybean canopy was positively correlated with the total dry weight of intercropped soybean. Sharma et al. (1994) conducted an experiment on soybean+maize intercropping system and reported that light interception by maize was not affected by soybean but a light interception by soybean was affected by maize. Light transmission to soybean decreased with increasing N. The plant height, as well as dry matter content of maize in soybean+maize intercropping system, was significantly increased with increasing N levels (Shivay and Singh 2000). The highest plant height (176.4 cm), leaf area index (LAI) (1.97), and dry matter accumulation (98.6 g/plant) of maize were recorded with 120 kg N/ha. In a similar experiment, Shivay et al. (2002) reported that increase in the dose of N also led to significant increase in LAI and net assimilation rate (NAR) at all the crop growth stages. Highest LAI was recorded at 60 DAS, whereas the maximum NAR was recorded at 60–90 DAS. Wandahwa et al. (2006) reported that the inorganic fertilizer (N) significantly reduced the growth and yield of soybean in the intercropping situation due to heavy growth of associated cereals.

Bhat et al. (2008) recorded maximum plant height (271 cm), stem girth (2.73 cm), leaves per plant (13.1), and dry weight per plant (145.6 g) of maize with 150 kg N/ha (through urea and *Azotobacter*). Kumar et al. (2008) reported that intercropping of black gram with maize had no significant effect on growth parameters of maize. Gao et al. (2010) reported that the roots of intercropped maize did not penetrate deeper in the soil than those of soybean but also extend below the soybean rows. The roots of soybean, however, were limited to the soybean rows only. Kushwaha and Chandel (1997) conducted an experiment on soybean+maize intercropping systems where the intercrop maize received 0, 50, 100, or 150 kg N and soybean were seed inoculated with *Bradyrhizobium japonicum* (Meena et al. 2017a, b). They observed that maize equivalent grain yield increased with up to 50 kg N in the intercrop and was higher than that of the sole crop receiving 120 kg N. Shivay et al. (2001) also reported that application of N significantly increased yield attributes, grain yield, maize equivalent yield, and economics of maize+legume intercropping systems. The highest grain yield was recorded with 120 kg N/ha.

The amount of fertilizer to be applied in a cereal intercropping system is varied with the site and types of component crops. Application of 75% recommended dose of fertilizer (RDF) to maize and 50% of soybean reported to significantly increase the productivity of the intercrops, maize equivalent yield, and profit from the system over supply of 50% RDF to maize and soybean (Meena et al. 2006). Although the cereal in an intercropping system was reported to positively respond to a higher dose of N (120 kg N/ha), the associated legume (pigeon pea) responded only up to the application of 80 kg N/ha (Satyam et al. 2008). The supply of biofertilizer (*Azotobacter*) along with 150 kg N/ha also reported increasing the productivity of maize intercropped with soybean (Satyam et al. 2008; Buragohain et al. 2017). These diverse types of N requirement for cereal+legume intercropping may be reported due to the difference in soil-available nutrient, types of intercrops grown, and their productivity level throughout the world. The maximum yield per hectare (5.46 Mg/ha) of maize in maize+soybean intercropping system was also recorded with 150 kg N/ha (through urea and *Azotobacter*) (Bhat et al. 2008). Sawargaonkar et al. (2008) conducted an experiment on soybean intercropped with maize at 75, 100, or 125% of the recommended rates of nitrogen (RDN) to maize. The application of 125 and 100% RDN resulted in higher maize grain equivalent yield and net monetary advantage over 75% RDN. However, application of 100% RDN registered a higher benefit/cost ratio than 125 and 75% RDN. Kumar et al. (2003) reported that intercropping of black gram with maize had no significant effect on yield parameters of maize. However, grain yield, stover yield, maize grain equivalent, N uptake, and economics significantly increased due to intercropping.

Marchiol et al. (1992) conducted an experiment on the soybean+maize intercropping system and reported that the protein content of the intercrop was greater than pure maize. But N application and soybean density did not produce any discernible changes in protein contents. In a field experiment during the monsoon seasons of 1992–1993 at Pantnagar, Uttar Pradesh, India, maize was grown alone and provided with 0 or 120 kg N/ha or intercropped with soybean. The intercrop received 0, 50, 100, or 150 kg N/ha, and soybean was seed inoculated with *Bradyrhizobium japonicum*. The protein content of maize grain improved with N application in the maize+soybean system (Kushwaha and Chandel 1997; Meena et al. 2017a, b). Shivay and Singh (2000) reported increases in N uptake of maize with increasing N levels, the highest being recorded with 120 kg N/ha. This was 11.2%, 44.8%, and

77.8% higher than the N uptake recorded with 80 and 40 kg N/ha and control, respectively. Duraisami et al. (2002) reported N at 120 and 240 kg/ha significantly improved the protein content in maize. Total N uptake was also the highest in maize (75 cm) + two rows of mung bean in comparison to their sole cropping. Intodia et al. (2007) also reported the similar result where the application of N up to 100 kg/ha brought about significant improvement in uptake of major (N, P, and K) and micronutrients (Zn, Mn, Fe and Cu) by grain and stover of maize.

11.7.1 Yield of Intercropped Legumes (Negative Impact)

In a field study in New Delhi, India, the soybean intercropped with pearl millet recorded lower pods per plant and seeds per pod as compared to soybean intercropped with maize and sorghum (Layek et al. 2014a, b). This may be attributed to stiff competition by pearl millet posed to soybean as observed in the tall and lanky plants, the fewer number of branches, reduced dry matter accumulation, etc. leading to reduced pods per plant (Layek et al. 2014a). Further, the overall competition might have also reduced seeds per pod. The maize and sorghum being widely spaced as well as non-tillering type had a relatively lesser competition to soybean as compared to pearl millet. The N levels too had a marginal impact on the yield parameters (Table 11.1). The higher values of pods per plant, seeds per plant, and seed index of soybean were observed when the intercropped cereal crops were

Table 11.1 Yield attributes, yield, and economics of soybean-based intercropping system in semiarid region of India

Treatment	Pods/ plant	Seeds/ pod	Grain yield (Mg/ha)	Biological yield (Mg/ha)	Net return ($/ha)
Sole soybean$_{30N}$	22.20	2.50	1.58	4.66	176.7
Soybean$_{30N}$ + maize$_{0N}$	18.80	2.30	1.27	4.15	152.6
Soybean$_{30N}$ + maize$_{30N}$	16.40	2.15	1.12	4.00	219.4
Soybean$_{30N}$ + maize$_{45N}$	15.60	2.14	1.09	4.01	239.4
Soybean$_{30N}$ + maize$_{60N}$	14.85	2.17	1.04	3.87	247.0
Soybean$_{30N}$ + sorghum$_{0N}$	18.45	2.30	1.20	3.99	138.6
Soybean$_{30N}$ + sorghum$_{30N}$	16.10	2.20	1.04	3.76	206.0
Soybean$_{30N}$ + sorghum$_{45N}$	15.45	2.17	1.01	3.70	226.4
Soybean$_{30N}$ + sorghum$_{60N}$	14.20	2.13	0.98	3.56	220.4
Soybean$_{30N}$ + pearl millet $_{0N}$	12.35	2.13	0.74	2.65	57.2
Soybean$_{30N}$ + pearl millet$_{20N}$	11.10	2.07	0.71	2.81	143.2
Soybean$_{30N}$ + pearl millet$_{30N}$	10.95	2.04	0.68	2.63	148.3
Soybean$_{30N}$ + pearl millet$_{40N}$	10.20	2.04	0.66	2.58	144.9
SEm±	0.55	0.05	0.04	0.18	502
LSD ($p = 0.05$)	1.65	0.15	0.10	0.52	1465

Data source: Layek et al. (2012)

grown with no N, while the increasing levels gradually declined the values of yield parameters. This may be mainly due to better growth of cereal intercrops with increasing levels of N, which in turn resulted in increased competition to soybean ultimately resulting in decreased yield parameters. The soybean grown with intercrops like pearl millet recorded the lowest yield and harvest index as compared to sole crop especially when the cereal counterparts are raised with a higher dose of N (Layek et al. 2012).

The yield parameters of soybean like pods per plant and seeds per pod were the highest in sole soybean. This is mainly due to unhindered growth in the sole crop as compared to soybean in intercropping systems. The soybean intercropped with pearl millet recorded lower pods per plant and seeds per pod as compared to soybean intercropped with maize and sorghum. This may be attributed to stiff competition by pearl millet posed to soybean as observed in the tall and lanky plants, the fewer number of branches, reduced dry matter accumulation, etc. leading to reduced pods per plant. Further, the overall competition might have also reduced seeds per pod. All intercropping systems with no N to intercropped cereals recorded lower values as compared to sole soybean. The intercropping of maize and sorghum with soybean supplied with all levels of N to intercrops (except no N) resulted in significantly higher net returns (Layek et al. 2012; Ram and Meena 2014). In soybean+pearl millet intercropping system, net returns with 50% RDN onward were statistically at par with sole soybean.

Similar results were also reported by Singh et al. (2007). The increase in yield of sole soybean was mainly due to the increased growth parameters and their positive influence on the yield parameters. The higher values of pods per plant, seeds per pod, and seed index in a sole crop of soybean resulted in increased seed yield in sole soybean. A similar type of result was also recorded by Maurya and Rathi (2000). While West and Griffith (1992) observed a 26% increase in productivity of maize and a 27% increase for soybean in a maize/soybean strip intercropping system, Ghaffarzadeh et al. (1994) recorded 20–24% increase in maize yields and 10–15% for soybean. Likewise, the biological yield also was the complementary effect of increased growth parameters, viz., increased number of branches per plant and dry matter accumulation per plant. The increased root biomass, leaf area index, crop growth rate, etc. might have positively affected the total biological yield of soybean in sole cropping. The increased biological yield further had a positive influence on the conversion of photosynthates into economic yield, i.e., seed yield. Thus, the harvest index too was higher with a sole crop of soybean. The higher yield of the sole crop in comparison to intercrop situation was also confirmed by Muoneke et al. (2007). As there is some competition to soybean from associated cereals (sorghum and pearl millet) in an intercropping system, the growth and yield parameters of soybean reduced as compared to sole soybean (Layek et al. 2012). However, Patra and Patra (2010) recorded a higher number of filled pods/plant in soybean due to intercropping with sesame as compared to sole cropping.

11.8 Nitrogen Transfer in Cereal+Legume Intercropping Systems

When cereals and legumes are grown in association in an intercropping system, the N nutrition of the associated cereal is improved due to the transfer of biologically fixed N from associated legume (Willey et al. 1983; Meena et al. 2015a). Although the cereals intercropped with legumes get a certain amount of N, it may not be sufficient for its optimum performance (Giller and Wilson 1991; Stern 1993). As the legumes have the capacity to fix atmospheric N in the soil, adapt diverse condition, and suit in different types of cropping systems (Ofori and Stern 1987), they can increase the productivity as well as sustainability of cereal+legume intercropping system (Jeyabal and Kuppuswamy 2001: Maingi et al. 2001). Further, as cereals need a large amount of N, it is likely that they may draw some quantity of N fixed by the soybean/legume through biological N fixation (BNF) (Senaratne et al. 1993). The information regarding the amount of N transfer to intercropped cereals from associated legumes and optimum dose of N in cereal+legume intercropping system is very meager. The extent of N drawn also depends on the degree of N made available to the intercropped cereal. It will be interesting to know how much N would be sufficient to cereal components to facilitate them to draw the biologically fixed N from the soybean without causing any severe depression in the performance of soybean (Layek et al. 2014a).

The optimum dose of N on 50% plant population basis to intercrops in soybean varied with crops (Table 11.2). The optimum dose of N for sorghum in intercropping was 54 and 56.7 kg/ha (against 60 kg/ha in sole cropping) and for pearl millet was 35.5 and 31.6 kg/ha (against 40 kg/ha for sole cropping) during 2009 and 2010, respectively. The pearl millet with short duration and lower productivity showed the lowest dose of N as optimum among the cereals. It may be due to higher amount of biologically fixed N (by soybean) transferred to pearl millet and sorghum as compared to maize in the intercropping. The highest contribution of nitrogen from soybean was observed in the case of pearl millet leading to the lesser amount of optimum dose.

The grain yield of sorghum and pearl millet in soybean+sorghum and soybean+pearl millet intercropping system increased significantly only up to 75% and 50% RDN. High grain yield of sorghum and pearl millet even with the supply of only 75% and 50% RDN, respectively, might be due to very less competition and supply of N from associated cereals through BNF (Giller and Wilson 1991). Thus, there is a possibility to minimize the demand for N fertilizers by 50% and 25% for pearl millet and sorghum, respectively, when they are intercropped with legume soybean (Layek et al. 2014a; Ram and Meena 2014). A good yield of cereal crops in cereal+legume intercropping with below RDN was also recorded by Ghosh et al. (2009).

Table 11.2 The optimum dose of N and yield of cereals in 2:1 soybean+cereal intercropping system

	Optimum dose (kg/ha)		Optimum grain yield (Mg/ha)		Linear regression equation	
Crop	2009	2010	2009	2010	2009	2010
Maize	66.6	70.7	1.74	1.96	$Y = -0.234x^2 + 32.60x + 591.8$	$Y = -0.241x^2 + 35.5x + 651.5$
					$R^2 = 0.991$	$R^2 = 0.991$
Sorghum	54.0	56.7	1.55	1.77	$Y = -0.320x^2 + 36.01x + 538.3$	$Y = -0.335x^2 + 39.4x + 607.8$
					$R^2 = 0.992$	$R^2 = 0.995$
Pearl millet	35.5	31.6	1.44	1.56	$Y = -0.634x^2 + 47.70x + 538.5$	$Y = -0.883x^2 + 57.2x + 636.3$
					$R^2 = 0.986$	$R^2 = 0.992$

Data source: Layek et al. (2014b)

Y yield (Mg/ha), *X* N dose (kg/ha), R^2 coefficient of determination

11.9 Indices for Measuring Intercrop Productivity and Efficiency

11.9.1 Equivalent Land Ratio (LER)

The LER is defined as the relative land area required under sole crop to produce the same yield as obtained under an intercropping system at the same management level (Willey et al. 1983; Fetene 2003). It is the summation of ratios of the yield of each component crop in an intercropping to its corresponding sole crop yield:

$$\mathrm{LER} = \mathrm{LER}_{\mathrm{legume}} + \mathrm{LER}_{\mathrm{cereal}}$$

$$\mathrm{LER}_{\mathrm{legume}} = \left(\mathrm{Yab} / \mathrm{Yaa}\right)$$

$$\mathrm{LER}_{\mathrm{cereal}} = \left(\mathrm{Yba} / \mathrm{Ybb}\right)$$

where Yab is the yield of legume “a” grown in association with intercrop cereal “b” and Yba is the yield of intercrop cereal “b” grown in association with legume “a.” The Yaa and Ybb represent the yields of legume “a” and cereal “b” grown in pure stand, respectively.

The LER indicates the efficiency of an intercropping system to use the limited available resources as against their pure stand (Mead and Willey 1980). When the value of LER is greater than one (1.0), the intercropping system is advocated (Ofori and Stern 1987). In contrast, when LER is lower than one (1.0), the intercropping is affecting the productivity of the intercrops grown in association (Caballero et al. 1995). The LER of 1.30 means there is a 30% advantage in yield in intercropping as against the cultivation of cereal and legume (intercrops) in isolation (sole cropping).

11.9.2 Crop Equivalent Yield (CEY)

It is the conversion of crop yields into one form to compare the crops grown under mixed cropping or intercropping or sequentially cropping (De Wit 1960). Here, the conversion is done in the form of soybean equivalent yield by considering the intercrop yield and market price of soybean and associated cereals.

$$\text{Soybean equivalent yield} = \frac{\text{Yield of intercrop}}{\text{Market price of soybean}} \times \text{market price of intercrop}$$

The higher soybean equivalent yield in the intercropping situation was also reported by Singh et al. (2008) and Gare et al. (2009). This may be mainly due to higher productivity of both soybean and maize in their intercropping systems than that of other intercropping combinations. It might also be due to the least adverse effect of intercrops on each other and higher response of maize to applied N up to 100% RDN. In a field experiment on clay loam during *Kharif* 1986 at Palampur, Himachal Pradesh, India, maize was intercropped with soybean in single rows (60 cm) or paired rows (30/90 cm). All intercropping treatments had land equivalent ratios of >1. Maize intercropped in single rows with soybean produced the highest maize equivalent yield (5.94 Mg/ha) among the intercropping treatments, while the equivalent maize yield of all the pure soybean cultivars was greater (6.15–8.78 Mg/ha) than the intercropping treatments (Kalia et al. 1992).

11.9.3 Aggressivity

Aggressivity (A) indicates the relative yield increase in crop "a" than crop "b" in an intercropping situation (McGilchrist 1965). The aggressivity (A) of a cereal+legume intercropping system can be derived from the following formula:

$$A_{cereal} = \{Yab / (Yaa \times Zab)\} - \{Yba / (Ybb \times Zba)\}$$

$$A_{legume} = \{Yba / (Ybb \times Zba)\} - \{Yab / (Yaa \times Zab)\}$$

Yab = yield of cereal "a" in cereal+legume intercropping system
Yaa = yield of cereal "a" in pure stand (sole cropping)
Zab = sown proportion of cereal "a" in intercropping
Yba = yield of legume "b" in cereal+legume intercropping system
Ybb = yield of legume "b" in pure stand (sole cropping)
Zba = sown proportion of legume "b" in intercropping

When the value of A becomes zero, none of the crops are considered as aggressive or both crops are equal in competition. If the value of A becomes positive, then cereal crop is considered as aggressive or dominant over intercropped legume. If the value of A becomes negative, then intercropped legumes are considered as aggressive or dominant over cereals.

In a cereal+legume annual intercropping system, the cereal component is known as a suppressing/aggressive crop and legume component as suppressed crop (Haynes 1980). For example, in intercropping systems of barley+faba bean (Strydhorst et al. 2008), maize+groundnut (Inal et al. 2007), and wheat+soybean, the barley, maize, and wheat are the aggressive crops, and the faba bean, groundnut, and soybean are the suppressed crops. In a soybean+maize/sorghum/pearl millet intercropping experiment conducted over 2 years' time (the year 2009 and 2010) in semiarid regions of India, the intercropped cereals showed positive aggressivity, while the soybean showed the negative aggressivity (Layek et al. 2014a). Among the intercrops, 100% RDN to intercropped pearl millet showed the highest aggressivity. This may be due to the higher competitive ability of pearl millet and better response to supplied N as compared to other intercrops like maize and sorghum (Layek et al. 2014b). As the positive aggressivity of intercropped cereals increased, the negative aggressivity increased in soybean. As the positive aggressivity of one crop in the intercropping system affects the performance of component crop (Ram and Meena

2014), it is reflected in the negative aggressivity. The results of aggressivity conformed that, in particular, cereals were the dominant species (A positive) in the soybean+cereal intercropping situation and among the cereals pearl millet was more aggressive than that of sorghum and maize. As the aggressivity value of soybean in this soybean+cereal intercropping was recorded to be negative; it could be called as the dominated crop. A similar type of result also confirmed for groundnut in groundnut+cereal intercropping by Ghosh (2004).

11.9.4 Relative Crowding Coefficient (RCC)

It is the measure of relative dominance of one component crop over the other in an intercropping system (De Wit 1960). It is estimated by the following formulae:

$$K = \left(K_{\mathrm{c}} \times K_{\mathrm{l}}\right)$$

where K = RCC of the intercropping system, K_c = RCC of intercropped cereal, and K_l = RCC of intercropped legume.

$$K_{\mathrm{c}} = \left\{\mathrm{Yab} \times \mathrm{Zba}\right\} / \left\{\left(\mathrm{Yaa} - \mathrm{Yab}\right) \times \mathrm{Zab}\right\}$$

$$K_{\mathrm{l}} = \left\{\mathrm{Yba} \times \mathrm{Zab}\right\} / \left\{\left(\mathrm{Ybb} - \mathrm{Yba}\right) \times \mathrm{Zba}\right\}$$

Yab = yield of cereal “a” in intercropping
Zba = sown proportion of legume “b” in intercropping
Yaa = yield of cereal “a” in sole cropping
Zab = sown proportion of cereal “a” in intercropping
Yba = yield of legume “b” in intercropping
Ybb = yield of legume “b” in sole cropping

When the value of the product of two coefficients ($K_c \times K_l$) is higher than one (>1), there is a yield advantage in the intercropping. However, if the value of K is one (1), there is no yield advantage/disadvantage in the system. If the value of K is less than one (<1), there is completion between intercrops and associated disadvantage in intercropping.

The relative crowding coefficient of legume intercropped with cereals like maize was higher in legume-based intercropping system, where the population of soybean is maintained fully and cereals were adjusted in between soybean/legume row by changing the crop geometry (Layek et al. 2014a). The pearl millet being more aggressive reduced K_s significantly as compared to other intercrops like maize and sorghum. This may be due to its tillering nature and also more responsive to applied N. In the experiment (Layek et al. 2014b), 2:1 soybean+pearl millet intercropping recorded lower values of RCC (K < 1); hence it is not recommended to grow pearl millet with soybean in intercropping system for the studied area. However, 2:1 soybean+maize and soybean+sorghum intercropping recorded greater values of RCC ($K > 1$). It indicates that these maize and sorghum are suitable for intercropping with soybean in the area as these offer less completion and gave higher system productivity (Layek et al. 2014a). The soybean+maize intercropping recorded the highest value of K in the year of deficient rainfall. Thus, cereal+legume intercropping (like soybean+maize) is advocated for adverse climatic condition (drought) to reduce the risk of total crop failure (Kitonyo et al. 2013; Meena et al. 2014).

11.9.5 Competition Ratio (CR)

Competitive ratio gives a good measure of the competitive ability of the component crops in an intercropping system (Willey and Rao 1980) and has some advantage over relative crowding coefficient (K) and aggressivity (A). It represents the ratio of individual land equivalent ratios (LERs) of the component crops and also takes into account the proportion of the crops sown in intercropping.

$$CR_{legume} = \left(LER_{legume} / LER_{cereal}\right)\left(Zba / Zab\right)$$

$$CR_{cereal} = \left(LER_{cereal} / LER_{legume}\right)\left(Zab / Zba\right)$$

Zab = sown proportion of legume in combination with intercrop cereal
Zba = sown proportion of cereal in combination with intercrop soybean

It gives better measure of the competitive ability of the crops as compared to y.

If the values of CR < 1, there is a positive benefit. It means there is limited competition between component crops and they can be grown as intercrops (Ghosh 2004). However, if the value is higher than one (CR > 1), there is a negative impact. In this condition, the competition between intercrops in the association is too high, and they are not recommended to grow as intercrops. The competition ratio (CR) of legume and intercrop cereal had an inverse relationship. A study was conducted in New Delhi, India, with different soybean+cereal intercropping with various doses of N to intercropped cereals. The soybean+maize intercropping supplied with no N (control) recorded the highest CR in soybean as compared to other intercropping and N management levels (Layek et al. 2014a). The increasing levels of N to associated cereal crops reduced the CR of soybean. The least CR of soybean was noticed when intercropped with pearl millet with 100% RDN. This decrease in CR of soybean may be due to competition exerted by the intercrops with increasing levels of N to them. As pearl millet is a more aggressive crop, it had exhibited stiffer competition leading to lesser CR of soybean as compared to other intercrops, viz., maize and sorghum (Layek et al. 2014a). The CR of intercrops was higher with pearl millet with 100% RDN and lowest when intercropped with maize with no N (Layek et al. 2014a). As the value of CR for soybean was less compared to intercropped cereals, the soybean is considered to be less competitive than cereals (Ghosh 2004; Meena et al. 2015). The CR values of pearl millet were higher than sorghum and maize in soybean+cereal intercropping, indicating that pear millet was more competitive than that of maize and sorghum (Layek et al. 2014a).The increasing levels of N enhanced their competitive capacity and increased their CR, while the lesser levels of N showed lesser competitive ability recording lesser CR (Dhima et al. 2007).

11.9.6 Actual Yield Loss (AYL)

The AYL is the proportionate yield loss or gain of component crops in an intercropping system in comparison to the respective pure stand. This index is based on the actual sown proportion of the component crops with its sole stand. It is calculated by the following formula (Banik 1996):

$$\mathrm{AYL} = \mathrm{AYL}_{\mathrm{cereal}} + \mathrm{AYL}_{\mathrm{legume}}$$

$\mathrm{AYL}_{\mathrm{cereal}}$ and $\mathrm{AYL}_{\mathrm{legume}}$ (partial actual yield loss of cereal and intercrop, respectively) represent the proportionate yield loss or gain of cereals and legumes in intercropping, about their yield in sole cropping.

$$\mathrm{AYL}_{\mathrm{cereal}} = \{\mathrm{Yab} / \mathrm{Zab}\} / \{(\mathrm{Yaa} / \mathrm{Zaa}) - 1\}$$

$$\mathrm{AYL}_{\mathrm{legume}} = \{\mathrm{Yba} / \mathrm{Zba}\} / \{(\mathrm{Ybb} / \mathrm{Zbb}) - 1\}$$

Yab = yield of s cereal “a” in intercropping
Zab = sown proportion of cereal “a” in intercropping
Yaa = yield of cereal “a” in sole cropping
Zaa = sown proportion of cereal “a” in sole cropping
Yba = yield of i legume “b” in intercropping
Zba = sown proportion of legume “b” in intercropping
Ybb = yield of legume “b” in sole cropping
Zbb = sown proportion of legume “b” in sole cropping

The AYL may have either positive or negative values indicating advantage or disadvantage in intercropping, respectively, when the yield is compared on per plant basis (Banik et al. 2000). The AYL index gives precise information on the inter- and intraspecific competition as well as the behavior of component crops in an intercropping system (Banik et al. 2000).

11.9.7 Intercropping Advantage (IA)

Intercropping advantage is calculated by comparing the advantage of intercropping in the monetary term (Banik et al. 2000).

$$\mathrm{IA}_{\mathrm{cereal}} = \left(\mathrm{AYL}_{\mathrm{cereal}}\right) \times \left(P_{\mathrm{cereal}}\right)$$

$$\mathrm{IA}_{\mathrm{legume}} = \left(\mathrm{AYL}_{\mathrm{legume}}\right) \times \left(P_{\mathrm{legume}}\right)$$

$$\mathrm{IA} = \mathrm{IA}_{\mathrm{cereal}} + \mathrm{IA}_{\mathrm{legume}}$$

where P_{cereal} is the commercial value of intercropped cereal (\$ Mg^{-1}) and P_{legume} is the commercial value of intercropped legume (\$ Mg^{-1}). Partial IA values for intercropped cereals were positive meaning intercrop cereals got a certain advantage due to intercropping with a legume. In a similar type of experiment, positive partial IA values for intercrops were recorded due to intercropping with cotton as reported by Muhammad Aasim et al. (2008) and Ram and Meena (2014).

11.9.8 Monetary Advantage Index (MAI)

The MAI is computed by using the following formula:

$$\mathrm{MAI} = \frac{\mathrm{LER} - 1}{\mathrm{LER}} \times \text{Value of combined intercrops}$$

where LER is a land equivalent ratio. The higher the MAI value, the better the cropping system.

11.10 Improving Productivity of Rice in Shifting Cultivation Area Through Legume-Based Intercropping: A Case Study

Shifting agriculture (also referred to as slash-and-burn agriculture) is prevalent among hilly areas of the northeast region (NER) of India where forest lands are slashed and burned to generate nutrients in situ to support the crop cultivation for a few years before the area is abandoned. Shifting cultivation or locally called *jhum* is practiced in about 0.88 million hectares in the NER. Generally, the farmers cultivate rice or maize crop in 1 or 2 years after burning the biomass and abandon it for several years for generating fertility.

Rice occupies a significant portion of this jhum area, but the productivity is very less (<1 Mg/ha). Plant-available nitrogen is generally low in shifting cultivation areas (Bruun et al. 2006; Yadav et al. 2017b), and the transfer of mineral N from burning of the aboveground plant biomass to the soil is also limited (Romanya et al. 2001). A participatory research was conducted in the farmers' *jhum* field of Sonidan village, Ri-Bhoi district, Meghalaya, India, to identify suitable rice varieties/lines and improved agronomic management practices for improving productivity and sustainable soil health. Five local rice varieties and six improved varieties were evaluated under both local and improved management practices. Soybean and groundnut were successfully grown as intercrops with rice. *Tephrosia* sp.

(leguminous shrub) was grown along the contour at suitable intervals for checking soil loss and improving soil fertility (Fig. 11.5). Improved package of practices was followed except for nutrient management practices. Rice and soybean and groundnut were sown in lines 30 cm apart across the slopes in 4:2 ratio (Fig. 11.7). Soybean and groundnut residues were incorporated to fertility improvement, reducing soil loss in the jhum field. Application of 50% recommended dose of nutrients (RDN) (30:30:20 kg N:P_2O_5:K_2O/ha) either through fertilizer or fertilizer+FYM intercropped with soybean and groundnut both recorded 40–60% enhancement in rice yield in *jhum* field (Fig. 11.6).

Fig. 11.5 *Tephrosia* sp. Grown as hedge row across the slope in between rice plots

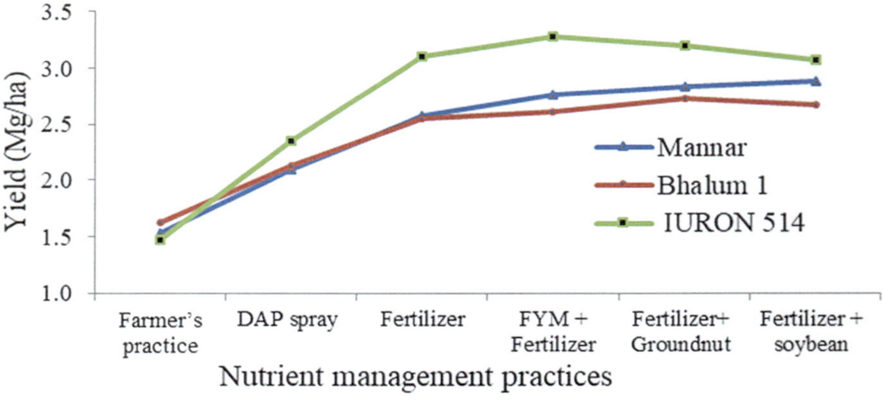

Fig. 11.6 Yield of rice as influenced by different nutrient management practices in jhum land (Layek et al. 2014c)

Rice + groundnut (4:2 ratio) in *jhum* field   Rice + soybean (4:2 ratio) in *jhum* field

Fig. 11.7 Intercropping of soybean and groundnut with rice in jhum land of Meghalaya, India

11.11 Constraints of Legume+Cereal Intercropping System

Although the benefits of cereal+legume intercropping systems are well documented, there are some constraints that need to be given due attention to popularize the cereal+legume intercropping system (Bationo et al. 2011; Mugendi et al. 2011; Maphumo 2011). In the acidic soil of different parts of the world, the low soil-available P and some micronutrients limit the growth of associated nitrogen-fixing *Rhizobium* (Sanchez et al. 1997) and consequently lower the BNF in soil (Fujita and Ofosu-Budu 1996; Giller 2001). The existing financial, technical, and infrastructural problems in the developing and underdeveloped countries limit the farmers to use mineral fertilizers, quality seeds, and other farm inputs, which in turn results in poor crop performance (Maphumo 2011). The poor market facility and lack of appropriate government policy and well-developed extension machineries limit the popularization of cereal+legume intercropping systems in farmers' land. Most of the cereal+legume intercropping researches are conducted in scientific farms with very little or no involvement of farming community particularly the resource-poor farmers (Matusso et al. 2014; Kumar et al. 2016).

11.12 Conclusion

The role of cereal+legume intercropping systems for improving the productivity and profitability and sustaining the soil health through improving physical, chemical, and biological soil parameters is well established. This system has huge potential to improve the food production from the marginal and degraded land of the developing countries especially in light of the changing climate. However, proper identification of location-specific cereals and legumes and their arrangement is necessary to optimally use the available resources. There is also a need to quantify the amount of N fertilizer to be added in the diverse cereal+legume intercropping systems for better nutrient productivity and profitability. Large-scale demonstration and scientific knowledge dissemination through extension machineries are also the need of the hour to popularize such intercropping system globally especially for marginal and resource-poor farming community for the sustainable development of agriculture.

11.13 Future Prospective

Although the history of intercropping is very old, it gets very little attention from researchers, farmers, and policy makers as a way for sustainable farming to provide diversified food. Participatory research on farmers' field involving small and marginal farmers, extension workers, and other related stakeholders is very much needed to create awareness on the role of legume in fixing nitrogen besides sustaining the productivity and improving soil quality and economic profitability. Proper supply of efficient strains of *Rhizobium* along with handling and storage of inoculation materials is very important to get the optimum result in legume intercropping. There is also a need to develop varieties, machineries, and research methodology specifically adopted for intercropping systems, so that it can be popular not only for the peasant farming but for commercial farming.

References

Ali RI, Awan TH, Ahmad M, Saleem MU, Akhtar M (2012) Diversification of rice-based cropping systems to improve soil fertility, sustainable productivity and economics. J Anim Plant Sci 22(1):108–112

Altieri MA (1994) Biodiversity and pest management in agroecosystems. Foods Products Press, New York

Anil L, Park RHP, Miller FA (1998) Temperate intercropping of cereals for forage: a review of the potential for growth and utilization with particular reference to the UK. Grass Forage Sci 53:301–317

Antoun H, Beauchamp CJ, Goussard N, Chabot R, Lalande R (1998) Potential of *Rhizobium* and *Bradyrhizobium* species as plant growth promoting rhizobacteria on non-legumes: effect on radishes (*Raphanus sativus* L.). Plant Soil 204:57–67

Baker CM, Blamey FPC (1985) Nitrogen fertilizer effects on yield and nitrogen uptake of sorghum and soybean, grown in sole cropping and intercropping systems. Field Crop Res 12:233–240

Balde AB, Scopel E, Affholder F, Corbeels M, Da Silva FAM, Xavier JHV, Wery J (2011) Agronomic performance of no-tillage relay intercropping with maize under smallholder conditions in Central Brazil. Field Crop Res 124(2):240–251

Banik P (1996) Evaluation of wheat (*Triticumaestivum*) and legume intercropping under 1:1 and 2:1 row replacement series system. J Agron Crop Sci 176:289–294

Banik P, Samsal T, Ghosal PK, Bagchi DK (2000) Evaluation of mustard (*Brassica campestris* var. toria) and legume intercropping under 1:1 and 1:2 row replacement series system. J Agron Crop Sci 185:9–14

Banik P, Midya A, Sarkar BK, Ghosh SS (2006) Wheat and chickpea intercropping systems in an additive series experiment: advantages and weed smothering. Eur J Agron 24:325–332

Bationo A, Kimetu J, Vanlauwe B, Bagayoko M, Koala S, Mokwunye AU (2011) Comparative analysis of the current and potential role of legumes in integrated soil fertility management in West and Central Africa. In: Bationo A, et al (eds) Fighting poverty in Sub-Saharan

Africa: the multiple roles of legumes in integrated soil fertility management. https://doi.org/10.1007/978-94-007-1536-3-6

Baudoin E, Nazaret S, Mougel C, Ranjard L, Moënne-Loccoz Y (2009) Impact of inoculation with the phytostimulatory PGPR *Azospirillum lipoferum* CRT1 on the genetic structure of the rhizobacterial community of field-grown maize. Soil Biol Biochem 41:409–413

Bedoussac L, Justes E (2010) Dynamic analysis of competition and complementarity for light and N use to understand the yield and the protein content of a durum wheat–winter pea intercrop. Plant Soil 330:37–54

Bedoussac L, Journet EP, Hauggaard-Nielsen H, Naudin C, Corre-Hellou G, Jensen ES, Prieur L, Justes E (2015) Ecological principles underlying the increase of productivity achieved by cereal-grain legume intercrops in organic farming: a review. Agron Sustain Dev 35:911–935

Beets WC (1982) Multiple cropping and tropical farming systems. Westview Press, Boulder

Behairy TG (1994) Effect of intercropping patterns on soybean growth and photosynthetic apparatus. Egypt J Physiol Sci 18:167–178

Bhat RA, Wani A, Beigh MAH, Dawson J (2008) Integrated nitrogen management on the growth and yield of maize (*Zea mays* L.) under conditions of Uttar Pradesh. Asian J Hort 3:229–231

Brooker RW, Bennett AE, Cong WF, Daniell TJ, George TS, Hallett PD, Hawes C, Iannetta PP, Jones HG, Karley AJ, Li L (2015) Improving intercropping: a synthesis of research in agronomy, plant physiology and ecology. New Phytol 206:107–117

Bruun TB, Mertz O, Elberling B (2006) Linking yields of upland rice in shifting cultivation to fallow length and soil properties. Agric Ecosyst Environ 113:139–149

Buragohain S, Sharma B, Nath JD, Gogaoi N, Meena RS, Lal R (2017) Impact of ten years of bio-fertilizer use on soil quality and rice yield on an inceptisol in Assam, India. Soil Res. https://doi.org/10.1071/SR17001

Caballero R, Goicoechea EL, Hernaiz PJ (1995) Forage yields and quality of common vetch and oat sown at varying seeding ratios and seeding rates of vetch. Field Crop Res 41:135–140

Cai H, You M, Lin C (2010) Effects of intercropping systems on community composition and diversity of predatory arthropods in vegetable fields. Acta Ecol Sin 30:190–195

Carlson JD (2008) Intercropping with maize in sub-arid regions. In: Community planning and analysis. Technical brief, April 16th, 6pp

Chai Q, Yu AZ, Chen GP, Huang P (2011) Soil evaporation under sole cropping and intercropping systems and the main driving factors. Chin J Eco Agric 6:014

Chalka MK, Nepalia V (2006) Nutrient uptake appraisal of maize intercropped with legumes and associated weeds under the influence of weed control. Indian J Agric Res 40:86–91

Davies B, Baulcombe D, Crute I, Dunwell J, Gale M, Jones J, Pretty J, Sutherland W, Toulmin C (2009) Reaping the benefits: science and the sustainable intensification of global agriculture. Royal Society, London

De Wit CT (1960) On competition. Verslag Land bouwkundige Onderzoekingen. Wageningen 66:1–81

Dhakal Y, Meena RS, De N, Verma SK, Singh A (2015) Growth, yield and nutrient content of mungbean (*Vigna radiata* L.) in response to INM in eastern Uttar Pradesh, India. Bangladesh J Bot 44(3):479–482

Dhakal Y, Meena RS, Kumar S (2016) Effect of INM on nodulation, yield, quality and available nutrient status in soil after harvest of green gram. Legum Res 39(4):590–594

Dhima KV, Lithourgidis AS, Vasilakoglou IB, Dordas CA (2007) Competition indices of common vetch and cereal intercrops in two seeding ratio. Field Crop Res 100:249–256

Dietrich JP, Schmitz C, Lotze-Campen H, Popp A, Müller C (2014) Forecasting technological change in agriculture – an endogenous implementation in a global land use model. Technol Forecast Soc Chang 81:236–249ed

Drechsel P, Heffer P, Magen H, Mikkelsen R, Singh H, Wichelns D (2015) Managing water and nutrients to ensure global food security, while sustaining ecosystem services. In: Drechsel et al. Managing water and fertilizer for sustainable agricultural intensification, 1st edn, IFA, IWMI, IPNI, IPI, Paris, pp 1–7, ISBN: 979-10-92366-02-0

Duraisami VP, Raniperumal, Mani AK (2002) Grain quality as influenced by fertilizer nitrogen, coir pith and biofertilizer in sole and intercropped sorghum-maize-soybean sequence. Mysore J Agric Sci 36:97–103

Dwivedi A, Dev I, Kumar V, Yadav RS, Yadav M, Gupta D, Singh A, Tomar SS (2015) Potential role of maize-legume intercropping systems to improve soil fertility status under smallholder farming systems for sustainable agriculture in India. Int J Life Sci Biotechnol Pharma Res 4(3):145

Eaglesham ARJ, Ayanaba A, Ranga Rao V, Eskew DL (1981) Improving the nitrogen nutrition of maize by intercropping with cowpea. Soil Biol Biochem 13:169–171

Elkoca E, Turan M, Donmez MF (2010) Effects of single, dual and triple inoculations with *Bacillus subtilis*, *Bacillus megaterium* and *Rhizobium leguminosarum* bv. Phaseoli on nodulation, nutrient uptake, yield and yield parameters of common bean (*Phaseolus vulgaris* l. cv. 'elkoca-05'). J Plant Nutr 33:2104–2119

Fetene M (2003) Intra-and inter-specific competition between seedlings of *Acacia etbaica* and a perennial grass (*Hyparrhenia hirta*). J Arid Environ 55:441–451

Fox RH, Myers RJK, Vallis I (1990) The nitrogen mineralization rate of legume residues in soil as influenced by their polyphenol, lignin, and nitrogen contents. Plant Soil 129:251–259

Fujita K, Ofosu-Budu KG (1996) Significance of intercropping in cropping systems. In: Ito O, Johansen C, Adu-Gyamfi JJ, Katayama K, Kumar Rao JVDK, Rego TJ (eds) Dynamics of roots and nitrogen in cropping systems of the semi- arid tropics, International agricultural series no. 3. Japan International Research Center for Agricultural Sciences, Tsukuba, pp 19–40

Fujita K, Ofosu-Budu KG, Ogata S (1992) Biological nitrogen fixation in mixed legume-cereal cropping systems. Plant Soil 141:155–176

Gan YT, Liang C, Hamel C, Cutforth H, Wang H (2011) Strategies for reducing the carbon footprint of field crops for semiarid areas: a review. Agron Sustain Dev 31:643–656

Ganeshamurthy AN, Ali M, Srinivasarao CH (2006) Role of pulses in sustaining soil health and crop production. Indian J Fert 1(12):29–40

Gao Y, Duan A, Qiu X, Sun J, Zhang J, Liu H, Wang H (2010) Distribution and use efficiency of photosynthetically active radiation in strip intercropping of maize and soybean. Agron J 102:1149–1157

Gare BN, More SM, Burli AV, Dodake SS (2009) Evaluation of yield stability in soybean based intercropping system under rainfed agriculture. Indian J Dryland Agric Res Dev 24:84–87

Ghaffarzadeh M, Prechac FG, Cruse RM (1994) Grain yield response of corn, soybean, and oat grown in a strip intercropping system. Am J Altern Agric 9:171–177

Ghanbari A, Dahmardeh M, Siahsar BA, Ramroudi M (2010) Effect of maize (*Zea mays* L.)-cowpea (*Vigna unguiculata* L.) intercropping on light distribution, soil temperature and soil moisture in arid environment. J Food Agric Environ 8(1):102–108

Ghosh PK (2004) Growth, yield, competition and economics of groundnut/cereal fodder intercropping systems in the semi-arid tropics of India. Field Crop Res 88:227–237

Ghosh PK, Mohanty M, Bandyopadhyay KK, Painuli DK, Misra AK (2006) Growth, competition, yields advantage and economics in soybean/pigeonpea intercropping system in semi–arid tropics of India: II. Effect of nutrient management. Field Crop Res 96:90–97

Ghosh PK, Bandyopadhyay KK, Wanjari RH, Manna MC, Misra AK, Mohanty M, Rao AS (2007) Legume effect for enhancing productivity and nutrient use-efficiency in major cropping systems–an Indian perspective: a review. J Sustain Agric 30:59–86

Ghosh PK, Tripathi AK, Bandyopadhyay KK, Manna MC (2009) Assessment of nutrient competition and nutrient requirement in soybean/sorghum intercropping system. Eur J Agron 31:43–50

Gianoli E, Ramos I, Alfaro-Tapia A, Valdez Y, Echegaray ER, Yabar E (2006) Benefits of a maize–bean–weeds mixed cropping system in Urubamba Valley, Peruvian Andes. Int J Pest Manag 52:283–289

Giller KE (2001) Nitrogen fixation in tropical cropping systems, 2nd edn. CABI, Wallingford

Giller KE, Wilson KJ (1991) Nitrogen fixation and tropical cropping systems. CABI, Wallingford

Gomes AA, Gomez KA (1983) Multiple cropping in the humid tropic of Asia, IDRC-176E. IDRC, Ottawa

Hauggaard-Nielsen H, Ambus P, Jensen ES (2001) Interspecific competition, N use and interference with weeds in pea–barley intercropping. Field Crop Res 70:101–109

Hauggaard-Nielsen H, Jornsgaard B, Kinane J, Jensen ES (2008) Grain legume–cereal intercropping: the practical application of diversity, competition and facilitation in arable and organic cropping systems. Renew Agric Food Syst 23:3–12

Haynes RJ (1980) Competitive aspects of the grass-legume association. Adv Agron 33:227–261

Herridge DF, Maecellos H, Felton WL, Turner GL, Peoples MB (1995) Chickpea increases soil N fertility in cereal systems through nitrate sparing and N_2 fixation. Soil Biol Biochem 27:545–551

Hinsinger P, Betencourt E, Bernard L, Brauman A, Plassard C, Shen J, Tang X, Zhang F (2011) P for two, sharing a scarce resource: soil phosphorus acquisition in the rhizosphere of intercropped species. Plant Physiol 156(3):1078–1086

Hulugalle NR, Lal R (1986) Soil water balance in intercropped maize and cowpea grown in a typical hydromorphic soil in western Nigeria. Agron J 77:86–90

IAASTD (2009) International assessment of agricultural knowledge, science and technology for development. Global Report. http://apps.unep.org/publications/pmtdocuments/-Agriculture%20at%20a%20crossroads%20-%20Synthesis%20report2009AgricultureatCrossroadsSynthesisReport.pdf

Ijoyah MO, Fanen FT (2002) Effects of different cropping pattern on performance of maize-soybean mixture in Makurdi, Nigeria. Sci J Crop Sci 1:39–47

Inal A, Gunes A, Zhang F, Cakmak I (2007) Peanut/maize intercropping induced changes in rhizosphere and nutrient concentrations in shoots. Plant Physiol Biochem 45:350–356

Innis DQ (1997) Intercropping and the scientific basis of traditional agriculture. Intermediate Technology Publications Ltd, London

Intodia SL, Intodia K, Intodia SK (2007) Effect of nitrogen, phosphorus and zinc fertilization on nutrient utilization by maize (*Zea mays* L.) on calcareous soils of south Rajasthan. Int J Chem Sci 5:209–216

Jat SK, Shivay YS, Parhar CM, Meena HN (2012) Evaluation of summer legumes for their economic feasibility, nutrient accumulation and soil fertility. J Food Legum 25(3):239–242

Jensen ES (1996) Grain yield, symbiotic N_2 fixation and interspecific competition for inorganic N in pea-barley intercrops. Plant Soil 182:25–38

Jeyabal A, Kuppuswamy G (2001) Recycling of organic wastes for the production of vermicompost and its response in rice–legume cropping system and soil fertility. Eur J Agron 15:153–170

Kalia RD, Singh RV, Singh R (1992) Performance of soybean intercropping with maize in different planting patterns under rainfed conditions of Himachal Pradesh. Haryana J Agron 8:78–80

Kariaga BM (2004) Intercropping maize with cowpeas and beans for soil and water management in Western Kenya. Proceedings of the 13th international soil conservation organization conference, July 2004, conserving soil and water for society, Brisbane

Kitonyo OM, Cheminingwa GN, Muthomi JW (2013) Productivity of farmer-preferred maize varieties intercropped with beans in semi-arid Kenya. Intl J Agron Agric Res 3:6–16

Kremen C, Miles A (2012) Ecosystem services in biologically diversified versus conventional farming systems: benefits, externalities, and trade-offs. Ecol Soc 17:40

Kumar K, Goh KM (2000) Crop residues and management practices: effects on soil quality, soil nitrogen dynamics, crop yield, and nitrogen recovery. Adv Agron 68:197–319

Kumar K, Reddy MD, Sivasankar A, Reddy NV (2003) Yield and economics of maize (*Zea mays*) and soybean (*Glycine max*) in intercropping under different row proportions. Indian J Agric Sci 73:69–71

Kumar RBP, Ravi S, Balyan JS (2008) Effect of maize (*Zea mays*) +black gram intercropping and integrated nitrogen management on productivity and economics of maize. Int J Plant Sci (Muzaffarnagar) 3:53–57

Kumar S, Sheoran S, Kumar SK, Kumar P, Meena RS (2016) Drought: a challenge for Indian farmers in context to climate change and variability. Progress Res An Int J 11:6243–6246

Kushwaha HS, Chandel AS (1997) Effect of soybean (*Glycine max*) intercropping under different nitrogen levels on yield, yield attributes and quality of maize (*Zea mays*). Indian J Agric Sci 67:249–252

Ladha JK, George T, Bohlool C (2013) Biological nitrogen fixation for sustainable agriculture: extended versions of papers presented in the symposium, role of biological nitrogen fixation in sustainable agriculture at the 13th congress of soil science Springer, Japan

Lado M, Ben-Hui M (2004) Organic matter and aggregate size interactions in infiltration, seal formation and soil loss. Soil Sci Soc Am J 68:935–942

Lal R, Hansen DO, Uphoff N, Slack S (2003) Food security and environmental quality in the developing world. CRC Press, Boca Raton. 464 pp

Layek J, Shivakumar BG, Rana DS, Munda S, Lakshman K (2012) Growth pattern, physiological indices and productivity of different soybean (*Glycine max*) based intercrops as influenced by nitrogen nutrition. Indian J Agron 57:349–356

Layek J, Anup D, Ramkrushna GI, Venkatesh A, Verma BC, Roy A, Panwar AS, Ngachan SV (2014a) Improving productivity of *jhum* rice through agronomic management practices. In: Book of abstracts. National seminar on shifting cultivation (*jhum*) in 21st century: fitness and improvement. 28–29 November, 2014, at CPGS, CAU, Umiam, Meghalaya, p 65

Layek J, Shivakumar BG, Rana DS, Munda S, Lakshman K, Das A, Ramkrushna GI (2014b) Soybean–cereal intercropping systems as influenced by nitrogen nutrition. Agron J 106:1–14

Layek J, Shivakumar BG, Rana DS, Munda S, Lakshman K, Panwar AS, Das A, Ramkrushna GI (2014c) Performance of soybean (*Glycine max*) intercropped with different cereals under varying levels of nitrogen. Indian J Agric Sci 85:1571–1557

Li L, Sun J, Zhang F, Guo T, Bao X, Smith FA, Smith SE (2006) Root distribution and interactions between intercropped species. Oecologia 147:280–290

Lithourgidis AS, Vasilakoglou IB, Dhima KV, Dordas CA, Yiakoulaki MD (2006) Forage yield and quality of common vetch mixtures with oat and triticale in two seeding ratios. Field Crop Res 99:106–113

Lithourgidis AS, Vlachostergios DN, Dordas CA, Damalas CA (2011) Dry matter yield, nitrogen content, and competition in pea–cereal intercropping systems. Eur J Agron 34:287–294

Liu Y, Wu L, Baddele JA, Watson CA (2011) Models of biological nitrogen fixation of legumes. A review. Agron Sustain Dev 31:155–172

Lupwayi NZ, Kennedy AC (2007) Grain legumes in northern plains: impacts on selected biological processes. Agron J 99:1700–1709

Maingi MJ, Shisanya AC, Gitonga MN, Hornetz B (2001) Nitrogen fixation by common bean (*Phaseolus vulgaris* L.) in pure and mixed stands in semi-arid south east Kenya. Eur J Agron 14:1–12

Malezieux E (2012) Designing cropping systems from nature. Agron Sustain Dev 32(1):15–29

Mao L, Zhang L, Li W, van der Werf W, Sun J, Spiertz H, Li L (2012) Yield advantage and water saving in maize/pea intercrop. Field Crop Res 138:11–20

Maphumo P (2011) Comparative analysis of the current and potential role of legumes in integrated soil fertility management in southern Africa. In: Bationo A et al (eds) Fighting poverty in Sub-Saharan Africa: the multiple roles of legumes in integrated soil fertility management. Springer, Dordrecht, pp 175–200. https://doi.org/10.1007/978-94-007-1536-3-8

Marchiol L, Miceli F, Pinosa M, Zerbi G (1992) Intercropping of soybean and maize for silage in northern Italy: effect of nitrogen level and plant density on growth, yield and protein content. Eur J Agron 1:207–211

Mashhadi HR, Zand E (2004) Canopy profile distribution of leaf area, light and nitrogen in some Iranian winter wheat cultivars released during the last 50 years. In: Proceedings of the 4th international crop science congress, Brisbane, Australia, 26th September–1st October 2004

Matusso JMM, Mugwe JN, Mucheru-Muna M (2014) Potential role of cereal-legume intercropping systems in integrated soil fertility management in smallholder farming systems of Sub-Saharan Africa. Res J Agric Env Manag 3:162–174

Maurya BM, Rathi KS (2000) Growth and development of soybean as influenced by intercropping with pigeonpea and phosphorus level. Gujarat Agric Univ Res J 26:1–5

McGilchrist CA (1965) Analysis of competition experiments. Biometrics 21:975–985
Mead R, Willey RW (1980) The concept of a 'land equivalent ratio' and advantages in yields from intercropping. Exp Agric 16:217–228
Meena OP, Gaur BL, Singh P (2006) Effect of row ratio and fertility levels on productivity, economics and nutrient uptake in maize (*Zea mays*)+soybean (*Glycine max*) intercropping system. Indian J Agron 51:178–182
Meena VS, Maurya BR, Meena RS, Meena SK, Singh NP, Malik VK (2014) Microbial dynamics as influenced by concentrate manure and inorganic fertilizer in alluvium soil of Varanasi, India. Afr J Microb Res 8(1):257–263
Meena VS, Maurya BR, Meena RS (2015) Residual impact of well-grow formulation and NPK on growth and yield of wheat (*Triticumaestivum* L.). Bangladesh J Bot 44(1):143–146
Meena RS, Yadav RS, Meena H, Kumar S, Meena YK, Singh A (2015a) Towards the current need to enhance legume productivity and soil sustainability worldwide: a book review. J Clean Prod 104:513–515
Meena RS, Meena VS, Meena SK, Verma JP (2015b) The needs of healthy soils for a healthy world. J Clean Prod 102:560–561
Meena HR, Meena RS, Lal R, Singh GS, Mitran T, Layek J, Patil SB, Kumar S, Verma T (2017a) Response of sowing dates and bio regulators on yield of clusterbean under current climate in alley cropping system in eastern U.P. Indian Leg Res. Accepted in press
Meena RS, Kumar V, Yadav GS, Mitran T (2017b) Response and interaction of *Bradyrhizobium japonicum* and *Arbuscular mycorrhizal* fungi in the soybean rhizosphere: a review. Plant Growth Regul. Accepted in press
Miao Q, Rosa RD, Shi H, Paredes P, Zhu L, Dai J, Gonçalves JM, Pereira LS (2016) Modeling water use, transpiration and soil evaporation of spring wheat–maize and spring wheat–sunflower relay intercropping using the dual crop coefficient approach. Agric Water Manag 165:211–229
Milder JC, Terhi M, Sara JS (2011) Performance and potential of conservation agriculture for climate change adaptation and mitigation in Sub-Saharan Africa. In: Ecoagriculture discussion paper no. 6. 2011. EcoAgriculture Partners, Washington, DC
Mugendi DN, Waswa BS, Mucheru-Muna MW, Kimetu JM, Palm C (2011) Comparative analysis of the current and potential role of legumes in integrated soil fertility management in East Africa. In: Bationo A et al (eds) Fighting poverty in Sub-Saharan Africa: the multiple roles of legumes in integrated soil fertility management. https://doi.org/10.1007/978-94-007-1536-3-8
Mugwe J, Mugendi DN, Mucheru-Muna M, Kungu J (2011) Soil inorganic N and N uptake by maize following application of legume biomass, tithonia, manure and mineral fertilizer in Central Kenya. In: Bationo A, et al (eds) Innovations as key to the green revolution in Africa. Springer, Dordrecht, pp 605–612. https://doi.org/10.1007/978-90-481-2543-2_62
Muhammad Aasim, Ejaz Muhammad Umer, Karim Abdul (2008) Yield and competition indices of intercropping cotton (*Gossypium hirsutum* L.) using different planting patterns. Tarim B I L Imleri Derg 14:326–333
Muoneke CO, Ogwuche MAO, Kalu BA (2007) Effect of maize planting density on the performance of maize/soybean intercropping system in a guinea savannah agroecosystem. Afr J Agric Res 2:667–677
Musa M, Leitch MH, Iqbal M, Sahi FUH (2010) Spatial arrangement affects growth characteristics of barley-pea intercrops. Int J Agric Biol 12:685–690
Nieder R, Benbi DK (2008) Carbon and nitrogen in the terrestrial environment. Springer, Dordrecht, p 430
Odendo M, Bationo A, Kimani S (2011) Socio- economic contribution of legumes to livelihoods in sub-saharan africa. In: Bationo A et al (eds) Fighting poverty in Sub-Saharan Africa: the multiple roles of legumes in integrated soil fertility management. Springer, Dordrecht, pp 27–46. https://doi.org/10.1007/978-94-007-1536-3-8
Ofori F, Stern WR (1987) Cereal-legume intercropping systems. Adv Agron 41:41–90

Ogindo HO, Walker S (2005) Comparison of measured changes in seasonal soil water content by rainfed maize-bean intercrop and component cropping in semi arid region in South Africa. Phys Chem Earth 30:799–808
Pandita AK, Shah MH, Bali AS (2000) Effect of row ratio in cereal-legume intercropping systems on productivity and competition functions under Kashmir conditions. Indian J Agron 45(1):48–53
Papastylianou I (2004) Effect of rotation system and N fertilizer on barley and common vetch grown in various crop combinations and cycle lengths. J Agric Sci (Camb) 142:41–48
Patra BC, Patra AP (2010) Productivity of soybean and sesame as influenced by intercropping in entisol of West Bengal. Environ Ecol 28:262–265
Peoples MB, Hauggaard-Nielsen H, Jensen ES (2009) The potential environmental benefits and risks derived from legumes in rotations. In: Emerich DW, Krishnan HB (eds) Nitrogen fixation in crop production. American Society of Agronomy, Madison
Qamar IA, Keatinge JDH, Mohammad N, Ali A, Khan MA (1999) Introduction and management of common vetch/barley forage mixtures in the rainfed areas of Pakistan. 3. Residual effects on following cereal crops. Aus J Agric Res 50:21–27
Ram K, Meena RS (2014) Evaluation of pearl millet and mungbean intercropping systems in Arid Region of Rajasthan (India). Bangladesh J Bot 43(3):367–370
Ramakrishnan PS (2007) Traditional forest knowledge and sustainable forestry: a north-east India perspective. For Ecol Manag 249(1):91–99
Ren Y, Liu J, Wang Z, Zhang S (2016) Planting density and sowing proportions of maize–soybean intercrops affected competitive interactions and water-use efficiencies on the Loess Plateau, China. Eur J Agron 72:70–79
Roldan A, Caravaca F, Hernandez MT, Garci C, Sanchez-Brito C, Velasquez M, Tiscareno M (2003) No-tillage, crop residue additions, and legume cover cropping effects on soil quality characteristics under maize in Patzcuaro watershed (Mexico). Soil Tillage Res 72:65–73
Romanya J, Casals P, Vallejo VR (2001) Short term effects of fire on soil nitrogen availability in Mediterranean grasslands and shrublands growing in old fields. For Ecol Manag 147:39–53
Sainju UM, Stevens WB, Caesar-TonThat T, Liebig MA (2012) Soil greenhouse gas emissions affected by irrigation, tillage, crop rotation, and nitrogen fertilization. J Environ Qual 41(6):1774–1786
Sanchez PA, Shepherd KD, Soule MJ, Place FM, Buresh RJ, Izac AMN, Uzo Mokwunye A, Kwesiga FR, Ndiritu CG, Woomer PL (1997) Soil fertility replenishment in Africa: an investment in natural resource capital. In: Buresh RJ, Sanchez PA, Calhoun F (eds) Replenishing soil fertility in Africa, Special Publication, No. 51. Soil Science Society of America, Madison, pp 1–46
Satyam B, Masthan SC, Reddy BB (2008) Economics of different levels of nitrogen application in maize based intercropping systems with legumes under rainfed conditions. Indian J Dryland Agric Res Dev 23:74–79
Savci S (2012) An agricultural pollutant: chemical fertilizer. Int J Env Sci Dev 3:73
Sawargaonkar GL, Shelke DK, Shinde SA, Kshirsagar S (2008) Performance of *kharif* maize based legumes intercropping systems under different fertilizer doses. Int J Agric Sci 4:152–155
Senaratne R, Liyanage NDL, Ratnasinghe DS (1993) Effect of K on nitrogen fixation of intercrop groundnut and the competition between intercrop groundnut and maize. Fertil Res 34:9–14
Seran TH, Brintha I (2010) Review on maize based intercropping. J Agron 9:135–145
Shah Z, Shah SH, Peoples MB, Schwenke GD, Herridge DF (2003) Crop residue and fertiliser N effects on nitrogen fixation and yields of legume–cereal rotations and soil organic fertility. Field Crop Res 83:1–11
Sharma AR, Behera UK (2009) Recycling of legume residues for nitrogen economy and higher productivity in maize (*Zea mays*)–wheat (*Triticum aestivum*) cropping system. Nutr Cycl Agroecosyst 83:197
Sharma RS, Agrawal KK, Jain KK (1994) Influence of spatial arrangement and nitrogen levels on light utilization and productivity in maize-soybean intercropping system. J Oilseeds Res 11:217–221

Sharma S, Aneja MK, Mayer J, Munch JC, Schloter M (2005) Characterization of bacterial community structure in rhizosphere soil of grain legumes. Microb Ecol 49:407–415

Shetty SVR, Ntare BR, Bationo A, Renard C (1995) Millet and cowpea in mixed farming systems of the sahel: a review of strategies for increased productivity and sustainability. In: Livestock and sustainable nutrient cycling in mixed farming systems of sub-Saharan Africa. Volume 2: technical papers. Proceedings, Addis Ababa, Ethiopia, 22–26 November 1993, pp 293–303

Shivay YS, Singh RP (2000) Growth, yield attributes, yields and nitrogen uptake of maize (*Zea mays* L.) as influenced by cropping systems and nitrogen levels. Ann Agric Res 21(4):494–498

Shivay YS, Singh RP, Pal M (2001) Productivity and economics of maize as influenced by intercropping with legumes and nitrogen levels. Ann Agric Res 22:576–558

Shivay YS, Singh RP, Pandey CS (2002) Physiological analysis of growth in maize (*Zea mays* L.) as influenced by cropping systems and nitrogen levels. Indian J Plant Physiol 7:126–130

Singh K, Singh H, Hooda RS, Pannu RK, Singh VP (2007) Nitrogen levels, times of application and weed management effects on yield and nutrients uptake by pearl millet. Res Crop 8:89–92

Singh KP, Ved P, Srinivas K, Srivastava AK (2008) Effect of tillage management on energy-use efficiency and economics of soybean (*Glycine max*) based cropping systems under the rainfed conditions in North-west Himalayan region. Soil Tillage Res 100:78–82

Singh KK, Ali M, Venkatesh MS (2009) Pulses in cropping systems. IIPR, Kanpur, p 39

Steiner KG (1982) Intercropping in tropical smallholder agriculture with special reference to West Africa. Schriftenreihe der GT2, N. 137, Eischbon

Stern WR (1993) Nitrogen fixation and transfer in intercrop systems. Field Crop Res 34:335–356

Strydhorst SM, King JR, Lopetinsky KJ, Harker KN (2008) Forage potential of strip intercropping: I. Yield advantage and interspecific interactions on nutrients. Field Crop Res 71:123–137

Subba Rao GV, Kumar Rao JVDK, Kumar J, Johansen C, Deb UK, Ahmed I et al (2001) Spatial distribution and quantification of rice fallow in South Asia—potential for legumes. ICRISAT, Hyderabad.; NRSA, Hyderabad and DFID, UK, p 315

Thiessen Martens JR, Entz MH, Hoeppner JW (2005) Legume cover crops with winter cereals in southern Manitoba: fertilizer replacement values for oat. Can J Plant Sci 85(3):645–648

Tscharntke T, Klein AM, Kruess A, Steffan-Dewenter I, Thies C (2005) Landscape perspectives on agricultural intensification and biodiversity–ecosystem service management. Ecol Lett 8:857–874

Tsubo M, Walker S, Ogindo HO (2005) A simulation model of cereal–legume intercropping systems for semi-arid regions: I. Model development. Field Crop Res 93:10–22

Ullah A, Bhatti MA, Gurmani ZA, Imran M (2007) Studies on planting patterns of maize (*Zea mays* L.) facilitating legumes intercropping. J Agric Res 45(2):113–118

Undie UL, Uwah DF, Attoe EE (2012) Effect of intercropping and crop arrangement on yield and productivity of late season maize/soybean mixtures in the humid environment of south southern Nigeria. J Agric Res 4(4):37

United Nations, Department of Economic and Social Affairs, Population Division (2017) World population prospects: the 2017 revision, key findings and advance tables. Working Paper No. ESA/P/WP/248

Vandermeer JH (1989) The ecology of intercropping. Cambridge University Press, Cambridge

Varma D, Meena RS, Kumar S (2017) Response of mungbean to fertility and lime levels under soil acidity in an alley cropping system in Vindhyan region, India. Int J Chem Stu 5(2):384–389

Veres A, Petit S, Conord C, Lavigne C (2013) Does landscape composition affect pest abundance and their control by natural enemies? A review. Agric Ecosyst Environ 166:110–117

Vesterager JM, Nielsen NE, Høgh-Jensen H (2008) Effect of cropping history and phosphorous source on yield and nitrogen fixation in sole and intercropped cowpea-maize systems. Nutr Cycl Agroecosyst 80:61–73

Wandahwa P, Tabu IM, Kendagor MK, Rota JA (2006) Effect of intercropping and fertilizer type on growth and yield of soybean (*Glycine max* L. Merrill). J Agron 5:69–73

Wang ZK, Zhao XN, Wu PT, He JQ, Chen XL, Gao Y, Cao X (2015) Radiation interception and utilization by wheat/maize strip intercropping systems. Agric For Meteorol 204:58–66

West TD, Griffith DR (1992) Effect of strip intercropping corn and soybean on yield and profit. J Prod Agric 5:107–110

Wezel A, Casagrande M, Celette F, Vian JF, Ferrer A, Peigne J (2014) Agroecological practices for sustainable agriculture. A review. Agron Sustain Dev 34(1):1–20

Willey RW, Rao MR (1980) A competitive ratio for quantifying competition between intercrops. Exp Agric 16:117–125

Willey RW, Natarajan M, Reddy MS, Rao MR, Nambiar PTC, Kannaiyan J, Bhatnagar VS (1983) Intercropping studies with annual crops. In: Better crops for food, vol 97. CIBA Foundation Symposium. pp 83–100

Yadav GS, Lal R, Meena RS, Datta M, Babu S, Das, Layek J, Saha P (2017a) Energy budgeting for designing sustainable and environmentally clean/safer cropping systems for rainfed rice fallow lands in India. J Clean Prod 158:29–37

Yadav GS, Lal R, Meena RS, Babu S, Das A, Bhomik SN, Datta M, Layak J, Saha P (2017b) Conservation tillage and nutrient management effects on productivity and soil carbon sequestration under double cropping of rice in North Eastern Region of India. Ecol India. http://www.sciencedirect.com/science/article/pii/S1470160X17305617

Yan F, Schubert S, Mengel K (1996) Soil pH changes during legume growth and application of plant material. Biol Fertil Soils 23:236–242

Yang F, Wang XC, Liao DP, Lu FZ, Gao RC, Liu WG, Yong T, Wu X, Du J, Liu J, Yang W (2015) Yield response to different planting geometries in maize-soybean relay strip intercropping systems. Agron J 107(1):296–304

Zhang F, Li L (2003) Using competitive and facilitative interactions in intercropping systems enhances crop productivity and nutrient-use efficiency. Plant Soil 248(1):305–312

Zhang G, Yang Z, Dong S (2011) Interspecific competitiveness affects the total biomass yield in an alfalfa and corn intercropping system. Field Crop Res 124:66–73

Zougmore R, Kambou FN, Ouattara K, Guillobez S (2000) Sorghum-cowpea intercropping: an effective technique against runoff and soil erosion in the Sahel (Saria, Burkina Faso). Arid Land Res Manag 14:329–334

Legume Green Manuring: An Option for Soil Sustainability

12

B. L. Meena, R. K. Fagodiya, K. Prajapat, M. L. Dotaniya, M. J. Kaledhonkar, P. C. Sharma, Ram Swaroop Meena, Tarik Mitran, and Sandeep Kumar

Contents

B. L. Meena (✉) · R. K. Fagodiya · K. Prajapat · M. J. Kaledhonkar · P. C. Sharma
ICAR – Central Soil Salinity Research Institute, Karnal, India
e-mail: bl.meena@icar.gov.in

M. L. Dotaniya
ICAR – Indian Institute of Soil Science, Bhopal, India

R. S. Meena
Department of Agronomy, Institute of Agricultural Sciences (BHU), Varanasi, UP, India

T. Mitran
Soil and Land Resources Assessment division, NRSC, ISRO, Hyderabad, Telangana, India

Carbon Management and Sequestration Centre, The Ohio State University, Columbus, OH, USA

S. Kumar
Department of Agronomy, CCS Haryana Agricultural University, Hisar, Haryana, India

R. S. Meena et al. (eds.), *Legumes for Soil Health and Sustainable Management*,
https://doi.org/10.1007/978-981-13-0253-4_12

Abstract

Overuse of nitrogen (N) fertilizer to enhance agricultural production is threatening the environment. The concentrations of reactive forms (e.g., NOx, N_2O, NO_3^-, NH_3) of N have increased to around 120% in the atmosphere as a result of different industrial units and use of chemical fertilizers in agriculture. The scenario compels to rethink about the role of biological nitrogen fixation (BNF). Green manuring with inclusion of legumes appears to be the most feasible option. Intensive agriculture with repeated tillage, use of high-analysis fertilizers, burning of agricultural residue, and non-incorporation of biodegradable solid waste from domestic and industrial sectors into soil mass have resulted in the decline of soil organic carbon (SOC). This in turn impaired soil health, decreased soil biodiversity, and aggravated the demand for essential plant nutrients, leading to the agricultural land becoming less productive and sometimes unfit for economic cultivation. The uncontrolled use and improper management of synthetic fertilizers, especially, the nitrogenous fertilizers, emit nitrate (NO_3^-) causing water pollution and nitrous oxide (N_2O), speeding up climate change process and oxides of N (NO_x) causing air pollution. The OC and soil nitrogen have a positive correlation. It suggests that soil nitrogen level can be improved with improving levels of soil organic matter (SOM). It will also help in reducing environmental damage due to overuse of nitrogen fertilizers. Green manuring with legumes has added advantage as legumes fix atmospheric nitrogen and are easily decomposable. Legume green manuring (LGM) improves SOC, nutrient availability, physicochemical and biological properties of soil, and crop productivity. Several legumes which were used for green manuring showed high N accumulation rate, i.e., 80–100 kg ha^{-1} in duration of 45–60 days of crop growth. Legume crop cultivation, say seed legumes in symbiotic association with *Rhizobium*, contributes around 10 Tg N $year^{-1}$, while forage legumes (cover crops) contribute 12 Tg N $year^{-1}$. Application of LGM is an important option to optimize the BNF and to ensure soil sustainability. The LGM may have a realistic and applicable potential in the area where soil properties are marginal for crop production.

Keywords

Legume green manuring · Nitrogen fixation · Legume crops · Soil health

Abbreviations

BNF	Biological nitrogen fixation
C	Carbon
cm	Centimeter
CO_2	Carbon dioxide
FYM	Farm yard manure
GLM	Green leaf manuring
ha	Hectare

IRRI	International Rice Research Institute
K	Potassium
kg	Kilogram
LGM	Legume green manuring
Mg	Megagram
mm	Millimeter
MWD	Mean weight diameter
N	Nitrogen
N_2O	Nitrous oxide
NH_3	Ammonia
NO_3^-	Nitrate
NOx	Oxides of nitrogen
NUE	Nitrogen use efficiency
OC	Organic carbon
OM	Organic matter
P	Phosphorous
pb	Bulk density
SOC	Soil organic carbon
SOM	Soil organic matter
Tg	Teragram
WHC	Water holding capacity
WUE	Water use efficiency

12.1 Introduction

The current world population is ~7.3 billion, and it will further increase to 8.5 and 9.7 billion during 2030 and 2050, respectively, and this is expected to be stabilized at ~11.2 billion by the end of the twenty-first century (UNDESA 2015). This additional three to four billion people will require extra food grain production, from shrinking land and water resources, to ensure food and nutritional security (Sulieman and Tran 2015). Increase in quality food grain production under intensive agricultural management practices would only be possible by doubling the use of energy and fertilizers consumption (Sulieman and Tran 2015). Food grain production increased manyfold at global level during the twentieth century, and a tremendous yield increase was observed due to the increase in the net cultivable area. Further, use of short-duration high-yielding varieties, synthetic fertilizers, and pesticides is inevitable (Sihag et al. 2015). This approach has resulted in gradual degradation of soil organic matter (SOM) because of the breakdown of stable soil aggregates and decomposition of organic matter (OM). Consequently, soil health is deteriorated in terms of reduction in water holding capacity (WHC) of soils, surface and groundwater pollution, and multiple nutrient deficiencies (Gill et al. 2008; Meena et al. 2013). Soluble nutrient is provided by synthetic fertilizers for crop production that are easily vulnerable to loss, if soils and irrigation water are not properly managed.

Moreover, higher application of synthetic fertilizers has led to an imbalance of the nutrient cycle, particularly N, illustrated by the growing accumulation of several reduced (NH_3) and oxidized (NOx, N_2O, NO_3^-) forms (Fagodiya et al. 2017), causing water pollution (NO_3^-), air pollution (NOx), and climate change (N_2O) (Galloway et al. 2003). There are severe concerns about sustainable soil productivity, and today most of the countries have moved into a post-green revolution phase and are facing the problem of stagnation or declining crop productivity. Hence, both farmers and researchers have opted for conservation agriculture practices, resources conservation, and use of green manuring into the farming system to enhance further food grain production while maintaining soil health (Meena and Majumdar 2016; Meena et al. 2016a, b, c).

It has been widely reported that leguminous green manure crops play an important role in soil health management (Whitbread et al. 2000) and recently received higher attention for improving soil fertility and agricultural sustainability (Ray and Gupta 2001; Fageria 2007). Green manuring is the practice of incorporation of undecomposed fresh/dry plant material into soils, both either in place or brought from a distance (Pieters 1927). In addition to this, green manure legume crops also fix atmospheric N biologically. Biological nitrogen fixation (BNF) is a microbiological process in which atmospheric N_2 is converted into a plant-usable form, which offers an economically attractive and ecologically sound option of reducing external inputs and improving internal resources (Sulieman and Tran 2014).

Legume green manuring (LGM) can enhance agriculture sustainability by improving nutrient retention (Dinnes et al. 2002), enhancing soil fertility (Fageria and Baligar 2005), by decreasing soil erosion (Smith et al. 1987), and reducing global warming (Robertson et al. 2000). LGM also has a major role to improve the SOC pool, thereby improving soil physicochemical and biological properties (Fageria and Baligar 2005; Fageria 2007). The incorporation of legume green manure crops into soil releases organic substances like organic acid, amino acids, sugars, vitamins, and mucilage (Shukla et al. 2011) during crop growth as well as after decomposition. These substances are capable to bind soil particles together and form better soil aggregation (MacRae and Mehuys 1985), resulting in increased hydraulic conductivity (MacRae and Mehuys 1985; Boparai et al. 1992), water holding capacity (WHC), water infiltration (Raimbault and Vyn 1991), and total pore space (Anderson et al. 1997) of the soil. Further, the green manure incorporation provides carbon (C) and energy to soil biota required for OM decomposition and nutrient recycling (Griffin and Garren 1976; Hu et al. 2006). In the process of LGM, soil pH is changed by addition of OM (Singh et al. 1992). The addition of organic amendments into soils, particularly green manure, has potential to control weeds and soil-borne diseases and to disrupt the life cycle of agriculture pest (Kumar et al. 2014; Varma et al. 2017). Green manure is important to small-scale farmers, for whom it is difficult to buy expensive mineral fertilizers (Meena et al. 2014). Therefore, legume green manure crops have great potential for sustainable food grain production. Keeping above facts in mind, the objective of this chapter is to provide information on the LGM for sustainable soil management and crop production.

12.2 Legumes as Green Manure

The green manuring practices are of two types: (a) in situ green manuring crops and (b) ex situ green leaf manuring (GLM) (Singh et al. 1991). In in situ green manuring, short-duration (~45 to 60 days) crops are grown and incorporated into soil at the same site. In ex situ green manuring, foliage and tender parts of green manuring crops collected from nearby forests, shrubs, and trees are incorporated into the soil at 15–30 days prior to the sowing of main crops (SSSA 1997).

There are many crops which can be used for green manuring, however the selection of the green manure crop depends upon several factors like the prevailing climatic conditions, cropping system practiced, availability of seed, and other factors including local habits and prejudices. The legume crops have an edge over nonlegume crops due to ability of fixing atmospheric N (Rao 2014). LGM is categorized under following categories:

(a) Grain legumes: pigeon pea (*Cajanus cajan*), green gram (*Vigna radiata*), soybean (*Glycine max*), or groundnut (*Arachis hypogaea*)
(b) Multipurpose perennial legume trees: subabul (*Leucaena leucocephala*), gliricidia (*Gliricidia sepium*), and kassod tree (*Cassia tora*)
(c) Non-grain legume crops: sunn hemp (*Crotalaria juncea*), dhaincha (*Sesbania rostrata*), centrosema (*Centrosema acutifolium*), stylo (*Stylosanthes guianensis*), and desmodium (*Desmodium ovalifolium*) (Palaniappan 1994)

The most commonly grown LGM crops are sunn hemp, dhaincha, berseem (*Trifolium alexandrinum*), and green gram. The leguminous crops, namely, cowpea (*Vigna unguiculata*), green gram, black gram (*Vigna mungo*), pigeon pea, chick pea (*Cicer arietinum*), black lentil (*Lens culinaris*), pea (*Pisum sativum*), lathyrus (*Lathyrus latifolius*), kidney beans (*Phaseolus vulgaris*), tephrosia (*Tephrosia purpurea*), groundnut, soybean, dhaincha, and sunn hemp; woody legumes, namely, subabul, gliricidia, karanj (*Pongamia glabra*), and delonix (*Delonix regia*), have the ability to fix the atmospheric N in their root nodules. Legume crops and tree species suitable for green manuring in different agroclimatic zones are mentioned in Table 12.1. Dry matter accumulation by these legume crops may vary from 1 to 10 Mg ha^{-1} $year^{-1}$ under ideal soil and environmental conditions, and the total N accumulation in the aboveground biomass ranges from 0.02 to 0.3 Mg ha^{-1} $year^{-1}$ (Lathwell 1990). Besides BNF, a green manure crop should have some desirable characteristics, viz., fast-growing habit, short duration, early onset of BNF, high N accumulation rate, high tolerance to biotic stresses (pest and disease), abiotic stresses (flood, drought, salinity, and adverse temperatures), wide range of ecological adaptability, timely release of nutrients, photoperiod insensitivity, high seed production, higher seed viability, and most importantly easiness in incorporation (Meena et al. 2015a).

Table 12.1 Leguminous green manure crops for different regions

(A) **In situ green manure crops**			
(a) **Tropical region**		(b) **Temperate region**	
Common name	**Scientific name**	**Common name**	**Scientific name**
Cluster bean	*Cyamopsis tetragonoloba*	Subterranean clover	*Trifolium subterraneum*
Cowpea	*Vigna unguiculata*	Ladino clover	*Trifolium repens*
Pueraria	*Pueraria phaseoloides*	Crimson clover	*Trifolium incarnatum*
Green gram	*Vigna radiata*	Faba bean	*Vicia faba*
Lablab	*Lablab purpureus*	Soybean	*Glycine max*
Dhaincha	*Sesbania aculeate, S. rostrata*	Red clover	*Trifolium pratense*
White lupin	*Lupinus albus*	Black lentil	*Lens culinaris*
Gray bean	*Mucuna cinerecum*	Alfalfa	*Medicago sativa*
Pigeon pea	*Cajanus cajan*	Barrel medic	*Medicago truncatula*
Sunn hemp	*Crotalaria breviflora*	Hairy vetch	*Vicia villosa*
Buffalo bean	*Mucuna aterrima*	Milk vetch	*Astragalus sinicus*
Jack bean	*Canavalia ensiformis*	Winter pea	*Pisum sativum*
Velvet bean	*Mucuna deeringiana*	Sweet clover	*Melilotus officinalis*
Stylo	*Stylosanthes guianensis*	Cura clover	*Trifolium ambiguum*
Desmodium	*Desmodium ovalifolium*	Purple vetch	*Vicia benghalensis*
Milk vetch	*Astragalus sinicus*	Common vetch	*Vicia sativa*
Zornia	*Zornia latifolia*	**(B) Ex situ green leaf manuring shrubs and trees**	
Jumby bean	*Leucaena leucocephala*	**Common name**	**Scientific name**
Kudzu	*Pueraria phaseoloides*	Subabul	*Leucaena leucocephala*
Adzuki bean	*Vigna angularis*	Gliricidia	*Gliricidia sepium*
Black gram	*Phaseolus mungo, P. trilobus*	Karanj	*Pongamia glabra*
Soybean	*Glycine max*	Milkweed	*Calotropis gigantea*
Alfalfa	*Medicago sativa*	Tephrosia	*Tephrosia purpurea*
Wild indigo	*Indigofera tinctoria*	Wild indigo	*Indigofera teysmannii*
Berseem	*Trifolium alexandrinum*	Sesbania	*Sesbania speciosa, S. rostrata*
Sunn hemp	*Crotalaria juncea, C. striata*	Kassod	*Cassia tora*

Modified from: Fageria and Baligar (2005) and Reddy (2016)

12.3 Biological Nitrogen Fixation in Legume Green Manuring

BNF is the process of conversion of atmospheric N_2 into ammonia (NH_3) or other molecules which are easily available to plants and other living organisms into the soil (Postgate 1998). The BNF and mineralization of leguminous green manure crop in soil are depicted in Fig. 12.1. Overall reaction of BNF is given below.

$$N_2 + 8H^+ + 8e^- \rightarrow 2NH_3 + H_2$$

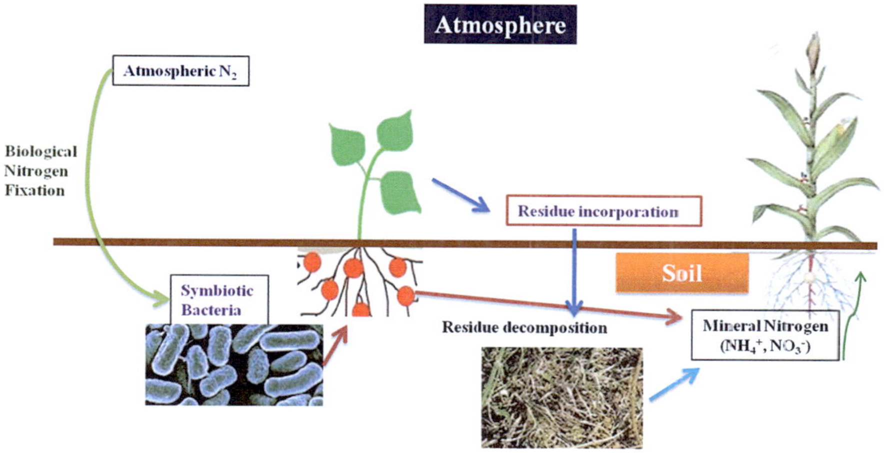

Fig. 12.1 N fixation and mineralization of leguminous green manure crop in soil

In many studies, it has been proven that the BNF is the most efficient process or way to supply the major amount of N required by legume crops or plant (Sulieman and Tran 2014) (Table 12.2). After sowing of the legume seed, soon after its germination, the inherent *Rhizobium* present in soil are externally added as seed inoculums and later on enters into the root hairs of legumes and moves through infection thread toward the main root. After invasion, these bacteria multiply rapidly in the root of legumes, which result into swelling of root cells to form nodules. The atmospheric N around the root hairs of legumes is "fixed" by binding it to other elements and converting it into a plant available ammonical form. The *Rhizobium* bacteria use carbohydrates as a source of hydrogen in the conversion of atmospheric N to ammonia (Fageria 2007), and this symbiotic association of legume with *Rhizobium* contributes 40% of world total N fixation (Ladha et al. 1992; Meena et al. 2015a). A detailed survey of the literature showed that on an average, grain legume crops could fulfill 50–80% of own N requirement through BNF. BNF through legume crop cultivation in agriculture adds 33 Tg N year^{-1}, of which symbiotic association of *Rhizobium* with seed legume crops, forage leguminous cover crops, non-*Rhizobium* N-fixing bacteria, cyanobacteria in rice, and endophytic N-fixing organisms in sugarcane contribute 10, 12, 4, 6, and 3 Tg N year^{-1}, respectively (Smil 1999).

Ladha et al. (1988) reported an average accumulation of 2.6 kg N ha^{-1} day^{-1} under different legume green manure crops. Incorporation of such high N-fixing legume crops at 45–65 days of growth stage into soils results in rice yield equal to the application of N fertilizers at 50–100 kg N ha^{-1}. Similarly, Ladha et al. (1988) reported that 45–60 days old dhaincha species could fix N equal to 200 kg N ha^{-1}. Dhaincha legume crop age of 55 days fixed about 303 kg N ha^{-1} (at 5.5 kg N day^{-1}) and 383 kg N ha^{-1} (at 6.96 kg N day^{-1}) without and with inoculation of *Azorhizobium*

Table 12.2 N accumulation in major leguminous green manure crops

Crop species	Growth duration (days)	N accumulation (kg ha^{-1})	Reference
Glycine max	45	115	Meelu et al. (1985)
Crotalaria juncea	45	169	
Cajanus cajan	45	33	
Sesbania aculeata	45	225	
Vigna radiata	45	75	
Dolichos lablab	45	63	
Indigofera tinctoria	45	45	
Sesbania rostrata	56	176	Furoc et al. (1985)
Sesbania aculeata	56	144	
Vigna unguiculata	45	75	
Vigna radiata	45	75	Morris et al. (1986)
Sesbania rostrata	60	219	Ladha et al. (1988)
Sesbania cannabina	60	171	
Sesbania aegyptiaca	57	39	Ghai et al. (1985)
Sesbania grandiflora	57	24	
Cluster bean	49	91	Singh et al. (1991)
Common vetch	Flowering	105–210	
Sweet clover	Flowering	150–300	
Milk vetch	Flowering	65–131	Watanabe (1984)

bacteria, respectively (Ladha et al. 1988). The N supplied by hairy vetch (*Vicia villosa*) and crimson clover (*Trifolium incarnatum*) in cover crop experiments ranged from 72 to 149 kg N ha^{-1} (Hargrove 1986; Ladha et al. 1988; Holderbaun et al. 1990). Thus, it is clear that the amount of N fixed by different legume species varies, and it depends on the legume species, its variety, the number of effective root nodules, type of soil, agronomical and water management practices, and prevailing climatic conditions and their interactions with other factors (Buresh and De Datta 1991; Fageria and Baligar 2005). Sharma and Ghosh (2000) evaluated dhaincha as an intercrop with direct-seeded rice as well as incorporated pure dhaincha before transplanting of rice under flood-prone lowland conditions and found that dhaincha accumulated 80–86 kg N ha^{-1} in pure stand and 58–79 kg N ha^{-1} when intercropped with direct-seeded rice in alternate rows at 50 days of growth.

12.4 Nutrient Composition of Legume Green Manuring

LGM crops incorporation into the soil improves essential plant nutrients for succeeding crop (Bhuiyan and Zaman 1996). Legume green manuring crops, having low C/N ratio (high N), may behave like the chemical nitrogenous fertilizer application, when compared to the other crops used for BNF. However, the mineral composition and N content of legume may vary considerably depending upon the species (Singh and Bhattacharyya 1989; Verma et al. 2015a), crop growth duration, and

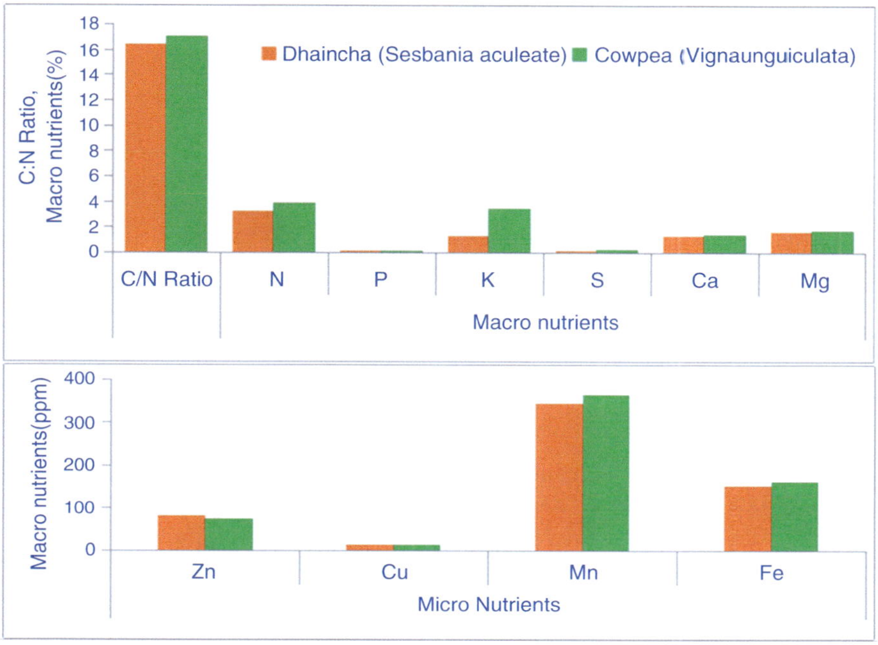

Fig. 12.2 Nutrient composition and supply of nutrient by 8-week-old green manuring crops. (Data source: Bhuiyan and Zaman 1996; Singh et al. 1992)

growth condition. With increased crop growth duration, carbonaceous content (carbohydrate and cellulose) increases and nitrogenous content (amino acid and protein) decreases leading to increase in C/N ratio of crops which results in slower decomposition of plant material (Ishikawa 1988). To overcome this problem, green manure crops should be incorporated at the flowering stage. The N and C content in roots, shoots, and leaves may also vary. In general, leaves contain lower C/N ratio as compared to stem and roots (Palm et al. 1988). In addition to N and C content, green manuring crops may also contain considerable amount of other nutrient and trace elements. A study conducted by Bhuiyan and Zaman (1996) under protected cultivation in greenhouse showed that cowpea has a higher mineral composition than dhaincha under wetland rice. The nutrient composition of dhaincha and cowpea is depicted in Fig. 12.2.

12.5 Decomposition and Mineralization

Incorporation of legume green manure into soil undergoes decomposition and mineralization process (Fageria 2007; Meena et al. 2018). Decomposition is a biological breakdown and transformation of complex organic compounds into simpler organic and inorganic molecules (Joffe 1955; Fox et al. 1990).

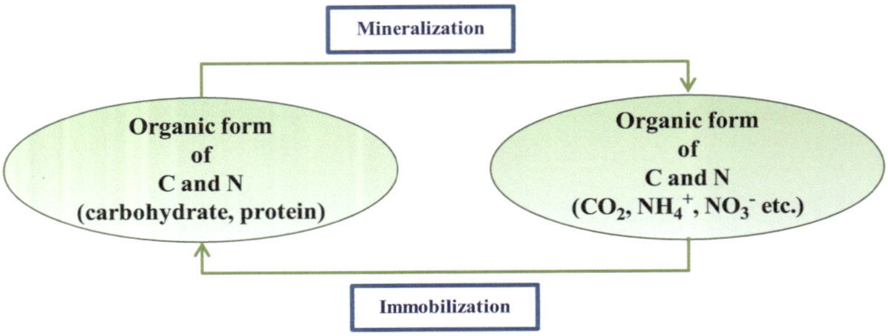

Fig. 12.3 Mobilization and immobilization of nutrient in soil

$$\text{Complex organic molecules} + O_2 = \text{Nutrients} + H_2O + CO_2 + \text{Energy}$$

There are mainly two processes, namely, mineralization and immobilization (Gil and Fick 2001; Dinnes et al. 2002) as shown in Fig. 12.3.

- Mineralization is the main process of conversion of organic carbon (OC) and N into the mineral form, i.e., CO_2 and NH_4^+ by microorganism. During this process, nutrients are released to soil solution and are made available to plants.
- Immobilization is the reverse process of mineralization in which the mineral form of C and N fixed into the organic form.

Decomposition and mineralization of green manures are affected by the type of soil, crops, crop growth stage at which it is incorporated, and prevailing climatic conditions (Fageria and Baligar 2005; Reddy 2016). Among these, the dominant factors are the quantity and quality of green manure crops. Soil factors, which affected the decomposition and mineralization of green manure, are the soil texture, structure, soil reaction, microbial activity, and the status of soil nutrients (Thonnissen et al. 2000; Dinnes et al. 2002; Dhakal et al. 2015). Decomposition and mineralization of OM mainly depend upon the availability of N in soil (Joffe 1955; Gil and Fick 2001). Due to low C/N ratio, lower lignin content, and high quantity of easily decomposable material, LGM generally leads to rapid and fast mineralization as compared to cereal residues (Janzen and Kucey 1988; Gil and Fick 2001). Decomposition rate of green manure legume was higher in sandy soils than the fine-textured soils (Verbene et al. 1990). Soybean residues incorporation into soils lost 68% of biomass within 1 month of incorporation (Broder and Wagner 1988). The incorporation of hairy vetch legume green manure rapidly released N within 15 days after incorporation (Varco et al. 1989). Legume residue incorporation into soils under field conditions led to that of the <30% of legume N recovered by a subsequent nonlegume cereal crop, and a major amount of it is retained in soil as organic forms of N (Ladd et al. 1983; Harris et al. 1994). Fractional C and N release was greater in dried and rewetted soil for green manuring legumes compared to continuously moist soil for all parts except for nodules (Franzluebbers et al. 1994). The

NH_4-N release and its accumulation increased significantly with a decrease in soil water; however there was a decrease in the release of NO_3-N (Brar and Sidhu 1995). Soil temperature also has an effect on the pattern of N release during decomposition of added green manure residue in the soil (Brar and Sidhu 1997). The suitable range of soil temperature and moisture are 20–30 °C and −0.01 to −0.05 MPa for faster decomposition of green manure into soil and subsequent release of nutrients (Cassman and Munns 1980; Sinha et al. 2009). Magid et al. (2001) reported higher N mineralization of both black medic and white sweet clover occurred at low temperature. Thonnissen et al. (2000) indicated that soybean and *Indigofera* decomposed rapidly and lost ~70% of the biomass in 1 month after incorporation.

12.6 Legume Green Manuring and Soil Properties

Modern agriculture, having intensive cropping system coupled with frequent tillage of the soil, higher doses of chemical fertilizers, and overlooking of the application of organic manures, led to the breakdown of soil aggregates and destruction of SOM. This further led to soil physicochemical and biological health degradation. Due to faulty management practices, the agricultural land is becoming less productive day by day. The adverse effects such as soil compaction, reduction in SOM, and reduced crop productivity have been recognized in many areas over several years (Unger and Kaspar 1994; Meena et al. 2016a). Green manuring with legumes also had a magnificent role of providing OM into the soil, thereby improving the physicochemical properties as well as biological properties of soil (Ebelhar et al. 1984; Fageria and Baligar 2005). The role of leguminous green manuring on soil physical, chemical, and biological properties is depicted in Fig. 12.4.

Besides the improvement of soil health, LGM helps in reducing insect-pest, disease incidence, and weed management (Kumar et al. 2014; Verma et al. 2015), acts as binding material in soil, and helps in improvement of soil structure (Schutter and Dick 2001). LGM between successive crop growth increased SOM (Pung et al. 2004) which stimulates the soil microbial activity and mineralization of plant nutrients (Eriksen 2005). Therefore, it enhances the soil quality and its fertility (Doran et al. 1988).

12.6.1 Soil Physical Properties

The continuous use of LGM results in buildup of OM in soil, which improves the soil physical properties (Table 12.3) and quantities of organic acid, amino acids, sugars, vitamins, and mucilage (Shukla et al. 2011). These organic substances are capable of binding the soil particles and better soil aggregation (MacRae and Mehuys 1985) which led to better hydraulic conductivity (Boparai et al. 1992) and improved drainage by increasing infiltration and percolation (Raimbault and Vyn 1991). The infiltration rates are mainly controlled by bulk density (*ρb*) and aggregate stability; higher infiltration rate reduces soil erosion (Martens and Frankenberger

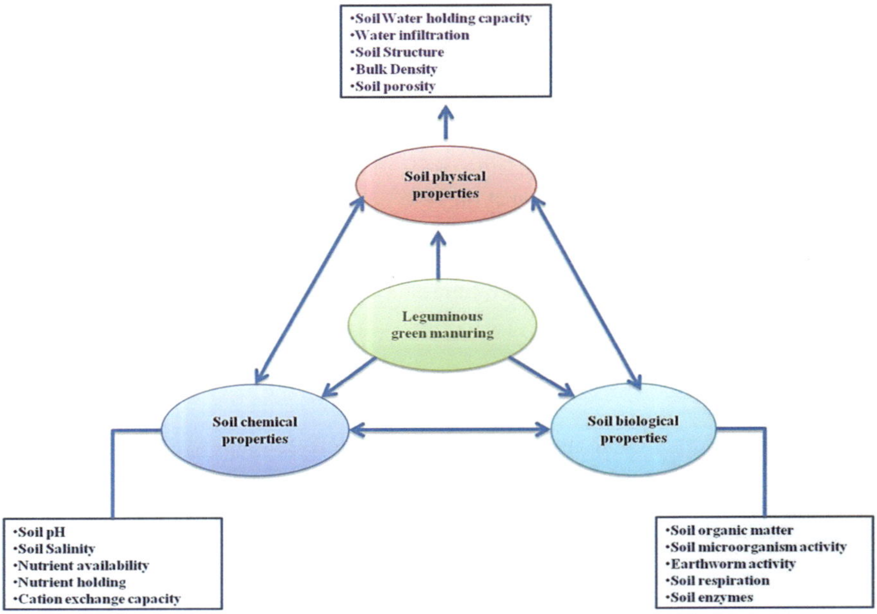

Fig. 12.4 Soil properties influenced by leguminous green manuring

Table 12.3 Physical properties of soil affected by LGM under rice-wheat system

	At rice harvest			At wheat harvest		
Treatments	Bulk density (Mg m^{-3})	MWD[a] of soil aggregates (mm)	Hydraulic conductivity (cm day^{-1})	Bulk density (Mg m^{-3})	MWD of soil aggregates (mm)	Hydraulic conductivity (cm day^{-1})
Fallow	1.52	0.519	3.65	1.55	0.531	3.74
S. rostrata	1.45	0.761	4.64	1.51	0.728	4.39
S. aculeata	1.46	0.713	4.55	1.51	0.699	4.28
Green gram	1.46	0.714	4.56	1.52	0.700	4.18

Data source: Mandal et al. (2003)
[a]Mean weight diameter

1992). The LGM improves total pore space by decreasing the soil *pb*, enhanced root development, soil water content, and nutrient use efficiency (Anderson et al. 1997).

The SOC is a more suitable index for crop production in arid and semiarid regions, as it has positive effects on the WHC (Dias-Zorita et al. 1999). The penetration of green manure roots, particularly dhaincha and sunn hemp in soil, holds it in place and prevents soil erosion (Schumann et al. 2000). Green gram and sesbania green manuring reduced the soil *pb* to the extent of 0.03–0.07 Mg m^{-3} and enhanced the mean weight diameter of soil aggregates in the upper layer of soil when compared to fallow soils during the growth of rice and wheat (Mandal et al.

2003). In sandy soils, green manure helps to hold more water by reducing drainage rate through improvement in soil physical properties (Selvi and Kalpana 2009; Yadav et al. 2017).

12.6.2 Soil Chemical Properties

Green manuring have effects on soil chemical properties, and particularly soil pH is reduced, (a) by addition of organic acids and generation of CO_2 during OM decomposition, resulting in reduced soil pH of furnishing proton by organic manure; (b) simultaneously soil pH is increased by reduction of H^+ by organic anions to H_2O and CO_2 during mineralization of organic manure (Singh et al. 1992; Buragohain et al. 2017). Reduction of organic substances may reduce Fe and Mn oxides causing soil pH to rise (Meena et al. 2017), because of oxidation of Fe and Mn consumed proton generated during decomposition of OM. The 12 year long-term experiments conducted by Yadav and Singh (1986) indicated that soil pH reduced with the time under green manuring. Soil pH after 12 years of uninterrupted sugarcane crop cultivation was 8.0 under green manuring and 8.5 under control treatments, respectively. Green manures also prevent nutrients from being washed out from the soil. The nutrients are drawn up by the green manure crop from deep soil layers and held inside the plant and recycled back to soil upon decomposition. Organic materials, acting as a slow-release source of N, are expected to more closely match N supply and N demand of crops, and this could reduce N losses (Becker et al. 1994). In flooded soils condition, the average N loss from applied green manure and split-applied urea were 14% and 35%, respectively (Becker et al. 1995). Thus, N from green manure crops contributes significantly lower to air and water pollution compared to urea application.

Green manure added around 50–60 kg N ha^{-1} to the soil for the succeeding crop of rice (Singh and Bhattacharyya 1989). The leaching losses of green manure N are normally expected to be smaller than inorganic fertilizers, because it must be mineralized before it can be lost via leaching (Singh et al. 1992), and make it available to succeeding main crop, thereby reducing needs of N fertilizer (Stute and Posner 1993). The NO_3^- leaching from nonlegume manuring ranged 29–94%; while, from legume manuring, it was 6–48%, showing that green manuring with legume crop of hairy vetch had more potential of reduction of NO_3^- leaching than nonlegume crops (rye, *Secale cereale*) (Sainju and Singh (1997). Green manuring enhanced the availability of phosphorus (P) from added rock phosphate in rice crop (Cavigelli and Thien 2003). Hundal et al. (1992)and Bah et al. (2006) evaluated the contributions of different green manures to P nutrition in rice crop in soils of arid tropical climate and found that the utilization of phosphorous fertilizers had markedly enhanced it from 3% to 39% in treatments having green manure.

The integrated effect of legume green manure crops with mineral fertilizers improved SOC, nutrient availability, intake of nutrients by crops, and yield of rice-wheat system (Kumar and Prasad 2008). The incorporation of dhaincha at flowering stage adds about 60–90 kg ha^{-1} N into soil (Pandey et al. 2008), and it also helps improve the soil chemical properties. Legume green manure crops prevent nutrient leaching, decrease weed growth, and reduce the harmful effect of agrochemicals and soil-borne phytopathogens (Kumar et al. 2010; Dhakal et al. 2016).

12.6.3 Soil Biological Properties

The decomposition of green manures serves major functions for microflora providing both C and energy for growth and formation of new cell material, which further multiplies its colony saprophytically on the decomposing OM (Ye et al. 2014). A large number of soil microorganisms exist in the soil as long as there is a C source for energy (Kumar et al. 2014). Soil-inhabiting microorganisms are very critical for decomposing organic residues and recycling soil nutrients (Akpor et al. 2006). The process of decomposition is of great significance because unless the energy and nutrients are released through microbial activity, the primary product cannot exist for a long time (Kumar et al. 2014).

The LGM has two main positive points from the microbiological point of view: (a) primarily it provides nutrient-rich OM for the microbial community which easily converts organically bound nutrients in plant residues to easily available nutrient form to the crops; (b) secondly it enhances the biodiversity of soil microorganisms. This microbial diversity can be increased by incorporating different legume green manure in crop rotation and cropping system programs (Schutter and Dick 2001; Eriksen 2005; Kumar et al. 2016). Since the begining of agriculture, it has been found that legume green manures and other organic amendments improve the soil tilth ability and fertility (MacRae and Mehuys 1985). The increase in beneficial microbial community and its activity is most often directly related to an overall increase in soil organic matter. Sikora and Stott (1996) and Griffin and Garren (1976) studied the colonization of *Aspergillus flavus* and *Aspergillus niger* fungi in different soil textures in deep-plowed, decomposing rye, a green manuring crop in the soils of groundnut field. The greater colonization of *A. flavus was reported,* in heavy textured soil adjacent to rye (*Secale cereal*) and groundnut crops i and *A. flavus* population was as high as 165 propagules g^{-1} soil adjacent to rye.

12.7 Effect of Legume Green Manuring on Crop Yield

The positive impact of legume green manuring on SOM and other soil properties associated with increase in nutrition to growing crop is well reflected on grain yield. It was found that the average yield of rice grain increased by 1.7 Mg ha^{-1} in green manuring treatments, over controlled plots. The increase in rice yield ranged from 0.5 to 3.3 Mg ha^{-1}. The average application of N @ 80 kg N ha^{-1}, green manure N shows an agronomic N use efficiency (NUE) of 20 kg rice grain increased kg^{-1} N applied (IRRI 1990). Besides this, green manures supply N constantly, and due to slow release of N from the green manure incorporation in soils, this would match with the requirement of N by plants resulting in improved crop performance

(Westcott and Mikkelson 1987). The yield response of high-yielding varieties of rice crop to different green manuring crops in India ranged from 0.65 to 3.1 Mg ha^{-1} (Singh et al. 1991). Ali and Narciso (1993) compared NUE in long-term fertilizer experiments conducted in India, Indonesia, and the Philippines by the International Network on Soil Fertility and Fertilizer Evaluation (INSURF). The NUE of lowland rice is higher in the case of green manuring when compared to mineral fertilizer N (Peoples et al. 1995). In general, the linear relationship of NUE and N application rate does not differ significantly among different N sources, but it tends to decrease more and more with a higher application rate of green manuring (Becker et al. 1995). The highest rice production was recorded in dhaincha green manuring and it was at on par with 20 Mg ha^{-1} of FYM (Misra et al. 1996). To find out the relationship between the rate of application of dhaincha green manure and rice yield, a field experiment was conducted at Bangladesh Rice Research Institute, Regional Station, Barisal. The rice grain yield showed a quadratic relation to the added dhaincha green manure, and maximum rice yield was obtained at ~6 Mg ha^{-1}.

The effect of dhaincha green manuring on crop growth and yield of direct seeded rice and transplanted rice under intermediate water stagnation condition was studied by Sharma and Das (1994) and Meena et al. (2016b). They found that the highest yield of the rice crop was observed when rice and dhaincha were grown at a 2:1 ratio as 20 cm row-to-row spacing. The increase in rice yield under dhaincha green manuring was attributed to greater panicle weight, which was probably due to a continued supply of N following decomposition of added dhaincha OM. The dhaincha green manuring intercropped with direct seeded rice and the conventional practice of before transplanting incorporation of dhaincha green manuring was compared under lowland flood-prone conditions It was found that the grain yield of direct seeded rice was significantly higher when 20 kg N ha^{-1} was applied at sowing and dhaincha was incorporated at 50 days of growth (Sharma and Ghosh 2000).

Bokhtiar et al. (2003) found that dhaincha and sunn hemp green manuring and supplemented urea increased yield of subsequent sugarcane by up to 57%. Besides this, there was a significant increase in SOM, total N, available P, and S of the soil. Comparison of the relative efficiency of green manures with inorganic N sources by Selvi and Kalpana (2009) revealed that the cowpea green manuring application at 0.17% (3.4 Mg ha^{-1}) and dhaincha green manuring at 0.36% (7.2 Mg ha^{-1}) on a dry matter basis would be more than enough to produce a yield of rice crop equal to that obtained with the application of the recommended dose of fertilizers, i.e., at 80-25-35 kg/ha N-P-K. Further, demonstration of rice grain yield response showed that the cowpea was a better green manure crop than dhaincha mainly for wetland rice. In addition, substantial residual effects of cowpea and dhaincha green crops were also observed and in some cases even up to two to three successive rice crops (Bhuiyan and Zaman 1996; Selvi and Kalpana 2009; Meena 2013).

Pooniya et al. (2012) conducted a field experiments with summer green manuring crops, namely, green gram, cowpea, and dhaincha, it was found that that the highest crop residue was added by dhaincha, i.e., 38.56 Mg ha^{-1}, which in turn led to the recycle of 180.5, 22.6, and 267.8 kg N, P, K ha^{-1}, respectively. Further dhaincha incorporation also led to a significantly higher yield of succeeding basmati rice crop, and it was 2.38%, 4.14%, and 10.82% higher over cowpea, green gram, and summer fallow, respectively.

12.8 Limitations of Legume Green Manuring

In spite of the wide range of literature which reveals associated benefits of legume LGM, its applicability still remains in the research farms. Their adaptation by farmers is still not a common practice due to lack of awareness and some limitations at farm levels. There are few cases to cite. Haryana Government promotes LGM in rice-wheat areas of state. Becker et al. (1995) reviewed several available literatures (Ladha et al. 1992; Ali and Narciso 1993; Garrity and Flinn 1988; Ashoka et al. 2017) and identified the following limitations of green manuring and the possible reasons behind it. They are listed below:

(a) Establishment and incorporation of green manuring crops are relatively costly.
(b) Narrow window period between the two crops for growing and incorporating green manure crops during most of the cropping season.
(c) Green manure crop, if not incorporated at proper growth stage and time, may lead to immobilization of N on a temporary basis.
(d) Being a high water requiring crop, it may not be suitable for dryland agriculture.
(e) Problems of decomposition of green manuring in the sowing of the following crop if proper moisture is not available, particularly in semiarid regions (Aase et al. 1996).
(f) No visible economic benefits are seen during initial few seasons of green legume manuring.
(g) Easy availability of fertilizers and their ease of application in comparison to green manuring.
(h) Prices of mineral fertilizers are relatively low when compared to the high price of land and labor.

12.9 Conclusion

The ever increasing human population poses a burden on soils to produce more food. The intensive use of agrochemicals is threatening the soil sustainability. Use of chemical fertilizers in intensive cropping systems may lead to degradation of natural resources, particularly soils. These degraded soils will not be fit for profitable agriculture. Therefore, incorporation of legume green manure crops into the soil is emphasized for crop production. In ancient times, also legumes were recognized as suitable crops for green manuring to improve soil health and crop productivity. In addition to fixing of atmospheric N, it helps in conservation of soil water and reduction of soil erosion. LGM is to be considered superior over the nonlegume crops due to a higher N content and lower C/N ratio, helping in easy decomposition of OM and mineralization for nutrient release at a faster rate. Besides this, it also reduced the N immobilization risk for succeeding crops. Therefore, practices of LGM have a large scope for inclusion, to make the farming system more sustainable.

12.10 Future Perspectives

Future research for legume green manures is needed on following topics:

(a) The studies across the globe have established the benefits of green manures on soil physicochemical and biological health. However, their application to the farm level is still at a limited stage because of lack of awareness and suitability under particular environments and cropping system.
(b) There is need to develop a location-specific cropping system with compatible legume green manure crop either partly or in the window period between two crops.
(c) Benefits of green legume manure need to be quantified regarding fertilizer saving, water saving, increase in crop productivity, and more importantly soil health improvement, so that the extra cost involved in raising the green manure crop may be justified to the ultimate adapter, the farmers.
(d) Efforts are to be directed to find out alternative techniques like brown manuring of legume crops by growing them as intercropped, which can save time as well as need of incorporation. Such easily acceptable techniques are to be devised.

References

Aase JK, Pikul JP, Prueger JP, Hatfield JR (1996) Lentil water use and fallow water loss in a semi-arid climate. Agron J 88:723–728

Akpor OB, Okoh AI, Babalola GO (2006) Culturable microbial population dynamics during decomposition of *Theobroma cacao* leaf litters in a tropical soil setting. J Biol Sci 6(4):768–774

Ali M, Narciso JH (1993) The perception and reality of GM use in rice an economic evaluation, IRRI Social Science Division Paper Series. The International Rice Research Institute, Los Bafios, p 42

Anderson IC, Buxton DR, Karlen DL, Cambardella C (1997) Cropping system effects on nitrogen removal, soil nitrogen, aggregate stability, and subsequent corn grain yield. Agron J 89:881–886

Ashoka P, Meena RS, Kumar S, Yadav GS, Layek J (2017) Green nanotechnology is a key for eco-friendly agriculture. J Clean Prod 142:4440–4441

Bah AR, Zaharah AR, Hussin A (2006) Phosphorus uptake from green manures and phosphate fertilizers applied in an acid tropical soil. Commun Soil Sci Plant Anal 37:2077–2093

Becker M, Ladha JK, Simpson IC, Ottow JCG (1994) Parameters affecting residue N mineralization in flooded soils. Soil Sci Soc Am J 58:1666–1671

Becker M, Ladha JK, Ali M (1995) Green manure technology: potential, usage, and limitations. A case study for lowland rice. Plant Soil 174:181–194

Bhuiyan NI, Zaman SK (1996) In: Rahmaned et al (eds) Biological nitrogen fixation associated with rice production. Kluwer Academic Publishers., The Netherlands, Dordrecht, pp 51–64
Bokhtiar SM, Gafur MA, Rahman ABMM (2003) Effects of *Crotalaria* and *Sesbania aculeata* green manures and N fertilizer on soil fertility and the productivity of sugarcane. J Agric Sci 140:305–309
Boparai BS, Singh Y, Sharma BD (1992) Effect of green manure (*Sesbania aculeata*) on physical properties of soil and growth of rice-wheat and maize-wheat cropping system. Int Agrophys 6:95–101
Brar DS, Sidhu AS (1995) Effect of soil water on patterns of nitrogen release during decomposition of added green manure residue. J Indian Soc Soil Sci 43(1):14–17
Brar DS, Sidhu AS (1997) Effect of temperature on pattern of nitrogen release during decomposition of added green manure residue in soil. J Res 34(3):275–278
Broder MW, Wagner GH (1988) Microbial colonization and decomposition of corn, wheat and soybean residue. Soil Sci Soc Am J 52:112–117
Buragohain S, Sharma B, Nath JD, Gogaoi N, Meena RS, Lal R (2017) Impact of ten years of biofertilizer use on soil quality and rice yield on an inceptisol in Assam, India. Soil Res. https://doi.org/10.1071/SR17001
Buresh RJ, De Datta SK (1991) Nitrogen dynamics and management of rice-legume cropping systems. Adv Agron 45:1–59
Cassman KG, Munns DN (1980) Nitrogen mineralization as affected by soil moisture, temperature and depth. Soil Sci Soc Am J 44:1233–1237
Cavigelli MA, Thien SJ (2003) Phosphorus bioavailability following incorporation of green manure crops. Soil Sci Soc Am J 67:1186–1194
Dhakal Y, Meena RS, De N, Verma SK, Singh A (2015) Growth, yield and nutrient content of mungbean (*Vigna radiata* L.) in response to INM in eastern Uttar Pradesh, India. Bangladesh J Bot 44(3):479–482
Dhakal Y, Meena RS, Kumar S (2016) Effect of INM on nodulation, yield, quality and available nutrient status in soil after harvest of green gram. Legum Res 39(4):590–594
Dias-Zorita M, Buschiazzo DE, Peinemann N (1999) Soil organic matter and wheat productivity in the semiarid Argentine Pampas. Agron J 91:276–279
Dinnes DL, Karlen DL, Jaynes DB, Kaspar TC, Hatfield JL, Colvin TS, Cambardella CA (2002) Nitrogen management strategies to reduce nitrate leaching in tile-drained Midwestern soils. Agron J 94:153–171
Doran JW, Fraser DG, Culik MN, Liebhardt WC (1988) Influence of alternative and conventional agricultural management on soil microbial process and nitrogen availability. Am J Alt Agric 2:99–106
Ebelhar SA, Giddens JE, Beaty ER (1984) Nitrogen from legume cover crops for no-till corn. Agron J 76:51–55
Eriksen J (2005) Gross sulphur mineralization-immobilization turnover in soil amended with plant residues. Soil Biol Biochem 37:2216–2224
Fageria NK (2007) Green manuring in crop production. J Plant Nutr 30(5):691–719
Fageria NK, Baligar VC (2005) Role of cover crops in improving soil and row crop productivity. Commun Soil Sci Plant Anal 36:2733–2757
Fagodiya RK, Pathak H, Kumar A, Bhatia A, Jain N (2017) Global temperature change potential of nitrogen use in agriculture: a 50-year assessment. Sci Report 7:44928
Fox RH, Myers RJ, Vallis I (1990) The nitrogen mineralization rate of legume residues in soil as influenced by their polyphenol, lignin, and nitrogen contents. Plant Soil 129:251–259
Franzluebbers K, Weaver RW, Juo ASR, Franzluebbers AJ (1994) Carbon and nitrogen mineralization from cowpea plants part decomposing in moist and in repeatedly dried and wetted soil. Soil Biol Biochem 26(10):1379–1387
Furoc RE, Dizon MA, Morris RA, Marqueses EP (1985) Effects of flooding regime and planting dates to N accumulation of three Sesbania species and consequently to transplanted rice. Paper presented at the 16th annual scientific convention of the crop science Society of Philippines, 8–10 May 1985, Central Luzon State University Munoz, Nueva Ecija, Philippines, 1985

Galloway JN, Aber JD, Erisman JW, Seitzinger SP, Howarth RW, Cowling EB, Cosby BJ (2003) The nitrogen cascade. Biol Sci 53:341–356
Garrity DP, Flinn JC (1988) Farm-level management systems for green manure crops in Asian rice environments. In: Sustainable agriculture: green manure in rice fanning. The International Rice Research Institute, Los Baiios, pp 111–129. IRRI, Philippines
Ghai K, Rao DLN, Batra L (1985) Comparative study of the potential of sesbania for green manuring. Trop Agric 62:52–56
Gil JL, Fick WH (2001) Soil nitrogen mineralization in mixtures of eastern gamagrass with alfalfa and red clover. Agron J 93:902–910
Gill MS, Pal SS, Ahlawat IPS (2008) Approaches for sustainability of rice (*Oryza sativa*)-wheat (*Triticum aestivum*) cropping system in Indo-Gangetic plains of India-a review. Indian J Agron 53(2):81–96
Griffin GJ, Garren KH (1976) Colonization of rye green manure and peanut fruit debris by *Aspergillus flavus* and *Aspergillus niger* group in field soils. Appl Environ Microbiol 32(1):28–32
Hargrove WL (1986) Winter legumes as nitrogen sorghum. Agron J 78:70–74
Harris GH, Hesterman OB, Paul EA, Peters SE, Janke RR (1994) Fate of legume and fertilizer nitrogen-15 in a long-term cropping systems experiment. Agron J 86:910–915
Holderbaun JF, Decker AM, Meisinger JJ, Mulford FR, Vough LR (1990) Fall-seeded legume cover crops for no-tillage corn in the humid east. Agron J 82:117–127
Hu C, Cao ZP, Ye ZN, Wu WL (2006) Impact of soil fertility maintaining practice on soil microbial biomass carbon in low production agro-ecosystem in northern China. Acta Ecol Sin 26:808–814
Hundal HS, Dhillon NS, Dev G (1992) Contribution of different green manures to P nutrition of rice. Indian J Soil Sci Soc 40:76–81
IRRI (1990) World rice statistics. The International Rice Research Institute, Los Bafios, Philippines. 320p
Ishikawa M (1988) Green manure in rice: the Japanese experience. In: Green manure in rice farming. International Rice Research Institute, Los Banos, pp 45–61
Janzen HH, Kucey RMN (1988) Carbon, nitrogen, and sulfur mineralization of crop residues as influenced by crop species and nutrient regime. Plant Soil 106:35–41
Joffe JS (1955) Green manuring viewed by a pedologist. Adv Agron 7:14–187
Kumar V, Prasad RK (2008) Integrated effect of mineral fertilizers and green manure on crop yield and nutrient availability under rice-wheat cropping system in Calciorthents. J Indian Soc Soil Sci 56(2):209–214
Kumar R, Srivastava S, Srivastava M, Sinha A (2010) Effects of organic amendments on soil mycoflora. Asian J Plant Path 4(2):73–81
Kumar R, Mahajan G, Srivastava S, Sinha A (2014) Green manuring: a boon for sustainable agriculture and pest management – a review. Agric Rev 35(3):196–206
Kumar S, Sheoran S, Kumar SK, Kumar P, Meena RS (2016) Drought: a challenge for Indian farmers in context to climate change and variability. Prog Res An Int J 11:6243–6246
Ladd JN, Amato M, Jackson RB, Butler JH (1983) Utilization by wheat crops of nitrogen from legume residues decomposing in soils in the field. Soil Biol Biochem 15:231–238
Ladha JK, Watanabe I, Saono S (1988) Nitrogen fixation by leguminous green manure and practices for its enhancement in tropical lowland rice. In: The International Rice Research Institute (ed) Sustainable agriculture: green manure in rice farming. IRRI, Los Banos, pp 165–183
Ladha JK, Pareek RP, Becker M (1992) Stem-nodulating legume-rhizobium symbiosis and its agronomic use in lowland rice. Adv Soil Sci 20:147–192
Lathwell DJ (1990) Legume green manures, Trop Soils Bulletin Number 90-01. Soil Management Collaborative Research Support Program, North Carolina State University, Raleigh
MacRae RJ, Mehuys RG (1985) The effect of green manuring on the physical properties of temperate area soils. Adv Soil Sci 3:71–94
Magid J, Henriksen O, Thorup-Kristensen K, Mueller T (2001) Disproportionately high N mineralisation rates from green manures at low temperatures – implications for modelling and management in cool temperate agroecosystems. Plant Soil 228:73–82

Mandal UK, Singh G, Victor US, Sharma KL (2003) Green manuring: its effect on soil properties and crop growth under rice-wheat cropping system. Eur J Agron 19:225–237
Martens DA, Frankenberger J (1992) Modification of infiltration rates in an organic amended irrigated soil. Agron J 84:707–717
Meelu OP, Furoc RE, Dizon MA, Morris RA, Marqueses FP (1985) Evaluation of different green manures on rice yield and soil fertility. Paper presented at the 16th annual scientific convention of the Crop Society of Philippines, 8–10 May, 1985, Central Luzon State University, Munoz, Nueva Ecija, Philippines
Meena RS (2013) Response to different nutrient sources on green gram (*Vigna radiata* L.) productivity. Indian J Ecol 40(2):353–355
Meena BL, Majumdar SP (2016) Improving yield of barley grown on coarse textured soil by compaction and sulphur fertilization. Ecol Env Cons 22:151–156
Meena BL, Rattan RK, Datta SP (2013) Efficacy of seed treatment in ameliorating iron deficiency in aerobic rice on a calcareous soil. J Indian Soc Soil Sci 61:147–152
Meena BL, Meena RL, Ambast SK, Pandey M (2014) Impact assessment of agriculture technological interventions in tsunami affected South Andaman- a case study. Bharatiya Krishi Anushandhan Patrika 28(3):141–148
Meena RS, Yadav RS, Meena H, Kumar S, Meena YK, Singh A (2015a) Towards the current need to enhance legume productivity and soil sustainability worldwide: a book review. J Clean Prod 104:513–515
Meena VS, Maurya BR, Meena RS (2015b) Residual impact of well-grow formulation and NPK on growth and yield of wheat (*Triticum aestivum* L.). Bangladesh J Bot 44(1):143–146
Meena BL, Majumdar SP, Meena VK, Dotaniya ML (2016a) Response of compaction with sulphur fertilization to nutrient content, uptake and economics of barley on highly permeable soil. Int J Agric Sci 34:1719–1722
Meena BL, Rattan RK, Datta SP, Meena MC (2016b) Effect of iron application on iron nutrition of aerobic rice grown in different soils. J Environ Biol 37:1377–1383
Meena RS, Bohra JS, Singh SP, Meena VS, Verma JP, Verma SK, Shiiag SK (2016c) Towards the prime response of manure to enhance nutrient use efficiency and soil sustainability a current need: a book review. J Clean Prod 112:1258–1260
Meena BL, Rattan RK, Datta SP (2017) Solubility relationships of iron and evaluation of its fertility status in degraded soils. Commun Soil Sci Plant Anal 48:1059–1067
Meena RS, Kumar V, Yadav GS, Mitran T (2018) Response and interaction of *Bradyrhizobium japonicum* and *Arbuscular mycorrhizal* fungi in the soybean rhizosphere: a review. Plant Growth Regul 84:207–223
Misra RD, Gupta VK, Pandey DS (1996) Management crop residues in rice. Int Rice Res Notes 21:71–72
Morris RA, Furoc RF, Dizon MA (1986) Rice response to a short duration green manure. II. Nitrogen recovery and utilization. Agron J 78:413–416
Palaniappan SP (1994) Green manuring: nutrient potential and management. In: HLS T (ed) Fertilizers, organic manure, recyclable waste and biofertilizers. Fertilizer development and Consultation Organization, New Delhi
Palm O, Weerakoon WL, DeSilva MAP, Thomas R (1988) Nitrogen mineralization of *Sesbania sesban* used as green manure for lowland rice in Sri Lanka. Plant Soil 108:210–209
Pandey DK, Pandey R, Mishra RP, Kumar S, Kumar N (2008) Collection of Dhaincha (*Sesbania* spp.) variability in Uttar Pradesh, biodiversity and agriculture (souvenir). Uttar Pradesh Biodiversity Board, Lucknow, pp 48–51
Peoples MB, Herridge DF, Ladha JK (1995) Biological nitrogen fixation: an efficient source of nitrogen for sustainable agricultural production. Plant Soil 174:3–28
Pieters JA (1927) Green manuring, principles and practice. Wiley, New York, p 365
Pooniya V, Shivaya YS, Rana A, Lata N, Radha P (2012) Enhancing soil nutrient dynamics and productivity of Basmati rice through residue incorporation and zinc fertilization. Eur J Agron 41:28–37

Postgate J (1998) Nitrogen fixation, 3rd edn. Cambridge University Press, Cambridge. Proc Indian Natn Sci Acad 80(2):359–378
Pung H, Aird PL, Cross S (2004) The use of Brassica green manure crops for soil improvement and soil borne disease management. 3rd Australian Soil borne diseases Symposium 8–11 February 2004. pp 1–2
Raimbault BA, Vyn TJ (1991) Crop rotation and tillage effects on corn growth and soil structural stability. Agron J 83:979–985
Rao DLN (2014) Recent advances in biological nitrogen fixation in agricultural systems. Proc Indian Natn Sci Acad 80:359–378
Ray SS, Gupta RP (2001) Effect of green manuring and tillage practices on physical properties of puddled loam soil under rice-wheat cropping system. J Indian Soc Soil Sci 49(4):670–678
Reddy PP (2016) Sustainable intensification of crop production. doi:https://doi.org/10.1007/978-981-10-2702-4
Robertson GP, Paul EA, Harwood RR (2000) Greenhouse gases in intensive agriculture: contributions of individual gases to the radioactive forcing of the atmosphere. Science 289:1922–1925
Sainju UM, Singh BP (1997) Winter cover crops for sustainable agricultural systems: influence on soil properties, water quality and crop yields. Hortic Sci 2:21–28
Schumann RA, Meyer JH, Van Antwerpen R (2000) A review of green manuring practices in sugarcane production. Proc S Afr Sugar Technol Assess 74:93–100
Schutter M, Dick R (2001) Shifts in substrate utilization potential and structure of soil microbial communities in response to carbon substrates. Soil Biol Biochem 33(11):1481–1491
Selvi RV, Kalpana R (2009) Potentials of green manure in integrated nutrient management for rice – a review. Agric Rev 30(1):40–47
Sharma AR, Das KC (1994) Effect of green manuring with dhaincha (*Sesbania aculeata*) on growth and yield of direct-sown and transplanted rice under intermediate deepwater conditions. J Agric Sci 122(3):359–364
Sharma AR, Ghosh A (2000) Effect of green manuring with *Sesbania aculeate* and nitrogen fertilization on the performance of direct-seeded flood-prone lowland rice. Nutr Cycl Agroecosyst 57:141–153
Shukla KP, Sharma S, Singh NK, Singh V, Tiwari K, Singh S (2011) Nature and role of root exudates: efficacy in bioremediation. Afr J Biotechnol 10(48):9717–9724
Sihag SK, Singh MK, Meena RS, Naga S, Bahadur SR, Gaurav, Yadav RS (2015) Influences of spacing on growth and yield potential of dry direct seeded rice (*Oryza sativa* L.) cultivars. Ecoscan 9(1–2):517–519
Sikora LJ, Stott DE (1996) Soil organic carbon and nitrogen. In: Methods for assessing soil quality. SSSA Special Publication 49. Soil Science Society of America, Madison, pp 157–167
Singh KN, Bhattacharyya HC (1989) Direct-seeded rice. Published by Mohan Primlani for Oxford and IBH publishing Co. Pvt. Ltd. New Delhi, pp 135–142
Singh Y, Khind CS, Singh B (1991) Efficient management of leguminous green manures in wetland rice. Adv Agron 45:135–189
Singh Y, Singh B, Khind CS (1992) Nutrient transformations in soils amended with green manures. Adv Soil Sci 20:237–309
Sinha A, Kumar R, Kamil D, Kapur P (2009) Release of nitrogen, phosphorus and potassium from decomposing *Crotalaria juncea* L. in relation to different climatic factors. Environ Ecol 27(4B):2077–2081
Smil V (1999) Nitrogen in crop production: an account of global flows. Glob Biogeochem Cycles 13:647–662
Smith MS, Frye WW, Varco JJ (1987) Legume winter cover crops. Adv Soil Sci 7:95–139
SSSA (1997) Glossary of soil science terms. Soil Science Society of America, Madison
Stute JK, Posner JL (1993) Legume cover option for grain rotations in Wisconsin. Agron J 85:1128–1132
Sulieman S, Tran L-SP (2014) Symbiotic nitrogen fixation in legume nodules: metabolism and regulatory mechanisms. Int J Mol Sci 15:19389–19393

Sulieman S, Tran L-SP (2015) Legume nitrogen fixation in a changing environment. Springer International Publishing, Cham, p 35. https://doi.org/10.1007/978-3-319-06212-9_3

Thonnissen C, Midmore DJ, Ladha JK, Olk DC, Schmidhalter U (2000) Legume decomposition and nitrogen release when applied as green manures to tropical vegetable production systems. Agron J 92:253–260

UN DESA (2015) United Nations, Department of Economic and Social Affairs, Population Division (UN DESA), world population prospects: the 2015 revision, key findings and advance tables. Working Paper No. ESA/P/WP.241, United Nations, Department of Economics and Social Affairs, New York

Unger PW, Kaspar TC (1994) Soil compaction and root growth: a review. Agron J 86:759–766

Varco JJ, Frye WW, Smith MS, Mackown CT (1989) Tillage effects on nitrogen recovery by corn from a nitrogen-15 labeled legume cover crop. Soil Sci Soc Am J 53:822–827

Varma D, Meena RS, Kumar S (2017) Response of mungbean to fertility and lime levels under soil acidity in an alley cropping system of Vindhyan Region, India. Int J Chem Stu 5(2):384–389

Verbene EL, Hassink J, Willingen P, Goot JJ, Van Veen JA (1990) Modeling organic matter dynamics in different soils. Neth J Agric Sci 38:221–238

Verma JP, Meena VS, Kumar A, Meena RS (2015a) Issues and challenges about sustainable agriculture production for management of natural resources to sustain soil fertility and health: a book review. J Clean Prod 107:793–794

Verma SK, Singh SB, Prasad SK, Meena RN, Meena RS (2015b) Influence of irrigation regimes and weed management practices on water use and nutrient uptake in wheat (*Triticum aestivum* L. Emend. Fiori and Paol.). Bangladesh J Bot 44(3):437–442

Watanabe I (1984) Use of green manure in northeast Asia. In: International Rice Research Institute (ed) Organic matter and rice. IRRI, Los Banos, pp 229–234

Westcott MP, Mikkelson DS (1987) Comparison of organic and inorganic nitrogen sources for rice. Agron J 79:937–943

Whitbread AM, Blair GJ, Lefroy RDB (2000) Managing legume leys, residues and fertilisers to enhance the sustainability of wheat cropping systems in Australia 1. The effects on wheat yields and nutrient balances. Soil Tillage Res 54(1/2):63–75

Yadav RL, Singh K (1986) Long term experiments with sugarcane under intensive cropping system and variation in soil fertility. Indian J Agron 31:322–325

Yadav GS, Lal R, Meena RS, Babu S, Das A, Bhomik SN, Datta M, Layak J, Saha P (2017) Conservation tillage and nutrient management effects on productivity and soil carbon sequestration under double cropping of rice in North Eastern Region of India. Ecol India. http://www.sciencedirect.com/science/article/pii/S1470160X17305617

Ye X, Liu H, Li Z, Wang Y, Wang Y, Wang H, Liu G (2014) Effects of green manure continuous application on soil microbial biomass and enzyme activity. J Plant Nutr 37(4):498–508

Legume Plant Growth-Promoting Rhizobacteria (PGPRs): Role in Soil Sustainability

13

R. Sammauria and S. Kumawat

Contents

R. Sammauria (✉) · S. Kumawat
Division of Agronomy, Rajasthan Agricultural Research Institute,
SKN Agriculture University, Durgapura-Jaipur, Rajasthan, India
e-mail: rakeshsammauria@sknau.ac.in

R. S. Meena et al. (eds.), *Legumes for Soil Health and Sustainable Management*,
https://doi.org/10.1007/978-981-13-0253-4_13

Abstract
The interactions of plants with different microorganisms in the rhizosphere have varied effects on plant health, productivity, and soil fertility. Numerous types of bacteria play certain vital roles in agriculture production systems by way of establishing different types of relationships that are commonly called as plant growth-promoting rhizobacteria (PGPRs). These PGPRs have multiple roles that have positive impact on plant physiological activities and defense mechanism against biotic and abiotic stresses. Such effects may include symbiotic and asymbiotic N fixation and mobilization; production of siderophores, phytohormones, and antibiotics; counteracting the effect on various pathogenic fungi; solubilization of soil phosphate; and production of growth-promoting indole acetic acid (IAA). Besides, under various stress situations (such as drought, salinity, and metal toxicity), these PGPRs not only halt the deleterious effects of such stresses but also make the plant grow and develop normally. PGPRs through production of ACC (1-aminocyclopropane-1-carboxylate) deaminase can successfully deter the detrimental stress effect of ethylene, enabling the plants to grow normally by alleviating the harmful effect of this. The toxic effects of higher concentration of heavy metals can also be counteracted through PGPRs. These favorable microorganisms reduce the chemical load in rhizosphere and also have multiple synergistic effects on the plant growth and efficacy of other microorganism. Recently, PGPRs have become the most important tool for use as biofertilizers because they provide sustainability of agroecosystems. As the understanding of these PGPRs is increasing, it has become quite evident that apart from legumes, these are also contributing cereals and other nonlegume host crops in a variety of ways. Eco-friendly PGPRs have now become a very important supplement for agriculture production and soil health with involvement of minimal cost. In the light of this background, various aspects of PGPRs biotechnology with special reference to legumes are reviewed and discussed in this chapter.

Keywords
BNF · PGPRs · Rhizobacteria · *Rhizobium*-legume symbiosis · Siderophores · Soil health and sustainable agriculture

Abbreviations

ACC	1-Aminocyclopropane-1-carboxylate
BNF	Biological N fixation
GA	Gibberellic acid
IAA	Indole acetic acid
ISR	Induction of systemic resistance
N	N
NH_3	Ammonia
PGPRs	Plant growth-promoting rhizobacteria
pH	Chemical reaction of soil (acidic/neutral/alkaline)
SOM	Soil organic matter

13.1 Introduction

The globe is under tremendous pressure for the producing of adequate amount of food to nourish the ever-increasing human population. Thus, it has redefined the importance of sustainable intensification of production and this has to be accomplished through low-cost environment-friendly technologies/interventions. Contemporary world of agriculture is poised with twin contrasting challenges of maintaining a reasonable level of production from existing degrading production systems to cater needs of huge populations of human and livestock ensuring sustainability of resources. Soil has its own quality as it is a vital renewable natural resource, but it is under tremendous pressure to produce more to suit the demands of increasing production. To fulfill this demand, widespread utilization of chemicals, excessive exploitation of water for the purpose of irrigation, and use of mechanization do take place. This has led to soil degradation and consequent poor soil health. Besides, there are some ill effects on soil health like reduction in the soil organic matter (SOM) and loss of biodiversity in the soil. This leads to degradation of soil health leading to negative impact on essential ecosystem functions, sustainability of agricultural production, and soil resilience capacity (Abrol et al. 2000; Meena et al. 2015a). Thus soil health is probably the biggest key for catering both the objectives of elevated production and sustainability of production systems (Saha and Mandal 2009; Meena et al. 2016a, b). The diverse use of agrochemicals (fertilizers, pesticides, and others) causes negative effects of variable magnitude to the soil environment (particularly of rhizosphere), when they are used excessively and indiscriminately. Chemical fertilizers have become the essential component of modern agriculture to satisfy the need of food for present population of the world. With continuous development of new generation technologies, it is possible to reduce the use of chemicals in agriculture without compromising the yield or quality of the crop production. The applied nitrogenous chemical fertilizer use efficiency is ~30 to 40% due to losses as leaching, volatilization, denitrification, etc. (Mia and Shamsuddin 2010), and about 75–90% of the added P is precipitated and fixed on metal-cation complexes, leading to low availability to crops (Sharma et al. 2013a, b). Therefore excessive amount of fertilizers are required for adequate nutrition that not only increases the cost of production but also aggravates chemical load in rhizosphere that prove highly detrimental to environment. For example, Brazil saves approximately US$ 7 billion per year (Hungria et al. 2013) and US$ 1.2 billion per year by 50% replacement of the N-fertilizer with *Azospirillum* sp. inoculation (Hungria et al. 2010), where 70% of the nitrogenous fertilizers are imported particularly for soybean crops, suggesting that the use of inoculants could reduce the use of chemical fertilizers worldwide.

As the result of the several research efforts on the differential response of plants to different soils, understanding and knowledge of the intricacy of the interactions among soil microorganism, roots, and plants has been established. As the seed germinates and roots are developed in the soil, the organic matter present in the soil facilitate the growth and development of active microbial populations in and around the root zone, and this phenomenon is called as "the rhizosphere effect" (Whipps 1990; Meena et al. 2017a). "Rhizosphere" may be defined as "the soil compartment influenced by the root" (Hiltner 1904). The elevated levels of microbial biomass and their activity have been observed in the rhizosphere, and the rhizospheric sphere may contain several-fold increased number of microbial organisms than non-rhizospheric soils. Dardanelli et al. (2010) reported 19–32 times counts of such microorganisms in rhizosphere than soil devoid of roots. The rhizobacteria which are able to colonize the rhizosphere and facilitating or inducing plant growth are frequently called as plant growth-promoting rhizobacteria (PGPRs) (Kloepper and Schroth 1978) or plant health-promoting rhizobacteria (PHPR) according to their mode of action. Today PGPRs are being used worldwide for agricultural benefits to develop and utilize effective, low-cost, eco-friendly technologies in the crop production program for combating increasing production cost and adverse effect of excessive use of synthetic fertilizer on soil health. PGPRs are capable of influencing plant growth both as direct and indirect modes, and thus, cumulative effect of these surely has positive effect on the growth and production of crop concerned and even of succeeding one. Direct effect may be observed as absorption or uptake of certain nutrients from the environment or by providing any compounds to plants, synthesized by the bacterium, whereas indirect effect may be observed in form of decreasing or blocking the harmful effects of one or more phytopathogenic organisms (Glick 1995), and there may be some complimentary effect of other favorable microorganisms too. Further, Yang et al. (2009) have described the term "induced systemic tolerance" (IST) for PGPR-induced physical and chemical changes in plants that result in enhanced capacity to tolerate various abiotic stresses. Inclusion of beneficial microorganisms is very crucial for the sustainable agriculture production strategies. The biological N fixation (BNF) through legume rhizobia symbiotic interaction can benefit not only the host crop, but it may also have positive effects on companion or even on subsequent crops (Lupwayi et al. 2004; Buragohain et al. 2017). Apart from symbiotic functioning, rhizobias have also been observed to function as nonsymbiotic PGPR. Through this mechanism, nonlegume crops such as rice or wheat have been reported to benefit to a great extent from rhizobia as endophytes (Biswas et al. 2000a; Chaintreuil et al. 2000). Worldwide, the land degradation has emerged as a serious threat to agricultural production systems. It is an urgent need to halt such degradation for sustaining their productive capacities and rehabilitate soil health. BNF may play a vital role in this regard. Symbiotic fixation of atmospheric N through the association of *Rhizobium* species and host legumes is an infinite source of N for agriculture. Such N obtained by BNF also improve the yield of crops raised after harvesting of legumes, and the extent of such benefit may be equivalent to the application of 30–80 kg of fertilizer-N/ha (Zahran 1999). Nonlegume crops also get benefit when grown in association with legume (Mandimba 1995; Ram and Meena 2014). Endosymbiotic interactions between legume plants and the genera *Bradyrhizobium*, *Azorhizobium*, *Rhizobium*, *Mesorhizobium*, and *Sinorhizobium* have been reported (Dardanelli et al. 2010). Rhizobia perform symbiotic relationships with plants of leguminous family and biologically convert atmospheric N into NH_3. That biologically fixed N not only can be utilized by host leguminous plants to which symbiotic relationship has been established but also by the companion plants or even by succeeding crops. Such symbiotic BNF is also important for protein-rich legumes and improves the soil health by enhancing fertility status and rhizospheric beneficial microbial activities.

Inclusion of legumes in cropping systems can partly fulfill the crop N requirement by substituting to N fertilizers through BNF. Symbiotic relationship between legumes and rhizobacteria has tremendous potential as through BNF about 20% of global crop demands of N can be met. Such BNF has been estimated to contribute ~45 to 70 million tonnes of fixed N to agriculture each year (Aliyu et al. 2013; Zahran 1999). Out of these, about half of such N is obtained from the cool and hot regions, whereas remaining half is received from the tropical regions. Fixation of atmospheric N through biological mode is done by the symbiosis of certain rhizobia bacteria and a host legume that is a significant source of N. Legume hosts vary in terms of their host specificity for bacterial strain with which the symbiotic relationship may be established. Apart from host specificity, there are many factors related to soil and environment that can influence upon the symbiotic relationship and its performance. Factors such as high/low soil temperature, nutrient deficiencies, low levels of soil moisture, low pH (<5.5), and low clay and organic matter have detrimental impact on the performance of rhizobia (Aliyu et al. 2013; Meena et al. 2014). The beneficial effects of various PGPRs are numerous influencing multiple of their processes and countering various stresses imposed by the environments of various growing situations. Besides, these also have pronounced positive impacts on various soil characteristics; those are responsible for marinating soil health dynamics. Thus, this is an attempt to synthesize the information obtained from evaluation of the effect of different bacterial species of PGPR on the varied aspects of plant growth and development and health of soil.

13.2 PGPRs' Current Need for Soil Sustainability

In today's scenario, the demand for quality produce is increasing and there is increasing awareness on food and environmental security. This has compelled the revival of interest in eco-friendly, sustainable, and agrochemical substituting or organic agriculture practices. One of the major concerns of contemporary global agricultural production systems are sustainability in agricultural systems along with conservation of environment and their production capacity. Further, the sustainability of agricultural systems in long run is a function of effective management of internal resources of production systems and their complex interrelationship. Thus, focus should be concentrated on the modern concept of rhizo-engineering based on favorable partitioning of the exotic biomolecules, which are responsible for the interaction between plant and microbes. In the future, the technological interventions owing to biotechnological and molecular advances will essentially improve the understanding of rhizosphere biology, and such understanding will open new avenues of utilizing and managing microbial populations for the benefit of agricultural production systems.

With the advancement of identification and multiplication of various strains, the multi strain bacterial association may play more effective role in improving plant growth than the single inoculation. Combined inoculation has highly synergistic effect in comparison to single inoculations. One of the effective technologies for improving plant growth at low temperature is the addition of ice-nucleating PGPR. Investigations on N fixation and phosphate solubilization by PGPRs is in progress, but little research could be done on potassium solubilization and its continuous mining under intensive production systems posing increasing toll on soil potassium reserves and more importantly maintaining soil solution potassium in close tandem of uptake with the pattern of exhaustive depletion patterns. The availability of multiple strains and microbial inoculants for more and more nutrients will surely expand the field of inoculants for the benefit of growers not only by availability of multiple options but by way of increasing the confidence of the farmers. Of late, the PGPRs have gained global significance and consent for agricultural advantages (Meena and Meena 2017). The microbial inoculants are very important agents of sustainability of agriculture as apart from supplying the nutrients to plants, they also improve the nutrient use efficiency. Further, they also increase efficiency of applied fertilizers, reduce abiotic and biotic stresses, and also act against the xenobiotic compounds.

13.3 Types of PGPRs

On the basis of mode of action and host PGPRs' relationship and localization, these PGPRs may be divided into several categories for better understanding of their role and importance (Fig. 13.1). Several researchers have made adequate attempts to categorize PGPRs accordingly. Broadly, these PGPRs have two distinct major groups on the basis of host relationship as (1) bacteria working under symbiosis and

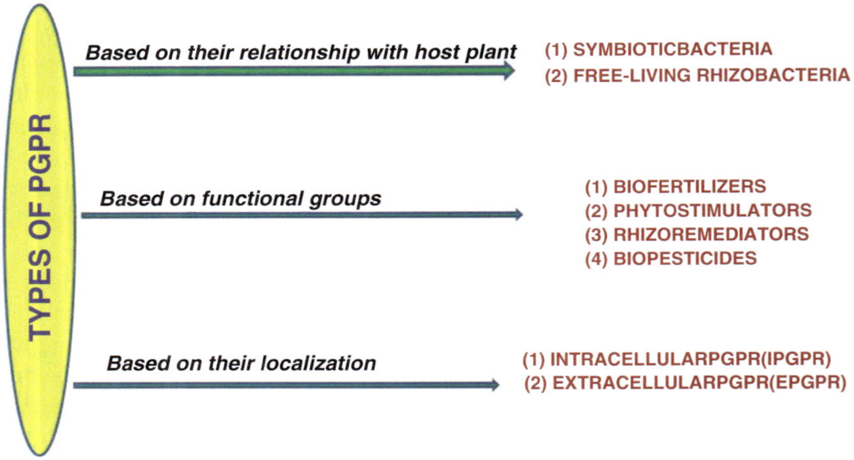

Fig. 13.1 Flow diagram of types of PGPRs

Table 13.1 *Rhizobium* species suitable for different crops

Group	Suitable *Rhizobium* species	Crops
Rhizobium (fast growing)	*Rhizobium leguminosarum*	Peas (*Pisum*), *Vicia*, lathyrus, lentil (*Lens*)
	Rhizobium trifolii	Berseem (*Trifolium*)
	Rhizobium phaseoli	Kidney beans (*Phaseolus*)
	Rhizobium meliloti	Melilotus, lucern (*Medicago*), fenugreek
	Rhizobium lupine	*Lupinus*
	Rhizobium spp.	Cowpea
Bradyrhizobium (slow growing)	*B. japonicum*	Soybean (*Glycine*), groundnut
	B. elkanii	Soybean (*Glycine*), groundnut
Mesorhizobium	*Mesorhizobium ciceri*	Chickpea
Azorhizobium (fast growing)	*A. caulinodans*	*Sesbania* (root and stem nodules)
Others	Miscellaneous group	Other legumes

(2) rhizobacteria not working in symbiosis (Khan 2005; Somers et al. 2004; Meena et al. 2017a) and have been classified into the functional groups on the basis of the inherent activities: (1) biofertilizers (improve access of nutrients to the plant), (2) phytostimulators (induce or improve the plant growth by synthesizing phytohormones), (3) rhizoremediators (degrade the organic pollutants), and (4) biopesticides (control plant diseases by way of either production of antimicrobial metabolites or by enabling plant tolerance under stress situation by the production of antimicrobial metabolites).

Localization of PGPRs is very important for having their effect, and on this basis, these have been classified as (1) intracellular PGPRs (iPGPRs) (the bacteria remain present within plant cells, cause deformity in plant cells in the form of nodules after infecting the roots, and reside in these forming nodules) and (2) extracellular PGPRs (ePGPRs) that live outside the cell, and these do not make nodules by infecting the plant cells; instead, they enhance plant growth by way of producing certain substances, improving nutrient availability, and imparting resistance toward disease(s). The PGPRs have their effects on plants by way of making association with the host plants, and a specific PGPR makes its association with its host plant in a specific and at a particular condition, and such behavior of one PGPR would be different than the other PGPR. Therefore, on the basis of stage of association with plant roots, the eggers have also been divided into three groups: (1) those that reside in the periphery of the root, but do not make direct contact with the roots, (2) those that colonize on the surface of the root, and (3) those that live in intercellular spaces of cells of the root cortex of host plants. Among these PGPRs, pipers are Gram-negative mostly and rod-shaped, with a few being Gram-positive rods, cocci, and pleomorphic type. In general term, iPGPRs include the members of Rhizobiaceae, which are able to form nodules on the root of leguminous plants (Table 13.1). Out of them, some of the agronomically important ePGPRs include genera such as *Enterobacter*, *Caulobacter*, *Serratia*, *Flavobacterium*, *Acinetobacter* sp., *Aeromonas*, *Agrobacterium*, *Alcaligenes* sp., *Phyllobacterium* sp., *Bacillus*,

Table 13.2 Various roles of *Rhizobium* for plant growth promotion

PGPRs	Role/mechanism	Plants	Reference(s)
R. leguminosarum bv. viciae 128C53K, *Mesorhizobium loti* MAFF303099	ACC deaminase activity	*Arabidopsis thaliana*	Contesto et al. (2008)
Rhizobium tropici (CIAT899), *Rhizobium*	Higher leghemoglobin concentrations, nitrogenase activity N_2 fixation efficiency, indole acetic acid, and cytokinin	*Phaseolus vulgaris L.*	Figueiredo et al. (2008)
Rhizobium	Indole-3-acetic acid	*Oryza sativa L.*	Biswas et al. (2000b)
R. leguminosarum (strain E11)	Indole-3-acetic acid	*Oryza sativa L.*	Dazzo et al. (2000)
Rhizobacteria	Indole-3-acetic acid and improve growth	*Brassica juncea*	Asghar et al. (2002)
Rhizobia	HCN production	Legumes	Thamer et al. (2011)
Rhizobia	Induction of plant stress resistance	Peanuts	El-Akhal et al. (2013)
Rhizobium	N fixation	Legumes, rice	Young and Haukka (1996) and Yanni et al. (2001)
Sinorhizobium	Chitinase and B glucanases production	Pigeon pea	Kumar et al. (2010)
Bradyrhizobium sp. 750	Heavy metal mobilization	Soybean	Dary et al. (2010)
Bradyrhizobium	Siderophore, P-solubilization, IAA, HCN, IAA	Groundnut	Deshwal et al. (2003) and Meena et al. (2017a)
Mesorhizobium	IAA, siderophore, ammonia, HCN, P-solubilization	Chickpea	Wani et al. (2008b)

Pseudomonas, Erwinia, Bacillus, Hyphomicrobium, Azotobacter, Azospirillum, and Acetobacter.

The family *Rhizobiaceae* comprises of six genera: *Rhizobium, Ensifer* (*Sinorhizobium*), *Mesorhizobium*, *Allorhizobium*, *Azorhizobium*, and *Bradyrhizobium*, which are collectively called as rhizobia (Table 13.2).

13.4 Role of PGPRs in Promoting Plant Growth

PGPRs promote plant growth through several mechanisms; those may include (i) abiotic stress mitigation in plants, (ii) nutrient fixation/mobilization to facilitate easy uptake of nutrients, (iii) release of plant growth-promoting substances, (iv) involvement in the production of siderophores, (v) volatile organic compounds production, and (vi) the production of protective enzyme like chitinase, glucanase, and ACC deaminase for curing the plant diseases. However, the mechanism of action of

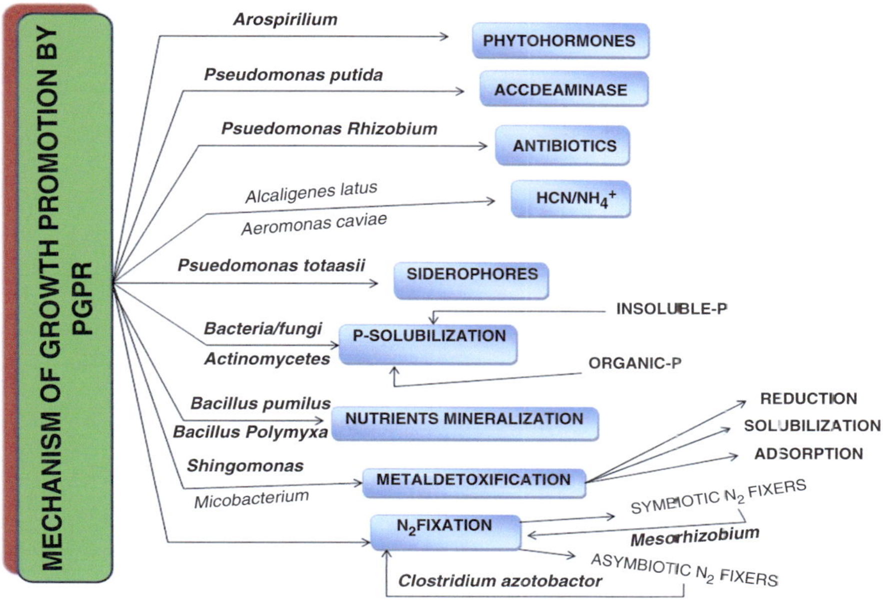

Fig. 13.2 Diverse mechanisms of plant growth promotion by PGPRs

different PGPRs may vary on the basis of type of host plants (Vejan et al. 2016) (Fig. 13.2). PGPRs enhance plant growth by synthesizing the precursors of phytohormones, many vitamins, enzymes, siderophores, and antibiotics and by inhibiting ethylene synthesis. They are also capable to fix atmospheric N (N_2 fixers) and solubilize inorganic P, leading to increase in availability of these nutrients in rhizosphere. They can also mineralize organic phosphate, thereby increasing its availability, and improve tolerance capacity to plant toward various abiotic stresses such as drought, salinity, and metal toxicity. Utilization of these microbes has numerous traits, including their role of bioremediation by way of imparting resistance to excess levels of metal or by promoting plant growth in the metal-contaminated soils, thus making them appropriate options for bioremediation. Rhizobia-host plant symbiosis improve the plant growth in several ways such as (1) biological N_2 fixation, (2) elevating the nutrient availability in the rhizosphere, (3) inducing the root surface area, (4) enhancing other beneficial symbioses of the host, (5) reducing or preventing the deleterious effects of phytopathogenic organisms, and (6) combining modes of action (Khan et al. 2009; Meena et al. 2016a, b).

In symbiotic N fixation, legume having symbiosis with rhizobia utilizes sugars or carbohydrates and hydrogen source to tap and fix N from the atmosphere that require involvement of high amount of ATP energy. Under sufficient N availability situations, such type of symbiotic N fixation does not take place. Therefore, applications of N fertilizers impede nodule formation and subsequent N fixation (Fig. 13.3). The initial process of symbiosis starts with moderate deficiency of N on periphery

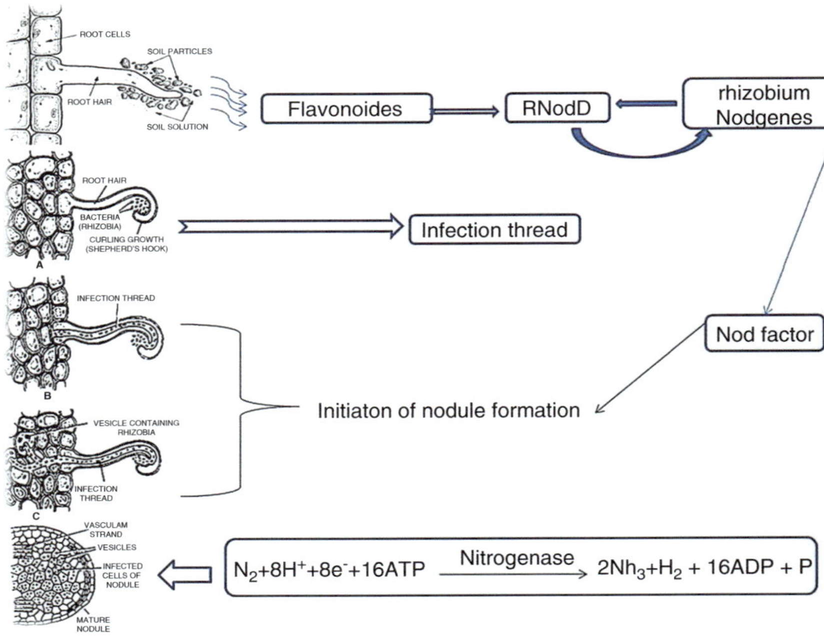

Fig. 13.3 Schematic view of symbiotic nodulation process and biological N fixation

of roots and exudates of such plant roots first attract the rhizobia present in that vicinity. At first, these bacteria come into contact and then enter through a susceptible location (root hair), move along inside, establish and multiply themselves, and then move to base of the root hair. After entry into the plant cell, the bacteria keep on multiplying regularly, resulting in increase in size of cell and ultimately forming the nodule. The fixation of atmospheric N is facilitated by a Mo-containing enzyme named nitrogenase; therefore, for symbiotic fixation of N, adequate supply of Mo to legume crops is very important. The presence of oxygen is detrimental to the functioning of nitrogenase enzyme; hence, atmospheric N is fixed by the reduction process. The bacteria and nitrogenase enzyme require N without oxygen for N fixation. This problem is solved with the help of a protein named leghemoglobin that binds to and transports oxygen and removes it from the reach of nitrogenase. The combination of the protein with oxygen makes inside of an oxygen-free nodule red or pink, which is why these colors indicate an active N-fixing nodule. Fixed N is utilized for making amino acids; those in turn synthesize the proteins. The quantity of N thus obtained is determined by various factors such as soil moisture, availability of N, and the presence of efficient and competitive rhizobial strains. Phosphorus and N are the key elements in symbiotic N_2-fixation because they affect nodulation and N_2-fixation process. P deficiency leads to reduce nodule mass and decreased N production (Tahir et al. 2009; Dhakal et al. 2016).

13.5 Nutrient Availability for Plant Uptake

PGPRs are capable to enhance the nutrient concentration in the rhizosphere by nutrient fixation; those can be utilized by plants as and when required, preventing their loss by way of leaching out. For instance, N is the scarcest element for plants because it is required for the synthesis of amino acids and proteins. It is not wrong to say that agricultural sustainability is basically a matter of N management. The BNF has been a very strong source of N to the soils, which is directly utilized by the plant, and it is less vulnerable to losses like denitrification, volatilization, and leaching. Approximately, 80% of this BFN is obtained from symbiosis including leguminous plants and different rhizobial species. The legume-*Rhizobium* symbiosis has paramount biological and agronomical importance and acts as a sizeable source of N and therefore assumes a significant role in agriculture (Dita et al. 2006; Meena et al. 2015b). Four families, namely, *Methylobacteriaceae*, *Bradyrhizobiaceae*, *Phyllobacteriaceae*, and *Rhizobiaceae*, have genera that are capable to stimulate nodule formation and fix N_2 in symbiotic relationship with legumes. In desert regions, rhizobia are very crucial, as by way of their symbiotic association with some of trees/shrubs like acacia they fix N, playing a very important role in soil fertility management under such extreme conditions.

Some PGPRs also have the properties of solubilizing the other very important plant nutrient, namely, phosphate; such solubility effect will ensure an enhanced availability of phosphorus that may be readily utilized by the plants. *Kocuria turfanensis* strain 2M4 isolated from rhizospheric soil was found to be a phosphate solubilizer, an IAA producer, and a siderophore producer (Vejan et al. 2016). Inoculation of plants with phosphate solubilizing microorganisms can enhance plant growth. A large number of strains of *Rhizobium* and *Bradyrhizobium* are capable to solubilize inorganic phosphate (Antoun et al. 1998; Meena et al. 2017b). For instance, Abd-Alla (1994) has demonstrated that strain TAL 1236 of *R. leguminosarum* bv. viciae releases phosphorus from organic compounds through the reaction of acid and alkaline phosphatase. In a field study, Chabot et al. (1996) have found that phosphate solubilization by strains of *R. leguminosarum* bv. phaseoli was highly instrumental in improving the growth of maize and lettuce under moderate fertile growing situations.

13.6 Phytohormones

These plant growth regulators are synthetic organic substances. They are present in minute quantity that also have great effect on various biochemical, physiological, and morphological processes in plants system, and this way they have got very important role in improving agricultural productivity. Among the PGPRs, N_2-fixing bacteria are solely known for their capacity of fixing N; however, they also have been reported to produce IAA. For instance, under in vitro conditions, species of *Bradyrhizobium*, *Rhizobium*, and *Mesorhizobium* also produce IAA, and for producing IAA two possible pathways have been reported: (1) the indole-3-pyruvic acid pathway, found in PGPRs, *Enterobacter cloacae*, *Rhizobium*, and *Bradyrhizobium*, and the (2) indole-3-acetamide (IAM) pathway, found in tumor-forming bacteria, such as *Pseudomonas syringae* pv. *savastanoi* and *Agrobacterium*, for which genes are plasmid-borne (Khan et al. 2009). Gibberellin (GA) is another key plant growth regulator that has got a key role in certain processes like seed germination and development, floral development, flower and fruit development, and steam and leaf growth. However, the most overwhelming physiological effect of GA has been shoot elongation.

Khan et al. (2014) demonstrated significant enhancement in various growth characteristics of tomato plants when treated with the gibberellin-producing *Sphingomonas* sp. LK11 strain. Cytokinins are responsible to induce cell division in plant, and the proliferation of root hairs, but block lateral root formation and primary root elongation. Another plant hormone ethylene is responsible to regulate many processes such as the ripening of fruits and the abscission of leaves (Vejan et al. 2016; Meena et al. 2016a, b).

13.7 Production of Siderophores

Siderophores are low-molecular-weight iron-binding protein compounds and show their presence in the process of chelating ferric iron (Fe^{3+}) from the soil. Under low availability of Fe, microbial siderophores can make Fe available to plants, stimulating their growth (Vejan et al. 2016). This mechanism of action is shown in Fig. 13.4.

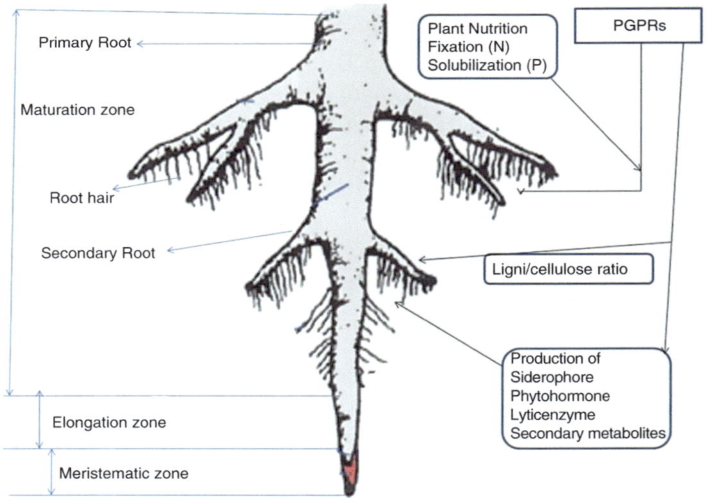

Fig. 13.4 The mechanism of action utilized by plant growth-promoting rhizobacteria (PGPRs) toward growth induction in plants. The flow and location of N fixation, phosphorus solubilization, and siderophore production are shown

In low Fe soils, symbiotic nodulation and subsequent N fixation is facilitated by the presence or generation of these siderophores, and this amply justifies the presence of symbiotic rhizobia in Fe-deficient soils. Various strains of *Mesorhizobium* demonstrated the formation of siderophores in Chrome Azurol S (CAS) agar medium while the supernatants of this strain yielded 2,3-dihydroxybenzoic acid (DHBA) and salicylic acid as phenolate-type siderophores. Other *Rhizobium and* bradyrhizobial strains inoculating green gram (*Vigna radiata* L. Wilczek), pigeon pea (*Cajanus cajan*), and pea (*Pisum sativum*) have exhibited the generation of siderophores using CAS solution assay and CAS agar plate (Khan et al. 2009; Meena et al. 2017a). Besides some isolates of *Rhizobium meliloti* obtained from *Mucuna pruriens*, a medicinal plant, few isolates could have inhibitory effect on the development of phytopathogens (*Macrophomina phaseolina*) and distinct positive impact on various growth parameters and symbiotic nodulation in plants of groundnut (*Arachis hypogaea*) infected with *M. phaseolina* (Arora et al. 2001). Moreover, the combined effect of PGPRs (*Bacillus* and *Pseudomonas*) and *Rhizobium* sp. improved the defense mechanism of plants by producing enzymes such as POX and polyphenol oxidase L-phenylalanine ammonialyase (PAL). It has also been found that the combined application of PGPRs and rhizobia enhance the systemic resistance of pigeon pea against the wilt caused by fusaria (Dutta et al. 2008). The stress owing to presence of heavy metals in soils and/or water is becoming great concern of agriculture production systems and some corrective mechanisms are essentially required to combat such stress. The various PGPRs have the capacity of producing adequate amounts of IAA under stress conditions owing to excess presence of zinc (Zn) and nickel (Ni). *Bradyrhizobium* (RM8) tolerant to Ni and Zn, *Rhizobium* sp. (RL9) tolerant to Zn, and *Rhizobium* sp. (RP5) tolerant to Zn and Ni are some examples under such metal-stressed conditions (Wani et al. 2008a, b).

13.8 Production of Volatile Organic Compound

PGPRs produce volatile organic compounds which promote the plant growth and increase systemic resistance against pathogens. The rhizobacteria can diminish the disease caused by foliar pathogens by activating plant-mediated resistance mechanism called as induced systemic resistance. Some other PGPR strains produce volatile organic compounds that are responsible to enhance plant biomass, resistance against disease, and stress tolerance toward abiotic factors. Some of these effects are observed directly or some others as indirectly. Volatile organic compound production is a unique characteristic of a larger range of microorganisms residing in soil (Dhakal et al. 2016). Different types of species that produce volatile compound differ in their identity and quantity and obviously, their effects are of various types (Vejan et al. 2016).

13.9 Rhizobia Symbiosis

Beijerinck (1888) was the first to observe the symbiotic effect of some of strains, and he named this strain as *Bacillus radicicola* which was isolated from plant root nodules. Frank called this bacterium *Rhizobium leguminosarum*. The word *rhizobia* was used to describe bacteria that belong to the genus of *Rhizobium*. However, recently it was established that a very small number of rhizobia belong to this genera and other genera came to be known such as *Bradyrhizobium* (Jordan 1982; Meena et al. 2017a), *Sinorhizobium* (Chen et al. 1988), and *Mesorhizobium* (Jarvis et al. 1997). At that time, rhizobia were thought to have a place only with the *Alphaproteobacteria*, namely, the order *Rhizobiales*, which also involved species belonging to nonlegume microsymbionts. For rhizobial symbiotic association, two distinct genes are necessary for a functional symbiosis, one for N fixation and another for nodulation (Fig. 13.5). Genes responsible for nodule formation, such as nod ABC encode enzymes regulates the biosynthesis and release of Nod factors are host specific lipochitooligosaccharides that have tendency of reacting with flavonoids present in the plant. Genes meant for N fixation involve the structural genes for the nitrogenase (nif HDK), and this enzyme is primarily responsible for fixation of atmospheric N. The symbiotic association of rhizobia-legumes combination is distinct as each strain of bacteria has its own host specificity having varied range. Figure 13.3 illustrates an early symbiosis stage with rhizobia inside infection threads in plant roots.

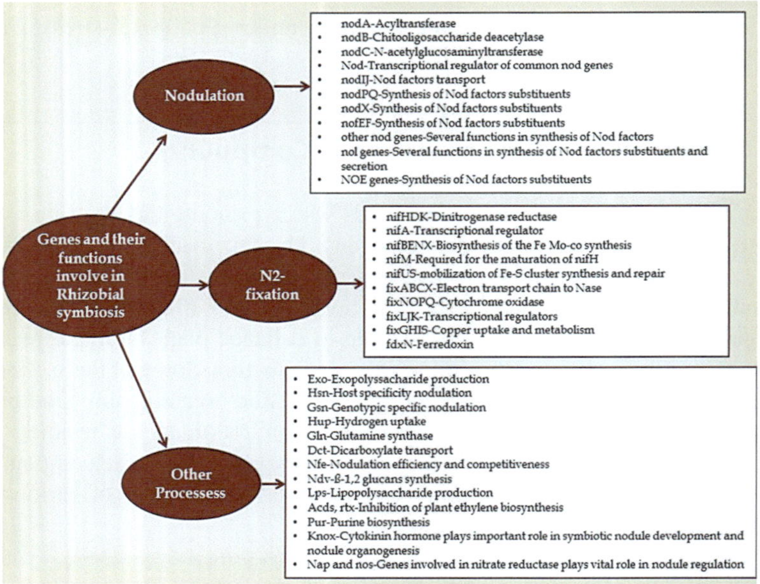

Fig. 13.5 Genes and functions of gene product in rhizobial symbiosis

13.10 Mesorhizobia Use in Agriculture

In 1997, the genera *Mesorhizobium* was reported (Jarvis et al. 1997). Thereafter some species have been reappointed which were previously assigned in the *Rhizobium* genus. Mesorhizobia are highly efficient fixers of N through symbiosis with host legumes as they are capable of doing so under any type of prevailing climatic situation covering a wide range of hosts from different habitats (Table 13.3).

Table 13.3 Species, type strains, original host legumes of *Mesorhizobium*, and their accession numbers for 16S rRNA and nodC genes

Species	Type strain	Host legume	Reference	Accession numbers	
				16S rRNA gene	*nodC*
M. abyssinica	AC98c^T	*Acacia abyssinica/A. tortilis*	Degefu et al. (2013)	GQ847896	GQ848002
M. albiziae	CCBAU 61158^T	*Albizia kalkora*	Wang et al. (2007)	DQ100066	DQ311092
M. alhagi	CCNWXJ12-2^T	*Alhagi sparsifolia*	Chen et al. (2010)	EU169578	EU722486
M. amorphae	ACCC 19665^T	*Amorpha fruticosa*	Wang et al. (1999)	AF041442	AF217261
M. australicum	WSM2073^T	*Biserrula pelecinus*	Nandasena et al. (2009)	AY601516	Mesau 05911
M. camelthorni	CCNWXJ 40-4^T	*Alhagi sparsifolia*	Chen et al. (2011)	EU169581	EU722491
M. caraganae	CCBAU 11299^T	*Caragana microphylla*	Guan et al. (2008)	EF149003	EU130405
M. chacoense	LMG 19008^T	*Prosopis alba*	Velázquez et al. (2001)	AJ278249	DQ450937
M. ciceri	UPM-Ca ^{T}	*Cicer arietinum*	Nour et al. (1994)	DQ444456	DQ450938
M. gobiense	CCBAU 83330^T	*Oxytropis glabra*	Han et al. (2008)	EF035064	EF050784
M. hawassense	AC99b^T	*Sesbania sesban*	Degefu et al. (2013)	GQ847899	GQ848005
M. huakuii	CCBAU 260 ^{T}	*Astragalus sinicus*	Chen et al. (1991)	D13431	Not available
M. loti	NZP 2213^T	*Lotus corniculatus*	Jarvis et al. (1997)	X67229	DQ450939
M. mediterraneum	UPM-Ca3 ^{T}	*Cicer arietinum*	Nour et al. (1995)	L38825	DQ450940
M. metallidurans	STM 2683^T	*Anthyllis vulneraria*	Vidal et al. (2009)	AM930381	gi:496153723
M. muleiense	CCBAU 83963^T	*Cicer arietinum*	Zhang et al. (2012)	HQ316710	HQ316752

(continued)

Table 13.3 (continued)

Species	Type strain	Host legume	Reference	Accession numbers	
				16S rRNA gene	*nodC*
M. opportunistum	WSM2075^T	*Biserrula pelecinus*	Nandasena et al. (2009)	AY601515	Mesop_6438
M. plurifarium	ORS 1032^T	*Acacia senegal*	de Lajudie et al. (1998)	Y14158	FJ745283
M. qingshengii	CCBAU 33460^T	*Astragalus sinicus*	Zheng et al. (2013)	JQ339788	JQ339881
M. robiniae	CCNWYC 115^T	*Robinia pseudoacacia*	Zhou et al. (2010)	EU849582	EU849563
M. sangaii	SCAU7^T	*Astragalus luteolus*	Zhou et al. (2013)	EU514525	JN129438
M. septentrionale	SDW014^T	*Astragalus adsurgens*	Gao et al. (2004)	AF508207	DQ450941
M. shangrilense	CCBAU 65327^T	*Caragana bicolor*	Lu et al. (2009)	EU074203	EU687487
M. shonense	AC39a^T	*Acacia abyssinica*	Degefu et al. (2013)	GQ847890	GQ847995
M. silamurunense	CCBAU 01550^T	*Astragalus membranaceus*	Zhao et al. (2012)	EU399698	EU418404
M. tamadayense	Ala-3^T	*Anagyris latifolia*	Ramirez-Bahena et al. (2012)	AM491621	AM491624
M. tarimense	CCBAU 83306^T	*Lotus frondosus*	Han et al. (2008)	EF035058	EF050786
M. temperatum	SDW018^T	*Astragalus adsurgens*	Gao et al. (2004)	AF508208	DQ450942
M. thiogangeticum	SJT^T	*(Clitoria ternatea)*	Ghosh and Roy (2006)	AJ864462	Not available
M. tianshanense	A-1BS^T	*Glycyrrhiza pallidiflora*	Chen et al. (1995)	AF041447	DQ450943

Modified: Laranjo et al. (2014)

Cicer arietinum (chickpea) is one of the most important legume host crops with which *Mesorhizobium* species establish symbiotic relationship. Field research conducted on chickpea crop in India and Turkey suggests that combined application of mesorhizobia and phosphate solubilizing bacteria is better in effect in comparison to sole applications of either. A current review in India revealed the beneficial outcomes of chickpea field combined applications with *Mesorhizobium* sp. and *Pseudomonas aeruginosa*, which represent for an appreciable increment of 32% in grain yield, compared to the uninoculated control (Verma et al. 2013). Most of the *Mesorhizobium* spp. work as root nodule symbionts, and they are capable to fix N with the help of mimosoid temperate legumes, like *Biserrula pelecinus*, *Astragalus* spp., *Cicer arietinum*, *Amorpha fruticosa*, *Caragana* spp., and *Lotus* spp. Recently, a *M. plurifarium* strain has been reported to cause nodulation even in the nonlegume *Parasponia andersonii*.

13.11 Role of PGPRs in Legume Nodulation and Productivity

Many researchers have established PGPRs' role to induce emergence of host plants and hence called them as emergence-promoting rhizobacteria (EPR) (Table 13.4). These effects may be of particular importance in soils falling under extreme climates or soils having poor structure or other problem soils. PGPRs have significant

Table 13.4 Beneficial effects of *Rhizobium* inoculation on selected crops

Host crops	*Rhizobium*	Site of colonization	Growing situation	Percent improvement in different attributes	References
Rice	*Bradyrhizobium*	Rhizosphere	Gnotobiotic	20 (total biomass)	Chaintreuil et al. (2000) and Bhattacharjee et al. (2008)
	Rhizobium leguminosarum bv. trifolii, *R. vietnamiensis*	Roots rhizosphere	Greenhouse and field nursery pot trial, field pot trial	15–22, 8–22 (grain yield)	Yanni et al. (1997, 2001)
				23 and 59 (shoot/root weight), 19 (yield)	Biswas et al. (2000a, b)
				13–22 (yield)	Tran Van et al. (2000)
	R. leguminosarum	Rhizosphere	Pot culture	18 (biomass)	Hussain et al. (2009)
			Glasshouse	25 (grain size)	
				43 (yield)	
Green gram	*Bradyrhizobium MRM6*	–	Pots	Increase herbicide tolerance (quizalofop-p-ethyl and clodinafop)	Ahemad and Khan (2011a, b, 2012)
	Bradyrhizobium sp. *(vigna) RM8*	–	Pots	82 (nodule numbers)	Wani et al. (2007) and Dhakal et al. (2016)
				120 (leghemoglobin)	
				34 (seed yield)	
				13 (grain protein)	
	R. phaseoli	–	Pots	In the presence of tryptophan, it mitigated the adverse effects of salinity and increase growth and yield	Zahir et al. (2010)
Pea, lentil	*Rhizobium strain MRP1, MRL3*	–	Pots	Increase nodulation and leghemoglobin content, amount of N and P in plant, seed yield and seed protein	Ahemad and Khan (2009, 2010, 2011b)
Mustard	*Sinorhizobium* sp. *Pb002*		Microcosms	Increased the efficiency of lead phytoextraction	Di Gregorio et al. (2006)

(continued)

Table 13.4 (continued)

Host crops	*Rhizobium*	Site of colonization	Growing situation	Percent improvement in different attributes	References
Chickpea	*Mesorhizobium* sp. *RC3*	–	Pots	71 (dry matter accumulation), 86 (number of nodules), 36 (seed yield) and 16 (grain protein)	Wani et al. (2008a)
Wheat	*R. trifolii*	Roots	Pot trials	24 (wheat shoot dry matter and grain yield)	Hilali et al. (2001)
Alfalfa	*Rhizobium leguminosarum* bv. phaseoli *CPMex46*	–	Growth chamber	Improved Cu and Fe translocation from root to shoot	Carrillo-Castaneda et al. (2003)
Maize	*R. trifolii*	Roots	Greenhouse and field	34 (yield), 11 (yield)	Riggs et al. (2001)
	Sinorhizobium spp.	–	Greenhouse	49–82 (yield)	Riggs et al. (2001)
	R. etli bv. phaseoli	Roots	Gnotobiotic	20–45 (total biomass)	Zamora and Matinez-Romero (2001)

Modified: Ahemad and Kibret (2014) and Mia and Shamsuddin (2010)

influence on development and nutrition of plants having both direct and indirect effects. Direct effect is exhibited in form of solubilization or asymbiotic N fixation, hormone production, etc., and indirect effect may be witnessed as alteration or modification of plant-microorganism relations, and interactions like the mycorrhizal or the *Rhizobium* symbioses (Requena et al. 1997; Ahmad et al. 2008; Dhakal et al. 2016) found that the stomatal conductance of plant leaf was more in PGPRs-inoculated plants than non-inoculated ones under limiting water conditions indicating that PGPR-inoculated plants are more capable in utilizing the available water more efficiently. This role of PGPRs is of great significance when water is progressively becoming a limiting factor in most of the production systems.

Results of an investigation revealed that different strains of PGPRs, identified from semiarid agroecosystem of desert region of Spain, exhibited differential performance in terms of fixation of N and content of N in plants (Requena et al. 1997; Ram and Meena 2014). Further, it was also observed that some of the rhizobacteria fix N as free-living fixers.

Field trials performed in Tunisia demonstrated that inoculation of different types of chickpea cultivars with highly effective *rhizobia* strains appears to hike in number of nodules and shoot dry weight (Ben Romdhane et al. 2007).

13.12 Role of PGPRs Under Stress Conditions

Ethylene has multiple roles in plant life. It may act as plant growth regulator as well as functions as stress hormone under certain conditions. The role of ethylene as stress hormone is evident by its elevated levels of production under stress situations caused by abiotic or biotic factors such as high salt concentration, deficit or excess of moisture, presence of toxic levels of heavy metals, and under pathogenic stress (Saleem et al. 2007).

The increasing levels of ethylene under stress conditions are an indication of cessation of growth as cell division and elongation, root elongation, and N fixation are almost stopped and the plant leads toward premature senescence (Khan et al. 2009). More recently, it was observed that rhizobia, by adopting certain methodologies, successfully lower down the level of ethylene thus reducing the inhibitory effect of ethylene on growth of plant, particularly nodulation. Two distinct methodologies have been reported for reduction in ethylene levels. The first one involves rhizobial production of an amino acid called as rhizobiotoxin that has inhibitory effect on biosynthesis of ethylene through inhibiting functioning of ACC synthatase enzyme (Okazaki et al. 2007). In second mechanism, rhizobacteria produces ACC deaminase enzyme restricting the production of ethylene. Duodu et al. (1999) describe the role and action of rhizobiotoxine in mutualistic symbiosis and found that for adequate number of symbiotic nodules in green gram (*Vigna radiata* L. Wilczek) owing to *Bradyrhizobium*, rhizobiotoxine is required. Rhizobiotoxin that facilitated reduced level of production of ethylene is confined only to slow-growing *Rhizobium* sp.; however, toxicity of ethylene levels is increased with faster-growing rhizobia, and this ACC deaminase has a very crucial role in this process. At first, this enzyme could be identified in *Pseudomonas* sp. strain ACP and the yeast, *Hansenula saturnus*. Since then, it has been found in many other PGPR strains and in many symbiotic N_2-fixing bacteria, like *R. hedysari, Rhizobium* spp., *Mesorhizobium, Rhizobium leguminosarum* bv. viciae, and *Bradyrhizobium* (Musarrat et al. 2009; Meena et al. 2017a). The production of ethylene is regulated by this enzyme through converting ACC into ketobutyrate and ammonia. This will result into more nodulation as found in plants inoculated with *S. meliloti* Rm11466 and also higher nodule formation occurred by Rm11466 due to combined inoculation with *S. meliloti* Rm5356 and Rm11466. (Ma et al. 2004). The ACC deaminase was also found in several soil bacteria like rhizobia. Genes that encode ACC deaminase found in PGPRs include *R. leguminosarum*, *R. gallicum*, *B. japonicum*, *Mesorhizobium loti*, and *R. radiobacter*.

The ACC deaminase genes (AcdS) found in *R. leguminosarum* bv. viciae are 64% similar with the gene found in PGPRs, *Pseudomonas putida* UW4 (Shah et al. 1998) pea roots. All these rhizobial strains had lower activity of ACC deaminase against those shown by *P. putida* UW4 (Fig. 13.6). The introduction of the ACC deaminase gene and leucine-responsive regulatory protein (LRP)-like gene from strain 128C53K, into a strain of *S. meliloti*, which is unable to produce this enzyme, made it 35–40% more promising and effective than the wild type at nodulating *M. sativa* (alfalfa) by enhanced nodule numbers and biomass (Ma et al. 2004). Klee

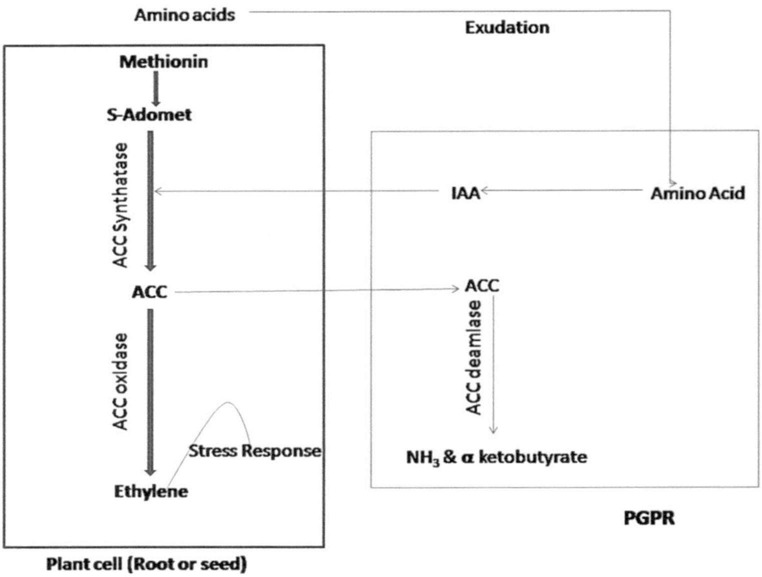

Fig. 13.6 Schematic representation of mechanisms through which the PGPRs attached to either a seed or plant root lower ethylene concentration and prevent ethylene inhibition of root elongation

et al. (1991) also had similar results that the expression of *Pseudomonas* sp. 6G5 acdS gene in tomato results in reduced formation of ethylene, and this causes delay in fruit maturity. In this context, by applying techniques used in genetics engineering, we are able to produce ACC deaminase from rhizobial strains (Table 13.5). This will be helpful to get better yield of leguminous crops under stressed conditions. Ahmad et al. (2013) and Meena et al. (2015c) found that under salt-affected environments, *Rhizobium* and *Pseudomonas* ACC-deaminase-producing strains are able to enhance the growth, physiology, and quality of mung beans.

Applications of PGPR containing ACC deaminase in relation to the nature of stress are described in Table 13.5.

13.13 Salinity Stress

Salt stress is responsible to inhibit the starting phase of *Rhizobium*-legume symbioses. For example, it was found that the presence of higher concentration of common salt could cause little curling or deformation in the root hairs of soybean when inoculated with *Bradyrhizobium japonicum,* diluting the effect of salt stress (Tu 1981). Under salt stress, rate of respiration in nodules is reduced and cytosolic proteins formation in nodules is also reduced (e.g., leghemoglobin). This led to reduction in the property of N_2-fixation (Zahran 1999; Meena and Yadav 2014).

Table 13.5 Plant tolerance to different abiotic stress by PGPRs

PGPRs	Crops	Stress tolerance mechanism	References
Rhizobium sp.	Sunflower	Production of EPS which affects the structure of rhizospheric soil	Nada et al. (2012)
Rhizobium and *Pseudomonas*	Wheat (*T. aestivum*)	Restricted Na^+ uptake	Nada et al. (2012)
Azospirillum sp.	Wheat (*T. aestivum*)	Increased water circulation	Nada et al. (2012) and Meena et al. (2015)
Rhizobium leguminosarum bv. viciae 128C53K	*Pisum sativum* L.	Symbiotic nodulation improved	Ma et al. (2003)
Pseudomonas putida	*Vigna radiata*	Salt	Mayak et al. (1999)
Rhizobium leguminosarum (LR-30), *Mesorhizobium ciceri* (CR-30 and CR-39), and *Rhizobium phaseoli* (MR-2)	Wheat	IAA produced by the consortia improved the growth, biomass and drought tolerance index	Hussain et al. (2014)
P. putida UW4	*Lycopersicon esculentum*	Salt	Yan et al. (2014)
ACC-deaminase-producing rhizobacteria	Wheat (*T. aestivum*)	Increased root-shoot length, biomass, and lateral root number	Shakir et al. (2012)
P. fluorescens ACC-5	*Pisum sativum*	Drought	Zahir et al. (2008)
Pseudomonas sp.	*Pisum sativum*	Drought	Arshad et al. (2008)
Azospirillum brasilense	Barley (*Hordeum vulgare*)	Salt	Omar et al. (2009)
Azospirillum	Wheat (*T. aestivum*)	Osmotic stress (20% PEG	Pereyra et al. (2006)
Pseudomonas fluorescens	Wheat (*T. aestivum*)	Temperature	Egamberdiyeva and Hoflich (2003)
Pantoea agglomerans,			
Mycobacterium sp.			
Rhizobium etli overexpressing trehalose-6-phosphate synthase gene	*Phaseolus vulgaris*	Trehalose act as a signaling molecule in upregulation of genes involved in stress tolerance, carbon and nitrogen metabolism	Suarez et al. (2008)
Bacillus isolate 23-B and *Pseudomonas* 6-P with *Mesorhizobium ciceri*	*Cicer arietinum*	Higher proline concentration, improved germination, root and shoot length, and fresh weight of the seedlings	Sharma and Khanna (2013)

(continued)

Table 13.5 (continued)

PGPRs	Crops	Stress tolerance mechanism	References
Gluconacetobacter diazotrophicus PAL5	Sugarcane	Inoculation activated the ABA-dependent signaling genes conferring drought resistance	Vargas et al. (2014)
Azospirillum sp.	Chickpea (*Cicer arietinum* L.)	Nutrient	Rokhzadi and Toashih (2011)
Azotobacter chroococcum		Deficiency	
Mesorhizobium ciceri			
Pseudomonas fluorescens			
Bacillus polymyxa	Zea maize L. (*Zea maize* cv. Felix)	Nutrient	Egamberdiyeva (2007)
Mycobacterium phlei		Deficiency	
Pseudomonas alcaligenes			
Aeromonas hydrophila	Soybean (*Glycine max*)	Temperature	Zhang et al. (1997)
Serratia liquefaciens			
Serratia proteamaculans			

Stress caused by soil salinity has detrimental effect on growth and development of plant. Soil salinity is also responsible for many negative impacts such as toxicity level of certain elements such as Na^+ and Cl^-, formation of ethylene production, osmotic stress, plasmolysis, nutrient imbalance, formation of reactive oxygen species, and obstruction with certain vital processes including photosynthesis. The formation and secretion is indicator of growth of plant and this is also produced by rhizobacteria. The production of ethylene is an indicator of inhibition of growth because as the production of ethylene elevates, it starts inhibiting IAA signal transduction. The ACC deaminase and IAA both works together synergistically. This will lead to decrease the suppressive effect of ethylene on auxin response factor synthesis; thus, IAA becomes free to exhibit its effect. This will result into enhanced plant growth curtailing the excess formation of ethylene (Gamalero et al.). In this context, Saravanakumar and Samiyappan (2007) also found the presence of increased salinity resistance in groundnut plants and better yield when inoculated with *Pseudomonas fluorescens* strain TDK1 containing ACC deaminase activity compared with groundnut plant inoculated with *Pseudomonas* strains lacking ACC deaminase activity. Mayak et al. (2004) also recorded similar findings when tomato seedlings inoculated with *Achromobacter piechaudii* having ACC deaminase activity in the presence of NaCl salt (up to 172 mM). Cheng et al. (2007) reported growth inhibition by salt when only wild-type *P. putida* UW4 inoculated with not an ACC deaminase protected canola plants against ethylene stress.

Today numerous findings are available on the co-inoculation of bacteria and fungi on plants under salt stress conditions. Their major focus is on the triple symbioses of rhizobia-AM fungi-legume (Gamalero et al.). For instance, when *Acacia cyanophylla* inoculated with *A. cyanophylla* and *Bradyrhizobium* sp. in a combination of endomycorrhizal fungi obtained from dune soils of coastal regions, the tolerance against salinity was increased. Furthermore, same findings were achieved by combination of *Glomus intraradices* and *Bradyrhizobium* sp. with *Acacia auriculiformis* or *Acacia mangium* seedlings under the situation of excess salt stress (Diouf et al. 2005). Enhanced plant tolerance to salt was seen when faba bean (*Vicia faba*) grown with *Glomus clarum* (a mycorrhizal fungi) in the presence of N-fixing rhizobia.

13.14 Drought Stress

Deficit of water and subsequent moisture stress and drought are global phenomena across all the globe and about half of the globe remains under drought every year. The drought or drought-like situations cause significant damage to BNF, and the use of PGPRs is a new possibility under such conditions.

Meena and Yadav (2013) found that combination of legume-*Rhizobium* in chickpea and pigeon pea plants is more sensitive to drought, salinity, and temperature than rhizobia growing alone. Root hairs are the primary site of infection and entry for rhizobia and these three stress factors restrict the development of rot hair. This will result into poor or no nodulation. Integrated inoculation of bean (*Phaseolus vulgaris* L.) with *Rhizobium tropici* and two strains of *Paenibacillus polymyxa* under drought stress resulted in improvement in growth traits including better nodulation (Figueiredo et al. 2008). Dodd et al. (2005) obtained increased seed yield (25–41%) and seed N accumulation and nodulation in drought-affected pea (*Pisum sativum* L.) inoculated with ACC deaminase bacteria *Variovorax paradoxus* 5C-2 compared with well-watered uninoculated plants.

13.15 Water Logging Stress

Ethylene production is an indicator of stress within the plant system. Excess water situations also cause stress and under waterlogged conditions in roots and stem, elevated levels of ethylene are observed. Under such conditions, ACC is formed in roots and then it is transported to stem where this transported ACC produce ethylene in the presence of oxidase enzyme. Grichko and Glick (2001) studied the effect of ACC deaminase PGPRs in tomato. They found that tomato showed important tolerance to flooding stress inoculated either with *Pseudomonas putida* UW4, *Enterobacter cloacae* CAL2, *P. putida* (ATCC17399/pRKACC), or *P. putida* (ATCC17399/pRK415).

13.16 Temperature Stress

Certain PGPRs have been found to combat the stress effect caused by temperature, and use of *Burkholderia phytofirmans* strain PsJN has been found effective in maintaining normal growth of potato (*Solanum tuberosum*) and growth and physiological activity of grapevine under temperature stress conditions. Bensalim et al. (1998), Barka et al. (2006), Cheng et al. (2007), and Meena et al. (2015b) have also reported positive impact of a psychrotolerant ACC deaminase bacterium *P. putida* UW4 at low temperature under salt stress.

13.17 Pathogenicity Stress

It is now well known that some ethylene synthesis inhibitors are capable to reduce the occurrence of the pathogenic infections in plants. ACC deaminase bacteria are found to be an antagonist against plant pathogen *Fusarium oxysporum* as reported by Yuquan et al. (1999). Furthermore, beneficial effects of ACC deaminase bacterium *Pseudomonas fluorescens* have been reported in suppressing the growth and development of pathogenic fungal species of *Fusarium oxysporum* and *Fusarium proliferatum* (Donate-Correa et al. 2005). The antagonism of *Burkholderia* sp. toward disease-causing pathogens, namely, *R. solani* and *Sclerotinia sclerotiorum*, has also been reported, and such effect is ascribed to the ACC deaminase produced by these endophytic bacteria (Pandy et al. 2005). While Rasche et al. (2006) recorded absence of antagonistic effect of ACC deaminase bacteria against bacterial pathogen *Erwinia carotovora* ssp. *atroseptica* (Eca), in another study, Rasche et al. (2006) reported that ACC deaminase bacteria were having antagonistic effect on two potato pathogens, namely, *Ralstonia solanacearum* and *Rhizoctonia solani*. Belimov et al. (2007) concluded that at lower concentration (inoculum size), bacterial ACC deaminase of *Pseudomonas brassicacearum* Am3 (pathogenic bacteria) could suppress the pathogenic effect of this bacteria and normal growth of tomato plants was obtained.

13.17.1 Heavy Metals Stress

Minute quantities of some metal elements are required by plant as essential micronutrients, but their supplies in excess or deficit have negative effect on plant growth and development. These metals are found in great amounts in various effluents particularly in urban areas, and application of this effluent water to plants poses the threat of toxicity of these metals to such plants or they may be accumulated in plants, which may damage the health of the consumers of these plants. Nie et al. (2002) communicated ACC deaminase genes in canola (*Brassica napus*) plants. The canola grown with bacterium *E. cloacae* CAL2 in the soil possessing toxic levels of arsenic and further it was observed that larger amounts of arsenate was accumulated in case of transgenic canola communicated with ACC deaminase genes.

Some examples of combating heavy metal stress through incculation with ACC deaminase rhizobacteria have been given in Table 13.6.

Table 13.6 Mitigation of heavy metal stress through ACC deaminase produced by PGPRs

Host species	PGPRs	Effect	References
Brassica napus	*Kluyvera ascorbata* SUD165	Plant could have normal growth under high levels of nickel (Ni^{2+}), lead (Pb^{2+}), zinc (Zn^{2+}), and chromate (CrO_4^{2-})	Burd et al. (1998)
Brassica juncea L.	*Kluyvera ascorbata* SUD165	Toxic effects of heavy metals like nickel, lead, and zinc (Ni^{2+}, Pb^{2+}, and Zn^{2+}) were alleviated by inoculation	Burd et al. (2000)
	Kluyvera ascorbata SUD165/26		
Brassica juncea L.	*Variovorax paradoxus*	Negated the detrimental effect of high concentration of cadmium (Cd^{2+})	Belimov et al. (2005)
	Rhodococcus sp.		
Brassica juncea L.	*Sinorhizobium* sp. *Pb002*	Reduce the effect of lead	Di Gregorio et al. (2006)
Brassica juncea L.	*Pseudomonas brassicacearum*	Imparted tolerance toward cadmium toxicity	Belimov et al. (2001) and Dadhich et al. (2015)
	Pseudomonas marginalis		
	Pseudomonas oryzihabitans, *Pseudomonas putida*	Stimulated root elongation even under high concentration of cadmium	
	Pseudomonas sp.		
	Alcaligenes xylosoxidans		
	Alcaligenes sp.		
	Variovorax paradoxus		
	Bacillus pumilus		
	Rhodococcus sp.		
Lycopersicum esculentum Mill	*Kluyvera ascorbata* SUD165	Stress/toxic effects of the heavy metals like nickel, lead, and cadmium (Ni^{2+}, Pb^{2+}, and Zn^{2+}) alleviated	Burd et al. (2000)
	Kluyvera ascorbata SUD165/26		
Phragmites australis	*Pseudomonas asplenii* ACa	Under high levels of copper (Cu^{2+}) and creosote, normal growth could be restored after inoculation	Reed et al. (2005)
Pisum sativum L.	*Pseudomonas brassicacearum* Am3, *Pseudomonas marginalis* Dp1	Normal nutrient uptake by roots, alleviating inhibitory effect of cadmium toxicity	Safronova et al. (2006)

(continued)

Table 13.6 (continued)

Host species	PGPRs	Effect	References
Pisum sativum L.	*Pseudomonas brassicacearum*, *Pseudomonas marginalis*	Imparted tolerance toward cadmium toxicity and caused normal root growth and elongation	Belimov et al. (2001)
	Pseudomonas oryzihabitans, *Pseudomonas putida*		
	Pseudomonas sp.		
	Alcaligenes xylosoxidans		
	Alcaligenes sp.		
	Variovorax paradoxus		
	Bacillus pumilus		
	Rhodococcus sp.		

Modified: Singh et al. (2015) and Saleem et al. (2007)

13.18 Conclusion

The PGPRs are a group of bacteria that are boon to the contemporary agriculture through their multifarious roles in sustainable crop production and soil health. The utilization of PGPRs is a unique and novel approach to harness the environment-friendly options for serving the variety of purposes from nutrition to stress tolerance, bioremediation of heavy metals, countering the disease-causing pathogens, and production of siderophores, antibiotics, phytohormones, and growth-promoting indole acetic acid. With the change of climate and pollution levels, the threat of multiple stresses are becoming a greater challenge, and sustainable cost-effective and eco-friendly options are urgently required for addressing such stresses. In this context, PGPRs are proving to be a very potent agent of imparting tolerance toward many of stresses. Ethylene production is a strong indicator of incidence of stress by any factor. PGPRs are capable of production of ACC deaminase that can inhibit the stress effect of ethylene, and this mechanism of dealing with ethylene is not only cost-effective but also compatible to the environment. Thus, the ethylene-lowering capability of nodule-forming bacteria is significant, and there is a need for further research efforts to increase understanding of intricate relationship of these PGPRs and plants. Besides, by the way of symbiotic and asymbiotic fixation of N and solubilizing the phosphates and extending complimentary effect toward other beneficial microorganism of soil, these PGPRs have become very strong tools of restoring and improving of soil health. Hence, use of such PGPRs may be a viable alternative to synthetic fertilizers for increasing the agronomic efficiency by reducing production cost and environmental pollutions.

13.19 Future Prospects

Some of the Potential Future Prospects of PGPR in Sustainable Agriculture May Be

- Proper identification and characterization of various PGPRs suitable for location-specific various purposes/objectives.
- Deciphering the role of any PGPR whether it is having singular or multiple roles and how their functions may be complimented with the suitable synergy of other inputs/items.
- Climate changes have posed serious challenge before agriculture fraternity globally, and potential of PGPRs must be fully utilized to mitigate or alter the climate change stress in the agricultural production systems.
- Pollution of water bodies is one another serious concern and mixing of various industrial and urban effluents making water sources unsuitable for use in crop production purpose. The bioremediation of such water bodies particularly contamination of heavy metals may be accomplished through use of PGPRs.
- The property of PGPRs to counter pathogenicity of various kinds of fungi must be utilized for the disease management under nonchemical or organic farming conditions with identification and development of proper protocol.
- Interrelationship or interaction of these PGPRs with other production factors and varied inputs must be understood to harness the potential adequately and amicably.
- Identification of suitable and stronger strains of different PGPRs suited to various farming situations across agroclimatic situations and clientele.
- Biotechnological tools must be utilized for development of superior strains for identified objectives.
- Soil health cards must have the population indicators of PGPRs and suitable methodologies to assess their concentration in soil must be developed.
- Adequate dissemination of technology among various stakeholders for widespread utilization by the end users.

References

Abd-Alla MH (1994) Use of organic phosphorus by Rhizobium leguminosarum bv. viciae phosphatases. Biol Fertil Soils 8:216–218

Abrol IP, Bronson KF, Duxbury JM, Gupta RK (2000) Long-term fertility experiments in rice-wheat cropping systems. Rice-wheat consortium paper series 6. Rice-wheat consortium for the Indo-Gangetic Plains. New Delhi, p 171

Ahemad M, Khan MS (2009) Toxicity assessment of herbicides quizalafop-p-ethyl and clodinafop towards Rhizobium pea symbiosis. Bull Environ Contam Toxicol 82:761–766

Ahemad M, Khan MS (2010) Comparative toxicity of selected insecticides to pea plants and growth promotion in response to insecticide-tolerant and plant growth promoting Rhizobium leguminosarum. Crop Prot 29:325–329

Ahemad M, Khan MS (2011a) Response of greengram [Vigna radiata (L.) Wilczek] grown in herbicide-amended soil to quizalafop-p-ethyl and clodinafop tolerant plant growth promoting Bradyrhizobium sp. (vigna) MRM6. J Agric Sci Technol 13:1209–1222
Ahemad M, Khan MS (2011b) Effect of tebuconazole-tolerant and plant growth promoting Rhizobium isolate MRP1 on pea-Rhizobium symbiosis. Sci Hortic 129:266–272
Ahemad M, Khan MS (2012) Productivity of greengram in tebuconazole-stressed soil, by using a tolerant and plant growth-promoting Bradyrhizobium sp. MRM6 strain. Acta Physiol Plant 34:245–254
Ahemad M, Mulugeta Kibret M (2014) Mechanisms and applications of plant growth promoting rhizobacteria: current perspective. J King Saud Univ Sci 26(1):1–20
Ahmad F, Ahmad I, Khan MS (2008) Screening of free-living rhizospheric bacteria for their multiple plant growth promoting activities. Microbiol Res 163:173–181
Ahmad M, Zahir ZA, Khalid M (2013) Efficacy of Rhizobium and Pseudomonas strains to improve physiology, ionic balance and quality of mung bean under salt-affected conditions on farmer's fields. Plant Physiol Biochem 63:170–176
Aliyu IA, Yusuf AA, Abaidoo RC (2013) Response of grain legumes to rhizobial inoculation in two savanna soils of Nigeria. Afr J Microbiol Res 7(15):1332–1342
Antoun H, Beauchamp CJ, Goussard N, Chabot R, Llande R (1998) Potential of Rhizobium and Bradyrhizobium species as plant growth promoting rhizobacteria on non-legumes: effect on radishes. Plant Soil 204:57–67
Arora NK, Kang SC, Maheshwari DK (2001) Isolation of siderophore producing strains of Rhizobium meliloti and their biocontrol potential against Macrophomina phaseolina that causes charcoal rot of groundnut. Curr Sci 81:673–677
Arshad M, Shaharoona B, Mahmood T (2008) Inoculation with Pseudomonas spp. containing ACC-deaminase partially eliminates the effects of drought stress on growth, yield, and ripening of pea (Pisum sativum L.). Pedosphere 18:611–620. https://doi.org/10.1016/S1002-0160(08)60055-7
Asghar HN, Zahir ZA, Arshad M, Khaliq A (2002) Relationship between in vitro production of auxins by rhizobacteria and their growth-promoting activities in Brassica juncea L. Biol Fertil Soils 35:231–237
Barka EA, Nowak J, Clément C (2006) Enhancement of chilling resistance of inoculated grapevine plantlets with a plant growth promoting rhizobacterium, Burkholderia phytofirmans strain PsJN. Appl Environ Microbiol 72:7246–7252
Beijerinck MW (1888) Die Bacterien der Papilionaceenknölchen. Bot Ztg 46:797–804
Belimov AA, Safronova VI, Sergeyeva TA, Egorova TN, Matveyeva VA, Tsyganov VE, Borisov AY, Tikhonovich IA, Kluge C, Preisfeld A, Dietz KJ, Stepanok VV (2001) Characterization of plant growth promoting rhizobacteria isolated from polluted soils and containing 1-aminocyclopropane-1-carboxylate deaminase. Can J Microbiol 47:242–252
Belimov AA, Hontzeas N, Safronova VI, Demchinskaya SV, Piluzza G, Bullitta S, Glick BR (2005) Cadmium-tolerant plant growth-promoting bacteria associated with the roots of Indian mustard (Brassica juncea L. Czern.). Soil Biol Biochem 37:241–250
Belimov AA, Dodd IC, Safronova VI, Hontzeas N, Davies WJ (2007) Pseudomonas brassicacearum strain Am3 containing 1- aminocyclopropane-1-carboxylate deaminase can show both pathogenic and growth-promoting properties in its interaction with tomato. J Exp Bot. https://doi.org/10.1093/jxb/erm010
Ben Romdhane S, Tajini F, Trabelsi M, Aouani M, Mhamdi R (2007) Competition for nodule formation between introduced strains of Mesorhizobium ciceri and the native populations of rhizobia nodulating chickpea (Cicer arietinum) in Tunisia. World J Microbiol Biotechnol 23:1195–1201
Bensalim S, Nowak J, Asiedu SK (1998) A plant growth promoting rhizobacterium and temperature effects on performance of 18 clones of potato. Am J Potato Res 75:145–152
Bhattacharjee RB, Singh A, Mukhopadhyay (2008) Use of N fixing bacteria as biofertiliser for non-legumes: prospects and challenges. Appl Microbiol Biotechnol 80:199–209
Biswas JC, Ladha JK, Dazzo FB (2000a) Rhizobia inoculation improves nutrient uptake and growth of lowland rice. Soil Sci Soc Am J 64:1644–1650

Biswas JC, Ladha JK, Dazzo FB, Yanni YG, Rolfe BG (2000b) Rhizobial inoculation influences seedling vigor and yield of rice. Agron J 92:880–886

Buragohain S, Sharma B, Nath JD, Gogaoi N, Meena RS, Lal R (2017) Impact of ten years of biofertilizer use on soil quality and rice yield on an inceptisol in Assam, India. Soil Res. https://doi.org/10.1071/SR17001

Burd GI, Dixon DG, Glick BR (1998) A plant growth-promoting bacterium that decreases nickel toxicity in seedlings. Appl Environ Microbiol 64:3663–3668

Burd GI, Dixon DG, Glick BR (2000) Plant growth-promoting bacteria that decrease heavy metal toxicity in plants. Can J Microbiol 46:237–245

Carrillo-Castaneda G, Munoz JJ, Peralta-Videa JR, Gomez E, Gardea-Torresdey GL (2003) Plant growth-promoting bacteria promote copper and iron translocation from root to shoot in alfalfa seedlings. J Plant Nutr 26:1801–1814

Chabot R, Antoun H, Cescas MP (1996) Growth promotion of maize and lettuce by phosphate-solubilizing Rhizobium legumioserum biovar phaseoli. Plant Soil 184:311–321

Chaintreuil C, Giraud E, Prin Y, Lorquin J, Ba A, Gillis M, de Lajudie P, Drefus B (2000) Photosynthetic Bradyrhizobia are natural endophytes of the African

Chen WX, Li GS, Qi YL, Wang ET, Yuan HL, Li JL (1991) Rhizobium huakuii sp. nov. isolated from the rootnodules of Astragalus sinicus. Int J Syst Bacteriol 41:275–280

Chen WX, Wang E, Wang SY, Li YB, Chen XQ (1995) Characteristics of Rhizobium tianshanense sp. nov., a moderately and slowly growing root-nodule bacterium isolated from an arid saline environment in Xinjiang, Peoples-Republic-of-China. Int J Syst Bacteriol 45:153–159

Chen WX, Yan GH, Li JL (1988) Numerical taxonomic study of fast-growing soybean rhizobia and a proposal that Rhizobium fredii be assigned to Sinorhizobium gen. nov. Int J Syst Bacteriol 38:392–397

Chen WM, Zhu WF, Bontemps C, Young JPW, Wei GH (2010) Mesorhizobium alhagi spnov. isolated from wild Alhagi sparsifolia in north-western China. Int J Syst Evol Microbiol 60:958–962

Cheng Z, Park E, Glick BR (2007) 1-aminocyclopropane-1-carboxylate deaminase from Pseudomonas putida UW4 facilitates the growth of canola in the presence of salt. Can J Microbiol 53:912–918

Contesto C, Desbrosses G, Lefoulon C, Bena G, Borel F, Galland M, Gamet L, Varoquaux F, Touraine B (2008) Effects of rhizobacterial ACC deaminase activity on Arabidopsis indicate that ethylene mediates local root responses to plant growth-promoting rhizobacteria. Plant Sci 175:178–189

Dadhich RK, Meena RS, Reager ML, Kansotia BC (2015) Response of bio-regulators to yield and quality of Indian mustard (*Brassica juncea* L. Czernj. and Cosson) under different irrigation environments. J App Nat Sci 7(1):52–57

Dardanelli MS, Carletti SM, Paulucci NS, Medeot DB, Rodriguez Caceres EA, Vita FA, Bueno M, Fumero MV, Garcia MB (2010) Benefits of plant growth-promoting rhizobacteria and rhizobia in agriculture. In: Plant growth and health promoting bacteria, vol 18. Springer

Dary M, Chamber-Pérez MA, Palomares AJ, Pajuelo E (2010) "In situ" phytostabilisation of heavy metal polluted soils using Lupinus luteus inoculated with metal resistant plant-growth promoting rhizobacteria. J Hazard Mater 177:323–330

Dazzo FB, Yanni YG, Rizk R, DeBruijn FJ, Rademaker J, Squartini A, Corich V, Mateos P, Martinez-Molina E (2000) Progress in multinational collaborative studies on the beneficial association between Rhizobium leguminoserum bv. Trifolii and rice. In: Ladha JK, Reddy PM (eds) The quest for nitrogen fixation in rice. IRRI, Los Banos, pp 167–189

Degefu T, Wolde-Meskel E, Liu B, Cleenwerck I, Willems A, Frostegard A (2013) Mesorhizobium shonense sp. nov., Mesorhizobium hawassense sp. nov. and Mesorhizobiumabyssinicae sp. nov., isolated from root nodules of different agroforestry legume trees. Int J Syst Evol Microbiol 63:1746–1753

de Lajudie P, Willems A, Nick G, Moreira F, Molouba F, Hoste B (1998) Characterizationof tropical tree rhizobia and description of Mesorhizobium plurifarium sp. nov. Int J Syst Bacteriol 48:369–382

Deshwal VK, Pandey P, Kang SC, Maheshwari DK (2003) Rhizobia as a biological control agent against soil borne plant pathogenic fungi. Ind J Exp Biol 41:1160–1164

Dhakal Y, Meena RS, Kumar S (2016) Effect of INM on nodulation, yield, quality and available nutrient status in soil after harvest of green gram. Leg Res 39(4):590–594

Di Gregorio S, Barbafieri M, Lampis S, Sanangelantoni AM, Tassi E, Vallini G (2006) Combined application of triton X-100 and Sinorhizobium sp. Pb002 inoculum for the improvement of lead phytoextraction by Brassica juncea in EDTA amended soil. Chemosphere 63:293–299

Diouf D, Duponnois R, Ba AT, Neyra M, Lesueur D (2005) Symbiosis of Acacia auriculiformis and Acacia mangium with mycorrhizal fungi and Bradyrhizobium spp. improves salt tolerance in greenhouse conditions. Funct Plant Biol 32:1143–1152

Dodd IC, Belimov AA, Sobeih WY, Safronova VI, Grierson D, Davies WJ (2005) Will modifying plant ethylene status improve plant productivity in water-limited environments? 4th international crop science congress. http://www.cropscience.org.au/icsc2004/poster/1/3/4/510_doddicref.htm. Accessed 17 June 2007

Donate-Correa J, Leon-Barrios M, Perez-Galdona R (2005) Screening for plant growth-promoting rhizobacteria in Chamaecytisusproliferus (tagasaste), a forage tree-shrub legume endemic to the Canary Islands. Plant Soil 266:261–272

Duodu S, Bhuvaneswari TV, Stokkermans TJW, Peters NK (1999) A positive role for rhizobitoxine in Rhizobium-legume symbiosis. Mol Plant-Microbe Interact 12:1082–1089

Dutta S, Mishra AK, Dileep Kumar BS (2008) Induction of systemic resistance against fusarial wilt in pigeon pea through interaction of plant growth promoting rhizobacteria and rhizobia. Soil Biol Biochem 40:452–461

Egamberdiyeva D (2007) The effect of plant growth promoting bacteria on growth and nutrient uptake of maize in two different soils. Appl Soil Ecol 36:184–189

Egamberdiyeva D, Hoflich G (2003) Influence of growth-promoting bacteria on the growth of wheat in different soils temperatures. Soil Biol Biochem 35:973–978

El-Akhal MR, Rincon A, Coba de la Peña T, Lucas MM, El Mourabit N, Barrijal S, Pueyo JJ (2013) Effects of salt stress and rhizobial inoculation on growth and N fixation of three peanut cultivars. Plant Biol 15:415–421. [CrossRef] [PubMed]

Figueiredo VB, Buritya HA, Martınez CR, Chanway CP (2008) Alleviation of drought stress in the common bean (Phaseolus vulgaris L.) by co-inoculation with Paenibacillus polymyxa and Rhizobium tropici. Appl Soil Ecol 40:182–188

Gao JL, Turner SL, Kan FL, Wang ET, Tan ZY, Qiu YH (2004) Mesorhizobium septentri-onale sp. nov. and Mesorhizobium temperatum sp. nov., isolated from Astragalusadsurgens growing in the northern regions of China. Int J Syst Evol Microbiol 54:2003–2012

Ghosh W, Roy P (2006) Mesorhizobium thiogangeticum sp. nov., a novel sulfur-oxidizingchemolithoautotroph from rhizosphere soil of an Indian tropical leguminousplant. Int J Syst Evol Microbiol 56:91–97

Glick BR (1995) The enhancement of plant growth by free-living bacteria. Can J Microbiol 41:109–117

Grichko VP, Glick BR (2001) Flooding tolerance of transgenic tomato plants expressing the bacterial enzyme ACC deaminase controlled by the 35S, rolD or PRB-1b promoter. Plant Physiol Biochem 39:19–25

Guan SH, Chen WF, Wang ET, Lu YL, Yan XR, Zhang XX (2008) Mesorhizobium caraganaesp. nov., a novel rhizobial species nodulated with Caragana spp. in China. Int J Syst Evol Microbiol 58:2646–2653

Han TX, Han LL, Wu LJ, Chen WF, Sui XH, Gu JG (2008) Mesorhizobium gobiense sp. nov. and Mesorhizobium tarimensesp. nov., isolated from wild legumes growing in desert soils of Xinjiang. China Int J Syst Evol Microbiol 58:2610–2618

Hilali A, Prevost D, Broughton WJ, Antoun H (2001) Effects deinoculation avec des souces de Rhizobium legumioserum biovar trifolii sur la crossance du dans deux sols du Maroc. Can J Microbiol 47:590–599

Hiltner L (1904) Uber neuere Erfahrungen and Probleme auf dem Gebiete der Bodenbakteriologie unter besonderer Ber€ucksichtigung der Gr€und€ungung und Brache. Arbeiten der Deutschen Landwirtschaftlichen Gesellschaft 98:59–78
Hungria M, Campo RJ, Souza EM, Pedrosa FO (2010) Inoculation with selected strains of Azospirillum brasilense and A.lipoferum improves yields of maize and wheat in Brazil. Plant Soil 331:413–425
Hungria M, Nogueira MA, Araujo RS (2013) Co-inoculation of soybeans and common beans with rhizobia and azospirilla: strategies to improve sustainability. Biol Fertil Soils 49:791–801
Hussain MB, Mehboob I, Zahir ZA, Naveed M, Asghar HN (2009) Potential of Rhizobium spp. for improving growth and yield of rice (Oryza sativa L.). Soil Environ 28(1):49–55
Hussain MB, Zahir ZA, Asghar HN, Asghar M (2014) Exopolysaccharides producing rhizobia ameliorate drought stress in wheat. Int J Agric Biol 16:3–13
Jarvis BDW, van Berkum P, Chen WX, Nour SM, Fernandez MP, Cleyet-Marel JC (1997) Transfer of Rhizobium loti, Rhizobium huakuii, Rhizobium ciceri, Rhizobium mediterraneum, and Rhizobium tianshanense to Mesorhizobium gen. nov. Int J Syst Bacteriol 47:895–898
Jordan DC (1982) Transfer of Rhizobium japonicum Buchanan 1980 to Bradyrhizobium gen.nov., a genus of slow-growing, root nodule bacteria from leguminous plants. Int J Syst Bacteriol 32:136–139
Khan AG (2005) Role of soil microbes in the rhizospheres of plants growing on trace metal contaminated soils in phytoremediation. J Trace Elem Med Biol 18:355–364
Khan MS, Zaidi A, Wani PA, Ahemad M, Oves M (2009) Functional diversity among plant growth-promoting rhizobacteria: current status. In: Microbial strategies for crop improvement. Springer, pp 105–160
Khan AL, Waqas M, Kang SM (2014) Bacterial endophyte Sphingomonas sp. LK11 produces gibberellins and IAA and promotes tomato plant growth. J Microbiol 52:689–695. [CrossRef] [PubMed]
Klee HJ, Hayford MB, Kretzmer KA, Barry GF, Kishore GM (1991) Control of ethylene synthesis by expression of a bacterial enzyme in transgenic tomato plants. Plant Cell 3:1187–1193
Kloepper JW, Schroth MN (1978) Plant growth-promoting rhizobacteria on radishes. In: Fourth international conference on plant pathogen bacteria, Angers, France 2, pp 879–882
Kumar H, Bajpai VK, Dubey RC (2010) Wilt disease management and enhancement of growth and yield of Cajanus cajan (L) var. Manak by bacterial combinations amended with chemical fertilizer. Crop Prot 29:591–598
Laranjoa M, Alexandrea A, Oliveira S (2014) Legume growth-promoting rhizobia: an overview on the Mesorhizobium genus. Microbiol Res 169:2–17
Lu YL, Chen WF, Wang ET, Han LL, Zhang XX, Chen WX (2009) Mesorhizoblum shangrilense sp nov., isolated from root nodules of Caragana species. Int J Syst Evol Microbiol 59:3012–3018
Ma W, Guinel FC, Glick BR (2003) Rhizobium leguminosarum biovar viciae 1-aminocyclopr opane-1-carboxylate deaminase promotes nodulation of pea plants. Appl Environ Microbiol 69:4396–4402
Ma W, Charles TC, Glick BR (2004) Expression of an exogenous 1-aminocyclopropane-1-carboxylate deaminase gene in Sinorhizobium meliloti increases its ability to nodulate alfalfa. Appl Environ Microbiol 70:5891–5897
Mandimba GR (1995) Contribution of nodulated legumes on the growth of Zea mays L. under various cropping systems. Symbiosis 19:213–222
Mayak S, Tirosh T, Glick BR (2004) Plant growth-promoting bacteria confer resistance in tomato plants to salt stress. Plant Physiol Biochem 42:565–572
Mayak S, Tivosh T, Glick BR (1999) Effect of wild type and mutant plant growth-promoting rhizobacteria on the rooting of mungbean cuttings. J Plant Growth Regul 18:49–53
Meena H, Meena RS (2017) Assessment of sowing environments and bio-regulators as adaptation choice for clusterbean productivity in response to current climatic scenario. Bangladesh J Bot 46(1):241–244

Meena RS, Yadav RS (2013) Groundnut yields as influenced by heat unit efficiency, fertility levels and varieties under different growing environment in hyper arid of Rajasthan. Indian J Ecol 40(1):110–114

Meena RS, Yadav RS (2014) Phonological performance of groundnut varieties under sowing environments in hyper arid zone of Rajasthan, India. J App Nat Sci 6(2):344–348

Meena VS, Maurya BR, Meena RS, Meena SK, Singh NP, Malik VK (2014) Microbial dynamics as influenced by concentrate manure and inorganic fertilizer in alluvium soil of Varanasi, India. Afr J Microb Res 8(1):257–263

Meena VS, Maurya BR, Meena RS (2015) Residual impact of well-grow formulation and NPK on growth and yield of wheat (*Triticum aestivum* L.). Bangladesh J Bot 44(1):143–146

Meena RS, Meena VS, Meena SK, Verma JP (2015a) The needs of healthy soils for a healthy world. J Clean Prod 102:560–561

Meena RS, Yadav RS, Meena H, Kumar S, Meena YK, Singh A (2015b) Towards the current need to enhance legume productivity and soil sustainability worldwide: a book review. J Clean Prod 104:513–515

Meena RS, Meena VS, Meena SK, Verma JP (2015c) Towards the plant stress mitigate the agricultural productivity: a book review. J Clean Prod 102:552–553

Meena H, Meena RS, Singh B, Kumar S (2016a) Response of bio-regulators to morphology and yield of clusterbean [*Cyamopsis tetragonoloba* (L.) Taub.] under different sowing environments. J App Nat Sci 8(2):715–718

Meena RS, Bohra JS, Singh SP, Meena VS, Verma JP, Verma SK, Shiiag SK (2016b) Towards the prime response of manure to enhance nutrient use efficiency and soil sustainability a current need: a book review. J Clean Prod 112:1258–1260

Meena RS, Kumar V, Yadav GS, Mitran T (2017a) Response and interaction of *Bradyrhizobium japonicum* and *Arbuscular mycorrhizal* fungi in the soybean rhizosphere: a review. Plant Growth Reg. Accepted in press

Meena RS, Meena PD, Yadav GS, Yadav SS (2017b) Phosphate solubilizing microorganisms, principles and application of microphos technology. J Clean Prod 145:157–158

Mia MAB, Shamsuddin ZH (2010) Rhizobium as a crop enhancer and biofertilizer for increased cereal production. Afr J Biotechnol 9(37):6001–6009

Musarrat J, Al Khedhairy AA, Al-Arifi S, Khan MS (2009) Role of 1-aminocyclopropane-1-carboxylate deaminase. In: Rhizobium–legume symbiosis. Microbial strategies for crop improvement. Springer, pp 63–83

Nada A, Jelena B, Branislava B (2012) Mitigating abiotic stress in crop plants by microorganisms. Proc Natl Sci Matica Srpska Novi Sad 123:17–26

Nandasena KG, O'Hara GW, Tiwari RP, Willems A, Howieson JG (2009) Mesorhizobium australicum sp nov and Mesorhizobium opportunistum sp nov., isolated from Biserrulapelecinus L. in Australia. Int J Syst Evol Microbiol 59:2140–2147

Nie L, Shah S, Burd GI, Dixon DG, Glick BR (2002) Phytoremediation of arsenate contaminated soil by transgenic canola and the plant growth-promoting bacterium Enterobacter cloacae CAL2. Plant Physiol Biochem 40:355–361

Nour SM, Fernandez MP, Normand P, Cleyet-Marel J-C (1994) Rhizobium ciceri sp. nov., consisting of strains that nodulate chickpeas (Cicer arietinum L.). Int J Syst Bacteriol 44:511–522

Nour SM, Cleyet-Marel J-C, Normand P, Fernandez MP (1995) Genomic heterogeneity ofstrains nodulating chickpeas (Cicer arietinum L.) and description of Rhizobiummediterraneum sp. nov. Int J Syst Bacteriol 45:640–648

Okazaki S, Sugawara M, Yuhashi KI, Minamisawa K (2007) Rhizobitoxine-induced chlorosis occurs in coincidence with methionine deficiency in soybeans. Ann Bot 100:55–59

Omar MN, Osman MEH, Kasim WA, Abd El-Daim IA (2009) Improvement of salt tolerance mechanisms of barley cultivated under salt stress using Azospirillum brasiliense. Tasks Veg Sci 44:133–147

Pandey P, Kang SC, Maheshwari DK (2005) Isolation of endophytic plant growth promoting Burkholderia sp. MSSP from root nodules of Mimosa pudica. Curr Sci 89:170–180

Pereyra MA, Zalazar CA, Barassi CA (2006) Root phospholipids in Azospirillum-inoculated wheat seedlings exposed to water stress. Plant Physiol Biochem 44:873–879
Ram K, Meena RS (2014) Evaluation of pearl millet and mungbean intercropping systems in Arid Region of Rajasthan (India). Bangladesh J Bot 43(3):367–370
Ramirez-Bahena MH, Hernandez M, Peix A, Velazquez E, Leon-Barrios M (2012) Mesorhizobial strains nodulating Anagyris latifolia and Lotus berthelotii in Tamadaya ravine (Tenerife, Canary Islands) are two symbiovars of the same species, Mesorhizo-bium tamadayense sp. nov. Syst Appl Microbiol 35:334–341
Rasche F, Velvis H, Zachow C, Berg G, Van Elsas JD, Sessitsch A (2006) Impact of transgenic potatoes expressing anti-bacterial agents on bacterial endophytes is comparable with the effects of plant genotype, soil type and pathogen infection. J Appl Ecol 43:555–566
Reed MLE, Warner BG, Glick BR (2005) Plant growth-promoting bacteria facilitate the growth of the common reed Phragmitesaustralis in the presence of copper or polycyclic aromatic hydrocarbons. Curr Microbiol 51:425–429
Requenaf N, Jimenez I, Toro M, Barea JM (1997) Interactions between plant-growth promoting rhizobacteria (PGPR), arbuscular mycorrhizal fungi and Rhizobium spp. in the rhizosphere of Anthyllis cytisoides, a model legume for revegetation in mediterranean semi-arid ecosystems. New Phytol 136:667–677
Riggs PJ, Chelius MK, Iniguez AL, Kaeppler SM, Triplett EW (2001) Enhanced maize productivity by inoculation with diazotrophicbacteria. Aust J Plant Physiol 28:829–836
Rokhzadi A, Toashih V (2011) Nutrient uptake and yield of chickpea (Cicer arietinum L.) inoculated with plant growth-promoting rhizobacteria. Aust J Crop Sci 5:44–48
Safronova VI, Stepanok VV, Engqvist GL, Alekseyev YV, Belimov AA (2006) Root-associated bacteria containing 1-aminocyclopropane- 1-carboxylate deaminase improve growth and nutrient uptake by pea genotypes cultivated in cadmium supplemented soil. Biol Fertil Soils 42:267–272
Saha N, Mandal B (2009) Soil health – a precondition for crop production. In: Microbial strategies for crop improvement. Springer, pp 161–181
Saleem M, Arshad M, Hussain S, Bhatti AS (2007) Perspective of plant growth promoting rhizobacteria (PGPR) containing ACC deaminase in stress agriculture. J Ind Microbiol Biotechnol. (2007 34:635–648
Saravanakumar D, Samiyappan R (2007) ACC deaminase from Pseudomonas Xuorescens mediated saline resistance in groundnut (Arachis hypogea) plants. J Appl Microbiol 102:1283–1292
Shah S, Li J, Moffat BA, Glick BR (1998) Isolation and characterization of ACC deaminase genes from two different plant growth promoting rhizobacteria. Can J Microbiol 44:833–843
Shakir MA, Asghari B, Arshad M (2012) Rhizosphere bacteria containing ACC deaminase conferred drought tolerance in wheat grown under semi-arid climate. Soil Environ 31:108–112
Sharma P, Khanna V, Kumar PI (2013a) Efficacy of aminocyclopropane-1-carboxylic acid (ACC)-deaminase-producing rhizobacteria in ameliorating water stress in chickpea under axenic conditions. Afr J Microbiol Res 7:5749–5757
Sharma SB, Sayyed RZ, Trivedi MH, Gobi TA (2013b) Phosphate solubilizing microbes: sustainable approach for managing phosphorus deficiency in agricultural soils. Springerplus 2:1–14
Singh RP, Shelke GM, Kumar A, Jha PN (2015) Biochemistry and genetics of ACC deaminase: a weapon to "stress ethylene" produced in plants. Front Microbiol 6:937. 00937
Somers E, Vanderleyden J, Srinivasan M (2004) Rhizosphere bacterial signalling: a love parade beneath our feet. Crit Rev Microbiol 30:205–240
Suarez R, Wong A, Ramirez M, Barraza A, OrozcoMdel C, Cevallos MA (2008) Improvement of drought tolerance and grain yield in common bean by over expressing trehalose-6-phosphate synthase in rhizobia. Mol Plant Microb Interact 21:958–966
Tahir MM, Abbasi MK, Rahim N, Khaliq A, Kazmi MH (2009) Effect of Rhizobium inoculation and NP fertilization on growth, yield and nodulation of soybean (Glycine max L.) in the sub-humid hilly region of Rawalakot Azad Jammu and Kashmir, Pakistan. Afr J Biotechnol 8(22):6191–6200

Thamer S, Schadler M, Bonte D (2011) Dual benefit from a belowground symbiosis: N fixing rhizobia promote growth and defense against a specialist herbivore in a cyanogenic plant. Plant Soil 341:209–219. [CrossRef]

Tran Van V, Bergo O, Ke SN, Balandreau J, Heulin T (2000) Repeated beneficial effects of rice inoculation with a strain of Burkholderia Vietnamiensis on early and late yield components in low fertility sulphate acid soils of Vietnam. Plant Soil 218:273–284

Tu JC (1981) Effect of salinity on Rhizobium-root hair interaction, nodulation and growth of soybean. Can J Plant Sci 61:231–239

Vargas L, Santa Brigida AB, Mota Filho JP, de Carvalho TG, Rojas CA (2014) Drought tolerance conferred to sugarcane by association with Gluconacetobacter diazotrophicus: a transcriptomic view of hormone pathways. PLoS One 9(12):e114744

Vejan P, Abdullah R, Khadiran T, Ismail S, Boyce AN (2016) Role of plant growth promoting Rhizobacteria in agricultural sustainability—a review. Molecules 21:573

Verma JP, Yadav J, Tiwari KN, Kumar A (2013) Effect of indigenous Mesorhizobium spp. and plant growth promoting rhizobacteria on yields and nutrients uptake of chickpea (Cicer arietinum L.) under sustainable agriculture. Ecol Eng 51:282–286

Vidal C, Chantreuil C, Berge O, Maure L, Escarre J, Bena G (2009) Mesorhizobium metallidurans sp nov., a metal-resistant symbiont of Anthyllis vulneraria growing onmetallicolous soil in Languedoc, France. Int J Syst Evol Microbiol 59:850

Wang ET, van Berkum P, Sui XH, Beyene D, Chen WX, Martínez-Romero E (1999) Diversity of rhizobia associated with Amorpha fruticosa isolated from Chinesesoils and description of Mesorhizobium amorphae sp. nov. Int J Syst Bacteriol 49:51–65

Wang FQ, Wang ET, Liu J, Chen Q, Sui XH, Chen WF (2007) Mesorhizobium albiziae sp.nov., a novel bacterium that nodulates Albizia kalkora in a subtropical region ofChina. Int J Syst Evol Microbiol 57:1192–1199

Wani PA, Khan MS, Zaidi A (2007) Effect of metal tolerant plant growth promoting Bradyrhizobiumsp. (vigna) on growth, symbiosis, seed yield and metal uptake by greengram plants. Chemosphere 70:36–45

Wani PA, Khan MS, Zaidi A (2008a) Chromium reducing and plant growth promoting Mesorhizobium improves chickpea growth in chromium amended soil. Biotechnol Lett 30:159–163

Wani PA, Khan MS, Zaidi A (2008b) Impact of zinc-tolerant plant growth promoting rhizobacteriaon lentil grown in zinc-amended soil. Agron Sustain Dev 28:449–455

Whipps JM (1990) Carbon economy. In: Lynch JM (ed) The rhizosphere. Wiley, Chichester, pp 59–97

Yan J, Smith MD, Glick BR, Liang Y (2014) Effects of ACC deaminase containing rhizobacteria on plant growth and expression of Toc GTPases in tomato (Solanum lycopersicum) under salt stress. Botany 92:775–781

Yang J, Kloepper JW, Ryu CM (2009) Rhizosphere bacteria help plants tolerate abiotic stress. Trends Plant Sci 14:1–4

Yanni YG, Rizk RY, Corich V, Squartini A, Ninke K, Philip Hollingsworth S, Orgambide G, de Bruijn F, Stoltzfus J, Buck-ley D, Schmidt TM, Mateos PF, Ladha JK, Dazzo FB (1997) Natural endophytic association between Rhizobium legumino- sarum bv. trifolii and rice roots and assessment of its potential to promote rice growth. Plant Soil 194:99–114

Yanni Y, Rizk R, Abd-El Fattah F (2001) The beneficial plant growth-promoting association of Rhizobium legumioserum bv. trifolii with rice roots. Aust J Plant Physiol 28:845–870

Young JPW, Haukka KE (1996) Diversity and phylogeny of rhizobia. New Phytol 133:87–94

Yuquan X, Rong S, Zhixing L (1999) Quickly screening a strain of Pseudomonas B8 with both ACC deaminase activity and antagonism against Fusarium oxysporum. http://www.wanfangdata.com.cn/qikan/periodical.articles/shjtdxxb/shjt99/shjt9902/990223.htm. Accessed 11 June 2007

Zahir ZA, Munir A, Asghar HN, Shaharoona B, Arshad M (2008) Effectiveness of rhizobacteria containing ACC deaminase for growth promotion of peas (Pisum sativum) under drought conditions. J Microbiol Biotechnol 18:958–963

Zahir ZA, Shah MK, Naveed M, Akhter MJ (2010) Substrate-dependent auxin production by Rhizobium phaseoli improves the growth and yield of Vigna radiata L. under salt stress conditions. J Microbiol Biotechnol 20:1288–1294

Zahran HH (1999) Rhizobium-legume symbiosis and N fixation under severe conditions and in an arid climate. Microbiol Mol Biol Rev 63:968–989

Zhang F, Dashti N, Hynes RK, Smith DL (1997) Plant growth-promoting rhizobacteria and soybean [Glycine max (L.) Merr] growth and physiology at suboptimal root zones temperatures. Ann Bot 79:243–249

Zhang JJ, Liu TY, Chen WF, Wang ET, Sui XH, Zhang XX (2012) Mesorhizobium muleiense sp nov., nodulating with Cicer arietinum L. Int J Syst Evol Microbiol 62:2737–2742

Zheng WT, Li Y, Wang R, Sui XH, Zhang XX, Zhang JJ (2013) Mesorhizobium qingshengii sp. nov., isolated from effective nodules of Astragalus sinicus. Int J Syst Evol Microbiol 63:2002–2007

Zhao CT, Wang ET, Zhang YM, Chen WF, Sui XH, Chen WX (2012) Mesorhizobium silamurunense sp. nov., isolated from root nodules of Astragalus species. Int J Syst Evol Microbiol 62:2180–2186

Zhou PF, Chen WM, Wei GH (2010) Mesorhizobium robiniae sp. nov., isolated from rootnodules of Robinia pseudoacacia. Int J Syst Evol Microbiol 60:2552–2556

Zhou S, Li Q, Jiang H, Lindström K, Zhang X (2013) Mesorhizobium sangaii sp. nov., isolated-from the root nodules of Astragalus luteolus and Astragalus ernestii. Int J Syst Evol Microbiol 63:2794–2799

Biofertilizers: A Sustainable Approach for Pulse Production

14

Subrata Nath Bhowmik and Anup Das

Contents

Abstract

Nutrient needs of plants can be met through a number of sources which include mineral fertilizers, organic manures, recycled wastes and by-products, biological nitrogen (N) fixation (BNF), natural minerals and to lesser extent nutrients recycled through irrigation waters, and precipitation. These supplement the soil nutrient reserves for nourishing the crops. Presently, soil management strategies are mainly dependent on inorganic chemical-based fertilizers, which caused a serious threat to human health and environment. The exploitation of beneficial microbes as a biofertilizer has become a paramount importance in agriculture for their potential role in food security and sustainable productivity. The eco-friendly

S. Nath Bhowmik (✉)
ICAR Research Complex for NEH Region, Tripura Centre, Lembucherra, Tripura, India

A. Das
Division of Crop Production, ICAR Research Complex for NEH Region, Umiam, India

R. S. Meena et al. (eds.), *Legumes for Soil Health and Sustainable Management*,
https://doi.org/10.1007/978-981-13-0253-4_14

approaches inspire a wide range of application of plant growth-promoting rhizobacteria (PGPRs), endo- and ectomycorrhizal fungi, cyanobacteria, and many other useful microscopic organisms. The interactions of these beneficial microbes with environment determine crop health in natural agroecosystem by providing numerous services to crop plants, viz., soil organic matter (SOM) decomposition, nutrient acquisition and recycling, weed control, water absorption, and biocontrol, thus enhancing soil fertility and maintaining soil heath in eco-friendly manner. Various complementing combinations of microbial inoculants for management of major nutrients such as N and phosphorus (P) are necessary for sustainable production. Biofertilizers also cut the cost of chemical fertilizers used in agriculture considerably. An estimated amount of US$ 1421–15,237 of chemical fertilizer in the form of urea per hectare per year can be substituted by biofertilizer. The present chapter highlights the broad canvas of biofertilizers that enhance N and P nutrition in varied crops with special reference to pulses in the form of several perspectives. The mode of action of these microorganisms within and the transformation of nutrients elucidated. In the Indian scenario, the use of biofertilizers faces various constraints, such as longevity, etc. that need to be overcome to achieve substantial fertilizer savings. One of the key issues that remain is the method of formulation of these biofertilizers. Some prospective solutions to tackle the issue are brought out in this chapter.

Keywords
Biofertilizer · Biological nitrogen fixation · N-fixing microorganism · Biological nitrogen fixation · Phosphate-solubilizing microorganism · PGPR

14.1 Introduction

The modern agriculture played a pivotal role in meeting the food demands of a burgeoning human population, which has also led to an escalating reliance on synthetic agrochemicals like fertilizers and pesticides (Santos et al. 2012). Synthetic fertilizers are industrially manufactured substances composed of known quantities of macronutrients (e.g., N, P, potassium (K), etc.) and micronutrients (e.g., zinc (Zn), boron (B), iron (Fe), etc.), and their indiscriminate and imbalance use causes air, ground, and water pollution by nitrate (NO3) leaching and surface water bodies pollution through eutrophication (Youssef and Eissa 2014). Considering this, recent efforts have been focused toward the production of "nutrient-rich high-quality food" to ensure biosafety (Bhardwaj et al. 2014; Buragohain et al. 2017). The innovative approach of farm production attracts the increasing demand of biological fertilizers as a viable option to agrochemicals (Raja 2013). Biological fertilizers or biofertilizers enrich the soil environment with micro- and macronutrients via N-fixation and P and K solubilization or mineralization and enhance plant growth-regulating/growth-promoting substances, production of antibiotics, and biodegradation of organic matter (OM) in the soil (Sinha et al. 2014).

14.1.1 What Are Biofertilizers?

Biofertilizers are carrier or liquid-based preparations containing microorganisms, in sufficient numbers which benefit plant growth and nutrition (Motsara et al. 1995). Depending upon their roles and association with plants, biofertilizer are of a different kind (Fig. 14.1). The use of biofertilizer can reduce the N, P and K requirement of crops from chemical fertilizer sources. They are largely acknowledged to be low-cost supplements to chemical fertilizers and devoid of harmful effect either on soil health or environment. The additional advantages of biofertilizers include longer shelf life of microbial cells causing no adverse effects to the ecosystem (Sahoo et al. 2014). When biofertilizers are used as seed or soil inoculants, they multiply and participate in nutrient cycling and contribute to crop growth and productivity (Motsara et al. 1995). In general, 60–90% of the total applied fertilizer is lost, and the remaining 10–40% is absorbed by plants (Bhardwaj et al. 2014). Thus, microbial inoculants have a high significance in integrated nutrient management systems (INMS) for sustaining agricultural productivity and promoting healthy environment (Adesemoye and Kloepper 2009).

India is the largest producer, consumer, and importer of pulses in the world, with 24% share in the global production (Reddy 2009). The major pulse crops grown in India are *Cicer arietinum*, *Cajanus cajan*, *Vigna radiata*, *Vigna mungo*, *Lens culinaris*, and *Pisum sativum* (Reddy and Reddy 2010). About 90% of the global *Cajanus*

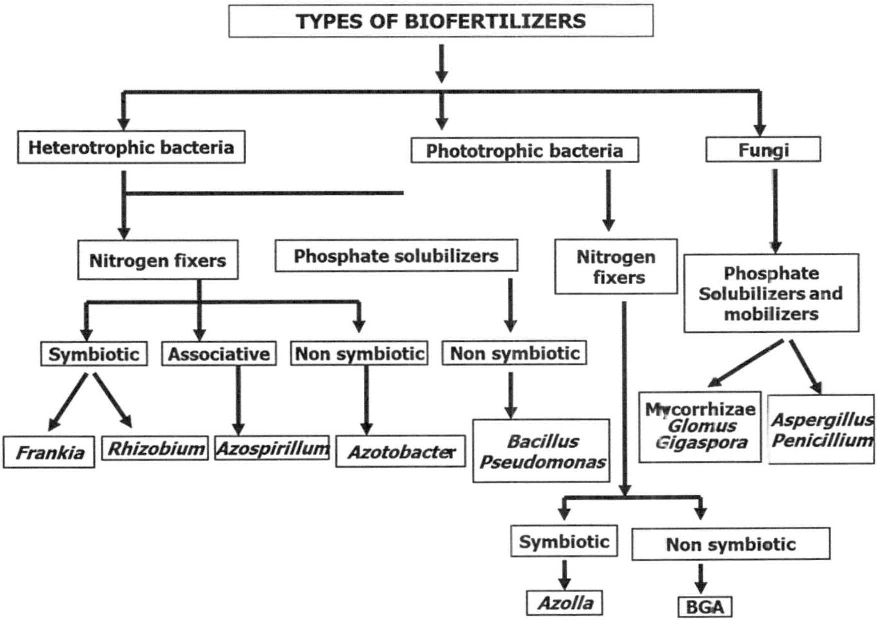

Fig. 14.1 Types of biofertilizer and their functions (Motsara et al. 1995)

cajan, 65% of *Cicer arietinum*, and 37% of *Lens culinaris* area fall in India, corresponding to 93%, 68%, and 32% of the global production, respectively (FAOSTAT 2009). The country's pulses production has been hovering around 14–15 million tones (Mt) from a near-stagnated area of 22–23 Mha, since 1990–1991 (Ali and Gupta 2012; Meena et al. 2015a). The country imports pulses ~2.5 to 3.5 Mt every year to meet the protein requirement of the nation. The strongly increasing trend in the import of pulses is a cause of concern, since an increase in demand from India has a cascading effect on international prices, thus putting tremendous pressure on the precious foreign exchange (Reddy 2009). By 2050, the domestic requirements of pulses would be ~26.50 Mt, demanding substantial stepping up in the production (Reddy 2004). This challenging task has to be accomplished under increasing production constraints, especially abiotic stresses, abrupt climatic changes, the emergence of new species/strains of insect pests and diseases, and multiple deficiencies of secondary and micronutrients in the soil (Ali and Gupta 2012).

This chapter is intended to cater to the needs of agriculturists and plant biologists whose work focuses in eco-friendly field management that reveals the potential application of biofertilizers for sustainable crop culture with special reference to pulse production. Further, the mode of actions of biofertilizers within and the transformation of nutrients elucidated. The use of biofertilizers faces various constraints, such as longevity, viable cell counts, etc. that need to be overcome to achieve substantial fertilizer savings. One of the key issues that remain is the method of formulation of these biofertilizers. Prospects and some of the key difficulties associated with the use of biofertilizers in sustainable agriculture with special reference to pulses are brought out in this chapter.

14.2 Beneficial Microbes and Pulse Production

Rhizosphere, the narrow zone of soil surrounding plant roots, contains ~1011 microbial cells per gram of root and >30,000 prokaryotic species that in general improve plant growth and productivity (Egamberdieva et al. 2008; Mendes et al. 2013). The collective genome of rhizosphere microbial community is larger compared to that of plants and is referred to as microbiome (Bulgarelli et al. 2013) whose interactions determine crop health in natural agroecosystem through numerous services being provided to crop plants, viz., nutrient acquisition, OM decomposition, nutrient recycling, water absorption, and pest control (Berg et al. 2013). Rhizosphere microbial communities as an option for synthetic fertilizers have become a subject of importance in sustainable agriculture and biosafety program. The agriculturally useful microbial populations encompass plant growth-promoting rhizobacteria (PGPR), N-fixing cyanobacteria, mycorrhiza, plant disease suppressive beneficial bacteria, stress-tolerant endophytes, and biodegrading microbes (Singh et al. 2011; Meena et al. 2015b). The term PGPR is currently applied to a wide spectrum of strains that have, in common, the ability to promote plant growth following inoculation onto seeds and subterranean plant parts (Kloepper et al. 1988; Bhowmik and Singh 2004).

14.2.1 Microorganisms for Nitrogen Fixation

For optimum pulses productivity, N is an essential plant nutrient (Dudeja et al. 2011). Since it is widely consumed by the majority of plants, most of the soils are deficient in it. Moreover, soil N is also lost due to leaching and volatilization (Brahmaprakash and Sahu 2012). Although air contains 78.09% N, plants cannot make use of it as such. The N-fixing bacteria synonymously called diazotrophs are a special type of microorganisms which can reduce atmospheric N into ammonia in the presence of nitrogenase enzyme. Microorganisms and plants assimilate N in their body parts in ammonical form for growth and development. On the basis of their mode of N-fixation, these bacteria are classified into three physiological groups, i.e., symbiotic, associative symbiotic, and free living (Fig. 14.2).

14.2.1.1 Chemical V/S Biological Nitrogen Fixation

In most of the agricultural systems, N is often the most limiting nutrient that dictates crop production. Despite its presence in large quantities in the atmosphere, plants cannot utilize N since it is in an inert form (Brahmaprakash and Sahu 2012). N is made available in the form of fertilizers which are chemical fixation of atmospheric N through the Haber-Bosch process (Motsara et al. 1995). This process requires high temperature (400–500 °C) and pressure (20 Mpa) and corresponds to energy inputs of about 875 m^3 of natural gas, 5.5 barrels of oil, or 2 mt of coal to fix one metric ton of ammonia (Dixon and Wheeler 1986). Dinitrogen is the most stable

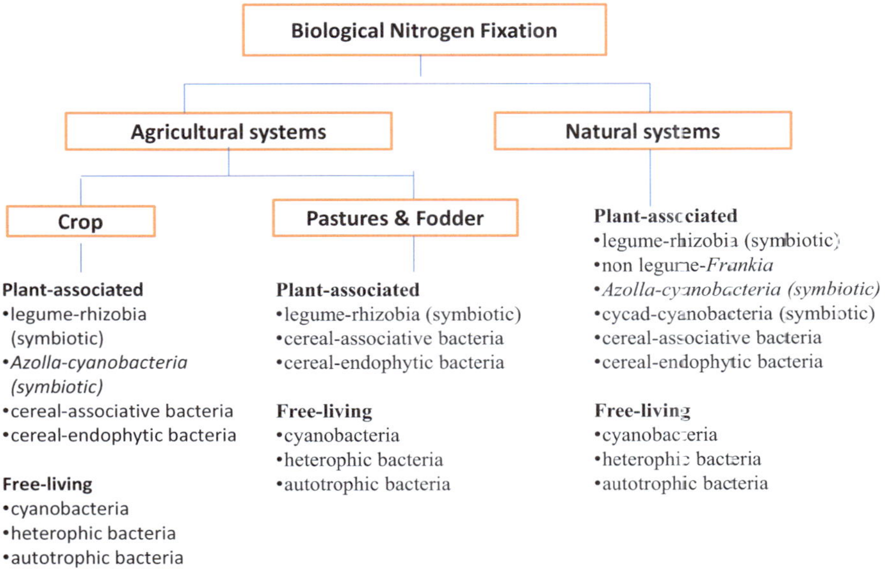

Fig. 14.2 Biological nitrogen-fixing agents in agricultural and terrestrial natural systems (Modified: Herridge et al. 2008)

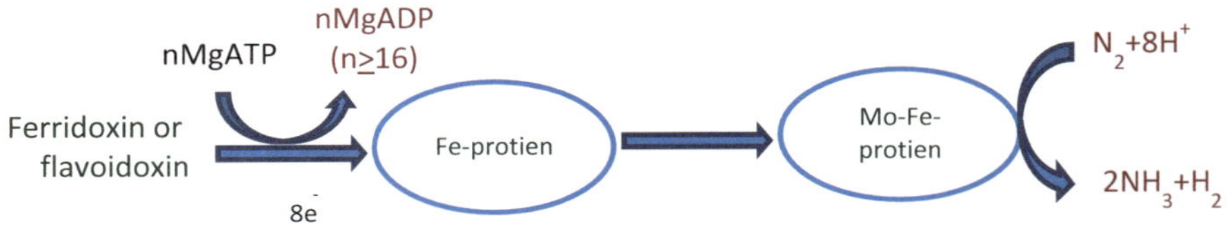

Fig. 14.3 Reduction of dinitrogen to ammonia by nitrogenase enzyme complex

diatomic molecule known, and two atoms are joined by a very stable triple bond. Very high amount of energy (945 kJ) is required to break this triple bond and therein rests one of the major challenges of dinitrogen fixation (Herridge et al. 2008).

Atmospheric dinitrogen can also be fixed biologically (by diazotrophs = *prokaryotes* that fix dinitrogen) to ammonia, which is utilized by plants. Prokaryotes are unicellular organisms lacking a well-defined nucleus (Stanier 1987). The ammonia may also be converted to nitrate by few microorganisms in the soil which is then available to plant (Madigan et al. 2009). The nitrate thus formed is amenable for denitrification reactions in deeper horizons of soil leading to the formation of N_2 gas which will escape to the atmosphere. This is the typical path of N cycle.

BNF is bacteria-mediated enzymatic fixation of N at ambient temperature and pressure (Pelczar et al. 1993). The magnitude of BNF in the biosphere is not easy to determine, but approximately it amounts to ~107 Mt/year compared to ~160 Mt/year of man-made N-fixation which is 1.5 times higher than the natural fixation (Galloway et al. 2008). In the global N cycle, every N atom in the atmosphere cycles once in a million year (Postgate 1989). BNF contributes 65% of N consumption in agriculture (Burris and Roberts 1993). All the bacteria fixing atmospheric N catalyze the reaction through nitrogenase enzyme. The nitrogenase enzyme has two components – (1) Mo-Fe protein, called dinitrogenase, and (2) Fe protein, called dinitrogenase reductase. First Mo-Fe protein takes part in reducing dinitrogen to ammonia, and second Fe protein assists Mo-Fe protein by providing electrons for reduction of dinitrogen. The mechanism of N fixation is the same in all N-fixing bacteria; the reduction of one molecule of dinitrogen requires 16 ATP in vitro and 20–30 ATP under field conditions, as it is less efficient (Fig. 14.3).

14.2.1.2 Symbiotic Nitrogen Fixation

Legume-*Rhizobium* symbiosis is an important aspect of symbiotic nitrogen fixation (SNF) which is optimally exploited to benefit agriculture for sustainability. Over a century ago, German scientists, Hellriegel and Wilfarth, experimentally demonstrated the N-fixation in legume nodule by nodule-inducing ferment (*Rhizobium*): the stage was set for the popularity of the *Rhizobium* inoculation technology world over. In this symbiosis, macro-symbiont is the legume plant, and micro-symbiont is the prokaryotic bacteria *(Rhizobium).* The macro-symbiont legume belongs to the Leguminosae family, divided into three subfamilies comprising of 700 genera and 14,000 species (Beringer et al. 1979). Only about 200 of these are cultivated by man.

Table 14.1 Cross-inoculation group and *Rhizobium*-legume association (Morel et al. 2012)

Rhizobia	Legume cross-inoculation group
Bradyrhizobium japonicum	Soybean group: *Glycine max*
Ensifer meliloti	Alfalfa group: *Melilotus* spp., *Trigonella* spp., *Medicago sativa*,
Rhizobium leguminosarum bv *trifolii*	Clover group (clover I, II, III, and IV): *Trifolium* spp.
Bradyrhizobium spp.	Cowpea group: *Arachis hypogea*, *Cajanus cajan*, *Vigna radiata*, *Vigna mungo*, *Vigna unguiculata*, *Vigna* spp., *Phaseolus lunatus*, *Acacia mearnsii*, *Enterolobium* spp., *A. mangium*, *Desmodium* spp., *Psophocarpus tetragonolobus*, *Albizia* spp., *Stylosanthes* spp., *Centrosema* sp., *Voandzeia subterranea*, *Lablab purpureus*, *Cyamopsis tetragonoloba*, *Calopogonium mucunoides*, *Pueraria phaseoloides*, *Macroptilium atropurpureum*
Rhizobium leguminosarum bv *phaseoli*	Bean group: *Phaseolus coccineus*, *Phaseolus vulgaris*
Rhizobium leguminosarum bv *viciae*	Pea group: *Pisum* spp., *Vicia* spp., *Vicia faba*, *Lens culinaris*
Rhizobium lupine	Group lupines
Mesorhizobium loti	Chickpea group: *Lotus corniculatus* L., *Cicer* spp.,
Rhizobium spp.	Crown vetch

Rhizobium is symbiotic bacterium which fixes atmospheric N in leguminous and certain nonlegumes like *Parasponia* plants by forming nodules (Bhardwaj et al. 2014; Meena et al. 2017a). Pink-colored nodules are considered most effective in N-fixation varying from 50 to 200 kg N/ha per season (Motsara et al. 1995). It is helpful for pulse legumes like *L. culinaris*, *C. arietinum*, *C. cajan*, *V. radiata*, *V. mungo*, *P. sativum*, etc., oil-seed legumes like *Glycine max* and *Arachis hypogea*, forage legumes like *Trifolium* and *Medicago sativa*, and tree legumes like *Leucaena*. *Rhizobium*, a symbiotic and crop-specific bio-inoculant, can be used for legume crop and trees, for example, *Rhizobium meliloti* for *Medicago sativa*, *Rhizobium trifolii* for *Trifolium*, *Rhizobium phaseoli* for temperate beans, *Rhizobium japonica* for *Glycine max*, *Rhizobium leguminosarum* for *Lens culinaris* and *Pisum sativum*, and *Rhizobium lupini* for *Cicer arietinum*. Hence, the use of right inoculant for the right legume is the farmer key for obtaining adequate benefit from the use of *Rhizobium* biofertilizers. Such information must be given by the manufacturer and clearly specified on the label. Plants mutually compatible with the same species of rhizobia are called "cross-inoculation groups" (Table 14.1). With regard to rhizobia, a considerable change in taxonomic status has come about during recent times. Sahgal and Johri (2003) outlined the current rhizobial taxonomy and enlisted 36 species distributed among 7 genera (*Azorhizobium*, *Allorhizobium*, *Mesorhizobium*, *Methylobacterium*, *Bradyrhizobium*, *Rhizobium*, and *Sinorhizobium*), based on the polyphasic taxonomic approach. *Rhizobium* is however limited by their specificity, and only certain legumes are benefited by this symbiosis. There is a need to develop an efficient symbiosis of host-specific rhizobial isolates and also to develop isolates with superior nodulation competitiveness that can overcome the limitations of low N-fixation, poor crop yield, and poor effectiveness under field conditions.

The genus *Mesorhizobium* has been described as in between *Rhizobium* and *Bradyrhizobium* and identified in Asia, Africa, Europe, Australia, South and North America, and the Arctic (Jarvis et al. 1997; Poinsot et al. 2001). Different species of mesorhizobia such as *M. temperadae*, *M. mediterraneum*, *M. tianshanense*, and *M. ciceri* have been reported to form nodules in *C. arietinum* (Dudeja and Narula 2008).

Rhizobium inoculation increases the crop yield in the range 10–35% (Motsara et al. 1995). However, variable crop responses to *Rhizobium* inoculation in the main field were noticed in the regime of different field conditions, crop types, inoculant, and soil conditions. Multilocational testing, All-India Coordinated Trials, and other field trials in *C. cajan*, *P. vulgaris*, *A. hypogaea*, *G. max*, and *C. arietinum* have been reported to increase in grain yields by 14–30% due rhizobial inoculation (Khurana and Dudeja 1994; Annapurna and Balasundaram 1995). One ton *Rhizobium* biofertilizer is reported to provide 100 t N/ha, with the application dose of 0.5 kg per ha (Motsara et al. 1995).

14.2.1.2.1 Response of Pulses to *Rhizobium*

Inoculation response of pulses is far below the desirable level, mostly inconsistent and dependent on many variables. Estimates of N-fixation by different pulse crops with respect to India are presented in Table 14.2. Most cultivated soils to pulses are known to harbor *V. unguiculata* group of rhizobia, and nodulation surveys indicate a need for inoculation every season for the majority of the pulses grown in India. The competition of inefficient native strains to efficient inoculants strains appears to be a bottleneck in realizing higher yields from *Rhizobium* inoculation. The yield increase due to inoculation in *C. cajan* reported to be varied from 1.2% to 20.3%, 8 to 47.8%, and 1.8 to 26.4% in 1992, 1993, and 1994, respectively, in different locations in India. The grain yield increase also appears to be an interaction of varieties and strains of *Rhizobium* (Brahmaprakash and Hegde 2005; Dhakal et al. 2016).

Rhizobial numbers are the primary determinant of the number of nodules formed. Hence, it is essential to evaluate the native rhizobial population in soil in order to predict legume response to rhizobial inoculants. Depending upon the overall nodule formation in *C. arietinum*, a rating index has been prepared indicating poor (1–10 nodules/plant), moderate (11–20 nodules/plant), good (21–30 nodules/plant), and very good nodulation (>30 nodules/plant). The multilocational trial for *C. arietinum*

Table 14.2 Estimates of nitrogen fixation in different pulses

Crop	N-fixed (kg/ha per year)
Cicer arietinum	26–63
Lens culinaris	35–100
Cajanus cajan	68–200
Vigna radiate	50–55
Pisum sativum	46

Modified: Wani and Lee (1991) and Bhata and vsh (2016)

nodulation by the native/indigenous rhizobia in India reveals that 55.7% of the locations had poor nodulation, 29.8% of the locations moderate, 12% good, and only 2.5% of the locations showed very good nodulation. Low population index (LPI) of native rhizobia dictates the need for seed inoculation with rhizobial inoculants. Moreover, rhizobial population dynamics in the soil is considerably influenced by soil moisture. The locations near to a canal with the provision of sprinkler irrigation showed better nodulation in *C. arietinum* as compared to sand dunes of Loharu/Badra region of Haryana state (Khurana et al. 1997a, b).

A complementary, coordinated effort on the part of plant breeders and microbiologists is now necessary to successfully select a high-yielding variety with elevated N-fixing abilities for sustainable agriculture. A multidisciplinary coordinated research project on pulse crops such as *C. cajan*, *C. arietinum*, and MuLLaRP (mung (*V. radiata*), urd (*V. mungu*), lentil (*L. culinaris*), *Lathyrus* (*Lathyrus sativus*), rajma (*Phaseolus vulgaris*), and pea (*P. sativum*)) is in operation in India funded by Indian Council of Agricultural Research (ICAR), in many different centers for research on improvement of several aspects of these pulses. The emphasis in this project is on crop improvement, crop production, and plant protection of these legumes. The response of pulse crops to *Rhizobium* inoculation under different agroclimatic conditions is given in Table 14.3. Depending on agroclimatic conditions, the variety planted and pest control measures undertook increased yields from 0% to 75% over controls under All-India Coordinated Trials of pulses (Rewari 1984, 1985). Further, field experiments have conclusively demonstrated the addition of N to soil due to pulses cultivation.

Dudeja and Khurana (1999) demonstrated an increase in grain yield up to 40% with a simultaneous savings of 40 kg of N as urea on inoculating local cultivar of *C. arietinum* with mesorhizobia over uninoculated control through multilocational testing. Inoculation technology in *C. arietinum* (Wani et al. 1995; Bhattari et al. 1997), *Lens culinaris* (Bhattari et al. 1997), and *Cajanus cajan* (Khurana and Dudeja 1994) recorded varied yields across the world despite numerous trials that indicate enhanced growth and grain yield in response to inoculation of many pulse crops (Table 14.4) and reports from regions such as South Asia of poor nodulation in >40% in farmers' field. Average values for % Ndfa (N derived from the air) through legume-rhizobia symbiosis for the pulses are 63 in experiments and 65 in farmer's field which is the second highest to *Vicia faba* and *Lupinus* (Herridge et al. 2008). Dudeja and Khurana (2001) determined the availability of N to the extent of 80 kg of N/ha or above made by *Mesorhizobium* and particularly by biological N-fixation by nodulating cultivar ICC 435 of *Cicer arietinum* and its non-nodulating mutant 435 M. The grain yield in their multilocational trial recorded to the tune of 65–401 kg/ha, corresponding to 9–33% increase in grain yield over the uninoculated control. Overall, mean depicted that an average increase of 120–250 kg of grain yield could be achieved by applying an efficient inoculant strain costing less than US$0.40 under different conditions (Dudeja et al. 2011).

Globally legume inoculants can be divided between the relatively sophisticated markets of North and South America, Europe, and Australia which are dominated by a small number of manufacturers producing consistently high-quality inoculants

Table 14.3 Response of pulse crops to *Rhizobium* inoculation under different agroclimatic conditions

Crop	Location	Increase (%) in grain yield over uninoculated
Vigna mungo	Pudukkottai, Tamil Nadu	4.0–21.0
	Dholi, Bihar	11.0–29.0
	Pantnagar, Uttar Pradesh	17.0–21.0
	Varanasi, Uttar Pradesh	0.14–2.32
Vigna radiata	Lam, Andhra Pradesh	1.4–75.0
	Delhi	10.0–49.0
	Pantnagar, Uttar Pradesh	4.0–15.0
	Dantiwada, Gujarat	0–3.5
	Hisar, Haryana	<1.0
	Kovilpatti, Tamil Nadu	12.5
	Jodhpur, Rajasthan	<1.0
Cajanus cajan	Hisar, Haryana	5.0–25.0
	Pantnagar, Uttar Pradesh	2.0–25.0
	Sardarkrushinagar, Gujarat	9.0–21.0
	Sehore, Madhya Pradesh	13.0–29.0
	Rahuri, Maharashtra	20.0–41.0
	Hyderabad, Telangana	<1.0
	Kovilpatti, Tamil Nadu	40–47.2
Cicer arietinum	Varanasi, Uttar Pradesh	4.0–19.0
	Hisar, Haryana	24.0–43.0
	Dholi, Bihar	25.0–42.0
	Delhi	18.0–28.0
	Sehore, Madhya Pradesh	20.0–41.0
	Dahod, Gujarat	33.0–67.0
	Badnagar, Maharashtra	8.0–12.0
	Kovilpatti, Tamil Nadu	4.0–8.2
	Indore, Madhya Pradesh	4.0–8.2
	Rewa, Madhya Pradesh	2.9–22.0
	Varanasi, Uttar Pradesh	0–13.5
Lens culinaris	Pantnagar, Uttar Pradesh	4.0–26.0
	Ludhiana, Punjab	No response
Cyamopsis tetragonoloba	Agra, Uttar Pradesh	11.2–16.6
	Jhansi, Uttar Pradesh	3.7–8.6
	Jodhpur, Rajasthan	<1.0
Macrotyloma uniflorum	Bangalore, Karnataka	<1.0
	Hyderabad, Telangana	<1.0
Vigna unguiculata	Hyderabad, Telangana	<1.0
Dolichos sp.	Hyderabad, Telangana	<1.0
Vigna aconitifolia	Jodhpur, Rajasthan	<1.0
Glycine max	Kovilpatti, Tamil Nadu	0.0–5.0
	Indore, Madhya Pradesh	0.0–1.6
	Rewa, Madhya Pradesh	0.0–5.0

Modified from: Rewari (1984, 1985)

Table 14.4 Summary of inoculation responses for pulses

Species	Number of trials	Significant yield enhancement due to inoculation compared to farmer practice (% of trials)
Vigna radiata	78	67
Cicer arietinum	87	49
Vigna mungo	33	64
Cajanus cajan	35	63
Lens culinaris	50	86

Modified: Peoples et al. (2009)

and the less sophisticated and more fragmented production of inoculants in Asia and Africa of variable quality (Singleton et al. 1997; Herridge 2008).

In Asia and Africa, limited local production and distribution of inoculants (Sattar et al. 1997) and poor inoculant quality (Singleton et al. 1997) are the major constraints. There may also be a lack of knowledge about inoculants by farmers and extension personnel (Hoa et al. 2002) and an inability to convince farmers of the merits of inoculation because of inconsistencies in crop response or an inability to demonstrate visible differences between inoculated and uninoculated treatments (Joshi 1994; Khurana and Dudeja 1994). Some of these limitations could be addressed with increased effort in training and education and especially with improved inoculant quality, emphasis in R&D on rhizobial strain selection and methods of inoculants production and application, private-sector investment, and quality control (Herridge 2008). However, economic restrictions that bear on resource-poor farmers may remain an insurmountable problem. The use of promiscuously nodulating *Glycine max* lines that nodulate and fix N with naturalized rhizobia and, therefore, requires no inoculation is an approach being evaluated in Africa; it represents a potential solution to the poor adoption of inoculant technology (Sanginga 2003). There seems no good reason why similar strategies should not apply to other legume crops.

BNF benefits not only the legumes themselves but also any intercropped or succeeding crop, reducing or removing the need for N fertilization. In soils with low mineral N content, N-fixing microorganisms provide ammonium into the legume biomass, allowing faster growth than their plant competitors. In contrast, if N is abundant, N-fixing microorganisms tend to be competitively excluded by non-fixing species because the N-fixation process is bio-energetically costly (Houlton et al. 2008). It means that there is a range of physiological and ecological situations that tend to constrain BNF in legume systems, mainly by the N demand of the plant and by the C: N stoichiometry of the ecosystem (Peoples et al. 2009; Varma and Meena 2016).

The type and size of rhizobial populations in the soil are decided by particular legume species grown most recently (Thies et al. 1995). For example, invariably high populations of *Bradyrhizobium* sp. in Alfisols of Karnataka State, India, were recorded when a host pulse was part of recent cropping systems (Table 14.5). However, the population of essential rhizobia dropped significantly when a particular host legume has been excluded from recent cropping system, thus necessitating inoculation of appropriate strains of rhizobia to ensure satisfactory and effective nodulation (Hegde 1994).

Table 14.5 Effect of cropping system on *Bradyrhizobium* population in Alfisols at six sites in Karnataka State, India

Cropping system	Rhizobia per g soil
Cereal-legume intercrop	7000
Cereal-legume rotation	34,000
Sole legume	63,000
Sole non-legume	87

Modified: Hegde (1994)

Table 14.6 Inoculation effect of pulses on the yield of subsequent crops

	Yield of subsequent crop (Mg ha^{-1})		
Previous crop	Inoculated	Uninoculated	Crop
Cajanus cajan	2.46	2.08	*Triticum aestivum*
Cicer arietinum	2.76	2.52	*Oryza sativa*
Vigna mungo	2.16	2.08	*Triticum aestivum*
Lens culinaris	2.56	2.26	*Oryza sativa*

Modified: Subba Rao and Tilak (1977)

The existence and persistence of root-nodule bacteria are much regulated by the presence or absence of host plant. Stressful edaphic factors like pH, temperature, aridity, excess of heavy metals, low clay content, soil salinity, and fallowing are often detrimental to the survival of most rhizobia in the absence of appropriate host (Giller 2001; Slattery et al. 2001; Howieson and Ballard 2004). Nearly at some stage, all farmers' fields shall encounter one or more such stresses described above. Despite the principle of rhizobial ecology that useful populations of rhizobia remain in the soil when appropriate legume has been a component of recent cropping system, rhizobial numbers are expected to decline drastically in the stressful soil. For instance, populations of *Rhizobium leguminosarum* bv. *viciae* were frequently <100 per g soil (0–10 cm) in acidic soils within 1 year of the growth of a well-nodulated crop of *P. sativum* (Marshall et al. 1993).

Under field conditions plant responses to *Rhizobium* have been quite variable and depended on soil conditions, quality of inoculums, and effectiveness of native rhizobia. *Rhizobium*-legume symbiosis could meet more than 80% N needs of the legume crop (Singh 2005). Experimental evidence indicated that 10–15% increase in the yield of pulse crops consequent to *Rhizobium* inoculation (Brahmaprakash et al. 2007). Legume yields may not always be augmented by biofertilization. N concentration in plant parts and grains may increase, or the benefits may be regarding residual N which will benefit the succeeding crops (Table 14.6) by increasing yields as a result of enrichment of soil N due to N-fixation and its conserving effect of legumes at some instances. *V. radiata*, *V. unguiculata*, *V. mungo*, and *C. arietinum* have been reported to leave about 30–40 kg N/ha for the subsequent crop, and *C. cajan* were found to be an exhaustive crop (Dudeja et al. 2011; Dhakal et al. 2015).

14.2.2 Multiple Microorganisms for Enhancing Pulse Productivity

Enhanced crop productivity can be achieved by inoculating rhizobia with other plant growth-promoting organisms such as phosphate solubilizers, *Bacillus*, *Pseudomonas*, arbuscular mycorrhizal fungi (AMF), etc.

14.2.2.1 Phosphate-Solubilizing Microorganisms

P is one of the essential nutrients for the crops, commonly present as inorganic forms such as compounds of calcium (Ca), iron (Fe), aluminum (Al), etc. and as organic forms such as compounds of phytins, phospholipids, and nucleic acids in the soil (Richardson 2001). A large portion of the available form of P present in the super phosphate and other chemical fertilizers is converted to insoluble forms soon after application to soil. So, the availability of P to crop plants is always a limiting factor. The dual inoculation of *Rhizobium* and phosphobacteria is an important practice for N and P nutrients for pulse crops (Balamurugan and Gunasekaran 1996). P is vital for nodule formation, better N-fixation, root development, and yield of legumes (Rooge et al. 1998; Nkaa et al. 2014; Meena et al. 2017b).

Tropical soil features P deficiency naturally. Further, most of the P gets fixed, making it unavailable to plant. It is estimated that 75% of P of superphosphate applied gets fixed, and only 25% is available for plant growth in tropical soil. There are some bacteria and fungi which can solubilize unavailable form of P to available form by the virtue of secretion of certain organic acids (lactic acid, succinic acid, acetic acid, fumaric acid) and are labeled as phosphate-solubilizing microorganisms (PSM). These can grow on insoluble tricalcium and rock phosphate. The cheaper source of rock phosphate (RP) like Mussoorie rock phosphate, Udaipur rock phosphate, etc., available in our country, can be used along with the PSM. A considerable amount of foreign exchange can be saved by application of PSM, as the raw material for the manufacture of superphosphate is imported. Inoculation of crops can solubilize applied phosphate varying from 40 to 50 kg P_2O_5/ha and increase crop yield by 10–20%. The commercially important PSM in biofertilizer industry is *Bacillus megaterium* var. phospaticum. This biofertilizer is recommended for all crops. One ton PSM inoculant provides 24 t P_2O_5/ha, with the application dose 0.5 kg/ha (Motsara et al. 1995; Khan et al. 2009).

14.2.2.1.1 Performance of PSM on Co-inoculation with Rhizobia

Efficient and economic use of P-fertilizers could be achieved by using PSM in pulse crops. P-uptake and P content were improved by the application of BioPhos (PSM inoculants) in many leguminous plants. The advantage of PSM to uninoculation in respect of nodulation, P-uptake, pod yield, and net utilization of pulse production was recognized. Seed inoculation with PSM noted the beneficial effect on the yield of crops was the saving by the use of 75:25 ratio of MRP: SSP (Agasimani et al. 1994). Additionally, application of PSM can substitute 25% of phosphate fertilizers. The combination of rock phosphate with PSM in field application has increased the yield of pulses. P solubilization and its release from rock phosphate by PSM were affected in higher P-uptake and dry matter yield in pulses (Sharma et al. 2013).

BNF depends significantly on the available form of phosphorous. Hence, co-inoculation of nitrogen (N) fixers and PSMs benefits the plant better (by providing both N and phosphorus (P)) than solo inoculation with either group of organisms. The dual inoculation of phosphobacterium (*Bacillus megaterium* var. *phosphaticum*) and *Rhizobium* significantly increased the plant height, nodule number, and nodule weight of black gram under field condition (Surendra et al. 1993; Balachandran and Nagarajan 2002). PSB can improve the competitive ability and symbiotic effectiveness of inoculated *Rhizobium* sp. in *Lens culinaris* (Kumar and Chandra 2008) and leghemoglobin content of nodules of *Cicer arietinum* (Tagore et al. 2013) under field conditions. This is because the P is an essential nutrient in the form of ATP for the enzyme nitrogenase of *Rhizobium* and plays a key role in the energetics of plant metabolism. Production of siderophores is the prominent factor for the biocontrol effect as suggested by Siddiqui et al. (2007). Gupta et al. (1992) demonstrated the possibility of saving of 10 kg P chemical fertilizer by entirely replacing with rock phosphate and PSB biofertilizer in the presence of *Rhizobium* in the field. Multilocational trials were conducted with 20 and 40 kg P_2O_5/ha as SSP with and without mesorhizobial and PSM inoculation during 1994–1996. Twenty percent increase in yield in *C. arietinum* was recorded by the use of SSP with biofertilizer inoculation (Khrurana et al. 1999). When no P source was used, inoculation with PSM inoculants enhanced the *C. arietinum* grain yield by 9.4%. Dual application of rock phosphate with PSM inoculants bettered the *C. arietinum* grain yield up to 34.8%, compared to only 24% and 9% with P source alone and without P, respectively. PSM inoculants on combination with DAP enhanced the grain yield of *C. arietinum* by 46.1%, compared to 30.7% with a P source alone. Utilization of PSM inoculants resulted in saving of about 20 kg P_2O_5/ha and hence is a better alternate source of P nutrient for pulses.

14.2.2.2 Arbuscular Mycorrhizal Fungi

Mycorrhiza is a symbiotic association of a special group of fungi with the roots of plants and that benefits the translocation and uptake of phosphorous in plants. Arbuscular mycorrhizal fungi (AMF) popularly known as VAM is the fungi that colonize roots of several crop plants, important in agriculture, horticulture, and tropical forestry. They are formed by the group of fungi that are usually present in all soils from the phylum *Glomeromycota*, including ten genera: *Paraglomus*, *Glomus*, *Gigaspora*, *Geosiphon*, *Scutellospora*, *Diversispora*, *Sclerocystis*, *Acaulospora*, *Entrophospora*, and *Archaeospora* (Schuessler et al. 2001; Redecker and Raab 2006). The photobiont is formed by more than 90% of all vascular flowering plant families with around 170 described species (Smith et al. 1997; Brundrett 2009). These are obligate symbionts and necessitate living host for their multiplication and growth. AMF presence improves the absorption surface area of roots and promotes plant growth by water uptake and improved P nutrition usually by absorption and translocation of soil solution P by their high-affinity P-uptake mechanisms (Muchovej 2001; Aggarwal et al. 2011). For instance, the P-depletion zone around non-mycorrhizal roots extends to only 1–2 mm, nearly the length of a root hair, whereas extraradical hyphae of AMF extend 8 cm or more beyond the

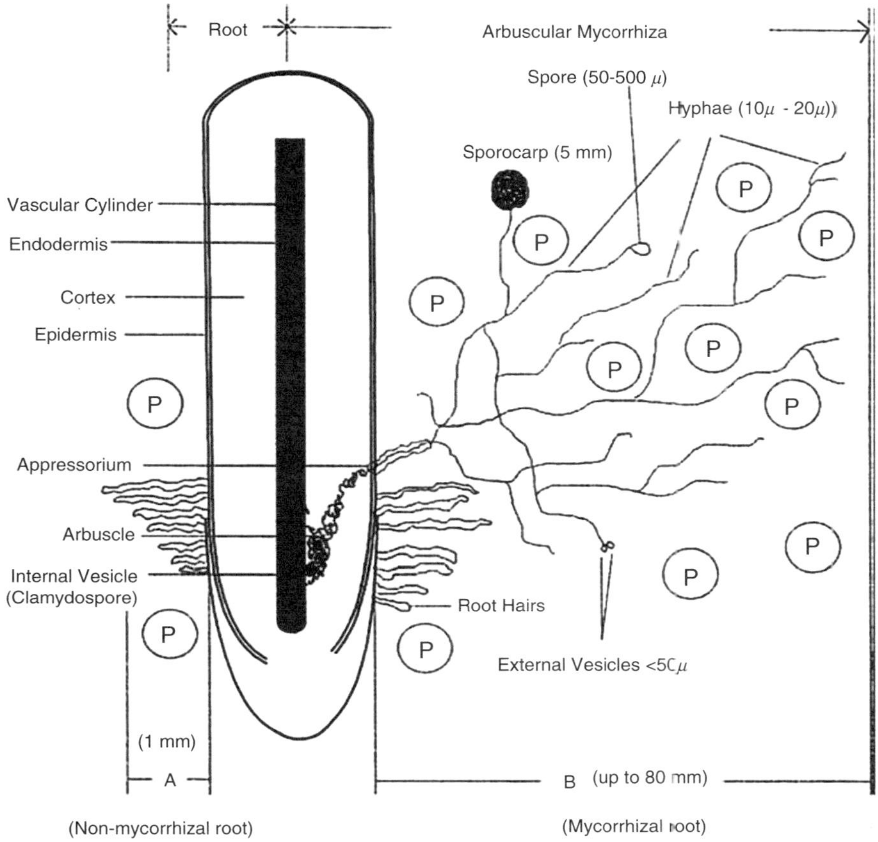

Fig. 14.4 Root colonized by AMF. Note the zone of P (or another nutrient) absorption by a non-mycorrhizal root (**a**) and by a mycorrhizal root (**b**) P-phosphate ion (Modified: Muchovej 2001)

root making the P in this greater volume of soil available to the host (Fig. 14.4). Additionally, increase in uptake of other nutrients, including N, S, B, Cu, K, Zn, Ca, Mg, Na, Fe, Mn, Al, and Si from the soil by AMF hyphae, has been documented (Clark and Zeto 2000). Indirectly BNF in legumes is expedited by AMF by facilitating the host plants with P and other immobile nutrients, such as copper (Cu) and zinc (Zn), essential for BNF (Li et al. 1991; Kothari et al. 1991; Clark and Zeto 2000). BNF can decline or even inhibited without AMF at low nutrient availability (Azcón et al. 1991; Muleta 2010). Table 14.7 enlists the direct or indirect effects of AMF on the growth and development of plants (Gosling et al. 2006; van der Heijden et al. 2008).

In the regime of low N and P availability, as encountered in several tropical soils, the possible transfer of nutrients from the mycorrhizal plant to another plant via AMF hyphal links developed inherently underground can take place (Heap and

Table 14.7 Effects of mycorrhizal fungi on crop productivity in organic farming systems

Direct effects	Indirect effects
Stimulation of plant productivity of various crops	Weed suppression
Enhanced seedling establishment	Stimulation of soil aggregation and soil structure
Nutrient acquisition (P, N, Cu, Fe, Zn)	Stimulation of N-fixation by legumes
Drought resistance	Suppression of some soil pathogens
Heavy metal/salt resistance	Stimulation of soil biological activity (phosphate solubilization)
	Increased soil carbon storage
	Reduction of nutrient leaching

Modified: Gosling et al. (2006) and van der Heijden et al. (2008)

Newman 1980). Research have established that AMF did increase N transfer from mycorrhizal legumes to another nonleguminous plant (Vankessel et al. 1985; Muleta 2010; Ram and Meena 2014).

14.2.2.2.1 Performance of AMF on Co-inoculation with Rhizobia

The occurrence of two symbionts in legume roots, *Rhizobium* bacteria and AM fungi which can sometime act synergistically, makes legumes a special case. This is well illustrated in *Stylosanthes guyanensis* grown in very low phosphate (2 ppm Olsen P) soil from the Brazilian cerrado (Mosse et al. 1976) where the combination of *Glomus fasciculatum* "E3" and rock phosphate greatly increased nodulation and N-fixation in addition to P-uptake and growth. Thus, the increased P-uptake by AMF stimulated the activity of *Rhizobium* which is well known to depend on an adequate supply of phosphate.

AMF are known to form a symbiotic relationship with a range of members of family Leguminosae (Pagano et al. 2007; Valsalakumar et al. 2007; Molla and Solaiman 2009). Studies reveal AMF colonizing *V. radiata* was of the trend *G. mosseae* (81%) followed by *G. microcarpum* (24%) and *G. margarita* (24%) and *Scutellospora* sp. (5%) (Valsalakumar et al. 2007). The range of distribution varied from a single species of AM fungus to three species belonging to two genera in one sample. In a study under five agroecological zones (AEZs) of Bangladesh, Molla and Solaiman (2009) surveyed the association of AMF in ten diverse leguminous crops and conclusively reported that *C. arietinum*, *V. radiata*, *L. culinaris*, and *V. mungo* were highly mycotrophic among all. In general, *Glomus* was most common, while *Sclerocystis* was the least prevalent genus in the study. The application of AMF inoculant in soils increased the growth and yields of legumes significantly under both greenhouse and field conditions. For instance, inoculation with AMF improved growth of *C. arietinum* and doubled P-uptake at low and intermediate levels of P in a pot experiment on sterilized low-P calcareous soil (Weber et al. 1992). Further, seed production in legumes significantly increased upon co-inoculation of mesorhizobia and AMF in multilocational field trials (Dudeja et al. 2011; Varma et al. 2017).

The majority of legumes have the ability to take part in a dual symbiotic interaction with rhizobia and AMF (Hazarika et al. 2000; Sprent 2001; Lodwig et al. 2003; Navazio et al. 2007). Rhizobia and AMF together play a key role in natural ecosystems and influence plant nutrition, productivity, resistance, and plant community structure (Cleveland et al. 1999; van der Heijden et al. 2006; Demir and Akkopru 2007). The tripartite symbiosis (between legumes-mycorrhiza and rhizobia) in a given ecosystem profoundly improved the overall performance of legumes (Subba Rao et al. 1986; Daniels-Hylton and Ahmad 1994; Barea et al. 2002; Xavier and Germida 2002; Zaidi et al. 2003; Stancheva et al. 2008). Generally, AMF improves phosphate nutrition of legume which, in turn, enhances plant growth and dinitrogen fixation. The effect of AM fungi on legumes often is duplicated by increasing P availability to non-AMF plants. Pot and field studies demonstrated that AMF inoculation enhanced BNF equivalent to that attainable by the application of phosphate fertilizer (75–100 kg P/ha). Recent studies have demonstrated that the two symbioses share some components of their developmental programs (Harrison 2005; Oldroyd et al. 2005; Navazio et al. 2007).

Positive effects of dual colonization of roots of legumes including pulses have been investigated by a number of workers. Most studies have concentrated on indirect relationships between AMF and rhizobia, in which a successful symbiosis has been measured as an increased uptake of nutrient elements, P in particular, by the plant (Ames and Bethlenfalvay 1987) and an increased dinitrogen fixation and nodule biomass (Fredeen and Terry 1988), but more direct non-nutritional effects of AM fungus-rhizobia interactions have also been reported (Bethlenfalvay et al. 1985).

Maximum nitrogenase activity and dry biomass of the legume resulted from co-inoculation of *Bradyrhizobium* sp. (*Vigna*) strain S-24 with *Scutellospora calospora* (Saxena et al. 1997). The nodulation competitiveness of the strain S-24 was significantly higher (60–63%) in the presence of *G. fasciculatum*, *G. mosseae*, and *S. calospora* when compared to nutrient with a single inoculation of S-24 (51%).

Jia et al. (2004) investigated the effects of the interactions between the microbial symbionts (*Rhizobium* and AMF) on N and P accumulation by broad bean (*Vicia faba*) and how increased N and P content influence biomass production, leaf area, and net photosynthetic rate. The AM fungus was found to promote biomass production and photosynthetic rates by increasing P/N accumulation, and an increase in P was consistently correlated with an increase in N accumulation and N productivity, expressed regarding biomass and leaf area. Photosynthetic N use efficiency, irrespective of the inorganic source of N (e.g., NO_3-), was enhanced by increased P supply due to AMF colonization. However, the presence of *Rhizobium* significantly declined AMF colonization irrespective of N supply; without *Rhizobium*, AMF colonization was higher in low-N treatments. The presence or absence of AMF did not have a significant effect on nodule mass, but high N with or without AMF led to a significant decline in nodule biomass. Furthermore, plants with the *Rhizobium* and AMF had higher photosynthetic rates per unit leaf area.

Geneva et al. (2006) reported that the dual inoculation of pea plants with *G. mosseae* or *G. intraradices* and *R. leguminosarum* bv. *viceae* strain D 293 significantly increased the plant biomass, photosynthetic rate, nodulation, and N-fixing activity in comparison with a single inoculation of *R. leguminosarum* bv. *viceae* strain D 293. In addition, the co-inoculation significantly increased the total P content in plant tissues, acid phosphatase activity, and percentage of root colonization. Among all microbial pairing, the mixture of *R. leguminosarum* and *G. mosseae* was most effective at low P level, while *G. intraradices* inoculated with *R. leguminosarum* was most effective at higher P level, as also reported for *L. culinaris* (Xavier and Germida 2002). In other greenhouse trial, Mehdi et al. (2006) reported that the effects of AMF (*G. mosseae* and *G. intraradices*), rhizobial (*R. leguminosarum* bv. *viciae*) strains, and P- (superphosphate and phosphate rock) fertilizers considerably increased the dry biomass of shoots and seeds, P and N contents (shoots and seeds) of *L. culinaris* cv. "Ziba" plants, and percent of root colonized by AMF. The rhizobial strain possessing P-solubilizing ability showed a more beneficial effect on plant growth and nutrient uptake than the strain without this activity, although both strains had similar N-fixing efficiency. Moreover, the P-uptake efficiency was increased when P-fertilizers were applied along with AM fungi and/or P-solubilizing rhizobial strains emphasizing the remarkable importance of dual inoculation in the improvement of plant growth responses as also reported by Zarei et al. (2006) and Meena et al. (2013) for rhizobium-mycorrhizae inoculated *L. culinaris* plants.

In another study, Singh et al. (1991) and Boby et al. (2008) demonstrated the effect of live yeast cells (*Saccharomyces cerevisiae*) on nodulation and dry biomass of shoot and roots of legumes such as *Glycine max*, *L. leucocephala*, *C. cajan*, *Phaseolus aureus*, *Phaseolus mungo*, and *V. unguiculata* in the presence of both AMF and *Rhizobium* strains. The results indicate that inoculation with live yeast cells remarks ably enhanced the measured plant parameters. Root infection (native AMF) and the formation of vesicles, arbuscules, and spores were also increased with yeast inoculation. The increase in the parameters, however, varied with legumes and the type of yeast culture.

Legumes generally have less extensive root systems than graminaceous plants, for example, and many are poor foragers for soil phosphate. In addition, many strains of *Rhizobium* found in culture collection are adapted to media containing 100 times more P than is usual in the soil solution (Munns and Mosse 1980) and can require at least 0.1% P in the shoot tissues to produce nodules (Mosse et al. 1976). Some indigenous rhizobia, on the other hand, can nodulate at lower concentrations of P to which they are adapted (Mosse 1977), an important consideration in dual inoculation of pulses in P-deficient tropical soils.

14.2.2.3 Plant Growth-Promoting Rhizobacteria and Biocontrol Agents

Rhizosphere microorganisms closely associated with roots with beneficial properties are called plant growth-promoting rhizobacteria (PGPR). PGPR includes a diverse group of soil bacteria that can improve host plant growth by interacting with other soil organisms, thereby either by promoting the growth of beneficial microbes such as rhizobia or phosphate solubilizers or plants directly or by inhibiting the growth of pathogenic bacteria (Vejan et al. 2016). Solo inoculation of rhizobia improved the yield by 12.4% compared to dual inoculation with PSB and PGPR by 22.1% increase in grain yield of *C. arietinum* in multilocational trials (Dudeja et al. 2011).

The effects of *G. intraradices*, *Rhizobium* sp., and *Pseudomonas striata* on the root-rot disease complex of *C. arietinum* caused by *Meloidogyne incognita* and *Macrophomina phaseolina* were observed. Dual inoculation of *G. intraradices* with *P. striata* plus *Rhizobium* to pathogen-inoculated plants caused a greater increase in plant biomass yield and a number of pods, chlorophyll, N, P, and potassium contents than by inoculation of *G. intraradices* plus *Rhizobium* or *G. intraradices* plus *P. striata* (Akhtar and Siddiqui 2008, 2009). Microbial inoculants (*Mesorhizobium*, PSB, and PGPR) with *Trichoderma viride* enhanced the productivity of *C. arietinum* (Dudeja et al. 2011; Yadav et al. 2017).

14.3 Formulations of Biofertilizers for Sustainable Agriculture

The success of inoculation technology depends on two factors such as the microbial strain and inoculants formulation. In practical terms, the formulation determines potential success of inoculants (Fages 1992). Inoculant (synonym for biofertilizer) is prepared containing beneficial microorganisms which enhance plant growth. As strains of same microbial species share many physiological properties, the technological progress thus developed for a particular strain is readily adaptable to another strain of some species with only minor modifications (Bashan 1998). Although formulation is vital in successful commercialization of biofertilizer, research in this area is meager. In addition to limited availability of published scientific information with regard to inoculant formulation, the information available is fragmented (Xavier et al. 2004). A survey of a bibliographic database of scientific literature shows that major emphasis was given to the development of improved strains through different approaches. Indeed many such strains have been constructed and granted patent in many developed countries but failed to appear on the commercial market, perhaps because of inappropriate formulation.

Development of improved formulations often rests with inoculant manufacturer's research and development facility which is primarily located in developed countries where target market exist, but they fail to consider the unique problems in applying these inoculants in developing countries. The most important characteristic common to most of biofertilizers is the unpredictability of their performance. To harness the benefits of biofertilizers in agriculture, the consistency of their performance must be improved (Wani and Lee 2002).

Bacteria introduced to soil may fail to establish in sufficient numbers in the rhizosphere because of competition from native numbers, and little is known about the factors controlling competitiveness of bacterial strains, more so under field conditions. Rhizosphere is the region of the soil under the influence of plant roots. Agricultural practices in developing countries and under semiarid conditions are two examples wherein biofertilizers may find their greatest challenges. Low-input agriculture is often followed by the farmers of developing countries in which fertilizers, pesticides, and agro-technical machinery are least utilized. Application or uses of biofertilizers in such systems require additional infrastructure, cost, labor, and technical knowledge. Drought phenomenon often reduces the population of beneficial microorganisms in the soil. However, ample scope resides in the development of microbial inoculants for contributing immensely in a stress situation (Bashan 1998; Trivedi et al. 2012). Most important constraints for adoption of biofertilization in India have been attributed to poor quality of inoculants produced, lack of knowledge about inoculation technology for extension personnel and farmers, effective inoculants delivery/supply system, and lack of committed policy to exploit biofertilizers successfully (Wani and Lee 1991; Bodake et al. 2009; Jangid et al. 2012)

14.3.1 Inoculant Formulations

Formulation step is a vital aspect for producing microbial inoculants and determines the success of a biological agent. Formulation typically consists of establishing viable microorganisms in a suitable carrier together with additives that aid in stabilization and protection of microbial cell during storage, transport, and at the target. The formulation should also be easy to handle and apply so that it is delivered to target in most appropriate manner and form, one that protects microorganisms from harmful environmental factors and maintains or enhances the activity of the organisms in the field. Therefore, several critical factors including user preference have to be considered before delivery of a final product (Xavier et al. 2004; Meena and Meena 2017).

14.3.1.1 Carriers for Inoculant Formulations

A suitable carrier plays a major role in formulating microbial inoculants. Carriers are inert materials, used for transporting microbes from the laboratory to land. Carrier is a delivery vehicle which is used to transfer live microorganism from an agar slant of laboratory to a rhizosphere. A good quality inoculant should be made of a superior carrier material. Hence, Smith (1992) has listed the characters of a superior-quality carrier material for microbial inoculants, which includes:

- High water-holding and water-retention capacity
- No heat of wetting
- Nearly sterile, chemically, physically, and uniformly
- Nontoxic in nature, easily biodegradable, and not polluting
- Nearly neutral pH or easily adjustable
- Supports growth and survival of microorganisms
- Amenable to nutrient supplement
- Rapid release of microorganisms in soil
- Manageable in mixing, curing, and packaging operations
- Available in powder or granular form in adequate quantities and at reasonable cost

Considering the above characteristics, it is clear that not a single universal carrier is available which fulfill all the desirable features, but good ones should have as many as possible. Peat was the carrier of choice and most commonly used for *Rhizobium* inoculants worldwide for decades. Despite its popularity, variability in quality which is dependent on source, the release of antibacterial agents due to sterilization often poses a threat to biofertilizer production (Chao and Alexander 1984), and its availability is restricted to a very few countries. All these factors have forced researchers to look for alternative carrier materials. Some of the alternative carriers researched for biofertilizers which include lignite and coal (Kandasamy and Prasad 1971; Dube et al. 1980) clays and inorganic soils (Chao and Alexander 1984), compost, farm yard manure, soybean meal (Iswaran et al. 1972), wheat bran (Jackson et al. 1991), press mud (Philip and Jauhri 1984), spent agricultural waste material (Sadasivam et al. 1986), and spent mushroom compost (Bahl and Jauhri 1986). Apart from these, many other synthetic and inert materials like vermiculite (Paau et al. 1991; Sparrow and Ham 1983), perlite, ground rock phosphate, calcium sulfate, polyacrylamide gels (Dommergues et al. 1979), and alginate (Bashan 1986; Jung et al. 1982) have also been evaluated. Most of the evaluated carriers are either naturally abundant resources or available waste materials. Little research has have been conducted with objectives of synthesizing a carrier with superior characteristics (Bashan 1998). Biofertilizers are commonly designed for seed treatment and direct application to the soil. Hence, delivery of biofertilizers to crops is vital to determine the formulation, for instance, powder for seed treatment granulated for soil application (Walter and Paau 1993; Verma et al. 2015a, b).

Principal drawbacks of solid carrier-based inoculants originate from a great variability in quality of carriers which is source dependent, and composition of carriers is undefined and complex. The quality of carrier material to a great extent affects the final product and causes trouble in inoculant dosage, storage conditions (Van Elsas and Heijnen 1990), and difference of inoculant effectiveness between different manufacturers and batches from the same manufacturer (Bashan 1998). Carrier-based bacterial inoculants often confront lower tolerance for physical stress during storage, particularly for temperature differences. Some types of peat can even reduce plant growth (Huber et al. 1989). They are often prone to contamination that can reduce the shelf life of the inoculants (Van Elsas and Heijnen 1990; Fages 1992; Olsen et al. 1994a, b). Adhesion of inoculant to seeds is overcome by adhesives. However, incorporation of adhesive during application is labor and cost intensive (Smith 1995; Datta et al. 2017). Solid carrier-based inoculant production involves a significant amount of cost; labor; energy-intensive processing such as mining, drying, and milling; and neutralization before its use in a commercial production.

The carrier-based inoculants produced in India generally have a short storability, poor quality, high contamination, and erratic field performance. High-quality biofertilizers would be expected to have a higher population of desired microorganisms and sufficient viability and remain uncontaminated for longer period of storage. The carriers used in India are nearly inert material and form clumps upon drying, which leads to significant loss of viability. Seed is not a favorable environment for most of the plant growth-promoting bacteria, as they are soil bacteria, yet seed inoculation is a common practice for microbial inoculation. The usual carrier-based biofertilizers do have some drawbacks for seed inoculation like seed coat damage, seed coat toxicity, the death of cells due to desiccation, and possible contact of microorganisms with agricultural chemicals. Today, advances in inoculant technology are concerned with improving quality, extending useful shelf life, and developing new formulations for use under less favorable conditions. Liquid inoculants and alginate-based granular formulations are two important new inoculant formulations which are an alternative to peat-/lignite-based ones.

14.3.1.2 Liquid Inoculants

Liquid inoculants are not the usual broth culture from a fermenter or water suspension of the carrier-based biofertilizers, as often made out to be. It is a special liquid formulation containing not only the desired microorganisms and their nutrients but also contains special cell protectant and amendments that promote cell survival in a package and after application to seed or soil. Various liquid media are being used to culture microorganisms. These media normally consist of carbon, N, and vitamin sources, which promote the growth of microorganisms. However, additives used in liquid inoculants improve quality of inoculants by increasing the population density and enhanced shelf life (Tittabutr et al. 2007).

Additives used in the preparation of liquid inoculants have been selected based on their ability to protect microbial cells in the package and on seeds extreme conditions such as high temperature, desiccation, and toxic condition of seeds and need chemicals. Most of the additives are high-molecular-weight polymers with good water solubility, non-toxicity, and complex chemical nature (Deaker et al. 2004; Ashoka et al. 2017) and can limit heat transfer and possess good rheological properties and high water activities (Mugnier and Jung 1985). Some of the polymers which are presently used in preparation of liquid inoculants include polyvinylpyrrolidone (PVP), methylcellulose, polyvinyl alcohol, polyethylene glycol, gum arabic, trehalose, glycerol, Fe-EDTA, sodium alginate, tapioca flour, etc. (Singleton et al. 2002; Tittabutr et al. 2007). Polyvinylpyrrolidone was known to bind toxic compounds present in seed exudates that are mobilized during inoculation and seed germination. PVP has a high water-binding capacity and appears to cause slow drying of an inoculant after application. PVP solution tends to coalesce into ridges on their seed coat as it dries, perhaps providing a thicker layer of protection than some other compounds. Its sticky consistency may also enhance cell and inoculants adherence seeds (Singleton et al. 2002). Sometimes seed-released compounds may bind iron in yeast extract, making it unavailable to cells. Supplementary iron may, therefore, replace Fe bound by seed exudates (Ali and Loynachan 1990). Glycerol has a high water-binding capacity and may protect cells from effects of desiccation by slowing the drying rate. Its flow characteristics appear to promote rapid and even coating on seeds (Al-Rashidi et al. 1982; Mary et al. 1985). Trehalose is widely reported to enhance cell tolerance to desiccation, osmotic, and temperature stress. It acts by stabilizing both enzymes and cell membranes, is a compatible osmoticum as well, and is readily manufactured by *Bradyrhizobium* given ideal conditions (Streeter 1985; Lippert and Galinski 1992). Addition of PVP in a medium was known to protect both fast- and slow-growing *Rhizobium* (Bushby and Marshall 1977). Bushby and Marshall (1977) and Vincent et al. (1962) have shown that addition of maltose (9%) and montmorillonite clay could protect *Rhizobium* against high temperature and desiccation. Polymers that are soluble in liquid inoculant formulations are convenient for batch processing of inoculants and make seed application a simpler process for farmers. Liquid *Rhizobium* inoculants prepared with PVP as an osmoprotectant had improved shelf life, nodulation, and N-fixation on par with lignite-based inoculants in cowpea.

Liquid inoculants can be produced by a simple fermentation process, packed directly from the fermenter aseptically, and stored. It minimizes the production cost by avoiding processing and sterilization of solid carrier material. The complete sterilization could be achieved with liquid formulations, and any contamination during the storage can be easily detected. Liquid inoculants could be produced with minimum labor, space, and energy, and also the quantity of inoculum required is less compared to carrier-based formulations, hence easier for farmers to handle. The first yardstick to measure the quality of biofertilizer is the viable cell density of desired microorganisms which essentially provides adequate number of microorganisms on each seed. The liquid inoculants developed were known to have a population of *Rhizobium* sp. and PSB up to the level of 108 cells/ml (Sridhar et al. 2004; Vithal 2004; Dayamani 2010; Velineni and Brahmaprakash 2011; Kumar et al. 2016). A higher population of microorganisms are often prescribed in formulation to compete with native *Rhizobium* and to offset death of cells due to biotic and abiotic stresses. Since the liquid biofertilizer has high cell count, each seed receives more than thousands of cells. Additives in liquid biofertilizer protect the cells on the inoculated seeds against toxicity, desiccation, and osmotic shock (Vithal 2004). The storage and transportation conditions are not congenial many a times for the bioinoculants as temperature in many parts of the country may reach up to 45 °C; in such condition the quality of biofertilizer will decline drastically. Liquid biofertilizers were known to have more than 1 year shelf life compared to carriers. Studies have revealed that these liquid inoculants can be stored without losing viability in high-temperature (45 °C) conditions also (Vithal 2004). Imposition of stress to bacteria results in an adaptive response. This necessitates changes in metabolic processes in cells, which are then reflected in an alteration of protein profiles (Saxena et al. 1996). Synthesis of additional 19 salt-stress proteins (SSPs) in *Rhizobium* (40–52 kDa) and synthesis of 19 heat-shock proteins (ranging from 8 to 60 kDa) in *Bradyrhizobium japonicum* at 43 °C have been reported (Munchbach et al. 1999); *Bradyrhizobium* sp. (*Arachis*) grown at room temperature in liquid inoculant synthesized 60 and 47 kDa proteins of higher intensity but the same proteins of lower intensity in yeast extract mannitol broth (YEMB). *Bradyrhizobium* sp. (*Arachis*) on exposure to heat stress showed the presence of bands of same proteins (60 and 47 kDa) in liquid inoculant. Similarly, under salt stress (0.05 M NaCl), *Bradyrhizobium* sp. (*Arachis*) grown in liquid inoculant synthesized the extra proteins of 66 kDa but not in YEMB (Brahmaprakash et al. 2007). This kind of mechanisms provides potential to grow at different types of soil as we know that performance of inoculants depends largely on soil conditions. The amount of inoculant needed for seed inoculation is less, and there is no need of any sticker material unlike carrier-based inoculants. Liquid inoculants can easily be adapted to advanced seeding equipments; since it can be sprayed onto seeds, it passes through seed drill and dries before it travels into the seed bin on the planter.

14.3.1.3 Field Response of Liquid *Rhizobium* Inoculants Formulation

Liquid rhizobial formulations are comparable to that of peat-based products performance under field conditions (Hynes et al. 2001). The field efficiency of the liquid inoculant was tested on farm trials under different agroclimatic conditions of India for two successive years in four different legumes like *C. cajan*, *A. hypogea*, *C. arietinum*, and *G. max*. Results of trials showed that liquid *Rhizobium* inoculants performed better than the carrier-based *Rhizobium* inoculants (Brahmaprakash et al. 2007; Meena et al. 2016). There is a need to improve formulation for better field performance. In this direction the research in progress indicates that solid carrier materials have been replaced with microbe-friendly liquid formulations. Further, a formulation containing a consortium is to be developed for field application.

14.3.1.4 Polymer Entrapped Inoculants Formulation

Polymers proved to be potential bacterial carriers in biofertilizers that offered substantial advantages over peat (Jung et al. 1982). These formulations encapsulate living cells, protect microorganisms against many environmental stresses, and release them to the soil, gradually but in large quantities, where the polymers are degraded by soil microorganisms. They can be dried stored at ambient temperatures for longer periods, can offer a consistent batch quality and a better defined environment for the bacteria, and can be manipulated easily according to the needs of specific bacteria. These inoculants can be amended with nutrients to improve short-term survival of bacteria upon inoculation, which is essential to the success of an inoculation process, especially with associative PGPB (Bashan 1998). However, a major limitation for the inoculation industry is that polymers are expensive compared to peat-based inoculants and require more handling by the industry (Fages 1992).

The encapsulation of microorganisms into a polymer matrix is still under research. At present there is no commercial bacterial product using this technology. The concept underlying immobilized microbial cells is to entrap beneficial microorganisms into a matrix. The formulation (bacterial matrix) is then fermented in a bacterial growth medium. Immobilized microbial cells are easy to produce, store, and handle during industrial operations. Encapsulated bacterial formulations in agriculture have two advantages (1) to temporarily protect the encapsulated microorganisms from the soil environment and microbial competition and (2) to release them gradually for the colonization of plant roots (Bashan 1986; Digat 1991; Bashan and Carrillo 1996).

14.3.1.5 Alginate-Based Formulations

Alginate is a commonly used polymer for encapsulation of microorganisms and is naturally occurring, composed of β-1,4-linked D-mannuronic acid and L-glucuronic acid. It is available from different macroalgae (De Lucca et al. 1990) as well as several bacteria (Smidsrod and Skjak-Braek 1990). Alginate cost has recently dropped because of its massive production in the Far East, making it potentially more attractive to the inoculant industry. The preparation of beads containing bacteria is fairly easy and involves a multistep procedure (Fages 1992; Digat 1991).

The main advantages of alginate preparations are their nontoxic nature, biodegradability, and their slow release of microorganisms into soil (Fages 1992; Kitamikado et al. 1990). This technology was used to encapsulate the plant-beneficial bacteria *Azospirillum brasilense* and *Pseudomonas fluorescens* (Fages 1992), which were later successfully used to inoculate wheat plants under field conditions. The bacteria survived in the field long enough, and their populations were comparable to the survival of bacteria originating from other carrier-based inoculants (Bashan et al. 1987). Furthermore, the addition of clay and skim milk to the beads significantly increased bacterial survival over alginate beads alone. Alginate mixed with perlite was used to entrap *Rhizobium* (Hegde and Brahmaprakash 1992). Colonization of wheat roots by beneficial cells released from the beads was superior to that achieved by direct soil inoculation. These studies provide clear evidence that alginate beads are efficient slow-release carriers for plant inoculants, thus providing a protective environment in the soil. Several other alginate-based preparations have been tried for the encapsulation of AM fungi (Ganry et al. 1982), ectomycorrhizal fungi (Marx and Kenney 1982; Le Tacon et al. 1985), *Frankia* inoculation (Sougoufara et al. 1989), and fungi used as biocontrol agents against soil-borne pathogens (Fravel et al. 1985; Lewis and Papavizas 1985).

Alginate preparations may have solved many of the problems associated with traditional peat inoculant. These inoculant formulations may solve the problems associated with tropical, low-input agriculture. In many parts of the tropical region, there is always a chance of prolonged dryness after sowing and microbial inoculation. Alginate encapsulated formulations are already desiccated due to lower water activity; microorganisms will be allowed at metabolic activities and are released into soil only after sufficient moisture is available, which always coincide with the germination of seeds. Considering the cost involved in the production of alginate formulations, attempts have been made to amend these formulations with material like rock phosphate, cement, bentonite clays, granite powder, gypsum, lignite, and talc by which cost of production can be minimized besides adding bulkiness formulation (Lewis and Papavizas 1985; Meena et al. 2015b).

14.3.2 AMF Inoculants

The AMF being an obligate symbiont, there are many constraints in its large-scale commercial production and application. The only method of production in association with host plant by pot culture, as the production of AMF in the artificial media, has met with little or no success. There are different types of AMF inoculum required for different purposes. The spores of AM fungi are used as inocula generally for experiments in vitro conditions. Large-scale production of spores is difficult (Bagyaraj et al. 2002).

14.3.2.1 Infected Root Inoculums

Large-scale production of the infected root is possible in aeroponic culture. Infected roots contain internal mycelium and external mycelia (may have spores).Infected roots colonize a host after 1 or 2 days of inoculation. Root inocula without spores should be used within a week. In vitro reproduction of some fungi on cultured tissue roots has been demonstrated (Napamornbodi et al. 1988). The production process is difficult and expensive. Other problems are (a) infected root introduced as inocula which act as an attractive nutrient source for several saprophytic and parasitic microorganisms, (b) short survival time, and (c) large quantities of inocula required.

14.3.2.2 Soil-Based Inoculums

Soil inoculum is produced using traditional pot-culture techniques containing all AM fungal structures that are highly infective. The success of good soil inoculums production depends on the selection of host plant and the ambient conditions under which a defined AM fungus can be mass multiplied (Bagyaraj et al. 2002; Meena et al. 2013a). AMF can be mass multiplied through soil pot culture using *Chloris gayana* as host in conjunction with *Azospirillum*, with a view to reducing the length of production period (Bhowmik and Singh 2004; Bhowmik et al. 2015), and the root pieces along with the soil substrate is used as an inoculant.

14.3.2.3 Peat-Based Inoculants (Nutrient Film Technique)

AMF inocula obtained from pot cultures were incorporated into peat and compressed into blocks. Lettuce plants are allowed to grow in the peat block for 2–5 weeks, and then the blocks are transferred to nutrient film technique (NFT) channels (Cooper 1985). The NFT channels slope and nutrient solution flow at 200 ml per minute. Plants are allowed growing NFT system for 8–10 weeks. During this time, mass reproduction of the fungus takes place. The peat blocks are allowed to dry, chopped, and used as AMF inoculums. The shelf lives of such peat-based inoculants are around 6 months (Bagyaraj et al. 2002).

14.3.3 Mixed Bacterial Inoculants

Recent studies show a promising trend in the field of inoculation technology. A consortium of microorganisms that interact synergistically is presently being devised. Mixed inoculation of the plant with *Azospirillum* with other microorganisms enhanced the growth of plant compared to single inoculation (Bashan and Holguin 1997). For instance, plant growth can be increased by dual inoculation with *Azospirillum* and phosphate-solubilizing bacteria (Alagawadi and Gaur 1992; Belimov et al. 1995). *Azospirillum* is also considered to be a *Rhizobium*-"helper" stimulating nodulation, nodule activity, and plant metabolism, all of which stimulate many plant growth variables and plant resistance to unfavorable conditions (Andreeva et al. 1993; Itzigsohn et al. 1993). The synergistic interaction between *Rhizobium* and AMF in legume plants is well established (Bagyaraj 1984).

Co-inoculation with N-fixing bacteria and arbuscular mycorrhizal fungi creates synergistic interactions that may result in a significant increase in growth, P content, enhanced mycorrhizal infection, and an enhancement in the uptake of mineral nutrients such as N, P, K, Cu, Zn, and Fe (Bagyaraj 1984; Li and Huang 1987; Singh et al. 1990; Li et al. 1992; Garbaye 1994).

14.3.4 Formulations for Microbial Consortia: Inoculants for the Future

Despite progress in research on mixed inoculants, microbial inoculants with multiple organisms are not yet produced commercially. Until now, the research on mixed microbial inoculation was only confined to the development and inoculation of each bacterium in a separate formulation. But development of new inoculant formulation like polymer-entrapped desiccated inoculants has opened new vistas in mixed microbial inoculants. In this direction concept of "microbial consortium" assumes greater importance for sustainable agriculture. Microbial consortium is a group of microbial species that work together to carry out an overall reaction process, in our case beneficial organisms that together help promoting plant growth.

Feasibility of production of microbial consortium using *Rhizobium* and PSB using lignite, liquid, and alginate granules has been tested (Nethravathi and Brahmaprakash 2005; Shanker and Brahmaprakash 2004). It was observed that microbial consortium developed using alginate encapsulation was able to conserve the viability of both the organisms used for more than 6 months. But in liquid formulations, fast-growing bacilli had outnumbered the slow-growing rhizobia. There is a need to exercise caution in selecting the bacterial strains and formulation in the development of microbial consortium. Care should be taken to avoid bacterial strains which have antagonistic interactions among themselves. Alginate encapsulation a promising inoculant formulation for microbial consortia as they are desiccated formulation, microorganisms will be in the metabolically inactive state. The development of microbial consortium may minimize cost, labor, and energy involved in the production of inoculants. But more and more single strains microbial inoculants must be registered before inoculation industry can contemplate the development and commercialization of multi-bacterial inoculants (Polonenko 1994).

Indian agriculture, since the 1960s, has progressed tremendously due to the introduction of high-yielding varieties responding to high fertilizer inputs leading to enhanced food grains production. This high-input agriculture has also led to undesirable effects on environment and overall sustainability of the farming system such as adverse effects of agrochemicals. Fertilizer contamination of ground water has led to, over a period, eutrophication of lake and river water; caused decrease in oxygen content, death of aquatic life, and nitrate pollution; and increased emission of gaseous dinitrogen and metal toxicity. The nitrate toxicity caused health hazards such as birth defects, impaired nervous system, cancer, and methemoglobinemia (blue baby syndrome) (Knobeloch et al. 2000).

Table 14.8 Substitution of chemical fertilizers by biofertilizers

Sl. no.	Biofertilizers	Substitutes/ha per year	References
1.	*Rhizobium*	108.6–217.3 kg of urea	Mahdi et al. (2010)
2.	*Azolla*	20–40 kg urea/10 Mg	Mahdi et al. (2010)
3.	*Azospirillum*	60 kg urea in maize	Fulcheri and Frioni (1994)
4.	BGA	54–65 kg urea	Goyal and Venkataraman (1971) and Venkataraman (1981)
5.	*Frankia*	195 kg urea	Silvester (1975)

Modified from Brahmaprakash and Sahu (2012)
Calculated based on kg N-fixed × 2.17/ha per year

In this context, every unit of chemical fertilizers getting substituted by biofertilizers adds to sustainability and in the long run reduces the hazardous load of chemicals in the ecosystem. A rough estimate of the chemical fertilizers that may be substituted by biofertilizers is presented in Table 14.8. The future of inoculant technology and its benefits for sustainable agriculture depend on improving inoculant quality and effectiveness. Hence, the challenge is to develop and popularize inoculant formulation with long shelf life and effective in its response once inoculated, be it seed or soil. There is a need for extensive formulation like freeze-dried and fluid bed-dried inoculants. More recently mixed microbial inoculants have become popular; hence, further research work is required in this area, and also appropriate regulations and quality control guidelines are needed.

14.4 Future Perspectives of Biofertilizers in Pulse Production

Application of different biofertilizers as an essential factor of pulse cultivation is the new promising area nowadays. These inoculants are already being successfully utilized in India and few developing countries for pulse production and are expected to grow with time (Muleta 2010; Dudeja et al. 2011; Brahmaprakash and Sahu 2012). Therefore, it is logical to anticipate that in the future, the widespread utilization of biofertilizers will offer various potent strategies for overall development of agriculture. Table 14.9 summarily indicates selected examples of commercially available biofertilizers that have been used to improve the pulse production efficiency worldwide. However, more extensive utilization of biofertilizers will require addressing few issues with more attention and necessary actions to resolve the issues (Manoharachary 2004; Dudeja and Duhan 2005; Verma et al. 2015a).

1. Selection of most efficient *V. radiata* and *V. mungo* rhizobia among the predominant ones may prove to be better way of selecting efficient rhizobia with better competitive ability as has been observed in the case of *C. arietinum*.

Table 14.9 Commercially available biofertilizer worldwide for pulse production

Serial number	Type of biofertilizer with examples	References
1.	Nitrogen-fixing biofertilizers	Vessey (2003) and Dudeja et al. (2011)
	Rhizobium	
	Bradyrhizobium	
	Mesorhizobium	
2.	Phosphate-solubilizing biofertilizers	Khan et al. (2009)
	Bacillus	
	Pseudomonas	
	Aspergillus	
3.	Phosphate-mobilizing biofertilizer	Mehrvarz et al. (2008) and Jha et al. (2012)
	Arbuscular mycorrhizal fungi	
4.	Plant growth-promoting biofertilizers	Mehrvarz et al. (2008)
	Pseudomonas sp.	
5.	Phospho-bacterium and mycorrhizae	Suja (2008)
6.	*Rhizobium* and phosphobacterins	Dutta and Bandyopadhyay (2009)

Modified from Mahanty et al. (2016)

2. Comprehensive study on rhizobia infecting *V. radiata*, *V. Mungo*, *C.* Cajan, *Cyamopsis tetragonoloba*, and V. *unguiculata* using molecular techniques for proper classification and exploration of its multiple functions is needed. Currently PCR-based techniques deciphered the endophytic occurrence of these rhizobia in cereals and increase fertilizer use efficiency in these crops.
3. Identification of host genes involved in efficiency and competitiveness, as rhizobial genes involved in nodulation, dinitrogen fixation, and competitiveness are expressed in the host, so complementary genes in host should also be identified for further improvement in dinitrogen fixation. This may also include a robust pulse breeding program to improve the susceptibility of host to allochthonous symbiotic microbial species.
4. Elaborating the use of biofertilizers from laboratory and greenhouse experiments to large-scale commercial use will require a number of advanced new approaches for the growth, storage, shipping, formulation, and application of these bacteria and AM fungi. Transferring technological know-how on biofertilizer production to the industrial level and for optimum formulation is highly deserved.
5. It is necessary to have requisite extension program to educate farmers and village level workers about the long-term benefit of biofertilizers. The ill effects of prolonged use of synthetic fertilizers should also be acknowledged to people side by side.
6. Invention of genetically modified organisms (GMOs) more efficacious in stimulating plant growth is required. Concurrently scientists will need to establish to both the public and regulatory agencies worldwide that GMOs do not present any new hazards or risk.

7. Establishing germination of AM fungi on agar media, multiplication and commercialization of efficient AM fungi; identification of marker genes and genetic mechanisms for effective "P" transport, for increased efficacy and adaptation to increase vigor yield, etc., establishing robust repository of AM fungi; and biotechnological application of AM fungi are the future lines proposed.
8. Investigation into the microbial persistence of biofertilizers in stressful soil environments is to be mandatory.
9. Assessment of biofertilizers on the basis of agronomic practices, soil, and socio-economic aspects under diverse agricultural production systems is desired for proper utility.
10. Quality control system for the production of biofertilizers and their use in the field guarantee and explore the benefits of plant-microorganism symbiosis. Hence, "Biofertilizer Act" and strict regulation for quality control in markets and application should be established.

14.5 Conclusion

Biofertilizers are low-cost inputs with high benefits in agriculture. There is a need to popularize this low-cost technology with the farming community to reap higher dividends. Although encouraging results have been obtained by pot and glasshouse experiments, the implementation of integrated nutrient management technologies by farmers is gloomy at field level. Incorporation of micronutrients is essential to increase pulse production in deficient soils of India. About 40% of the pulse-growing regions have low to the medium population of native *Rhizobium*. In these circumstances, productivity of pulse may be increased by 10–12% via seed inoculation with low-cost rhizobial biofertilizer. Biofertilizers supplementing P nutrition in agriculture may be vital in saving the much needed foreign exchange if we succeed in making the "fixed" P available to crops. AMF inoculant is promising to improve the supply of phosphate and micronutrients like zinc for a variety of pulse crops, while phosphate solubilizers are the best option in rainfed areas of poor P availability. Shortage of adequate quantity of quality biofertilizer is one of the chief limitations in the popularization of biofertilizers. Therefore, efforts are needed to ensure availability of critical inputs like quality biofertilizers, etc. at the state level. A combined effort between soil chemists, microbiologists, and agronomists is required to facilitate judicious use of inorganic and microbiological inputs to realize better yields while ensuring the agriculture remains sustainable.

References

Adesemoye AO, Kloepper JW (2009) Plant–microbes interactions in enhanced fertilizer-use efficiency. Appl Microbiol Biotechnol 85:1–12

Agasimani CA, Mudalgiriyapla, Sreenivasa MN (1994) Response of groundnut to phosphate solubilising microorganisms. Groundnut News 6:5

Aggarwal A, Kadian N, Tanwar A, Yadav A, Gupta KK (2011) Role of arbuscular mycorrhizal fungi (AMF) in global sustainable development. J Appl Nat Sci 3:340–351
Akhtar MS, Siddiqui ZA (2008) Biocontrol of a root-rot disease complex of chickpea by *Glomus intraradices*, *Rhizobium* sp. and *Pseudomonas striata*. Crop Prot 27:410–417
Akhtar MS, Siddiqui ZA (2009) Effects of phosphate solubilizing microorganisms and Rhizobium sp. on the growth, nodulation, yield and root-rot disease complex of chickpea under field condition. Afr J Biotechnol 8:3479–3488
Alagawadi AR, Gaur AC (1992) Inoculation of *Azospirillum brasilense* and phosphate solubilizing bacteria on yield of sorghum [*Sorghum bicolor* (L.) Moench] in dry land. Trop Agric 69:347–350
Ali M, Gupta S (2012) Carrying capacity of Indian agriculture: pulse crops. Curr Sci 102:874–881
Ali FS, Loynachan TE (1990) Inhibition of *Bradyrhizobium japonicum* by diffusates from soybean seed. Soil Biol Biochem 22:973–976
Al-Rashidi RK, Loynachan TE, Frederick LR (1982) Desiccation tolerance of four strains of *Bradyrhizobium japonicum*. Soil Biol Biochem 14:489–493
Ames RN, Bethlenfalvay GJ (1987) Localized increase in nodule activity but no competitive interaction of cowpea rhizobia due to pre-establishment of vesicular arbuscular mycorrhiza. New Phytol 106:207–215
Andreeva IN, Red'kina TV, Ismailov SF (1993) The involvement of indole acetic acid in the stimulation of *Rhizobium*-legume symbiosis by *Azospirillum brasilense*. Russ J Plant Physiol 40:901–906
Annapurna K, Balasundaram VR (1995) Microbiological report on All India Coordinated Soybean Improvement Programme. ICAR, New Delhi
Ashoka P, Meena RS, Kumar S, Yadav GS, Layek J (2017) Green nanotechnology is a key for eco-friendly agriculture. J Clean Prod 142:4440–4441
Azcón R, Rubio R, Barea JM (1991) Selective interactions between different species of mycorrhizal fungi and *Rhizobium meliloti* strains, and their effects on growth, N2-fixation (15N) and nutrition of *Medicago sativa* L. New Phytol 117:399–404
Bagyaraj DJ (1984) Biological interaction with VA mycorrhizal fungi. In: Powell CL, Bagyaraj DJ (eds) VA mycorrhiza. CRC Press, Boca Raton, pp 131–153
Bagyaraj DJ, Mehrotra VS, Suresh CK (2002) Vesicular arbuscular mycorrhizal biofertilizer for tropical forest plants. In: Kannaiyan S (ed) Biotechnology of biofertilizers. Narosa Publishing House, New Delhi
Bahl N, Jauhri S (1986) Spent mushroom compost as a carrier for bacterial inoculant production. In: Proceedings of the international symposium on scientific and technological aspects of cultivating edible fungi. The Pennsylvania State University, University Park
Balachandran D, Nagarajan P (2002) Dual inoculation of *Rhizobium* and phosphobacteria with phosphorus on black gram cv. Vamban 1. Madras Agric J 89:691–693
Balamurugan S, Gunasekaran S (1996) Effect of combined inoculation of *Rhizobium* sp., and phosphobacteria at different levels of phosphorus in groundnut. Madras Agric J 83:503–505
Barea JM, Azcon R, Azcon-Aguilar C (2002) Mycorrhizosphere interactions to improve plant fitness and soil quality. Anton Leeuw 81:343–351
Bashan Y (1986) Alginate beads as synthetic inoculants carriers for the slow release of bacteria that affect plant growth. Appl Environ Microbiol 51:1089–1098
Bashan Y (1998) Inoculants for plant growth promoting bacteria for use in agriculture. Adv Biotechnol 16:729–770
Bashan Y, Carrillo A (1996) Bacterial inoculants for sustainable agriculture. In: Pérez-Moreno J, Ferrera-Cerrato R (eds) New horizons in agriculture: agroecology and sustainable development. Proceedings of the 2nd international symposium on agroecology, sustainable agriculture and education. Colegio de Postgraduados en ciencias agricolas, Montecillo
Bashan Y, Holguin G (1997) *Azospirillum*-plant relationships: environmental and physiological advances (1990–1996). Can J Microbiol 43:103–121
Bashan Y, Levanony H, Ziv-Vecht O (1987) The fate of field-inoculated *Azospirillum brasilense* Cd in wheat rhizosphere during the growing season. Can J Microbiol 33:107

Belimov AA, Kojemiakov AP, Chuvarliyeva CV (1995) Interaction between barley and mixed cultures of nitrogen fixing and phosphate-solubilizing bacteria. Plant Soil 173:29–37
Berg G, Zachow C, Müller H, Phillips J, Tilcher R (2013) Next-generation bio products sowing the seeds of success for sustainable agriculture. Agronomy 3:648–656
Beringer JE, Berwin AVB, Johnston Schulman JB, Hopwood DA (1979) The *Rhizobium*-legume symbiosis, in the cell as a habitat. University Press, Cambridge
Bethlenfalvay GJ, Brown MS, Stafford AE (1985) The glycine-glomus-rhizobium symbiosis: antagonistic effects between mycorrhizal colonization and nodulation. Plant Physiol 79:1054–1059
Bhardwaj D, Ansari MW, Sahoo RK (2014) Biofertilizers function as key player in sustainable agriculture by improving soil fertility, plant tolerance and crop productivity. Microb Cell Factories 13:66. http://www.microbialcellfactories.com/content/13/1/66
Bhata S, Umesh KB (2016) Estimating positive externalities of nitrogen fixation by pulses. Agric Econ Res Rev 29(2):201–209. https://doi.org/10.5958/0974-0279.2016.00048.3
Bhattari S, Maskey SL, Kama RL (1997) On-farm experiments on rhizobial inoculants in Nepal. In: Rupela OP, Johansen E, Herridge DF (eds) Extending nitrogen fixation research to farmers' fields. ICRISAT, Hyderabad
Bhowmik SN, Singh CS (2004) Mass multiplication of AM inoculum: effect of plant growth promoting rhizobacteria and yeast in rapid culturing of *Glomus mosseae*. Curr Sci 86:705–709
Bhowmik SN, Yadav GS, Datta M (2015) Rapid mass multiplication of *Glomus mosseae* inoculum as influenced by some biotic and abiotic factors. Bangladesh J Bot 44:209–214
Boby VU, Balakrishna AN, Bagyaraj DJ (2008) Interaction between *Glomus mosseae* and soil yeasts on growth and nutrition of cowpea. Microbiol Res 163:693–700
Bodake HD, Gaikwad SP, Shirke VS (2009) Study of constraints faced by the farmers in adoption of bio-fertilizer. Int J Agric Sci 5:292–294
Brahmaprakash GP, Hegde SV (2005) Nitrogen fixing in pigeonpea. In: Ali M, Shivkumar (eds) Advances in pigeonpea research. Indian Institute of Pulses Research, Kanpur
Brahmaprakash GP, Sahu PK (2012) Biofertilizers for sustainability. J Indian Inst Sci 92:37–62
Brahmaprakash GP, Girisha HC, Vithal N, Laxmipathy R, Hegde SV (2007) Liquid *Rhizobium* inoculants formulations to enhance biological nitrogen fixation in food legumes. J Food Legum 20:75–79
Brundrett MC (2009) Mycorrhizal associations and other means of nutrition of vascular plants: understanding the global diversity of host plants by resolving conflicting information and developing reliable means of diagnosis. Plant Soil 320:37–77
Bulgarelli D, Schlaeppi K, Spaepen S, Loren V, van Themaat E, Schulze-Lefert P (2013) Structure and functions of the bacterial microbiota of plants. Annu Rev Plant Biol 64:807–838
Buragohain S, Sharma B, Nath JD, Gogaoi N, Meena RS, Lal R (2017) Impact of ten years of biofertilizer use on soil quality and rice yield on an inceptisol in Assam, India. Soil Res. https://doi.org/10.1071/SR17001
Burris RH, Roberts GP (1993) Biological nitrogen fixation. Annu Rev Nutr 13:317–335
Bushby HVA, Marshall KC (1977) Water status of rhizobia in relation to their susceptibility to desiccation and to their protection by montmorillonite. J Gen Microbiol 99:19–28
Chao WL, Alexander M (1984) Mineral soils as carriers for *Rhizobium* inoculants. Appl Environ Microbiol 47:94–97
Clark RB, Zeto SK (2000) Mineral acquisition by arbuscular mycorrhizal plants. J Plant Nutr 23:867–902
Cleveland CC, Townsend AR, Schimel DS, Fisher H, Howarth RW, Hedin LO, Perakis SS, Latty EF, von Fischer JC, Elseroad A, Wasson MF (1999) Global patterns of terrestrial biological nitrogen (N2) fixation in natural ecosystems. Glob Biogeochem Cycles 13:623–645
Cooper AJ (1985) Crop production recirculating nutrient solution. Sci Hortic 3:35–38
Daniels-Hylton KDM, Ahmad MH (1994) Inoculation response in kidney beans (*Phaseolusvulgaris* L.) to vesicular-arbuscular mycorrhizal fungi and rhizobia in non-sterilized soil. Biol Fertil Soils 18:95–98

Datta R, Baraniya D, Wang YF, Kelkar A, Moulick A, Meena RS, Yadav GS, Ceccherini MT, Formanek P (2017) Multi-function role as nutrient and scavenger of free radical in soil. Sustain MDPI 9(8):402. https://doi.org/10.3390/su9081402

Dayamani KJ (2010) Formulation and determination of effectiveness of liquid inoculants of plant growth promoting rhizobacteria. PhD thesis, University of Agricultural Sciences, Bangalore, India

De Lucca AJ, Connick WJ Jr, Fravel DR, Lewis JA, Bland JM (1990) The use of bacterial alginates to prepare biocontrol formulations. J Ind Microbiol 6:129–134

Deaker R, Roughley RJ, Kennedy IR (2004) Legume seed inoculation technology- a review. Soil Biol Biochem 36:75–88

Demir S, Akkopru A (2007) Using of arbuscular mycorrhizal fungi (AMF) for biocontrol of soil borne fungal plant pathogens. In: Chincholkar SB, Mukerji KG (eds) Biological control of plant diseases. Haworth, Philadelphia, pp 17–37

Dhakal Y, Meena RS, De N, Verma SK, Singh A (2015) Growth, yield and nutrient content of mungbean (*Vigna radiata* L.) in response to INM in eastern Uttar Pradesh, India. Bangladesh J Bot 44(3):479–482

Dhakal Y, Meena RS, Kumar S (2016) Effect of INM on nodulation, yield, quality and available nutrient status in soil after harvest of green gram. Legum Res 39(4):590–594

Digat B (1991) A new encapsulation technology for bacterial inoculants and seed bacterization. In: Keel C, Koller B, Défago G (eds) Plant growth-promoting rhizobacteria – progress and prospects. IOBC/WPRS Bulletin, Zurich

Dixon ROD, Wheeler CT (1986) Nitrogen fixation in plants. Chapman and Hall, New York

Dommergues YR, Diem HG, Divies C (1979) Polyacrylamide entrapped *Rhizobium* as an inoculant for legumes. Appl Environ Microbiol 37:779–981

Dube JN, Mahere DP, Bawat AF (1980) Development of coal as a carrier for rhizobial inoculants. Sci Cult 46:304

Dudeja SS, Duhan JS (2005) Biological nitrogen fixation research in pulses with special reference to mungbean and urdbean. Indian J Pulses Res 18:107–118

Dudeja SS, Khurana AL (1999) Long term multiplication field evaluation of chickpea rhizobia in India. In: Proceedings of international symposium on long term fertilization trials as a basis for sustainable land use and quantification of matter cycles. UFZ-Bericht, Halle

Dudeja SS, Khurana AL (2001) Integrated management of N and P availability in chickpea through the use of Rhizobium and phosphate solubilizers. In: Proceedings of XIV international plant nutrition colloquium. Hannover

Dudeja SS, Narula N (2008) Molecular diversity of root nodule forming bacteria. In: Khachatourians GG, Arora DK, Rajendran TP, Srivastava AK (eds) Agriculturally important microorganisms, vol 2. Academic World International, Bhopal

Dudeja SS, Singh NP, Sharma P, Gupta SC, Chandra R, Dhar B, Bansal RK, Brahmaprakash GP, Potdukhe SR, Gundappagol RC, Gaikawad BG, Nagaraj KS (2011) Biofertilizer technology and pulse production. In: Singh A et al (eds) Bioaugmentation, biostimulation and biocontrol, soil biology. Springer, Berlin

Dutta D, Bandyopadhyay P (2009) Performance of chickpea (*Cicer arietinum* L.) to application of phosphorus and biofertilizer in laterite soil. Arch Agron Soil Sci 55:147–155

Egamberdieva D, Kamilova F, Validov S, Gafurova L, Kucharova Z, Lugtenberg B (2008) High incidence of plant growth stimulating bacteria associated with the rhizosphere of wheat grown on salinated soil in Uzbekistan. Environ Microbiol 10:1–9

Fages J (1992) An industrial view of *Azospirillum* inoculants: formulation and application technology. Symbiosis 13:15–26

FAOSTAT (2009) Online interactive database on agriculture. FAOSTAT. www.fao.org

Fravel DR, Marois JJ, Lumsden RD, Connick WJ (1985) Encapsulation of potential biocontrol agents in an alginate-clay matrix. Phytopathology 75:774–777

Fredeen AL, Terry N (1988) Influence of vesicular–arbuscular mycorrhizal infection and soil phosphorous level on growth and carbon metabolism of soybean. Can J Bot 66:2311–2316

Fulcheri E, Frioni D (1994) *Azospirillum* inoculation of maize: effects on yield on a field experiment in Central Argentina. Biophysics 244:686–691

Galloway J, Raghuram N, Abrol YP (2008) A perspective on reactive nitrogen in a global, Asian and Indian context. Curr Sci 94:1375–1381

Ganry F, Diem HG, Dommergues YR (1982) Effect of inoculation with *Glomus mosseae on* nitrogen fixation by field grown soybeans. Plant Soil 68:321–329

Garbaye L (1994) Helper bacteria: a new dimension to the mycorrhizal symbiosis. New Phytol 128:197–210

Geneva M, Zehirov G, Djonova E, Kaloyanova N, Georgiev G, Stancheva I (2006) The effect of inoculation of pea plants with mycorrhizal fungi and *Rhizobium* on nitrogen and phosphorus assimilation. Plant Soil Environ 52:435–440

Giller KE (2001) Nitrogen fixation in tropical cropping systems. CAB International, Wallingford

Gosling P, Hodge A, Goodlass G, Bending GD (2006) Arbuscular mycorrhizal fungi and organic farming. Agric Ecosyst Environ 113:17–35

Goyal SK, Venkataraman GS (1971) Effects of algalization on high yielding rice varieties. II. Response of soil types. Phykos 10:32–38

Gupta SB, Vyas MK, Patil SK (1992) Effect of phosphorus solubilizing bacteria and thiram at different levels of phosphorus on soybean and soil micro flora. J Indian Soc Soil Sci 40:854–856

Harrison MG (2005) Signaling in the arbuscular mycorrhizal symbiosis. Annu Rev Microbiol 59:19–42

Hazarika DK, Das KK, Dubey LN, Phookan AK (2000) Effect of vesicular arbuscular mycorrhizal (VAM) fungi and *Rhizobium* on growth and yield of green gram (*Vigna radiata* (L.)Wilczek.). J Mycol Plant Pathol 30:424–426

Heap AJ, Newman EL (1980) Links between roots by hyphae of vesicular arbuscular mycorrhizas. New Phytol 85:169–171

Hegde SV (1994) Population of cowpea rhizobia in farmers' fields in southern Karnataka: influence of cropping system, locations, and N-level. In: Rupela OP, Kumar Rao JVDK, Wani SP, Johansen C (eds) Linking biological nitrogen fixation research in Asia. ICRISAT, Hyderabad

Hegde SV, Brahmaprakash GP (1992) A dry granular inoculants of *Rhizobium* for soil application. Plant Soil 144:309–311

Herridge DF (2008) Inoculation technology for legumes. In: Dilworth MJ, James EK, Sprent JI, Newton WE (eds) Nitrogen fixation: origins, applications and researchprogress. Nitrogen-fixing leguminous symbioses, vol 7. Springer, The Netherlands

Herridge DF, Peoples MB, Boddey RM (2008) Marschner review: global inputs of biological nitrogen fixation in agricultural systems. Plant Soil 311:1–18

Hoa NTL, Thao TY, Lieu P, Herridge DF (2002) N2 fixation of groundnut in the eastern region of south Vietnam. In: Herridge D (ed) Inoculants and nitrogen fixation of legumes in Vietnam. ACIAR proceedings. ACIAR, Canberra

Houlton BZ, Wang YY, Vitousek PM, Eield CB (2008) A unifying framework for dinitrogen fixation in the terrestrial biosphere. Nature 454:327–330

Howieson J, Ballard R (2004) Optimising the legume symbiosis in stressful and competitive environments within southern Australia – some contemporary thoughts. Soil Biol Biochem 36:1261–1273

Huber DM, El-Nasshar L, Moore HW, Mathre DE, Wagner JE (1989) Interaction between a peat carrier and bacterial seed treatments evaluated for biological control of the take – all diseases of wheat (*Triticum aestivum* L.). Biol Fertil Soils 8:166–171

Hynes RK, Jans DC, Bremer E, Lupwayi NZ, Rice WA, Clayton GW, Collins MM (2001) *Rhizobium* sp. population dynamics in the pea rhizosphere of rhizobial inoculants strain applied in different formulations. Can J Microbiol 47:595–600

Iswaran V, Sen A, Apte R (1972) Plant compost as a substitute for peat for legume inoculants. Curr Sci 41:299

Itzigsohn R, Kapulnik Y, Okon Y, Dovrat A (1993) Physiological and morphological aspects of interactions between *Rhizobium meliloti* and alfalfa (*Medicago saliva*) in association with *Azospirillum brasilense*. Can J Microbiol 39:610–615

Jackson AM, Whipps JM, Lynch JM (1991) Production, delivery systems and survival in soil of four fungi with disease biocontrol potential. Enzym Microb Technol 13:636–642

Jangid MK, Khan IM, Singh S (2012) Constraints faced by the organic and conventional farmers in adoption of organic farming practices. Indian Res J Ext Educ Spec Issue 2:28–32

Jarvis BDW, van Berkum P, Chen WX, Nour SM, Fernandez MP, Cleyet-Marel JC, Gillis M (1997) Transfer of *Rhizobium loti, Rhizobium huakuii, Rhizobium ciceri, Rhizobium mediterraneum* and Rhizobium tianshanense to Mesorhizobium gen. nov Int J Syst Bacteriol 47:895–898

Jha MN, Kumar P, Chourasia SK (2012) Hope, hype and reality of biofertilizer. Fertil Technol 121:448–480

Jia Y, Gray VM, Straker CJ (2004) The influence of *Rhizobium* and arbuscular mycorrhizal fungi on nitrogen and phosphorus accumulation by *Vicia faba*. Ann Bot 94:251–258

Joshi PK (1994) Field response of groundnut to *Bradyrhizobium* inoculation. In: Rupela OP, Kumar Rao JVDK, Wani SP, Johansen C (eds) Linking biological nitrogen fixation research in Asia. ICRISAT, Hyderabad

Jung G, Mugnier J, Diem HG, Dommergues YR (1982) Polymer-entrapped *Rhizobium* as an inoculant for legumes. Plant Soil 65:219–231

Kandasamy R, Prasad NN (1971) Lignite as a carrier of rhizobia. Curr Sci 40:496

Khan MZ, Zaidi A, Wani PA, Oves M (2009) Role of plant growth promoting rhizobacteria in the remediation of meta contaminated soils. Environ Chem Lett 7:1–19

Khrurana AL, Dudeja SS, Singh M (1999) Increasing the efficiency of phosphatic fertilizer with phosphate solubilizing bacteria to improve soil fertility and chickpea (*Cicer arietinum*) productivity. In: Proceedings of international symposium on long term fertilization trials as a basis for sustainable land use and quantification of matter cycles. UFZ-Bericht, Halle

Khurana AL, Dudeja SS (1994) On-farm experience in the use of rhizobial inoculants on pigeonpea in India. In: Rupela OP, Kumar Roo JVDK, Wani SP, Johansen C (eds) Linking biological nitrogen fixation research in Asia. ICRISAT, Hyderabad

Khurana AL, Namdeo SL, Patel BI, Dudeja SS (1997a) On-farm experiments on rhizobial inoculants – problems and possible solutions. In: Rupela OP, Johansen C, Herridge DF (eds) Extending nitrogen fixation researchto farmers' fields. ICRISAT, Hyderabad

Khurana AL, Namdeo SL, Patel BJ, Dudeja SS (1997b) On-farm experiments on rhizobial inoculants: problems and possible solutions. In: Rupela OP, Johansen C, Herridge DF (eds) Extending nitrogen fixation research to farmers' fields. Proceeding of managing legumes nitrogen fixation in cropping systems of Asia, ICRISAT Asia Center

Kitamikado M, Yamaguchi K, Tseng CH, Okabe B (1990) Methods designed to detect alginate-degrading bacteria. Appl Environ Microbiol 56:2939–2940

Kloepper JW, Lifshitz R, Schroth MN (1988) Pseudomonas inoculants to benefit plant production. In: ISI atlas of science. Institute for Scientific Information, Philadelphia

Knobeloch L, Salna B, Hogan A, Postle J, Anderson H (2000) Blue babies and nitrate-contaminated well water. Environ Health Perspect 108:675–678

Kothari SK, Marschner H, Romheld V (1991) Contribution of the VA mycorrhizal hyphae in acquisition of phosphorus and zinc by maize grown in a calcareous soil. Plant Soil 131:177–185

Kumar R, Chandra R (2008) Influence of PGPR and PSB on *Rhizobium leguminosarum* Bv. viciae strain competition and symbiotic performance in lentil. World J Agric Sci 4:297–301

Kumar S, Sheoran S, Kumar SK, Kumar P, Meena RS (2016) Drought: a challenge for Indian farmers in context to climate change and variability. Progress Res Int J 11:6243–6246

Le Tacon F, Jung G, Mugnier J, Michelot P, Mauperin C (1985) Efficiency in a forest nursery of an ectomycorrhizal fungus inoculum produced in a fermentor and entrapped in polymeric gels. Can J Bot 63:1664–1668

Lewis JA, Papavizas GC (1985) Characteristics of alginate pellets formulated with *Trichoderma* and *Gliocladium* and their effect on the proliferation of the fungi in the soil. Plant Pathol 34:571–577

Li CY, Huang LL (1987) Nitrogen fixing (acetylene reducing) bacteria associated with ectomycorrhizas of Douglas –fir. Plant Soil 98:425–428

Li XL, Marschner H, George E (1991) Acquisition of phosphorus and copper by VA–mycorrhizal hyphae and root-to-shoot transport in white clover. Plant Soil 136:49–57

Li CY, Massicote HB, Moore LVH (1992) Nitrogen-fixing *Bacillus* sp. associated with Douglas -fir tuberculate ectomycorrhizae. Plant Soil 140:35–40

Lippert K, Galinski EA (1992) Enzyme stabilization by ectoine type compatible solutes: protection against heating, freezing and drying. Appl Microbiol Biotech 37:61–65

Lodwig EM, Hosie AHF, Bourdés A, Findlay K, Allaway D, Karunakaran R, Downie JA, Poole PS (2003) Amino-acid cycling drives nitrogen fixation in the legume–Rhizobium symbiosis. Nature 422:722–726

Madigan MT, Martinko JM, Dunlap PV, Clark DP (2009) Brock biology of microorganisms, 12th edn. Pearson Benjamin, Cummings

Mahanty T, Bhattacharjee S, Goswami M, Bhattacharyya P, Das B, Ghosh A, Tribedi P (2016) Biofertilizers: a potential approach for sustainable agriculture development. Environ Sci Pollut Res. https://doi.org/10.1007/s11356-016-8104-0

Mahdi SS, Hassan GI, Samoon SA, Rather HA, Dar Showkat A, Zehra B (2010) Biofertilizers in organic agriculture. J Phytol 2:42–54

Manoharachary C (2004) Biodiversity, taxonomy, ecology, conservation and biotechnology of arbuscular mycorrhizal fungi. Indian Phytopathol 57:01–06

Marshall DI, Bateman JD, Brockwell J (1993) Validation of a serial-dilution, plant-infection test for enumerating *Rhizobium leguminosarum* bv. *viciae* and its application for counting rhizobia in acid soils. Soil Biol Biochem 25:261–268

Marx DH, Kenney DS (1982) Production of ectomycorrhizal fungus inoculum. In: Schenck NC (ed) Methods and principles of mycorrhizal research. The American Phytopathological Society, St. Paul

Mary P, Ochin D, Tailliez R (1985) Rates of drying and survival of *Rhizobium meliloti* during storage at different relative humidities. Appl Environ Microbiol 50:207–211

Meena H, Meena RS (2017) Assessment of sowing environments and bio-regulators as adaptation choice for clusterbean productivity in response to current climatic scenario. Bangladesh J Bot 46(1):241–244

Meena KK, Meena RS, Kumawat MS (2013) Effect of sulphur and iron fertilization on yield attribute, yield, nutrient uptake of mungbean (*Vigna radiata* L.). Indian J Agric Sci 83(4):108–112

Meena VS, Maurya BR, Verma R, Meena RS, Jatav GK, Meena SK, Meena R, Meena SK (2013a) Soil microbial population and selected enzyme activities as influenced by concentrate manure and inorganic fertilizer in alluvium soil of Varanasi. Bioscience 8(3):931–935

Meena RS, Yadav RS, Meena H, Kumar S, Meena YK, Singh A (2015a) Towards the current need to enhance legume productivity and soil sustainability worldwide: a book review. J Clean Prod 104:513–515

Meena RS, Meena VS, Meena SK, Verma JP (2015b) Towards the plant stress mitigate the agricultural productivity: a book review. J Clean Prod 102:552–553

Meena H, Meena RS, Singh B, Kumar S (2016) Response of bio-regulators to morphology and yield of clusterbean [*Cyamopsis tetragonoloba* (L.) Taub.] under different sowing environments. J App Nat Sci 8(2):715–718

Meena RS, Kumar V, Yadav GS, Mitran T (2017a) Response and interaction of *Bradyrhizobium japonicum* and *Arbuscular mycorrhizal* fungi in the soybean rhizosphere: a review. Plant Growth Reg. Accepted in press

Meena RS, Meena PD, Yadav GS, Yadav SS (2017b) Phosphate solubilizing microorganisms, principles and application of microphos technology. J Clean Prod 145:157–158

Mehdi Z, Nahid S-R, Alikhani HA, Nasser A (2006) Responses of lentil to co-inoculation with phosphate-solubilizing rhizobial strains and arbuscular mycorrhizal fungi. J Plant Nutr 29:1509–1522

Mehrvarz S, Chaichi MR, Alikhani HA (2008) Effects of phosphate solubilizing microorganisms and phosphorus chemical fertilizer on yield and yield components of barley (Hordeum vulgare L.). Am Eurasian J Agric Environ Sci 3:822–828

Mendes R, Garbeva P, Raaijmakers JM (2013) The rhizosphere microbiome: significance of plant beneficial plant pathogenic and human pathogenic microorganisms. FEMS Microbiol Rev 37:634–663

Molla MN, Solaiman ARM (2009) Association of arbuscular mycorrhizal fungi with leguminous crops grown in different agro-ecological zones of Bangladesh. Arch Agron Soil Sci 55:233–245

Morel MA, Braña V, Castro-Sowinski S (2012) Legume crops, importance and use of bacterial inoculation to increase production. In: Goyal A (ed) Crop plant. ISBN: 978-953-51 0527-5, http://www.intechopen.com/books/crop-plant/legume-crops-importance-and-useofbacterial-inoculation-to-increase-production. Last visited on April, 2017

Mosse B (1977) Plant growth responses to vesicular-arbuscular mycorrhiza. X. Responses to Stylosanthes and maize to inoculation in unsterile soils. New Phytol 78:277–288

Mosse B, Powell CL, Hayman DS (1976) Plant growth responses to vesicular-arbuscular mycorrhiza. IX. Interactions between VA mycorrhiza, rock phosphate and symbiotic nitrogen fixation. New Phytol 76:331–342

Motsara MR, Bhattacharyya P, Srivastava B (1995) Biofertilizer technology, marketing and usage-A source book-cum-glosarry. Fertilizer Development and Consultation Organisation, New Delhi

Muchovej RM (2001) Importance of mycorrhizae for agricultural crops. http://www.phc.eu/files/publicaties/Ag_Mycorrhizae_Public_document.pdf. Last accessed April 2017

Mugnier J, Jung G (1985) Survival of bacteria and fungi in relation to water activity and solvent properties of water in biopolymer gels. Appl Environ Microbiol 50:108–114

Muleta D (2010) Legume responses to arbuscular mycorrhizal fungi inoculation in sustainable agriculture. In: Khan MS et al (eds) Microbes for legume improvement. Springer, Wien. https://doi.org/10.1007/978-3-211-99753-6_12

Munchbach M, Nocker A, Narberhaus F (1999) Multiple small heat shock proteins in rhizobia. J Bacteriol 181:83–90

Munns DN, Mosse B (1980) Mineral nutrition of legume crops. In: Summerfield RJ, Bunting AH (eds) Advances in legume science. HMSO, London

Napamornbodi O, Rajanasiriwong W, Thamsurakul S (1988) Production of VAM fungi, *Glomus intraradices* and *G. mosseae* in tissue culture. In: Mycorrhiza for green Asia. University of Madras, Tamil Nadu

Navazio L, Moscatiello R, Genre A, Novero M, Baldan B, Bonfante P, Mariani P (2007) A diffusible signal from arbuscular mycorrhizal fungi elicits a transient cytosolic calcium elevation in host plant cells. Plant Physiol 144:673–681

Nethravathi CS, Brahmaprakash GP (2005) Alginate based composite biofertilizer of *Bradyrhizobium japonicum* and *Bacillus megaterium* for soybean (*Glycine max* (L) Merrill). J Soil Biol Ecol 25:1–13

Nkaa FA, Nwokeocha OW, Ihuoma O (2014) Effect of phosphorus fertilizer on growth and yield of cowpea (*Vigna unguiculata*). J Pharm Biol Sci 9:74–82

Oldroyd GED, Harrison MJ, Udvardi M (2005) Peace talks and trade deals: keys to long-term harmony in legume-microbe symbioses. Plant Physiol 137:1205–1210

Olsen PE, Rice WA, Collins MM (1994a) Biological contaminants in North American legume inoculants. Soil Biol Biochem 27:699–701

Olsen PE, Rice WA, Bordeleau LM, Biederbeck VO (1994b) Analysis and regulation of legume inoculants in Canada: the need for an increase in standards. Plant Soil 161:127–134

Paau AS, Graham LL, Bennett M (1991) Progress in formulation research for PGPR and biocontrol inoculants. In: Keel C, Koller B, Défago G (eds) Plant growth-promoting rhizobacteria-progress and prospects. IOBC/WPRS Bulletin, Zurich

Pagano MC, Cabello MN, Scotti MR (2007) Phosphorus response of three native Brazilian trees to inoculation with four arbuscular mycorrhizal fungi. J Agric Technol 3:231–240

Pelczar MJ Jr, Chan ECS, Krieg NR (1993) Microbiology, 5th edn. Tata McGraw-Hill Publishing Company Limited, New Delhi

Peoples MB, Brockwell J, Herridge DF, Rochester IJ, Alves BIR, Urquiaga S, Boddey RM, Dakora FD, Bhattarai S, Maskey SL, Sampet C, Rerkasem B, Khans DF, Hauggaard-Nielsen H, Jensen BS (2009) The contributions of nitrogen-fixing crop legumes to the productivity of agricultural systems. Symbiosis 48:1–17

Philip K, Jauhri KS (1984) Pressmud: a potential carrier for *Rhizobium* and *Azotobacter* 1. Comparative analytical studies of various carrier materials. Z Mikrobiol 139:5–4

Poinsot V, Bélanger E, Laberge S, Yang GP, Antoun H, Cloutier J, Treilhou M, Dénarié J, Promé JC, Debellé F (2001) Unusual methyl-branched a, ß-unsaturated acyl chain substitutions in the Nod factors of an Arctic *Rhizobium, Mesorhizobium* sp. Strain N33 (*Oxytropicarctobia*). J Bacteriol 183:3721–3728

Polonenko DR (1994) Commercial opportunities for multiorganism inoculants. In: Germida JJ (ed) Proceedings of the BIOREM 4th annual general meeting. BIOREM, Montreal

Postgate JR (1989) Trends and perspectives in nitrogen fixation research. Adv Microb Physiol 30:1–22

Raja N (2013) Biopesticides and biofertilizers: ecofriendly sources for sustainable agriculture. J Biofertil Biopestic. doi:1000e112:1000e112

Ram K, Meena RS (2014) Evaluation of pearl millet and mungbean intercropping systems in arid region of Rajasthan (India). Bangladesh J Bot 43(3):367–370

Reddy AA (2004) Consumption pattern, trade and production potential of pulses. Econ Polit Wkly 39:4854–4860. http://ssrn.com/abstract=1537541

Reddy AA (2009) Pulses production technology: status and way forward. Econ Polit Wkly 44:73–82

Reddy AA, Reddy GP (2010) Supply side constrains in production of pulses in India: a case study of lentil. Agric Econ Res Rev 23:129–136

Redecker D, Raab P (2006) Phylogeny of the Glomeromycota (arbuscular mycorrhizal fungi): recent developments and new gene markers. Mycologia 98:885–895

Rewari RB (1984) Summarised results of microbiology trials. All-India Co-ordinated Research Project on Improvement of Pulses, New Delhi

Rewari RB (1985) Summarised results of microbiology trials. All-India Co-ordinated Research Project on Improvement of Pulses, New Delhi

Richardson AE (2001) Prospects for using soil microorganisms to improve the acquisition of phosphorous by plants. Aust J Plant Physiol 28(9):897–906

Rooge RB, Patil VC, Ravikrishnan (1998) Effect of phosphorus application with phosphate solubilisation organisms on the yield, quality and P uptake of soybean. Legum Res 21:85–90

Sadasivam KV, Tyagi RK, Ramarethinam S (1986) Evaluation of some agricultural wastes as carriers for bacterial inoculants. Agric Wastes 17:301–306

Sahgal M, Johri BN (2003) The changing face of rhizobial systematics. Curr Sci 84:43–48

Sahoo RK, Ansari MW, Pradhan M, Dangar TK, Mohanty S, Tuteja N (2014) Phenotypic and molecular characterization of efficient native Azospirillum strains from rice fields for crop improvement. Protoplasma. https://doi.org/10.1007/s00709-013-0607-7

Sanginga N (2003) Role of biological nitrogen fixation in legume based cropping systems; a case study of West Africa farming systems. Plant Soil 252:25–39

Santos VB, Araujo SF, Leite LF, Nunes LA, Melo JW (2012) Soil microbial biomass and organic matter fractions during transition from conventional to organic farming systems. Geoderma 170:227–231

Sattar MA, Khanam D, Ahmad S, Haider MR, Podder AK, Bhuiyan MA (1997) On-farm experiments on rhizobial inoculants in Bangladesh: results, problems, and possible solutions. In: Rupela OP, Johansen C, Herridge DF (eds) Extending nitrogen fixation research to farmers' fields. ICRISAT, Hyderabad

Saxena D, Mohammed A, Khanna S (1996) Modulation of protein profiles in *Rhizobium* sp under salt stress. Can J Microbiol 42:617–620

Saxena AK, Rathi SK, Tilak KVBR (1997) Differential effect of various endomycorrhizal fungi on nodulating ability of green gram by *Bradyrhizobium* sp. (*Vigna*) strain S24. Biol Fertil Soils 24:175–178

Schuessler A, Schwarzott D, Walker C (2001) A new fungal phylum, the Glomeromycota: phylogeny and evolution. Mycol Res 105:1413–1421

Shanker S, Brahmaprakash GP (2004) Development of composite biofertilizer containing *Bradyrhizobium+Bacillus megaterium* in different formulations. J Soil Biol 24:63–70

Sharma SB, Sayyed RZ, Trivedi MH, Gobi TA (2013) Phosphate solubilizing microbes: sustainable approach for managing phosphorus deficiency in agricultural soils. Springer Plus 2:587. http://www.springerplus.com/content/2/1/587

Siddiqui ZA, Baghel G, Akhtar MS (2007) Biocontrol of *Meloidogyne javanica* by *Rhizobium* and plant growth-promoting rhizobacteria on lentil. World J Microbiol Biotechnol 23:435–441

Silvester WB (1975) Ecological and economical significance of the non-legume symbioses. In: Newton WE, Nyman CJ (eds) 1st international symposium onnitrogen fixation. Washington State University Press, Washington, DC

Singh MS (2005) Effect of Rhizobium, FYM and chemical fertilizers on legume crops and nutrient status of soil-a review. Agric Rev 26:309–312

Singh CS, Amawate JS, Tyagi SP, Kapoor A (1990) Interaction effect of *Glomus fasciculatum* and *Azospirillum brasilense* on yields of various genotypes of wheat (*Triticum aestivum*) in pots. Zentralbl Mikrobiol 145:203–208

Singh CS, Kapoor A, Wange SS (1991) The enhancement of root colonisation of legumes by vesicular-arbuscular mycorrhizal (VAM) fungi through the inoculation of the legume seed with commercial yeast (*Saccharomyces cerevisiae*). Plant Soil 131:129–133

Singh JS, Pandey VC, Singh DP (2011) Efficient soil microorganisms: a new dimension for sustainable agriculture and environmental development. Agric Ecosyst Environ 140:339–353

Singleton PW, Boonkerd N, Carr TJ, Thompson JA (1997) Technical and market constraints limiting legume inoculant use in Asia. In: Rupela OP, Johansen C, Herridge DF (eds) Extending nitrogen fixation researchto farmers' fields. ICRISAT, Hyderabad

Singleton P, Keyser H, Sande E (2002) Development and evaluation of liquid inoculants. In: Herridge D (ed) Inoculants and nitrogen fixation of legume in Vietnam. ACIAR Proceedings

Sinha RK, Valani D, Chauhan K, Agarwal S (2014) Embarking on a second green revolution for sustainable agriculture by vermiculture biotechnology using earthworms: reviving the dreams of Sir Charles Darwin. Int J Agric Health Saf 1:50–64

Slattery JJ, Coventry DR, Slattery W (2001) Rhizobial ecology as affected by the soil environment. Aust J Exp Agric 41:289–298

Smidsrod O, Skjak-Braek G (1990) Alginate as immobilization matrix for cells. Trends Biotechnol 8:71–78

Smith RS (1992) Legume inoculant formulation and application. Can J Microbiol 38:485–492

Smith RS (1995) Inoculant formulations and applications to meet changing needs. In: Tikhonovich IA, Provorov NA, Romanov VI, Newton WE (eds) Nitrogen fixation: fundamentals and applications. Kluwer Academic Publishers, Dordrecht

Smith SE, Read DJ, Harley JL (1997) Mycorrhizal symbiosis, 2nd edn. Academic, London

Sougoufara B, Diem HG, Dommergues YR (1989) Response of field -grown *Casuarina equisetifolia* to inoculation with *Frankia* strain ORS 021001 entrapped in alginate beads. Plant Soil 118:133–137

Sparrow SD, Ham GE (1983) Survival of *Rhizobium phaseoli* in six carrier materials. J Agron 75:181–184

Sprent JI (2001) Nodulation in legumes. Royal Botanic Gardens, London

Sridhar V, Brahmaprakash GP, Hegde SV (2004) Development of a liquid inoculant using osmoprotectants for phosphate solubilizing bacteria. Karnataka J Agric Sci 17:251–257

Stancheva I, Geneva M, Djonova E, Kaloyanova N, Sichanova M, Boychinova M, Georgiev G (2008) Response of alfalfa (*Medicago sativa* L) growth at low accessible phosphorus source to the dual inoculation with mycorrhizal fungi and nitrogen fixing bacteria. Gen Appl Plant Physiol 34:319–326

Stanier RY (1987) General microbiology, 5th edn. Macmillan, London

Streeter JG (1985) Accumulation of *alpha*, *alpha*-trehalose by *Rhizobium* bacteria and bacteroids. J Bacteriol 164:78–84

Subba Rao NS, Tilak KVBR (1977) Souvenir bulletin. Directorate of Pulses Development, Government of India

Subba Rao NS, Tilak KVBR, Singh CS (1986) Dual inoculation with *Rhizobium* sp. and *Glomus fasciculatum* enhances nodulation, yield and nitrogen fixation in chickpea (*Cicer arietinum* Linn). Plant Soil 95:351–359

Suja G (2008) Strategies for organic production of tropical tuber crops. In: Venkateswarlu B, Balloli SS, Ramakrishna YS (eds) Organic farming in rain fed agriculture: opportunities and constraints. Central Research Institute for Dryland Agriculture, Hyderabad, pp 135–143

Surendra ST, Pathan MA, Gupta KP, Khandkar UR (1993) Effect of phosphate solubilising bacteria at different levels of phosphate on black gram. India J Agron 38:131–133

Tagore GS, Namdeo SL, Sharma SK, Kumar N (2013) Effect of *Rhizobium* and phosphate solubilizing bacterial inoculants on symbiotic traits, nodule leghemoglobin, and yield of chickpea genotypes. Int J Agron 2013:1–8. Article ID 581627, 8 pages. https://doi.org/10.1155/2013/581627

Thies JE, Woomer PL, Singleton PW (1995) Enrichment of *Bradyrhizobium* spp. populations in soil due to cropping of the homologous host legume. Soil Biol Biochem 27:633–637

Tittabutr P, Payakapong W, Teaumroong N, Singleton PW, Boonkerd N (2007) Growth, survival and field performance of bradyrhizobial liquid inoculants formulations with polymeric additives. Sci Asia 33:69–77

Trivedi P, Pandey A, Palni LMS (2012) Bacterial inoculants for field applications under mountain ecosystem: present initiatives and future prospects. In: Maheshwari DK (ed) Bacteria in agrobiology: plant probiotics. Springer, Berlin. https://doi.org/10.1007/978-3-642-27515-9_2

Valsalakumar N, Ray JG, Potty VP (2007) Arbuscular mycorrhizal fungi associated with green gram in south India. Agron J 99:1260–1264

van der Heijden MGA, Streitwolf-Engel R, Riedl R, Siegrist S, Neudecker A, Ineichen K, Boller T, Wiemken A, Sanders IR (2006) The mycorrhizal contribution to plant productivity, plant nutrition and soil structure in experimental grassland. New Phytol 172:739–752

van der Heijden MGA, Rinaudo V, Verbruggen E, Scherrer C, Bàrberi P, Giovannetti M (2008) The significance of mycorrhizal fungi for crop productivity and ecosystem sustainability in organic farming systems. In: 16th IFOAM organic world congress, Modena

Van Elsas JD, Heijnen CE (1990) Methods for the introduction of bacteria into soil: a review. Biol Fertil Soils 10:127–133

Vankessel C, Singleton PW, Hoben HJ (1985) Enhanced N-transfer from a soybean to maize by vesicular arbuscular mycorrhizal (VAM) fungi. Plant Physiol 79:562–563

Varma D, Meena RS (2016) Mungbean yield and nutrient uptake performance in response of NPK and lime levels under acid soil in Vindhyan region, India. J App Nat Sci 8(2):860–863

Varma D, Meena RS, Kumar S (2017) Response of mungbean to fertility and lime levels under soil acidity in an alley cropping system in Vindhyan Region, India. Int J Chem Stud 5(2):384–389

Vejan P, Abdullah R, Khadiran T, Ismail S, Boyce AN (2016) Review role of plant growth promoting Rhizobacteria in agricultural sustainability—a review. Molecules 21:573. https://doi.org/10.3390/molecules21050573

Velineni S, Brahmaprakash GP (2011) Survival and phosphate solubilizing ability of *Bacillus megaterium* in liquid inoculants under high temperature and desiccation stress. J Agric Sci Technol 13:795–802

Venkataraman GS (1981) Blue green algae for rice production—a manual for its production'. F.A.O. Soil Bulletin. 46, p 102

Verma JP, Meena VS, Kumar A, Meena RS (2015a) Issues and challenges about sustainable agriculture production for management of natural resources to sustain soil fertility and health: a book review. J Clean Prod 107:793–794

Verma JP, Jaiswal DK, Meena VS, Meena RS (2015b) Current need of organic farming for enhancing sustainable agriculture. J Clean Prod 102:545–547

Vessey JK (2003) Plant growth promoting rhizobacteria as biofertilizers. Plant Soil 255:571–586

Vincent JM, Thompson JA, Donovan KO (1962) Death of root nodule bacteria on drying. Aust J Agric Res 13:258

Vithal N (2004) Development of liquid inoculant formulations for *Bradyrhizobium* sp.: (*Arachis*), *Azospirillum lipoferum* and *Azotobacter chroococcum*. Ph.D. thesis, University of Agricultural Sciences, Bangalore

Walter JF, Paau AS (1993) Microbial inoculant production and formulation. In: Metting FB Jr (ed) Soil microbial ecology. Marcel Dekker, New York

Wani SP, Lee KK (1991) Role of biofertilizers in upland crop production. In: Tandon HLS (ed) Fertilizers of organic manures, recycle waste and biofertilizers. Fertilizer Development and Consultation Organization, New Delhi

Wani SP, Lee KK (2002) Biofertilizers for sustaining cereal crop production. In: Kannaiyan S (ed) Biotechnology of biofertilizers. Narosa publishing House, New Delhi

Wani SP, Rupela OP, Lee KK (1995) Sustainable agriculture in the semi-arid tropics through biological nitrogen fixation in grain legumes. Plant Soil 174:29–49

Weber E, George E, Beck DP, Saxena MC, Marschner H (1992) Vesicular-arbuscular mycorrhiza and phosphorus uptake of chickpea grown in Northern Syria. Exp Agric 28(433):442

Xavier LJC, Germida JJ (2002) Response of lentil under controlled conditions to co-inoculation with arbuscular mycorrhizal fungi and rhizobia varying in efficacy. Soil Biol Biochem 34:181–188

Xavier IJ, Holloway G, Leggett M (2004) Development of rhizobial inoculant formulations. In: Proceedings of the great plains inoculant forum. Plant Management Network, Saskatoon, Saskatchewan

Yadav GS, Babu S, Meena RS, Debnath C, Saha P, Debbaram C, Datta M (2017) Effects of godawariphosgold and single supper phosphate on groundnut (*Arachis hypogaea*) productivity, phosphorus uptake, phosphorus use efficiency and economics Indian J Agric Sci 87(9):1165–1169

Youssef MMA, Eissa MFM (2014) Biofertilizers and their role in management of plant parasitic nematodes – a review. J Biotechnol Pharm Res 5:1–6

Zaidi A, Khan MS, Amil M (2003) Interactive effect of rhizotrophic microorganisms on yield and nutrient uptake of chickpea (*Cicer arietinum* L.). Eur J Agron 19:15–21

Zarei M, Saleh-Rastin N, Alikhani HA, Aliasgharzadeh N (2006) Responses of lentil to co-inoculation with phosphate-solubilizing rhizobial strains and arbuscular mycorrhizal fungi. J Plant Nutr 29:1509–1522

Role of Soil Phosphorus on Legume Production

15

Tarik Mitran, Ram Swaroop Meena, Rattan Lal,
Jayanta Layek, Sandeep Kumar, and Rahul Datta

Contents

T. Mitran (✉)
Soil and Land Resources Assessment division, NRSC, ISRO, Hyderabad, Telangana, India

Carbon Management and Sequestration Centre, The Ohio State University, Columbus, OH, USA

R. S. Meena
Department of Agronomy, Institute of Agricultural Sciences (BHU), Varanasi, UP, India

R. Lal
Carbon Management and Sequestration Center, SENR/FAES, The Ohio State University, Columbus, OH, USA

J. Layek
Division of Crop Production, ICAR Research Complex for NEH Region, Umiam, Meghalaya, India

Carbon Management and Sequestration Centre, The Ohio State University, Columbus, OH, USA

S. Kumar
Department of Agronomy, CCS Haryana Agricultural University, Hisar, Haryana, India

R. Datta
Department of Geology and Pedology, Mendel University, Brno, Czech Republic

R. S. Meena et al. (eds.), *Legumes for Soil Health and Sustainable Management*,
https://doi.org/10.1007/978-981-13-0253-4_15

Abstract

Legumes play a significant role in sustainable agriculture through their ability to improve soil fertility and health. Legumes, with a mutual symbiotic relationship with some bacteria in soil, can improve nitrogen (N) amount through biological N-fixation (BNF). But to maximize such functions, legumes need more phosphorus (P) as it is required for energy transformation in nodules. Besides, P also plays a significant role to root development, nutrient uptake, and growth of legume crops. But most of the agricultural soils have inadequate amounts of P to support efficient BNF as it exists in stable chemical compounds which are least available to plants. The deficiency of P causes significant yield reduction in leguminous crops. The mineral P sources are nonrenewable, unlike N. So there is a need to enhance P use efficiency (PUE) for better legume productivity and soil sustainability. Improving the PUE of applied fertilizer requires enhanced P acquisition from the soils by crops for growth and development. It is necessary to better exploit soil P resources through increasing labile soil P using leguminous crops in a rotation cycle. Moreover, incorporation of legumes in cropping system with better P management under P-deficient conditions could be a promising tool for improving legume productivity. Endowed with inherent potential PUE, deep root system, root exudate-mediated P-solubilization, and nutrient-rich residues, legumes can improve soil fertility and enhance the soil profile and efficient nutrient cycling. The data obtained from various research studies show that agriculturally important legumes can fix 40–60 million metric tons of N annually. In view of this importance of P, this chapter emphasizes on the PUE and its role in legume production for food security programs, soil sustainability, and management.

Keywords

Biological nitrogen fixation · Legume · Nodulation · Phosphorus

Abbreviations

ADP	Adenosine diphosphate
ATP	Adenosine triphosphate
BNF	Biological nitrogen fixation
INM	Integrated nutrient management
MAP	Monoammonium phosphate
N	Nitrogen
OM	Organic matter
P	Phosphorus
PSB	Phosphorus solubilizing bacteria
PUE	Phosphorus use efficiency
SIFS	Sustainable integrated farming systems
SOM	Soil organic matter
SOP	Soluble organic phosphorus

15.1 Introduction

Legume crops is a major part of sustainable integrated farming systems (SIFS) as they fix atmospheric nitrogen (N) (Korir et al. 2017; Suzaki et al. 2015). The practice of including legumes in cropping system plays a key role to increase soil fertility through symbiotic N-fixation. Legumes induce N-fixing bacteria in some genera, viz., *Rhizobium*, *Sinorhizobium*, *Mesorhizobium*, and *Bradyrhizobium* (Berg 2009; Fenchel 2011; Meena et al. 2017, 2018). Cultivation of leguminous crops can be an alternative source of nutrients as it is a renewable and eco-friendly source of N (Oldroyd and Dixon 2014). But some biotic and abiotic factors disturb the symbiotic relationship between legumes and bacteria with negative effects on its productivity (Udvardi and Poole 2013) and stressful events such as drought, low and high pH, salinity, extreme temperatures, heavy metal problems (Zahran 1999; Dimkpa et al. 2009; Xie et al. 2009; Meena et al. 2015b). Among the nutrients, deficiency of phosphorus (P) in soil has an adverse impact on legume production as it is required for energy transformation in nodules and enhanced N-fixation (Rotaru and Sinclair 2009; Udvardi and Poole 2013; Yadav et al. 2017). The P is a primary nutrient essential for plant growth and development and important for regulation of various enzymatic activities and constituent for energy transformation (Schulze et al. 2006). Some molecules which contain P include nucleic acids, proteins, lipids, sugars, and adenylate and are required for the functioning of plant cells (Zhang et al. 2014). The P also plays a significant role in many metabolic processes including energy generation, respiration, membrane synthesis and its integrity, nucleic acid synthesis, photosynthesis, activation or inactivation of enzymes, signaling, and carbohydrate metabolism (Vance et al. 2003; Zhang et al. 2014). Therefore, P- deficient soil and low availability impose major restrictions on the vegetative and reproductive growth development of crop (Vance et al. 2003; Zhang et al. 2014). The P constraint directly decreases photosynthesis through its negative effects on vegetative crop growth of leaf area development and photosynthetic ability per unit leaf area (Vance et al. 2003; Sulieman et al. 2013). Likewise, inadequate supply of P can also affect carbon (C) absorption and distribution between plant shoots and its underground parts (Zhang et al. 2014). The P also plays a crucial role in the development of the symbiotic relationship between legumes and bacteria as a certain amount of P is required to carry out biological nitrogen fixation (BNF) (Oliveira et al. 2002; Rotaru and Sinclair 2009). There is considerable evidence that nodulated legumes require more P than nonsymbiotic plants grown solely on a mineral N source (Rotaru and Sinclair 2009; Sulieman and Schulze 2010).

A large amount of P is required for metabolic pathways of energy transfer that takes place during nodule functioning (Hernandez et al. 2009; Cabeza et al. 2014a, b). But most of the agricultural soils have inadequate amounts of P to support efficient BNF (Brown et al. 2013). The inadequacy of P in soil is mainly due to its retention as adsorbed P on the surface of soil particles and associated with amorphous aluminum (Al) and iron (Fe) oxides (Mitran and Mani 2017). About 90% of the inorganic P fertilizers are used in agriculture crop production produced from high-grade rock phosphates which expected to be depleted shortly within 30–50 years (Abrol and Palaniappan 1988; Cordell and Drangert 2009). So there will be possibilities of less vegetative growth and production of legumes as P availability expected to decrease shortly as the growth of the N-fixing legumes severely affected under P-deficient condition due to poor nodule functioning (Sulieman and Tran 2015; Dhakal et al. 2016). So there is a need to improve P resources to better legume crop productivity and soil sustainability through increasing PUE in legumes. There are some adaptive strategies which can also help to conserve the supply of P under the deficient condition and enhance N-fixation. The objective of this chapter is to evaluate the potential role of P in legume productivity as well as pointing out some adaptive strategies to improve PUE in the deficient soil and enhance BNF and productivity of legumes.

15.2 Importance of Phosphorus in Legumes

The P is a vital component of adenosine diphosphate (ADP) and adenosine triphosphate (ATP) the "energy unit" (Cabeza et al. 2014a, b; Nesme et al. 2014). These are high-energy phosphate compounds that control most processes in legume crops including respiration, photosynthesis, nucleic acid synthesis, and protein and plant cell formation through nutrient transport (Sawyer 1947; Nesme et al. 2014; Meena et al. 2014). ATP formed during photosynthesis has P in its structure and processes from the beginning of seedling growth to the formation of grain and maturity (Nesme et al. 2014). The specific growth factors that have been associated with P in legume crops are the following (Fig.15.1):

- It's essential for commercial seed productions.
- It promotes the root growth of leguminous crops.
- It helps to early maturity in legumes.

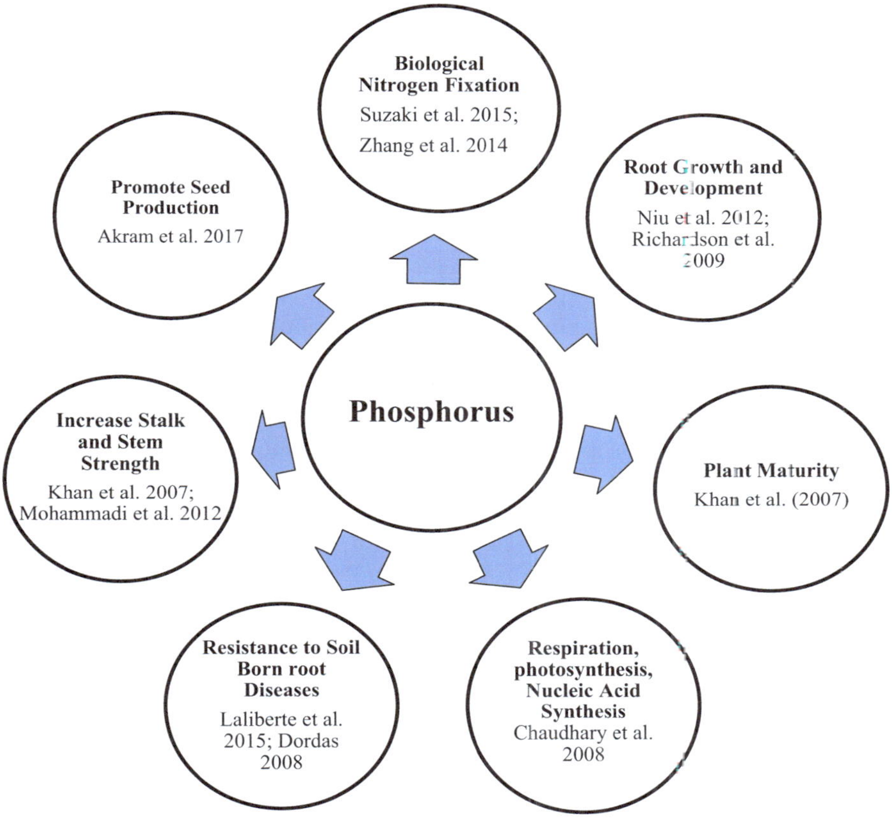

Fig. 15.1 Role of P in legumes

- It enhances the stalk strength in vegetative stage of legumes.
- It promotes the resistance to soil born root rot diseases.
- It stimulates root development in legumes.

15.3 Impact of Phosphorus Deficiency in Legumes

The BNF takes place in root nodules which are the outgrowths induced by N-fixing rhizobial bacteria (Fenchel 2011). However, such symbiotic relationship is dramatically affected by various biotic and abiotic factors (Schulze 2004) and stressful events such as drought, low and high pH, salinity, extreme temperatures, heavy metal problems.

Low P availability is affecting the legume production in most of the soils (Lopez-Arredondo et al. 2014; Schulze et al. 2006). The supply and availability of P are very important as it's a major component for N transformation and regulation of enzymatic activities to enhance BNF (Vance et al. 2003; Zhang et al. 2014). The P plays key roles in metabolic processes related to the aboveground organs, glycolysis, including energy generation, nucleic acid synthesis, respiration, and photosynthesis of legume crops (Chaudhary et al. 2008). The limited availability of P in soil leads to poor plant growth and development of legume crops. P deficiency has some negative effects on BNF, nodule formation, and photosynthetic ability in leaf and hence reduces photosynthesis (Sulieman and Schulze 2010; Sulieman et al. 2013; Yadav et al. 2017). The legume crops have more demand for P for optimal N-fixation compared to non-modulating plants like cereals because of P having a crucial role in nodule energetic transformations (Roatru 2009). A number of researchers observed a significant correlation between P concentration in nodule and N-fixation (Schulze et al. 2006; Rotaru and Sinclair 2009; Cabeza et al. 2014a, b). The metabolic pathways such as N-fixation that occur in bacteroids, as well as the ammonium assimilation into amino acids and ureides that occur in the plant cell fraction of nodules, require a large amount of P in energy transfer during nodule functioning (Sulieman and Tran 2015). In the absence of optimum supply of P, the growth of the legumes severely retards, and nodules are not sufficient to support the requirements for plant growth and development (Hernandez et al. 2009; Sulieman and Tran 2015). Studies revealed that up to 20–25% of total plant P was estimated to be allocated to nodule fraction (Jebara et al. 2005; Kouas et al. 2005). Tang et al. (2001) observed that under the P-deficient situation, even much higher P is preferentially partitioned to the nodules for maintaining N-fixation. The efficient P allocation and proper usage of available P in nodules during P limitations are very much essential for the optimal symbiotic interaction between the host plant and its rhizobial partner (Kouas et al. 2005; Al-Niemi et al. 1997; Meena et al. 2017). Hence, the P allocation rate may play an important role in the determination of the symbiotic efficiency as well as the degree of legume adaptability under deficient nutritional conditions (Sulieman and Tran 2015).

15.4 Phosphorus Cycle in Legume-Cultivated Soil

P can be applied to the soil in the form of manures, fertilizers, plant residues, and agricultural wastes, municipal and industrial by-products, etc. (Fig. 15.2). The native sources of P in soil are primary P minerals (apatite) and secondary clay minerals, i.e., calcium (Ca), Fe, Al-phosphates, which also play a significant role in maintaining the buildup of available P in soil through dissolution and desorption process (Mitran and Mani 2017). Within the soil, organic forms of phosphate such as living soil biomass, soil organic matter (SOM), and soluble organic P (SOP) can be made available to plants by bacteria that break down organic matter (OM) to inorganic forms of P; this process is known as mineralization (Meena et al. 2018). Processed plant and animal products, such as manure or compost, have been reported to have lower P use efficiency than that of water-soluble P mainly applied in the form of fertilizers (McLaughlin and Alston 1986; Nachimuthu et al. 2009; Oberson et al. 2010). Soil solution P ($H_2PO_4^-$ and HPO_4^{-2}) can be immobilized to

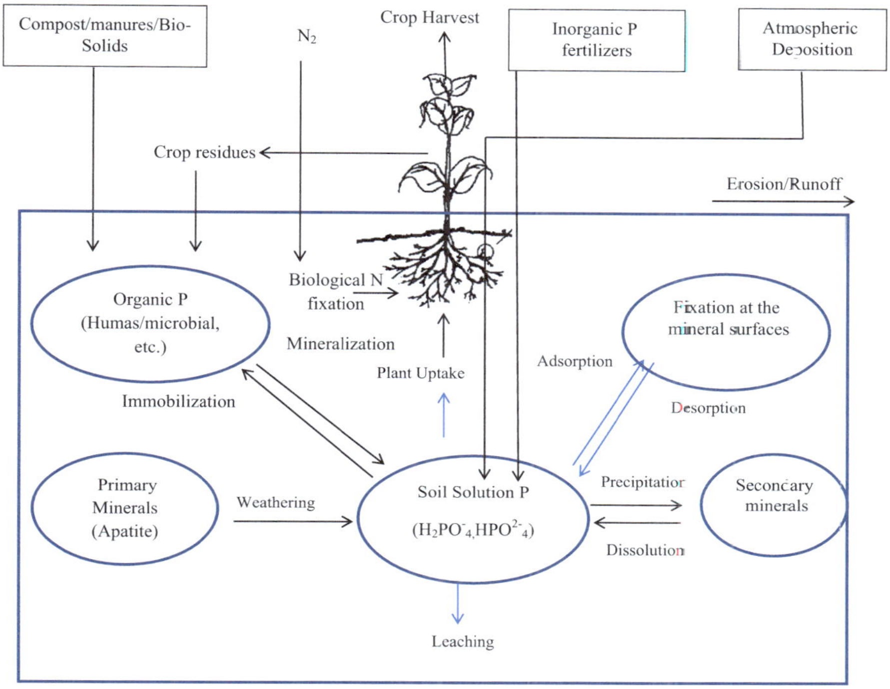

Fig. 15.2 Phosphorus cycle in legume-cultivated soils

organic P or adsorbed on the surface of soil particles and associated with amorphous Al and Fe oxides and become unavailable to plants (Ohel et al. 2004). The inherent soil properties and climate condition also affect the crop growth and response of crops to applied P fertilizers. Climatic parameters such as rainfall, temperature, etc. and soil attribute like soil temperature, aeration, salinity, etc. also affect the rate of P mineralization (NRCS, USDA). The long-term application of P inputs (inorganic P fertilizer, manure, compost) have effects on an available P due to release and erosion losses resulting eutrophication in water bodies and low land agriculture (Ulen et al. 2007; Meena et al. 2015c). Although P doesn't readily leach out from the root zones; the potential for P loss is mainly associated with erosion and runoff (Farkas et al. 2013). But integrated nutrient management (INM) in legume field has the potential to increase PUE and decrease soil P losses and efficiently uptake by the crops (Ali et al. 2002; Mitran and Mani 2017; Dhakal et al. 2016). At the same time, it should be aimed at replenishing SOM content, optimizing soil biological activity, and minimizing erosion and water runoff that support to increase PUE (Schroder et al. 2010; Spiess 2011). There are several mechanisms by which legumes can adapt to low P availability such as by activating high-affinity orthophosphate ion transporters for taking up P or by releasing organic acids which solubilize P bound to Ca and by releasing phosphatase enzymes to hydrolyze organic P compounds

(George et al. 2011; Richardson et al. 2011). The legume crops are colonized by phosphorus solubilizing bacteria (PSB) and able to access the P in plant available from within rhizospheric zone (Morel and Plenchette 1994; Meena et al. 2018). The enhanced uptake of P promotes biological nitrogen fixation and enriches N content in the soil which in turn influences growth and yield of legume crops.

15.5 Sources of Available Phosphorus for Leguminous Crops

A number of phosphatic fertilizers (Table 15.1) are available based on their solubility (Ghosal and Chakraborty 2012). The available phosphate can be defined by their solubility either in water or in neutral or alkaline ammonium citrate (Ghosal and Chakraborty 2012). It varies from country to country; some are using water to extract available P from fertilizer or by dissolving it in citrate or both. These definitions are not always adequate for evaluation of fertilizer availability for alkaline and calcareous soils. In calcareous soil, where pH is in the higher range, water solubility of P is hindered (Leytem and Mikkelsen 2005). Some of the highly water-soluble phosphate fertilizers are monocalcium phosphates, phosphoric acids, ammonium ortho- and polyphosphates, etc., whereas calcium metaphosphates, di- and tricalcium phosphates, and basic slag are not soluble in water but are citrate soluble (MacKay et al. 1990; Yadav et al. 2017). Apatites are major components of source rock phosphate that are insoluble even in ammonium citrate (Chien et al. 2011). Phosphatic fertilizers are either ordinary superphosphate (approximately 16% P_20_5) or concentrated superphosphate (43–46% P_20_5 approximately); both are predominantly monocalcium phosphates (Ca $[H_2P0_4]_2$) with relatively small amounts of iron and aluminum phosphates and dicalcium phosphate ($CaHP0_4$). Orthophosphoric acid as a phosphate (55% P_20_5) fertilizer is very effective in calcareous and alkaline soils where the Ca content is large enough to prevent undesirable acidification. The

Table 15.1 Sources of phosphate fertilizer

P fertilizer	P_20_5%
Water soluble	
Superphosphate (ordinary)	16–20
Superphosphate (concentrated)	40–45
Monoammonium phosphate	61
Diammonium phosphate	46
Insoluble	
Apatite (rock phosphate)	20–30
Bone meal	18–20
Citrate soluble	
Dicalcium phosphate	35–40
Basic slag	3–5

Data sources: Ghosal and Chakraborty (2012) and Chien et al. (2011)

solubility of ammonium phosphate fertilizers is higher than superphosphate fertilizers. The N and P content of fertilizer grade monoammonium phosphate (MAP) and diammonium phosphate (DAP) is approximately 12% and 18% N and 61% and 46% P_2O_5, respectively. These fertilizers are industrially attractive having a high nutrient content, the low tendency for caking, and low hygroscopicity. Whereas the nitric phosphate fertilizers are highly hygroscopic and citrate soluble which contains 4–13% P and 14–20% N. The nitric phosphate fertilizers are effective in neutral, alkaline, and calcareous soils as a P source to plants is a function of the ratio of water- to citrate-soluble phosphate. The nitric phosphates with a low water solubility are considered unsuitable in calcareous, neutral, and alkaline soils (Venkateswarlu et al. 1970; Sharma and Singh 1976; Bijay et al. 1976).

15.6 Phosphorus Use Efficiency in Legumes

The PUE is low in agriculture soils. When P is applied to the soil through a source of fertilizer or organic manure, it undergoes several biochemical reactions which remove phosphate ions from the soil solution (Kruse et al. 2015). It is measured that only 15–30% of applied fertilizer P is taken up by crops in the year of its application (Swarup 2002; Syers et al. 2008). However, the remaining 70–90% becomes part of the soil P pool, which is fixed but subsequently released to the crop over the following months and years (Roberts and Johnston 2015). Improving the PUE for growth in legume crops requires enhanced P acquisition from the soil and enhanced use of P in processes that lead to faster growth and a greater allocation of biomass to the harvestable parts (Kruse et al. 2015). In biomass calculations, measurements are often restricted to the aboveground portion of plant parts in leguminous crops. The PUE is the amount of total biomass produced per unit of P uptake (Hammond et al. 2009; Varma and Meena et al. 2016). Intraspecies and large genotypic differences for PUE are well known for different legumes such as cowpea (*Vignaunguiculata* L.; Sanginga et al. 2000), soybean (*Glycine max* L.; Furlani et al. 2002; Jemo et al. 2006), faba bean (*Vicia faba* L.; Daoui et al. 2012), and common bean (*Phaseolus vulgaris* L.; Vadez et al. 1999).

15.7 Role of Phosphorus in Legume Production

15.7.1 Growth, Root Development, and Nutrient Uptake in Legumes

Continuous cultivation of crops or following mono-cropping sequence without field fallowing shows a severe deficiency of most of the major and micro nutrients especially N, P, and zinc (Abbasi et al. 2008). The major nutrient demand for N in a deficient soil is normally achieved by the use of chemical fertilizers. However, the high cost of mineral N fertilizers and their unavailability at the time of requirement are the two major constraints responsible for low fertilizer N inputs. This

Table 15.2 Potential roles of P on N-fixation in legumes

Sl. no.	Role	References
1	Increasing top and root growth of legume plants	Zahran (1999) and Zafar et al. (2011)
2	Decreasing time needed for active nodule development	Tang et al. (2001)
3	Increasing the size and number of nodules	Hayat et al. (2008), Korir et al. (2017) and Kasturikrishna and Ahlawat (1999)
4	Increasing the amount of N assimilated nodules per unit weight	Sulieman et al. (2013) and Schulze et al. (2006)
5	Act as ingredients for *Rhizobium* bacteria to convert atmospheric N to ammonium	Berg (2009), Fenchel (2011) and Suzaki et al. (2015)
6	Promotes translocation of photosynthate from leaves to root and the movement of N-containing compound from nodules to another plant part	Zahran (1999) and Vance et al. (2003)
7	Controlling key enzyme reactions and regulate metabolic pathways	Rotaru and Sinclair (2009), Zhang et al. (2014) and Hernandez et al. (2007)

emphasizes the importance of developing an alternative means to meet the demand of nutrients (especially N and P) in plants through the use of beneficial bacteria in the ecosystem that is sustainable ergonomically, environmentally friendly, and affordable (Souza et al. 2015; Meena et al. 2016). As most of the nutrients are poorly available or may deficient, the efficient utilization of such from the soil by root is a major concern (Buerkert et al. 2001). The rate of root growth, an extension of root hairs, and the plasticity of root architecture are very much important for effective exploration of soil and interception of nutrients (Richardson et al. 2009). The recent studies indicated that P enhanced root system which provides greater root-soil contact and eventually higher uptake of P and other important and low mobility nutrients and absorption of higher concentration of mineral nutrients (Zafar et al. 2011) (Table 15.2). Almost all the legumes required P in relatively large amounts for growth and have been reported to promote leaf area, biomass, yield, nodule number, and nodule mass (Kasturikrishna and Ahlawat 1999). P supplement in legumes has great potential for promoting growth and higher yield, increases nodule number, as well as enhances symbiotic establishment for increased N-fixation (Ndakidemi et al. 2006). Several studies have reported the important role of P in growth and production of legumes in many tropical soils (Buerkert et al. 2001; Ohyama 2010; Kisinyo et al. 2012). The low availability of P in the bulk soil limits plant uptake. So there is a need to study how beneficial bacteria and P application can affect the uptake of nutrients in leguminous crops (Ndakidemi et al. 2011; Olivera et al. 2004) reported that the application of P significantly increased root and shoot P concentration (six- and fourfold, respectively) and nodule biomass (fourfold) in common bean (*Phaseolus vulgaris* L). Makoi et al. (2013) reported that *Rhizobium* inoculation significantly increases the uptake of P, potassium (K), magnesium (Mg), zinc (Zn), Fe, and Ca in different plant organs. Weisany et al. (2013) reported that the leguminous crops take up small amounts of nutrients relatively in the early season, but as they grow, the nutrient uptake increases. The *Bradyrhizobium* inoculants have been developed and are primarily used for supplying N to plants, and inoculation enhances the uptake of P, K, S, Mn, Fe Ca, Mg, B, Cu, Mo, and Zn in leguminous plants. A number of researchers have reported that the application of P fertilizers and inoculation with *Bradyrhizobium* significantly enhanced nodulation, shoot biomass, and grain yield and improve symbiotic nitrogen fixation of mash bean crop (Zaman et al. 2008; Vance 2001; Meena et al. 2017).

15.7.2 N-Fixation in Legumes

The atmospheric N gas concentration is ~80% and mostly unusable by living organisms. All the living organisms including plants, animals, and microorganism need N for the synthesis of proteins, nucleic acid, amino acid, and other necessary nitrogenous compound necessary for life (Ohyama 2010). The N deficiency in the soil causes death of plants, animals, and microorganisms as they are not able to use atmospheric N. BNF is the process that changes inert N to biologically useful NH_3 to the plants. This process is mediated in nature only by the bacteria. Legumes have a mutual symbiotic relationship with some N-fixing bacteria in the soil which can improve levels of N in the plant root zone (Ghosh et al. 2007; Peoples et al. 1989; Dhakal et al. 2016). In a natural ecosystem and a cropping system, legume can fix N in the soil in the range of 30–180 kg/ha (Frankow-Lindberg and Dahlin 2013). A common soil bacterium, *Rhizobium*, invades the root and multiplies within the cortex cells. During development of the bacteria, plant provides all the essential nutrients and energy for the bacteria (Fenchel 2011; Suzaki et al. 2015). After a couple of weeks of infection, small nodules are visible depending on legume species and germination conditions. Hayat et al. (2008) observed less than 100 nodules per plant in beans and several hundred nodules per plant in soybean and may have 1000 or more nodules on a well-developed peanuts plant.

Peanut nodules are white or gray in color and not able to fix atmospheric N usually. With the progress of growing period, the nodules become pink or reddish in color, indicating N-fixation has started. The pink or red color is caused by leg hemoglobin which contains both iron and molybdenum that controls oxygen flow to the bacteria. P is one of the important ingredients for *Rhizobium* to convert atmospheric N to ammonium (NH_4) which can be used by plants. P influences nodule development through its basic functions in plants as an energy source when 16 molecules of ATP are converted to ADP as each molecule of N is reduced to NH_3 (Berg 2009). The translocation of photosynthate from leaves to root and the movement of N-containing compound from nodules to other plant part are vital to an efficient symbiotic system (Zahran 1999; Meena et al. 2017). The number of researchers across the worlds has reported increased N-fixation in legumes by adding phosphate to the P-deficient soil (Ahlawat and Ali 1993; Bekere and Hailemariam 2012).

Hayat et al. (2008) observed 26% and 30% higher nodules in green gram (*Vigna radiata* (L) and black gram (*Vingna mungo* (L) crop, respectively, due to P fertilization over non-fertilized beans. The significant role of P in the symbiotic N-fixation process could be summarized by the following:

- Increase top and root growth of legume plants.
- Enhance early formation of active nodules for benefitting from hosting legumes.
- Increase the size and number of nodules.
- Improve the amount of N assimilated nodules per unit weight.
- Total amount of N increasing in the harvested portion of the host legume plants.
- Rhizobia bacteria in surrounding of soil, it helps in improving the root of density of crop plants.

The P supplements and *Rhizobium* inoculation is important to the soil fertility because of its potential for excellent N-fixation by increasing nodulation in legumes (Zhang et al. 2014; Bedoussac et al. 2015; Suzaki et al. 2015). The incorporation of legumes in cereal-based cropping system significantly enriches the N content in soil by BNF from the atmosphere and improved subsequence crop yield and productivity of soils (Liu et al. 2011; Zhang et al. 2014; Bedoussac et al. 2015; Ram and Meena 2014). Among all the essential nutrients required by plants, N is one of the most crucial elements, and deficiency of it causes significant yield reduction in the agricultural crop in all types of soil (Shah et al. 2003; Bedoussac et al. 2015). Hence, application of nitrogenous fertilizers is essential for optimum crop productivity for most of the crops. Due to continuous removal of N by intense cereal mono-cropping system, soil's capacity to supply the quantities of N required for optimum yield is declining rapidly (Layek et al. 2014a; Bedoussac et al. 2015). Continuous application of costly N fertilizers cannot subside the effect alone. Therefore, N fertilizer must be supplemented with rotations utilizing legumes break crops which can increase supply and availability of N through BNF (Layek et al. 2014b). Cultivation of various varieties or cultivars of grain legumes for BNF has become one of the most attractive strategies for the development of sustainable agricultural systems (Hardarson 1993; Shiferaw et al. 2004). The legume residues to subsequent crops can fix N through the decomposition and mineralization process (Hara 2001; Fatima et al. 2007; Shu-Jie et al. 2007; Dhakal et al. 2016). Because of relatively high N content and low C:N ratio, legume residues can supply more mineral N to the succeeding crops than that of cereal residues (Lynch et al. 2016). However, the N in leguminous crop residues is only partially available to plants during the first growing season (Wagger 1989; Stevenson and Kessel 1997) and gradually transferred from the labile pool to more stabilize C pools in soil (Hassink and Dalenberg 1996). Hence, legumes are playing a significant role for sustaining soil health by solubilizing insoluble P in soil, improving the soil physical environment, increasing soil microbial activity, and restoring organic matter (Ghosh et al. 2007; Layek et al. 2014a; Bedoussac et al. 2015).

15.7.3 Productivity of Legumes

The P is involved in various functions in growth and metabolism in legumes (Hernandez et al. 2007). It is frequently a major limiting nutrient for plant growth including legumes in most of the tropical soils. Thus, application of an optimum dose of P fertilizer has a significant influence on improving growth and productivity of legume crops. Along with synthetic fertilizers, PSB could also play an important role in increasing P availability by solubilizing the fixed P and supplying it to plants in a more available form (Khan et al. 2007). Srinivasarao et al. (2007) reported that among the kharif (rainy season) pulses, pigeon pea (*Cajanus cajan*) having dominant deep-rooted system performs extremely well under rainfed conditions and responds significantly to applied P in all type of soils with low available P status. They have also reported that application of 80 kg P_2O_5 ha^{-1} in pigeon pea significantly increased seed yield by 29.2% over control in Northern Indian soils, whereas in Central India, the soil produces maximum yield when applied with P at the rate of 90 kg P_2O_5 ha^{-1} which 54.6% higher was over control (Table 15.3). In a study, Singh and Ahlawat (2007) reported that application of 30 kg ha^{-1} P_2O_5 increased seed yield of pigeon pea approximately up to 1300 kg ha^{-1}, but *Rhizobium* inoculation with this P level increased the yield up to 1800 kg ha^{-1}. A similar result has also been reported by other researchers (Singh and Ahlawat 2007; Meena et al. 2014). Srinivasarao et al. (2007) reported that response of black gram to applied P at a different region of India varies from 60 to 90 kg P_2O_5 ha^{-1}. Dhillon and Vig (1996) suggested that if the available P status in the soil was low to medium, the response of green gram to applied P was found up to 40 kg P_2O_5 ha^{-1} while it was only 20 kg P_2O_5 ha^{-1} in soil testing high in available P. They have also found that the degree of response of lentil to applied P depended to a great extent on available P status of the soil. As per All India Coordinated Research Project (AICRP 1999) report, the response of chickpea to applied P was observed up to 60 kg P_2O_5 ha^{-1} (Table 15.3).

But the degree of response varied from region to region. Similarly, growth attributes of cowpea (*Vigna unguiculata*) such as plant height, leaf area, the number of branches, and the number of leaves were significantly increased by the application

Table. 15.3 Impact of phosphorus fertilization on yield of legumes

SL. no.	Legume crop	P dose (P_2O_5 kg ha^{-1})	Yield response	References
1	Pigeon pea (*Cajanus cajan*)	80–90	29.2–54.6% increment in yield over control	Srinivasarao et al. (2003)
2	Pigeon pea (*Cajanus cajan*)	30	Yield up to 1300 kg ha^{-1}	Singh and Ahlawat (2007)
3	Black gram (*Vigna mungo*)	60–90	Optimum yield	Srinivasarao et al. (2007)
4	Green gram (*Vigna radiata*)	40	Optimum yield	Dhillon and Vig (1996)
5	*Phaseolus* beans (*Phaseolus vulgaris*)	150	62% increase in seed yield	Ruschel et al. (1982)

of phosphorus fertilizer (Krasilnikoff et al. 2003; Nyoki et al. 2013). Ndakidemi and Dakora (2007) attributed this to the fact that phosphorus is required in large quantities in the shoot and root tips where metabolism is high, and cell division is rapid. P application has significantly improved yield and yield attributes of cowpea varieties, as it is utilized the applied P fertilizer judiciously in growth and development processes. This is in conformity with the findings of several workers (Okeleye and Okelana 2000; Natare and Bationo 2002; Ndakidemi and Dakora 2007; Singh et al. 2011) who also discovered a significant increase in yield of cowpea in response to phosphorus application. Application of phosphorus did not only increase cowpea yield but rather enhanced nodulation and phosphorus content of leaf and stem over the without application of P (Agboola and Obigbesan 2001). In Kenya, fertilizing Phaseolus beans with 150 kg/ha of P increased seed yield by 62% and increased nitrogen fixation from an average of 8–60 kg/ha. In an experiment with green gram in Pakistan, increasing the P fertilizer rate from 25 to 35 kg P/ha resulted in an increase in N fixation from 20 to 48 kg/ha (Ruschel et al. 1982).

15.8 Benefits of Phosphorus Supplementation in Legumes

Grain legumes are being popularized throughout the globe at an increasing level due to their vast use in different situations including human food, animal feed, as well as industrial demands (Zhang et al. 2011; Bedoussac et al. 2015). Considering the increasing needs for human consumption of plant proteins (pulses) and the economic constraints of applying fertilizer in legumes, there is a major role for grain legumes in cropping systems, especially in regions where affordability of fertilizer is difficult (Ndakidemi and Dakora 2003). Grain legumes such as soybean (*Glycine max*), cowpea, and common bean (*Phaseolus vulgaris*) have the potential to grow in different agroecological zones (Yagoub et al. 2012). Legumes are economically important crops used in a wide range of products like tortillas, chips, doughnuts, bread, spreads, and types of snacks or liquid form of yogurt and milk and thus play a significant role in the sustainability of agricultural systems (Das and Ghosh 2016; Meena et al. 2015d). BNF is becoming more attractive, environmentally friendly, and economically viable N inputs and acts as a substitute of inorganic fertilizers for resource-poor farmers (Bekere and Hailemariam 2012). Most tropical soils experience low N, which is the major constraint in crop production. Small-scale agriculture which is practiced in most sub-Saharan Africa covers the majority of the people, of which chemical fertilizers are unaffordable because of increasing prices in each year (Tadele 2017). Intercropping of cereals and legumes and crop rotation with legumes has found to be alternative sources and means of improving the fertility of the soil and boost crop productivity and farmer's income (Ndakidemi and Dakora 2003; Zhang et al. 2011; Layek et al. 2014a; Meena et al. 2015a). Several studies have shown that BNF incorporates residual N in the soil which adds OM nutrients for the next cropping season to cereal crops as well as other legumes (Zahran 1999; Lithourgidis et al. 2006, 2011). The BNF is therefore considered to have economic and ecological environmental benefits (Ndakidemi and Dakora 2003; Bedoussac et al. 2015). The nutrient supply in crop production is one of the key components to higher yields (Gehl et al. 2005). The per capita consumption of fertilizer in Tanzania is standing at 8 kg ha^{-1} as compared with 52 kg ha^{-1} for South Africa and Zimbabwe and 27 kg ha^{-1} for Malawi (Walter 2007; Gyaneshwar et al. 2002). The combined application of bacterial inoculants and P fertilizer to field legume plants significantly increased biomass production and grain yield as compared with the single use of N and P or rhizobial strains alone (Ndakidemi et al. 2006). From the economic analysis, the increase in grain yield with inoculation translated into a significantly higher marginal rate of return and profit for soybean and common bean farmers in Tanzania (Ndakidemi et al. 2006). In view of increasing price of fertilizers, it seems the cost of nutrients will be increasing in most cropping systems (Komareka et al. 2017). Evidently, legumes will remain the component of the farming system in remote areas comprised of poor farmers due to their capacity to fix N. Research efforts should be directed in assessing the optimum combinations between organic and inorganic fertilizers along with legume incorporation in cropping system that will offer immediate economic returns to the resource-poor farmers who cannot afford the full package of inorganic fertilizers (Chhonkar 2002; Yadav et al. 2013).

15.9 Adaptive Strategies to Overcome P Deficiency for Better N-Fixation and Legume Productivity

There is a need to develop some adaptive strategies which can help to conserve the supply of P under the deficient condition and enhance legume productivity (Veneklaas et al. 2012; Meena et al. 2015d). The adaptive response of nodule metabolism to P deficiency is crucial to improving symbiotic efficiency under P-deficient situations (Esfahani et al. 2014). There are a number of adaptive strategies (Fig. 15.3) such as P-homeostasis in nodule, increasing P acquisition, upgrading N-fixation per unit of nodule mass, and consumption per unit of nodule mass which compensate for the reduction in the number of nodules (Vance et al. 2003; Lopez-Arredondo et al. 2014; Sulieman and Tran 2015). However, the molecular mechanism is including maintenance of the P-homeostasis in nodules for rhizobia-legume symbiosis emerging as a main adaptive strategy for P-deficient soil (Sulieman and Tran 2015).

The main concept of such strategies is to conserve more P concentration in the nodule which can maintain a high rate of N-fixation (Graham 1992; Nogales et al. 2002; Dhakal et al. 2016). There are several ways to P stabilization in the symbiotic tissues such as including higher P allocation to nodules, the formation of a strong P sink in nodules, direct P acquisition via nodule surface and P remobilization from organic-P containing products (Sulieman and Tran 2015). Several studies have shown that symbiotic N-fixation could continue without any disturbance if total plant P is estimated to be allocated toward nodule up to 20% (Jebara et al. 2005; Tajini et al. 2009). Nodules represent a preferential strong sink for P incorporation during P starvation among the other plant parts (Le Roux et al. 2008; Hernandez

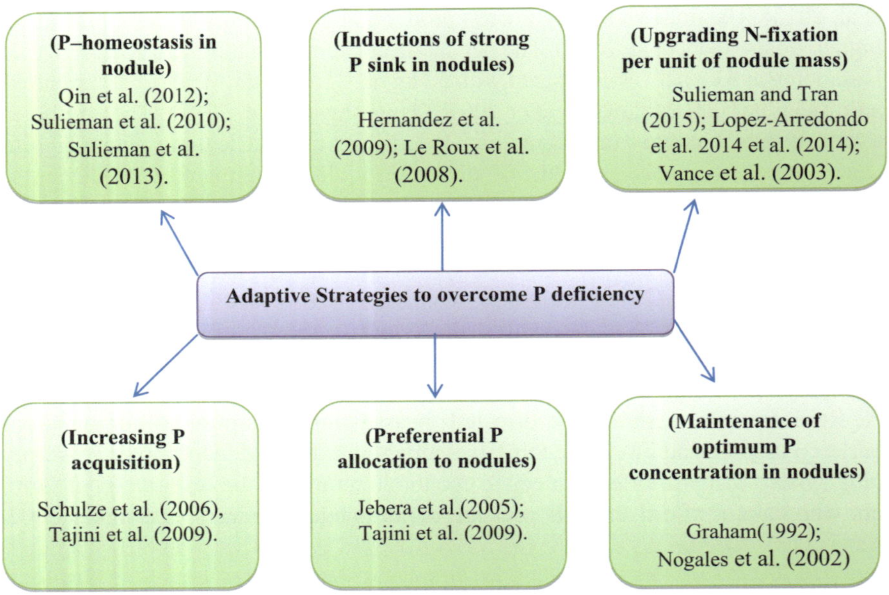

Fig. 15.3 Adaptive strategies to overcome P deficiency for better N-fixation and legume productivity

et al. 2009). Formation of cluster root and mycorrhizas also plays a key role in N-fixation by increasing root surface area and exudation of an organic acid and hence enhanced P acquisition during low P supply (Schulze et al. 2006; Tajini et al. 2009). Remobilization of organic P within the plant by encoding acid phosphatase (Qin et al. 2012; Zhang et al. 2014) is also an important biochemical and physiological adaptive strategy to P deficiency.

15.10 Conclusions

Legumes are becoming integral parts of the farming system because of its capabilities of atmospheric N-fixation through a mutualistic symbiotic relationship with a group of soil microflora. The BNF that occurs in bacteroids, as well as the ammonium assimilation into amino acids and ureides that occur in the plant cell fraction of nodules, requires a large amount of P in energy transfer during nodule functioning. Deficiency of P in soil at this crucial stage directly affects root growth, photosynthesis, sugar translocation, and many more functions which in turn directly or indirectly disturb N-fixation. So, therefore, P supplement and rhizobium inoculation is an important practice to enhance the soil N-fixation by increasing nodulation in legumes. But the mineral P sources are nonrenewable, and high-grade rock phosphates are expected to be depleted shortly. As the mineral P sources are nonrenewable, and solubility of P in soil is low and only 15–30% of applied fertilizer P is taken up by crops in the 1styear of its application. The efficiency of P fertilizer requires enhanced acquisition by plants from the soil which can be achieved by growing some legumes which are capable to grown in P deficient soils. Hence, developments of some adaptive strategies which can help to conserve the supply of P under the deficient condition and enhance fixation of N in legumes are needed for better productivity. Now a days, the molecular mechanism including maintenance of the P-homeostasis in nodules for rhizobia–legume symbiosis emerging as a main adaptive strategy to enhance P utilization in P-deficient soils.

15.11 Future Prospects

Worldwide production of grain legumes is increasing significantly due to their vast use in different situations including human food, animal feed, as well as industrial demands. Considering the increasing needs for human consumption of plant products and the economic constraints of applying fertilizer, there is a greater role for grain legumes in cropping systems, especially in regions where affordability of fertilizer is in question. Furthermore, in continuous removal of N by cereal monocropping systems, the capacity of the soil to supply sufficient quantities of N required for optimum yield is declining rapidly. Application of costly nitrogenous fertilizers continuously cannot subside the effect alone. So, therefore, N fertilizer must be supplemented with rotations utilizing legumes break crops which can increase supply and availability of N. BNF by various varieties or cultivars of grain legumes have become one of the most attractive strategies for the development of sustainable agricultural systems. Nevertheless, grain legumes have the ability to enhance the levels of SOM in cropping systems. Legumes can also play an important role in enhancing soil C sequestration. Besides N-fixation and high protein feed, legumes can also have considerable additional benefits such as positive impacts on biodiversity and soil quality. There is a great need for a strong focus on developing the role of legumes and their contribution to both a sustainable intensification of production and the livelihoods of smallholder farmers in many parts of the world.

Acknowledgment The first author is greatly thankful to Science and Engineering Research Board and Indo-US Science and Technology Forum of India for providing SERB INDO-US fellowship and Carbon Management and Sequestration Center, the Ohio State University, USA, for necessary help and support.

References

Abbasi M, Majeed KA, Sadiq A, Khan SR (2008) Application of Bradyrhizobium japonicum and phosphorus fertilization improved growth, yield, and nodulation of soybean in the sub-humid hilly region of Azad Jammu and Kashmir, Pakistan. Plant Prod Sci 11(3):368–376

Abrol IP, Palaniappan SP (1988) Green manure crops in irrigated and rainfed lowland rice-based cropping system in South Asia. In: Proceeding of symposium on sustainable agriculture: green manure in rice farming, 25–29 May 1987, Los Banos, Laguna, Philippines, Manila p 71–82

Agboola AA, Obigbesan GO (2001) Effect of different sources and levels of P on the performance and P uptake of Ife-Brown variety of cowpea. Ghana J Agric Sci 10(1):71–75

Ahlawat IPS, Ali M (1993) Fertilizer management in pulses. In: Fertilizer management in food crops. Fertilizer Development and Consultation Organization, India, p 114–138

Akram M, Hussain S, Hamid A, Majeed S, Chaudary SA (2017) Interactive effect of phosphorus and potassium on growth, yield, quality and seed production of chili (*Capsicum annuum* L.). J Hortic Sci 4:192

Ali M, Ganeshmurthy AN, Srinivasarao C (2002) Role of plant nutrient management in pulse production. Fert News 47(11):83–90

Al-Niemi TS, Kahn ML, McDermott TR (1997) Metabolism P in the bean-Rhizobium tropici symbiosis. Plant Physiol 113:1233–1242

Bedoussac L, Journet EP, Hauggaard-Nielsen H, Naudin C, Corre-Hellou G, Jensen ES, Prieur L, Justes E (2015) Ecological principles underlying the increase of productivity achieved by cereal-grain legume intercrops in organic farming: a review. Agron Sustain Dev 35:911–935

Bekere W, Hailemariam A (2012) Influences of inoculation methods and phosphorus levels on nitrogen fixation attributes and yield of soybean (*Glycine max* L.) at Haru, Western Ethiopia. Am J Plant Nutr Fert Technol 2(2):45–55

Berg G (2009) Plant–microbe interactions promoting plant growth and health: perspectives for controlled use of microorganisms in agriculture. Appl Microbiol Biotechnol 84:11–18

Bijay S, Mundal HS, Sekhon GS (1976) Evaluation of nitric phosphates differing in water solubility of their phosphorus fraction. J Agric Sci 87:325–330

Brown LK, George TS, Barrett GE, Hubbard SF, White PJ (2013) Interactions between root hair length and arbuscular mycorrhizal colonisation in phosphorus deficient barley (Hordeum vulgare). Plant Soil 372:195–205

Buerkert A, Bationo A, Piepho HP (2001) Efficient phosphorus application strategies for increased crop production in sub-Saharan West Africa. Field Crops Res 72(1):1–15

Cabeza RA, Liese R, Lingner A, Stieglitz I, Neumann J, Salinas-Riester G, Pommerenke C, Dittert K, Schulze J (2014a) RNA-seq transcriptome profiling reveals that *Medicago truncatula* nodules acclimate N2fixation before emerging P deficiency reaches the nodules. J Exp Bot 65:6035–6048

Cabeza R, Koester B, Liese R, Lingner A, Baumgarten V, Dirks J, Schulze J (2014b) An RNA sequencing transcriptome analysis reveals novel insights into molecular aspects of the nitrate impact on the nodule activity of Medicago truncatula. Plant Physiol 164(1):400–411

Chaudhary MI, Adu-Gyamfi JJ, Saneoka H, Nguyen NT, Suwa R, Kanai S, El-Shemy HA, Lightfoot DA, Fujita K (2008) Effect of phosphorus deficiency on nutrient uptake nitrogen fixation and photosynthetic rate in mashbean, mungbean and soybean. Acta Physiol Plant 30:537–544

Chhonkar PK (2002) Organic farming myth and reality. In: Proceedings of the FAI seminar on fertilizer and agriculture meeting the challenges, held at New Delhi, India (December)

Chien SH, Prochnow LI, Tu S, Snyder CS (2011) Agronomic and environmental aspects of phosphate fertilizers varying in source and solubility: an update review. Nutr Cycl Agroecosyst 89:229–255

Cordell D, Drangert JO (2009) The story of phosphorus: global food security and food for thought. Glob Environ Chang 19:292–305. [18] ENV.B.1/ETU/2009/0025 Plant research International, Wageningen UR, report 357, 123

Daoui K, Karrou M, Mrabet R, Fatemi Z, Draye X, Ledent JF (2012) Genotypic variation of phosphorus use efficiency among Moroccan faba bean (*Vicia faba* major) under rainfed conditions. J Plant Nutr 35:34–48

Das PK, Ghosh A (2016) Approach for enriched nutrition through pulse consumption. [online]. Int J Bio Sci 3(2):95–99

De Souza PA, Ponette-González AG, De Mello WZ, Weathers KC, Santos IA (2015) Atmospheric organic and inorganic nitrogen inputs to coastal urban and montane Atlantic Forest sites in southeastern Brazil. Atmos Res 160:126–137

Dhakal Y, Meena RS, Kumar S (2016) Effect of INM on nodulation, yield, quality and available nutrient status in soil after harvest of green gram. Leg Res 39(4):590–594
Dhillon NS, Vig AC (1996) Response of lentil to P in relation to organic carbon and Olsen P in soil. J Indian Soc Soil Sci 44:433–436
Dimkpa C, Weinand T, Asch F (2009) Plant–rhizobacteria interactions alleviate abiotic stress conditions. Plant Cell Environ 32:1682–1694
Dordas CA (2008) Dry matter, nitrogen and phosphorus accumulation, partitioning and remobilization as affected by N and P fertilization and source–sink relations. Eur J Agron 30(2., February 2009):129–139
Esfahani MN, Sulieman S, Schulze J, Yamaguchi-Shinozaki K, Shinozaki K, Tran LS (2014) Approaches for enhancement of N2 fixation efficiency of chickpea (*Cicer arietinum* L.) under limiting nitrogen conditions. Plant Biotechnol J 12:387–397
Farkas C, Beldring S, Bechmann M, Deelstra J (2013) Soil erosion and phosphorus losses under variable land use as simulated by the INCA-P model. Soil Use Manag 29:124–137
Fatima Z, Zia M, Chaudhary MF (2007) Interactive effect of Rhizobium strains and p on soybean yield, nitrogen fixation and soil fertility. Pak J Bot 39(1):255–264
Fenchel T (2011) Bacterial ecology. In: Encyclopedia of life sciences. Wiley, Chichester. https://doi.org/10.1002/9780470015902.a0000339.pub3. Accessed 15 Sept 2011
Frankow-Lindberg BE, Dahlin AS (2013) N_2 fixation, N transfer, and yield in grassland communities including a deep-rooted legume or non-legume species. Plant Soil 370(1–2):567–581
Furlani MC, Furlani PR, Tanaka RT, Mascarenhas HAA, Delgado MDP (2002) Variability of soybean germplasm in relation to phosphorus uptake and use efficiency. Sci Agric 59:529–536
Gehl RJ, Schmitt JP, Maddux LD, Gordon BW (2005) Corn yield response to nitrogen rate and timing in sandy irrigated soils. Agron J 97:1230–1238
George TS, Fransson AM, Hammond JP, White PJ (2011) Phosphorus nutrition: rhizosphere processes, plant response and adaptation. In: Bünemann EK, Oberson A, Frossard E (eds) Phosphorus in action. Soil Biol 26:245–274
Ghosal PK, Chakraborty T (2012) Comparative solubility study of four Phosphatic fertilizers in different solvents and the effect of soil. Res Environ 2(4):175–179
Ghosh PK, Bandyopadhyay KK, Wanjari RH, Manna MC, Misra AK, Mohanty M, Subba RA (2007) Legume effect for enhancing productivity and nutrient use-efficiency in major cropping systems – an Indian perspective: a review. J Sustain Agric 30(1):59–86
Graham PH (1992) Stress tolerance in *Rhizobium* and *Bradyrhizobium* and nodulation under adverse soil conditions. Can J Microbiol 38:475–484
Gyaneshwar P, Kumar GN, Parekh LJ, Poole PS (2002) Role of soil microorganisms in improving P nutrition of plants. Plant Soil 245:83–93
Hammond JP, Broadley MR, White PJ, King GJ, Bowen HC, Hayden R, Meacham MC, Mead A, Overs T, Spracklen WP (2009) Shoot yield drives phosphorus use efficiency in *Brassica oleracea* and correlates with root architecture traits. J Exp Bot 60:1953–1968
Hara GW (2001) Nutritional constraints on root nodule bacteria affecting symbiotic nitrogen fixation: a review. Aust J Exp Agric 41:417–433
Hardarson G (1993) Methods for enhancing symbiotic nitrogen fixation. Plant Soil 152:1–17
Hassink J, Dalenberg JW (1996) Decomposition and transfer of plant residue 14C between size and density fraction in soil. Plant Soil 179:159–169
Hayat R, Ali S, Siddique MT, Chatha TH (2008) Biological nitrogen fixation of summer legumes and their residual effects on subsequent rainfed wheat yield. Pak J Bot 40(2):711–722
Hernandez G, Ramirez M, Valdes-Lopez O, Tesfaye M, Graham MA, Czechowski T, Schlereth A, Wandrey M, Erban A, Cheung F (2007) Phosphorus stress in common bean: root transcript and metabolic responses. Plant Physiol 144:752–767
Hernandez G, Valdes-Lopez RM, Goffard N, Weiller G, Aparicio-Fabre R, Fuentes SI, Erban A, Kopka J, Udvardi M, Vance CP (2009) Global changes in the transcript and metabolic profiles during symbiotic nitrogen fixation in phosphorus-stressed common bean plants. Plant Physiol 151:1221–1238

Jebara M, Aouani ME, Payre H, Drevon J (2005) Nodule conductance varied among common bean (Phaseolus vulgaris) genotypes under phosphorus deficiency. J Plant Physiol 162:309–315

Jemo M, Abaidoo R, Nolte C, Tchienkoua M, Sanginga N, Horst W (2006) Phosphorus benefits from grain-legume crops to subsequent maize grown on acid soils of southern Cameroon. Plant Soil 284:385–397

Kasturikrishna S, Ahlawat IPS (1999) Growth and yield response of pea (Pisumsativum L.) to moisture stress, phosphorus, sulphur and zinc fertilizers. Indian J Agron 44:588–596

Khan MS, Zaidi A, Wani PA (2007) Role of phosphate –solubilising micro-organisms in sustainable agriculture-a review. Agron Sustain Dev 27:29–43

Kisinyo PO, Gudu SO, Othieno CO, Okalebo JR, Opala PA, Maghanga JK, Agalo DW, Ng'etich WK, Kisinyo JA, Osiyo RJ, Nekesa AO, Makatiani ET, Odee DW, Ogola BO (2012) Effects of lime, phosphorus and rhizobia on Sesbania sesban performance in a Western Kenyan acid soil. Afric J Agric Res 7(18):2800–2809

Komareka AM, Drogueb S, Chenounec R, Hawkinsa J, Msangia S, Belhouchettecd H, Flichman G (2017) Agricultural household effects of fertilizer price changes for smallholder farmers in central Malawi. Agric Syst 154:168–178

Korir H, Mungai NW, Thuita M, Hamba Y, Masso C (2017) Co-inoculation effect of rhizobia and plant growth promoting Rhizobacteria on common bean growth in a low phosphorus soil. Front Plant Sci 8:141

Kouas S, Labidi N, Debez A, Abdelly C (2005) Effect of P on nodule formation and N fixation in bean. Agron Sustain Dev 25:389–339

Krasilnikoff G, Gahoonia T, Erik-Nelson N (2003) Variation in phosphorus uptake by genotypes of cowpea (Vigna unguiculata (L.) Walp) due to differences in root and root hair length and induced rhizosphere processes. Plant Soil 251:83–91

Kruse J, Abraham M, Amelung W, Baum C, Bol R, Kühn O, Lewandowski H, Niederberger J, Oelmann Y, Rüger C, Santner J, Siebers M, Siebers N, Spohn M, Vestergren J, Vogts A, Leinweber P (2015) Innovative methods in soil phosphorus research: a review. J Plant Nutr Soil Sci 178:43–88

Laliberté E, Lambers H, Burgess TI, Wright SJ (2015) Phosphorus limitation, soil-borne pathogens and the coexistence of plant species in hyperdiverse forests and shrub lands. New Phytol 206:507–521

Layek J, Shivakumar BG, Rana DS, Munda S, Lakshman K, Das A, Ramkrushna GI (2014a) Soybean–cereal intercropping systems as influenced by nitrogen nutrition. Agron J 106:1–14

Layek J, Shivakumar BG, Rana DS, Munda S, Lakshman K, Panwar AS, Das A, Ramkrushna GI (2014b) Performance of soybean (*Glycine max*) intercropped with different cereals under varying levels of nitrogen. Indian J Agric Sci 85:1571–1557

Le Roux MR, Kahn S, Valentine AJ (2008) Organic acid accumulation inhibits N_2-fixation in P-stressed lupin nodules. New Phytol 177:956–964

Leytem AB, Mikkelsen RL (2005) The nature of phosphorus in calcareous soils. Better Crops 89(2):11–13

Lithourgidis AS, Vasilakoglou IB, Dhima KV, Dordas CA, Yiakoulaki MD (2006) Forage yield and quality of common vetch mixtures with oat and triticale in two seeding ratios. Field Crops Res 99:106–113

Lithourgidis AS, Vlachostergios DN, Dordas CA, Damalas CA (2011) Dry matter yield, nitrogen content, and competition in pea–cereal intercropping systems. Eur J Agron 34:287–294

Liu Y, Wu L, Baddele JA, Watson CA (2011) Models of biological nitrogen fixation of legumes. A review. Agron Sustain Dev 31(155):172

López-Arredondo DL, Leyva-González MA, González-Morales SI, López-Bucio J, Herrera-Estrella L (2014) Phosphate nutrition: improving low-phosphate tolerance in crops. Annu Rev Plant Biol 65:95–123

Lynch MJ, Mulvaney MJ, Hodges SC, Thompson TL, Thomason WE (2016) Decomposition, nitrogen and carbon mineralization from food and cover crop residues in the central plateau of Haiti. Springerplus 5(1):973

Mackay AD, Caradus JR, Pritchard MW (1990) Variation in aluminium tolerance in white clover. Plant Soil 123:101–105
Makoi JHJR, Bambara S, Ndakidemi PA (2013) Rhizobium inoculation and the supply of molybdenum and lime affect the uptake of macroelements in common bean (*P. vulgaris* L.) plants. Am J Crop Sci 7(6):784–793
McLaughlin MJ, Alston AM (1986) The relative contribution of plant residues and fertiliser to the phosphorus nutrition of wheat in a pasture/cereal system. Aust J Soil Res 24:517–526
Meena RS, Yadav RS, Meena VS (2014) Response of groundnut (*Arachis hypogaea* L.) varieties to sowing dates and NP fertilizers under western dry zone of India. Bangladesh J Bot 43(2):169–173
Meena VS, Maurya BR, Meena RS (2015a) Residual impact of well-grow formulation and NPK on growth and yield of wheat (*Triticum aestivum* L.). Bangladesh J Bot 44(1):143–146
Meena RS, Meena VS, Meena SK, Verma JP (2015b) Towards the plant stress mitigate the agricultural productivity: a book review. J Clean Prod 102:552–553
Meena RS, Dhakal Y, Bohra JS, Singh SP, Singh MK, Sanodiya P (2015c) Influence of bioinorganic combinations on yield, quality and economics of mungbean. Am J Exp Agri 8(3):159–166
Meena RS, Yadav RS, Meena H, Kumar S, Meena YK, Singh A (2015d) Towards the current need to enhance legume productivity and soil sustainability worldwide: a book review. J Clean Prod 104:513–515
Meena RS, Bohra JS, Singh SP, Meena VS, Verma JP, Verma SK, Shiiag SK (2016) Towards the prime response of manure to enhance nutrient use efficiency and soil sustainability a current need: a book review. J Clean Prod 112:1258–1260
Meena RS, Meena PD, Yadav GS, Yadav SS (2017) Phosphate solubilizing microorganisms, principles and application of microphos technology. J Clean Prod 145:157–158
Meena RS, Vijayakumar V, Yadav GS, Mitran T (2018) Response and interaction of Bradyrhizobium japonicum and arbuscular mycorrhizal fungi in the soybean rhizosphere. Plant Growth Regul 84:207–223
Mitran T, Mani PK (2017) Effect of organic amendments on rice yield trend, phosphorus use efficiency, uptake, and apparent balance in soil under long-term rice-wheat rotation. J Plant Nutr 40(9):1312–1322
Mohammadi K, Sohrabi Y, Heidari G, Khalesro S, Majidi M (2012) Effective factors on biological nitrogen fixation. Afr Agric Res 7(12):1782–1788
Morel C, Plenchette C (1994) Is the isotopically exchangeable phosphate of a loamy soil the plant available P? Plant Soil 158:287–297
Nachimuthu G, Guppy C, Kristiansen P, Lockwood P (2009) Isotopic tracing of phosphorus uptake in corn from P-33 labelled legume residues and P-32 labelled fertilisers applied to a sandy loam soil. Plant Soil 314:303–310
Natare BR, Bationo A (2002) Effects of phosphorus on yield of cowpea cultivars intercropped with pearl millet on Psammentic Paleustalf in Niger. Fert Res 32:143–147
Ndakidemi PA, Dakora FD (2003) Legume seed flavonoids and nitrogenous metabolites as signals and protectants in early seedling development. Funct Plant Biol 30:729–745
Ndakidemi PA, Dakora FD (2007) Yield components of nodulated cowpea (Vignaunguiculata) and maize (*Zea mays* L) plants grown with exogenous phosphorus in different cropping systems. Aust J Exp Agric 47:587–590
Ndakidemi PA, Dakora FD, Nkonya EM, Ringo D, Mansoor H (2006) Yield and economic benefits of common bean (Phaseolus vulgaris) and soybean (*Glycine max*) inoculation in northern Tanzania. Aust J Exp Agric 46:571–577
Ndakidemi PA, Bambara S, Makoi JHJR (2011) Micronutrient uptake in common bean (*Phaseolus vulgaris* L) as affected by Rhizobium inoculation, and the supply of molybdenum and lime. Plant Omics J 4(1):40–52
Nesme T, Colomb B, Hinsinger P, Watson CA (2014) Soil phosphorus management in organic cropping systems: from current practices to avenues for a more efficient use of P resources. In: Organic farming, prototype for sustainable agricultures. Springer, Berlin, p 23–45

Niu YF, Chai RS, Jin GL, Wang H, Tang CX, Zhang YS (2012) Responses of root architecture development to low phosphorus availability: a review. Ann Bot 112:391–408
Nogales J, Campos R, Ben Abdelkhalek H, Olivares J, Lluch C, Sanjuan J (2002) *Rhizobium tropici* genes involved in free-living salt tolerance are required for the establishment of efficient nitrogen-fixing symbiosis with *Phaseolus vulgaris*. Mol Plant-Microbe Interact 15:225–232
Nyoki D, Patrick A, Ndakidemi R (2013) Economic benefits of Bradyrhizobium japonicas inoculation and phosphorus supplementation in cowpea (Vignaunguiculata (L.)) grown in northern Tanzania. Am J Res Comm 1(11):173–189
Olivera M, Tejera N, Iribarne C, Ocana A, Lluch C (2004) Growth, nitrogen fixation and ammonium assimilation in common bean (Phaseolus vulgaris): effect of phosphorus. Physiol Plant 121(3):498–505
Oberson A, Tagmann HU, Langmeier M, Dubois D, Mader P, Frossard E (2010) Fresh and residual phosphorus uptake by ryegrass from soils having different fertilization histories. Plant Soil 334:391–407
Ohel F, Frossard E, Fliessbach A, Dubois D, Oberson A (2004) Basal organic phosphorus mineralization in soils under different farming systems. Soil Biol Biochem 36:667–675
Ohyama T (2010) Nitrogen as a major essential element of plants. In: Ohyama T, Sueyoshi K (eds) Nitrogen assimilation in plants. Research Signpost, Kerala, pp 1–17
Okeleye KA, Okelana MAO (2000) Effect of phosphorus fertilizer on nodulation, growth, and yield of cowpea (*Vignaunguiculata*) varieties. Indian J Agric Sci 67(1):10–12
Oldroyd GE, Dixon R (2014) Biotechnological solutions to the nitrogen problem. Curr Opin Biotechnol 26:19–24
Oliveira ALM, Urquiaga S, Dobereiner J, Baldani JI (2002) The effect of inoculating endophytic N2-fixing bacteria on micropropagated sugarcane plants. Plant Soil 242:205–215
Peoples MB, Faizah AW, Rekasem B, Herridge DF (1989) Methods for evaluating nitrogen fixation by nodulated legumes in the field. ACIAR Monograph No. 11:22–45
Qin L, Zhao J, Tian J, Chen LY, Sun ZA, Guo YX, Lu X, Gu M, Xu GH, Liao H (2012) The high-affinity phosphate transporter *GmPT5* regulates phosphate transport to nodules and nodulation in soybean. Plant Physiol 159(4):1634–1643
Ram K, Meena RS (2014) Evaluation of pearl millet and mungbean intercropping systems in Arid Region of Rajasthan (India). Bangladesh J Bot 43(3):367–370
Richardson AE, Barea JM, McNeill AM, Prigent-Combaret C (2009) Acquisition of phosphorus and nitrogen in the rhizosphere and plant growth promotion by microorganism. Plant Soil 321:305–339
Richardson AE, Lynch JP, Ryan PR, Delhaize E, Smith FA, Smith SE, Harvey PR, Ryan MH, Veneklaas EJ, Lambers H, Oberson A, Culvenor RA, Simpson RJ (2011) Plant and microbial strategies to improve the phosphorus efficiency of agriculture. Plant Soil 349:121–156
Roberts TL, Johnston AE (2015) Phosphorus use efficiency and management in agriculture. Resour Conser Recycl 105:275–281
Rotaru V, Sinclair TR (2009) Interactive influence of phosphorus and iron on nitrogen fixation by soybean. Environ Exp Bot 66(1):94–99
Ruschel AP, Vose PB, Matsui E, Victoria RL, Saito SMT (1982) Field evaluation of N2-fixation and N-utilization by Phaseolus bean varieties determined by 1S/V isotope dilution. Plant Soil 65:397–407
Sanginga N, Lyasse O, Singh BB (2000) Phosphorus use efficiency and nitrogen balance of cowpea breeding lines in a low P soil of the derived savanna zone in West Africa. Plant Soil 220:119–128
Sawyer CN (1947) Fertilization of lakes by agricultural and urban drainage. New England Water Works. Assoc J 61:109–127
Schroder JJ, Cordell D, Smit AL, Rosemarin A (2010) Sustainable use of phosphorus. EU Tender
Schulze J (2004) How are nitrogen fixation rates regulated in legumes. J Plant Nutr Soil Sci 167:125–137
Schulze J, Temple G, Temple SJ, Beschow H, Vance CP (2006) Nitrogen fixation by white lupin under phosphorus deficiency. Ann Bot 98:731–740

Shah Z, Shah SH, Peoples MB, Schwenke GD, Hrridge D (2003) Crop residue and fertilizer N effects on nitrogen fixation and yields of legume-cereal rotations and soil organic fertility. Field Crops Res 83:1–11

Sharma PD, Singh TA (1976) Evaluation of nitric phosphates of varying water solubility and granule size as source of phosphorus for wheat. Mysore J Agric Sci 10:568–574

Shiferaw B, Bantilan MCS, Serraj R (2004) Harnessing the potential of BNF for poor farmers: technological policy and institutional constraints and research need. In: Serraj R (ed) Symbiotic nitrogen fixation: prospects for enhanced application in tropical agriculture. Oxford & IBH, New Delhi, p 3

Shu-Jie M, Yun-Fa Q, Xiao-Zeng H, An M (2007) Nodule formation and development in soybean (*Glycine max* L.) in response to phosphorus supply in solution culture. Pedosphere 17(1):36–43

Singh U, Ahlawat IPS (2007) Phosphorus management in pigeon pea (*Cajanuscajan*) wheat (*Triticumaestivum*) cropping system. Indian J Agron 52(1):21–26

Singh A, Baoule AL, Ahmed HG, Aliyu U, Sokoto MB (2011) Influence of phosphorus on the performance of cowpea (*Vigna unguiculata*) varieties in the sudan savannah of Nigeria. Agric Sci 2:313–317

Spiess E (2011) Nitrogen, phosphorus and potassium balances and cycles of Swiss agriculture from 1975 to 2008. Nutr Cycl Agroecosys 91:351–365

Srinivasarao C, Masood A, Ganeshamurthy AN, Singh KK (2003) Potassium requirements of pulse crops. Better Crops Int 17(1):8–11

Srinivasarao C, Ganeshamurtthy AN, Ali M, Singh RN (2007) Effect of phosphorus levels on zinc, iron, copper and manganese removal by chickpea genotypes in Typic Ustochrept. J Food Legumes 20:45–48

Stevenson FC, Kessel CV (1997) Nitrogen contribution of pea residue in a hummocky terrain. Soil Sci Soc Am J 61:494–503

Sulieman S, Schulze J (2010) Efficiency of nitrogen fixation of the model legume *Medicago truncatula* (Jemalong A17) is low compared to *Medicago sativa*. J Plant Physiol 167:683–692

Sulieman S, Tran LSP (2015) Phosphorus homeostasis in legume nodules as an adaptive strategy to phosphorus deficiency. Plant Sci 239:36–43

Sulieman S, Van Ha C, Schulze J (2013) Growth and nodulation of symbiotic *Medicago truncatula* at different levels of phosphorus availability. J Exp Bot 64:2701–2712

Suzaki T, Yoro E, Kawaguchi M (2015) Leguminous plants: inventors of root nodules to accommodate symbiotic bacteria. Int Rev Cell Mol Biol 316:111–158

Swarup A (2002) Lessons from long term fertilizer experiments in improving fertilizer use efficiency and crop yields. Fert News 47(12):59–73

Syers JK, Johnston AE, Curtin D (2008) Efficiency of soil and fertilizer phosphorus use—reconciling changing concepts of soil phosphorus behaviour with agronomic information. FAO Fertilizer and Plant Nutrition Bulletin 18. FAO, United Nations, Rome

Tadele Z (2017) Raising crop productivity in Africa through intensification. Agronomy. 2017 7(1):22. https://doi.org/10.3390/agronomy7010022

Tajini F, Suriyakup P, Vailhe H, Jansa J, Drevon JJ (2009) Assess suitability of hydroaeroponic culture to establish tripartite symbiosis between different amf species, beans, and rhizobia. BMC Plant Biol 9:73

Tang C, Hinsinger P, Drevon JJ, Jaillard B (2001) Phosphorus deficiency impairs early nodule functioning and enhances proton release in roots of *Medicago truncatula* L. Ann Bot 88:131–138

Udvardi M, Poole PS (2013) Transport and metabolism in legume-rhizobia symbioses. Annu Rev Plant Biol 64:781–805

Ulen B, Bechmann M, Folster J, Jarvie HP, Tunney H (2007) Agriculture as a phosphorus source for eutrophication in the north-west European countries, Norway, Sweden, United Kingdom and Ireland: a review. Soil Use Manag 23:5–15

Vadez V, Lasso JH, Beck DP, Drevon JJ (1999) Variability of N2-fixation in common bean (Phaseolus vulgaris L.) under P deficiency is related to P use efficiency: N2-fixation tolerance to P deficiency. Euphytica 106:231–242

Vance CP (2001) Symbiotic nitrogen fixation and phosphorus acquisition. Plant nutrition in a world of declining renewable resources. Plant Physiol 127:390–397
Vance CP, Uhde-Stone C, Allan DL (2003) Phosphorus acquisition and use: critical adaptations by plants for securing a nonrenewable resource. New Phytol 157:423–447
Varma D, Meena RS (2016) Mungbean yield and nutrient uptake performance in response of NPK and lime levels under acid soil in Vindhyan region, India. J App Nat Sci 8(2):860–863
Veneklaas EJ, Lambers H, Bragg J, Finnegan PM, Lovelock CE, Plaxton WC, Price CA, Scheible WR, Shane MW, White PJ, Raven JA (2012) Opportunities for improving phosphorus-use efficiency in crop plants. New Phytol 195:306–320
Venkateswarlu J, Reddy KS, Ramesam M (1970) Availability of phosphorus in fertilizers with different water soluble phosphates. J Indian Soc Soil Sci 18:303–306
Wagger MG (1989) Time of desiccation effects on plant composition and subsequent nitrogen release for several winter annual cover crops. Agron J 81:236–241
Walter D (2007) Tanzania: the challenge of moving from subsistence to profit. Business for Development. OECD publication for development
Weisany W, Raei Y, Allahverdipoor KH (2013) Role of some of mineral nutrients in biological nitrogen fixation. Bull Environ Pharmacol Life Sci 2(4):77–84
Xie X, Zhang H, Paré PW (2009) Sustained growth promotion in arabidopsis with long-term exposure to the beneficial soil bacterium *Bacillus subtilis* (GB03). Plant Signal Behav 4:948–953
Yadav A, Gupta R, Garg VK (2013) Organic manure production from cow dung and biogas plant slurry by vermicomposting under field conditions. Int J Recycl Org Waste Agric 2:21. https://doi.org/10.1186/2251-7715-2-21
Yadav GS, Babu S, Meena RS, Debnath C, Saha P, Debbaram C, Datta M (2017) Effects of godawariphosgold and single supper phosphate on groundnut (*Arachis hypogaea*) productivity, phosphorus uptake, phosphorus use efficiency and economics. Indian J Agri Sci 87(9):1165–1169
Yagoub SO, Ahmed WMA, Mariod AA (2012) Effect of urea, NPK and compost on growth and yield of soybean (*Glycine max* L) in semi-arid region of Sudan. Agron Int Sch Res Netw 2012:678124., 6 pages. https://doi.org/10.5402/2012/678124
Zafar M, Abbasi M, Rahim N, Khaliq A, Shaheen A, Jamil M, Shahid M (2011) Influence of integrated phosphorus supply and plant growth promoting Rhizobacteria on growth, nodulation, yield and nutrient uptake in Phaseolus vulgaris. Afr J Biotechnol 10(74):16793–16807
Zahran HH (1999) Rhizobium-legume symbiosis and nitrogen fixation under severe conditions and in an arid climate. Microbiol Mol Biol Rev 63(4):968–989
Zaman M, Nguyen M, Blennerhassett J, Quin B (2008) Reducing NH_3, N_2O and N losses from a pasture soil with urease or nitrification inhibitors and elemental S-amended nitrogenous fertilizers. Biol Fert Soils 44:693–705
Zhang G, Yang Z, Dong S (2011) Interspecific competitiveness affects the total biomass yield in an alfalfa and corn intercropping system. Field Crops Res 124:66–73
Zhang Z, Liao H, Lucas WJ (2014) Molecular mechanisms underlying phosphate sensing, signaling, and adaptation in plants. J Integr Plant Biol 56:192–220

Grain Legumes: Impact on Soil Health and Agroecosystem

16

Nirmali Gogoi, Kushal Kumar Baruah,
and Ram Swaroop Meena

Contents

Abstract

Legumes are one of the richest sources of proteins, minerals, and fibers for animals and human being. They also have a great role in maintaining soil fertility through biological nitrogen fixation (BNF). Legumes help in solubilizing insoluble phosphorus (P) in soil, improving the soil physical environment, and increasing soil microbial activity and also have smothering effect on weed. Due to these positive roles in improving soil health and excellent adaptability to marginal environment, legumes are now considered as one of the important components of a cropping system. To reduce poverty, hunger, malnutrition, and

N. Gogoi · K. K. Baruah
Department of Environmental Science, Tezpur University, Napaam, Assam, India

R. S. Meena (✉)
Department of Agronomy, Institute of Agricultural Sciences (BHU), Varanasi, UP, India

R. S. Meena et al. (eds.), *Legumes for Soil Health and Sustainable Management*,
https://doi.org/10.1007/978-981-13-0253-4_16

environmental degradation, legume crop can be a substitute for cereal crop in marginal lands. Rediscoveries in genetics and genomics now open up new opportunities for improving productivity and quality in grain legume research. The carryover of nitrogen (N) derived from legume grain either in crop senescence or in intercropping system for succeeding crop is important. The necessitate of the interdisciplinary study on grain legumes to address their important role on soil health. Thus, the maximum beneficial effect in modern agriculture as the optimization of fertilizer N use is an essential not only to maintain and restore soil organic carbon (SOC) but also to minimize the nitrate pollution from agricultural source.

Keywords
Grain legumes · Nitrogen yield · Protein yield · Biological nitrogen fixation

Abbreviations

AM	Arbuscular mycorrhizal fungi
BNF	Biological nitrogen fixation
CED	Chronic energy deficiency
ISFM	Integrated soil fertility management practices
PEM	Protein energy malnutrition
PGPR	Plant growth-promoting rhizobacteria
SMB	Soil microbial biomass
SOC	Soil organic carbon

16.1 Introduction

Among the cultivated crops of the world, grain legume occupies an important position. The pods (matured, ripen, or unripen) of the grain legumes (family, Fabaceae) are used as human food as well as animal feed. In terms of production, grain legumes rank third after cereal and oilseed, but its importance is more in terms of agriculture and the environment due to the supplement of protein to human and livestock and the ability to fix atmospheric nitrogen (N) (Mantri et al. 2013). With the expanding world's population, from the current 7.5 billion to 11 billion by the end of the twenty-first century, about 70% more food will be needed (UN 2017; Alexandratas 2009). The cultivation of grain legume will play an important role in the food security of this growing world population. Among the grain legumes, the main sources of dietary protein for vegetarians come from chickpea (*Cicer arietinum*), common bean (*Phaseolus vulgaris*), grass pea (*Lathyrus sativus*), lentil (*Lens culinaris*), mung bean (*Vigna radiata*), urad bean (*Vigna mungo*), pea (*Pisum sativum*), pigeon pea (*Cajanus cajan*), and soybean (*Glycine max*). Though rich in protein and known as poor man's meat, some grain legumes like soybean and groundnut are also good sources of vegetable oil (Bellaloui et al. 2013; Meena and Yadav 2015). The ability to fix N biologically makes the legume crop an important candidate for cropping sequence to maintain the N fertility in agricultural soil and thus to improve soil physical condition and sustain the environmental balance (Courty et al. 2015). Cultivation of legume can significantly mitigate the agricultural contribution to climate change by reducing the energy use, emission of greenhouse gasses, and maintaining positive soil carbon balance. The presence of high soluble and insoluble fiber, oligosaccharide, and phenolics and essential nutrients such as vitamins, antioxidants, and bioactive compounds in food legume can provide several health benefits to human and the livestock (Shimelis and Rakshit 2005; Meena et al. 2015a).

Legumes are cultivated in diverse climates ranging from semiarid to subtropical and temperate region. Being shorter in crop duration, any changes in climatic parameters lead to drastic reduction of legume yield (Fang et al. 2010). They are more sensitive to various abiotic and biotic stresses than cereals and have higher cultivation risk and lower yield over competing cereal crops. Environmental factors such as water stress, temperature stress, salinity, high CO_2 concentration, and heavy metal pollution affect its growth, yield, and the quality of the produce (Wani et al. 2007; Varma and Meena 2016). Farmers also use their marginal lands to grow grain legume leading to reduced productivity. In a modeling approach, Cooper et al. (2009) predicted that 3 °C rise in temperature will reduce current average peanut production in Zimbabwe by 33% and pigeon pea in Kenya by 19%, due to shorter growing period and early maturity. Legume crops are slow grower at early stages of growth and susceptible to weed competition due to low soil N uptake at this period, which can reduce the yields by 25–40% (Pandey et al. 1998). To cope with the changing climate, legume breeding for stress resistance is very essential. Several grain legume genotypes have been identified with the ability to decrease the stomatal conductance with the soil drying – making them a perfect candidate to grow under water-limited situation (Zaman-Allah et al. 2011; Devi et al. 2009). Therefore, it is the need of the hour to give emphasis on enhancement of the grain legume production through agronomic and molecular breeding approach. Compared to natural ecosystem, soil health in an agroecosystem has to face many challenges owing to rapid disturbances from various agricultural operations during cultivation. Therefore, a holistic approach is needed to maintain the soil physical, chemical, and biological characteristics of an agroecosystem. Reduced soil disturbances in terms of tillage practices and keeping an organic soil cover are proved helpful in this regard (Meena et al. 2015b). This can be achieved by introducing a legume crop in an agroecosystem since they can grow under reduced tillage and be used as an organic cover.

Legumes can be grown in marginal land with less availability of macro- and micronutrient. Due to the presence of nitrogen-fixing ability, legumes can support their own growth and development at even soil with less fertility. With the process of growth, they accumulate good amount of biomass through photosynthetic carbon (C) fixation. These biomass finally enrich the soil with C by net exudation. Thus, the legume can maintain the SOC component. Once the soil is enriched with C, it improves the soil physicochemical properties. For succeeding crop, legumes can thus improve the soil quality through their biomass incorporation – this is the basic hypothesis of this chapter. In this chapter we discussed about the impact of legume cultivation on soil health (in terms of both biological and chemical health) and the role of legume crop in agroecosystem – with reference to modern agriculture.

16.2 Role of Grain Legumes in Food Security

With the superior grain composition and multi-nutritional benefits, grain legumes may help to reduce the malnutrition and to meet the dietary demands of the increasing global population. Food security is achieved when all people have access to enough food to live a healthy and active life. In most of the cases, malnutrition is caused by undernutrition diet with inadequate protein and calories. Protein energy malnutrition (PEM) and chronic energy deficiency (CED) are two most common nutrient deficiency diseases in India. Legumes are rich sources of plant protein and play a significant role in food security of the society. Malnutrition can be overcome by production of enough legumes which is cheap compared to animal protein, and poor ones can easily purchase it for their dietary protein need. The agricultural production has to increase by 70% by 2050 to deal with an estimated increase of 40% in the world's population (Burinsma 2009). To cope with the increasing food demand, it is necessary to adopt sustainable and improve technologies for ensured developments in food productivity and thereby food security (Gruhn et al. 2000; Landers 2007; Ashoka et al. 2017). Along with the human dietary need, legumes are also essential for intensive animal and milk production, where grain crops are used as a major feed source and forage legumes are needed to maintain animal health as medicine (Wattiaux and Howard 2001). These make legumes as an integral component of the modern agriculture.

Food legumes are the best sources of dietary proteins more particularly in developing countries and provide 20–40% of dietary protein requirements (Kudapa et al. 2013). They are the rich sources of carbohydrates, vitamins, and minerals (Wang et al. 2011). Essential nutrients (macro and micro), vitamins, dietary minerals, good quality dietary fibers, antioxidants, and other bioactive compounds are the important sources of grain legumes (Prakash and Gupta 2011; Wang et al. 2011). But, compared to cereal crops, the yield of grain legumes is substantially low due to its shorter life cycle and requirement of higher photosynthate to convert to protein. They have numerous health benefits such as lowering and preventing some forms of cardiovascular diseases, obesity, certain cancer, and diabetes mellitus (Goni and Valentin-Gamazo 2003) due to their high soluble and insoluble fiber, oligosaccharide, and phenolic contents. In many countries, grain legumes serve as a vital part of the daily diet and thus deliver a larger share of plant protein in human diet. Legume accounts for 27% of world primary crop production and contributes 33% nutritional protein needs of human diet (Vance et al. 2000). Legume seeds are rich in protein, containing 20–30% protein with high level of lysine (Duranti and Gius 1997), one of the essential amino acids which cannot be synthesized by mammals. Because of this higher protein content and increasing price of animal protein, the cultivation of legume in modern agriculture is highly essential. Large variation exists in the protein content of food legumes in different studies and across different region and ranges from 26% to 57% in soybean (Iqbal et al. 2006), 21% to 29% in common bean (Costa et al. 2006), 16% to 32% in pea (Costa et al. 2006), 22% to 36% in faba bean (*Vicia faba*) (Iqbal et al. 2006), 19% to 32% in lentil (Costa et al. 2006), 16% to 28% in chickpea (Iqbal et al. 2006), 16% to 31% in cowpea (Duranti 2006), 21% to 31% in mung bean (Duranti 2006; Dhakal et al. 2015), and 16% to 24% in pigeon pea (Duranti 2006). Genotypes from the same crop species, environmental conditions, and crop husbandry practices adopted during cultivation play an important role in protein content of grain legume. Major storage proteins present in grain legumes are globulins (70%) and albumins (20%), whereas prolamins and glutelins are some minor proteins (Duranti 2006). Legumin and vicilin are the major protein fractions of globulin and albumin. All food legumes contain more vicilin, and the relative proportion of legumin and vicilin varies with genotype.

In addition to the digestible proteins, many essential amino acids such as lysine, leucine, valine, isoleucine, and phenylalanine (Javaid et al. 2004) are also found in grain legumes. Among the grain legumes, soybean and peanut contain an excellent source of vegetable oils and contribute more than 35% to global processed vegetable oil production. The higher vitamins and mineral contents along with the antioxidant property increase market demand of the vegetable oil. The carbohydrate content in grain legumes ranges from 30% in soybean to 63% in chickpea. Legume starch has a higher proportion of amylopectin than amylose. However, the amylose content of legume starches tends to be slightly higher than that of cereal starches (Arab et al. 2010). Grain legumes are also the vital sources of minerals such as P, potassium (K), calcium (Ca), magnesium (Mg), and zinc (Zn). Some essential fatty acids such as omega fatty acids (omega-3 and omega-6) are not synthesized in the human body, so they must be obtained through nutrition or as supplements. Replacing animal products in the diet with plant products such as soybean provide benefits in cardiovascular health (Sirtori et al. 2009) through lowered cholesterol (Harland and Haffner 2008). Consumption of both soybean and lupin was found to decrease cholesterol in animals and humans, and it also helped in managing diabetes (Bertoglio et al. 2011).

16.3 Necessity of Grain Legumes in Modern Agriculture

Cultivation of crop and raising of livestock for food, fiber, biofuel, medicine, and other day-to-day needs of human life are agriculture. Crops provide the major part of human nutrition, fodder, and the most important requirement, medicine. Thus, agriculture is essential for survival, food, growth, health, productivity, and development of world economic system (Ram and Meena 2014). Legume-based farming brings the sustainability to the farming system. Legumes deserve a prominent role in the present cropping systems of both developed and developing agriculture

(Dhakal et al. 2016). Along with the human dietary need, legumes are also essential for intensive animal and milk production, where grain crops are used as a major feed source and forage legumes are needed to maintain animal health as medicine (Wattiaux and Howard 2001). The primary goal of incorporation of legumes in cropping system is to enhance the soil fertility (Meena et al. 2015b) and provide fodder for the animals and for the direct consumption as food by human. The lipids from grain legume also have the possibility to use as biodiesel (Jensen et al. 2012), one of the renewable sources of energy for clean environment.

The BNF ability of the grain legume makes them suitable to include in the cropping system as N is the most limiting nutrient for crops. It reduces the N demand and thereby decreases the production cost and environmental pollution since nitrogenous fertilizers are one of the prime causes of agricultural pollution. Moreover, the ability of the legumes to convert the unavailable form of phosphorus (P) to available form through releasing some organic acids by the roots also brings P sufficiency in a cropping system (Jensens 1996). In developing countries, with a crop livestock production, the nutrient deficiency is a common phenomenon in the cattle as they are mostly fed with cereal crop residues where the N content is below the threshold level. For an efficient digestion, about 1.0–1.2% of N content in livestock feed is necessary to support optimum growth of the microbes in the cattle rumen (Van Soest 1994). The N-rich legume residue can help to remediate the nutrient deficiency problem in livestock.

Besides the positive effects on soil fertility, grain legumes also reduce incidence of pests, diseases, and weeds. Therefore, with the developed agronomic practices like reduced tillage and organic farming, the production of grain legume is escalating (Meena et al. 2016). Crop rotation has a great influence on the yield performance of the crops in a cropping sequence, and helps to imorove the agro-economic and soil environmental sustainability (Reckling et al. 2014) (Fig. 16.1). For example, about 15–25% increase of cereal yield is reported by Kirkegaard et al. (2008), when grown in rotation with grain legume and thereby can reduce the need of agrochemicals. Hence, this is necessary to move toward the organic farming to achieve sustainability in agriculture (Verma et al. 2015a, b). Incorporation of legume crop and intercrop system is a good way of organic agriculture as diseases and pest attacks are disturbed without the application of chemicals. Another important issue of crop production in present day is escalating the costs of fertilizers. The cost of composite fertilizer is reported to increase by 113% between 2000 and 2007 (Huang 2007). Legumes have the ability to transfer fixed nitrogen to the coexisting crops; when legumes are grown with other crops, the weed competitions become less. For example, when they are cultivated with cereal, weeds are found in less number as cereals are a good competitor of weeds. The availability of P, K, Ca, and Mg is higher in`the intercropping systems than the monocultures (Vandermeer 1992; Li et al. 2007). In conservation agriculture, legumes are also used in rotation as a cover crop. When legumes are grown as intercrop, it not only increases the total productivity of the system but also plays an important role in efficient use of resources (Ghosh et al. 2007; Veronica et al. 2005; Varma et al. 2017). Results of legumes in intercropping systems are shown in Table 16.1.

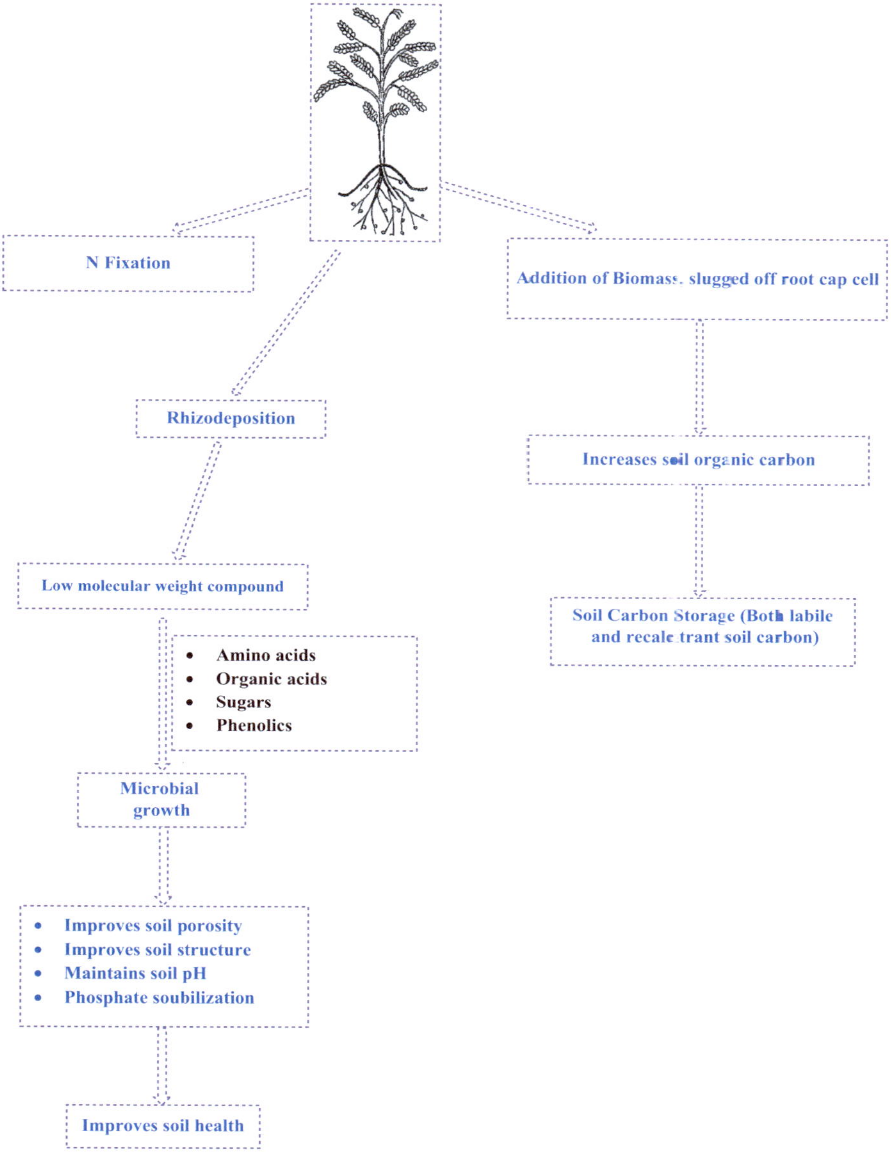

Fig. 16.1 Impact of legume on soil health

The reduced requirement of tillage in legume cultivation has positive influence in farm economic performance along with increased C sequestration due to the reduced disturbances in soil (Reckling et al. 2014). The decreased need of fertilizer application and agrochemicals helps in lowering the greenhouse gas emission and potential global warming. The emissions of greenhouse gasses and N deposition to terrestrial

Table 16.1 Various intercropping systems with legume in India

Intercropping system	Location	References
Sorghum intercropped with		
Green gram	Rajasthan	Laddha and Totawat (1997)
Soybean	Bhopal	Ghosh et al. (2005)
Pigeon pea	Hyderabad	Tobita et al. (1994)
Cowpea	New Delhi	
Black gram	New Delhi	
Groundnut	Junagarh, Gujarat	Ghosh (2004)
Maize intercropped with		
Groundnut	Junagarh, Gujarat	Ghosh (2004)
Black gram	Nainital	Singh (2000)
Soybean	West Bengal	Mandal et al. (2014)
Pearl millet intercropped with		
Pigeon pea	Hyderabad	Ghosh et al. (2008)
Groundnut intercropped with		
Pigeon pea	Hyderabad	Ghosh et al. (2008)

ecosystems are responsible for eutrophication and soil acidification (Clark et al. 2013). Agricultural emissions of both N and P compounds are a significant source of freshwater nutrients and are detrimental to biodiversity in aquatic ecosystems through eutrophication (Nemecek et al. 2008). The gaseous emissions of N compounds are dominated by ammonia, of which more than 93% comes from agriculture. According to findings of Pappa et al. 2011, the emission of nitrous oxide and leaching of nitrate from arable soils are high after the cultivation of grain legume and during the early stages of crop growth. But with the application of proper strategy, for example, using catch and cover crops (e.g., cereal-legume intercropping), it can be reduced substantially (Justus and Kopke 1995; Ram and Meena 2014). In temperate climate, when grain legumes are grown during summer with a fallow winter period, it also leads to nitrate leaching which can be minimized by growing cover crops. Thus, by recycling the nutrients on and between the farms, the cultivation of grain legume can potentially reduce the loss of nutrients and able to fulfill the basic requirements of modern agriculture in terms of resource utilization and effect on the environment and biodiversity.

Besides using as food and fodder, legumes can also be used in liquid form for producing milks, yogurt, and food formula for infant (Garcia et al. 1998). Legume can be milled to flour to make various chips and snacks. Other uses of legumes are production of biodegradable plastics (Paetau et al. 1994), oils, gums, dyes, and inks (Morris 1997).

16.4 Impact of Legume on Soil Biological Properties

The specially developed nodule structures of grain legumes support the atmospheric N fixation process with the help of the enzyme nitrogenase. In addition to the nitrogen storage in proteins, some legumes also have an extra layer of store of glycoprotein in their leaf cells (in between palisade and spongy mesophyll) (Klauer and Francesch 1997). After screening of the legume species for the presence of this paravenial layer, *Lansing* and Franceschi (2000) found that 39 legume species bear this potentially important structure of protein.

P is another essential element for plant growth to supply adequate energy within the cell. In the cell, the vacuole can store a substantial amount of phosphorus to provide the required energy transfer during later growth stages. In the soil solution, this important nutrient element usually makes complexes with calcium, iron, and aluminum and makes it unavailable for plant uptake, though the soil may have large amount of phosphorus (Sinclair and Vadez 2002; Meena et al. 2017a). In this regard, the cultivation of grain legume can improve the situation by following ways:

1. The release of available P is highly dependent on soil characteristics (Jones et al. 2003) such as pH. The organic acids (such as malate, citrate, oxalate, tartrate, and acetate) released by the roots of grain legume (Shen et al. 2002; Nwoke et al. 2008; Nuruzzaman et al. 2006) decrease the soil pH in the rhizosphere which helps in conversion of unavailable P to available form.
2. Grain legumes also release enzyme phosphatase into the soil which helps in breakdown of organic materials containing P (Gilbert et al. 1999; Helal 1990).

Soil biological properties such as soil microbial biomass (SMB) are generally used as an early indicator of changes in soil physicochemical properties because of soil management in agricultural ecosystems (Brookes 1995; Trasar-Cepeda et al. 1998; Suman et al. 2006) (Fig. 16.2). During the process of BNF, hydrogen gas is produced which in turn encourages the bacterial growth in the legume rhizosphere leading to higher microbial biomass C in the soil. The soil microbial C (C_{mic}) and N (N_{mic}) contribute 1–7% of total soil C (C_{org}) and up to 5% of total soil N (N_{tot}), respectively (McGill et al. 1986; Sørensen 1987; Anderson and Domsch 1989; Insam et al. 1989; Sparling 1992), which is among the most labile C and N pools in soils (Jenkinson and Ladd 1981). Consequently, size and activity of the SMB can influence nutrient availability and yield of the agroecosystems. The nodule-rhizosphere interaction of the leguminous plants results in enhanced microbial activity in the soils of legume crops. Alvey et al. (2003) reported that the introduction of legume crop rotations had a significant influence on the microbial community structure and increased microbial diversity. Similar results have been achieved in intercropping experiments in which bacterial biomass and activity varied from those in monocropping systems (Latati et al. 2014; Li et al. 2009; Qiang et al. 2004; Song et al. 2006; Tang et al. 2014; Wang et al. 2007). The ability of the leguminous rhizospheric fauna to capture atmospheric N and enhanced root exudation results in higher C:N ratio, and it has been found by Liang et al. (2014) that legume species,

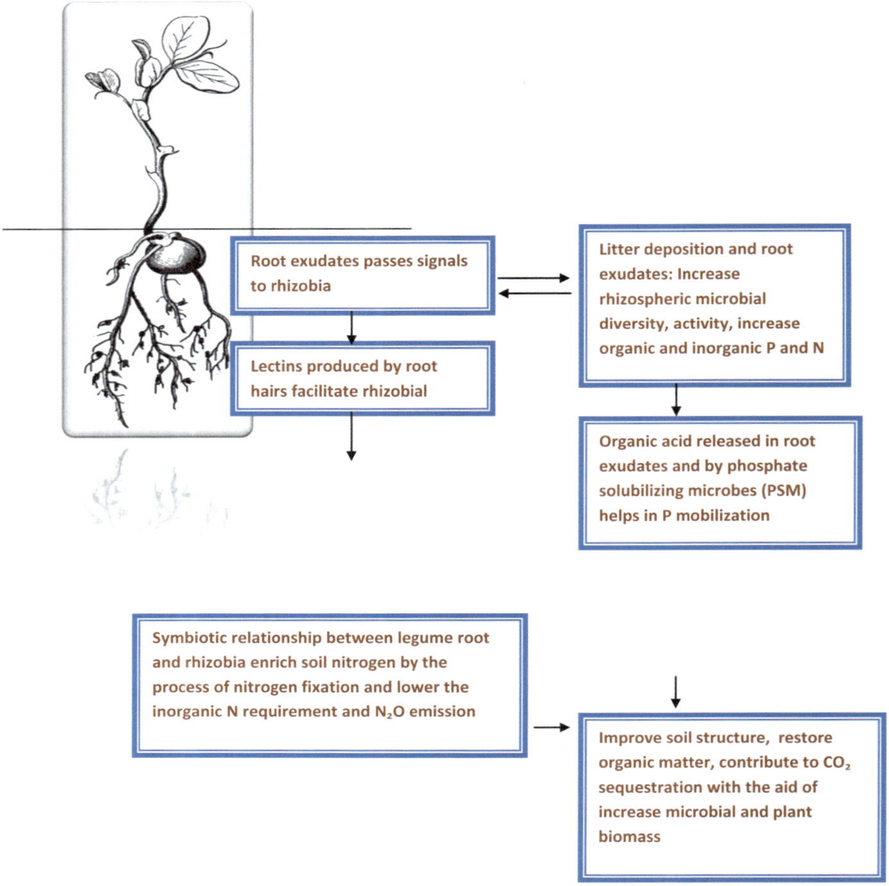

Fig. 16.2 Impact of legume on soil biological property

even with small variations in C:N ratio and lignin and cellulose contents, triggered ample divergence in soil microbial properties (Meena et al. 2014). The production and exudation of lectins by legumes have shown to be capable of influencing the mobility of plant growth-promoting rhizobacteria (PGPR) and improving root colonization and the phyto-beneficial activity of these PGPR (Schelud'ko et al. 2009). Legumes are known for their tripartite symbiosis (mycorrhiza-legume-*Rhizobium*) (Hay- man 1986) and have been shown to be responsible for colonization of specific arbuscular mycorrhizal (AM) fungi, mainly due to their special nutritional requirements associated with their root nodule activity (Scheublin et al. 2004; Vandenkoornhuyse et al. 2002; Meena et al. 2017b). The dual symbiosis of AM fungi and *Rhizobium* bacteria on legume plants enhances plant growth and yield under several environmental conditions. It is due to the higher dependency of the legume plants on mycorrhiza to achieve their maximum growth. The hyphae of the mycorrhiza have the ability to access a greater volume of soil and can absorb and transport fairly large amounts of low-diffusing nutrients like P to their host plant and help in nodule formation (Zahran 1999). Though the AM fungi don't possess specificity in symbiotic relationship, they differ in their ability to enhance nutrient uptake by the host plant. Therefore, the combination of different AM fungal strains or species is important since the compatibility of such interactions may be relevant to N fixation and to nutrient and water uptake by the legume plants (Vinicius Ide 2013). Legumes also appear to promote AM colonization in low-input systems. Previous studies largely showed that AM results in an increased flow of nutrients, plant productivity, and ecosystem sustainability (Gianinazzi and Wipf 2010). Legumes are also used extensively as a cover crops to reduce soil N loss and erosion in agricultural fields. Short-term management (e.g., 1 year) of legumes has shown the influence on microbial population of the cultivated soils. However, all soil properties and processes are not sensitive to short-term management with legume cover crops (Liang et al. 2014; Meena et al. 2014), while soil enzyme activities, microbial biomass, and respiration are sensitive toward the termination strategies of cover crops.

16.5 Soil Processes

The residual N supply obtained from introduction of legumes in crop rotation through symbiotically fixed N depends on climate, crop management practices, and the species of the legume grown (Heichel and Barnes 1984; Meena et al. 2015a, b, c, d). A cropping system with leguminous crops and sufficient N fertilizer also enhances SOC concentration (Varvel 1994). A study on *Mucuna* with maize resulted in a decline in runoff and erosion, an upsurge in soil organic matter content and in the production of maize grains, and an improvement of soil water regime (Blanchart et al. 2006). In a legume-nonlegume crop sequence, the amount of N returned to the soil for nonlegume succeeding crop depends on the following factors:

1. The quantity of legume residue returned to the soil
2. The content of the symbiotically fixed N in the residues
3. The availability of the legume residue N to the subsequent crop (Heichel 1987)

Drinkwater et al. (1998) documented a significant increase in C and N retention under legume-based cropping systems and suggested the contribution of narrow C:N organic residues combined with the relatively higher temporal diversity on the same. It was also reported that crop rotations, which include legumes, are able to maintain higher organic matter levels than continuous cropping systems with nonleguminous row crops (Campbell et al. 1991; Campbell and Zentner 1993; Stevenson 1982). Inclusion of legumes into crop rotations is justified by their natural capability to exploit atmospheric N, and this additional source of N is likely to avoid interspecific struggle between crops and legumes for N acquirement (Carof 2006; Hauggaard-Nielsen et al. 2008) and to make ample N contents available for the following crop through increased soil N content after destruction of the legume cover crop.

The N-rich legume residue also encourages the activities of earthworm in the soil, and thus, it improves the soil porosity promoting higher water and air movement (Meena et al. 2015). For example, growing legume has a positive effect on soil structure due to its continuous network of residual root channels and macropores which leads to improve soil water-holding capacity (Jensen et al. 2012). The higher protein content in the legume facilitates the decomposition of crop residue by encouraging the microbial growth in the soil (Dhakal et al. 2016) and their conversion into soil building organic matter because most of the crop residues are rich in C. Improvements in both soil humus and organic C content are reported after legume cultivation as they supply biomass and organic C and N in the soil (Lemke et al. 2007). Additionally, the reduced tillage used during cultivation of legume crops helps in buildup of organic C (Alpmann et al. 2013). The quantity of organic C buildup depends on the soil, climatic condition, and species of grain legume. Higher organic carbon sequestration has been documented in a mixture of grasses and legumes than the monocultivation of the same (Lopez-Bellido et al. 2009; Yadav et al. 2017).

Through the process of BNF, the grain legume can save some 150–200 kg ha^{-1} of N per year compared to other cereal or rapeseed crops (Peyraud et al. 2009). When inoculated with proper strains of *Rhizobium* bacteria, legume can supply up to 90% of their own N. Shortly after the germination of the seed, the bacteria penetrate the root to form the nodule where the N present in the soil air is bound and supply it to the aboveground plant during photosynthesis. The bacteria produce ammonia with the help of hydrogen acquired from the plant carbohydrate synthesized during photosynthesis. Though variable results were obtained regarding the savings of N fertilizer from different sites, Bues et al. (2013) had reported that an average of 21 kg ha^{-1} of nitrogen fertilizers can be saved in 3–6 years of rotations with grain legume. Some of the N fixed by legume is recycled – mostly during decomposition of aboveground and belowground crop residues (Meena et al. 2015). N cycling is mediated by soil organisms, and the rate and the pattern of nitrogen released from crop residues are regulated by soil microbial activity, residue quality (rhizodeposition), and soil environment. For example, in alkaline soil, legume can help in maintenance of plant soil microbial activity by reducing soil pH where the organic acid released from legume facilitates the process. The highest maize yield was reported by Ghosh and Singh (1994), while growing after cowpea (fodder) compared to the maize grown after maize (fodder). This enhancement in yield is primarily because of enrichment of soil N by leguminous cowpea (Tables 16.2 and 16.3).

Table 16.2 Grain yield of maize crop and total nitrogen content in soil as influenced by preceding summer crops and nitrogen applied to maize crop (Adopted from Ghosh and Singh 1994; Bues et al. 2013)

Treatment	Grain yield (kg ha^{-1})	Total N (%)	
		After summer crop	After maize harvest
Summer crop			
Black gram	3920	0.069	0.068
Green gram	4208	0.071	0.069
Cowpea (fodder)	4404	0.075	0.070
Cowpea (grain)	3594	0.071	0.070
Maize (fodder)	3477	0.065	0.066
Fallow	3946	0.068	0.068
$LSD_{(0.05)}$	506		0.0008
N to maize (kg ha^{-1})			
0	2790	–	0.063
30	3775	–	0.066
60	4451	–	0.067
90	4684	–	0.070
$LSD_{(0.05)}$	279	–	0.0008

Table 16.3 Yield potentiality of legumes

Grain legume	Yield (kg/ha)	References
Soybean (*Glycine max*)	1000	Masuda and Goldsmith (2009)
Pea (*Pisum sativum*)	182	Cousin (1997)
Pigeon pea (*Cajanus cajan*)	657	Singh (2013)
Lentil (*Lens culinaris*)	667	Singh (2013)
Rice bean (*Vigna umbellata*)	907–1089	Khadka and Acharya (2009)
Cowpea (*Vigna unguiculata*)	300	Ehlers and Hall (1997)
Faba bean (*Vicia faba*)	5112–5737	Song et al. (2006)
Common bean (*Phaseolus vulgaris*)	729	El-Al et al. (2011)
Groundnut (*Arachis hypogaea*)	310	Ramana et al. (2002)
Chickpea (*Cicer arietinum*)	792	Singh (2013)
Mung bean (*Vigna radiata*)	346	Singh (2013)
Black gram (*Vigna mungo*)	733–900	*agritech.tnau.ac.in*

16.6 Greenhouse Gas Emission

The enhanced N_2O emissions from agricultural and natural ecosystems are believed to be caused by increasing soil N availability due to increased use of fertilizer, BNF, and N deposition (IPCC 2013). The potentiality of N_2O emission from arable soil under agriculture is drastically reduced due to legume cultivation through the savings in fertilizers (N and P) as the estimated CO_2 emission from fertilizer production is about 300 Tg per year (Jensen et al. 2012). N_2O production in soil occurs mainly by two microbial processes:

(i) Nitrification in aerobic conditions
(ii) Denitrification in anaerobic conditions

Both the incidence and intensity of these processes are strongly affected by soil mineral N and the availability of soluble C, water and oxygen contents, temperature, pH, and soil texture (Conen et al. 2000; Gu et al. 2013; Smith et al. 1998). In agricultural fields, cover crops are frequently used as catch crops to mitigate nitrate leaching and erosion during the autumn and winter fallow periods (Thorup-Kristensen et al. 2003). When legume cover crops are used either alone or in mixture, they provide an additional N green manure effect for the subsequent crop (Tribouillois et al. 2015; Dhakal et al. 2016) and are responsible for the modification of mineral N availability in the soil, either reducing it during plant growth or increasing it after incorporation into the soil. They can also affect soil water content through increased transpiration compared to bare soil. Studying alternative crop emissions, Jeuffroy et al. (2013) observed that legume crops emit about five to seven times less GHG per unit area compared to other crops. Results of N_2O fluxes from different crops demonstrated that pea emitted 69 kg N_2Oha^{-1}, far less emissions than winter wheat (368 kg N_2Oha^{-1}) and rape (534.3 kg N_2Oha^{-1}). The company of legumes in the cereal-based crop rotation instead reduces the amount of synthetic N required by the following cereal crop and consequently decreases the N_2O emissions associated with synthetic N fertilizers (Jensen and Hauggaard-Nielsen 2003; De AntoniMigliorati et al. 2015). Tillage is another factor associated with N_2O emission from agricultural fields. There is a general tendency to observe higher emissions under conventional tillage (Plaza-Bonilla et al. 2014; Yadav et al. 2017) which can be minimized with the inclusion of legume as legume needs very low tillage compared to the conventional tillage used for cereal crops and is reported to increase carbon sequestration in the soil.

16.7 Crop-Legume Intercropping

Intercropping is a mixed cropping system of cultivating two or more crops in the same space at the same time (Andrews and Kassam 1976; Sanchez 1976) in a definite row arrangement. Four different types of intercropping, namely, mixed intercropping, row intercropping, strip intercropping, and relay intercropping, are in use. Due to higher density of crops under intercropping, particularly with the inclusion of legumes, microbial diversity of the soil increases which brings stability to the agroecosystem (Ram and Meena 2014). Crop-legume intercropping plays an important role in improving soil fertility, water and radiation use efficiency, weed, pest and disease control, and profit maximization for farmers. Success stories of pulse as an intercrop have already been documented by many researchers. For example, intercropping soybean with corn gives higher economic return with more crude protein compared to the pure stand. Rhizobia and legume are found in a symbiotic association, where both of them are benefited. Rhizobia receive food and shelter from the legume, and in return legume gets fixed N ammonia and is utilized in biosynthesis of amino acid and nucleotides. Crop plant when grown with legume in nutrient poor soil better yield is achieved compared to the plant grown alone. Cereal legume intercropping has higher capacity to restore soil mineral N through its ability to biologically fix atmospheric nitrogen (Fujita et al. 1992; Giller 2001; Meena et al. 2017b). Intercropping falls under organic farming, as here disease and pest are controlled biologically, while soil fertility is maintained organically. The use of biochar as organic amendment in intercropping was found to enhance legume N fixation and increased yield compared to single crop and facilitate N transfer from legume to coexisting crops (Ling Liu et al. 2017).

According to some researchers, legume plants are weak suppressors of weed as they grow slow at early development or lose leaves in the ripening stage (Hauggaard-Nielsen 2001; Jensen et al. 2005). But when cereal crops and legumes are grown together, the weed suppression ability increases. Disease risk minimization is another benefit obtained from crop-legume intercropping. Common bacterial blight and fungal rust can be controlled by intercropping (Boudreau and Mundt 1992; Fininsa 1996) with legume. Viral diseases such as cassava mosaic disease of cassava plant and whitefly attack can be reduced by intercropping cassava with green gram.

Cultivating crop repeatedly in the same piece of land reduces soil fertility, and the addition of chemical fertilizer is not the solution as it increases the price of the produce along with its effects on the ecosystem. In this situation, crop rotation is one of the adaptation options through which the soil fertility can be maintained. Intercropping cereals with legume is a main component of integrated soil fertility management practices (ISFM) (Sanginga and Woomer 2009; Mucheru-Muna et al. 2010; Meena et al. 2015b). Cereal legume intercropping is being practiced in agriculture for last decades. In this regard, right choice of both cereal and gain legume crop is very important; otherwise profit may shift to loss as maximum utilization of soil nutrient will be hampered. For example, combination of two crops having different ripening period reduces crop yield rather than increases yield. So cereal legume intercropping does not automatically improve crop yield, but the correct combination of crop is important. In rotation cropping system, legumes are mainly used as green manure. Though some other crops can also be grown as green manure, but due to N-fixing ability (Table 16.4), the legume crops are preferred the most. Green manuring in maize field with *Sesbania rostrata* + 30 kg N ha^{-1} gives same yield as application of 90 kg N ha^{-1}, indicating 60 kg of N is saved through green manuring (Tiwari et al. 2004). Sometimes, legume green manure crop can supply entire N need for the next crop. Legume litters contain K, P, and other nutrients which are recycled to the soil. In intercropping system, N is transferred to the coexisting crop. Intercropping of peanut with rice crop which transfers N from peanut to rice is prominent especially in N-poor soil (Chu et al. 2004; Meena et al. 2015). In maize and cowpea, when intercropping has been done at low N level, the N content of intercropped maize was found to be higher than sole maize crop, which shows the

Table 16.4 Nitrogen-fixing ability of legumes

Grain legume	N-fixing ability (kg ha^{-1})	Reference
Soybean (*Glycine max*)	71–108	
Pea (*Pisum sativum*)	90–128	Jensen (1996)
Pigeon pea (*Cajanus cajan*)	120–170	Adu-Gyamfi et al. (1997)
Lentil (*Lens culinaris*)	8–14	Cowell et al. (1989)
Rice bean (*Vigna umbellata*)	13–30	
Cowpea (*Vigna unguiculata*)	14–35	Okereke and Ayama (1992)
Faba bean (*Vicia faba*)	23–79	Danso et al. (1987)
Common bean (*Phaseolus vulgaris*)	20–60	Silva et al. (1993)
Groundnut (*Arachis hypogaea*)	150–200	Toomsan et al. (1995)
Chickpea (*Cicer arietinum*)	64–103	Fatima et al. (2008)
Mung bean (*Vigna radiata*)	19–54	Hayat et al. (2008)
Black gram (*Vigna mungo*)	16–79	Hayat et al. (2008)

transfer of fixed N from cowpea to maize (Francis 1986). Thus, with intercropping, food quality can be enhanced by increasing protein content of cereal and other crops, and the food security can be fulfilled to some extent. Intercropping in upland rice with soybean at the ratio of 4:2 was found beneficial to increased productivity along with soil fertility improvement (Hazarika et al. 2006). The cereal crops such as rice, wheat, and maize are cultivated extensively, these crops alone cannot contribute to all nutritional needs of the animals. Therefore, the diversification of crops by growing various valuable crops is necessary, which will provide all the dietary requirements of the human population including other animals. Thus, crop diversification with legume has advantage of N nutrition to the plant, along with breakage effect on disease cycle and pest (Voisin et al. 2014).

Legume can reduce disease and pest attack, increase production of coexisting crop with higher protein availability, and thus help in food security. In rotation, legume brings diversification in the cropping sequence which affects the associated diversity of wild flora, fauna, and soil microbes (Collette et al. 2011; Meena et al. 2014) with the potentiality of a dynamic and more sustainable agriculture (Peoples et al. 2009). By providing nectar and pollen, the mass flowering of grain and forage legumes contributes in the maintenance of wild and domesticated bees (Kopke and Nemeck 2010). Though there are controversial reports on the effects of legume on honey bees' population where it is argued that because of the regular disturbances in soil, use of biocides, and dense covering on the soil, the crop fields are not the foraging place for honey bees (Power and Stout 2011; Jeanneret et al. 2006). The diversification of cereal-dominated cropping systems with legumes enables pesticide savings, especially of specific fungicides in rotations (Von Richthofen et al. 2006; Kirkegaard et al. 2008).

16.8 Soil Erosion and Legume

The physical removal of soil by agents which provide the kinetic energy to move soil from one location to another is called soil erosion. Topsoil is the layer of soil where plants grow as it has the highest fertility than the other soil layers due to the presence of organic matter content, soil microorganism, and mineral nutrients. The primary causes of soil erosion are wind, water, grazing animals, and anthropogenic activity. Natural soil erosion is a slow process, and it is not a major problem as natural soil-forming processes can replenish it. Soil erosion is becoming a matter of concern as it is accelerated by anthropogenic activity. The use of land in different purposes indicates soil loss, so revegetation can help to reduce soil loss. Legumes are known to use as cover crop to control soil erosion. For example, legume shrubs (*Colutea arborescens*, *Dorycnium pentaphyllum*, and *Medicago strasseri*) grown as cover crops were found to reduce runoff and soil loss (García-Estringana et al. 2013). Hedgerow with leguminous species is planted for erosion control which also adds N to the soil. Bhatt and Bujarbaruah (2006) reported that on an average, pruning of the leguminous hedgerow species can add 20–80, 3–14, and 8–38 kg of N, P, and potassium (K) per hectare per year, respectively.

Organic matters are the integral component of topsoil and function as a main indicator of soil quality and fertility (Franzluebbers 2002; Verma et al. 2017). It has direct impact on plant growth and productivity. Cover crops are planted for soil erosion control, soil fertility, and quality management as subsequent cropping in the same land reduces the soil quality by removing soil organic matter. In conventionally tilled legume-based rotation, use of cover crop was found to be effective to mitigate SOC and soil organic nitrogen (SON) losses, increasing N use efficiency of the crop system while maintaining optimum productivity (Daniel Plaza-Bonilla et al. 2016). Soil erosion can be significantly reduced by crop and soil management practices, such as minimal tillage, contour ridging, mulching, fertilizer, intercropping, narrow plant spacing, and planting cover crop of grasses or legume (Howeler 1987 and 1994; Ruppenthal et al. 1997; Yadav et al. 2017). Annual legumes when grown as cover crop have the advantage of providing adequate cover within short duration of 6 weeks from planting and can be effectively used to control soil erosion faster.

16.9 Agronomic Use Efficiency

In natural ecosystem, plant follows ecological succession, and better adapted plant replaces the pre-existing one. But in managed ecosystem (like the agricultural land), cultivation of crops can be done according to the necessity of human being. For a sustainable production of crop, the management of soil is very important (Meena et al. 2015c). Soil fertility is generally maintained by application of chemical fertilizers. NPK are the main nutrients applied in field during crop cultivation. Testing of soil is essential before application of fertilizer to find out which element is less in soil for crop production. It was found that in most intensive crop production systems, 50–75% of N applied to field is not used by crops and N is lost by leaching into the soil causing environmental pollution, such as surface and groundwater pollution (Hodge 2000; Asghari and Cavagnaro 2011). Contaminated water with nitrate is not potable, and at higher concentration, it can cause serious health problems (Umar and Iqbal 2007). Well-grown grain legumes are self-sufficient in their N requirement and even can contribute to N economy of the entire cropping system by adding fixed N to the soil pool, using little or none from the soil reserves of N (Walley et al. 2007). Studies are in progress on whether increasing water use efficiency (WUE) and nutrient use efficiency (NUE) in food legumes is possible through agronomic means. Grain yield per unit of water use, evapotranspiration, or growing-season rainfall is termed as crop WUE of plant. Increasing WUE is associated with increasing grain yield and water use after flowering (Loss et al. 1997; Siddique et al. 2001). For example, late planting reduces the WUE with decreasing grain yield. Early planting is preferred to give better yield and higher WUE. Exceptions are there in field pea, where too early plantation leads to the development of black spot disease (Siddique et al. 1998). Use of herbicides or manual weeding increases the water use efficiency and crop yield by increasing NUE and the economic yield per unit of nutrient applied (Verma et al. 2017). NUE is declining gradually with time, and the nutrients lost from the agricultural system have detrimental effects on adjacent ecosystems (Cloern et al. 2007). Therefore, it is necessary to increase fertilizer use efficiency and apply minimum fertilizer as possible. During the process of domestication and breeding, the genetic diversity of some important crops has been reduced (Warschefsky et al. 2014). Genes from the crop with higher nutrient utilization ability can be used in genetic engineering for improving NUE of other crops. Performing organic farming can minimize the detrimental effects on environment and reduce the environmental risk. Legume can be utilized for better NUE. For example, legumes such as lupin have the capacity to utilize P from partially available sources than other crop species (Braum and Helmke 1995). Depending on the environmental conditions, the legume can add maximum possible N to the system leading to high crop yield. For example, legumes are reported to be sensitive to stress and stop fixing N on exposure to drought (Sinclair et al. 1987).

Low-Input Sustainable Agriculture (LISA) was replaced by Sustainable Agriculture Research and Education (SARE) program through an act passed in the US Congress during 1985. The main focus of this program was to maintain high land productivity to using the techniques that minimize the use of pesticides, fertilizers, and off-farm purchases through appropriate rotations; biological weed, pest, and disease control; integration of livestock with crops; and minimum tillage systems. Lower-input in sustainable systems do not mean practicing of only organic system; rather, it requires a farmer to understand more about the biological effects of a crop or management systems and how to use this information cheaply and effectively in farm programs (Meena et al. 2015b), e.g., integrated pest management.

For the healthy growth of food legumes, formation of adequate nodule is necessary (Dhakal et al. 2016) even in cool and dry conditions where rhizobia are not available in soil. Under that situation, the inoculation of rhizobia is essential. Cultivated legumes are mostly slow grower at early stage and prone to weed competition. Pandey et al. (1998) reported that weed can reduce the legume yields by 25–40%. Weed control in legume crop can minimize the loss of grain yield. Herbicides are becoming noneffective due to the development of herbicide-resistant weed variety. Manual weeding is also becoming increasingly expensive due to shortage of labor. Paolini et al. (2003) and McDonald et al. (2007) found an increased weed infestation with the increasing density and competitive ability of lentil, which enhances the cost for weed suppression relative to mechanical and chemical pest control mechanism.

16.10 Future Perspectives

Among the diverse species of legumes, only very few have received the attention of the researchers. Therefore, this is necessary to explore the other legume species (both wild and cultivated) for their multiple benefits. The explored valuable qualities of grain legumes should cross into the germplasm to produce higher nutritious food for human and livestock. To obtain it, research objective aiming on this area is required. Research emphasis focusing on the use of legume and their rhizobia for value-added future exploitation including the opportunities such as use as a source of pharmaceutical drugs against various diseases is very much essential. In this regard, to practically realize the benefits of rhizobia to its fullest, in-depth studies on the rhizobial manipulation are a must involving the agricultural biotechnologist (Meena et al. 2017b). There is also an urgent need to assess the overall socioeconomic and environmental significances which may arise from the widespread adaptation of legume-based agriculture so that it helps the farmers in decision-making. With the escalating rate of climate change, this is also important to breed legume cultivars for various abiotic stress resistances.

16.11 Conclusion

This chapter gives an overview of different aspects of legume growth, productivity, and their impact on soil health. Legumes are an important ingredient of human diet especially for the large vegetarian population of the world. In the era of green revolution with major focus on staple foods like rice, wheat, and potato, cultivation of legumes was relegated to the marginal land with least of inputs. This, coupled with the increasing population, resulted in reducing per capita availability of legumes to the common people. Cutting-edge technologies on legume culture need to be developed in order to face the challenges of climate change. Genomics, transgenics, molecular breeding, quality improvement, and biotic and abiotic stress management of different legume crops need more attention. Legumes can be considered as smart food for high nutritional value having low water footprint, low carbon footprint, and ability to sustain soil health. Agribusiness opportunities of legume crops are an emerging area which can help the small landholders of countries like India. These crops can be a good source of study for soil N dynamics and soil N_2O production and emission. The leguminous intercrop can increase soil available N for the subsequent crop. Legume as intercrop may reduce the N loss and can improve soil N availability for the subsequent crop. Legumes grown in an ecosystem can also be a good source of carbon sink in the form of biomass and in soil as well. Well-designed studies on legume crops and their impacts on soil C dynamics and carbon storage are needed for climate resilient agriculture.

References

Adu-Gyamfi JJ, Ito O, Yoneyama T, Katayama K (1997) Nitrogen management and biological nitrogen fixation in sorghum/pigeonpea intercropping on Alfisols of the semi-arid tropics. Soil Sci Plant Nutr 43(1):1061–1066

Alexandratos N (2009) World food and agriculture to 2030/50 highlights and views from mid-2009, Expert Meeting on How to feed the World in 2050, Food and Agriculture Organization of the United Nations, Economic and Social Development Department, 24–26 Jun 2009

Alpmann D, Braun J, Schäfer BC (2013) Analyseeiner Befragunguntererfolgreichen KörnerleguminosenanbauernimkonventionellenLandbau, Erste Ergebnisseausdem Forschungsprojekt LEGUAN, In: ImFokus: HeimischeKörnerleguminosenvom AnbaubiszurNutzung, DLG Wintertagung, 15–17 Jan 2013

Alvey S, Yang CH, Buerkert A, Crowley DE (2003) Cereal/legume rotation effects on rhizosphere bacterial community structure in West African soils. BiolFertil Soils 37:73–82

Anderson TH, Domsch KH (1989) Ratios of microbial biomass carbon to total organic carbon in arable soils. Soil BiolBiochem 21:471–479

Andrews DJ, Kassam AH (1976) The importance of multiple cropping in increasing world food supplies, ASA Special Publication, vol 27. American Society of Agronomy, Madison, pp 1–10

Arab EAA, Helmy IMF, Bareh GF (2010) Nutritional evaluation and functional properties of chickpea (*Cicer arietinum* L.) flour and the improvement of spaghetti produced from its. J Am Sc 6(10):1055–1072

Asghari HR, Cavagnaro TR (2011) Arbuscular mycorrhizas enhance plant interception of leached nutrients. Funct Plant Biol 38:219–226

Ashoka P, Meena RS, Kumar S, Yadav GS, Layek J (2017) Green nanotechnology is a key for eco-friendly agriculture. J Clean Prod 142:4440–4441

Bellaloui N, Mengistu A, Kassem MA (2013) Effects of genetics and environment on fatty acid stability on soybean seed. Food NutrSci 4:165–175

Bertoglio JC, Calvo MA, Hancke JL, Burgos RA, Riva A, Morazzoni P, Ponzone C, Magni C, Durant M (2011) Hypoglycemic effect of lupin seed γ-conglutin in experimental animals and healthy human subjects. Fitoterapia 82(7):933–938

Bhatt BP, Bujarbaruah KM (2006) Agroforestry in North East India: opportunities and challenges (© 2005), ICAR Research Complex for NEH Region, Umiam, Meghalaya

Blanchart E, Villenave C, Viallatoux A, Barthès B, Girardin C, Azontonde A, Fellera C (2006) Long-term effect of a legume cover crop (*Mucuna pruriens var. utilis*) on the communities of soil macrofauna and nematofauna, under maize cultivation, in southern Benin. Eur J Soil Biol 42:S136–S144

Boudreau MA, Mundt CC (1992) Mechanisms of alterations in bean rust epidemiology due to intercropping with maize. Phytopathology 82:1051–1060

Braum SM, Helmke PA (1995) White lupin utilizes soil phosphorus that is unavailable to soybean. Plant Soil 176:95–100

Brookes PC (1995) The use of microbial parameters in monitoring soil pollution by heavy metals. BiolFertil Soils 19:269–279

Bues A, Preissel S, Reckling M, Zander P, Kuhlman T, Topp K, Watson CA, Stoddard F, Bokem M (2013) The environmental role of protein crops in the new Common Agricultural Policy (D4). European Parliament, Directorate General for Internal Policies, Policy Department B: Structural and Cohesion Policies, Agricultural and Rural Development

Campbell CA, Zentner RP (1993) Soil organic matter as influenced by crop rotation and fertilization. Soil SciSoc Am J 57:1034–1040

Campbell CA, Bowren KE, Schnitzer M, Zentner RP, Townley-Smith L (1991) Effect of crop rotations and fertilization on soil organic matter and some biochemical properties of a thick Black Chernozem. Can J Soil Sci 71:377–387

Carof M (2006) Fonctionnement de peuplementsen semis direct associant du blétendred'hiver (Triticumaestivum L.) à différentesplantes de couvertureenclimattempéré. INAPG (AgroParisTech)

Chu GX, Shen QR, Cao JL (2004) Nitrogen fixation and N transfer from peanut to rice cultivated in aerobic soil in an intercropping system and its effect on soil N fertility. Plant Soil 263:17–27

Clark CM, Yongfei B, Bowman William D, Cowles Jane M, Fenn Mark E, Gilliam Frank S, Phoenix Gareth K, Ilyas S, Stevens Carly J, Sverdrup Harald U, Throop Heather L (2013) Nitrogen deposition and terrestrial biodiversity. In: Levin SA (ed) Encyclopedia of biodiversity, vol 5, 2nd edn. Academic, Waltham, pp 519–536

Cloern J, Krantz T, Duffy JE (2007) Eutrophication. In: Cleveland CJ (ed) Encyclopedia of earth. Environmental information coalition. National Council for Science and the Environment, Washington, DC. Available at http://www.eoearth.org/article/Eutrophication Last revised December 18, 2007

Collette L, Hodgkin T, Kassam A, Kenmore P, Lipper L, Nolte C, et al. (2011) Save and grow: a policymaker's guide to the sustainable intensification of smallholder crop production. Food and Agriculture Organization of the United Nations (FAO), Rome. isbn:978-92-5-106871-7

Conen F, Dobbie KE, Smith KA (2000) Predicting N_2O emissions from agricultural land through related soil parameters. Glob Chang Biol 6:417–426

Cooper P, Rao KPC, Singh P, Traore PS, Rao K, Dixit P, Twomlow SJ (2009) Farming with current and future climate risk: advancing a "hypothesis of hope" for rainfed agriculture in the semi-arid tropics. J SAT Agric Res 7:1–19

Costa GEA, Queiroz-Monici KS, Reis SMPM, Oliveira AC (2006) Chemical composition, dietary fibre and resistant starch contents of raw and cooked pea, common bean, chickpea and lentil legumes. Food Chem 94:327–330

Courty PE, Smith P, Koegel S, Redecker D, Wipf D (2015) Inorganic nitrogen uptake and transport in beneficial plant root-microbe interactions. Crit Rev Plant Sci 34(1–3):4–16

Cousin R (1997) Peas (*Pisum sativum* L.). Field Crop Res 53:111–130

Cowell LE, Bremer E, Van Kessel C (1989) Yield and N, fixation of pea and lentil as affected by intercropping and N application. Can J Soil Sci 69:243–251

Danso SKA, Zapata F, Hardarson G (1987) Nitrogen fixation in fababeans as affected by plant population density in sole or intercropped systems with barley. Soil BiolBiochem 19(4):411–415

De AntoniMigliorati M, Bell M, Grace PR, Scheer C, Rowlings DW, Liu S (2015) Legume pastures can reduce N_2O emissions intensity in subtropical cereal cropping systems. Agric Ecosyst Environ 204:27–39

Devi MJ, Sinclair TR, Vadez V, Krishnamurthy L (2009) Peanut genotypic variation in transpiration efficiency and decreased transpiration during progressive soil drying. Field Crop Res 114:280–285

Dhakal Y, Meena RS, De N, Verma SK, Singh A (2015) Growth, yield and nutrient content of mungbean (*Vigna radiata* L.) in response to INM in eastern Uttar Pradesh, India. Bangladesh J Bot 44(3):479–482

Dhakal Y, Meena RS, Kumar S (2016) Effect of INM on nodulation, yield, quality and available nutrient status in soil after harvest of green gram. Leg Res 39(4):590–594
Drinkwater LE, Wagoner P, Sarrantonio M (1998) Legume-based cropping systems have reduced carbon and nitrogen losses. Nature 396:262–264
Duranti M (2006) Grain legume proteins and nutraceutical properties. Fitoterapia 77(2):67–82
Duranti M, Gius C (1997) Legume seeds: protein content and nutritional value. Field Crop Res 53(1–3):31–45
Ehlers JD, Hall AE (1997) Cowpea (*Vigna unguiculata* L. Walp.). Field Crop Res 53:187–204
El-Al HA, El-Hwat N, El-Hefnawy N, Medany M (2011) Effect of sowing dates, irrigation levels and climate change on yield of common bean (*phaseolus vulgaris* L.). AmEur J Agri Environ Sci 11(1):79–86
Fang X, Turner NC, Yan G, Li F (2010) Flower numbers, pod production, pollen viability and pistil production are reduced and flower and pod abortion increased in chick pea (*Cicer arietinum* L.) under terminal drought. J Exp Bot 61:335–345
Fatima Z, Bano A, Sial R, Aslam M (2008) Response of chickpea to plant growth regulators on nitrogen fixation and yield. Pak J Bot 40(5):2005–2013
Fininsa C (1996) Effect of intercropping bean with maize on bean common bacterial blight and rust diseases. Int J Pest Manag 42:51–54
Francis CA (1986) Multiple cropping systems. Macmillan, New York
Franzluebbers AJ (2002) Soil organic matter stratification ratio as an indicator of soil quality. Soil TillRes 66:95–106
Fujita K, Ofosu-Budu KG, Ogata S (1992) Biological nitrogen fixation in mixed legume-cereal cropping systems. Plant Soil 141:155–176
Garcia MC, Marina ML, Laborda F, Torre M (1998) Chemical characterization of commercial soybean products. Food Chem 62:325–331
Garcia-Estringanaa P, Alonso-Blázqueza N, Marquesa MJ, Bienesa R, González-Andrésc F, Alegrea J (2013) Use of Mediterranean legume shrubs to control soil erosion and runoff in central Spain. A large-plot assessment under natural rainfall conducted during the stages of shrub establishment and subsequent colonisation. Catena 102:3–12
Ghosh PK (2004) Growth, yield, competition and economics of groundnut/cereal fodder intercropping systems in the semi-arid tropics of India. Field Crop Res 88:227–237
Ghosh PK, Singh NP (1994) Soil-nitrogen status under summer legumes-maize (*Zea mays*) sequence. Indian J AgricSci 641(12):856–857
Ghosh PK, Manna MC, Bandyopadhyay KK, Tripathi AK, Wanjari RH, Hati KM, Misra AK, Acharya CL, Subba Rao A (2005) Interspecific interaction and nutrient use in soybean/sorghum intercropping system. Agron J 98(4):1097–1108
Ghosh PK, Bandyopadhyay KK, Wanjari RH, Manna MC, Misra AK, Mohanty M, SubbaRao A (2007) Legume effect for enhancing productivity and nutrient use-efficiency in major cropping systems–an Indian perspective: a review. J Sustai Agri 30(1):59–86. https://doi.org/10.1300/J064v30n01_07
Gianinazzi S, Wipf D (2010) Des champignons au service des plantes. PHM–Revue Hortic 521:9–11
Gilbert GA, Knight JD, Vance CP, Allan DL (1999) Acid phosphatase activity in phosphorus-deficient white lupin roots. Plant Cell Environ 22:801–810
Giller KE (2001) Nitrogen fixation in tropical cropping systems, vol 423, 2nd edn. CABI, Wallingford
Goni I, Valentin-Gamazo C (2003) Chickpea flour ingredient slows glycemic response to pasta in healthy volunteers. Food Chem 81(4):511–515
Gruhn P, Goletti F, Yudelman M (2000) Integrated nutrient management, soil fertility, and sustainable agriculture: current issues and future challenges. International food policy research Institute, Washinton, DC
Gu J, Nicoullaud B, Rochette P, Grossel A, Hénault C, Cellier P, Richard G (2013) A regional experiment suggests that soil texture is a major control of N_2O emissions from tile-drained winter wheat fields during the fertilization period. Soil BiolBiochem 60:134–141

Harland JI, Haffner TA (2008) Systematic review, meta-analysis and regression of randomised controlled trials reporting an association between an intake of circa 25 g soya protein per day and blood cholesterol. Atherosclerosis 200(1):13–27

Hauggaard-Nielsen H, Ambus P, Jensen ES (2001) Interspecific competition, N use and interference with weeds in pea-barley intercropping. Field Crop Res 70:101–109

Hauggaard-Nielsen H, Jørnsgaard B, Kinane J, Jensen ES (2008) Grain legume-cereal intercropping: the practical application of diversity, competition and facilitation in arable and organic cropping systems. Renew Agric Food Syst 23:3–12

Hayat R, Ali S, Ijaz SS, Chatha TH, Siddique MT (2008) Estimation of N_2-fixation of mung bean and mash bean through xylem ureide technique under rainfed conditions. Pak J Bot 40(2):723–734

Hayman DS (1986) Mycorrhizae of nitrogen fixing legumes. World J MicrobioBiotechno 2(1):121–145

Hazarika UK, Munda GC, Bujarbaruah KM, Das A, Patel DP, Prasad K, Kumar R, Panwar AS, Tomar JMS, Bordoloi JS, Sharma M, Gogoi G (2006) Nutrient management in organic farming. Technical Bulletin No 30, ICAR Research Complex for NEH Region, Meghalaya

Heichel GH (1987) Legume nitrogen: symbiotic fixation and recovery by subsequent crops. In: Heisel Z (ed) Energy in plant nutrition and pest control. Energy World Agric Elsevier, New York

Heichel GH, Barnes DK (1984) Opportunities for meeting crop nitrogen needs from symbiotic nitrogen fixation. In: Bezdicek DF (ed) Organic farming, Spec. Pub., vol 46. ASA, CSSA, SSSA, Madison

Helal HM (1990) Varietal difference in root phosphatase activity as related to the utilization of organic phosphorus. In: Bassam N (ed) Genetic aspects of plant mineral nutrition. Kluwer Academic Publishers, Dordrecht

Hodge I (2000) Agri-environmental Relationships and the choice of policy mechanism. World Econ 23(2):257–273

Howeler RH (1987) Soil conservation practices in cassava-based cropping systems. In: Thay TH, Mokhtaeruddin AM, Zahari AB (eds) Proc. Intern. Conf. streep land Agric. In Humid Tropics, held in Kuala Lumpur, Malaysia. Aug 16–21, 1987. p 490–517

Howeler RH (1994) Integrated soil and crop management to prevent environmental degradation in cassava-based cropping system in Asia. In: Bottema JWT, Stoltz DR (eds.) Upland agriculture in Asia. Proceeding of workshop held in Bogor, Indonesia. April 6–8, 1993. pp 195–224 http://agritech.tnau.ac.in/agriculture/pulses_index.html. Accessed on 27 Mar 2017

Huang WY (2007) Impact of rising natural gas prices on U.S. ammonia supply, Outlook Report No. (WRS-0702). USDA Economic Research Service, Washington, DC

Insam H, Parkinson D, Domsch KH (1989) The influence of macroclimate on soil microbial biomass levels. Soil BiolBiochem 21:211–221

Intergovernmental Panel on Climate Change (2013) Climate change 2013: the physical science basis. Climate change 2013: the physical science basis. In: Stocker TF, Qin D, Plattner G-K, Tignor M, Allen SK et al (eds) Contribution of working group I to the fifth assessment report of the intergovernmental panel on climate change. Cambridge University Press, Cambridge, UK. Available:: http://www.climatechange2013.org/images/uploads/WGI_AR5_SPM_brochure.pdf. Accessed on 4 Apr 2017

Iqbal A, Khalil IA, Ateeq N, Khan MS (2006) Nutritional quality of important food legumes. Food Chem 97:331–335

Javaid A, Ghafoor A, Anwar R (2004) Seed storage protein electrophoresis in ground nut for evaluating genetic diversity. Pak J Bot 36(1):25–29

Jeanneret P, Baumgartner D, Freiermuth R, Gaillard G (2006) Méthoded'évaluation de l'impact des activitésagricolessur la biodiversitédans les bilansécologiques. SALCA-BD, Rapport Agroscope 67

Jenkinson DS, Ladd JN (1981) Microbial biomass in soil measurement and turnover. In: Paul EA, Ladd JN (eds) Soil biochemistry, vol 5. Marcel Dekker, New York, pp 415–471

Jensen ES (1996) Grain yield, symbiotic N_2 fixation and interspecific competition for inorganic N in pea-barley intercrops. Plant Soil 182:25–38

Jensen ES, Hauggaard-Nielsen H (2003) How can increased use of biological N_2 fixation in agriculture benefit the environment? Plant Soil 252:177–186

Jensen ES et al. (2005) Intercropping-the practical application of diversity, competition and facilitation in arable and organic cropping systems; In: Researching Sustainable Systems. Proceedings of the First Scientific Conference of the International Society of Organic Agricultural Research (ISOFAR) p 22–25

Jensen ES, Peoples MB, Boddey RM, Gresshoff PM, Hauggaard-Nielsen H, Alves BJR, Morrison MJ (2012) Legumes for mitigation of climate change and the provision of feedstock for biofuels and biorefineries-a review. Agron SustainDev 32:329–364

Jeuffroy M, Baranger E, Carrouée B, Ed C, Gosme M, Hénault C, Schneider A, Cellier P (2013) Nitrous oxide emissions from crop rotations including wheat, rapeseed and dry pea. Biogeosci Discuss 9(7):9289

Jones DL, Dennis PG, Owen AG, van Hees PAW (2003) Organic acid behavior in soils- misconceptions and knowledge gaps. Plant Soil 248(1):31–41

Justus M, Köpke U (1995) Strategies to reduce nitrogen losses via leaching and to increase Precrop effects when growing Faba beans. Biol Agric Hortic: Inter J Sustain Prod Syst 11(1–4):145–155

Khadka K, Acharya BD (2009) Cultivation practices of ricebean. Local Initiatives for Biodiversity, Research and Development (LI-BIRD), Pokhara

Kirkegaard J, Christen O, Krupinsky J, Layzell D (2008) Break crop benefits in temperate wheat production. Field Crop Res 107(3):185–195

Klauer SF, Francesch VR (1997) Mechanism of transport of vegetative storage proteins to the vacuole of the paraveinal mesophyll of soybean leaf. Protoplasma 200(3):174–185

Köpke U, Nemecek T (2010) Ecological services of faba bean. Field Crop Res 115(3):217–233

Kudapa H, Ramalingam A, Nayakoti S, Chen X, Zhuang W, Liang X, Kahl G, Edwards D, Varshney RK (2013) Functional genomics to study stress responses in crop legumes: progress and prospects. Funct Plant Biol 40(12):1221–1233

Laddha KC, Totawat KL (1997) Effects of deep tillage under rainfed agriculture on production of sorghum (So$z~rn biocolor L. Moench) intercropped with green gram (Vigna radiata L. 'Wilczek) in western India. Soil Tillage Res 43:241–250

Landers PS (2007) The food stamp program: history, nutrition education, and impact. J Am Diet Assoc 107(11):1945–1951

Lansing AJ, Franceschi VR (2000) The paraveinal mesophyll: a specialized path for intermediary transfer of assimilates in legume leaves. Aust J Plant Physiol 27:757–767

Latati M, Blavet D, Alkama N, Laoufi H, Drevon JJ, Gérard F, Pansu M, Ounane SM (2014) The intercropping cowpea-maize improves soil phosphorus availability and maize yields in an alkaline soil. Plant Soil 385:181–191

Lemke RL, Zhonga Z, Campbell CA, Zentner R (2007) Can pulse crops play a role in mitigating greenhouse gases from north American agriculture? Agron J 99(6):1719–1725

Li L, Li SM, Sun JH, Zhou LL, Bao XG, Zhang HG, Zhang FS (2007) Diversity enhances agricultural productivity via rhizosphere phosphorus facilitation on phosphorous-deficient soils. Proc Natl Acad Sci 104:11192–11196

Li H, Shen J, Zhang F, Marschner P, Cawthray G, Rengel Z (2009) Phosphorus uptake and rhizosphere properties of intercropped and monocropped maize, faba bean, and white lupin in acidic soil. BiolFertil Soils 46:79–91

Liang S, Grossman J, Shi W (2014) Soil microbial responses to winter legume cover crop management during organic transition. Eur J Soil Biol 65:15–22

Liu L, Wang Y, Yan X, Li J, Jiao N, Hu S (2017) Biochar amendments increase the yield advantage of legume-based intercropping systems over monoculture. AgricEcosyst Environ 237:16–23

Lopez-Bellido RJ, Fontána JM, López-Bellidob FJ, López-Bellidoa L (2009) Carbon sequestration by tillage, rotation, and nitrogen fertilization in a Mediterranean vertisol. Agron J 102(1):310–318

Loss SP, Siddique KHM, Tennant D (1997) Adaptation of faba bean (Vicia faba L.) to dryland Mediterranean-type environments. III. Water use and water-use efficiency. Field Crop Res 54:153–162

Mandal MK, Banerjee M, Banerjee H, Alipatra A, Malik GC (2014) Productivity of maize (*zea mays*) based intercropping system during kharif season under red and lateritic tract of West Bengal. Bioscan 9(1):31–35
Mantri N, Basker N, Ford R, Pang E, Pardeshi V (2013) The role of micro-ribonucleic acids in legumes with a focus on abiotic stress response. Plant Genome 6(3):1–14
Masuda T, Goldsmith PD (2009) World soybean production: area harvested, yield, and long-term projections. Int Food Agribusiness Manag Rev 12(4):143–162
McDonald GK, Hollaway KL, McMurray L (2007) Increasing plant density improves weed competition in lentil (Lens culinaris). Aust J ExpAgric 47:48–56
McGill WB, Cannon KR, Robertson JA, Cook FD (1986) Dynamics of soil microbial biomass and water-soluble organic C in Breton L. after 50 years of cropping to two rotations. Can J Soil Sci 66:1–19
Meena RS, Yadav RS (2015) Yield and profitability of groundnut (*Arachis hypogaea* L) as influenced by sowing dates and nutrient levels with different varieties. Leg Res 38(6):791–797
Meena VS, Maurya BR, Meena RS, Meena SK, Singh NP, Malik V K (2014) Microbial dynamics as influenced by concentrate manure and inorganic fertilizer in alluvium soil of Varanasi, India
Meena RS, Meena VS, Meena SK, Verma JP (2015a) The needs of healthy soils for a healthy world. J Clean Prod 102:560–561
Meena VS, Maurya BR, Meena RS (2015b) Residual impact of well-grow formulation and NPK on growth and yield of wheat (*Triticum aestivum* L.). Bangladesh J Bot 44(1):143–146
Meena RS, Meena VS, Meena SK, Verma JP (2015c) Towards the plant stress mitigate the agricultural productivity: a book review. J Clean Prod 102:552–553
Meena RS, Yadav RS, Meena H, Kumar S, Meena YK, Singh A (2015d) Towards the current need to enhance legume productivity and soil sustainability worldwide: a book review. J Clean Prod 104:513–515
Meena RS, Bohra JS, Singh SP, Meena VS, Verma JP, Verma SK, Shiiag SK (2016) Towards the prime response of manure to enhance nutrient use efficiency and soil sustainability a current need: a book review. J Clean Prod 112:1258–1260
Meena RS, Meena PD, Yadav GS, Yadav SS (2017a) Phosphate solubilizing microorganisms, principles and application of microphos technology. J Clean Prod 145:157–158
Meena RS, Kumar V, Yadav GS, Mitran T (2017b) Response and interaction of *Bradyrhizobium japonicum* and *Arbuscular mycorrhizal* fungi in the soybean rhizosphere: a review. Plant Growth Reg., Accepted in press African J Microb Res 8(1):257–263
Morris JB (1997) Special-purpose legume genetic resources conserved for agricultural, industrial, and pharmaceutical use. Econ Bot 51(3):251–263
Mucheru-Muna M, Pypers P, Mugendi D, Kung'u J, Mugwe J, Merckx R, Vanlauwe B (2010) Staggered maize–legume intercrop arrangement robustly increases crop yields and economic returns in the highlands of Central Kenya. Field Crop Res 115:132–139
Nemecek T, Richthofen JV, Dubois G, Casta P, Charles R, Pahlf H (2008) Environmental impacts of introducing grain legumes into European crop rotations. Eur J Agron 28:380–393
Nuruzzaman M, Lambers H, Bolland MDA, Veneklaas EJ (2006) Distribution of carboxylates and acid phosphatase and depletion of different phosphorus fractions in the rhizosphere of a cereal and three grain legumes. Plant Soil 281(1):109–120
Nwoke OC, Diels J, Abaidoo R, Nziguheba G, Merckx R (2008) Organic acids in the rhizosphere and root characteristics of soybean (Glycine max) and cowpea (*Vigna unguiculata*) in relation to phosphorus uptake in poor savanna soils. Afri J Biotech 7(20):3620–3627
Okereke GU, Ayama N (1992) Sources of nitrogen and yield advantages for monocropping and mixed cropping with cowpeas (Vigna unguiculata L.) and upland rice (*Oryza sativa* L.). BiolFertil Soils 13:225–228
Paetau I, Chen CZ, Jane JJ (1994) Biodegradable plastic made from soybean products. 1.Effect of preparation and processing on mechanical-properties and water-absorption. J EnviroPolyDegrad 2(3):211–217

Pandey AK, Prasad K, Singh P, Singh RD (1998) Comparative yield loss assessment and crop-weed association in major winter crops of mid hills of N-W Himalayas. Indian J Weed Sci 30:54–57

Paolini R, Colla G, Saccardo F, Campiglia E (2003) The influence of crop plant density on the efficacy of mechanical and reduced-rate chemical weed control in lentil (Lens culinarisMedik.). Ital J Agron 7:85–94

Pappa VA, Rees RM, Walker RL, Baddeley JA, Watson CA (2011) Nitrous oxide emissions and nitrate leaching in an arable rotation resulting from the presence of an intercrop. AgricEcosyst Environ 141(1–2):153–161

Peoples MB, Brockwell J, Herridge DF, Rochester IJ, Alves BJR, Urquiaga S, Boddey RM, Dakora FD, Bhattarai S, Maskey SL, Sampet C, Rerkasem B, Khan DF, Hauggaard-Nielsen H, Jensen ES (2009) The contributions of nitrogen-fixing crop legumes to the productivity of agricultural systems. Symbiosis 48(1):1–17

Peyraud JL, Gall AL, Lüscher A (2009) Potential food production from forage legume-based-systems in Europe: an overview. Irish J Agric Food Res 48(2):115–135

Plaza-Bonilla D, Álvaro-Fuentes J, Arrúe JL, Cantero-Martínez C (2014) Tillage and nitrogen fertilization effects on nitrous oxide yield-scaled emissions in a rainfed Mediterranean area. AgricEcosyst Environ 189:43–52

Plaza-Bonilla D, Nolot JM, Passot S, Raffaillac D, Justes E (2016) Grain legume-based rotations managed under conventional tillage need cover crops to mitigate soil organic matter losses. Soil Till Res 156:33–43

Power F, Stout JC (2011) Organic dairy farming: impacts on insect–flower interaction networks and pollination Eileen. J Appl Ecol 48:561–569

Prakash D, Gupta C (2011) Role of phytoestrogens as nutraceuticals in human health. Pharmacologyonline 1:510–523

Qiang C, Peng H, Gaobao H (2004) Effect of intercropping on soil microbial and enzyme activity in the rhizosphere. Acta Pratacult Sin 14:105–110

Ram K, Meena RS (2014) Evaluation of pearl millet and mungbean intercropping systems in arid region of Rajasthan (India). Bangladesh J Bot 43(3):367–370

Ramana S, Biswas AK, Kundu S, Saha JK, Yadava RBR (2002) Effect of distillery effluent on seed germination in some vegetable crops. Bioresour Technol 82(3):273–275

Reckling M, Preissel S, Zander P, Topp K, Watson C, Murphy-Bokern D, Stoddard FL (2014) Legume futures report 1.6 -effects of legume cropping on farming and food systems, legume-supported cropping systems for Europe. www.legumefutures.de. Accessed on 04 Apr 2017

Ruppenthal M, Leihner DE, Steinmuller N, El-Sharkawy MA (1997) Losses of organic matter and nutrients by water erosion in cassava-based cropping systems. Exp Agric 33:487–498

Sanchez PA (1976) Properties and management of soils in the tropics. Wiley, New York, pp 478–532

Sanginga N, Woomer PL (2009) Integrated soil fertility management in Africa: principles, practices and development process (eds.). Tropical Soil Biology and Fertility Institute of the International Centre for Tropical Agriculture, Nairobi 263

Schelud'ko AV, Makrushin KV, Tugarova AV, Krestinenko VA, Panasenko VI, Antonyuk LP, Katsy EI (2009) Changes in motility of the rhizobacterium Azospirillum brasilense in the presence of plant lectins. Microbiol Res 164:149–156

Scheublin TR, Ridgway KP, Young JPW, van der Heijden MGA (2004) Nonlegumes, legumes, and root nodules harbor different arbuscular mycorrhizal fungal communities. Appl Environ Microbiol 70:6240–6246

Shen H, Yan X, Zhao M, Zheng S, Wang X (2002) Exudation of organic acids in common bean as related to mobilization of aluminium- and iron-bound phosphates. Environ Exp Bot 48(1):1–9

Shimelis EA, Rakshit SK (2005) Proximate composition and physico-chemical properties of improved dry bean (*Phaseolus vulgaris* L.) varieties grown in Ethiopia. LWT Food Sci Technol 38:331–338

Siddique KHM, Loss SP, Pritchard DL, Regan KL, Tennant D, Jettner RL, Wilkinson D (1998) Adaptation of lentil (Lens culinarisMedik.) to Mediterranean-type environments: effect of time of sowing on growth, yield, and water use. Aust J Agric Res 49:613–626

Siddique KHM, Regan KL, Tennant D, Thomson BD (2001) Water use and water use efficiency of cool season grain legumes in low rainfall Mediterranean-type environments. Eur J Agron 15:267–280

Silva PMD, Tsai SM, Bonetti R (1993) Response to inoculation and N fertilization for increased yield and biological nitrogen fixation of common bean (*Phaseolus vulgaris* L.). Plant Soil 152:123–130

Sinclair TR, Vadez V (2002) Physiological traits for crop yield improvement in low N and P environments. Plant Soil 245:1–15

Sinclair TR, Muchow RC, Bennett JM, Hammond LC (1987) Relative sensitivity of nitrogen and biomass accumulation to drought in field-grown soybean. Agron J 79:986–991

Singh VP (2000) Planting geometry in maize (*Zea mays*) and blackgram (*Phaseolus mungo*) intercropping system under rainfed low hill valley of Kumaon. Indian J Agron 45(2):274–278

Singh RP (2013) Status paper on pulses. Government of India, Ministry of Agriculture (Department of Agriculture & Cooperation), Directorate of pulses Development, 6th floor, Vindhyanchal Bhavan, Bhopal-462004 (M.P)

Sirtori CR, Galli C, Anderson JW, Sirtori E, Arnoldi A (2009) Functional foods for dyslipidaemia and cardiovascular risk prevention. Nutr Res Rev 22:244–261

Smith KA, Thomson PE, Clayton H, Mctaggart IP, Conen F (1998) Effects of temperature, water content and nitrogen fertilisation on emissions of nitrous oxide by soils. Atmos Environ 32:3301–3309

Song YN, Zhang FS, Marschner P, Fan FL, Gao HM, Bao XG, Sun JH, Li L (2005) Effect of intercropping on crop yield and chemical and microbiological properties in rhizosphere of wheat (*Triticum aestivum* L.), maize (*Zea mays* L.), and faba bean (*Vicia faba* L.). BiolFertil Soils 43:565–574

Sørensen LH (1987) Organic matter and microbial biomass in a soil incubated in the field for 20 years with 14C-labelled barley straw. Soil Biol Biochem 19:39–42

Sparling GP (1992) Ratio of microbial biomass carbon to soil organic carbon as a sensitive indicator of changes in soil organic matter. Aust J Soil Res 30:195–207

Stevenson FJ (1982) Origin and distribution of nitrogen in soil. In: Stevenson FJ (ed) Nitrogen in agricultural soils. ASA, Madison

Suman A, Lal M, Singh AK, Gaur A (2006) Microbial biomass turnover in Indian subtropical soils under different sugarcane intercropping systems. Agron J 98(3):698–704

Tang X, Bernard L, Brauman A, Daufresne T, Deleporte P, Desclaux D, Souche G, Placella SA, Hinsinger P (2014) Increase in microbial biomass and phosphorus availability in the rhizosphere of intercropped cereal and legumes under field conditions. Soil Biol Biochem 75:86–93

Thorup-Kristensen K, Magid J, Jensen LS (2003) Catch crops and green manures as biological tools in nitrogen management in temperate zones. Agron B-A Acad Press 79:227–302

Tiwari RC, Sharma PK, Khandelwal SK (2004) Effect of green manuring through Sesbaniacannabina and Sesbaniarostrata and nitrogen application through maize (Zea mays) in maize– wheat (*Triticumaestivum*) cropping system. Indian J Agron 49(1):15–17

Tobita S, Ito O, Matsunaga R, Rao TP, Rego TJ, Johansen C, Yoneyama T (1994) Field evaluation of nitrogen fixation and use of nitrogen fertilizer by sorghum/pigeonpea intercropping on an Alfisol in the Indian semi-arid tropics. Biol Fertil Soils 17:241–248

Toomsan B, McDonagh JE, Limpinuntana V, Giller KE (1995) Nitrogen fixation by groundnut and soyabean and residual nitrogen benefits to rice in farmers' fields in Northeast Thailand. Plant Soil 175:45–56

Trasar-Cepeda C, Leiros C, Gil-Sotres F, Seoane S (1998) Towards biochemical quality index for soils: an expression relating several biological and biochemical properties. BiolFertil Soils 26:100–106

Tribouillois H, Cruz P, Cohan JP, Justes É (2015) Modelling agroecosystem nitrogen functions provided by cover crop species in bispecific mixtures using functional traits and environmental factors. Agric Ecosyst Environ 207:218–228

Umar AS, Iqbal M (2007) Nitrate accumulation in plants, factors affecting the process, and human health implications. A review. Agron Sustain Dev 27:45–57

United Nations, Department of Economic and Social Affairs, Population Division (2017) World population prospects: the 2017 revision, key findings and advance tables. Working paper No. ESA/P/WP1248

Van Soest PJ (1994) Nutritional ecology of the ruminant, 2nd edn. Cornell University Press, Ithaca

Vance CP, Graham PH, Allan DL (2000) Biological nitrogen fixation: phosphorus – a critical future need?, nitrogen fixation: from molecules to crop productivity. Curr Plant Sci Biotech Agric 38:509–514

Vandenkoornhuyse P, Husband R, Daniell TJ, Watson IJ, Duck JM, Fitter AH, Young JPW (2002) Arbuscularmycorrhizal community composition associated with two plant species in a grassland ecosystem. Mol Ecol 11:1555–1564

Vandermeer JH (1992) The ecology of intercropping. Cambridge University Press, Cambridge, UK

Varma D, Meena RS (2016) Mungbean yield and nutrient uptake performance in response of NPK and lime levels under acid soil in Vindhyan region, India. J Appl Nat Sci 8(2):860–863

Varma D, Meena RS, Kumar S (2017) Response of mungbean to fertility and lime levels under soil acidity in an alley cropping system in Vindhyan Region, India. Int J Chem Stu 5(2):384–389

Varvel GE (1994) Rotation and nitrogen fertilization effects on changes in soil carbon and nitrogen. Agronomy J86:319–325

Verma JP, Jaiswal DK, Meena VS, Meena RS (2015a) Current need of organic farming for enhancing sustainable agriculture. J Clean Prod 102:545–547

Verma SK, Singh SB, Prasad SK, Meena RN, Meena RS (2015b) Influence of irrigation regimes and weed management practices on water use and nutrient uptake in wheat (*Triticum aestivum* L. Emend. Fiori and Paol.). Bangladesh J Bot 44(3):437–442

Veronica NR, Estela HBB, Rocío CO, Luisa AA (2005) Allelopathic potential of beans (*Phaseolus spp.*) and other crops. Allelo J15:197–210

Vinicius IF, Rosario A, Fernanda LM, Ricardo A (2013) Different interaction among glomus and rhizobium species on *Phaseolus vulgaris* and *Zea mays* plant growth, physiology and symbiotic development under moderate drought stress conditions. Plant Growth Regul 70:265–273

Voisin AS, Guéguen J, Huyghe C, Jeuffroy MH, Magrini MB, Meynard JM, Mougel C, Pellerin S, Pelzer E (2014) Legumes for feed, food, biomaterials and bioenergy. Eur Agron Sustain Dev 34:361–338

Von Richthofen JS, Pahl H, Casta P, Dubois G, Lafarga A, Nemecek T, Pedersen JB (2006) Economic impact of grain legumes in European crop rotations. Grain Leg 45:16–19

Walley FL, Clayton GW, Miller PR, Carr PM, Lafond GP (2007) Nitrogen economy of pulse crop production in the Northern Great Plains. Agron J 99:1710–1718

Wang D, Marschner P, Solaiman Z, Rengel Z (2007) Growth, P uptake and rhizosphere properties of intercropped wheat and chickpea in soil amended with iron phosphate or phytate. Soil BiolBiochem 39:249–256

Wang S, Melnyk JP, Rong Tsao R, Marcone MF (2011) How natural dietary antioxidants in fruits, vegetables and legumes promote vascular health. Food Res Int 44:14–22

Wani PA, Khan MS, Zaidi A (2007) Effect of metal tolerant plant growth promoting *Bradyrhizobium* sp. (vigna) on growth, symbiosis, seed yield and metal uptake by greengram plants. Chemosphere 70:36–45

Warschefsky E, Penmetsa RV, Cook DR, von Wettberg EJ (2014) Back to the wilds: tapping evolutionary adaptations for resilient crops through systematic hybridization with crop wild relatives. Am J Bot 101:1791–1800

Wattiaux MA, Howard TM (2001) Technical dairy guide: nutrition and feeding. University of Wisconsin. http://babcock.cals.wisc.edu/de/html/ch6/nutrition_eng_ch6.html. Accessed on 04 Apr 2017

Yadav GS, Lal R, Meena RS, Babu S, Das A, Bhomik SN, Datta M, Layak J, Saha P (2017) Conservation tillage and nutrient management effects on productivity and soil carbon sequestration under double cropping of rice in North Eastern Region of India. Ecol Indian. http://www.sciencedirect.com/science/article/pii/S1470160X17305617

Zahran HH (1999) Rhizobium-legume symbiosis and nitrogen fixation under severe conditions and in an arid climate. Microbiol Mol Biol Rev 63(4):968–989

Zaman-Allah M, Jenkinson DM, Vadez V (2011) Chickpea genotypes contrasting for seed yield under terminal drought stress in the field differ for traits related to the control of water use. Funct Plant Biol 38(4):270–281

MIX
Papier aus verantwortungsvollen Quellen
Paper from responsible sources
FSC® C105338

FSC
www.fsc.org

If you have any concerns about our products,
you can contact us on
ProductSafety@springernature.com

In case Publisher is established outside the EU,
the EU authorized representative is:
Springer Nature Customer Service Center GmbH
Europaplatz 3, 69115 Heidelberg, Germany

Printed by Libri Plureos GmbH
in Hamburg, Germany